DISCRIMINATION OF CHIRAL COMPOUNDS USING NMR SPECTROSCOPY

THE WILEY BICENTENNIAL–KNOWLEDGE FOR GENERATIONS

Each generation has its unique needs and aspirations. When Charles Wiley first opened his small printing shop in lower Manhattan in 1807, it was a generation of boundless potential searching for an identity. And we were there, helping to define a new American literary tradition. Over half a century later, in the midst of the Second Industrial Revolution, it was a generation focused on building the future. Once again, we were there, supplying the critical scientific, technical, and engineering knowledge that helped frame the world. Throughout the 20th Century, and into the new millennium, nations began to reach out beyond their own borders and a new international community was born. Wiley was there, expanding its operations around the world to enable a global exchange of ideas, opinions, and know-how.

For 200 years, Wiley has been an integral part of each generation's journey, enabling the flow of information and understanding necessary to meet their needs and fulfill their aspirations. Today, bold new technologies are changing the way we live and learn. Wiley will be there, providing you the must-have knowledge you need to imagine new worlds, new possibilities, and new opportunities.

Generations come and go, but you can always count on Wiley to provide you the knowledge you need, when and where you need it!

WILLIAM J. PESCE
PRESIDENT AND CHIEF EXECUTIVE OFFICER

PETER BOOTH WILEY
CHAIRMAN OF THE BOARD

DISCRIMINATION OF CHIRAL COMPOUNDS USING NMR SPECTROSCOPY

THOMAS J. WENZEL
Charles A. Dana, Professor of Chemistry, Bates College
Lewiston, Maine 04240, USA

WILEY-INTERSCIENCE
A JOHN WILEY & SONS, INC., PUBLICATION

Reproduced from Lovely, AE and Wenzel TJ, *Organic Letters*, 2006, 8, 2823-2826 with permission from the American Chemical Society.

© 2007 by John Wiley & Sons, Inc. All rights reserved

Published by John Wiley & Sons, Inc., Hoboken, New Jersey
Published simultaneously in Canada

No part of this publication may be reproduced, stored in a retrieval system, or transmitted in any form or by any means, electronic, mechanical, photocopying, recording, scanning, or otherwise, except as permitted under Section 107 or 108 of the 1976 United States Copyright Act, without either the prior written permission of the Publisher, or authorization through payment of the appropriate per-copy fee to the Copyright Clearance Center, Inc., 222 Rosewood Drive, Danvers, MA 01923, (978) 750-8400, fax (978) 750-4470, or on the web at www.copyright.com. Requests to the Publisher for permission should be addressed to the Permissions Department, John Wiley & Sons, Inc., 111 River Street, Hoboken, NJ 07030, (201) 748-6011, fax (201) 748-6008, or online at http://www.wiley.com/go/permission.

Limit of Liability/Disclaimer of Warranty: While the publisher and author have used their best efforts in preparing this book, they make no representations or warranties with respect to the accuracy or completeness of the contents of this book and specifically disclaim any implied warranties of merchantability or fitness for a particular purpose. No warranty may be created or extended by sales representatives or written sales materials. The advice and strategies contained herein may not be suitable for your situation. You should consult with a professional where appropriate. Neither the publisher nor author shall be liable for any loss of profit or any other commercial damages, including but not limited to special, incidental, consequential, or other damages.

For general information on our other products and services or for technical support, please contact our Customer Care Department within the United States at (800) 762-2974, outside the United States at (317) 572-3993 or fax (317) 572-4002.

Wiley also publishes its books in a variety of electronic formats. Some content that appears in print may not be available in electronic formats. For more information about Wiley products, visit our web site at www.wiley.com.

Wiley Bicentennial Logo: Richard J. Pacifico

Library of Congress Cataloging-in-Publication Data:

Wenzel, Thomas J.
 Discrimination of chiral compounds using NMR spectroscopy / Thomas J. Wenzel.
   p. cm.
 Includes bibliographical references and index.
 ISBN: 978-0-471-76352-9 (alk. paper)
 1. Chemical tests and reagents. 2. Nuclear magnetic resonance spectroscopy.
 3. Enantiomers–Derivatives. 4. Chirality. 5. Derivatization. I. Title.
 QD77.W489 2007
 543$'$.66–dc22

2006033572

Printed in the United States of America

10 9 8 7 6 5 4 3 2 1

To Brad and Erica

CONTENTS

Preface	xxi
Acknowledgments	xxiii

1. Introduction — 1

   1.1. Chiral Derivatizing Agents — 1
   1.2. Chiral Solvating Agents — 2
   1.3. Overview of Chiral Reagents and Methodologies — 4
   1.4. Future Prospects — 6

2. Aryl-Containing Carboxylic Acids — 8

   2.1. Introduction — 8
   2.2. α-Methoxy-α-trifluoromethylphenylacetic Acid (MTPA–Mosher's Reagent) — 11
      2.2.1. Analysis of Secondary Alcohols — 13
      2.2.2. Analysis of Secondary Diols and Polyols — 19
      2.2.3. Analysis of Primary Alcohols — 25
      2.2.4. Analysis of Tertiary Alcohols — 26
      2.2.5. Analysis of Secondary Amines — 26
      2.2.6. Analysis of Primary Amines — 28
      2.2.7. Use as a Chiral Solvating Agent — 29
      2.2.8. Use of MTPA Derivatives with Paramagnetic Lanthanide Chelates — 30
      2.2.9. Use of MTPA Derivatives with Diamagnetic Lanthanide Chelates — 34

	2.2.10. Preparation of MTPA Derivatives	35
	2.2.11. Liquid Chromatography–NMR Spectroscopy of MTPA Derivatives	35
	2.2.12. Database Methods with MTPA	35
2.3.	α-Methoxyphenylacetic Acid (O-methyl Mandelic Acid-MPA)	38
	2.3.1. Analysis of Secondary Alcohols	39
	2.3.2. Analysis of Diols	42
	2.3.3. Analysis of Primary Alcohols	43
	2.3.4. Analysis of Amines	44
	2.3.5. Analysis of Sulfoxides	45
	2.3.6. Variable-temperature Method for Assigning Absolute Stereochemistry	46
	2.2.7. Barium(II) Method for Assigning Absolute Stereochemistry	47
	2.3.8. Use of MPA Derivatives with Lanthanide Chelates	48
	2.3.9. Use as a Chiral Solvating Agent	49
	2.3.10. Preparation of MPA Derivatives–The "Mix and Shake" Method	50
2.4.	Mandelic Acid (2-Hydroxy-2-phenyl Acetic Acid) (MA)	51
2.5.	O-Acetyl Mandelic Acid (2-Acetoxy-2-phenyl Acetic Acid) (O-AMA)	53
2.6.	Tetrahydropyranyl-protected Mandelic Acid [(2R)-2-phenyl-2-[(2S)-tetrahydro-2-pyranyloxy] Ethanoic Acid]	56
2.7.	O-Nitromandelic Acid	56
2.8.	2-Phenylpropionic Acid (2-PPA)	57
2.9.	2-Methoxy-2-phenylpropionic Acid	58
2.10.	3-Phenylbutanoic Acid (3-PBA)/2-Phenylbutanoic Acid (2-PBA)	58
2.11.	α-Methyl-α-methoxy(pentafluorophenyl) (Acetic Acid)	60
2.12.	α-Cyano-α-fluorophenylacetic Acid (CFPA)/α-Cyano-α-fluoronaphthylacetic Acid (CFNA)/α-Cyano-α-fluoro-p-tolylacetic Acid (CFTA)	60
2.13.	N-Boc Phenylglycine (BPG)	62
2.14.	1,5-Difluoro-2,4-dinitrobenzene and Derivatives	63
	2.14.1. N-(5-Fluoro-2,4-dinitrophenyl)-1-phenylethylamide	64
	2.14.2. 1-Fluoro-2,4-dinitrophenyl-5-(S)-alanine Amide (Marfey's Reagent)	65
2.15.	2-Fluoro-2-phenylacetic Acid/2-Fluoro-2-(1-naphthyl)propionic Acid/2-Fluoro-2-(2-naphthyl)propionic Acid	66
2.16.	α-Methoxy-α-(1-naphthyl) Acetic Acid (1-NMA)/α-Methoxy-α-(2-naphthyl) Acetic Acid (2-NMA)	67
2.17.	2-(1-Naphthyl)-2-phenylacetic Acid	70
2.18.	2-Methoxy-2-(1-naphthyl) Propionic Acid (MαNP)	71
2.19.	1-Methoxy-2,3-dihydro-1H-cyclopenta[a]naphthalene-1-carboxylic Acid	72

2.20.	O-Aryl Lactic Acids	72
2.21.	α-(2-Anthryl)-α-methoxyacetic Acid (2-AMA)/α-(9-Anthryl)-α-methoxyacetic Acid (9-AMA)	74
2.22.	(α-[1-(9-Anthryl)]-2,2,2-trifluoroethoxy)acetic Acid	78
2.23.	2-Methoxy-2-(9-phenanthryl)propionic Acid	78
2.24.	Summary	79
	2.24.1. Analysis of Primary Alcohols	79
	2.24.2. Analysis of Secondary Alcohols	79
	2.24.3. Analysis of Tertiary Alcohols	80
	2.24.4. Analysis of Primary Amines	80
	2.24.5. Analysis of Secondary Amines	81

3. Other Carboxylic Acid-based Reagents — 82

3.1.	Camphanic Acid	82
3.2.	Menthoxyacetic Acid (MAA)	86
3.3.	2-(2,3-Anthracenedicarboximido)cyclohexane Carboxylic Acid	87
3.4.	3β-Acetoxy-Δ5-etiocholenic Acid	88
3.5.	Naproxen	89
3.6.	2-*tert*-Butyl-2-methyl-1,3-benzodioxole-4-carboxylic Acid	90
3.7.	*endo*-3-Benzamidonorbornane-2-carboxylic Acid	90
3.8.	Coumarin-based Reagents	91
3.9.	2,2-Diphenylcyclopropane Carboxylic Acid	91
3.10.	Amide Derivatives of Kemp's Triacid	91
3.11.	Amino Acids	92
	3.11.1. N-Boc-phenylalanine	92
	3.11.2. Isotopically Substituted Amino Acids	92
3.12.	(R)-(−)-(2,3,5,7-Tetranitro-9-fluorenyloximino)propanoic Acid	93
3.13.	2-Chloropropanoic Acid (2-Cl-PA)	93
3.14.	Chlorofluoroacetic Acid	94
3.15.	Perfluoropropoxy Propionic Acid/Perfluoro Isopropropoxy Propionic Acid	94
3.16.	2-Methylbutyric Acid	94
3.17.	Lactic Acid and Derivatives	95
	3.17.1. Trifluorolactic Acid	95
	3.17.2. (S)-O-Acetyllactylchloride	95
3.18.	Lasalocid	96
3.19.	Axial Chiral Carboxylic Acids	96
	3.19.1. (+)-1-[2-Carboxy-6-(trifluoromethyl)phenyl]pyrrole-2-carboxylic Acid	96
	3.19.2. 2′-Methoxy-1,1′-binaphthyl-2-carboxylic Acid (MBNC)	97
	3.19.3. 2′-Methoxy-1,1′-binaphthalene-8-carboxylic Acid	98
	3.19.4. 2′-Octylcarbamoyl-1,1′-binaphthyl-2-dicarboxylic Acid	100
	3.19.5. 2-(2′-Methoxy-1′-naphthyl)-3,5-dichlorobenzoic Acid (MNCB)	100

4. Hydroxyl- and Thiol-Containing Reagents — 102

- 4.1. 2,2,2-Trifluorophenylethanol (TFPE) — 103
- 4.2. 2,2,2-Trifluoro-1-(9-anthryl)ethanol (Pirkle's Alcohol (TFAE)) — 109
  - 4.2.1. Analysis of Sulfoxides — 109
  - 4.2.2. Analysis of Lactones — 110
  - 4.2.3. Analysis of Lactams — 111
  - 4.2.4. Analysis of Oxaziridines — 112
  - 4.2.5. Analysis of Axial Chiral Compounds — 112
  - 4.2.6. Analysis of Compounds that are Chiral by Virtue of Slow Rotation — 113
  - 4.2.7. Analysis of Metal Complexes — 115
  - 4.2.8. Analysis of Cyclophosphazenes — 116
  - 4.2.9. Analysis of Phosphine Oxides — 117
  - 4.2.10. Analysis of Calixarenes — 118
  - 4.2.11. Analysis of Other Substrates — 119
  - 4.2.12. Use of Lanthanide Chelates with TFPE and TFAE — 122
  - 4.2.13. Use as a Chiral Derivatizing Agent — 123
- 4.3. Other Anthryl-based Reagents — 123
  - 4.3.1. Analogs of TFAE — 123
  - 4.3.2. Ethyl-2-(9-anthryl)-2-hydroxyacetate — 126
  - 4.3.3. 2-(2,3-Anthracenedicarboximido)-1-cyclohexanol — 127
- 4.4. 2,2,2-Trifluoro-1-(1-pyrenyl)ethanol — 127
- 4.5. 2-(Trifluoromethyl)benzhydrol — 127
- 4.6. 1-Phenylethanol — 128
- 4.7. Methyl Mandelate (MM) — 129
- 4.8. Octahydro-8,9,9-trimethyl-5,8-methano-2H-1-benzopuran-2-ol (Noe's Reagent) — 130
- 4.9. Menthol — 132
- 4.10. Borneol — 135
- 4.11. (−)-10-Mercaptoisoborneol — 135
- 4.12. (R)-(−)-Pantolactone — 136
- 4.13. trans-Bis(hydroxydiphenylmethyl)-2,2-dimethyl-1,3-dioxacyclopentane — 136
- 4.14. 2-Hydroxymethyl-4,6-dimethyl-2-phenyl-1,3-dioxane — 137
- 4.15. 2-Butanol — 137
- 4.16. Assignment of Absolute Configuration Using Glycosidation Shifts — 138
  - 4.16.1. Analysis of Di- and Polysaccharides — 141
  - 4.16.2. β-D- and β-L-Fucofuranoside/Arabinofuranoside — 142
- 4.17. Linear Dextrins — 144
- 4.18. Axial Chiral (Atropoisomeric) Alcohols — 144
  - 4.18.1. 2,2′-Dihydroxy-1,1′-binaphthalene (BINOL) — 145
  - 4.18.2. 8,8′-Dihydroxy-1,1′-binaphthalene — 147

	4.18.3.	Alkyl-Substituted Binaphthols	147
	4.18.4.	2′-Hydroxy-1,1′-binaphthyl-2-yl[(3,5-dinitrobenzoyl)amino](phenyl)acetate	148
	4.18.5.	1,6-Di(o-chlorophenyl)-1,6-diphenylhexa-2,4-diyne-1,6-diol	148
	4.18.6.	4,4′6,6′-Tetrachloro-2,2′-bis(hydroxydiphenylmethyl)biphenyl	148
4.19.	Diol and Dithiol Reagents		149
	4.19.1.	Butane-2,3-diol/Butane-2,3-thiol	149
	4.19.2.	2,2-Dimethyl-1,3-propanediol	150
	4.19.3.	1,1,2-Triphenyl-1,2-ethanediol	151
	4.19.4.	Ethylene Dithiol	151
	4.19.5.	α,α,α′,α′-Tetraphenyl-1,3-dioxolane-4,5-dimethanols) (TADDOL)	151

5. Amine-based Reagents — 153

5.1.	Primary Amines		154
	5.1.1.	1-Phenylethylamine (PEA)/1-(1-Naphthyl)ethylamine (NEA)/1-(9-Anthryl)ethylamine (AEA)	154
	5.1.2.	Fluorinated Aryl Amines	161
	5.1.3.	9-(1-Amino-2,2-dimethylpropyl)-9,10-dihydroanthracene	161
	5.1.4.	(1S,2S)-1-Phenyl-2-amino-3-methoxy-1-propanol	162
	5.1.5.	Phenylglycinol	163
	5.1.6.	1-(3-Aminopropyl)-(5R,8S,10R)terguride	163
	5.1.7.	5-Amino-4-aryl-2,2-dimethyl-1,3-dioxans	164
	5.1.8.	Amino Acids	164
		5.1.8.1. Peptides	164
		5.1.8.2. Phenylglycine methyl ester (PGME)/Phenylglycine dimethyl-amide	165
		5.1.8.3. L-Cysteine	168
		5.1.8.4. L-Proline	169
	5.1.9.	(1R,2R)-1-(1′,8′-Naphthalimide)-2-aminocyclohexane Derivatives	169
	5.1.10.	1-(1-Naphthyl)-2,2-dimethylpropylamine	170
	5.1.11.	1-Methoxy-2-aminopropane	170
	5.1.12.	(+)-2-Amino-1-methoxymenth-8-ene	171
5.2.	Secondary Amines		171
	5.2.1.	Ephedrine	171
	5.2.2.	2-Methyl Piperidine	173
	5.2.3.	(N-Methyl)-α-isosparteinium cation	174
	5.2.4.	N-Boc-1-(1-naphthyl)ethylamine	174
	5.2.5.	N-Methyl-D-(−)-glucamine	175
	5.2.6.	(S)-2-(Diphenylmethyl)pyrrolidine	175

5.3. Tertiary Amines 175
    5.3.1. Quinine/Cinchonidine/Quinidine 175
    5.3.2. 2,8-Dimethyl-6H,12H-5,11-methanodibenzo[**b,f**][1,5]
        diazocine (Troger's Base) 180
    5.3.3. Brucine 180
    5.3.4. Strychnine 181
    5.3.5. (R)- and (S)-Tetrahydroisoquinoline 181
5.4. Diamine Reagents 182
    5.4.1. 1,2-Diphenyl-1,2-diaminoethane 182
    5.4.2. N,N'-Substituted 1,2-diphenyl-1,2-diaminoethane 183
5.5. Databases Using Amine Chiral Solvating Agents 184
    5.5.1. N,α-Dimethylbenzylamine (DMBA) 184
    5.5.2. Bis-1,3-methylbenzylamine-2-methylpropane (BMBA-pMe) 187
    5.5.3. (R-α-Methoxy-α-trifluoromethylphenylacetic Acid
        (MTPA)/BMBA-pMe 189

6. **Miscellaneous Organic-based Chiral Derivatizing
and Solvating Agents** **190**

    6.1. Amides 191
        6.1.1. N-(3,5-Dinitrobenzoyl)-1-phenylethylamine (DNB-PEA) 191
        6.1.2. N-(3,5-Dinitrobenzoyl)-L-leucine (DNB-Leu) 193
        6.1.3. N-(3,5-Dinitrobenzoyl)phenylglycine 195
        6.1.4. N-(3,5-Dinitrobenzoyl)-4-amino-3-methyl-1,2,3,4-
            tetrahydrophenanthrene (Whelk-O-1) 195
        6.1.5. N-1-(1-Naphthyl)ethyltrifluoroacetamide 196
        6.1.6. 1-(1-Naphthyl)ethyl Urea Derivatives
            of Amino Acids (NEU-AA) 197
        6.1.7. Bis Allyl Amide Derivatives 198
            6.1.7.1. N,N'-Diallyl-L-tartardiamide
                bis(4-$tert$-butylbenzoate) 198
            6.1.7.2. (S,S)-$trans$-Acenaphthene-1,2-dicarboxylic
                acid bis allylamide 198
        6.1.8. (R)-Phenylglycinol-N-3,5-dinitrobenzoyl-O-
            triethoxysilylpropylcarbamate 198
        6.1.9. N-Methyl Amide of (R)-N-Acetyl-4-methoxy Phenyl
            Glycine 199
        6.1.10. N-(n-Butylamide) of (S)-2-Phenylcarbamoyloxypropionic
            Acid 199
        6.1.11. 2,2'-Oxybis[N-(1-phenylethyl)acetamide] 200
        6.1.12. Tetrapeptide Species 200
        6.1.13. Tetraamidic Selector 200
    6.2. Lactams 201
        6.2.1. Dihydropyrimidone 202
        6.2.2. Amino Acid Isocyanurate Derivatives 202

6.3.	Aldehydes	203
	6.3.1. 15-Formyl-14-hydroxy-2,8-dithia[9](2,5) pyridinophane	203
	6.3.2. 2′-Methoxy-1,1′-binaphthalene-8-carbaldehyde	203
	6.3.3. 2-Hydroxy-2′-nitrobenzoate-3-aldehyde-1,1′-binaphthalene	204
	6.3.4. 2-Hydroxy-2′-substituted-3-aldehyde-1,1′-binaphthalene	205
6.4.	Ketones	205
	6.4.1. l-Menthone	205
	6.4.2. (S)-(+)-2-Propylcyclohexanone	206
6.5.	Isocyanates	206
	6.5.1. 1-Phenylethyl Isocyanate (PE-I)	206
	6.5.2. 1-(1-Naphthyl)ethyl Isocyanate (NE-I)	208
	6.5.3. Phenylethy Isothiocyanate (PE-IS)/Naphthylethyl Isothiocyanate (NE-IS)	209
	6.5.4. α-Methoxy-α-(trifluoromethyl)benzyl Isocyanate	210
	6.5.5. 2,3,4,6-Tetra-O-acetyl-β-D-glucopyranosyl Isothiocyanate	210
	6.5.6. (S)-2-Chloro-2-fluoroethanoyl Isocyanate	211
6.6.	Miscellaneous Reagents	211
	6.6.1. 2,2-Dimethoxypropane	211
	6.6.2. (−)-Chloromethylmenthyl Ether	212
	6.6.3. 2-Oxazolidones	213
	6.6.4. 5(R)-Methyl-1-(chloromethyl)-2-pyrrolidinone	214
	6.6.5. 5-Methyl-5-phenylpyrroline N-oxide	215
	6.6.6. (S)-2[(R)-Fluoro(phenyl)methyl]oxirane	215
	6.6.7. (S)-Triazine Selector	216
	6.6.8. 2-Deuterio-2,3-dihydro-2-methyl-6-nitrobenzothiophene-1-oxide	216
	6.6.9. 3a-Benzhydryl-3,3a,4,5-tetrahydro-2H-cyclopenta[b]furan	217
	6.6.10. Camphor-10-sulfonic Acid (CSA)	217
	6.6.11. Menthyl Chloroformate	219
	6.6.12. (N-Methylphenylsulfoximidoyl)methyl Lithium	220
	6.6.13. Coumarin Dimer	221
	6.6.14. Di-O-benzoyl Tartrate/Di-O-p-toluoyl Tartrate	221
	6.6.15. Dibenzoyl-L-tartaric Acid Anhydride	222
	6.6.16. DNA	223
	6.6.17. 5,8-Bis(aminomethyl)-1,12-dimethylbenzo[c] phenanthrene	223
	6.6.18. 2′-Methoxy-1,1′-binaphthalene-2-carbohydroxymoyl Chloride (MBCC)	223
6.7.	Self-discrimination of Chiral Compounds	224
6.8.	High-throughput Optical Purity Measurements	227

7. Reagents Incorporating Phosphorus, Selenium, Boron, and Silicon Atoms 229

7.1. Phosphorus-containing Reagents 230
    7.1.1. Phosphorus(V) Reagents 230
        7.1.1.1. Phosphinic Amides (Phos1) 230
        7.1.1.2. Phosphinothioic Acids (Phos2) 231
        7.1.1.3. cis-2-Chloro-3,4-dimethyl-5-phenyl-1,3,2-oxazophospholidin-2-one (Phos3) 234
        7.1.1.4. (2-Chloro-4(R),5(R)-dimethyl-2-oxo-1,3,2-dioxaphospholane (Phos4) 236
        7.1.1.5. L-Menthylphenylchlorophosphine Oxide (Phos5) 236
        7.1.1.6. 2-Chloro-5,5-dimethyl-4-phenyl-1,3,2-dioxaphosphorinane-2-oxide (Phos6) 236
        7.1.1.7. O,O-Di-(2-(S)-(N,N-diethyl-2-hydroxypropyl) phosphonate (Phos7) 237
        7.1.1.8. 2-Chloro-3-phenyl-1,3,2-diazaphosphabicyclo[3.3.0]octane-2-oxide (Phos8) 238
        7.1.1.9. Methyl Phosphonic Dichloride (Phos9)/Methyl Phosphonothioic Dichloride (Phos10) 239
        7.1.1.10. O,O-Diphenyldithioic Acid (Phos11) 239
        7.1.1.11. 1,1'-Binaphthyl-2,2'-diylphosphoric Acid (Phos12) 240
    7.1.2. Phosphorus(III) Reagents 242
        7.1.2.1. Diazaphospholidines (Phos13–Phos20) 242
        7.1.2.2. 1,3,2-Dioxaphospholanes (Phos21–Phos27) 246
        7.1.2.3. Dimenthyl Chlorophosphite (Phos28) 248
        7.1.2.4. Dichloromenthylphosphine (Phos29) 248
        7.1.2.5. Phosphorus Trichloride (Phos30) 249
        7.1.2.6. Hexaalkylphosphorus Triamides (Phos31) 249
        7.1.2.7. α-1-Phenylethylamino Trisaminoamine (Phos32) 249
        7.1.2.8. [5] HELOL Phosphite (Phos33) 249
    7.1.3. Ionic Reagents—TRISPHAT, BINPHAT, and BINTROP 250
    7.1.4. Configurational Analysis of Phosphates 260
        7.1.4.1. Oxygen Isotope Methods 260
        7.1.4.2. ROESY Studies 263
    7.1.5. Summary 264
7.2. Selenium-containing Reagents 264
    7.2.1. 4-Methyl-5-phenyloxazolidine-2-selone 264
    7.2.2. (R,R)-N,N'-Dimethylcyclohexyl-1,2-diazaselenophospholine 266
    7.2.3. 2-Phenylselenopropionic Acid 267
    7.2.4. 1,2,3-Selenadiazoles 267
7.3. Boron-containing Reagents 268
    7.3.1. 2-(1-Methoxyethyl)phenylboronic Acid 268
    7.3.2. Phenylboronic Acid 269
    7.3.3. Boric Acid [$B(OH)_3$] 270

CONTENTS   xv

       7.3.4. Camphanylboronic Acid   270
       7.3.5. 4S,5R-4-Methyl-5-phenyl-1,3,2-oxazaborolidine
             (Ephedrine Borane)   271
       7.3.6. (S)-(+)-N-Acetylphenylglycine Boronic Acid   271
       7.3.7. 1-Benzamido-1-(2-methylphenyl)methaneboronic Acid   271
       7.3.8. 2-Formylphenylboronic Acid   272
       7.3.9. 2,2'-(1,2-Phenylene)bis[(4R,5R)-4,5-diphenyl-
             1,3,2-dioxaborolane   272
       7.3.10. (1-Chloroalkyl)boronates   273
   7.4. Silyl-containing Reagents   274
       7.4.1. Menthylphenylsilylacetic Acid   274
       7.4.2. (Benzylmethylphenylsilyl)methylamine   274
       7.4.3. Monochlorosilanes   275

8. Host Compounds as Chiral NMR Discriminating Agents   **278**

   8.1. Cyclodextrins   279
       8.1.1. Introduction   279
       8.1.2. Native Cyclodextrins   280
            8.1.2.1. Lanthanide Coupling to Native Cyclodextrins   286
       8.1.3. Neutral Cyclodextrin Derivatives   287
            8.1.3.1. Hexakis(2,3,6-tri-O-methyl)-α-cyclodextrin
                    (TM-α-CD)/Heptakis(2,3,6-tri-O-methyl)-
                    β-cyclodextrin (TM-β-CD) – (Permethyl CDs)   287
            8.1.3.2. Heptakis(2,6-di-O-methyl)-β-cyclodextrin
                    (2,6-DM-β-CD)   289
            8.1.3.3. Benzylated and Benzoylated Cyclodextrins   290
            8.1.3.4. Carbamoylated Cyclodextrins   290
            8.1.3.5. Octakis(3-O-butanoyl-2,6-di-O-pentyl)-
                    γ-cyclodextrin (BP-CD)   291
            8.1.3.6. Hydroxyethyl and Hydroxypropyl
                    Cyclodextrins (HE-CD, HP-CD)   291
            8.1.3.7. Heptakis(2,3-di-O-acetyl)-β-CD (2,3-DA-β-CD)   291
            8.1.3.8. Mono(2-acetylamino-2-deoxy)-, Mono
                    (3-acetylamino-3-deoxy)- and Mono
                    (6-acetylamino-6-deoxy)-β-cyclodextrin
                    (AA-β-CD)   292
            8.1.3.9. Mono-3,6-anhydro-β-cyclodextrin   292
            8.1.3.10. Heptakis(3-O-acetyl-2,6-di-O-pentyl)-
                    β-cyclodextrin   292
            8.1.3.11. Octakis(2,6-di-O-pentyl-3-O-
                    trifluoroacetyl)-γ-cyclodextrin   292
            8.1.3.12. Cyanoethylated-β-cyclodextrin   292
            8.1.3.13. Per-O-octyl-α-cyclodextrin   293
            8.1.3.14. Biphenyl-capped-β-cyclodextrin   293
            8.1.3.15. Heptakis(2,3,6-tri-O-acetyl)-β-CD (TA-β-CD)   293

8.1.4. Anionic Cyclodextrin Derivatives ... 293
  8.1.4.1. Carboxymethylated Cyclodextrins (CM-CD) ... 293
  8.1.4.2. Heptakis(6-carboxymethylthio-6-deoxy)-
      β-cyclodextrin (6-CMT-β-CD) ... 296
  8.1.4.3. Sulfobutylether-β-cyclodextrin (SBE-β-CD) ... 296
  8.1.4.4. Heptakis(2,3-di-*O*-acetyl-6-sulfo)-β-CD
      (HDAS-β-CD) ... 297
  8.1.4.5. Sulfated Cyclodextrins (S-CD) ... 298
8.1.5. Cationic Cyclodextrin Derivatives ... 298
  8.1.5.1. Amino-substituted Cyclodextrins ... 298
  8.1.5.2. Bis(6-trimethylammonio)-β-cyclodextrin
      (TMA-β-CD) ... 299
  8.1.5.3. Mono(6-xylylenediamine)-β-cyclodextrin (X-β-CD) ... 299
8.1.6. Summary ... 299
8.2. Crown Ethers ... 300
  8.2.1. Introduction ... 300
  8.2.2. (18-Crown-6)-2,3,11,12-tetracarboxylic Acid
      (18-C-6-TCA) ... 301
    8.2.2.1. Amide Derivatives of (18-Crown-6)-2,3,11,12-
        tetracarboxylic Acid ... 303
  8.2.3. (18-Crown-6)-2,3,11,12-derivatives ... 303
  8.2.4. Crown Ethers Incorporating a 2,2'-Dihydroxy-1,1'-
      binaphthalene Unit ... 305
  8.2.5. Glycoside-derived Crown Ethers ... 305
  8.2.6. Crown Ether-like Compounds with
      Nitrogen-heterocyclic Units ... 306
  8.2.7. Triazole-18-crown-6 Ether ... 308
  8.2.8. Crown Ether-like Compounds with Phenol Units ... 308
  8.2.9. Aza Crown Ethers ... 309
8.3. Calixarenes and Calixresorcarenes ... 311
  8.3.1. Introduction ... 311
  8.3.2. Calix[4]resorcarenes ... 312
  8.3.3. Calixarenes ... 315
  8.3.4. Summary ... 319
8.4. Receptor Compounds for Carboxylic Acids ... 320
8.5. Cyclic Peptides ... 323
8.6. Miscellaneous Receptor Compounds ... 324
  8.6.1. Cyclic Tetraamidic Reagent ... 324
  8.6.2. Cyclosophoraoses (Cyclic(1-2)-β-D-glucans) ... 325
  8.6.3. Glycophane Receptor ... 325
  8.6.4. Cyclophane Receptors ... 326
  8.6.5. Cryptophane Receptor ... 326
  8.6.6. Receptor for Secondary Amines ... 327
  8.6.7. Spirobisindane Receptor ... 327
  8.6.8. Receptors for Diols ... 328
  8.6.9. 1,1'-Binaphthalene-based Receptors ... 328

CONTENTS xvii

8.7.	Enzymes		329
	8.7.1.	Calmodulin	329
	8.7.2.	Chymotrypsin	330

9. Chiral Discrimination with Metal-based Reagents — **331**

9.1. Introduction — 331
9.2. Lanthanide Complexes — 332
   9.2.1. Introduction — 332
   9.2.2. Peak Broadening with Lanthanide Shift Reagents — 335
   9.2.3. Catalytic Properties of Lanthanide Ions — 337
   9.2.4. Application of Lanthanide tris β-diketonates — 338
      9.2.1.1. Discrimination of Prochiral Nuclei — 339
      9.2.1.2. Analysis of Deuterium-Substituted Compounds — 341
      9.2.1.3. Analysis of Compounds Chiral by Virtue of Slow Rotation — 342
      9.2.4.1. Analysis of Compounds Chiral by Virtue of Geometrical Constraints — 344
   9.2.5. Miscellaneous Applications of Lanthanide tris β-diketonates — 346
   9.2.6. Analysis of Chiral Metal Complexes — 348
   9.2.7. Assignment of Absolute Stereochemistry — 351
   9.2.8. Pr(III) Complex of Tetraphenylimidodiphosphinate—$Pr(tpip)_3$ — 354
   9.2.9. Binuclear Lanthanide–silver Reagents — 355
      9.2.9.1. Analysis of Organic Salts — 358
   9.2.10. Aqueous Lanthanide Shift Reagents — 359
      9.2.10.1. Complexes of Propylenediaminetetraacetate (pdta) — 360
      9.2.10.2. Complexes of N,N,N',N'-Tetrakis(pyridylmethyl) propylene Diamine (TPPN) — 361
      9.2.10.3. Complexes with Other Ligands — 362
      9.2.10.4. Complexes with Tetraazocyclodecane Macrocyclic Ligands — 364
9.3. Transition Metal Complexes — 365
   9.3.1. Palladium Complexes — 365
      9.3.1.1. Bridged Dimers with Amine Ligands — 365
      9.3.1.2. Miscellaneous Palladium Complexes — 372
   9.3.2. Platinum Complexes — 374
      9.3.2.1. Complexes with Diphosphine Ligands — 374
      9.3.2.2. Complexes with Amine Ligands — 375
      9.3.2.3. Analysis of Geometrical Isomers — 378
   9.3.3. Rhodium Complexes — 378
      9.3.3.1. Rhodium Dimer—$[Rh_2(MTPA)_4]$ — 378
      9.3.3.2. Miscellaneous Rhodium Complexes — 384

9.3.4. Cobalt Complexes	385
9.3.4.1. Complexes with Porphyrin Ligands	385
9.3.4.2. Analysis of Geometrical Isomers	387
9.3.4.3. Analysis of DNA	389
9.3.5. Zinc Complexes	389
9.3.6. Ruthenium Complexes	392
9.3.7. Nickel Complexes	394
9.3.8. Copper Complexes	394
9.3.9. Silver Complexes	395
9.3.10. Tin Complexes	396
9.3.11. Titanium Complexes	397
9.3.12. Tungsten Complexes	397
9.3.13. Molybdenum Complexes	397
9.3.14. Mercury Complexes	397
9.3.15. Osmium Complexes	398
9.3.16. Uranium Complexes	398

10. Chiral NMR Discrimination with Highly Ordered Systems   **399**

10.1. Introduction	399
10.2. Liquid Crystals	400
10.2.1. Introduction	400
10.2.2. Early Studies with Liquid Crystals	401
10.2.3. ^{2}H NMR Analysis of Labeled Compounds	403
10.2.4. Natural Abundance ^{2}H NMR Analysis	410
10.2.5. ^{13}C NMR Analysis	412
10.2.6. ^{1}H NMR Analysis	414
10.2.7. ^{19}F NMR Analysis	414
10.2.8. Analysis Using Other Nuclei	415
10.2.9. Miscellaneous Studies with Liquid Crystals	415
10.3. Micelles and Gels	416
10.4. Ionic Liquids	419
10.5. Polymers	420
10.5.1. Cellulose-carbamate Polymers	420
10.5.2. Poly 3- and 4-((S-(α-Methylbenzylcarbamoyl) phenyl Isocyanate	421
10.5.3. Poly(ester/β-sulfoxides)	421
10.5.4. Chirasil-val	422
10.5.5. Poly(N-1-anthrylmaleimide)	422
10.5.6. Poly [(1-6)-2,5-anhydro-3,4-di-O-methyl-D-glucitol]	423

10.6. Aggregation of Achiral Compounds into Chiral Assemblies	423
10.7. Solid-state NMR Spectroscopy	425
10.7.1. Tri-*O*-thymotide Clathrates	425
10.7.2. Cholic Acid Channels	426
10.7.3. Analysis of Polymer-bound Compounds	427
References	**428**
Index	**537**

PREFACE

Nuclear magnetic resonance spectroscopy represents one of the most common methods employed for the analysis of chiral compounds. I used over 30 different keywords or combination of keywords as well as numerous CAS numbers in trying to identify articles that describe the development and application of chiral NMR discriminating agents. My search of the literature identified about 3000 publications in which NMR spectroscopy had been employed for the determination of optical purity or assignment of absolute configuration. Many of these reports described new chiral reagents or expanded the applicability of the existing chiral reagents. Many others involved the use of known reagents as tools to analyze a compound of interest to the investigator.

In organizing the manuscript, I had to make a choice between identifying compound classes and then describing particular chiral NMR reagents suitable for analyzing them and identifying chiral NMR reagents and describing the range and types of compounds for which they could be used for analysis. I opted for the latter organization, as too many compounds that were analyzed in the literature were polyfunctional. These polyfunctional substrates often did not have a particular, identifiable functional group that was solely responsible for interaction with the chiral reagent, or bound simultaneously through two sites, thereby resisting easy classification.

In identifying NMR reagents, I opted for a broad coverage so as to show the extensive range of systems that have been used for the analysis of chiral compounds. For those reagents that are the focus of only one or a handful of studies, I generally incorporated a brief discussion of all of the articles I was able to identify. For those chiral discriminating agents that have been extensively applied, which in some cases extend to hundreds of compounds, a discussion of every application was neither desirable nor feasible. Instead, with these reagents, I focus on aspects that demonstrate the overall utility of the reagent. This required a judgment on my part as to which articles best exemplify the utility of a particular reagent and no doubt reflects some of my own biases as to what is most interesting or important.

This text should be especially useful to the investigator who would like to identify a suitable reagent to analyze the optical purity or assign the absolute configuration of a particular compound. It should also be of use to investigators involved in the development of new chiral NMR discriminating agents who will benefit from the thorough review of prior work in this field.

THOMAS J. WENZEL

ACKNOWLEDGMENTS

This project would not have been possible without the considerable help of others. I thank Sylvia Deschaine for typing the first draft of the references. Courtney O'Farrell's effort in proofreading and reformatting the references is deeply appreciated. Monique Brown, Lily Conover, and Shawna-Kaye Lester's assistance in collecting copies of literature articles, including more than 1000 interlibrary loan requests they processed, was vital to the success of this project. The countless hours that Katelyn Provencher and Madeline Weber spent drawing the compound structures and figures are greatly appreciated. I especially thank Ann Lovely, Courtney O'Farrell, Katelyn Provencher, and Madeline Weber for their help in getting the compounds and figures into final form. The support from a Howard Hughes Medical Institute Award to Bates College in the form of a course reduction and funds for student stipends to assist with the project is most appreciated. Finally, I thank family, friends, and colleagues who provided encouragement throughout the project.

1

INTRODUCTION

Nuclear magnetic resonance spectroscopy is one of the most common methods used to determine optical purity and assign the absolute configuration of chiral compounds. The strategy that has been most exploited, as first recognized by Raban and Mislow in 1965,[1] is to use an optically pure chiral reagent to distinguish a pair of enantiomers through the formation of nonequivalent diastereomeric complexes. The optically pure probe molecule functions as either a derivatizing or solvating agent. Furthermore, the association of an optically pure compound with a prochiral molecule with nuclei that are enantiotopic by internal comparison (e.g., the methyl groups of 2-propanol) renders these nuclei nonequivalent such that distinct resonances are often observed in the NMR spectrum. Classifying chiral metal compounds as derivatizing or solvating agents is sometimes difficult. What is important is whether the substrate molecule undergoes fast or slow exchange with the metal center. Strategies based on different packing orders for a pair of enantiomers, such as those that occur in liquid crystals or solid-state systems, have also been used for chiral analysis in NMR spectroscopy.

1.1. CHIRAL DERIVATIZING AGENTS

Chiral derivatizing agents form a covalent bond with a reactive moiety of the substrate. Many chiral derivatizing agents are available for the analysis of carboxylic acids, alcohols, and amines, although strategies for preparing derivatives of many other functional groups will be described as well throughout the text. There are two

Discrimination of Chiral Compounds Using NMR Spectroscopy, by Thomas J. Wenzel
Copyright © 2007 John Wiley & Sons, Inc.

potential concerns with the application of chiral derivatizing agents. One is the possibility of kinetic resolution. Kinetic resolution refers to a situation in which one enantiomer reacts faster than the other with the chiral derivatizing agent. If the reagents are not allowed to react for a long enough time, the proportion of the two diastereomers will not be equivalent to the proportion of the two enantiomers in the original mixture. Kinetic resolution is significant when determining optical purity, but is not significant if the chiral derivatizing agent is being used to assign the absolute configuration of an optically pure substrate such as a natural product.

A second concern with chiral derivatizing agents is that no racemization occurs during the derivatization reaction. This can be significant whether it happens to the substrate or the chiral derivatizing agent. With some chiral derivatizing agents for which unacceptable levels of racemization did occur, further study was undertaken to develop reaction conditions that minimize or eliminate racemization. When pertinent, these studies are described in the text.

A general understanding is that chiral derivatizing agents should be optically pure. A method for using chiral derivatizing agents that are less than 100% optically pure has been described.[2] The purity of the reagent must first be accurately measured using an appropriate method. A set of equations was provided in the report to determine the optical purity of an unknown from the known purity of the chiral reagent.

Many chiral derivatizing agents incorporate moieties that produce specific and predictable shifts in the resonances of the substrate. In such cases, the shifts in the spectrum of an optically pure substrate in the derivatives with the (*R*)- and (*S*)- enantiomers of the chiral derivatizing agent can be used to assign the absolute configuration. In other situations, moieties on the substrate may cause specific and predictable shifts in resonances of the chiral derivatizing agent. If so, these can be used to assign absolute configurations as well.

Another procedure that is often used with chiral derivatizing or solvating agents is to look for the presence of specific trends in the shifts that correlate with the absolute configuration of the substrate. The assumption is that, if the trends are consistent among a series of compounds with known configurations, then they will be consistent for an unknown compound with a similar structure. Empirical trends such as these have been observed in many situations and are described where appropriate throughout the text.

An alternative, although much less used, derivatizing strategy involves a self-coupling reaction of a chiral molecule. The self-coupling of two chiral molecules leads to the formation of a mixture of *meso* (*R*,*S*) and *threo* [(*S*,*S*)/(*R*,*R*)] derivatives.[3] Assuming that these species exhibit distinct resonances in the NMR spectrum, the size of the different resonances depends on the optical purity of the compound. The utility of the method was demonstrated for several examples.

1.2. CHIRAL SOLVATING AGENTS

Chiral solvating agents associate with the substrate through noncovalent interactions. These can involve a mix of dipole–dipole, ion-pairing, and π–π interactions. Because of this, the choice of solvent is often an important parameter when using a particular chiral solvating agent. Organic-soluble chiral solvating agents are often more effective in nonpolar solvents that cannot effectively solvate the polar groups of the chiral solvating agent and substrate. Water-soluble chiral solvating agents, which are often organic host compounds, usually rely on hydrophobic effects to promote the interaction of a hydrophobic portion of the substrate within the hydrophobic cavity of the solvating agent. Steric effects are also important in the recognition properties of many chiral solvating agents.

Chiral solvating agents generally undergo fast exchange with substrates. With fast exchange, the NMR spectrum is a weighted average of the proportion of bound and unbound substrate. Resonances of the substrate double with the presence of chiral recognition. If slow exchange and enantiomeric discrimination occur, and not all of the substrate is bound to the solvating agent, three resonances are observed for a particular nucleus in the NMR spectrum. One is for the unbound substrate. The other two are for the bound forms of the (R)- and (S)-isomers of the substrate. Sometimes the resonances of the substrate or chiral solvating agent are broadened, which occurs if the system has an intermediate rate of exchange. In such cases, it may be possible to speed up the exchange to acceptable levels by warming the sample.

Most chiral solvating agents are used to determine optical purities. There are instances, though, in which the interaction is understood with enough specificity to be able to assign the absolute configuration based on the relative magnitudes of the shifts in the spectrum, much like observed with certain chiral derivatizing agents. Unlike chiral derivatizing agents, when measuring optical purities with a chiral solvating agent, it is not necessary to have 100% optical purity for the chiral reagent. What is needed is sufficient recognition to cause nonequivalence in the spectra of the enantiomers so that the peaks can be accurately integrated.

Chiral recognition with a chiral solvating agent can occur from two mechanisms. One is that the chiral solvating agent–substrate complexes are diastereomers and may have different chemical shifts. The other is that the two enantiomers often have different association constants with the chiral solvating agent, such that the time-averaged solvation environments are different. In many cases, both mechanisms likely contribute to some extent to the nonequivalence that is observed in the NMR spectrum.

Enantiomeric discrimination with chiral solvating agents is often concentration and temperature dependent. Higher concentrations of the chiral solvating agent generally favor formation of the distinct diastereomeric complexes and enhance the discrimination. Lowering the temperature of the solution usually raises the association constant of the chiral solvating agent–substrate complex, thereby enhancing the enantiomeric discrimination.

A diverse variety of chiral derivatizing and solvating agents have been developed as described in the ensuing chapters. Also, published review articles have described different aspects of chiral NMR solvating agents,[4,5] chiral derivatizing agents,[6] the use of NMR spectroscopy to assign absolute configurations,[7] the use of chiral fluorine-containing reagents for the determination of optical purity,[8] and the use of NMR spectroscopy for chiral analysis.[9-11]

1.3. OVERVIEW OF CHIRAL REAGENTS AND METHODOLOGIES

An important family of reagents that are described in Chapter 2 is aryl-containing carboxylic acids, the most well known of which is α-methoxy-α-trifluoromethylphenylacetic acid (MTPA). These are mostly used as chiral derivatizing agents for the assignment of absolute configurations of substrates such as alcohols and amines. Shielding by the aromatic ring of the derivatizing agent in the resulting diastereomeric complexes is used to make the assignment. Although MTPA is the most well known and commonly applied of these reagents, as will be discussed in Chapter 2, there are other reagents that are recommended for the analysis of certain classes of compounds.

Chapter 3 describes other carboxylic acids that have been used as either chiral derivatizing or solvating agents. Certain of these reagents such as camphanic acid have proven useful for distinguishing the *pro-(R)* and *pro-(S)* positions of α-deuterated primary alcohols. Several reagents based on axially chiral systems are also discussed.

Hydroxy-containing compounds, as described in Chapter 4, have been widely exploited for chiral analysis in NMR spectroscopy. This includes the application of 2-(9-anthryl)-2,2,2-trifluoroethanol, one of the most versatile chiral solvating agents ever developed. Shielding by the anthryl group of this reagent can also be used to assign the absolute configuration of certain classes of substrates. Alcohol reagents are also used as chiral derivatizing agents, especially in the analysis of carboxylic acids. Certain diols and glycosides have been used as effective chiral derivatizing agents for ketones and secondary alcohols, respectively. Axially chiral compounds such as 2,2'-dihydroxy-1,1-binaphthalene have been used as effective chiral derivatizing or solvating agents with suitable substrates.

Primary, secondary, and tertiary amines have been used as chiral derivatizing and solvating agents as described in Chapter 5. 1-Phenylethylamine, the first compound ever used as a chiral NMR solvating agent, and 1-(1-naphthyl)ethylamine are especially important chiral amines that have been used extensively to analyze carboxylic acids and other compounds as well. Phenylglycine methyl ester hydrochloride is another important reagent for assigning the absolute configuration of carboxylic acids. Quinine, a tertiary amine, has a variety of functional groups that influence its association with and chiral discrimination properties toward a number of compound classes. Some amine reagents have been exploited as effective chiral derivatizing agents for the analysis of aldehydes and ketones. Certain diamine reagents have proven to be useful reagents for chiral analysis by NMR spectroscopy.

As described in Chapter 5, chemical shift data measured with the chiral solvating agents N,α-dimethylbenzylamine and bis-1,3-methylbenzylamine-2-methylpropane have been used to construct ^{13}C and ^1H NMR databases for all of the configurations of particular structural motifs. The pattern of the shift data for the known configurations that best matches that of an unknown can be used to determine the stereochemistry. The method is especially well suited to the assignment of structural motifs within complex natural products.

Chapter 6 describes a collection of chiral reagents that encompass a variety of compound classes. These include reagents with amide, lactam, aldehyde, ketone, isocyanate, and heterocyclic ring functionalities. Many of these reagents have been studied on a limited basis and apply to specific types of substrates, although some of the amide compounds represent soluble analogs of widely applicable chiral liquid chromatographic phases and are effective with a variety of compound classes. Certain of the reagents described in Chapter 6 are used as chiral solvating agents and associate through combinations of dipole–dipole and π–π interactions. Others such as the ketones, aldehydes, and isocyanates are utilized as chiral derivatizing agents for particular classes of substrates.

Reagents specifically designed to incorporate phosphorus, selenium, boron, and silicon atoms are described in Chapter 7. Many of the phosphorus and selenium reagents specifically use ^{31}P and ^{77}Se NMR spectroscopy to facilitate the analysis of optical purity. The majority of the systems are used as chiral derivatizing agents. In the case of the phosphorus-, boron-, and silicon-containing reagents, the reactions usually involve addition of the substrate at the heteroatom to form the diastereomeric complexes. The primary selenium-containing reagent that has been studied incorporates the selenium atom as a spectroscopic probe rather than a reactive center. An important set of chiral cationic and anionic phosphorus-based reagents that form ion pairs with ionic substrates is also described in this chapter. The anionic reagents are especially useful in the analysis of cationic metal complexes, although organic cations can be analyzed as well.

Another versatile strategy for effective chiral recognition, as described in Chapter 8, is through the use of host–guest complexation. Cyclodextrins have been the most widely studied family of host compounds for chiral NMR applications. Cyclodextrins can be derivatized either selectively or indiscriminately at the different hydroxyl groups, providing a range of host compounds of varying solubility and chiral recognition properties. In the aggregate, these cyclodextrin derivatives have the potential to function as chiral discriminating agents for a broad array of substrates.

Crown ethers are another common group of host compounds that function rather selectively for primary amines, although recent work has shown how the only commercially available chiral crown ether for NMR studies is also an effective chiral solvating agent for secondary amines. Calixarenes and calixresorcarenes are less studied in NMR applications, but offer interesting potential for future development and applications. There are also many specialized receptor compounds that have been described that exhibit chiral recognition toward a specific compound or class of compounds.

The use of metal complexes as chiral discriminating agents in NMR spectroscopy is an area that has received considerable attention. The importance of paramagnetic lanthanide shift reagents within the entire field of chiral NMR analysis cannot be underemphasized. Although the use of chiral lanthanide shift reagents is mostly described in Chapter 9, the utilization of lanthanide species as a means of enhancing the chiral discrimination of other NMR reagents shows up throughout the text. The utilization of lanthanide shift reagents has diminished as more investigators have obtained access to high-field NMR spectrometers. One reason is that the enhanced dispersion caused by the addition of a paramagnetic lanthanide is often no longer necessary. The other is that the line broadening caused by the paramagnetic ions is more pronounced at higher field strengths.

Chiral reagents based on diamagnetic metal complexes of palladium, platinum, rhodium, and silver have significant applications as well. These metals are especially effective at bonding to soft Lewis bases, thereby broadening the scope of compound classes amenable to chiral analysis by NMR spectroscopy. The exceptionally large shielding of substrate nuclei caused by the porphyrin rings of metal complexes of cobalt, zinc, and ruthenium has been exploited in the development of chiral discriminating agents. In addition, a number of more specialized reagents involving other metal systems have been described in the literature.

One of the more intriguing developments in recent years, which is described in Chapter 10, involves the use of liquid crystals for chiral NMR discrimination. Liquid crystals become ordered in an applied magnetic field. Pairs of enantiomeric compounds often adopt a different packing order relative to the magnetic field when dissolved in a chiral, ordered liquid crystal. The different packing order can lead to different dipolar coupling or quadrupolar splitting for the enantiomers, which causes distinct resonances. As no specific interactions need to occur between the liquid crystal and the substrate, this method is potentially amenable to any chiral substrate including aliphatic hydrocarbons.

Solid-state NMR studies potentially offer some of the same potential as liquid crystals, but have been used to far less extent. Studies of micelles and ionic liquids are more limited in scope. Organic-soluble polymers have been used for chiral NMR discrimination, but often these studies have been aimed at understanding specific geometrical aspects of the association between the polymer and the substrate. Relatively recent work has begun to explore the ability of specially designed achiral compounds to aggregate in solution into assemblies with chiral pockets.

1.4. FUTURE PROSPECTS

Although it might be tempting to consider the area of chiral discrimination by NMR spectroscopy a mature field, a number of new developments in this area over the past 5 years show that it is an area of continued discovery and growth. As will become apparent from the ensuing chapters, we currently have an extensive range of reagents that make it possible to determine the optical purity and assign the absolute configuration of many classes of compounds. But that has not deterred investigators from developing new and more effective reagents or further refining and expanding the applications of existing ones. Further development of database approaches based on existing or newly collected NMR data is likely to occur. The desire to better understand fundamental aspects of noncovalent interactions, which is at the core of how chiral solvating agents function, and to design more selective chiral recognition systems means that we will continue to see new reagents for chiral analysis by NMR spectroscopy.

Finally, further improvements in instrumental design may ultimately have profound implications for the analysis of chiral compounds using NMR spectroscopy. A way to study chiral systems via NMR spectroscopy without the need of chiral reagents has been described.[12] A nuclear magnetic moment in the x-direction combined with a strong magnetic field induces, through odd-parity coupling in a freely tumbling molecule, an electric dipole moment in the y-direction. The electronic magnetic moment in the x-direction is of opposite signs for D- and L-enantiomers. A $\pi/2$ pulse causes a coherent precession of nuclear spins, and in favorable cases the rotating electric polarization at the NMR frequency can be detected.

The method would involve a simultaneous observation of rotating magnetization and electric polarization at some angular frequency following the $\pi/2$ pulse. The peaks obtained can, in theory, be assigned to particular chiral centers. The magnitude of the precessing electron polarization is quite small, so actually performing this experiment will prove challenging. However, if it was possible to perform the measurement in dilute solutions, it would provide a new way to study chiral systems by NMR spectroscopy.

2

ARYL-CONTAINING CARBOXYLIC ACIDS

2.1. INTRODUCTION

In 1973, Dale and Mosher published a landmark paper on the use of NMR spectroscopy as a means of assigning absolute configuration.[13] Although the Mosher reagent (α-methoxy-α-trifluoromethylphenylacetic acid–MTPA) was applied in this first report to secondary alcohols, the general strategy has since been exploited across a broad range of reagents that can be used with more classes of compounds than with just secondary alcohols. Mosher and Mosher-like reagents, as presented here, are optically pure compounds that are used to prepare derivatives of the substrate under study.

The basis of the Mosher method is predicated on two general factors of the derivative(s) that are illustrated in Figure 2.1 for a secondary alcohol. The first involves the adoption of a preferred conformation of the chiral derivatizing agent–substrate derivative[14] and the necessity that this be the same for the two diastereomeric derivatives and independent of L_1 and L_2.[15] The second involves shielding of the nuclei of the substrate caused by an aromatic ring within the chiral derivatizing agent. This shielding will cause certain resonances of the substrate to shift to lower frequency in the NMR spectrum. Without a preferred conformation, rapid rotation about the discriminating agent–substrate bond will average out the shielding of the substrate protons and make any distinction far less pronounced. The exact nature of the preferred conformation is not that important as long as there is a preferred

---

Discrimination of Chiral Compounds Using NMR Spectroscopy, by Thomas J. Wenzel
Copyright © 2007 John Wiley & Sons, Inc.

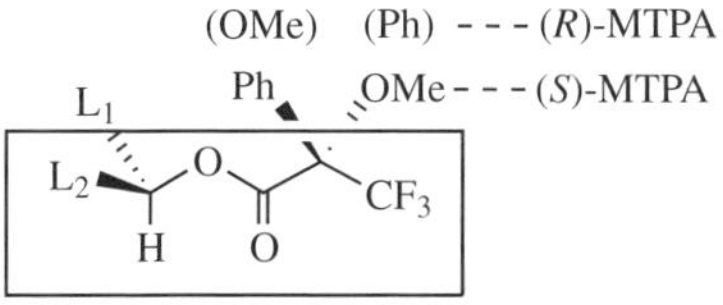

INTRODUCTION

Figure 2.1. Conformational model for the (R)- and (S)-MTPA derivatives of a secondary alcohol. The box represents the MTPA or "Mosher plane."

conformation, its structure is known, and the aromatic ring is in a position to induce specific and differential shielding of substituent groups on the substrate.

The conformation in Figure 2.1 has the Cα–OMe, C=O, and C$_1'$–H bonds approximately coplanar (the so-called "Mosher plane") and is designated as *syn-periplanar* (*sp*). In the (R)-MTPA derivative in Figure 2.1, the largest shift will occur for resonances of L_1, which will shift to lower frequencies because of shielding from the phenyl ring. The chemical shifts of L_2 are likely to be quite similar to those of the underivatized substrate. In the (S)-MTPA derivative, the resonances of L_2 are shielded relative to their position in the NMR spectrum of the underivatized substrate. The resonances of L_1 will have chemical shifts similar to those of the underivatized substrate. Subtracting the chemical shifts of the (R)-derivative from those of the (S)-derivative ($\Delta\delta^{RS}$) will cause negative values for the hydrogen resonances of L_1 and positive values for the hydrogen resonances of L_2.

For example, the $\Delta\delta^{RS}$ values reported for the three methyl resonances (positions 8, 9, and 10) of the (R)- and (S)-MTPA and (R)- and (S)-α-methoxyphenylacetic (MPA) acid derivatives of menthol are shown in Figure 2.2.[15] The $\Delta\delta^{RS}$ values are positive for the C10 methyl resonance and negative for the C8 and C9 resonances. The sign of $\Delta\delta^{RS}$ indicates whether a particular hydrogen is at the L_1 or L_2 position. Identification of which groups are at L_1 and L_2 provides a means to assign the absolute configuration of the substrate.

There are two cautionary notes about the use of MTPA derivatives. One is that beginning with the preliminary papers by Mosher, many reports using MTPA derivatives calculated $\Delta\delta^{SR}$ values, whereas reports with other similar chiral derivatizing agents are usually reported as $\Delta\delta^{RS}$ values. The other is that conversion of the (R)-isomer of the MTPA acid into its corresponding acid chloride results in the (S)-isomer of the acid chloride.

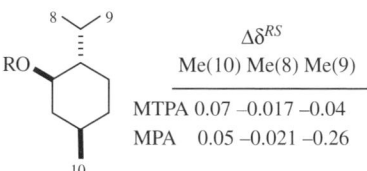

Figure 2.2. $\Delta\delta^{RS}$ values for the methyl resonances of the MTPA and MPA derivatives of (−)-menthol.

The greater the $\Delta\delta^{RS}$ values for the resonances of the substrate. The more certain the assignment of absolute configuration. The magnitude of the $\Delta\delta^{RS}$ values depends on the degree of shielding caused by the aromatic ring and the population of the preferred conformation. Replacement of the phenyl ring of MTPA with larger aromatic rings, such as naphthyl and anthryl groups, enhances the shielding. These larger rings also enhance the population of the *sp* conformation for secondary alcohols. Incorporating substituent groups onto the aromatic ring often either does not have a significant effect on the $\Delta\delta^{RS}$ values[15] or provides some anomalous $\Delta\delta^{RS}$ values due to undesirable alteration of the conformational preference.[16]

Two other variables that have the potential to influence the conformer distribution are solvent and temperature. In a computational study of derivatives of secondary alcohols with phenyl-, naphthyl-, and anthryl-containing derivatizing agents,[15] the similarity of the dipole moments of the *sp* and *anti-periplanar* (*ap*) rotamers indicated that the solvent does not have that much effect on the conformer populations. Decreasing the temperature has the effect of increasing the population of the *sp* conformer for derivatives of secondary alcohols. The improvement of $\Delta\delta^{RS}$ was much larger for MPA than for 1-(1-naphthyl)methoxyacetic acid (1-NMA) and 1-(9-anthryl)methoxyacetic acid (9-AMA), likely because the larger aromatic rings in 1-NMA and 9-AMA already enhance the population of the *sp* relative to the *ap* conformer.[15]

In 1973, when Dale and Mosher reported their pioneering work, most NMR spectrometers were low-field instruments with hydrogen resonant frequencies of 60–100 MHz. As such, proton NMR spectra of many substrates were often second order and definitive assignment of positive or negative $\Delta\delta^{RS}$ values for many protons was not possible. An early advantage of MTPA was the presence of the trifluoromethyl group, which enabled the potential use of ^{19}F NMR spectroscopy to distinguish the absolute configurations. Certain trends were observed that allowed ^{19}F shifts of the diastereomeric CF_3 group to be correlated with absolute configuration. Although these trends were applicable to many compounds, sufficient number of anomalous observations in this one-point analysis have since been noted and urge caution in its application.

In 1991, Kakisawa and co-workers described what many investigators now refer to as the modified Mosher method, which involves the utilization of high-field and two-dimensional NMR techniques to assign many more of the hydrogen resonances of L_1 and L_2 of the substrate.[17–19] Successful application of the method requires observation of a consistent trend wherein the $\Delta\delta^{RS}$ values of L_1 must all show one direction (i.e., negative), whereas those of L_2 are all opposite (i.e., positive). Obviously, if more resonances can be assigned that show the expected negative and positive trends for L_1 and L_2, then the more certain the assignment of absolute configuration. Any inconsistencies in the trends for $\Delta\delta^{RS}$ values of L_1 and/or L_2 bring into question the assignment of absolute configuration.

As already mentioned, there are now a number of reagents that seem to consistently give larger $\Delta\delta^{RS}$ values for substrates than MTPA. Many reports of new reagents compare their effectiveness only to that of MTPA and not to that of the wider range of possible alternatives. Specific details are discussed in the following section for each reagent. These larger values occur because of either greater population of one conformer and/or greater shielding from a larger aromatic ring. In the event that the conformational preference is quite high, and the shielding quite significant, it is possible to have a situation in which only one derivative of the substrate needs to be prepared. Knowing the absolute configuration of the chiral derivatizing agent allows one to position the most shielded group of the substrate *syn* to the aromatic ring, thereby facilitating an assignment of the stereochemistry. Such single derivatization approaches must always be applied with caution, and using $\Delta\delta^{RS}$ values from the derivatives of both enantiomers of the chiral derivatizing agent is likely to provide more reliable results.

Another facet to consider is the ease with which the compounds can be derivatized with the reagent and the extent to which any racemization of the reagent or substrate is avoided. Many investigations have found that (*S*)-MTPA seems less reactive than (*R*)-MTPA. Similarly, many investigators of other Mosher-like reagents report achieving derivitization under milder conditions with shorter reaction times. Specific details are discussed as appropriate in the following sections.

Application of these to polyfunctional substrates requires caution. If the stereogenic centers are far enough removed from each other, it is likely that each aromatic ring of the derivatizing agent only causes shielding of an independent set of hydrogen atoms, allowing multiple stereogenic centers to be simultaneously assigned. When the stereogenic centers are in close proximity to each other, the shifts may reflect shielding from more than one aromatic ring, complicating a straightforward analysis of the positive or negative nature of $\Delta\delta^{RS}$. Examples of the application of Mosher and Mosher-like reagents to polyfunctional substrates are described in the following sections.

2.2. α-METHOXY-α-TRIFLUOROMETHYLPHENYLACETIC ACID (MTPA–MOSHER'S REAGENT) (2.1)

2.1

MTPA, which is often referred to in the literature as Mosher's reagent, has been the most widely used chiral derivatizing reagent for the assignment of absolute configuration by NMR spectroscopy. Its use is described in hundreds of literature reports. The majority of these involve the study of secondary alcohols, but as will be described, there are several other functional groups for which MTPA can be reliably used to assign absolute configuration. It is important to acknowledge that for certain applications in which MTPA might be used, there are now other commercially available reagents that will yield more reliable assignments of absolute configuration. Nevertheless, MTPA is still widely used.

The initial applications of MTPA in NMR spectroscopy by Dale and Mosher in the late 1960s involved its use for the determination of enantiomeric purity.[20–22] When employing chiral derivatizing agents for the determination of enantiomer purity, it is important that no racemization occurs during the derivatization step, that no kinetic resolution occurs such that one enantiomer of the substrate reacts in preference to the other, and that substantial discrimination is observed in the NMR spectrum. In their initial work, Dale and Mosher compared the effectiveness of MTPA to that of several other chiral carboxylic acids including MPA.[21] In addition to the absence of racemization with MTPA, the ^{19}F signal of the CF_3 group provided an advantage when determining enantiomeric purity because of the low field strength spectrometers that were primarily available to researchers at that time. In a subsequent study of secondary alcohols and primary amines, it was noted that MTPA exhibited complete stability toward racemization over long reaction times.[20] Enantiomeric discrimination was noted in the ^{1}H and ^{19}F NMR spectra. Even though the ^{19}F signals were broadened slightly because of long-range coupling to the hydrogen atoms of the methoxy group, the ^{19}F signals of the diastereomers exhibited baseline resolution and were useful for determining enantiomeric purity.

Later on, these investigators appreciated that the conformational preference of MTPA,[13,23] mandelic acid,[13] and MPA[13] could be used as a means to assign absolute configurations based on differential shielding and deshielding by the chiral derivatizing agent. In an early study of 19 ester and 6 amide derivatives with MTPA, 18 of the derivatives showed a clear trend in the ^{19}F signals that correlated with their absolute configuration.

Even at this early time, Dale and Mosher noted how esters prepared from mandelic acid and MPA usually caused greater $\Delta\delta^{RS}$ values in the ^{1}H NMR spectrum than MTPA.[13] Nevertheless, the absence of racemization in derivatives with MTPA[13] and the ease and utility of measuring differences and noting trends in the ^{19}F NMR spectra[23] were advantages of MTPA over the other reagents.

Figures 2.2[24] and 2.3[25] provide comparative $\Delta\delta^{RS}$ values in the ^{1}H NMR spectra of menthol derivatives (MTPA and MPA) and 1-octen-3-ol derivatives (MTPA, MPA,

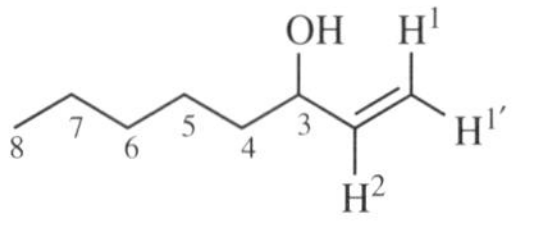

Position	MTPA	MPA	PPA
1	0.07	0.19	0.43
1'	0.03	0.12	0.20
2	0.09	0.14	0.18
3	0.04	0.02	0.03
4	0.03	0.07	0.14
4'	0.03	0.06	0.13
8	0.03	0.04	0.10

Figure 2.3. $\Delta\delta^{RS}$ values for the ^{1}H resonances of the MTPA, MPA, and α-phenylpropionic acid (PPA) derivatives of 1-octen-3-ol.

and α-phenylpropionic acid (2-PPA)), respectively. The larger $\Delta\delta^{RS}$ values with MPA are apparent. A thorough comparison of the effectiveness of reagents such as MPA and 9-AMA to that of MTPA for secondary alcohols is described in the section on those reagents.[26,27]

It was also found that temperature had a significant effect on the $\Delta\delta^{RS}$ values of the MPA derivatives but not those of MTPA derivatives.[28] Lowering the temperature of MPA derivatives of secondary alcohols further enhanced the population of the preferred *sp* conformer over that of the *ap* conformer, thereby increasing the differential shielding effects of the phenyl ring on L_1 and L_2. In fact, changes in the $\Delta\delta^{RS}$ values with temperature of the MPA derivatives ($\Delta\delta_{T_1T_2}$ values) can be used as a diagnostic tool to assign absolute configurations. The computational studies of the MTPA derivatives indicated that there are three preferred rotamers (one *ap* form and two *sp* forms) instead of two as observed with MPA.[27-29] Shielding and deshielding effects of the different rotamers of the MTPA derivatives and a lower population of the most preferred *sp* conformer reportedly account for the smaller $\Delta\delta^{RS}$ values of MTPA compared to those of MPA with secondary alcohols. Lowering the temperature of the MTPA derivatives had a much smaller effect on the $\Delta\delta^{RS}$ values than that observed for MPA because there was less redistribution of MTPA into the preferred *sp* conformer.[28]

2.2.1. Analysis of Secondary Alcohols

The optical purity of 1-*O*-alkyl-3-*O*-benzyl-*sn*-glycerols (**2.2**) has been determined through the use of MTPA derivatives.[30] The method was more effective for determining the optical purity than the use of chiral lanthanide shift reagents. The absolute stereochemistry of steroids such as **2.3** has been assigned using the MTPA derivatives of the C24 hydroxy group.[31] The orientation of the MTPA group relative to the steroid side chain clearly differentiated among the C21 methyl and C26, C27, and C28 hydrogen resonances.

The absolute configuration of the carbinol center of nitrosamine **2.4** was assigned on the basis of $\Delta\delta^{RS}$ values with MTPA.[32] Differential shielding of the 4- and 6-positions on the pyridine ring and the *N*-methyl and *N*-methylene resonances in the (*R*)- and (*S*)-MTPA derivatives were useful in making the assignment. The absolute configuration of the secondary carbinol site in the metal porphyrin **2.5** was assigned on the basis of differences in shifts in the ^{19}F NMR spectrum and $\Delta\delta^{RS}$ values in the ^{1}H spectrum of the MTPA derivatives.[33]

MTPA esters have been used to assign the stereochemistry of a series of 2-substituted cyclopentanols.[34] Both ^{1}H and ^{19}F NMR data were used, and the analysis was complicated because of the conformational flexibility of the cyclopentanol ring. Even with the additional conformers, there were still trends that suggested an orientation of the MTPA group relative to the cyclopentanol ring as illustrated in Figure 2.4. Similarly, analysis of the $\Delta\delta^{RS}$ values for the 2-substituted cyclopentanols of structure **2.6** was complicated due to the mobility of the cyclopentanol ring.[35] Nevertheless, there were clear trends in the values that did permit the assignment of the stereochemistry. The distribution of positive and negative $\Delta\delta^{RS}$ values obtained with the MTPA derivative of 5-deoxy-1,2-dimethoxyarabinose (**2.7**) allowed the assignment of absolute configuration.[36]

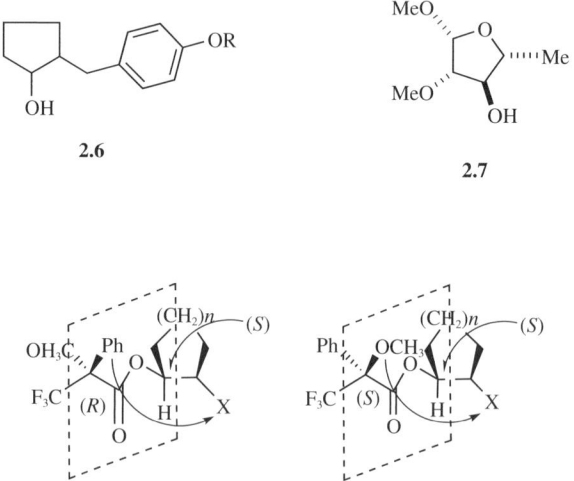

Figure 2.4. Configurations of the MTPA derivatives with 2-substituted cyclopentanols. [Reproduced with permission.[34]]

The utility of using MTPA to assign the absolute configurations of 2- and 3-substituted cyclohexanols has been explored in detail.[37] For the 3-substituted derivatives, the carbon atoms juxtaposed to the phenyl ring were shielded relative to those juxtaposed to the methoxy group, enabling the development of an empirical rule for assigning the stereochemistry. However, the 2-methyl- and 2-phenylcyclohexanol did not follow the rule as steric interactions were believed to distort the preferred conformation. A similar inconsistency in the trends was observed for 3-methylcyclopentanol.

A common strategy when using MTPA to assign the absolute configuration of an unknown is to compare the $\Delta\delta^{RS}$ values to those obtained with known compounds of similar structure. For example, the shifts of the pentakis-(S)-MTPA ester of amphidinolide C (**2.8**) were compared to the MTPA ester of C1–C10 and C17–C29 segments of known configurations.[38] By comparing the shifts of the knowns to those of the unknown, the reliability of the assignment was improved.

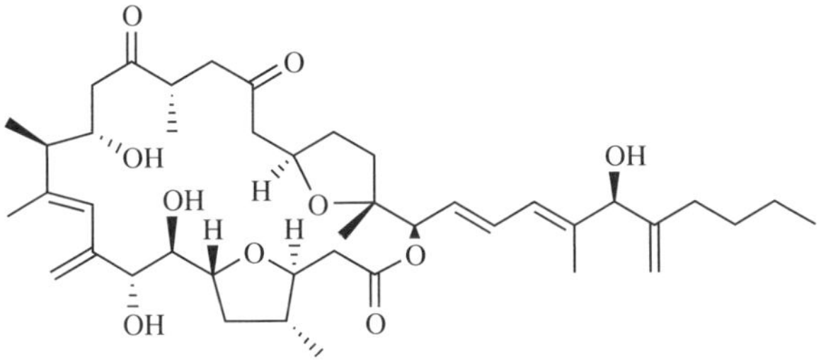

2.8

Another strategy for increasing the reliability of stereochemical assignments obtained with chiral derivatizing agents is to use nuclear Overhauser effect (NOE) data. For example, NOE correlations of the methoxy and phenyl groups of the MTPA with hydrogens of menthol and 1-phenylethanol differ for the configuration of the alcohol.[39] The presence of an NOE between the methoxy hydrogens of the MTPA and the hydrogen atoms of the alcohol in the respective diastereomers could be used to assign the absolute configuration.

MTPA has been used to determine enantiomeric purity and assign the absolute configurations of α-hydroxyphosphonates (**2.9**) using both ^{1}H and ^{31}P NMR data.[40–42] The phenyl ring of the (R)-MTPA adopted a specific orientation relative to the phosphonate, as shown in Figure 2.5,[41,42] that consistently deshielded the ^{31}P resonance of the (S)-alcohol relative to that of the (R)-alcohol. Nonequivalence of the ^{31}P resonances for the two enantiomers was typically in the range of 0.30–1.51 ppm.[41]

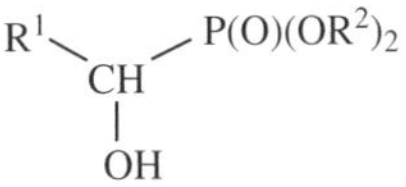

$$R^1\text{–CH(OH)–P(O)(OR}^2)_2$$

2.9

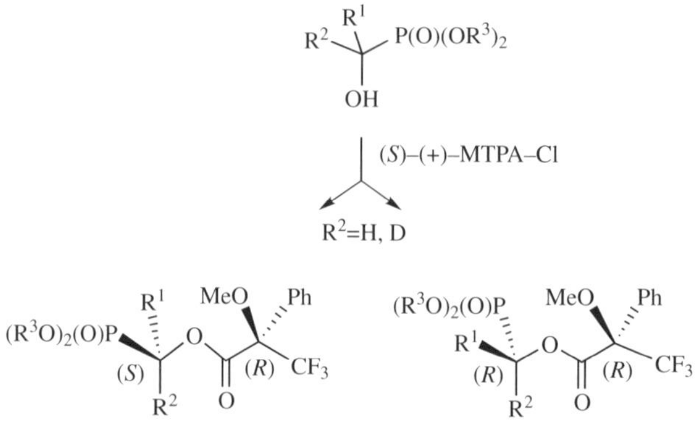

Figure 2.5. Conformational model of Mosher esters derived from α-hydroxyphosphonates. [Reproduced with permission from Elsevier.[41]]

One must always be careful of anomalous behavior that may complicate the assignment of absolute configuration using MTPA derivatives. A series of butenolide compounds, an example of which is illustrated as **2.10**, exhibited some anomalous $\Delta\delta^{RS}$ values for the protons at the C2' position.[43] X-ray crystallographic data showed a clear presence of the Mosher plane; however, there were some derivatives in which the furanone ring adopted different orientations.[43]

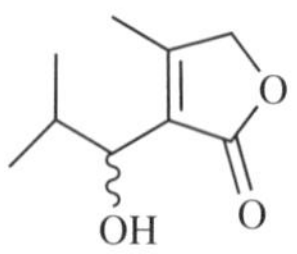

2.10

The absolute configuration predicted using the MTPA derivative of a pyrrolo[2,1-*b*]quinazoline-vasicinone alkaloid (**2.11**) was found to be in error.[44] For **2.11**, the heteroatom adjacent to the carbinol hydrogen altered the conformation of the MTPA group from its predicted orientation, thereby altering the $\Delta\delta^{RS}$ values.

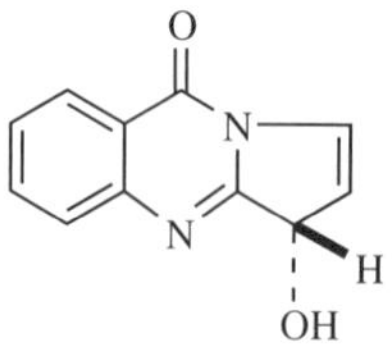

2.11

A series of secondary carbinols with an aromatic ring on the α-carbon exhibited irregular $\Delta\delta^{RS}$ values for the hydrogen resonances of the β-carbon.[45] All other carbons showed the expected trend based on the preferred *sp* conformation. The anomalous results for the β-carbon were caused by additional shielding from the aromatic substituent group at the α-position.

Attempts to apply MTPA to soulattrolide (**2.12**) were inconclusive solely on the basis of ^{1}H NMR data.[46] The differences in shifts between the (*R*)- and (*S*)-MTPA derivatives were too small without enough diagnostic hydrogens to make a definitive assignment. Incorporating the use of molecular mechanics to determine the geometry of the preferred conformation enabled an assignment of the absolute configuration.

2.12

A possible occurrence in larger molecules is the presence of steric compression (Fig. 2.6) that alters the shielding caused by the phenyl ring of the MTPA away from the expected *sp* conformation. The compounds bullatin (**2.13**) and 30(*S*)-bullatin (**2.14**), which are examples of annonaceous acetogenins, exhibit anomalously positive $\Delta\delta^{RS}$ values for H11 because of such steric compression.[47] The presence of steric compression and the potential influence on the assignment of absolute configurations were identified and analyzed in a series of reports.[17–19,48–50] Initially attempting to use the expected trends in the ^{19}F NMR data, the analysis of a marine natural product indicated the wrong configuration.[50] Using ^{1}H NMR data instead, the correct configuration was obtained. For other natural products, it was noted that predictions of $\Delta\delta^{RS}$ trends for the MTPA derivatives worked reliably for compounds with equatorial hydroxyl groups, whereas those with axial hydroxyl groups were prone to steric compression and anomalies in the predicted trends.[17,48]

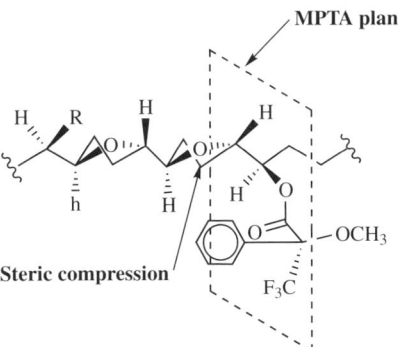

Figure 2.6. Model showing the presence of steric compression in derivatives with MTPA. [Reproduced with permission from Elsevier.[47]]

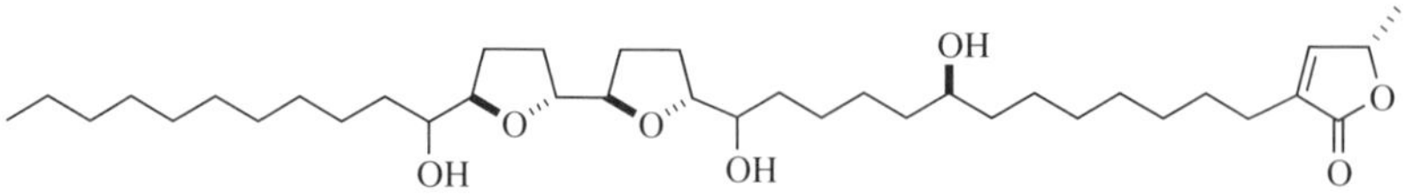

2.13

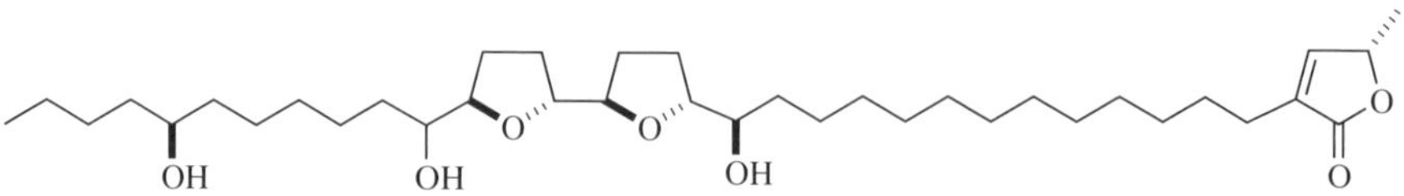

2.14

The most important factor in determining the reliability of assignments of stereochemistry based on ^{19}F NMR data with MTPA esters was the difference in steric bulkiness of substituents on the β- and β′-carbons.[18] Steric repulsion between the phenyl ring of MTPA and the β-substituent was essential in causing the characteristic chemical shift difference between the ^{19}F and methoxy signals. It was also found that using benzene-d_6 as the solvent had an influence on the predicted trends of $\Delta\delta^{RS}$ values due to shielding effects of the solvent, such that chloroform-d or methanol-d_4 were recommended for studies with MTPA.[18,19] This ultimately led these authors to propose what is now often called the modified Mosher method.[17–19]

The modified Mosher method relies on using high-field NMR and two-dimensional techniques to assign and obtain accurate shifts for as many hydrogen resonances as possible in the (*R*)- and (*S*)-MTPA derivatives. The larger number of resonances increases the reliability of assignments of absolute configuration or indicates when anomalous results are obtained.

A review article describing the modified Mosher method, including a description of some of the anomalies that can occur, has been published.[19] In particular, secondary carbinols with a hydroxy group in the axial position are prone to steric compression. Also, the presence of aromatic rings such as furan, phenyl, and naphthyl substituted β to the carbonyl group in compounds such as **2.15** provides additional shielding contributions that influence the magnitude and sign of the $\Delta\delta^{RS}$ values for certain of the hydrogen resonances.[19]

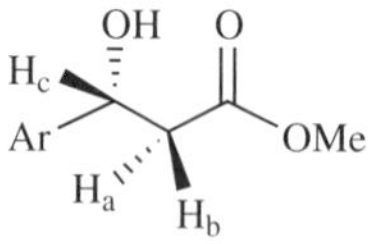

2.15

A review of prior literature findings of the use of MTPA derivatives of α- and β-(hetero)aromatic alcohols has been undertaken.[51] Results on compounds with single aryl, 2- and 3-furyl, pyrrole, thiophene, and 2-, 3-, and 4-pyridine substituents were reported. In particular, the correlation of shifts of the methoxy resonance of the

MTPA with absolute configuration were examined. Of the 50 examples found in the literature at that time, 46 showed the expected trend, 3 showed no difference between the two, and 1 exhibited a contrary result.[51] These findings suggest that the MTPA can be applied quite reliably for determining the absolute configuration of α- and β-(hetero)aromatic alcohols.

2.2.2. Analysis of Secondary Diols and Polyols

Compounds with more than one secondary hydroxyl group can be analyzed with reagents like MTPA, but sometimes present additional considerations if two or more MTPA moieties exert shielding or deshielding influences over the same hydrogen atoms. In compounds where the two secondary hydroxyl groups are far removed from each other and hydrogen resonances near both sites can be assigned, it is often possible to prepare the bis-MTPA derivative and assign the stereochemistry of both sites from the same spectra.[52,53] For example, the absolute configuration about both the C13 and C38 secondary alcohol sites in **2.16** was assigned on the basis of shifts with MTPA.[54]

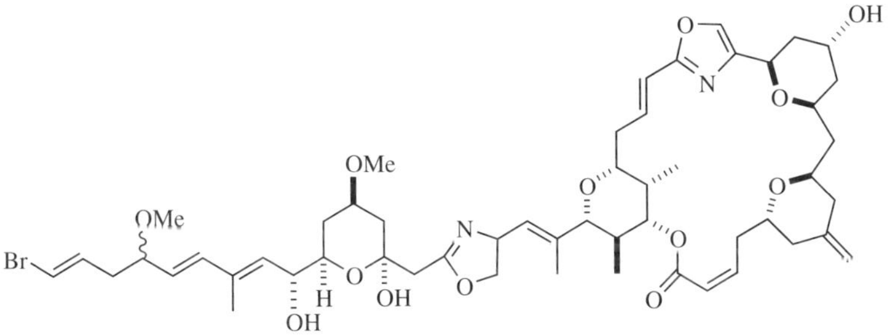

2.16

The stereochemistries about C3 and C8 of **2.17** were assigned on the basis of shifts of the bis-MTPA ester, but the presence of the MTPA group at C8 did influence the magnitude of some of the shifts at C3, as noted from the data for the analogous compound with a hydrogen instead of a hydroxyl group at C8.[55] The bis-MTPA adducts of monodeuterated α,α'-dimethyl derivatives of 1,4-benzenedimethanol (**2.18**),[56] which can exist in either a (R,R)-, (S,S)-, or meso-(R,S)-form, showed conformity to the expected rules and the absolute configurations of the two sites could be assigned.[56] It was also possible to unequivocally assign the (R) and (S) ends of the meso-isomer based on the shifts of the C-methyl groups.

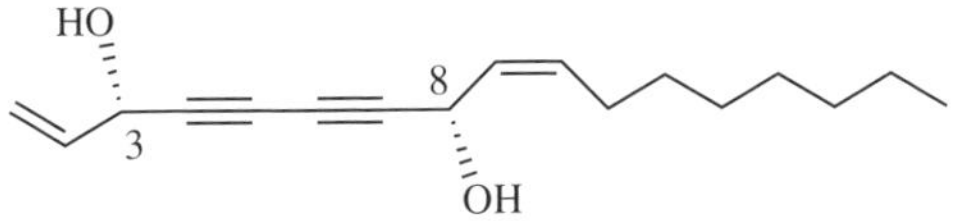

2.17 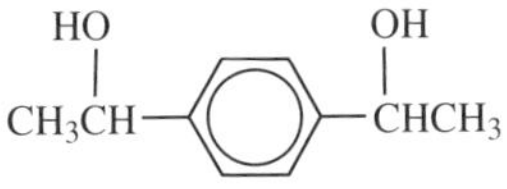   **2.18**

In a study of compounds such as **2.19** with up to four chiral centers,[57] the changes caused by derivatives with MTPA generally occurred as predicted by the model. In the specific case of **2.19**, there were some anomalies in the $\Delta\delta^{RS}$ values of the hydrogen atoms at the C4 and C5 positions because of opposing effects of the two MTPA units. These could be rationalized and the stereochemistry assigned, but they do point out the need for caution when assigning the absolute configuration of nearby stereogenic centers in diols.

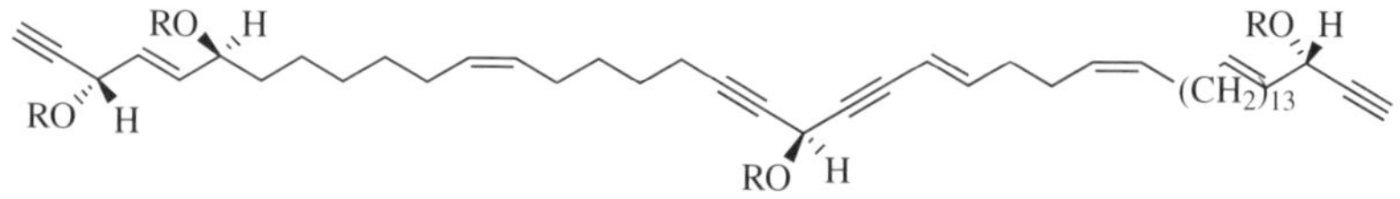

2.19

Similarly, the influence of the MTPA units in the bis-MTPA derivatives of **2.20** occurred cooperatively such that larger shielding and much larger $\Delta\delta^{RS}$ values (0.25 – 0.60 ppm in some cases) were observed relative to what typically occurs with MTPA dertivatives of secondary alcohols.[58] With an understanding of the orientation of the MTPA groups in these derivatives, the $\Delta\delta^{RS}$ values could be used to assign the absolute configurations.[58]

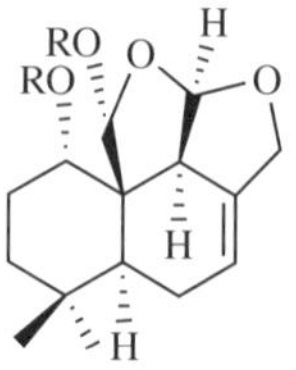

2.20

The stereochemistry at the C24 and C25 diol moiety of **2.21** was determined through the use of the C26 MTPA derivative of the corresponding 22,23-dihydro compound (**2.22**).[59] Model 2,3-dimethylpentane-1,2-diols of known configurations were used to confirm the stereochemical assignment.

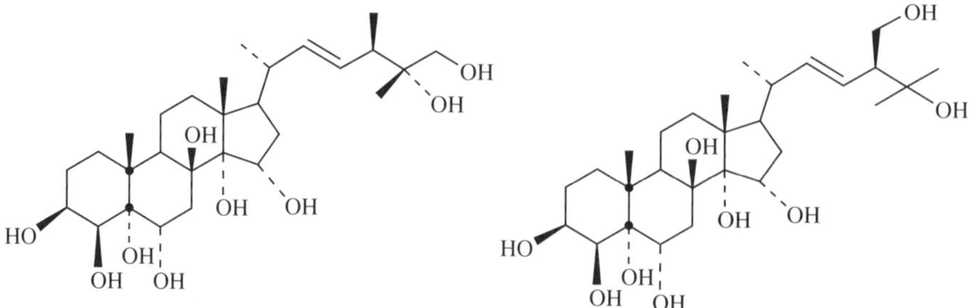

2.21

2.22

The ^{1}H NMR spectrum of the (S)-MTPA diester of **2.23** showed four doublets ((S,S)-diol, (R,R)-diol, and each stereogenic center in the (R,S)-diol) for a nonstereoselective process.[60] A corresponding stereoselective synthesis showed the (S,S)-diol as the major product.

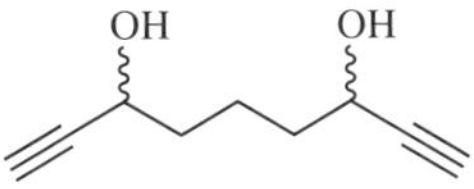

2.23

The application of MTPA in assigning the configuration of adjacent secondary hydroxyl groups has been reported for a number of compounds.[61–67] The nature of the $\Delta\delta^{RS}$ values obtained for the bis-MTPA esters of *syn*- and *anti*-1,2-glycols is quite different, but does show specific trends that can be used to reliably assign the stereochemistry of these units.[66]

The situation is more complicated if the 1,2-glycol has aromatic groups such as phenyl, furyl, or methoxy phenyl rings.[67] In these cases, the aromatic ring of the glycol also causes shielding and has the potential to influence the orientation of the MTPA group. Relatively straightforward assignments predicated on the established MTPA geometry occurred with the *anti*-glycols. The same was not true for mono- and bis-aromatic *syn*-glycols. In these cases, the combined effects caused anomalies in the $\Delta\delta^{RS}$ values that complicated the assignment.[67] These complications did not arise in *syn*-glycols that had simple alkyl substituent groups.

The suitability of using $\Delta\delta^{RS}$ values with bis-MTPA esters to assign the stereochemistry of *anti*-2,3-octanediol and *syn*-2,3-octanediol has been demonstrated.[64,65] The method was confirmed by comparison to the corresponding mono-MTPA derivative.

Consistent trends in the $\Delta\delta^{RS}$ values of vicinal hydroxyl groups were observed in a study of a series of dihydroalkylhexenones (**2.24**) from *Lannea edulis*.[61] The potential for the effects of the adjacent MTPA groups to either reinforce or cancel each other out was noted, and any analysis of the stereochemistry of vicinal diols must be done with care.

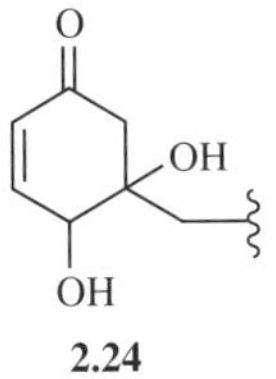

2.24

The stereochemistry about the secondary carbinol in diol **2.25** was assigned on the basis of $\Delta\delta^{RS}$ values of the bis-MTPA ester.[63] It was possible to distinguish and assign the stereochemistry of all four isomers of **2.26** using the bis-MTPA esters.[62] The four diastereomeric MTPA derivatives were separated by high-performance liquid chromatography prior to the NMR analysis.

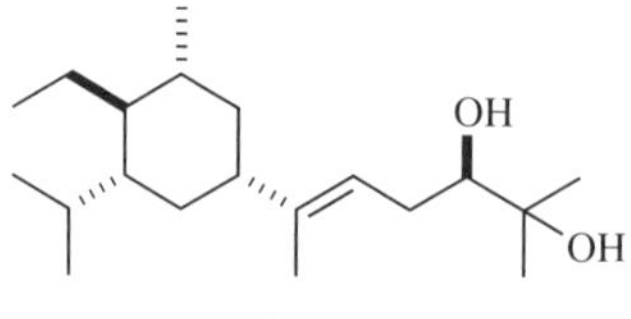

2.25

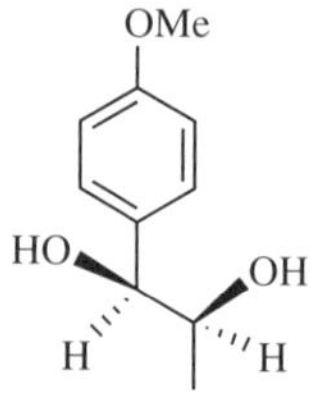

2.26

The trends in $\Delta\delta^{RS}$ values for MTPA derivatives of a series of cis-dihydrodiol metabolites of polycyclic arenes such as **2.27** and heteroarenes **2.28** were successfully used to assign absolute configurations.[68] Splitting of the methoxy resonance of the MTPA was conveniently used to determine enantiomeric purity, whereas there was a consistent trend in the $\Delta\delta^{RS}$ values that was dependent on the (R)- or (S)-configuration of the benzylic center.[68]

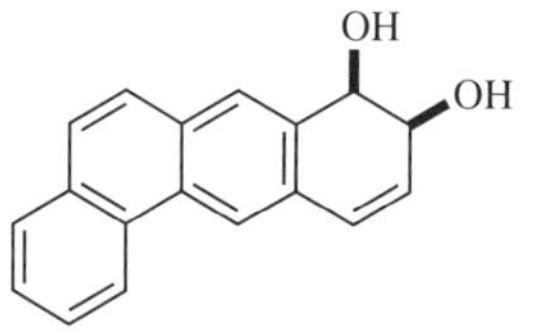

2.27

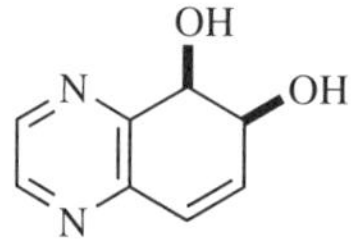

2.28

Application of MTPA in the analysis of cis-dihydrodiols (**2.29**) could not be performed directly on the compounds because they are aromatized during the derivatization step.[69,70] Derivatization with 4-phenyl-1,2,4-triazoline-3,5-dione provided the cycloadduct **2.30**. The bis-MTPA esters of **2.30** were then useful in determining the enantiomeric purity and assigning the absolute configurations. The utility of MTPA was compared to derivatives obtained from (+)- and (−)-2-(1-methoxyethyl)benzene boronic acid for these monocyclic cis-diols.[69] The $\Delta\delta^{RS}$ values in the NMR spectrum of the cis-diols were larger with the MTPA than with the boronic derivative. Assignments of the absolute configuration were also correlated with X-ray crystallographic data.

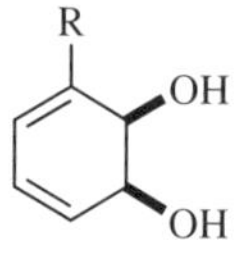

2.29

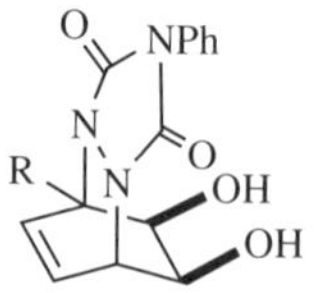

2.30

A detailed analysis of the use of MTPA to assign the stereochemistry of 1,3-diols has been reported.[71] Using compounds with established configurations as models, the trends in $\Delta\delta^{RS}$ values for the syn- and anti-configurations were examined. For acyclic syn-1,3-diols, there was a systematic trend of the $\Delta\delta^{RS}$ values that agreed with expected models of MTPA conformation. For acyclic anti-1,3-diols, the $\Delta\delta^{RS}$ values were irregularly arranged. Coupling of these trends with the circular dichroism exciton chirality method,[72] which works for *anti*-diols but not *syn*-glycols,[73] allowed a reliable assignment of absolute configuration. For cyclic 1,3-diols, the $\Delta\delta^{RS}$ values agreed with predicted trends.[71]

Alternatively, the relative configuration of 1,3-diols (*threo* or *erythro*) can be assigned by conversion to the corresponding acetonides by reaction with acetone and analysis of the shifts in the ^{13}C NMR spectrum.[73] Protection of one of the alcohol groups allowed for the preparation of mono-MTPA derivatives that could be used to assign the stereochemistry.

Another means of identifying the relative stereochemistry of diols or polyols is to use a series of databases established by Kishi and co-workers (see Sections 2.2.12 and 5.5). Having established the relative stereochemistries of the contiguous dipropionate motif shown in the box in **2.31** through the use of the appropriate database, it was then possible to assign the absolute stereochemistry at one site using the corresponding MTPA derivative and then assign all of the other absolute configurations. The utility of this method was demonstrated with oasomycin (**2.31**)[74] and caylobolide A (**2.32**).[75]

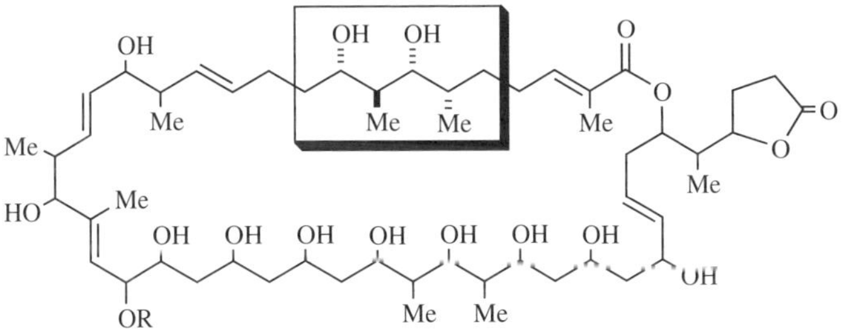

2.31

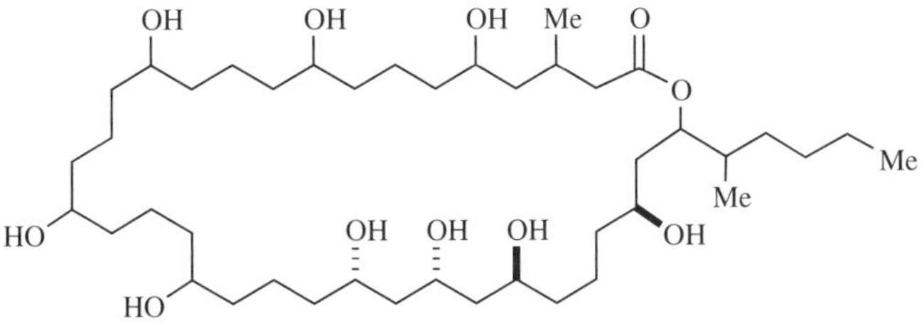

2.32

Another strategy with polyol compounds is to selectively derivatize sites, which is most often accomplished by blocking specific sites before derivatizing the desired site(s) with MTPA. For rostratin A (**2.33**), it was found that lowering the reaction temperature of the reagents in pyridine-d_5 to $-5°$C led to selective derivatization of the C5(5′) sites with no evidence of reaction at the C8(8′) sites.[76] The reaction was performed directly in the NMR tube.

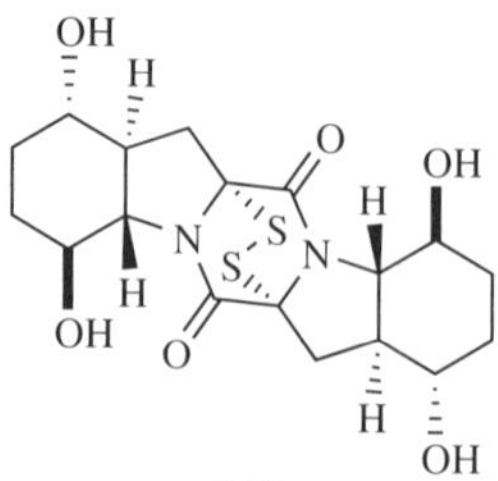

2.33

A family of compounds that have been examined using MTPA derivatives are the annonaceous acetogenins, compound **2.34** being just one example.[77–81] These consist of tetrahydrofuran rings separated by alkyl spacers and usually have several secondary hydroxyl groups. A detailed study on the use of MTPA to assign the stereochemistry of secondary carbinol centers in annonaceous acetogenins has been undertaken.[81] ^{1}H and ^{19}F NMR data were used in the analysis. Assignments of unknown compounds were validated on the basis of a thorough analysis of model compounds with known configurations. The authors urged particular caution in using only ^{19}F NMR data for the determination as this only provides a single point of comparison. Some of the trends in the ^{19}F NMR data depended on quite subtle structural features of the compounds.[81]

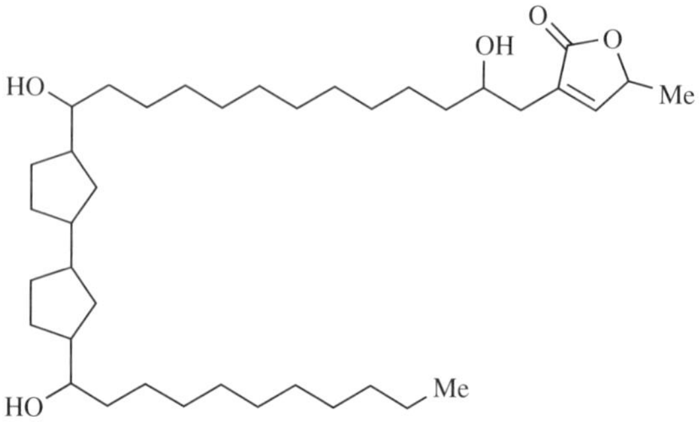

2.34

A strategy for assigning the stereochemistry of sites in annonaceous acetogenins is to prepare cyclic formaldehyde acetals out of the hydroxyl groups that are in 1,2-, 1,4-, and 1,5-configurations.[78] This is advocated as the MTPA moieties of nearby hydroxyl sites in *per*-MTPA derivatives often interfere in the analysis through combined shielding and deshielding effects. After preparation of the acetal, isolated hydroxyl groups were converted to the MTPA derivatives for assignment of stereochemistry. With the absolute stereochemistry of the remote sites known, the relative configurations of adjacent sites could be determined from coupling constants and used to assign the stereochemistry of the other chiral centers.[78]

Other investigators have used *per*-MTPA derivatives of annonaceous acetogenins to directly assign absolute configurations. For example, the absolute configuration of nonadjacent bis-tetrahydrofurans was assigned using their *per*-MTPA derivatives.[77] The shift data in the *per*-MTPA derivative were compared to studies of a series of mono-tetrahydrofuran species with known configurations. The importance of using model compounds as a guide was stressed because of the potential for more than one MTPA moiety to influence the shifts of resonances of particular hydrogen atoms.[77]

per-MTPA esters have been used to assign the configurations of vicinal diols in annonaceous acetogenins.[80] The strategy first involved assigning the relative configuration as *threo* or *erythro* based on coupling constants. Assignment of the absolute stereochemistry based on $\Delta\delta^{RS}$ values with MTPA was not straightforward because both MTPA units contributed to the shifts of adjacent hydrogen resonances. This was especially so for diol units with a *threo* arrangement. In combination with NOESY data, the authors provided guidelines for determining whether the effects of the two MTPA units are additive or canceling.

A procedure for assigning the absolute configuration of polyols including annonaceous acetogenins using only (*R*)-MTPA *per*-esters has been described.[79] The (*R*)-MTPA ester was used because the (*R*)-MTPA acid chloride was found to react faster and under milder conditions than the (*S*)-MTPA acid chloride. In this case, both enantiomers of the substrate were used, and there were consistent trends in the shifts that correlated with the absolute stereochemistry.

2.2.3. Analysis of Primary Alcohols

The utility of MTPA esters for assigning the stereochemistry of primary alcohols that are chiral at the C2 position has been investigated. In one report, 14 primary alcohols with known stereochemistry were analyzed.[82] The $\Delta\delta^{RS}$ values of the methylene resonances were diagnostic of the absolute configuration, although differences in $\Delta\delta^{RS}$ values of at least 0.1 ppm were judged necessary to make reliable predictions. Assignments were unreliable if the compound had a conjugated group or a consecutive chiral center at C3.[82]

Similarly, the ^{1}H coupling patterns of the α-methylene group of MTPA esters of a series of 2-substituted-1-propanols were found to correlate with the absolute configuration.[83] The (*R*)-MTPA–(*S*)-substrate derivative showed a doublet for the α-methylene resonance, whereas the same resonance in the (*S*)-MTPA–(*S*)-substrate diastereomer appeared as a multiplet. Of the 21 substrates studied, 17 showed a similar trend, whereas 4 did not, which suggests that considerable caution must be used when applying this method to a compound with an unknown configuration.[83]

The use of MTPA to assign the absolute configuration of primary alcohols chiral at the C2 position has been applied to 2,6-dimethylheptanol[84] and **2.35**.[85] The utility of using shifts in the ^{19}F NMR spectra of MTPA esters of a series of sulfonamide esters (**2.36**) to assign absolute configurations of primary alcohols and amines with a chiral center at the β-position was explored.[86] The ^{19}F data did show a specific trend for the (*R*)- and (*S*)-MTPA derivatives that correlated with the absolute configuration at the β-site.

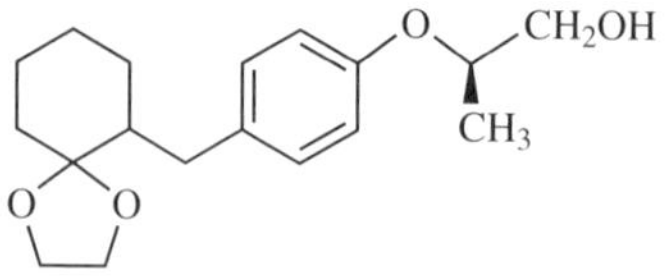

2.35 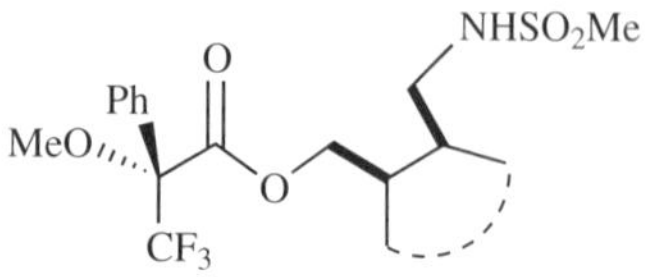 **2.36**

2.2.4. Analysis of Tertiary Alcohols

The use of MTPA has been extended to assign the absolute configuration of tertiary alcohols such as (1S,2R,5R)- and (1S,2R,5R)-pinan-2-ol (**2.37**).[87] The preferred conformation of the (R)- and (S)-MTPA relative to the substrate was ascertained through the use of NOE data. It was then possible to assign the stereochemistry using the predicted sign of the $\Delta\delta^{RS}$ values for different hydrogen atoms of the substrate.

2.37

2.2.5. Analysis of Secondary Amines

MTPA is an effective chiral derivatizing agent for amines. A thorough analysis of the use of MTPA for assigning the absolute configuration of secondary amines has been undertaken.[88,89] The first article in this set examined the use of MTPA with cyclic piperidines (**2.38**) and pyrrolidines (**2.39**).[88] Derivatives of amines were conveniently prepared using the acid chloride of MTPA. Because of slow rotation about the resulting amide bond, the derivatives consisted of a pair of rotamers, each observable in the NMR spectrum. These usually had unequal populations, so the two forms were easily distinguished. Each rotamer also had information that was diagnostic in assigning the absolute configuration. The conformational analysis of these derivatives, which was crucial to the success of stereochemical assignments, is more complex than that for esters. There was not a simple pattern of $\Delta\delta^{RS}$ values with these amine ring systems comparable to the MTPA plane that occurs with esters. Both rotation about the amide bond and conformational changes of the rings had to be taken into account. Several conformational schemes were presented that could be reliably applied to the assignment of the absolute configuration of substituted piperidines and pyrrolidines. An interesting observation was the exceptionally large $\Delta\delta^{RS}$ values (on the order of 0.75–1.00 ppm for piperidines) for several substrate protons. For MTPA esters, $\Delta\delta^{RS}$ values as high as 0.3 ppm are relatively rare. The $\Delta\delta^{RS}$ values for pyrollidines are smaller but still quite significant. The piperidines consistently showed the largest $\Delta\delta^{RS}$ values for the hydrogen nuclei at position C3 rather than C2, which was rationalized based on the spatial relationships in the preferred conformation of the MTPA amide.

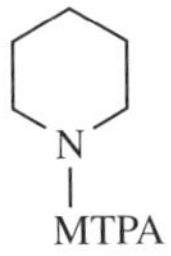

2.38 (piperidine-N-MTPA) **2.39** (pyrrolidine-N-MTPA)

The second article described the application of the conformational rules for configurational assignment to a large number of secondary amines, several of which were of greater structural complexity.[89] Again, the $\Delta\delta^{RS}$ values with the MTPA amides were significantly larger than those with esters. One especially noteworthy example is the $\Delta\delta^{RS}$ value of H2 in **2.40**, which was a remarkable 2.44 ppm. The method was also used with N-methyl tertiary amines. These were first demethylated to form the corresponding secondary amine, which was then analyzed as its MTPA derivative. Steric bulk around the amine can inhibit derivatization with MTPA, and it was observed that the (S)-MTPA acid chloride reacted more slowly than the (R)-MTPA acid chloride. For these derivatives, the shielding by the phenyl ring of the MTPA was so large that it was possible to confidently assign absolute configurations using only one isomer of the MTPA.[89] MTPA is clearly the reagent of choice for the analysis of secondary amines.

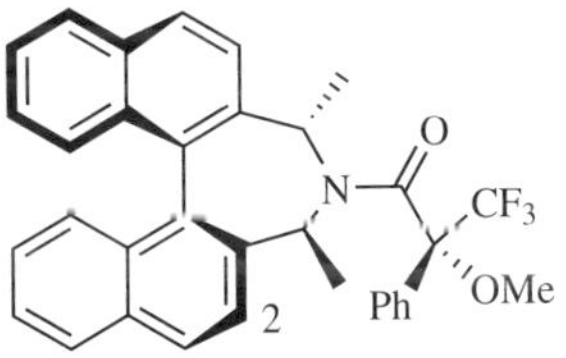

2.40

In a study of a series of compounds such as timolol, acetutolol, metaprolol, and propranolol, which are variants of structure **2.41** that have both a hydroxyl and a secondary amine group, it was found that the acylation of the amine with MTPA occurred in preference to the hydroxyl group.[90]

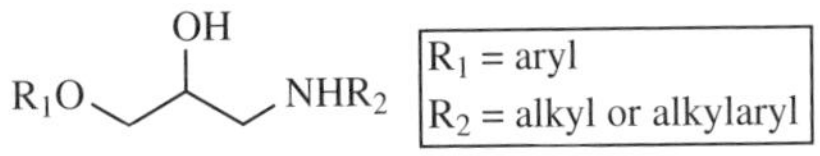

R_1O—CH(OH)—CH$_2$—NHR$_2$

R_1 = aryl
R_2 = alkyl or alkylaryl

2.41

Even with the presence of established trends, care must always be taken when assigning stereochemistries using MTPA derivatives. For example, the MTPA amide of isoanabasane (**2.42**) produced a ratio of *syn-* and *anti-*forms that was unexpected compared to prior studies of amides.[91] In particular, the derivative with (S)-MTPA showed an unexpected predominance of the *syn-*conformer. Effects of the pyridyl and phenyl rings likely accounted for the unexpected conformer populations.

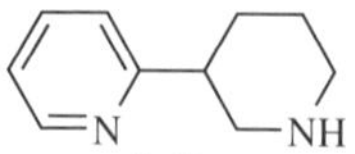

2.42

2.2.6. Analysis of Primary Amines

The enantiomeric purities[92–94] and absolute configurations[95] of primary amines can be determined by analyzing the corresponding MTPA derivatives. A detailed analysis of the conformational preferences of MTPA amides of primary amines with a chiral center at the α-carbon has been undertaken. Computational studies indicated that three main conformers (*sp*, *ap*1, and *ap*3) are preferred because of restricted rotation about the Cα—CO and Cα—phenyl bonds. The preference for the *sp* conformer with MTPA amides was greater than that with MTPA esters, and the shielding of the phenyl ring in the *sp* conformer of the MTPA amide accounted for exceptionally large $\Delta\delta^{RS}$ values when compared to MTPA esters. In fact, the $\Delta\delta^{RS}$ values with MTPA derivatives of primary amines were larger than those with MPA, such that MTPA is the reagent of choice for the analysis of primary amines.[96]

The utility of MTPA for assigning the absolute stereochemistry of the α-carbon of amino acid esters and acyclic amines has been demonstrated.[97] In all cases, the α-proton of the amino acid, methoxy group of the MTPA, and amide NH showed $\Delta\delta^{RS}$ values that correlated with absolute configuration.

The NMR shift data of the MTPA diamide of **2.43**, which has a primary and a secondary amine, were compared to those of a known compound to assign the absolute stereochemistry.[97,98] The bis-MTPA amide consisted of a mixture of four stereoisomers, all of which could be distinguished in the NMR spectrum.

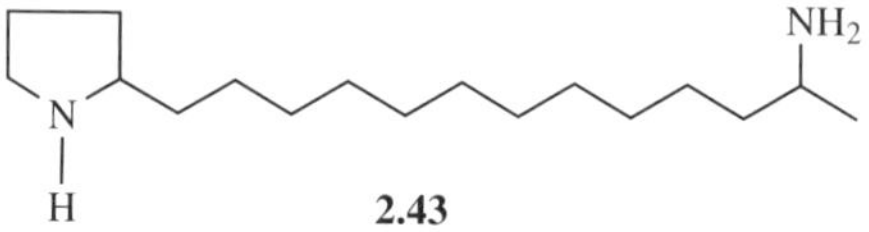

2.43

An analysis of the absolute configurations of the two chiral centers in the amino alcohol **2.44** was facilitated through the use of MTPA derivatives.[99] The amino alcohol moiety was first converted to its corresponding oxazolidone and the coupling constants used to determine whether the adjacent chiral centers had a *threo* or *erythro* configuration. The relative configuration was further confirmed by measurement of NOE data. The shifts in the bis-MTPA derivative were then used to assign the absolute configuration.

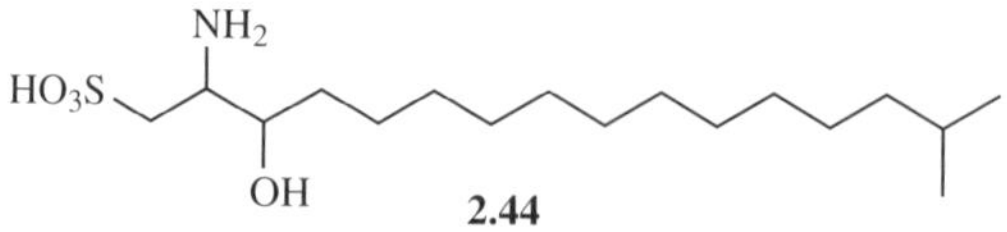

2.44

The utility of MTPA amides for assigning the absolute configuration of compounds with structure **2.45**, which has a chiral center β to the amine, was examined.[86]

A specific trend in the ^{19}F signals with the (R)- and (S)-MTPA derivatives that correlated with the absolute configuration at the β-site was found.

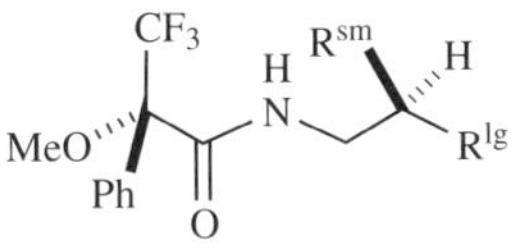

2.45

2.2.7. Use as a Chiral Solvating Agent

MTPA has been used as a chiral NMR solvating agent for the determination of enantiomeric purity of amines. Addition of one of the optically pure isomers of MTPA to an amine in a solvent, such as chloroform-d, leads to the formation of diastereomeric salts that often exhibit different chemical shifts. The acid form of MTPA can be added to a neutral amine or the sodium salt of MTPA to a chiral ammonium ion.[100] The more non-polar the solvent. The larger the chiral discrimination in such systems.

MTPA caused larger chiral discrimination than 2-phenylpropionic acid in the NMR spectra of a variety of amines.[100] Chiral discrimination with MTPA was even noted in methanol-d_4.[100] Other studies have used MTPA to determine the enantiomeric purity of piperidines.[101–103] For trans-4-(4-fluorophenyl)-3-hydroxymethyl-1-methylpiperidine (**2.46**), the discrimination with MTPA was larger in benzene-d_6 than that in chloroform-d.[103]

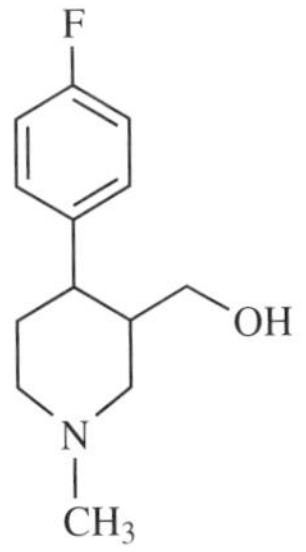

2.46

The utility of MTPA, mandelic acid, and N-(3,5-dinitrobenzene)phenylglycine for determining the optical purity of a series of mono- and diamines was compared.[104] The results varied with different substrates such that no one reagent of the three was consistently preferable to the others. The utility of MTPA as a chiral solvating agent for determining the enantiomeric purity of 8-benzyl-5,6,7,8-tetrahydroquinoline (**2.47**) was compared to mandelic acid, lactic acid, and β-cyclodextrin.[105] The enantiomeric discrimination in the ^{15}N NMR spectrum was similar using MTPA or mandelic acid, but much less with lactic acid or β-cyclodextrin.

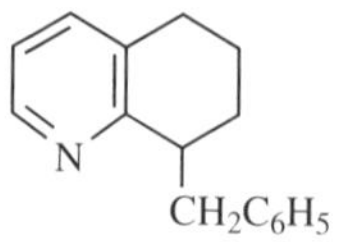

2.47

MTPA induced chiral discrimination in the ^{1}H NMR spectra of several tertiary amines in benzene-d_6 and chloroform-d.[106] In some cases, protons in the same compounds exhibited different senses of nonequivalence in the presence of MTPA. Also, cooling the sample to −45°C led to large increases in the enantiomeric discrimination in the ^{1}H NMR spectrum, presumably because of greater association of the MTPA with the substrate.

2.2.8. Use of MTPA Derivatives with Paramagnetic Lanthanide Chelates

Paramagnetic lanthanide shift reagents have been used with MTPA derivatives to enhance the distinction of the diastereomers. Binding of the paramagnetic lanthanide ion to the derivative occurs through the oxygen atoms of the carbonyl and methoxy groups of the MTPA as seen in Figure 2.7, provided there are no other hard Lewis base functional groups in the substrate. Addition of a lanthanide species such as Eu(fod)$_3$ or Pr(fod)$_3$ (fod = 6,6,7,7,8,8,8-heptafluoro-2,2-dimethyl-3,5-octanedione) causes differential shifts in the ^{1}H or ^{13}C NMR spectra of the two MTPA–substrate diastereomers, thereby enhancing the discrimination in the spectrum.[107–112] The methoxy resonance of the MTPA moiety is conveniently monitored for the determination of enantiomeric purity.

A more significant finding was that the magnitude of the lanthanide-induced shifts often correlated with the absolute configuration of the substrate.[112–118] Using the (R)-MTPA derivative of secondary carbinols, a consistent observation was that the methoxy resonance of the diastereomer with the (R)-isomer of the substrate exhibited the larger lanthanide-induced shift.[113,118] The difference in the lanthanide-induced shift between the (R)- and (S)-MTPA–substrate derivatives correlated with the difference in the steric bulk of the two substituents of the alcohol. The greater the

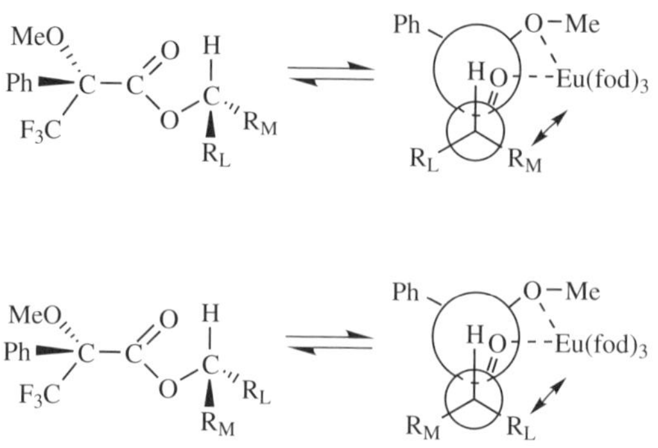

Figure 2.7. Binding of Eu(fod)$_3$ to MTPA derivatives.

difference in the steric bulk of the substituents (R_L and R_M in Figure 2.7 indicating the large- and medium-sized substituents), the greater the difference in the lanthanide-induced shift and more certain the assignment. Several of the compounds had additional heteroatoms and these did not influence the europium binding or disrupt the observed trends of the lanthanide-induced shift. Of the 32 secondary carbinols examined, only 4 produced lanthanide-induced shifts that were counter to the trend and these were primarily with cyclic compounds. If only one enantiomer of the substrate is available, as for a natural product, it was recommended that the lanthanide-induced shifts of both the (R)- and (S)-MTPA derivatives be examined.[113]

Similarly, the influence of Eu(fod)$_3$ on the ester methyl and carbinol methine resonances of (R)-MTPA derivatives of twenty-eight 2- and 3-hydroxycarboxylic acid methyl esters was examined.[116] The resonances of the (S)-isomer of the substrate consistently shifted further than those of the (R)-isomer. There was an evidence that the europium also coordinated to some extent at the carboxymethyl carbonyl group, but this did not alter the observed trend of the lanthanide-induced shift with configuration.

In a study of the MTPA esters of 12 primary carbinols with a chiral center at C2, there was a systematic trend in the lanthanide-induced shift that correlated with the stereochemistry of the chiral center.[114] The difference in lanthanide-induced shifts for the two diastereomers depended on the difference in steric bulk of the two substituent groups attached at the stereocenter.

The correlation of lanthanide shift data with absolute configuration was also studied for MTPA derivatives of 11 axial chiral biaryl compounds, of which **2.48** and **2.49** represent some of the several structures examined.[117] Specific trends were observed for the mono- and disubstituted MTPA derivatives that correlated with the stereochemistry of the compound.

2.48   **2.49**

The absolute configuration of 2-hydroxy-1,1'-binaphthyl was assigned on the basis of Eu(fod)$_3$ shift data of the MTPA derivatives.[119] The findings in this study resulted in a revision of an earlier configurational assignment.

The use of Eu(fod)$_3$ shift data was evaluated with a series of MTPA amide derivatives of amino acid esters.[115] Addition of Eu(fod)$_3$ caused larger shifts of the methoxy resonance of the MTPA than methyl resonance of the derivative of leucine methyl ester, indicating that the europium bound in a bidentate manner to the carbonyl and methoxy oxygen atoms of MTPA. A wide range of amino acids were examined and the methoxy resonance of the (S,S)-MTPA–amino acid pair consistently exhibited larger lanthanide-induced shifts than the (R,S) pair, unless the substituent on the α-carbon atom is a benzyl group or its analogs. Changing the steric bulk of the ester group on the amino acid and the presence of other chiral centers in the amino acid did not alter the general trend of relative lanthanide-induced shifts for the two stereoisomers. If only one enantiomer of the amino acid was available, the absolute configuration could be assigned on the basis of the differential lanthanide-induced shifts in the (R)-MTPA and (S)-MTPA derivatives.[115]

The utility of using lanthanide shift data to assign the absolute configurations of MTPA derivatives of cyclopentanols and cyclohexanols has been examined.[120–123] As the separation of the methoxy signal in the ^{1}H NMR spectrum was insufficient to perform the analysis, the ^{19}F signal of the trifluoromethyl group was used instead.[120,121] With these compounds, an unambiguous correlation between the differences in lanthanide-induced shifts in the ^{19}F spectrum and the absolute configuration was not observed.[121] There were some consistent trends among analogous series of compounds such as the 2- and 3-substituted cyclohexanols.[124] Greater distinction was observed when the hydroxyl group was in an equatorial rather than in an axial position. The lanthanide-induced shifts in the spectra of the cyclopentanol derivatives did not follow a particular trend.[121]

A detailed study of the interaction of Eu(fod)$_3$ with the MTPA esters of cis- and trans-4-tert-butylcyclohexanol, which are in a fixed conformation, was undertaken to gain additional insights into the binding of europium to the derivatives.[122,123] It was found that the europium most likely bonded solely to the carbonyl group of the MTPA rather than in a bidentate manner involving the methoxy oxygen atom. This produced two rotamers, which complicated some of the predictions of the relative lanthanide-induced shifts for the two configurations of the stereocenter.

An examination of the lanthanide-induced shifts in the ^{19}F NMR spectrum of the MTPA derivative of 2-chlorocyclohexanol pointed out further limitations of this method.[125] For 2-chlorocyclohexanol, the accepted rule did not work as the lanthanide-induced shifts of the MTPA derivative of the (1R,2R)-derivative were less than those of the (1S,2S)-derivative. Additional exceptions to the rule were noted in a study of cis-3-tert-butylcyclohexanol, trans-carveol, cis-3-methylcyclohexanol, and α- and β-tetralol.[126] This method must therefore be applied with considerable caution when assigning the absolute configurations of cyclic carbinols.

The utilization of Eu(Fod)$_3$ shift data was evaluated for a series of MTPA derivatives of cyclic and acyclic allylic and acetylenic secondary alcohols (**2.50**).[127] The steric bulk of the substituent groups had a significant effect on the differences in lanthanide-induced shifts. Almost all of the compounds except for the cyclooctenols showed a consistent trend of the lanthanide-induced shifts of the alcohol, although another concern is that a few of the MTPA esters underwent an elimination reaction upon addition of the europium chelate, which is a Lewis acid.[127]

$$R-\underset{H}{C}=\underset{H}{C}\diagdown\overset{R_1}{\underset{}{\diagup}}CHOH$$

$$R-C\equiv C\diagdown\overset{R_1}{\underset{}{\diagup}}CHOH$$

2.50

The utility of adding Eu(fod)$_3$ to (R)-MTPA derivatives of terpenoid, steroidal, and cycloaliphatic secondary alcohols has been reported.[128] In all cases, the ^{1}H methoxy signal of the (R)-carbinol exhibited larger europium-induced shifts than that of the (S)-carbinol.

The europium-induced shifts of the methoxy resonances of a series of MTPA derivatives with eight spiro alcohols and diols, of which **2.51** and **2.52** are representatives, were examined for consistency with the absolute configuration.[129] The geometry of the europium binding with the derivatives could be used to justify the differences in lanthanide-induced shifts. For the diol derivatives, it was necessary to have two essentially equivalent hydroxyl groups for the method to work.

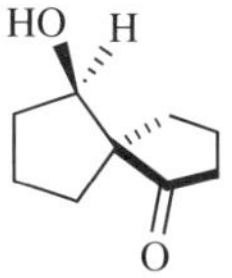

2.51

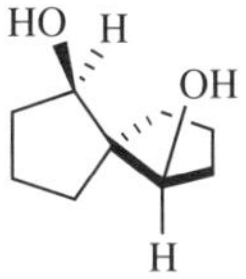

2.52

The Eu(fod)$_3$-induced shifts in the methoxy (^{1}H) and trifluoromethyl (^{19}F) resonances of (R)-MTPA derivatives of bicyclic carbinols such as **2.53** correlated with the absolute configuration.[130] The consistent trend was that the resonances of the (R)-isomer of the substrate exhibited larger lanthanide-induced shifts. The only inconsistencies resulted if the compound had two groups of similar steric bulk.

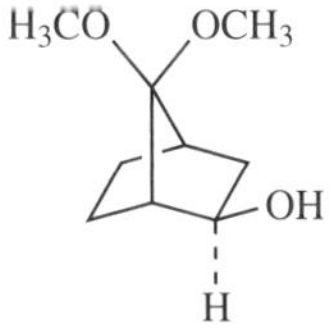

2.53

Application of the relative magnitude of lanthanide-induced shifts to MTPA–substrate derivatives for the purpose of assigning absolute configuration should be done with caution. Comparison to known model compounds is likely warranted. The use of lanthanide shift data along with other evidence such as $\Delta\delta^{RS}$ values to further confirm stereochemical assignments is likely warranted as well, especially if there are inconsistencies in the $\Delta\delta^{RS}$ values.[131–133] Utilization of only the methoxy ^{1}H resonance or ^{19}F signal is based on one comparison, a method that is inherently riskier than using $\Delta\delta^{RS}$ values for several hydrogen resonances of a substrate. Also, paramagnetic lanthanide chelates cause broadening in the spectra that is enhanced by the higher field strength of the instrument. In some cases, this broadening may be significant enough to overcome the beneficial effect of adding the lanthanide species.

2.2.9. Use of MTPA Derivatives with Diamagnetic Lanthanide Chelates

The addition of Ba(II) perchlorate to MPA derivatives of secondary alcohols causes a conformational realignment that leads to characteristic changes in the $\Delta\delta^{RS}$ values. These changes can be used to assign the absolute configuration of the alcohol.[134] Although the addition of Ba(II) to MTPA esters does not have a similar effect, it has been found that the addition of La(hfa)$_3$ (hfa = 1,1,1-5,5,5-hexafluoro-2,4-pentanedione) to MTPA esters caused a conformational change that could be used to confirm the configurational assignment.[135] Binding of the lanthanum to the MTPA derivative likely involves a chelate bond at the oxygen atoms of the carbonyl and methoxy groups of the MTPA moiety as shown in Figure 2.8. The conformational realignment reverses the position of the phenyl ring and hence the shift order of the hydrogen resonances. La(hfa)$_3$ is especially suited to this application because the ligand has only one signal that rarely overlaps with others in the ^{1}H NMR spectrum.

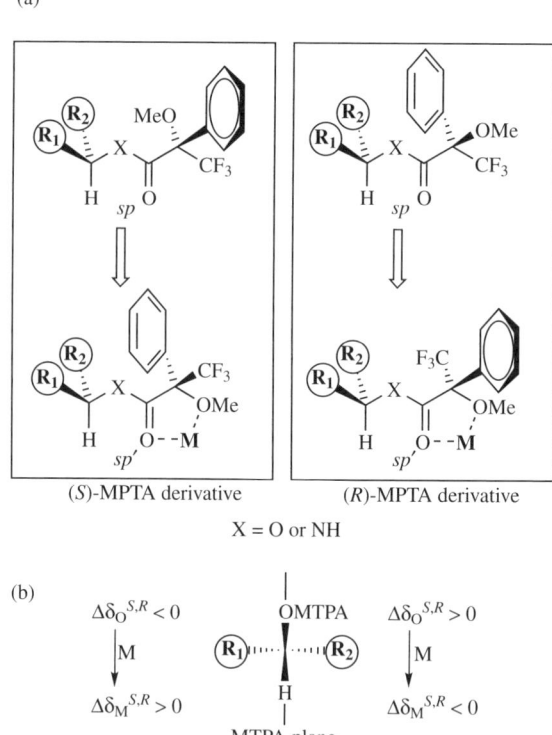

Figure 2.8. (a) Conformational changes of (R)- and (S)-MTPA esters and amides on chelation of a metal ion. (b) The signs of $\Delta\delta_O^{SR}$ and $\Delta\delta_M^{SR}$ on the right and left sides of the MTPA plane. [Reproduced with permission from Elsevier.[135]]

The method requires a substrate with several hydrogen resonances to monitor. The reversal in shift order provides an indication whether the absolute configuration based on the expected structure for MTPA is as thought. The method is predicated on examining the changes caused by La(hfa)$_3$ in the spectra of both the (R)- and (S)-MTPA derivatives.[135]

2.2.10. Preparation of MTPA Derivatives

Procedures have been described for preparing MTPA derivatives of secondary alcohols in pyridine-d_5 directly in an NMR tube using the acid chloride of MTPA.[136] Alternatively, polystyrene-supported carbodiimide resins have been used to facilitate the derivatization and analysis of secondary alcohols[137] and amines.[138] Sample clean-up prior to NMR analysis is easily achieved by filtering out the resin. Still another procedure is to covalently attach the MTPA to a solid resin.[139] The so-called "mix and shake" method also works with MPA and is described in more detail in the section on MPA.

2.2.11. Liquid Chromatography–NMR Spectroscopy of MTPA Derivatives

A microscale liquid chromatography–NMR technique for analyzing microgram range samples of compounds as their MTPA ester derivatives has been described.[140] Using a stopped-flow LC–NMR method, it was possible to obtain the spectra of the derivatives and assign the absolute configurations.

2.2.12. Database Methods with MTPA

An interesting strategy for assigning the stereochemistry of particular structural motifs found within compounds, and especially natural products, is the use of diagnostic databases. One method of constructing a database involves a comparison of ^{13}C and ^{1}H NMR shifts for a particular structural motif. Figure 2.9 shows the structural motif for two contiguous propionate units, and **2.54** is a model compound that contains this structural unit.[141] The eight possible diastereomers of **2.54** were synthesized and the chemical shifts of the ^{13}C and ^{1}H resonances were determined for each individual diastereomer.

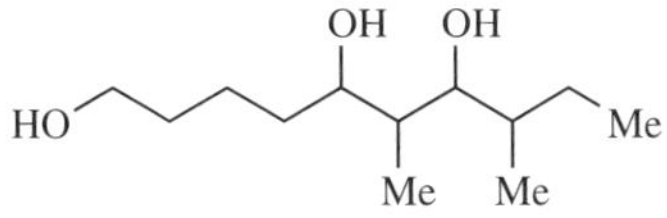

2.54

The chemical shift of a particular carbon or hydrogen resonance in one of the diastereomers was then subtracted from the average value of all eight isomers.

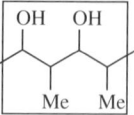

Figure 2.9. Structural motif for two contiguous propionate units.

Figure 2.10 shows an example of the plot of these differences for all eight possible stereoisomers. An important feature is that each of the eight diastereomers shows a different pattern. For the assignment of the stereochemistry of an unknown, the individual ^{1}H and ^{13}C chemical shifts are measured and subtracted from the average value available from the database. A plot similar to that in Figure 2.10 is constructed and the pattern of the eight known configurations that best matches the pattern of the unknown provides the actual stereochemistry. The validity of this method was shown by assigning the stereochemistry of the two contiguous propionate units of oasomycins A and B (**2.31**) and comparing the findings to those obtained by the preparation of MTPA esters.[74] In developing the database, model compounds with other substituents

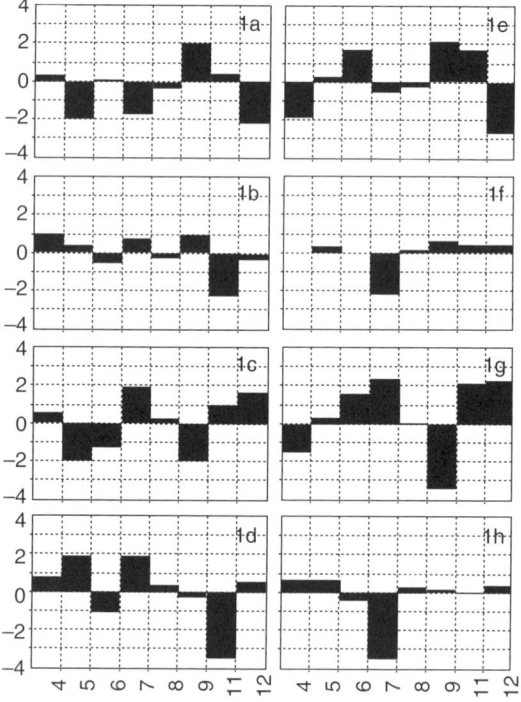

Figure 2.10. Representative plot of the differences in carbon chemical shifts between the average and the values of each stereoisomer (100 MHz, CD$_3$OD). The x and y axes represent carbon number and $\Delta\delta(\delta_{1a-h} - \delta_{ave})$ in ppm, respectively. [Reproduced with permission from American Chemical Society.[141]]

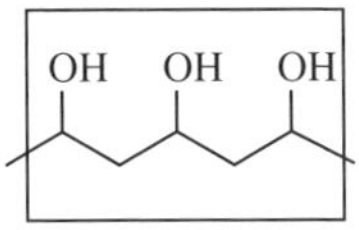

Figure 2.11. 1,3,5-Triol motif.

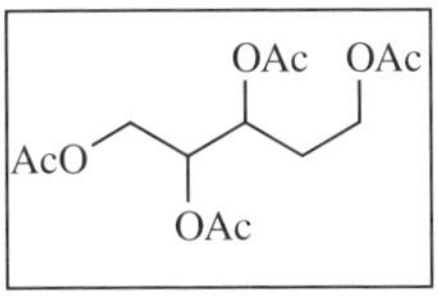

Figure 2.12. 1,2,3,5-Tetraol peracetate motif.

on the side chain were examined to gauge or incorporate their effects. This enabled the development of an empirical formula for incorporating adjustments for a variety of other functional groups.[141] The database for the motif in Figure 2.9 was also constructed in methanol-d_4, dimethylsulfoxide-d_6, and chloroform-d. Even though the resonances of the eight model diastereomers had different chemical shifts in methanol-d_4 and dimethylsulfoxide-d_6, the different profiles were almost identical, recommending the use of either of these two solvents.

In order to predict the stereochemistry of the entire C21–C38 portion of the desertomycin/oasomycin class of natural products, the ^{13}C database described above could be applied to the C29–C33 positions and two new databases, one for a 1,3,5-triol unit (Fig. 2.11) and another for a 1,2,3,5-tetraol peracetate unit (Fig. 2.12), were prepared.[142] The conclusions reached on the stereochemistry of the C21–C38 portion of these compounds were then confirmed by a stereoselective synthesis.[143]

Further analysis showed that the center carbon of the 1,3,5-triol database exhibited a distinctive shift that was dependent on the 1,3- and 3,5-relative configurations, but independent of any units outside of the structural motif.[144] Therefore, using 1,3-diol databases, it was possible to use additivity increments to predict the chemical shift of the central carbon of the 1,3,5-triol unit. This method was then applied to the assignment of stereochemistry of 1,3,5-triol units in several natural products. This particular database method was applied to the assignment of the 1,3,5-triol unit of caylobolide (**2.32**) by measuring the ^{13}C chemical shifts in dimethylsulfoxide-d_6.[75]

Finally, by constructing two additional databases for the structural motifs shown in Figures 2.13 and 2.14, and using the three databases already mentioned, it was

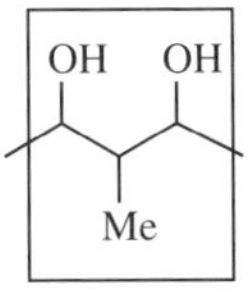

Figure 2.13. 2-Methyl-1,3-diol motif.

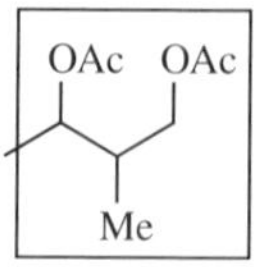

Figure 2.14. 2-Methyl-1,3-diol peracetate motif.

possible to assign the complete stereochemistry of the desertomycin/oasomycin natural products.[145]

2.3. α-METHOXYPHENYLACETIC ACID (O-METHYL MANDELIC ACID-MPA) (2.55)

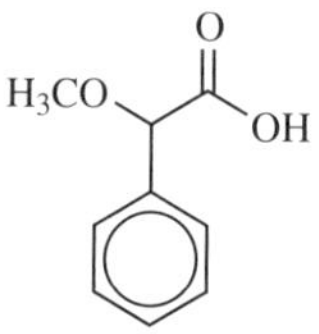

2.55

α-Methoxyphenylacetic acid, also referred to in some articles as *O*-methyl mandelic acid, was first used by Raban and Mislow in 1966 to determine the optical purity of propan-2-ol-1-d_3 and 2,2-dimethylpropan-1-ol-1-*d* by analyzing the ^{1}H NMR spectrum.[146] Dale and Mosher compared the effectiveness of MPA with seven other acids for determining the enantiomeric purity of secondary alcohols.[21] MPA was found to be especially effective among the group of reagents that was evaluated, although at that time the reaction conditions led to considerable racemization of the MPA during the derivatization.[21] As early as 1968, it was noted that the shifts of MPA derivatives of secondary alcohols correlated with absolute configurations, although it was considered too early to fully generalize these findings.[147] In 1972, the use of MPA derivatives in assigning the absolute configuration of amines due to predictable shielding from the phenyl group was first noted.[148]

MPA, MTPA, and mandelic acid were evaluated in Dale and Mosher's classic paper in 1973 on the use of these reagents to assign absolute configurations of secondary alcohols.[13] Although MTPA became the derivatizing agent of choice, it was noted in this early report that the distinction in the ^{1}H NMR spectra of the diastereomers with MPA and mandelic acid was actually greater. Many subsequent comparisons of MPA to MTPA have shown that MPA usually produced derivatives with secondary alcohols that have larger $\Delta\delta^{RS}$ values. One early drawback to the use of MPA was a tendency for racemization to occur during the derivatization process. Better procedures for preparing derivatives with MPA that eliminated the racemization were described in 1986.[149] MPA has been extensively studied for its utility in determining optical purity and assigning absolute configurations.[13,20,146,147,150]

The assignment of absolute configurations with MPA is predicated on the derivatives having a preferred conformation.[13] The nature of the conformation of derivatives with MPA and MTPA has been thoroughly examined.[27,29] Computational methods indicate that derivatives with MPA have two favored conformations (*sp* and *ap*), whereas those with MTPA have three (*sp*1, *sp*2, and *ap*).[14,27] Furthermore, the *sp* conformation with secondary alcohols is more predominant with MPA than with MTPA, which accounts for the substantially larger $\Delta\delta^{RS}$ values.[13] Also, the shielding effects of the phenyl ring in the two MPA rotamers are quite comparable and therefore additive, whereas the shielding and deshielding effects with the three MTPA rotamers tend to counterbalance each other. It is important to note that the preferred conformation with MPA gives the reverse sense of positive and negative $\Delta\delta^{RS}$ values when compared to MTPA.

The nature of the substituent groups in particular substrates has the ability to influence the relative populations of the preferred rotamers with MPA and MTPA, influencing the magnitude of the $\Delta\delta^{RS}$ values and validity of the assigned absolute configurations.[29] It was noted that the shifts of the CαH hydrogen in MPA derivatives of secondary alcohols and primary amines were especially susceptible to the population of the different conformers, such that the more the *sp* rotamer was favored the lower the frequency at which the CαH resonated.[29] This can be put to practical advantage in assigning absolute stereochemistry.

2.3.1. Analysis of Secondary Alcohols

Several other studies have compared the effectiveness of MPA and MTPA for determining the absolute configuration of secondary alcohols.[151–153] For a series of hydroxyiridals, an example of which is **2.56**, after protecting the primary hydroxyl group, the MPA derivative at the secondary hydroxyl group had consistently larger $\Delta\delta^{RS}$ values than the MTPA derivative, thereby providing more confidence in the assignment of absolute configuration.[151] Furthermore, it was found that MPA reacted faster and under milder conditions than MTPA. The MTPA derivative of sarcoglaucol-16-one (**2.57**) was not suitable for assigning the absolute configuration, whereas the MPA derivatives had sufficient $\Delta\delta^{RS}$ values to enable the assignment.[152] Both the ^{1}H and ^{13}C NMR spectra of a variety of secondary alcohols including 1-iodo-3-hydroxy-1*E*-undecene (**2.58**) had significantly larger $\Delta\delta^{RS}$ values with the MPA than with MTPA derivatives.[153] Differential shielding was observed up to nine bonds removed (2.7 Hz for the methyl resonance of **2.58** at 500 MHz) from the chiral center in the MPA derivative.[153]

2.56

2.57

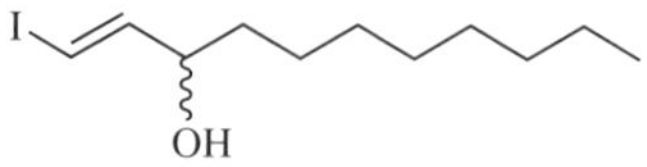

The absolute configuration of lactone **2.59** was assigned on the basis of $\Delta\delta^{RS}$ values with MPA.[154] Distinct shielding of the methoxy methyl group of **2.60** in its derivative with MPA could be used to assign the absolute configuration.[155]

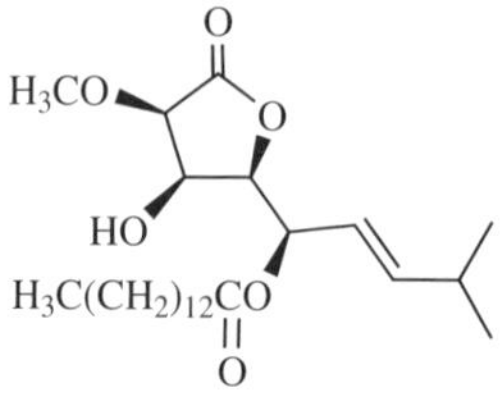

MPA has been used to assign the stereochemistry of viridoxins (**2.61**).[156] Even though the recommended method for preparing the derivatives was used, racemization on the order of 30% with the (S)-MPA and 15% with the (R)-MPA was observed. Even with the racemization, it was still possible to assign the absolute configuration from the ^{1}H NMR spectrum.

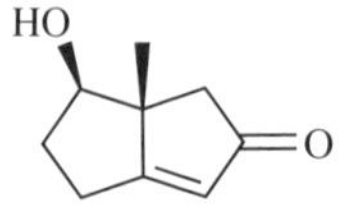

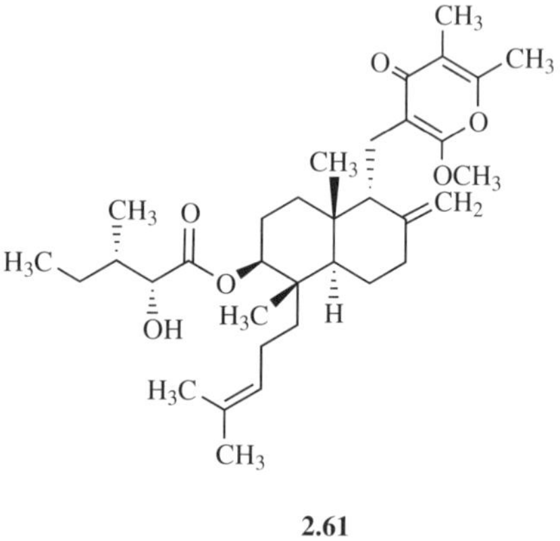

The absolute configurations of a series of aromatic–heteroaromatic carbinols, one of which is **2.62**, were assigned on the basis of $\Delta\delta^{RS}$ values of the MPA derivatives.[157] The $\Delta\delta^{RS}$ values were positive for hydrogen resonances on one of the rings and negative on the other. Derivatives with MTPA provided very little differentiation and were not suitable for assigning the absolute configuration.

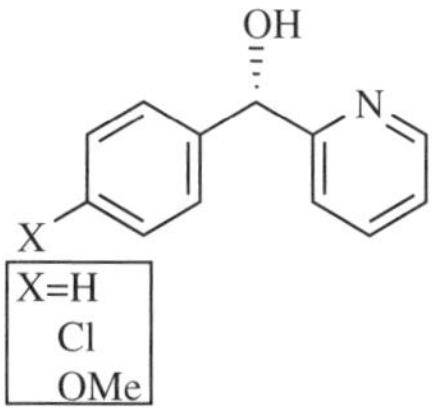

α-METHOXYPHENYLACETIC ACID (O-METHYL MANDELIC ACID-MPA)

2.62

X=H
Cl
OMe

The aldehyde resonance of **2.63** exhibited specific shifts due to shielding from the phenyl group of its MPA derivative that could be used to determine the stereochemistry.[158]

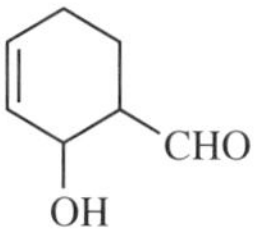

2.63

The MPA derivatives of a series of kamahines (**2.64**) were expected to have a *syn–syn*-coplanar structure of the carbonyl proton of the secondary carbinol, ester carbonyl, and methoxy oxygen.[159] Although the positive and negative nature of the $\Delta\delta^{RS}$ values showed a general consistency with the expected conformation, X-ray and computational data indicated that the conformation was slightly distorted from the ideal. This suggests that applications of MPA to hemiacetal esters require due caution in making stereochemical assignments.

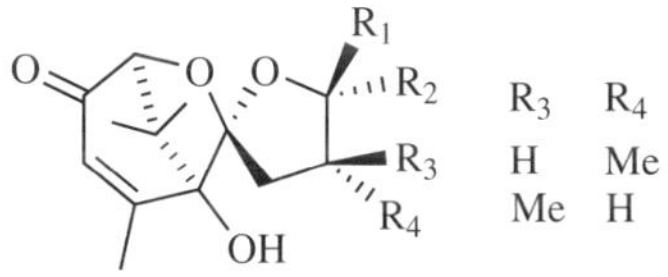

R$_3$	R$_4$
H	Me
Me	H

2.64

Assigning the absolute configurations of cyanohydrins (**2.65**) is not just an extension of the procedure for secondary alcohols because of complicating effects of the cyano group.[160] A thorough analysis of the $\Delta\delta^{RS}$ values in the ^{1}H and ^{13}C NMR spectra of MPA derivatives was performed for cyanohydrins with known configurations. The *sp* conformation predominated and the $\Delta\delta^{RS}$ values for the ^{1}H resonance and ^{13}C resonance of the cyano group correlated with the absolute configuration. Values with 2-(9-anthryl)methoxyacetic acid (9-AMA) were also measured and were larger than those with MPA.[160]

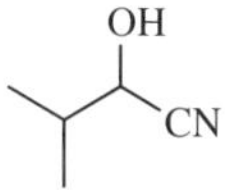

2.65

Several reports have used MPA derivatives to assign the absolute configuration of α-hydroxyphosphonates (**2.66**) with a variety of R groups.[161–163] Both ^{1}H and ^{31}P NMR data were used in confirming the assignment. In particular, following the model from secondary alcohols, the phenyl ring of the (S)-MPA derivative of the (R)-hydroxy phosphonate caused shielding of the phosphorus atom[161,162] with nonequivalence in the ^{31}P NMR spectrum of approximately 0.2–0.6 ppm.[161] MPA has been used to assign the absolute configuration of N,N-disubstituted γ-amino-β-hydroxyphosphonates (**2.67**).[164] The $\Delta\delta^{RS}$ values in the ^{1}H and ^{31}P NMR spectra were used to assign the absolute configuration.

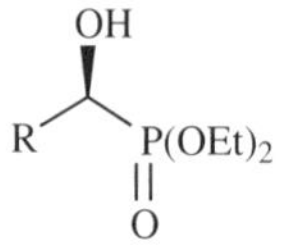

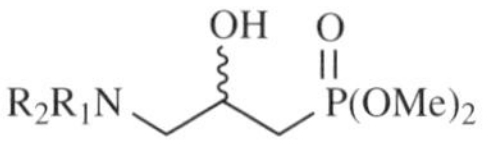

2.3.2. Analysis of Diols

Diols can present a challenge if the two phenyl rings are close enough to each other to simultaneously shield the same protons. If the two hydroxyl groups are far enough apart, as in **2.68**,[165] the absolute configuration about each secondary carbinol center can be assigned on the basis of fully independent shielding effects.[165] Similarly, MPA was also applied to assign the stereochemistry at C11 in the triol **2.69**.[166] The bis-MPA ester of the corresponding diol (**2.70**) was first tested as a model to show that the two MPA units at the primary hydroxyls did not interfere in the assignment. The stereochemistry at C11 could then be assigned in the tris-MPA ester.

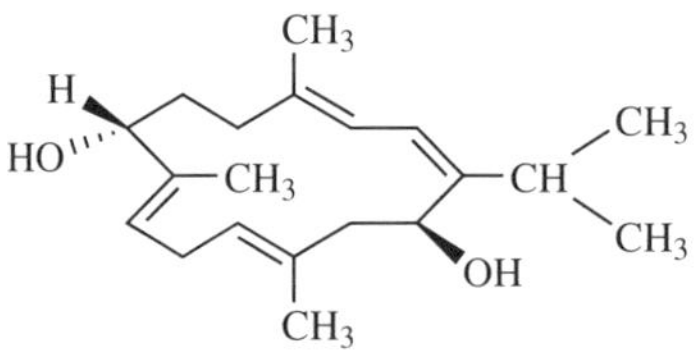

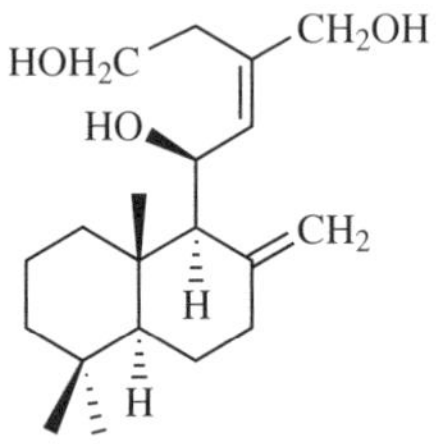

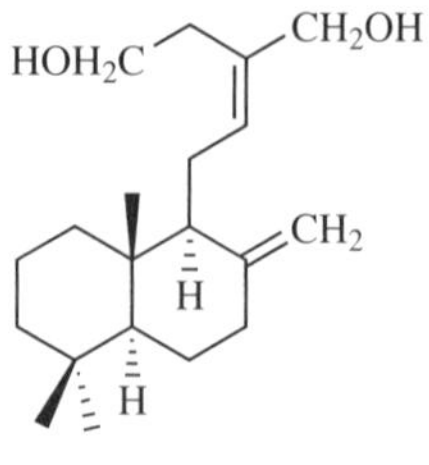

An analysis of the bis-MPA esters of the corresponding diol (**2.71**) of 7,7′-spirobi[3]ferrocenophane-6,6′-dione was successfully used to determine enantiomeric purity and assign the stereochemistry.[167]

2.71

Using bis-MPA esters of 2R,4R-(−)-pentanediol as a model, a scheme was developed to determine the configuration of three chiral carbinol stereocenters in natural products.[168] The first step involved selectively blocking two of the hydroxyl groups and using the MPA ester to assign the configuration of the third. Then, shift data on the tris-MPA ester could be used to assign the stereochemistry of the remaining two.

The analysis of absolute configurations of 1,2-diols is complicated by the combined effects of the two phenyl rings in the bis-MPA derivatives. A detailed analysis of the use of bis-MPA esters for the assignment of absolute stereochemistry of 1,2-diols has been performed.[169] By studying compounds of primary and secondary 1,2-diols with known configurations, and incorporating the analysis of deuterated derivatives, coupling constants, circular dichroism studies, and variable temperature studies, it was possible to determine that the effects of the more populated rotamer dominated the shifts in the spectrum. A graphical model that involved comparing the signs of the $\Delta\delta^{RS}$ values for L_1 and L_2 to that of the methylene proton with the higher $\Delta\delta^{RS}$ value was presented for assigning the absolute configuration.

In a follow-up report, the use of variable-temperature NMR to reliably assign the absolute configuration of 1,2-diols using only one MPA enantiomer was described.[170] Lowering the temperature led to an increase in the population of the preferred *sp* conformer and systematic changes in the shifts of the methylene protons. The $\Delta\delta_{T_1T_2}$ values for either the bis-(*R*)-or bis-(*S*)-MPA derivative could be used to assign the absolute configuration.

A detailed analysis on the use of MPA to assign the absolute configuration of *syn*- and *anti*-1,2-, 1,3-, 1,4-, and 1,5-diols has been reported.[171,172] The $\Delta\delta^{RS}$ values in these different compounds reflected the composite shielding and deshielding effects of the two MPA moieties. By examining the $\Delta\delta^{RS}$ values for known configurations of all of the diols, and seeing a reliability of the correlations over a wide range of substrates, the investigators were able to construct tables of the expected signs of the $\Delta\delta^{RS}$ values for the hydrogen resonances of the different stereoisomers. The procedure for assigning the absolute configuration of a diol with unknown stereochemistry is to prepare the bis-(*R*) and bis-(*S*) esters of the diol and compare the signs of the $\Delta\delta^{RS}$ values to the various possibilities.

2.3.3. Analysis of Primary Alcohols

MPA has been used to assign the stereochemistry of *cis*- and *trans*-6-chloro-9-[2-hydroxymethyl)cyclopentyl]purine (**2.72**)[173] and an α-substituted primary alcohol.[174]

Differential shielding of protons on each side of the hydroxymethyl group could be used to unequivocally assign the absolute configuration.

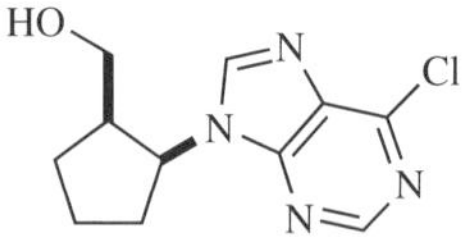

2.72

2.3.4. Analysis of Amines

Soon after MPA was used to analyze secondary alcohols, it was applied in the analysis of chiral amines. As early as 1968, MPA was used to determine the enantiomeric purity of a primary amine (1-(1-naphthyl)ethylamine — NEA) and secondary amine (*N*-methyl-1-(1-naphthyl)ethylamine).[175] The acid chloride of MPA was used to prepare the derivative and no appreciable racemization occurred during the derivatization. The optical purity of deoxyephedrine (**2.73**) was determined through the use of MPA amides.[176] In this case, though, (*S*)-*cis* and (*S*)-*trans* rotational isomers were observed. Warming up the sample sped up the rotation so that it was fast on the NMR timescale and facilitated the analysis of optical purity. The MPA amide was used to determine the enantiomeric purity of 1-ferrocenyl-2-aminopropane.[177]

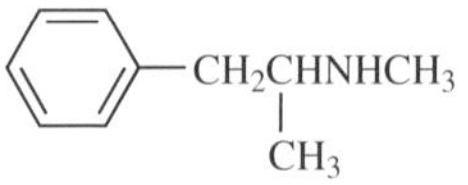

2.73

MPA derivatives can also be used to assign the absolute configuration of α-chiral primary amines.[178] The amide derivatives contrast with those of the esters as the *ap* conformation is preferred (Fig. 2.15).[178] Three reaction methods were described for the preparation of the amide derivative using either the free amine or the amine salt.[178]

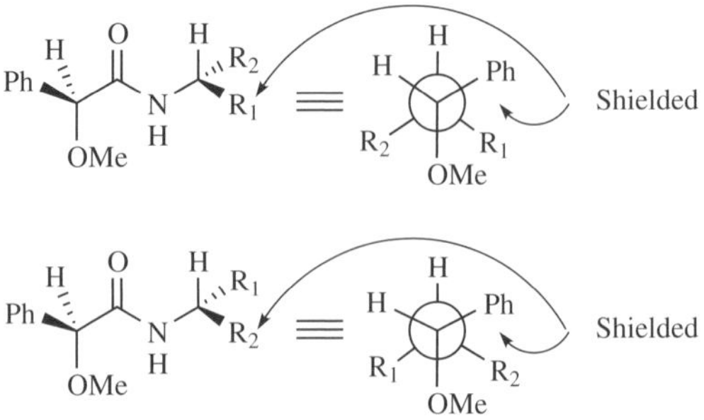

Figure 2.15. Conformational model of the MPA derivatives of primary amines.

No racemization was observed in these reactions. The *ap* conformation provided systematic signs of the $\Delta\delta^{RS}$ values that correlated with absolute configuration.

The $\Delta\delta^{RS}$ values of MPA derivatives of primary amines were very sensitive to solvent polarity.[179] The use of nonpolar solvents stabilized the less polar *ap* conformer causing a substantial enhancement in the $\Delta\delta^{RS}$ values.

A detailed computational analysis of the conformational preference of MPA and MTPA amides with primary amines has been reported.[94,179] The MPA derivative has two preferred conformers (*ap* and *sp*) with a preference for the *ap*, whereas the MTPA derivative has three preferred conformers (*sp*, *ap*3, and *ap*1) with a prefernce for the *sp*. Even though computations showed that the *ap* conformer of the MPA derivative has a higher population than the *sp* conformer of the corresponding MTPA derivative, the position of the phenyl ring in the *sp* conformer with MTPA produced much greater shielding such that the $\Delta\delta^{RS}$ values with MTPA were much larger than those with MPA.[94] MTPA is therefore recommended for the analysis of primary amines.

The analysis of the absolute configuration of 1,2-amino alcohols is complicated by the combined effects of the two phenyl rings in the bis-MPA derivatives. The analysis is further complicated by the *sp*-conformer preference of the ester functionality and *ap* preference of the amide group.[180] A consideration of these conformers led to a clear prediction of the sign of $\Delta\delta^{RS}$ values that correlated with the absolute configuration of *anti*- and *syn*-1,2-aminoalcohols. The study of model compounds with known configurations showed that the trends in $\Delta\delta^{RS}$ values did not change with rotation about the bonds.

2.3.5. Analysis of Sulfoxides

A procedure to assign the absolute configuration of sulfoxides using a covalent complex with MPA has been described.[181] The method involved the conversion of the sulfoxide to a sulfoximine, as illustrated in Figure 2.16. The sulfoximine was then condensed with racemic MPA. The $\Delta\delta^{RS}$ values of the sulfoximine derivative were

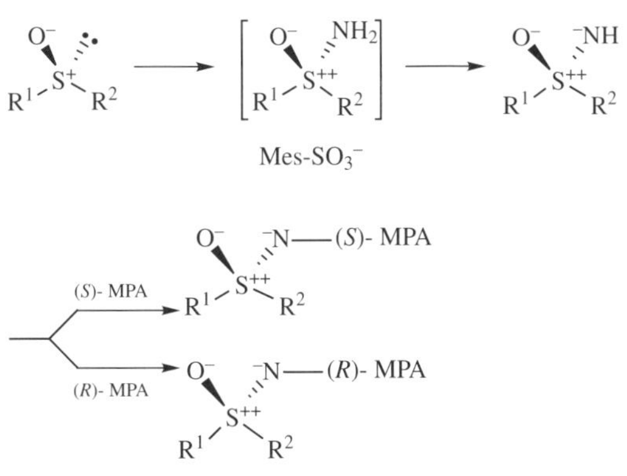

Figure 2.16. Reaction sequence for the analysis of absolute configurations of sulfoxides using MPA.

positive or negative depending on which side of the MPA plane a particular hydrogen atom was positioned. The $\Delta\delta^{RS}$ values could be used to assign the absolute stereochemistry.

2.3.6. Variable-temperature Method for Assigning Absolute Stereochemistry

The influence of rotamer population on the $\Delta\delta^{RS}$ values of MPA derivatives with secondary alcohols led to an interesting and useful observation of the effect of temperature.[27,28,182–185] Lowering the probe temperature with MPA derivatives led to an enhanced population of the *sp* rotamer, which increased the shielding on one substituent group and deshielded the other. Typical probe temperatures were in the range of 175–220 K. A 100 K decrease in temperature with the MPA derivatives of secondary alcohols approximately doubled the $\Delta\delta^{RS}$ values.[182] The effect of decreasing the temperature on $\Delta\delta^{RS}$ values with MTPA, 2-(2-naphthyl)methoxyacetic acid (2-NMA), 2-(1-naphthyl)methoxyacetic acid (1-NMA), and 9-AMA was far less dramatic.[27,182] In the case of 2-NMA, 1-NMA, and 9-AMA, it is likely that the larger aromatic ring already caused a much larger population of the *sp* conformer such that reducing the temperature did not lead to as much of a change in the conformational preference.[182] For MTPA, lowering the temperature did not lead to as great a preference for the *sp* conformer as with MPA, and the more complicated shielding and deshielding effects of the three preferred conformations reduced the effect of lowering the temperature.[27]

The effect of temperature on the chemical shift values of the MPA derivatives of secondary alcohols led to an alternative procedure for assigning absolute stereochemistry.[15,28,182,184] The derivative was prepared with either (*R*)- or (*S*)-MPA. The chemical shifts were recorded at ambient and reduced temperatures, and values of $\Delta\delta_{T_1T_2}$ were determined. The hydrogen resonances on the substituent shielded by the phenyl group showed positive $\Delta\delta_{T_1T_2}$ values, whereas the hydrogen resonances on the other substituent exhibited negative values. An advantage of this method is that only one MPA derivative needs to be prepared. The method was tested on a range of secondary alcohols. Values of $\Delta\delta_{T_1T_2}$ of a few hundredths to approximately 0.2 ppm were obtained and were sufficiently large so that absolute configurations could be reliably assigned.[182,184]

The method was successfully applied to a series of allylic alcohols with cyclopentanol (**2.74**) and cyclohexanol units.[185] Comparing the shifts in methylene chloride-d_2 at 22°C versus −80°C caused upfield shifts of 0.41–0.76 ppm for the shielded hydrogens that were consistent with the model. The H-vinyl resonance was found to be especially suited to the analysis and most diagnostic of the configuration.

2.74

The absolute configuration of a series of acryloins (**2.75**) was assigned on the basis of the $\Delta\delta_{T_1T_2}$ values of the MPA derivatives.[186] In a study of 3-hydroxycarotenoids, the $\Delta\delta^{RS}$ values with MPA and other agents, such as 9-AMA, 1-NMA, and 2-NMA, could be used to assign the absolute configuration. However, the values of $\Delta\delta_{T_1T_2}$ with the MPA derivative of zeaxanthin (**2.76**) at ambient temperature and 215 K did not show the expected trend as all the values were positive.[183]

2.75

2.76

2.2.7. Barium(II) Method for Assigning Absolute Stereochemistry

The addition of Ba(II) to MPA esters caused a systematic change in the $\Delta\delta^{RS}$ values that could be used to assign the absolute configuration.[134] The Ba(II) bound in a chelate manner to the methoxy and carbonyl oxygen atoms of the derivative as shown in Figure 2.17, increasing the population of the *sp* conformer and, for the example

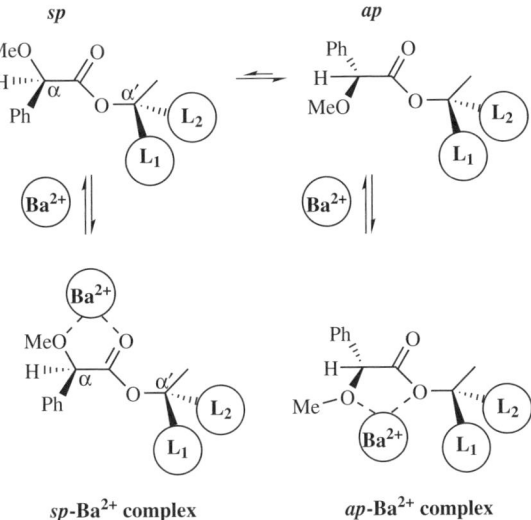

Figure 2.17. Formation of Ba(II) complexes for MPA esters. [Reproduced with permission from American Chemical Society.[135]]

shown in the figure, increasing the shielding of L_1 by the phenyl ring of MPA.[134] The process involved recording the spectrum of the MPA derivative in acetonitrile-d_3, adding solid barium(II)perchlorate to obtain a saturated solution, and recording a second spectrum. Values of $\Delta\delta^{Ba}$ were calculated and were positive for the resonances of one substituent of the substrate and negative for the other. One criterion for the successful utilization of this method is that the substrate cannot have another functional group that will complex with Ba(II).

The addition of Ba(II) has also been used successfully with MPA amide derivatives of α-chiral primary amines.[187,188] Similar to the situation with secondary alcohols, addition of Ba(II) led to an enhancement of the *sp* conformer similar to that shown in Figure 2.17 for the amide derivative. As MPA amides of primary amines show a preference for the *ap* conformer, addition of Ba(II) reversed the original conformer equilibrium and led to $\Delta\delta^{Ba}$ values that were characteristic of the absolute configuration of the amine. Ba(II) was more effective than Mg(II) and Ca(II).

Despite these encouraging reports on the use of the barium method with MPA derivatives, there have been some cautions provided about the general applicability of the method for secondary alcohols and primary amines.[189] For the case of *syn*- and *anti*-γ-oxo-α-amino acids (4-oxo-α-amino acids) (**2.77**), investigators noted that the $\Delta\delta^{Ba}$ values were inconsistent with the proposed model. Furthermore, it was noted that the sign of $\Delta\delta^{Ba}$ for L_1 and L_2 was often the same and basing the assignment on comparative magnitudes of $\Delta\delta^{Ba}$ values was riskier. Also, the diastereomeric ester resonances often shifted in the same direction on the addition of Ba(II). The presence of mobile side chains and another aromatic ring that contributed additional shielding were two particular situations that led to complications in the $\Delta\delta^{Ba}$ values.[189] One other possibility with the 4-oxo-α-amino acids was that the γ-oxo functionality might provide another site for the Ba(II) to complex, although there was no evidence of this observed in the NMR spectrum.

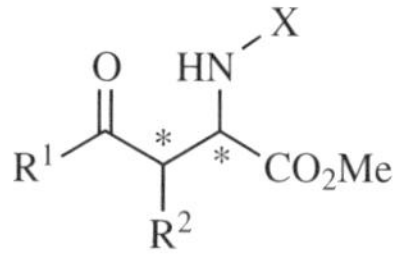

2.77

2.3.8. Use of MPA Derivatives with Lanthanide Chelates

It has also been shown that Eu(dpm)$_3$ (dpm = 2,2,6,6-tetramethyl-3,5-heptanedione or dipivaloylmethane) can be added to MPA derivatives of amines to enhance the discrimination of the methoxy resonances of the two diastereomers.[190] The availability of high-field NMR combined with enhanced broadening of lanthanides at higher fields likely obviates the need for utilization of lanthanide species with MPA derivatives.

2.3.9. Use as a Chiral Solvating Agent

MPA can be mixed with amines to form diastereomeric salts for the purpose of determining optical purity. For heterocyclic β-dimethylamine esters and amines of general structure **2.78** in chloroform-*d*, the extent of discrimination was greater with MPA than with mandelic acid.[191]

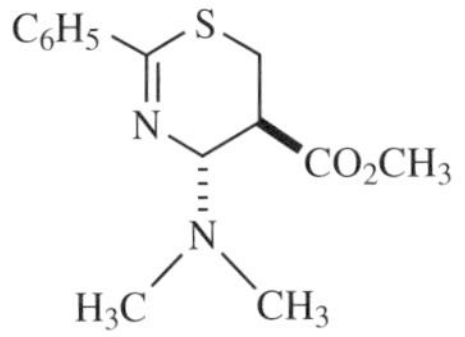

2.78

When coupled to (*R*)-MPA as a salt, the enantiotopic nuclei of the *meso*-isomer of *meso*-di-(*trans*-2-aminocyclohexyl)amine (**2.79**) became diastereotopic and certain of them were resolved.[192] Two-dimensional hetereonuclear multiple quantum correlation spectra and heteronuclear multiple-bond correlation spectra were used to detect proton–carbon coupling and enabled the distinction of the *meso*-isomer from the pair of enantiomers by analyzing differences in the C1/C1′ signals and differences in the H1/H1′ hydrogens coupled to them.[192]

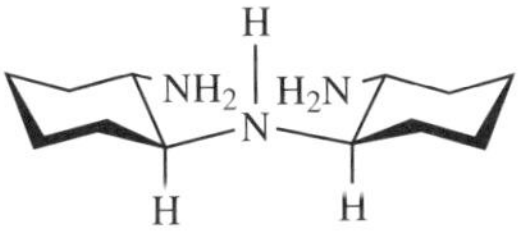

2.79

The enantiomeric purity and absolute configurations can be determined for sulfoxides that form an associated complex with MPA.[193–199] The association of MPA with acyclic sulfoxides occurs as shown in Figure 2.18.[193,197] The location of the phenyl ring was specific enough to cause characteristic shielding and nonequivalence of 0.01–0.03 ppm in the spectrum. The direction and magnitude of the shifts could be used to assign the absolute configuration of the sulfoxide. MPA has been used as a chiral solvating agent to determine the optical purity and absolute configuration of dialkylsulfoxides such as **2.80**.[200] Resolution of the α-sulfinyl diastereotopic

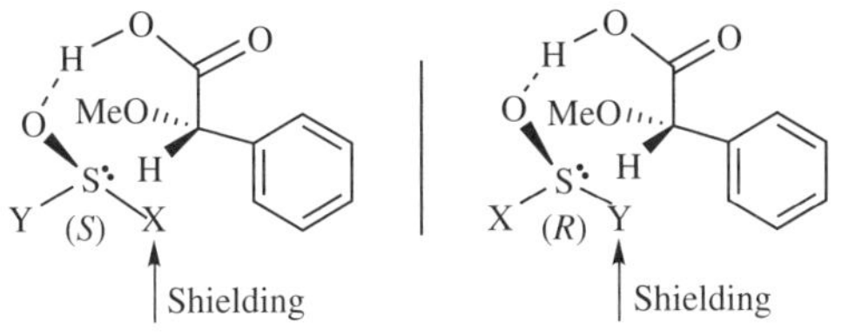

Figure 2.18. Binding model for the interaction of (*S*)-MPA with the two enantiomers of a sulfoxide. [Reproduced with permission from Elsevier.[197]]

hydrogen atoms occurred in the presence of MPA. A derivative of **2.80** with a single deuterium at the α-sulfinyl position was used to assign the resonances of the diastereotopic hydrogen atoms.[193]

$$\text{CH}_3(\text{CH}_2)_6\text{CH}_2-\overset{\overset{\text{O}}{\|}}{\text{S}}-(\text{CH}_2)_5-\overset{\overset{\text{O}}{\|}}{\text{C}}-\text{OMe}$$
2.80

The shifts in the NMR spectra of the compound (R)-dibutyl-d_9-sulfoxide with (S)-MPA were used as a model to confirm the assignment of absolute configuration of biologically produced long-chain sulfoxides of unknown configurations.[196] The analysis was performed in chloroform-d, although it was essential that the chloroform be free of DCl to avoid racemization of the sulfoxide.

Shifts with MPA have been used to assign the absolute configuration of glucopyranosyl sulfoxides such as **2.81**.[195,198] The binding of MPA to the two epimers is comparable to that in Figure 2.18. The structure of the associated complexes of these glucopyranosyl sulfoxides with MPA was confirmed by X-ray crystallographic data.[195,198] Only one enantiomer of the MPA was necessary to assign the absolute configuration.

2.81

In a study of alkyl sulfoxides, it was found that the $\Delta\delta^{RS}$ values in the ^{1}H and ^{13}C NMR spectra were larger in benzene-d_6 than those in chloroform-d.[199] A comparison of the results to 2-NMA showed that larger $\Delta\delta^{RS}$ values were observed with 2-NMA than with MPA.

2.3.10. Preparation of MPA Derivatives–The "Mix and Shake" Method

A convenient method for preparing MPA, MTPA, and Boc-phenylglycine derivatives, dubbed the "mix and shake method," has been devised.[139] The derivatizing reagent was covalently attached to a carboxypolystyrene resin through an anhydride linkage. One equivalent of the substrate, enough resin to contain two equivalents of the derivatizing agent, and a small amount of chloroform-d were added to an NMR tube. The anhydride linkage was hydrolyzed in the presence of a primary amine or secondary alcohol to form the desired derivative. The resin floats in chloroform, so adding extra chloroform to raise the resin out of the recording area of the probe allowed spectra to be obtained without removing the resin from the NMR tube. By using a 2:1 mixture of (R)- and (S)-MPA resin, it was possible to get distinguishable resonances for the (R)-MPA and (S)-MPA derivatives of the substrate for the purpose of determining $\Delta\delta^{RS}$ values. The reactions were found to be quantitative with no kinetic resolution for the compounds that were studied.[139] Barium(II) could also be added to the NMR tube for the purpose of monitoring $\Delta\delta^{Ba}$ values.[139]

The "mix and shake" method was applied to the analysis of the absolute configurations of a series of *cis-* and *trans*-3-hydroxy-β-lactams (**2.82**).[201] The samples had to sit overnight to achieve complete derivatization. The derivatives did provide $\Delta\delta^{RS}$ values that could be used to assign absolute configurations, and the findings from the NMR studies correlated with configurations obtained from X-ray crystallographic structures and chiral liquid chromatographic separations.[201]

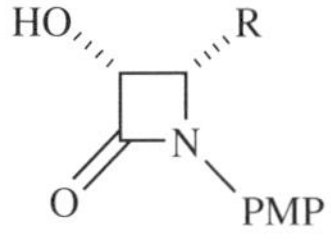

2.82

2.4. MANDELIC ACID (2-HYDROXY-2-PHENYL ACETIC ACID) (MA) (2.83)

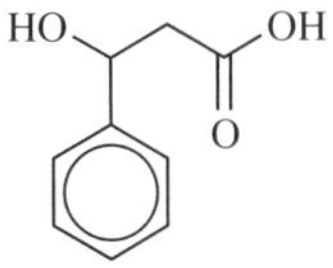

2.83

Mandelic acid (MA) was employed early on as a chiral discriminating agent in NMR spectroscopy for determining optical purity[202,203] and assigning absolute configurations of secondary alcohols.[13] The enantiomeric purity of a variety of secondary carbinols was determined by derivatization with MA and analysis of the ^{1}H and ^{13}C NMR spectra.[204] No racemization or self-condensation of the MA occurred during the derivatization.

The utility of MA, *O*-acetylmandelic acid (*O*-AMA), and MPA for assigning the absolute configuration of secondary alcohols has been compared.[205] The preferred conformation in which the methine proton, carbonyl group, and methoxy oxygen are all *syn* and coplanar allows predictable values of $\Delta\delta^{RS}$ to be correlated with absolute configuration. The $\Delta\delta^{RS}$ values for a variety of secondary alcohols were generally in the order MA > *O*-AMA > MPA.[205] The mandelic acid derivative was prepared by deacetylation of *O*-AMA, and the gain in discrimination likely arises from a higher preference for the *sp* conformer brought on by hydrogen bonding between the carbonyl oxygen and the hydroxyl group. This is similar to the early work of Dale and Mosher in which $\Delta\delta^{RS}$ values with MA were greater than those with MPA or MTPA.[13] It is important to note that derivatives with MA cause a reverse sense of configuration when compared to MTPA.

MA has been used to determine the enantiomeric purity of ethanol-1-*d* through the use of ^{2}H NMR spectroscopy.[206] Natural systems sometimes selectively incorporate ^{2}H into the C1 group of ethanol, resulting in the chiral mono [1-^{2}H]ethanol.[207] The

^{1}H NMR spectrum of the mandelate ester shows distinct resonances of the *pro-(R)* and *pro-(S)* positions. The selectivity of the natural process could be conveniently monitored by ^{2}H NMR spectroscopy.

MA has been used as a chiral solvating agent to determine the optical purity of amines. Addition of MA to amines leads to the formation of organic-soluble diastereomeric salts. Examples of substrates examined with MA include porphyrin-like ligands (**2.84**).[208] Distinct resonances were observed for the four stereoisomeric species in benzene-d_6 and the signals of the *meso* methine protons attached to C5 (C15) and C10 were especially diagnostic. MA has been used to discriminate *N,N*-dimethyl-α-ferrocenylethylamine (**2.85**)[209] and polyamine alkaloids such as **2.86**.[210]

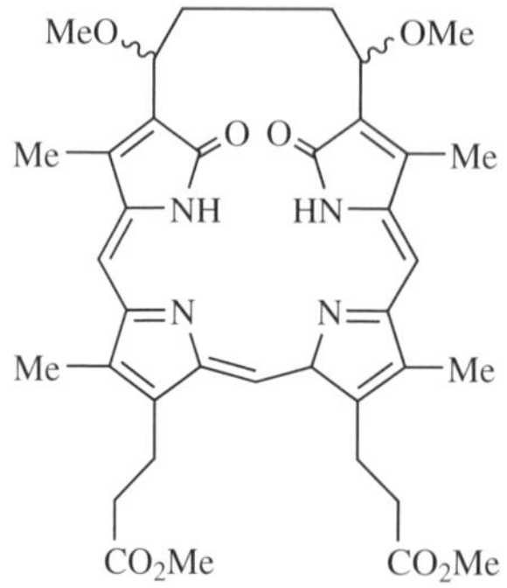

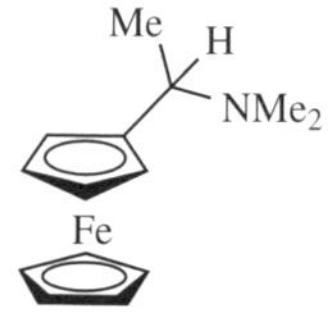

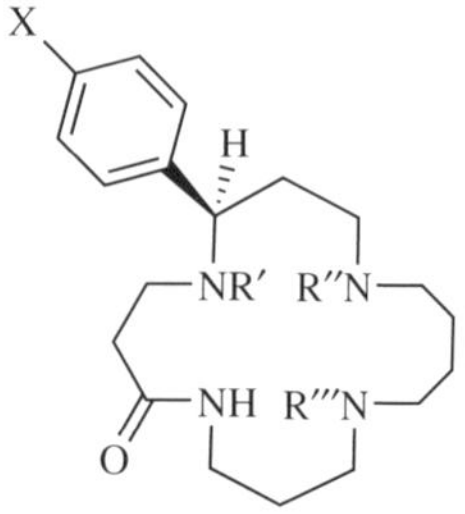

The utility of MA for the determination of optical purity of various monoamine and diamines was compared to that of MTPA and *N*-(3,5-dinitrobenzoyl)phenylglycine.[104] The extent of enantiomeric discrimination caused by the three reagents varied from substrate to substrate such that none offered consistent advantages over the others.

The chiral discrimination in the ^{15}N NMR spectrum of **2.47** with MA was compared to that with lactic acid, MTPA, β-cyclodextrin, and 2,2,2-trifluoro-1-phenylethanol.[105] MA and 2,2,2-trifluoro-1-phenylethanol produced the largest distinction. Larger shifts were observed in less polar solvents, and the extent of discrimination increased at lower probe temperatures.

A detailed analysis of the diastereomeric salts of MA with 1-phenylethylamine, ephedrine (**2.87**), and pseudoephedrine (**2.88**) has been reported.[211] The solution phase studies in dimethylsulfoxide-d_6 were compared to X-ray structures of the crystalline salts. The preferred conformations were stabilized through intramolecular hydrogen bonding between the hydroxyl group and the amine moiety of the substrate.[211]

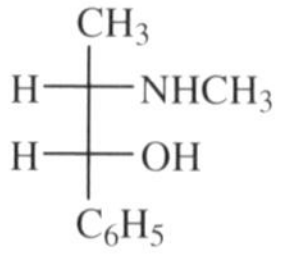

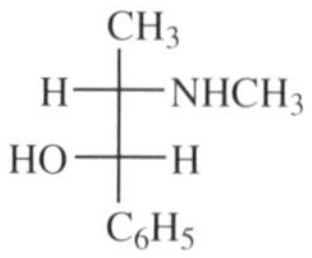

(S)-MA has been used as a chiral solvating agent for phosphine oxides such as **2.89**.[212] A comparison of the shifts in the associated substrate–MA complex to model compounds of known configuration allowed the assignment of the absolute stereochemistry of **2.89**.

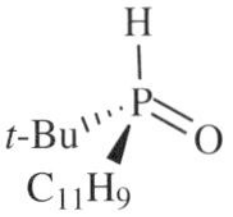

2.89

The enantiomeric Yb(III) complexes of ligands of structure **2.90** could be distinguished in the ^{1}H NMR spectrum in the presence of hydroxyl carboxylate anions such as mandelate, lactate, tartrate, malate, gluconate, and trifluorolactate.[213] The $\lambda\lambda\lambda\lambda$ and $\delta\delta\delta\delta$ enantiomers showed nonequivalent resonances in the presence of the anionic additive. The lactate and mandelate showed the largest discrimination in the ^{1}H NMR spectrum, although the ratio of the two metal enantiomers varied with the identity of the anion.

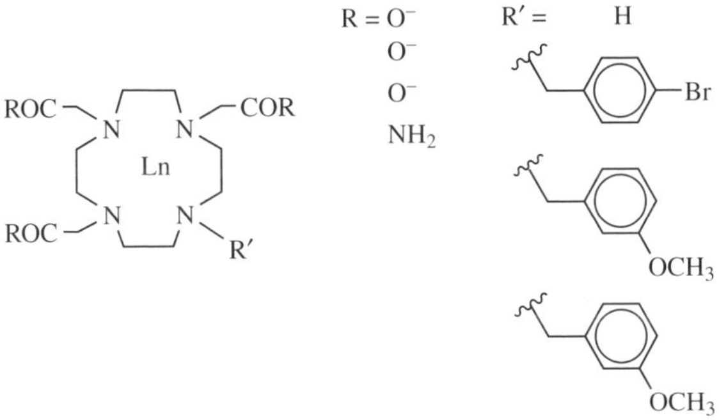

2.90

2.5. O-ACETYL MANDELIC ACID (2-ACETOXY-2-PHENYL ACETIC ACID) (O-AMA) (2.91)

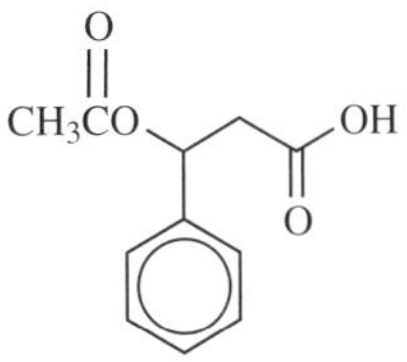

2.91

O-AMA has been used in a variety of studies to determine optical purity and assign absolute configurations. A relatively common application has been to use O-AMA to form diastereomeric salts with amines for the purpose of determining enantiomeric purity.

The extent of discrimination in the spectra of heterocyclic β-dimethylamine esters and amines in chloroform-d was compared for mandelic acid, MPA, and O-AMA.[191] The O-AMA gave larger discrimination than the other reagents for certain substrates. In a study of amines and β-amino alcohols, O-AMA functioned as a more effective chiral solvating agent than MTPA or mandelic acid.[214] The salts with O-AMA were also more soluble in chloroform-d and benzene-d_6. Lowering the temperature of the sample enhanced the population of the preferred conformer, increasing both shifts and extent of enantiomeric discrimination. It was also possible to correlate the chemical shifts in the spectra of the salts with the absolute configuration of the amines.[214] O-AMA has been used to determine the enantiomeric purity of macrocyclic polyamine alkaloids (**2.86**)[210] and albuterol (**2.92**).[215]

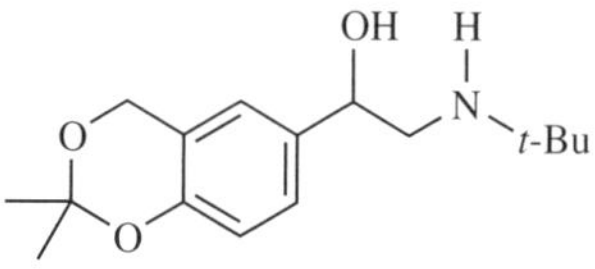

2.92

The utility of O-AMA as a reagent to assign absolute stereochemistries of secondary alcohols has been compared on several occasions to other common chiral derivatizing agents. The effectiveness of O-AMA for a series of mono-2-deuterated isopentenyl alcohols (**2.93**) was compared to that of MTPA and MPA.[216] Six common NMR solvents covering a range of polarities were tested. The best results were obtained with the O-AMA derivative in benzene-d_6 with homonuclear decoupling of the hydrogen at the 1-position. In a comparison of various secondary alcohols, the $\Delta\delta^{RS}$ values with O-AMA were larger than those with MPA but smaller than those achieved with MA.[205]

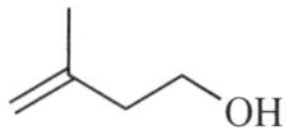

2.93

A comparison of the $\Delta\delta^{RS}$ values in the ^{13}C NMR spectra of 1-octen-3-ol derivatives of O-AMA, MTPA, and MPA resulted in the order MPA > O-AMA > MTPA.[25] Similarly, a comparative study of a series of squalene diols (**2.94**) found that the $\Delta\delta^{RS}$ values ranked in the order 9-AMA > MPA/O-AMA > MTPA.[217] The derivative with MPA was reportedly the most stable of those that were studied.

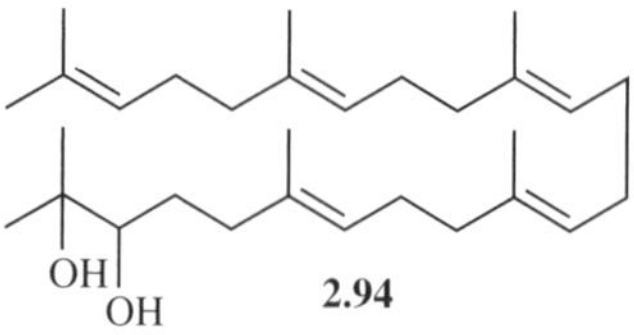

2.94

O-AMA has been used to determine the optical purity of chiral polymers such as **2.95** produced through an aldol polymerization process.[218,219] Certain resonances of the diastereomeric derivatives obtained with O-AMA were distinguishable in the ^{1}H NMR spectrum. Studies on the monomeric species were performed to confirm the findings on the polymers.[219]

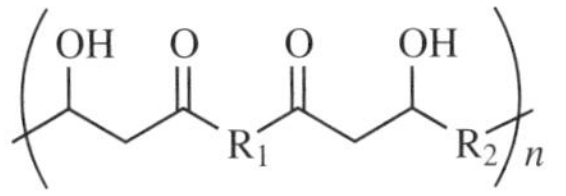

2.95

O-AMA has been used to assign the absolute configurations of crowded and conformationally locked cyclitol derivatives including the protected myo-inositol **2.96**.[220–222] Derivatization of the tetraol derivative (**2.97**) resulted in incomplete reaction and many partially derivatized compounds. Blocking certain hydroxyl groups by conversion to the ketal derivatives allowed preparation of the di-O-AMA derivatives of **2.96**. The use of O-AMA avoided any racemization and provided the desired derivatives of the protected compounds in good yield.[220] The two O-AMA groups produced a characteristic shielding effect in the NMR spectra of aliphatic and cyclic derivatives.[221] Comparison to model derivatives with only one O-AMA group was used to confirm the conclusions. The O-AMA derivative was found to work better for these compounds than MTPA, MPA, and MA.[221] Further study of the X-ray structure of the bis-O-AMA derivative of **2.96** showed that there were two conformations for the O-AMA groups, with one in the expected *sp* conformation and the other in an unexpected *ap* conformation. Presumably, steric crowding accounted for the unexpected conformation, meaning that studies of sterically crowded and conformationally locked alcohols should be done with caution.[222]

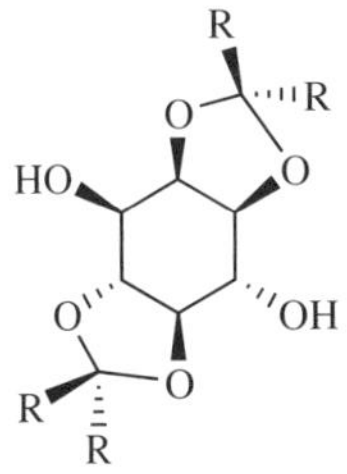

2.96 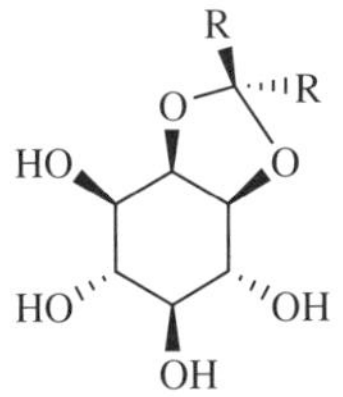   **2.97**

O-AMA was utilized in the analysis of the optical purity of [1,1,2,3,3-^{2}H$_5$]-propane-1,3-diol by analyzing the ^{1}H NMR spectrum of the monoester derivative.[223] The absolute configuration and optical purity of C1-oxygenated *E*-crotyl silanes (**2.98**) were assigned on the basis of the ^{1}H NMR spectrum of the (*R*)-O-AMA derivatives.[224] Characteristic shielding and deshielding from the (*R*)-O-AMA group was used to distinguish the (*S*)- and (*R*)-isomers of the substrate.

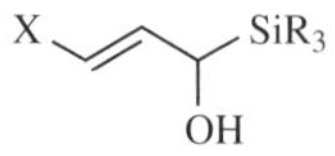

2.6. TETRAHYDROPYRANYL-PROTECTED MANDELIC ACID [(2R)-2-PHENYL-2-[(2S)-TETRAHYDRO-2-PYRANYLOXY] ETHANOIC ACID] (2.99)

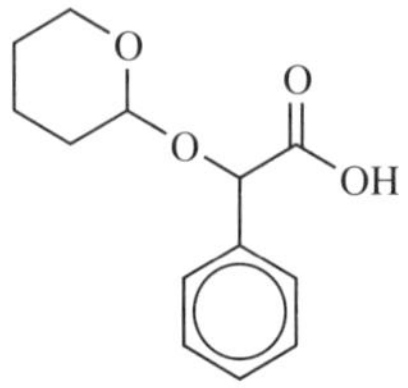

2.99 has been used to assign the absolute stereochemistries and determine the enantiomeric purities of a series of bicyclic and aliphatic secondary alcohols.[225] Unlike MPA, which required more stringent reaction conditions to avoid racemization of the reagent, **2.99** could be used to prepare the ester derivatives under mild conditions. The conformation of the resulting derivative was comparable to those with MPA and other mandelic acid derivatives.[225]

2.99 has also been used to prepare the amide of [1-(2,6-dimethylphenoxy)-2-aminopropane (**2.100**).[226] The derivative with **2.99** reportedly had a *syn*-coplanar arrangement of the NH and carbonyl groups, which was opposite to the *ap* conformation observed with the corresponding MPA derivatives.[226]

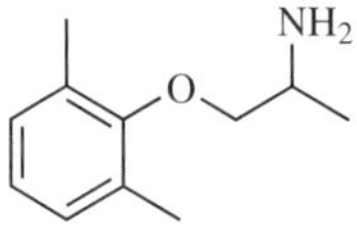

2.7. *O*-NITROMANDELIC ACID (2.101)

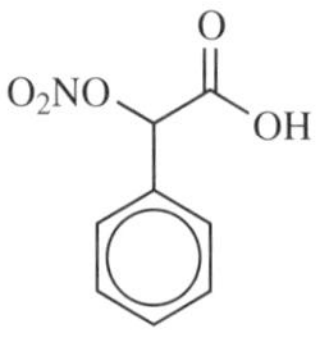

2.101 is readily prepared from optically pure mandelic acid and was examined as a chiral solvating agent to determine the optical purity of several diamines, a sulfoximine (**2.102**), and some alcohols.[227] The extent of enantiomeric discrimination in the ^{1}H NMR spectra was comparable to or better than that with mandelic acid, MTPA, and *N*-(3,5-dinitrobenzoyl)phenylglycine.

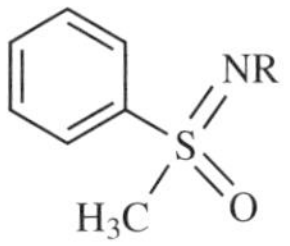

2.102

2.8. 2-PHENYLPROPIONIC ACID (2-PPA) (2.103)

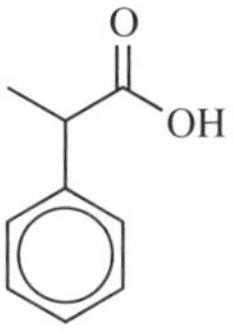

2.103

The use of 2-PPA as a chiral derivatizing agent is of considerable historical significance because it was the original reagent used by Raban and Mislow in the first application of NMR spectroscopy for chiral discrimination.[1] Reaction of the acid chloride of optically pure 2-PPA with 1-(*o*-fluorophenyl)ethanol caused the methyl resonance in the ^{1}H NMR spectrum to split into two doublets, the areas of which could be used to determine optical purity.

Since then, the reagent has been used to assign the absolute configuration of secondary alcohols[228–234] and thiols.[235,236] 2-PPA was used to assign the absolute configuration of thiolactones such as **2.104**.[235] The thiolactone ring was first opened to the corresponding mercapto alcohol, the alcohol converted to an acetate, and the thiol group converted to its thioester using 2-PPA. The shifts show characteristic trends that correlated with the absolute configuration. An analogous procedure has been used to determine the stereochemistry of cyclic lactones in which the lactone ring was opened and the corresponding 2-PPA esters analyzed.[234]

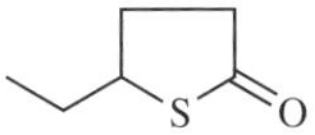

2.104

The absolute configurations of thiols were assigned by preparing thiol ester derivatives with (*R*)- and (*S*)-2-phenylpropionyl chloride and comparing $\Delta\delta^{RS}$ values for the two substituent groups.[237] The method was demonstrated on a variety of chiral thiols.

The differential shielding that occurred in the ^{13}C NMR spectra of the 2-PPA derivative of axial and equatorial cyclohexanols such as 3-trifluoromethylcyclohexanol was used to assign absolute configurations of these compounds.[231]

2-PPA derivatives of sulfenyl fluorohydrins (α-fluoro-α′-sulfenyl alcohols) with secondary hydroxyl groups (2.105) caused $\Delta\delta^{RS}$ values on the order of 0.1–0.1 ppm, which were sufficiently large to reliably assign the absolute stereochemistry.[228] The validity of the assignments based on the preferred conformational preference of the derivatives was confirmed for a few of the substrates by X-ray crystallographic data.

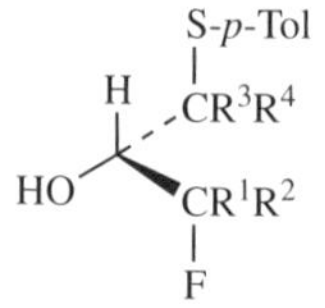

2.105

The $\Delta\delta^{RS}$ values in the ^{1}H and ^{13}C NMR spectra of 1-octen-3-ol were compared for derivatives with 2-PPA, MTPA, and MPA.[25] As seen by the ^{1}H NMR data in Figure 2.3, 2-PPA often provided the largest $\Delta\delta^{RS}$ values.

2.9. 2-METHOXY-2-PHENYLPROPIONIC ACID (2.106)

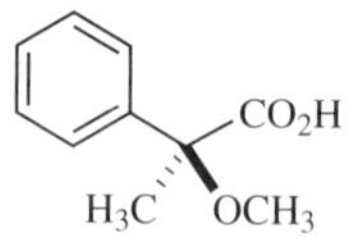

2.106

2.106 has been used as a chiral derivatizing agent for secondary alcohols and amines.[238] The derivatives had a preferred conformation such that the $\Delta\delta^{RS}$ values correlated with absolute configuration. The $\Delta\delta^{RS}$ values were larger than those with MTPA for many, but not all, of the substrates examined. The effectiveness of this reagent was not compared to that of other Mosher-like systems.

2.10. 3-PHENYLBUTANOIC ACID (3-PBA – 2.107)/ 2-PHENYLBUTANOIC ACID (2-PBA – 2.108)

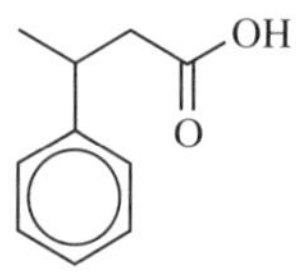

2.107

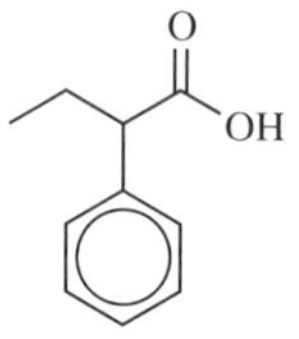

2.108

3-PHENYLBUTANOIC ACID

The (R)- and (S)-enantiomers of 3-PBA have been shown to cause differential shielding in the ^{1}H NMR spectra of norborneols such as **2.109**.[239] Racemization was not observed under the conditions used for the esterification. The $\Delta\delta^{RS}$ values and effectiveness of the reagent were not compared to those of other reagents.

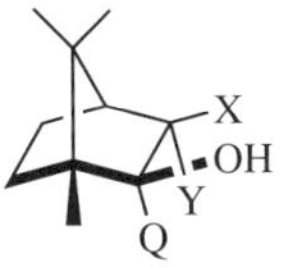

2.109

2-PBA was examined quite early on for its use in assigning the stereochemistry of secondary alcohols and α-chiral primary amines.[240] Assuming a preferred sp conformation, $\Delta\delta^{RS}$ values could be used to assign absolute configurations. The reagent has been applied in limited instances to assign the absolute configuration of secondary alcohols in natural products **2.110**[241] and **2.111**.[242] Reaction of 2-phenylbutanoic anhydride with **2.112** provided the corresponding ester, and the signal of the H7 cis to the oxygen atom at C6 was separated by 0.15 ppm.[243]

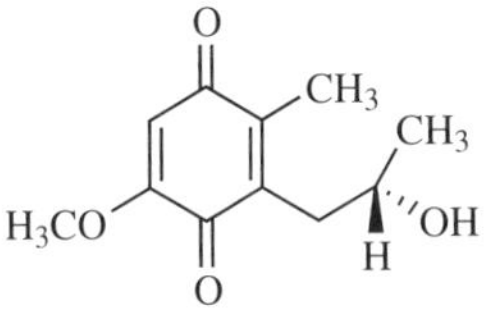

2.110

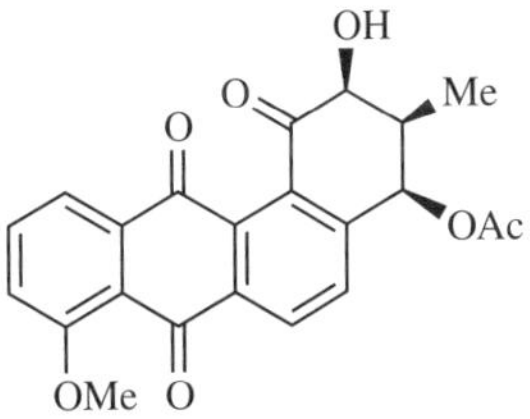

2.111

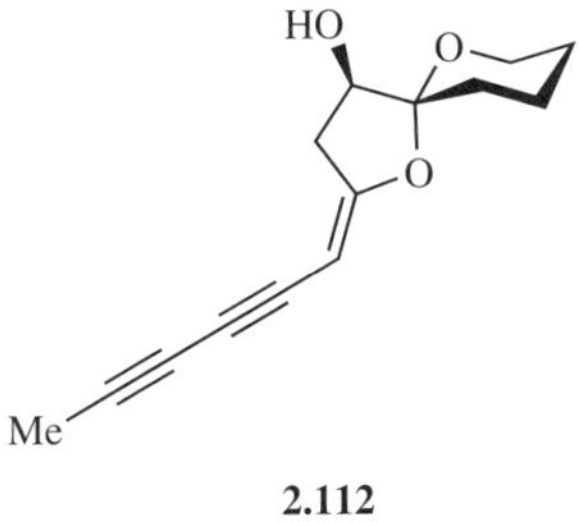

2.112

2.11. α-METHYL-α-METHOXY(PENTAFLUOROPHENYL) (ACETIC ACID) (2.113)

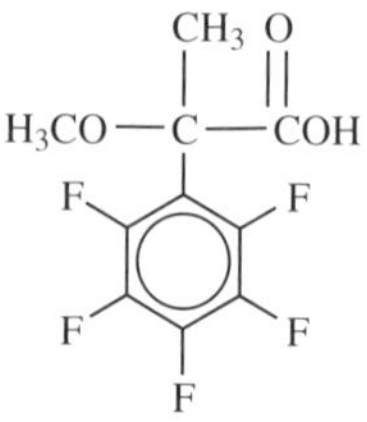

2.113

The utility of **2.113** for the analysis of chiral β-arylethyl amines such as 1-phenylethylamine (PEA) has been explored.[244,245] The derivatives were volatile enough to permit separation by gas chromatography, and discrimination was also noted in the NMR spectra of the diastereomeric products. In the derivative with PEA, the pentafluorophenyl group deshielded the methyl group of the PEA moiety, whereas the phenyl group of the PEA shielded the methyl group of the chiral reagent.

2.12. α-CYANO-α-FLUOROPHENYLACETIC ACID (CFPA) (2.114a)/ α-CYANO-α-FLUORONAPHTHYLACETIC ACID (CFNA) (2.114b)/ α-CYANO-α-FLUORO-*p*-TOLYLACETIC ACID (CFTA) (2.114c)

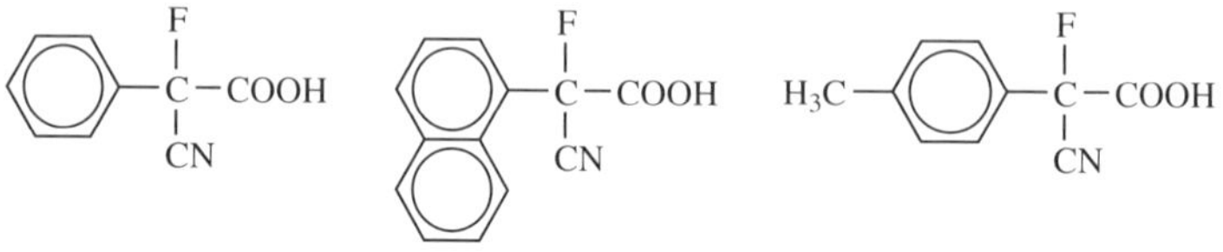

2.114a   **2.114b**   **2.114c**

A series of related reports examined the effectiveness of CFPA[246,247] and the corresponding 1-naphthyl[248] and p-tolyl (CFTA)[249–253] derivatives as chiral derivatizing agents for alcohols and amines. Both ^{1}H and ^{19}F NMR spectra were used for the analysis of enantiomeric purity. Nonequivalence in the spectra of the CFPA and CFTA derivatives was relatively comparable. Compared to the MTPA derivatives, discrimination in the ^{19}F NMR spectra of the diastereomers with CFPA and CFTA was three to six times greater, and in the ^{1}H NMR spectra about two times larger.[246–249] The effectiveness of these reagents was ascribed to the bulkiness of the substituent groups, which enhanced the population of the preferred conformer thereby increasing the shielding caused by the aromatic ring.[247] Another advantage was that these reagents reacted with hindered amines much faster than MTPA, reducing the likelihood of the occurrence of any kinetic resolution. For example, derivatization of 3,3-dimethylbutan-2-ol with CFPA was 500 times faster than that with MTPA.[254]

The ^{19}F signal of CFPA and CFTA shifts considerably upon derivatization such that it is easily distinguished from unreacted starting material.[246] As a result of the greater shielding and conformational preference, the reagent can be used to determine the enantiomeric purity of compounds with remotely disposed chiral centers.[246,247] For example, the enantiomeric purities of compounds **2.115** and **2.116**[246] were successfully determined through the use of CFPA derivatives, whereas the analysis with MTPA was unsuccessful. CFTA was easier to prepare in pure form than CFPA and CFNA because of an enzymatic resolution procedure.[249,252]

2.115

2.116

The utility of CFTA in assigning the absolute configurations of secondary alcohols[250,253] and α-amino esters[251] has been explored. Computational and X-ray crystallographic data showed that derivatives of secondary alcohols with CFTA adopt a conformation in which the C—F group, carbonyl, and carbinyl CH are *syn-periplanar* (Fig. 2.19).[250] The stereochemistry of the secondary hydroxyl site can be assigned on the basis of the positive and negative nature of the $\Delta\delta^{RS}$ values. Alternatively, it was expected and observed that the ^{19}F signal in the derivative of (*R*)-CFTA with the (*S*)-isomer of the substrate should resonate at lower frequency than the ^{19}F signal in the (*R*)-CFTA derivative with the (*R*)-isomer of the substrate.[250] This distinction was

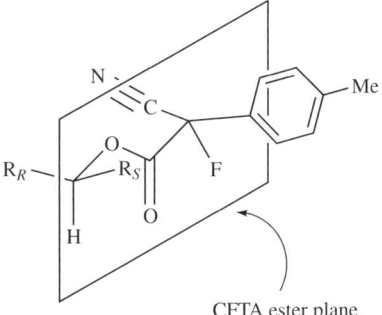

Figure 2.19. Conformational model of the CFTA esters of secondary alcohols.

expected because the (*R,S*)-ester will have a higher preference for the *sp* conformer than the (*R,R*)-ester. The application of CFTA in assigning the absolute stereochemistry of a series of benzhydrols such as **2.117** through the use of $\Delta\delta^{RS}$ values in the ^{1}H spectra has been described.[253] The $\Delta\delta^{RS}$ values of the CFTA derivatives were much larger than those with MTPA. The absolute configuration of monodeuterated benzyl alcohols (**2.118**) was assigned by analyzing the derivatives with CFTA.[253] The signs of $\Delta\delta^{RS}$ for the ^{1}H and ^{2}H resonances were opposite for **2.118** as would be expected.

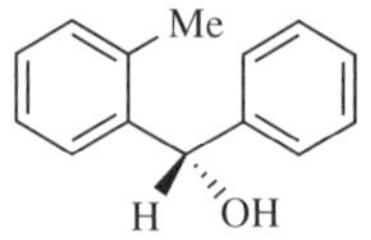

2.117 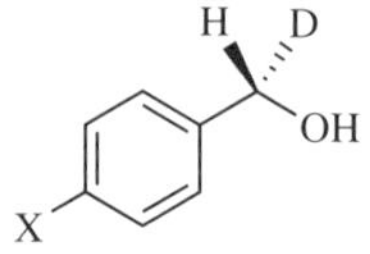   **2.118**

The *ap* conformation was preferred for CFTA amide derivatives of α-amino esters, which is opposite to that observed for CFTA esters.[251] For the α-amino esters, the ^{19}F signal in the (*R*)-CFTA–substrate diastereomer occurred 2–6 ppm to higher field than that of the (*S*)-CFTA–substrate diastereomer. A greater preference for the *ap* conformation in the (*R,S*)-diastereomer compared to that in the (*S,S*)-diastereomer was invoked to explain this trend.

2.13. *N*-BOC PHENYLGLYCINE (BPG) (2.119)

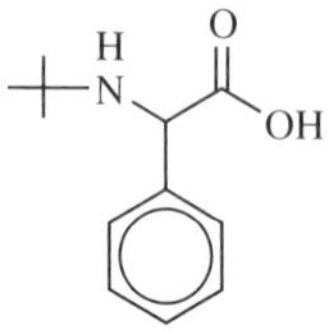

2.119

BPG is an effective chiral derivatizing agent for the analysis of α-substituted primary amines.[255,256] The corresponding amide derivative exhibited a preference for the *ap* over the *sp* conformer. The phenyl ring was coplanar with the Cα—H bond and caused differential shielding of the two substituent groups of the substrate. The $\Delta\delta^{RS}$ values could therefore be used to assign the absolute configuration. Comparisons on a variety of α-chiral primary amines showed that the $\Delta\delta^{RS}$ values with BPG were two to three times larger than those with MTPA and MPA. The conformational preference of the BPG derivatives also reversed the signs of the $\Delta\delta^{RS}$ values relative to MPA and MTPA. The larger $\Delta\delta^{RS}$ values with BPG resulted from the presence of only two preferred conformers and a more favorable orientation of the phenyl ring that produced greater shielding.

BPG was used to assign the absolute stereochemistry of the C3 site in 1,1-dimethyl-1,2,3,4-tetrahydro-6,7-dimethoxyquinolinium (**2.120**) on the basis of the $\Delta\delta^{RS}$ values.[255] The "mix and shake" method described earlier for the preparation of MPA derivatives has also been adapted to use with BPG derivatives of amines.[139]

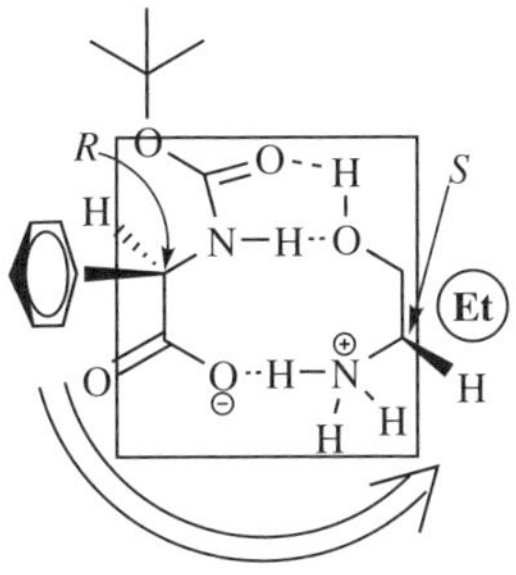

Figure 2.20. Binding model of (R)-BPG with a β-amino alcohol.

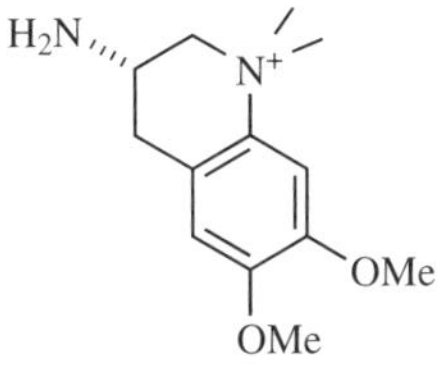

2.120

BPG has been used as a chiral derivatizing agent to assign the absolute configurations of β-amino alcohols and their ethers.[257] The substrate and BPG were simply mixed together in chloroform-d to form the corresponding salt. Ion pairing of the ammonium and carboxylate ions and hydrogen bonding between the hydroxyl group of the substrate and the HN group of the BPG stabilized the structure shown in Figure 2.20 for the (R)-isomer of BPG. The $\Delta\delta^{RS}$ values could be reliably used to assign the absolute stereochemistry. If the other R group of the β-amino ether was too large, the second association was inhibited, and the $\Delta\delta^{RS}$ values diminished considerably.

2.14. 1,5-DIFLUORO-2,4-DINITROBENZENE AND DERIVATIVES (2.121)

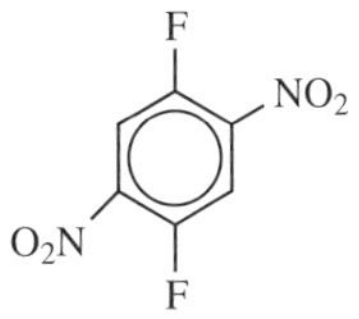

2.121

Although the reagents in this section are atypical in appearance from other Mosher-like reagents, the scheme by which they can be used to assign the absolute configurations of secondary alcohols is rather analogous to the other reagents discussed in this chapter. The method involves a two-step derivatization process in which **2.121** is

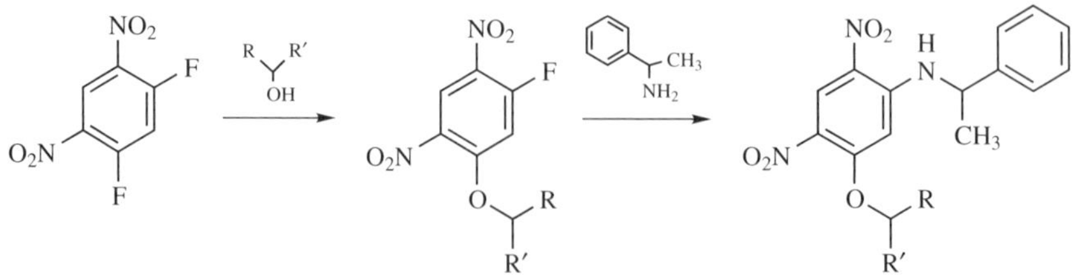

Figure 2.21. Reaction sequence of a secondary alcohol and 1-phenylethylamine with **2.121**.

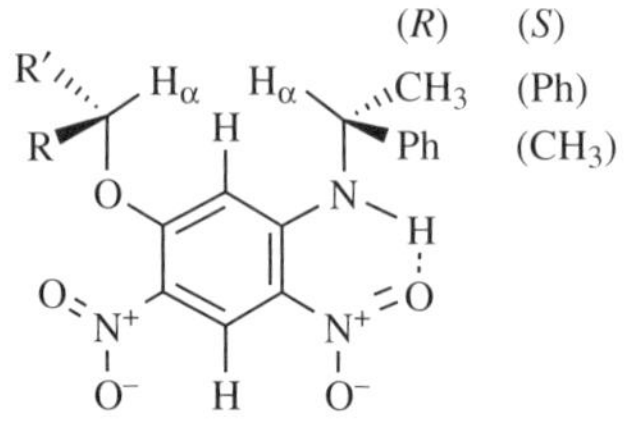

Secondary alcohol—DPEA

Figure 2.22. Conformation of the secondary alcohol/1-phenylethylamine derivative of **2.121**.

first reacted with the secondary alcohol and then with 1-phenylethylamine (PEA) as depicted in Figure 2.21.[258] The resulting derivative has a planar structure in which the phenyl ring of the PEA produces differential shielding of the substituent groups of the secondary alcohol (Fig. 2.22). The validity of the structure shown in Figure 2.22 was supported by NOE data for some of the derivatives. Measurements using the (R)- and (S)-PEA derivatives led to characteristic positive and negative $\Delta\delta^{RS}$ values for the two substituent groups of the secondary alcohol. For many cyclic and acyclic alcohols, the $\Delta\delta^{RS}$ values with this system were often several tenths of a ppm and were not only greater than those with MTPA, but with MPA and NMA as well.[258] **2.121** has not been studied that extensively to date, but the substantial $\Delta\delta^{RS}$ values for those substrates that have been studied suggest its potential utility.

2.14.1. *N*-(5-Fluoro-2,4-dinitrophenyl)-1-phenylethylamide

The 1-phenylethylamide derivative of **2.121** is an effective chiral derivatizing agent for amines.[259,260] Primary amines reacted by replacing the fluorine atom. Hydrogen bonding between the NH hydrogen and the oxygen of the adjacent nitro group caused a preferred conformation as illustrated in Figure 2.23.[259] This preferred conformation can be used to assign the stereochemistry of the α-carbon of primary amines based on the $\Delta\delta^{RS}$ values, which were larger than those observed using MTPA. The utility of this reagent was demonstrated with compounds such as isoleucinol and phenyl alaninol,[259] and then applied to the analysis of the absolute stereochemistry of the primary amine group in the 3-amino-2-hydroxydecanoic acid portion of microginin (**2.122**).[259,260]

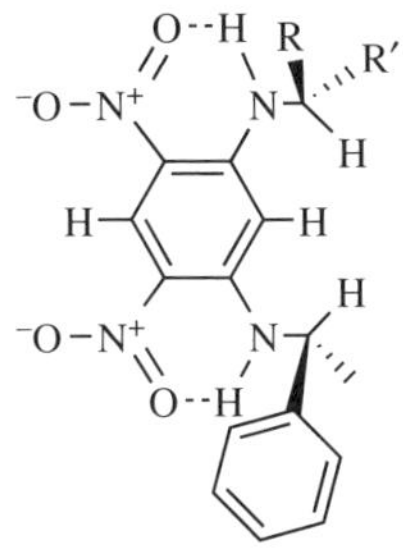

Figure 2.23. Conformation of primary amino derivatives of the (R)-enantiomer of N-(5-fluoro-2,4-dinitrophenyl)-1-phenylethylamide.

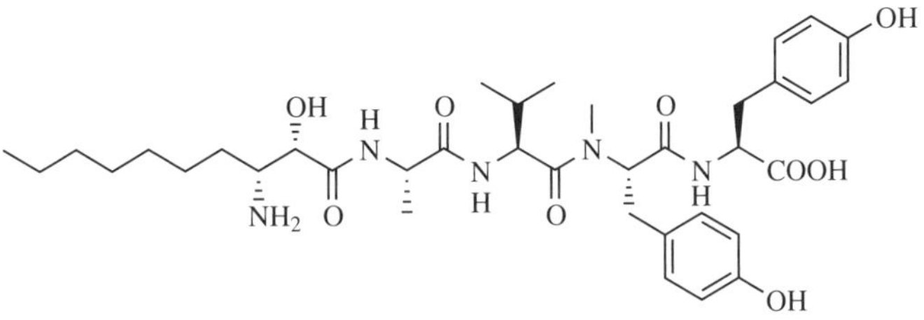

2.122

2.14.2. 1-Fluoro-2,4-dinitrophenyl 5-(S)-alanine Amide (Marfey's Reagent)

1-Fluoro-2,4-dinitrophenyl-5-(S)-alanine amide, widely known as Marfey's reagent, also undergoes substitution reactions with amines (**2.123**). The reagent has been widely used for the chromatographic separation of chiral amines. An NMR study of the diastereomers produced with six chiral benzylic amines such as PEA and similar compounds noted no discrimination of the diastereomers in chloroform-d and only a small amount in dimethylsulfoxide-d_6.[261] No distinction was observed in the NMR spectra of the diastereomers produced from aliphatic amines.

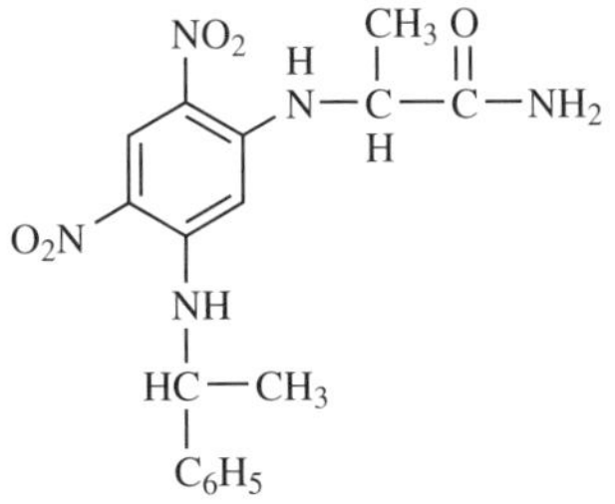

2.123

2.15. 2-FLUORO-2-PHENYLACETIC ACID/2-FLUORO-2-(1-NAPHTHYL)PROPIONIC ACID/2-FLUORO-2-(2-NAPHTHYL)PROPIONIC ACID (2.124)

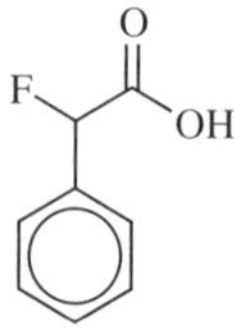

2.124

Substituted 2-fluoroacetic acids with aromatic groups, such as phenyl, phenoxy, phenylsulfide, benzyl, and phthaloyl, have been used to analyze chiral alcohols and amines.[247,262–264] The ^{19}F NMR spectrum was convenient to monitor. The most effective discrimination was observed with bulky substituent groups (phenyl or phthalimide). Two advantages of these reagents over MTPA were that the ^{19}F spectra were sharper and the ^{1}H resonance of the CH group was at much higher frequency and less prone to interference with resonances of the substrate.[262]

The acid form of 2-fluoro-2-phenylacetic acid has been applied to determine the enantiomeric purity of alcohols and amines, using almost exclusively differentiation of the ^{19}F signals for the two diastereomers.[265–273] In some cases, the ^{19}F shifts were shown to correlate with absolute configuration.[266,267,272] Computational methods indicated that the preferred conformation had a reference plane containing six atoms that included Cα (FPAA group), CO_2, Cα (chiral alcohol), and Hα.[271]

After protecting the amine group of aromatic-containing β-aminoalcohols such as **2.125** and derivatizing the alcohol with **2.124**, the diastereomeric ^{19}F resonances were nonequivalent by about 2.5 ppm.[272] Reaction of these substrates with MTPA was unsuccessful because of too much steric hindrance.

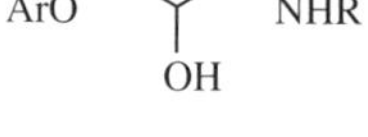

2.125

Analogous reagents with a 1-naphthyl or 2-naphthyl group, as well as the corresponding 2-fluoropropionic acid derivative, were used to determine the enantiomeric purity of 1,2-diglycerides (**2.126**) and other alcohols.[274] Substantial differences were observed in the chemical shifts of the ^{19}F resonances for the diastereomers. There was no discussion about whether the shifts could be used to determine the absolute configuration.

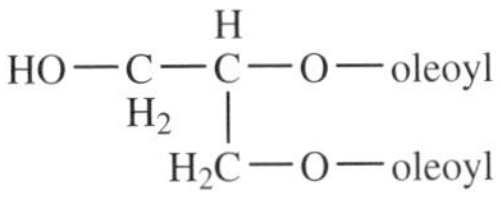

$$\text{HO} - \underset{H_2}{C} - \underset{|}{\overset{H}{C}} - O - \text{oleoyl}$$
$$H_2C - O - \text{oleoyl}$$

2.126

2.16. α-METHOXY-α-(1-NAPHTHYL) ACETIC ACID (1-NMA) (2.127)/ α-METHOXY-α-(2-NAPHTHYL) ACETIC ACID (2-NMA) (2.128)

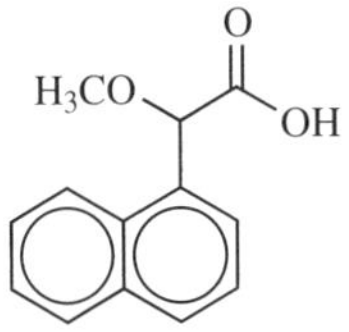

2.127 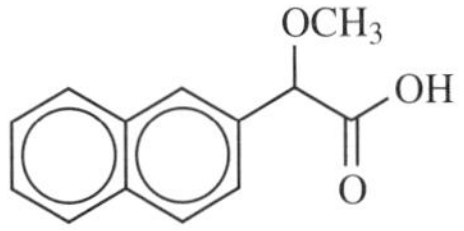   2.128

The use of 2-NMA and 1-NMA for the assignment of absolute configurations of secondary alcohols has been explored with a variety of compounds. Several of these reports compare the effectiveness of 2-NMA or 1-NMA to that of MTPA[275–277] and/or MPA.[183,199,276,278,279]

The absolute configuration of 10-nonacosanol (**2.129**) was assigned on the basis of shifts with the 2-NMA derivative, whereas the shifts with MTPA were not sufficient to perform the analysis.[280] The 2-NMA induced characteristic $\Delta\delta^{RS}$ values on all the ^{1}H resonances of the C9 group and up to eight carbons away from the chiral center for the C19 group, demonstrating the considerable shielding that occurred from the naphthyl ring.

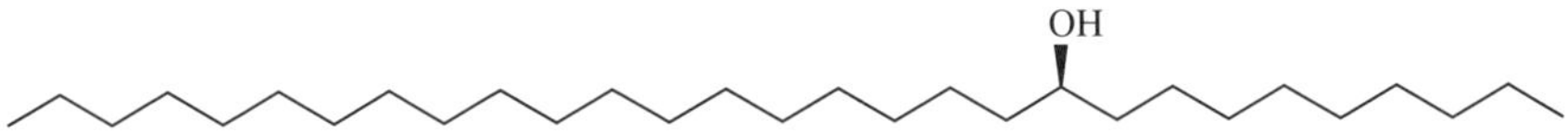

2.129

The absolute stereochemistries of the methyl esters of 8-hydroxytetradecanoic and 8-hydroxyhexadecanoic acid could not be determined with MTPA esters because of overlapping resonances, whereas the larger shifts with 2-NMA derivative did facilitate assignment of the configuration.[64] Similarly, the 1-NMA derivative of sarcophytol (**2.130**) gave $\Delta\delta^{RS}$ values on the order of three times greater than those with MTPA.[277]

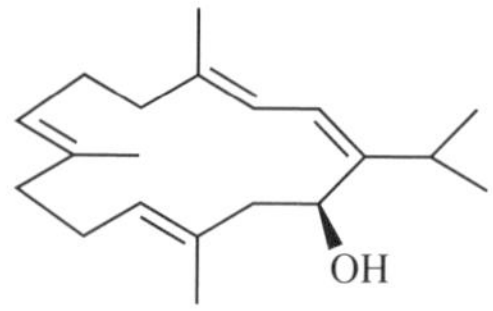

2.130

The use of 2-NMA on annonaceous acetogenins (**2.34**) provided $\Delta\delta^{RS}$ values that were 2–17 times larger than those observed with MTPA.[275] Some *per*-2-NMA derivatives of polyfunctional acetogenins showed additive effects from the naphthyl rings, although it was still possible to rationalize these effects and accurately assign the absolute configuration.

The 2-NMA derivative of the methyl ester of hydroxyfuranoic acid (**2.131**) provided improved long-range anisotropic effects in the spectrum when compared to the MPA derivative.[278] The assignment of some portions of the spectrum of the MPA derivative were complicated due to spectral overlap, whereas the greater shielding and shifts with the 2-NMA derivative facilitated the assignment. A similar finding was observed when comparing the 2-NMA and MPA derivatives of a series of alkyl sulfoxides.[199] The $\Delta\delta^{RS}$ values were larger in the 2-NMA derivative and extended further down the chain when compared to those with MPA. Both ^{1}H and ^{13}C NMR spectra were used in the analysis. It was also found that the $\Delta\delta^{RS}$ values of 2-NMA were two to three times larger in benzene-d_6 than in chloroform-d.[199]

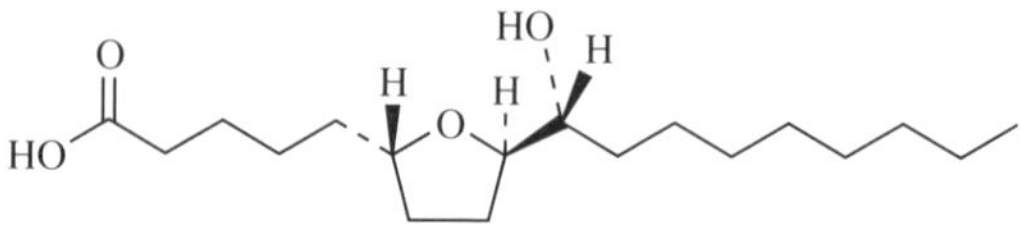

2.131

In a study of 3-hydroxycaretonoids such as 3-hydroxy-β-ionine (**2.132**), the $\Delta\delta^{RS}$ values obtained with 1-NMA and 2-NMA were greater than those with MPA, although even larger values were obtained with 9-AMA.[183]

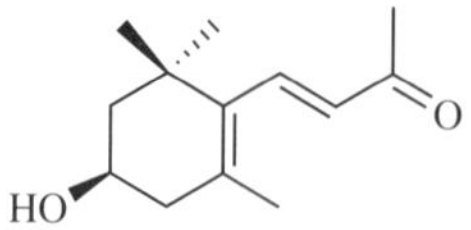

2.132

The stereochemistry of a series of acyclic β- (**2.133**) or γ-methyl-substituted secondary alcohols (**2.134**), which have two chiral centers, was assigned on the basis of $\Delta\delta^{RS}$ values of their 2-NMA esters.[281] The $\Delta\delta^{RS}$ values of a set of model compounds with *syn*- and *anti*-configurations were studied and found to show characteristic patterns. The absolute configuration could be unequivocally assigned, provided suitable thresholds of $\Delta\delta^{RS}$ values were exceeded.[281]

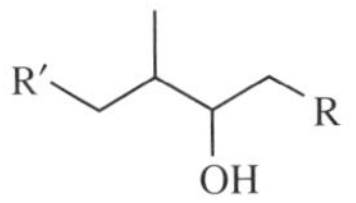

R'—CH(OH)—CH(R)   R'—C(OH)—CH(R)

2.133 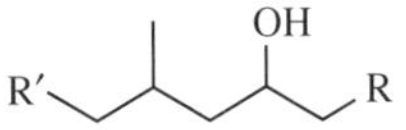   2.134

In some cases, the $\Delta\delta$ values with 2-NMA are large enough that only one derivative is needed to reliably assign the absolute configuration. Methyl 9-hydroxystearate (**2.135**) and 2-octanol are examples where this was found to be the case.[282,283] For 2-octanol, there were some anomalous $\Delta\delta$ values. It was better to calculate the $\Delta\delta$ values by subtracting chemical shifts in the spectrum of the 1-NMA or 2-NMA derivative from those in the acetate derivative. The $\Delta\delta$ values on the *syn*-side of the alcohol showed a characteristic pattern that could be used to reliably assign the absolute configuration.[283]

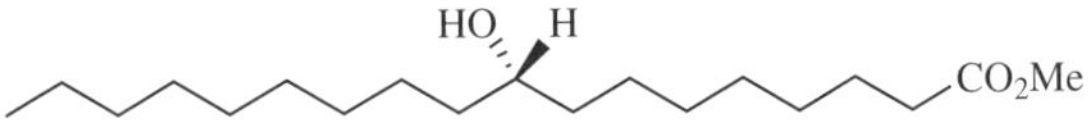

HO H
  \ /
  C —(CH₂)ₙ—CO₂Me

2.135

In a study of the methyl ester of 6-hydroxyoctadecanoic acid, it was noted that the desired *sp* conformer of the (*R*)-2-NMA derivative was more populated at lower temperatures (266 K versus 298 K), thereby increasing the shielding of the protons of one of the substituent groups and decreasing it on the other, further improving the reliability of the stereochemical assignment.[284]

The utility of 2-NMA, MPA, and MTPA for the assignment of absolute configurations of acyclic tertiary alcohols has been investigated.[276] Eleven model alcohols were examined. The derivatives with MTPA and MPA could not be used to assign the stereochemistry. The largest $\Delta\delta^{RS}$ values were observed with 2-NMA, but 2-NMA and MPA underwent racemization during the derivatization. Therefore, racemic 2-NMA was used to carry out the derivatization and the diastereomers were separated by column chromatography prior to the NMR analysis. Except for the β-methylene protons adjacent to the tertiary hydroxyl group, the $\Delta\delta^{RS}$ values showed a consistent positive or negative character that correlated with the absolute configuration. The preferred conformation shown in Figure 2.24 was confirmed by the smaller shifts of the α-methyl group compared to those of the β-hydrogen.

1-NMA and 2-NMA have been used as chiral solvating agents to determine the enantiomeric purity of chiral sulfoxides such as (*S*)-oxide-13-thiaoleate (**2.136**).[279] The enantiomeric discrimination in the ¹H NMR spectra with 1-NMA was larger than that observed with 2-NMA and MPA.

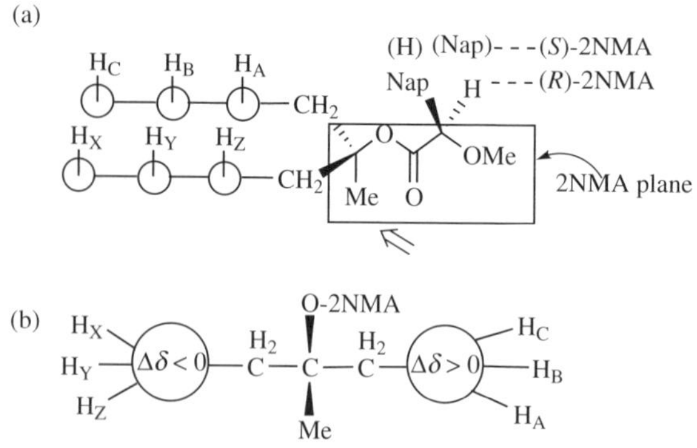

Figure 2.24. (a) The 2-NMA plane of a 2-NMA ester is shown. $H_{A,B,C}$ and $H_{X,Y,Z}$ are on the right and left sides of the plane, respectively. (b) A model to determine the absolute configurations of tertiary alcohols is illustrated. This model is a view of the 2-NMA ester drawn in (a) from the direction indicated by the outlines arrow. [Reproduced with permission.[276]]

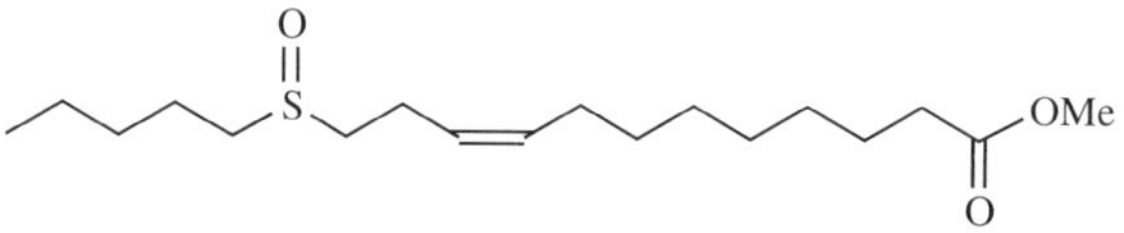

2.136

A method for the large-scale preparation of the optically pure forms of 1-NMA and 2-NMA has been reported.[285]

2.17. 2-(1-Naphthyl)-2-phenylacetic Acid (2.137)

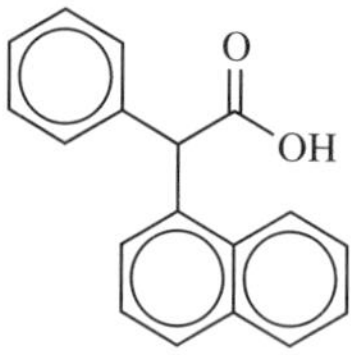

2.137

2.137 was examined as a chiral derivatizing agent for a series of secondary alcohols and α-chiral primary amines.[286] An X-ray structure of the **2.137** derivative with (1R,2S,5R)-2-isopropyl-5-methylcyclohexanol showed that the methine hydrogen atom of **2.137** and carbonyl group were aligned in an *ap* conformation with the plane of the naphthyl ring orthogonal to the phenyl plane. As a result, the phenyl ring was the one responsible for the shielding of the substituent groups of the substrate. The $\Delta\delta^{RS}$ values could be used to predict absolute configurations, although the magnitude of the $\Delta\delta^{RS}$ values was relatively comparable to those of the MTPA derivatives because of the multiconformational composition.[286]

2.18. 2-METHOXY-2-(1-NAPHTHYL) PROPIONIC ACID (MαNP) (2.138)

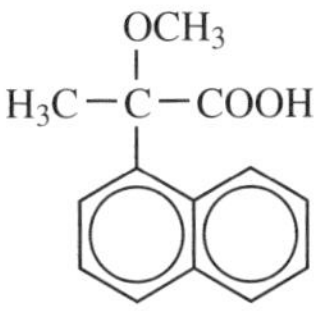

2.138

MαNP has been examined in a series of reports for the assignment of absolute stereochemistries of secondary alcohols.[287–296] The $\Delta\delta^{RS}$ values with MαNP were about four times larger than those with MTPA and two times larger than those with MPA,[287] and comparable to those with 1-NMA and 2-NMA. The larger $\Delta\delta^{RS}$ values compared to MTPA and MPA were ascribed to the greater shielding of the naphthyl compared to the phenyl ring. The *sp* conformation of the methoxy and oxygen atom of the ester carbonyl group was preferred and facilitated the assignment of absolute configuration based on the signs of the $\Delta\delta^{RS}$ values.[287] The preferred conformation was verified through X-ray crystallographic data.[287,288,293]

An advantage of MαNP was that it was less prone to racemization during the derivatization step when compared to 1-NMA, 2-NMA, MPA, and other acetic acid derivatives.[287,289,290] In particular, the addition of the methyl group at the chiral center prevents the epimerization of the reagent. As with other naphthyl and anthryl reagents, the $\Delta\delta$ values were often large and consistent enough to assign the configuration using only one enantiomer of MαNP. In several studies, the derivatives of a racemic mixture of the secondary alcohol with one enantiomer of MαNP were prepared and separated by liquid chromatography. The absolute configuration of each of the enantiomers could then be assigned by analyzing the ^{1}H NMR spectrum.[288,291–293,295] In cases where the compound had multiple chiral centers, the liquid chromatographic separation resulted in additional peaks, and assignment of absolute configurations of each diastereomer was achievable through the shifts in the ^{1}H NMR spectrum.[295]

MαNP has been used successfully with 1,2,3,4-tetrahydro-4-phenanthrenol (**2.139**),[288] citronellol (**2.140**),[290] menthol (**2.141**),[292,296] precursor alcohols to phthalides (**2.142**),[293] *meta*-substituted diphenyl methanols such as **2.143**,[294] and *o*-, *m*-, and *p*-substituted fluorinated diphenyl methanols.[297] For the *meta*-substituted diphenylmethanols, the aromatic resonances on one ring had positive $\Delta\delta^{RS}$ values, whereas those on the other aromatic ring had negative $\Delta\delta^{RS}$ values. The $\Delta\delta^{RS}$ values were greater than 0.3 ppm for some of the resonances, which were larger than those observed with MTPA or MPA.[298]

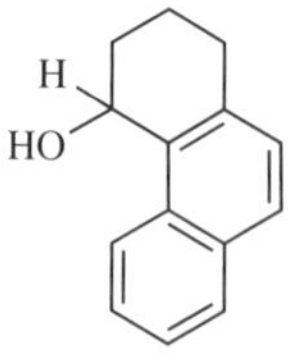

2.139 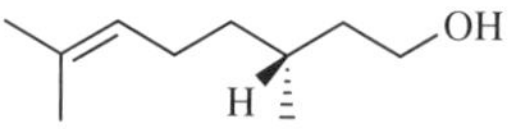 **2.140** 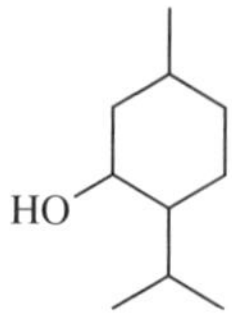 **2.141**

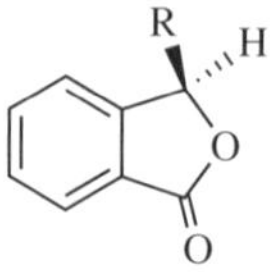

2.142 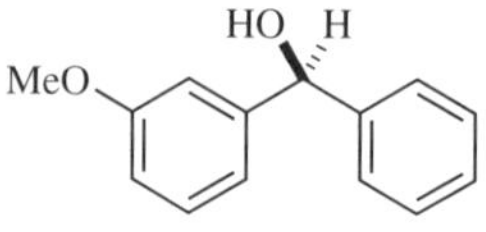 **2.143**

2.19. 1-METHOXY-2,3-DIHYDRO-1*H*-CYCLOPENTA[a]NAPHTHALENE-1-CARBOXYLIC ACID (2.144)

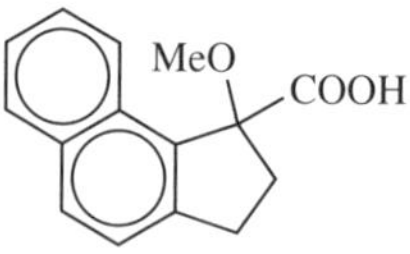

2.144

2.144 was shown to be an effective chiral derivatizing agent for the assignment of absolute configurations of secondary alcohols.[299] The naphthyl ring system caused larger shifts than those observed with MTPA or MPA. Also, the reagent has more restricted rotation than MTPA and MPA. Estimations of the populations of the *syn*- and *anti*-conformers were incorporated into molecular dynamic calculations to simulate the esterification shifts, facilitating the assignment of absolute configurations.

2.20. *O*-ARYL LACTIC ACIDS (2.145)

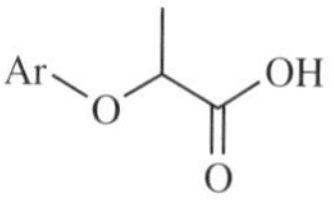

2.145

A variety of *O*-aryl lactic acid derivatives have been studied as chiral derivatizing agents. These reagents are easily prepared by reacting phenols with ethyl (*S*)-lactate under Mitsunobu conditions followed by hydrolysis of the ester.[300] For example, the *ortho*-fluorophenyl derivative was shown to be useful for the determination of the optical purity of primary, secondary, and tertiary alcohols.[301] No racemization was noted during the esterification reaction. The preferred conformation places the ester carboxyl linkage and ether methine in a common plane, allowing the phenyl ring to shield one substituent of the alcohol.[301] The α-hydrogen of the acid and that of the secondary alcohol are eclipsed with the carbonyl group of the ester.

The *sp* conformation of the ester carboxylate group was confirmed by X-ray structural data on the *para*-fluorophenyl lactic acid derivative of neomenthol.[302] The shielding in the NMR spectrum was specific enough, given the *sp* conformer that the absolute configuration of the secondary alcohol could be reliably assigned. NOE data on derivatives of hydroxyl substituted methylene vinyl ethers also confirmed the preference for the *sp* conformation and could be used to aid in the assignment of absolute stereochemistry.[302]

In comparison to most other MTPA-like reagents, larger shifts in the ^{1}H NMR spectra were actually observed at sites more remote from the chiral center.[301] Presumably, extension of the aromatic ring in the *O*-aryl lactic acid derivatives relative to many other MTPA-like reagents accounts for this observation. Even though the *O*-aryl bond can rotate, the $\Delta\delta^{RS}$ values showed specific trends that correlated with the absolute configuration.[301]

In a study of the effectiveness of a range of *O*-aryl lactic acid derivatives with secondary alcohols, the β-naphthyl derivative [(2-β-naphthyloxy)propionic acid] was especially effective in discriminating the enantiomers of substrates.[303]

In a set of aryl lactic acid derivatives that contain a fluorine atom on the aromatic ring (**2.146**, **2.147**, X = CN, Br, I, HgCl), the ^{19}F NMR resonance was convenient to monitor in the determination of the enantiomeric purity of alcohols and amines. Whereas the cyano group reduced the differentiation in the ^{19}F signals of derivatives with **2.146**, the bromo, iodine, and mercurychloride derivatives lead to noticeable increases.[304,305] With the mercury derivative, it was also possible to use the ^{199}Hg NMR spectrum to determine enantiomeric purity.[304] Reaction of the four isomeric menthols with (*R*)-2-(4-fluorophenoxy)propionic acid produced eight diastereomers, and separate signals for each of the eight were observed in the ^{19}F NMR spectrum.[305] This reagent also recognized the planar chirality in **2.148**.[305]

2.146   **2.147**   **2.148**

O-Aryl lactic acids have been used for the determination of the enantiomeric purity and absolute configurations of α-chiral primary amines and α-amino acid esters.[300] In contrast to the ester derivatives that favor the *sp* conformation, the amide derivatives prefer an *ap-Z* conformation in which the CαOAr and carbonyl groups are close to *anti-periplanar* as seen in Figure 2.25.[300] X-ray data for one of the amide derivatives provided support for the preference of the *ap* conformation. Reagents with

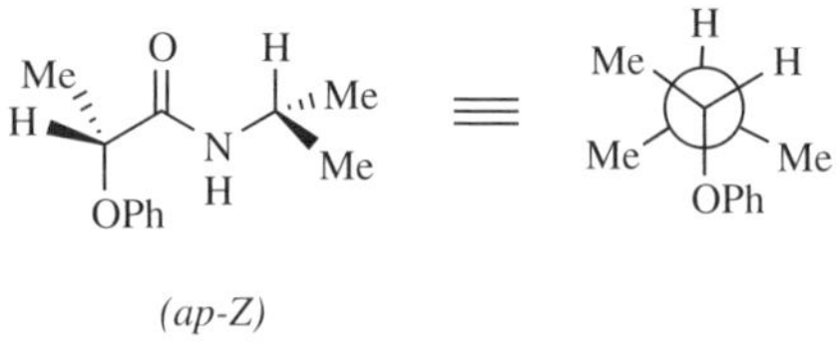

Figure 2.25. The *ap-Z* conformational preference of amide derivatives with *O*-aryl lactic acids.

four different aromatic rings (phenyl, 2-naphthyl, 2-methyl-4-chlorophenyl, and 2,4-dichlorophenyl) were examined and the largest discrimination occurred with the naphthyl derivative. The derivatives with an electron-withdrawing group were less effective as the electron-deficient aromatic ring produced less shielding. The extension of the aromatic ring from the chiral center of the lactic acid causes larger discrimination for hydrogens that are more remote from the chiral center.[300] For example, with the 2-methyl-4-chlorophenyl derivative of 2-aminobutane, the hydrogens of the methyl group at C1 were nonequivalent by 0.09 ppm, whereas those of the methyl group at position 4 had a discrimination of 0.15 ppm.[300] Similarly, the methine resonance of the derivative of valine methyl ester was distinct by 0.09 ppm, whereas the diastereotopic methyl resonances were nonequivalent by 0.17 and 0.18 ppm. The discrimination with the lactic acid derivatives was larger than that observed for the corresponding MTPA amides and similar in magnitude to those with MPA and *O*-AMA.[300]

O-Aryl lactic acids can also be used as chiral solvating agents for determining the enantiomeric purity of primary and secondary amines and amino alcohols.[306] Mixing the *O*-aryl lactic acid with the amine in chloroform-*d* caused the formation of diastereomeric salts. The enantiomeric discrimination with the *O*-(2-naphthyl) lactic acid derivative was larger than that occurred with MTPA or *O*-AMA. For those compounds that were studied, the resonances of the derivative with the (*R*)-enantiomer of the amine were consistently at lower frequency than those of the (*S*)-enantiomer. Without further study, it is difficult to know how generally this can be applied to the assignment of absolute configurations of amines.

2.21. α-(2-ANTHRYL)-α-METHOXYACETIC ACID (2-AMA) (2.149)/ α-(9-ANTHRYL)-α-METHOXYACETIC ACID (9-AMA) (2.150)

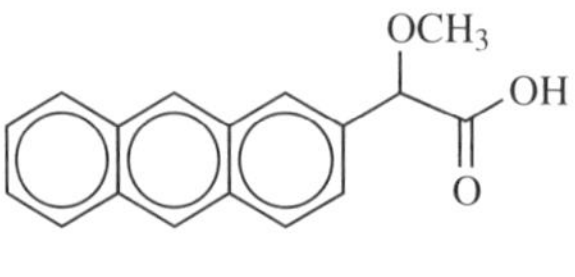

2.149

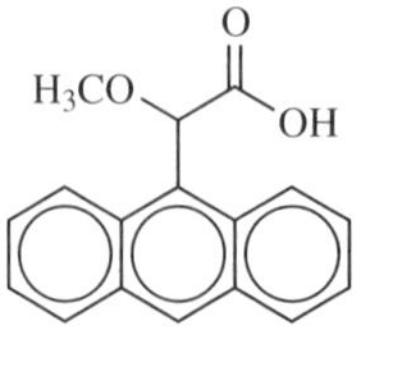

2.150

The anthryl-containing 2-AMA and 9-AMA generally produce larger $\Delta\delta^{RS}$ values than MTPA and other similar reagents for secondary alcohols. In many cases, the shifts were so large and the *sp* conformer so favored that only one diastereomer (using either (R)-AMA or (S)-AMA) was needed.[14,307–309] In linear alcohols, differential shielding was readily observed for hydrogens three to four bonds away from the stereocenter and in some cases up to seven bonds away.[308] Comparing sets of protons at comparable positions removed from the stereocenter makes it possible to assign the substituent group of the substrate adjacent (*syn*) to the anthryl group. Lowering the temperature with the 9-AMA derivative did not lead to much change in the $\Delta\delta^{RS}$ values because of the strong preference for the *sp* conformer that already existed.[15,308] The situation was more complicated with cyclic secondary alcohols, especially if the hydroxyl group was in the axial position.[308] In this case, it was better to incorporate the shifts of several sets of protons. Typically, one set of hydrogen atoms was shielded, whereas the others did not shift much.

A procedure for determining the optical purity and assigning the absolute configuration of both enantiomers of a variety of secondary alcohols using a single 9-AMA derivative and liquid chromatography–NMR has been described.[310] The method was successfully applied in a deuterated liquid chromatographic solvent composed of a mixture of acetonitrile, deuterium oxide, and formic acid, and the trend of $\Delta\delta^{RS}$ correlated with that observed in chloroform-*d*. If the substrate was enantiomerically pure, the derivatization was done using a 3:1 mixture of (R)- and (S)-9-AMA, and the absolute configuration could be assigned.

After protection of several of the hydroxyl groups of 2-(2,3,7,13-tetrahydroxyoctadecanoylamino)ethanesulfonic acid (**2.151**), the mono-2-AMA derivatives at C7 and C13 enabled an assignment of the stereochemistry of these sites.[309]

2.151

For a series of secondary squalene diols (**2.94**),[217] the $\Delta\delta^{RS}$ values for the hydrogen at the β-position were up to 1 ppm for 9-AMA, approximately 0.4 ppm for MPA and acetoxyphenylacetic acid, and about 0.15 ppm for MTPA. The only reservation was that the 9-AMA esters were reportedly not as stable as those with MPA. The 9-AMA caused significant $\Delta\delta^{RS}$ values for hydrogen atoms further down the chain than the other reagents, especially MTPA.

Application of 2-AMA to methyl 9-hydroxystearate (**2.135**) and 10-nonacosanol (**2.129**) caused sufficient shielding of the methyl resonances nine bonds away to cause separate signals for the two diastereomers.[307] 9-AMA produced larger values of $\Delta\delta^{RS}$ for cyclic secondary alcohols when compared to 2-AMA, whereas the reverse trend was observed for long-chain aliphatic alcohols.[15,307] The position of the 9-anthryl group in derivatives of cyclic alcohols (Fig. 2.26) justifies the larger shielding with 9-AMA compared to that with 2-AMA.[15] Similarly, 2-AMA is in a position to extend further than 9-AMA in derivatives of linear alcohols as seen in Figure 2.27.[15]

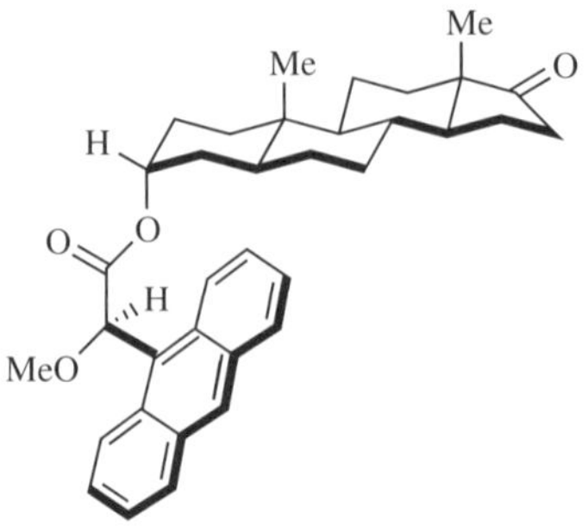

Figure 2.26. Position of the anthryl group in 9-AMA derivatives of cyclic alcohols.

In other comparative studies, 9-AMA produced larger $\Delta\delta^{RS}$ values than 1-NMA, 2-NMA, and MPA for a series of 3-hydroxycarotenoids (**2.132**).[183] Unlike the NMR studies, attempts at using the exciton chirality method for these hydroxyl carotenoids were unsuccessful for assigning the absolute configurations.[183] Similarly, the $\Delta\delta^{RS}$ values for a series of sesquiterpenes such as **2.152** were three to four times larger (about 0.1–0.7 ppm) with 9-AMA compared to those with MPA.[311]

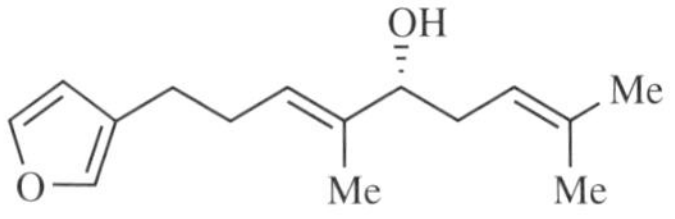

2.152

The stereochemistry of long-chain alcohols used in the synthesis of mono- and dideuterated tridecanoic acids was assigned on the basis of NMR data with the (S)-9-AMA derivative.[312] The enantiomers were resolved by liquid chromatography and then assigned by comparison to known compounds. The use of MTPA was unsuccessful in this analysis.

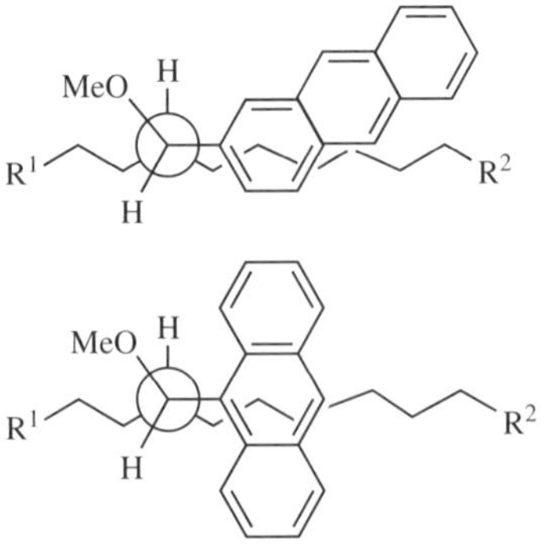

Figure 2.27. Position of the anthryl group in 9-AMA derivatives of linear alcohols.

The utility of MPA and 9-AMA for the assignment of the absolute configurations of *syn-* and *anti-*1,2-, 1,3-, 1,4-, and 1,5-diols has been explored.[171,172] The results were quite comparable for the two reagents, although the $\Delta\delta^{RS}$ values for the 9-AMA derivatives were typically larger than those with MPA. Models for monoalcohols were not directly applicable to these polyalcohols, as the two aromatic rings in the disubstituted derivatives resulted in a combination of shielding and deshielding effects for different hydrogen atoms. There was a systematic correlation of effects observed for a wide range of substrates that leads to distinctive patterns for the signs of the $\Delta\delta^{RS}$ values. Therefore, preparing the bis-(*R*)-9-AMA and bis-(*S*)-9-AMA esters and comparing the measured $\Delta\delta^{RS}$ values to those observed for known model compounds allowed the determination of the absolute configuration of both stereocenters.[172]

The utility of 9-AMA as a reagent for the stereochemical assignment of acyclic β-chiral primary alcohols has been examined.[313,314] Unlike the derivatives with (*R*)- and (*S*)-MPA and MTPA, which had very similar spectra for β-chiral primary alcohols, there were distinct trends of the positive and negative $\Delta\delta^{RS}$ values on each side of the asymmetric center in the derivative with 9-AMA that allowed the absolute configuration to be assigned. Computational studies were used to determine the conformational preference, which is shown in Figure 2.28. The conformational preference was supported by low-temperature NMR spectra as well. If the primary alcohol was highly hindered, the primary hydroxyl group on a cyclic compound or the chiral carbon at the γ-position, the method was not as reliable or required more caution.[313,314] The absolute configuration of *cis-* and *trans-*6-chloro-9-[2-(hydroxymethyl)cyclopentyl]-9*H*-purine (**2.153**) assigned on the basis of the $\Delta\delta^{RS}$ values with the MPA derivative was shown to be incorrect, whereas the $\Delta\delta^{RS}$ values of the 9-AMA derivative were reliable.[315] With primary alcohols of cyclic systems, the importance of corroborating the NMR data with theoretical calculations was emphasized.[315]

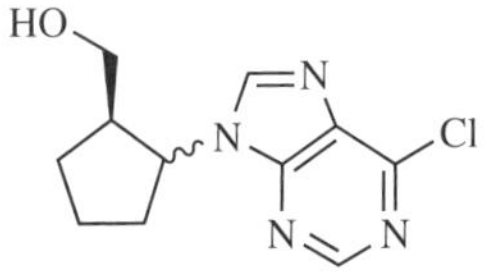

2.153

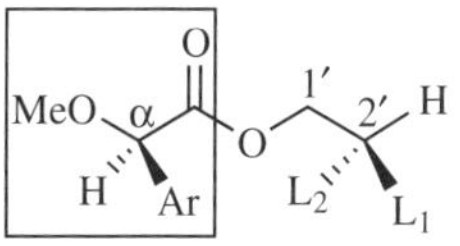

Figure 2.28. Conformational model of the 9-AMA derivatives of β-chiral primary alcohols.

2.22. (α-[1-(9-ANTHRYL)]-2,2,2-TRIFLUOROETHOXY]ACETIC ACID (2.154)

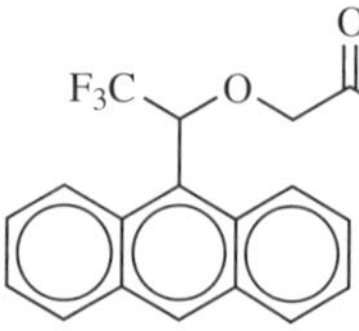

2.154

2.154 has been used to determine the optical purity and assign the absolute configurations of alcohols, amines, and thiols.[316] Shielding by the anthryl group was comparable in scope to other Mosher-like compounds, facilitating the assignment of absolute configurations. The distinction between the two diastereomers was larger for the anthryl derivative than for the corresponding phenyl and naphthyl reagents.[316] The optical purity of methyl-L-lactate[317] and absolute configuration of **2.155** were determined on the basis of derivatives with **2.154**.[318]

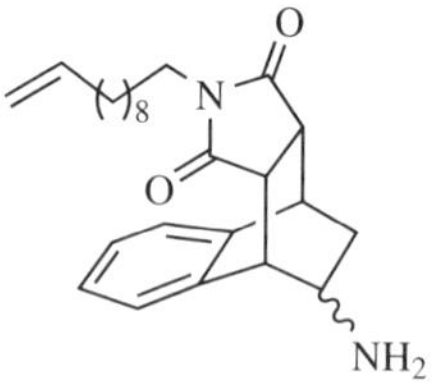

2.155

2.23. 2-METHOXY-2-(9-PHENANTHRYL)PROPIONIC ACID (2.156)

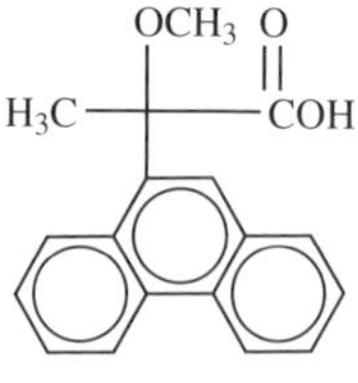

2.156

The derivatives of 3-octanol with **2.156** were separated by liquid chromatography and the absolute configurations were assigned on the basis of a preferred *sp* conformation of the methoxy and carbonyl groups in the derivatives.[319] The reagent was also used to assign the absolute configuration of 6-methyl-5-hepten-2-ol.[287,320] No comparison of the effectiveness of **2.156** to similar reagents was provided in these reports.

2.24. SUMMARY

The variety of reagents inspired by the original work of Dale and Mosher provides investigators with a range of choices when trying to determine optical purity or assign the absolute configurations of alcohols and amines. Commercial availability of the reagent is a crucial factor as very few laboratories are willing to spend the time and money to synthesize and purify the chiral reagent. Familiarity with a particular reagent often plays an important role as well, which explains the widespread use of MTPA relative to other similar reagents.

A number of rather thorough comparative studies of these reagents have been undertaken.[14–16,26,179,321] For certain classes of compounds, the use of MTPA is warranted. For other classes of compounds, there are reagents that are markedly superior to MTPA. Using a more optimal reagent is important as it enhances the likelihood that the absolute configuration will be reliably assigned. Reliable assignments can only be made when the magnitudes of the $\Delta\delta^{RS}$ values are above the experimental error and when the $\Delta\delta^{RS}$ values are all positive for the resonances of the hydrogen atoms on one side of the stereogenic center and negative for those on the other.[26] Attempts at assigning absolute configurations are inherently unreliable when identical signs of the $\Delta\delta^{RS}$ values occur for both substituents, when one of the substituents shows both positive and negative signs of $\Delta\delta^{RS}$, or when $\Delta\delta^{RS}$ values are only available for a single substituent of the substrate.[26] Assigning absolute configurations in polyfunctional substrates must always be approached with caution because of the possibility that resonances will show shielding or deshielding effects from two or more of the attached groups.[26] A summary guide to using these reagents, including experimental details such as reaction and NMR conditions, has been published.[321] A comprehensive review primarily focused on the use of aryl methoxy acids for the assignment of absolute configuration has also been published.[322]

2.24.1. Analysis of Primary Alcohols

The largest $\Delta\delta^{RS}$ values for the analysis of the absolute configuration of primary alcohols are often achieved with 9-AMA.[321] For a reliable assignment, it is best to prepare derivatives with the (R)- and (S)-9-AMA and use $\Delta\delta^{RS}$ values.

2.24.2. Analysis of Secondary Alcohols

MTPA, MPA, and 9-AMA can all be used to assign the absolute configuration of secondary alcohols. In general, the derivatives with 9-AMA show the largest $\Delta\delta^{RS}$ values of the three, whereas those with MTPA have the smallest $\Delta\delta^{RS}$ values.[321] Typical $\Delta\delta^{RS}$ values with 9-AMA are two to three times larger than those with MPA.[16]

Using MPA, there are also two ways of assigning the absolute configuration on the basis of a derivative with only a single MPA enantiomer. One is to lower the probe temperature to enhance the population of the *sp* conformer. The $\Delta\delta_{T_1 T_2}$ values from the spectra at ambient temperature and $-70°C$ should show different signs for the L_1

and L_2 groups. The second is to add Ba(II) to the sample, which also causes a higher population of the *sp* conformer. The $\Delta\delta^{Ba}$ values from the spectra before and after the addition of Ba(II) will have different signs for the L_1 and L_2 groups. This method can only be reliably used if there are no other functional groups that complex with barium.

With 9-AMA, the upfield shifts caused by the shielding of the anthryl group are so large that measuring the shifts in the spectrum of either the (*R*)- or (*S*)-9-AMA derivative can be used to reliably identify L_1 and L_2 and assign the configuration.[321]

Three other reagents that deserve mention for use with secondary alcohols are mandelic acid, MαNP, and CFTA. The use of mandelic acid goes back to Mosher's original papers, and in those cases where comparative data are provided, the $\Delta\delta^{RS}$ values are usually larger than those obtained with MPA and MTPA derivatives. Direct preparation of MA derivatives may be more difficult, but preparing the *O*-AMA derivative and hydrolyzing it to the MA derivative is an alternative procedure. MαNP, which recently has become commercially available, causes $\Delta\delta^{RS}$ values that are typically four times larger than those with MTPA and twice as large as those with MPA. The shielding is large enough that absolute configurations can be reliably assigned using only a single derivative. Finally, the presence of the methyl group on the chiral carbon reduces the likelihood that the reagent will racemize in the derivatization step. CFTA produces $\Delta\delta^{RS}$ values that are about two times as large as those with MTPA in the ^{1}H NMR spectrum and three to six times greater in the ^{19}F spectrum. The reagent is also expected to be less prone to racemization in the derivatization step than acetic acid derivatives with a hydrogen atom on the chiral carbon.

2.24.3. Analysis of Tertiary Alcohols

The utilization of these reagents to assign the absolute configuration of tertiary alcohols is limited in scope.[87,276,323] Comparative studies with 2-NMA, MPA, and MTPA indicate that the largest and most reliable $\Delta\delta^{RS}$ values are observed with 2-NMA.[276,323] The 2-NMA does racemize during the derivatization step which may necessitate that the two diastereomers be separated by liquid chromatography prior to the NMR analysis.

2.24.4. Analysis of Primary Amines

MTPA, MPA, and BPG can all be used to assign the absolute configuration of primary amines in which the stereogenic center is at the α-position.[321] In comparative studies of these reagents, the largest $\Delta\delta^{RS}$ values are usually observed for the BPG derivatives. The derivatives of amines with 9-AMA give rather small $\Delta\delta^{RS}$ values.[179] Unlike with secondary alcohols, the *ap* conformer is preferred for the derivatives of primary amines with MPA. The magnitude of the $\Delta\delta^{RS}$ values for primary amines is especially sensitive to solvent and is optimized in low polarity solvents, especially for derivatives with MPA.[179,321]

The MTPA derivatives of primary amines show three preferred conformers (*sp*, *ap*3, and *ap*1) with a preference for the *sp* form. This form orients the phenyl ring of the MTPA in such a way so as to produce large shielding effects, such that the $\Delta\delta^{RS}$ values of MTPA are usually larger than those with MPA.[94,179]

The absolute configuration of primary amines can be assigned on the basis of a single derivative with either (R)- or (S)-MPA and the addition of barium(II). Complexation of the Ba(II) enhances the population of the *sp* conformer and causes different signs of the $\Delta\delta^{Ba}$ values for the L_1 and L_2 substituents.

The CFTA reagent is also effective for assigning the absolute configuration of amines. Either ^{1}H or ^{19}F NMR data can be employed. The bulkiness of the substituent groups enhances the population of the preferred conformation, thereby increasing the $\Delta\delta^{RS}$ values. The CFTA reacts much faster with hindered amines than MTPA, which reduces the likelihood of kinetic resolution.

2.24.5. Analysis of Secondary Amines

The MTPA derivatives of secondary amines produce substantial $\Delta\delta^{RS}$ values such that MTPA is the reagent of choice for secondary amines.[88,89] The shielding of the MTPA derivatives of secondary amines is so pronounced that absolute configurations can reliably be assigned on the basis of a single derivative with only (R)- or (S)-MTPA. As the (S)-MTPA acid chloride is often less reactive than the (R)-MTPA acid chloride, the use of (R)-MTPA acid chloride is recommended in single-derivative analyses.

3

OTHER CARBOXYLIC ACID-BASED REAGENTS

A variety of other carboxylic acids have been used as either chiral derivatizing or solvating agents, primarily for the analysis of alcohols or amines. One relatively common application involves the analysis of the *pro*-(R) and *pro*-(S) positions of monodeuterated primary alcohols through preparation of suitable ester derivatives. Other reagents either have predictable shielding or provide derivatives with spectra that exhibit empirical trends that correlate with absolute configuration. A number of axial chiral compounds with carboxylic acid functional groups have been developed and used for chiral analysis.

The majority of these reagents have been applied only in limited capacity, but serve to show the range of carboxylic-acid-containing compounds that have been used as chiral discriminating agents in NMR spectroscopy.

3.1. CAMPHANIC ACID (3.1)

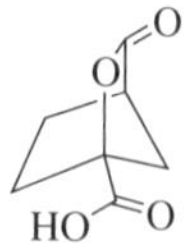

3.1

Camphanic acid has been used frequenty for the analysis of the *pro*-(R) and *pro*-(S) positions of α-deuterated primary alcohols. In the first report of this method, which

---

Discrimination of Chiral Compounds Using NMR Spectroscopy, by Thomas J. Wenzel
Copyright © 2007 John Wiley & Sons, Inc.

was done at relatively low field strengths, the distinction of the *pro*-(R) and *pro*-(S) positions for camphanate esters of substrates such as geraniol (**3.2**) and benzyl alcohol was enhanced through the addition of a paramagnetic lanthanide chelate [(Eu (dpm)$_3$—dpm = 2,2,6,6-tetramethyl-3,5-octanedione]. The lanthanide-induced shifts[324] also exhibited characteristic trends that correlated with the absolute configuration.

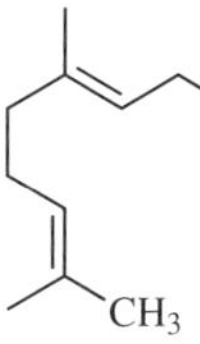

3.2

The combination of camphanate esters with Eu(III) shift reagents was subsequently used in the analysis of a variety of deuterated alcohols.[325–329] These include α-deuterated straight chain alcohols such as *n*-heptanol and *n*-octanol,[325] benzyl alcohol derivatives such as 3,5-dimethoxy-(7S)-[7-^{2}H]benzyl alcohol (**3.3**),[326] and 2-phenylethanol derivatives (**3.4**).[329] In this last instance, the analysis of the 2-phenylethanol derivative enabled the determination of the optical purity of the starting oxirane (**3.5**) that was converted to the alcohol by reaction with phenyl lithium.

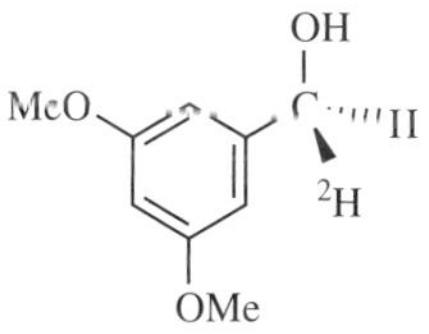

3.3 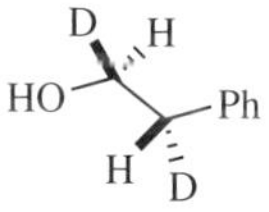   **3.4**    **3.5**

The combination of europium shifts with camphanic acid has also been applied to the analysis of [1-^{2}H]-3-phthalimido-1-propanols (**3.6**).[328] In these compounds, the order of shifts of the *pro*-(R) and *pro*-(S) positions was contrary to the empirical rule, which was ascribed to the presence of the phthalimide group.

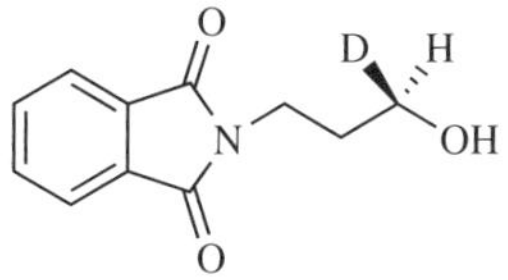

3.6

The optical purity and absolute configuration of 2- and/or 3-deuterated 1-phenyl-3-alkylpropanols (**3.7**) were determined on the basis of the europium-induced shifts in the ^{1}H NMR spectrum at 400 MHz.[327] The nonequivalence between the diastereotopic resonances was between 0.6–0.7 ppm at C2 and 0.2–0.3 at C3. Stereospecifically labeled compounds were used to confirm the reliability of the method.

3.7

Some studies have observed suitable enantiomeric discrimination of the *pro*-(R) and *pro*-(S) positions in deuterated and tritiated geraniol (**3.2**)[330,331] or [1,2-^{2}H$_2$]-ethylamine[332] derivatives without the addition of a europium shift reagent. The ^{3}H NMR of the camphanate derivative of geraniol monotritiated at the C8 position could be used to assign the absolute configuration.[330] The method was then used to determine the stereospecificity of enzymatic tritiation reactions. For the ethylamine, the resulting *N*-[1,2-^{2}H$_2$]ethylcamphanamide was analyzed.

The use of natural abundance ^{2}H NMR spectroscopy for studying the stereoselectivity of deuterium substitution into natural systems such as glucose, galactose, or mannitol by enzymes has been described.[333] The discrimination can occur in natural systems in which the CHD group is already adjacent to a chiral center or in a prochiral system to which a chiral center is added by derivatization of the alcohol into the corresponding esters with camphanic acid.

Camphanic acid has been used to analyze the *pro*-(R) and *pro*-(S) hydrogens of other monodeuterated or monotritiated amines through the analysis of the corresponding camphanamides.[334–341] This includes the analysis of the camphanamide of [2-^{2}H$_1$]glycine by ^{1}H NMR[334,337] or [2-^{3}H$_1$]glycine by ^{3}H NMR spectroscopy.[338]

The two prochiral hydrogen atoms at the C1′ positions of spermidine (**3.8**) were distinguishable in the camphanamide derivative.[335] The primary amines were blocked with Boc groups and the secondary amine was converted to the camphanamide. The spectra could be used to distinguish between the single and double displacement mechanisms of the spermidine synthase reaction.

3.8

A detailed analysis of the discrimination of the *pro*-(R) and *pro*-(S) positions of camphanamides of primary amines has been undertaken.[336] For α-deuterated primary amines, the spectra of the camphanamides in benzene-d_6 exhibited nonequivalence of 0.12–0.21 ppm for the H$_S$ and H$_R$ resonances. The distinction of the diastereotopic methylene protons was caused by differential shielding of the amide carbonyl group, such that the resonance of the *pro*-(S) position, which was closer to the amide carbonyl group, occurred at a higher frequency than the resonance of the *pro*-(R) position.

Camphanamides of α-arylethyl amines with phenyl, 1-naphthyl, and 9-anthryl groups show discrimination of the *pro*-(R) and *pro*-(S) hydrogen resonances.[339]

Hydrogen bonding in the *syn*-oriented O—C(1)—C(6)—NH unit and steric repulsion between the C1'-methyl and amide carbonyl groups could be used to rationalize the distinction of the *pro*-(R) and *pro*-(S) hydrogens, enabling the assignment of the absolute stereochemistry of these amines.

Camphanate esters have also been used to determine the optical purity of primary (**3.9**)[342] and secondary alcohols.[343–346] The optical purity of diisopropyl hydroxyl phosphonates with a hydroxyl group α, β, or γ to the phosphorus atom (**3.10**) was determined by the ^{1}H and ^{31}P NMR spectra of the corresponding camphanate ester.[346] The addition of Eu(fod)$_3$ (fod = 6,6,7,7,8,8,8-heptafluoro-2,2-dimethyl-3,5-octane-dione) to the (−)-camphanate esters of secondary alcohols consistently caused the camphanic methyl resonance of the (S)-alcohol to appear at lower frequency than that of the (R)-alcohol.[345] The method could be used to assign absolute configurations provided secondary alcohols were compared to model compounds with closely related structures.

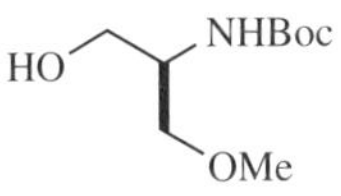

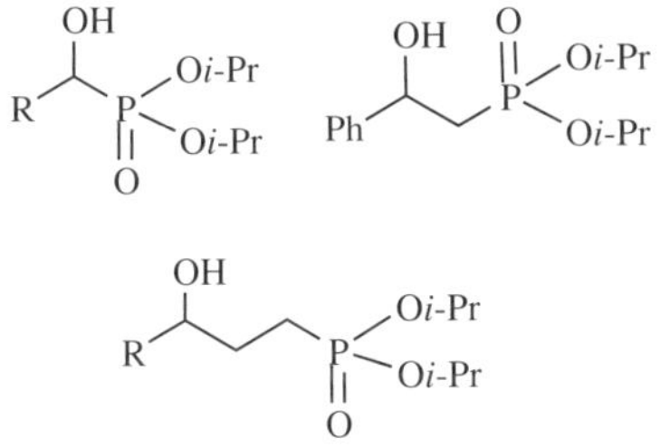

Hypericin (**3.11**) has a helix inversion barrier that causes the compound to be chiral. The diastereomers created by preparing the camphanate esters of the 3-hydroxy group had separate signals that could be used to analyze the inversion barrier of the compound.[347]

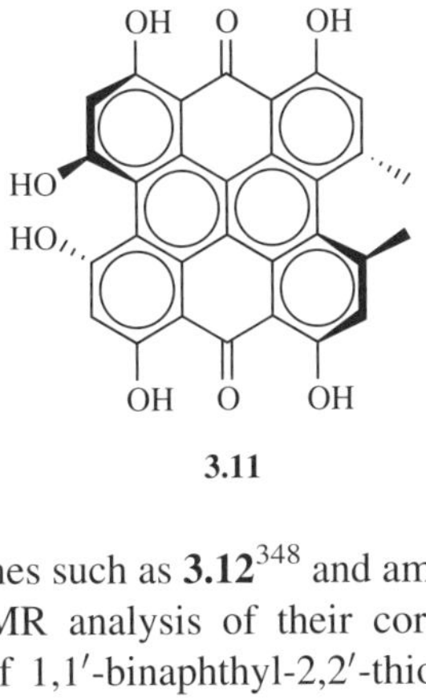

The optical purity of amines such as **3.12**[348] and amino acid methyl esters[349] has been determined by ^{1}H NMR analysis of their corresponding camphanamides. Finally, the optical purity of 1,1'-binaphthyl-2,2'-thiol (**3.13**) was determined by conversion to its corresponding dicamphanate thiol ester.[350] The three methyl resonances exhibited nonequivalence of 0.11, 0.35, and 0.93 ppm in the diastereomeric products.

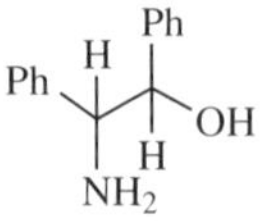

3.12

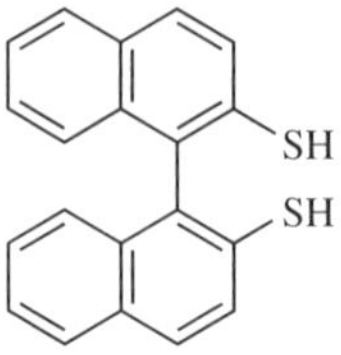

3.13

3.2. MENTHOXYACETIC ACID (MAA) (3.14)

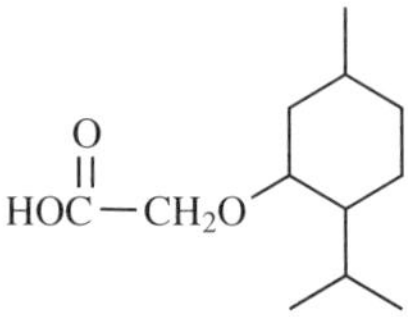

3.14

MAA has been a versatile reagent for the analysis of diols, especially those resulting from the epoxidation of polycyclic aromatic compounds. For example, the ^{1}H NMR spectrum of the bis-MAA esters of 4,5-dihydrodiols of benzo[a]pyrene (**3.15**) in benzene-d_6 can be used to determine optical purity and assign the stereochemistry.[351] The methylene resonances of the C(O)CH$_2$—O— unit of the menthoxy group are diagnostic. Using (−)-MAA, each of the methylene groups of the derivative of the (−) form of **3.15** appeared as two singlets, whereas in the (+) form they appeared as an AB quartet. Circular dichroism spectra were used to confirm the assignment.

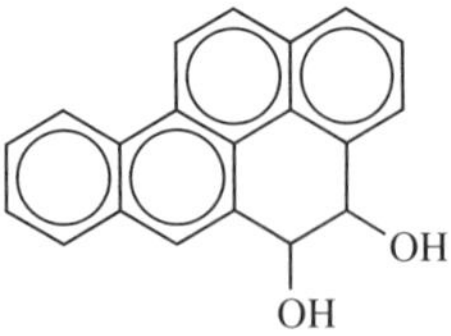

3.15

An analogous procedure has been used to assign the stereochemistry of 1,2,3,4-tetrahydrobenz[a]anthracene 3,4-diol (**3.16**),[352] trans-3,4-dihydroxy-3,4-dihydrobenzo[c]phenanthrene (**3.17**),[353] trans-3,4-dihydroxy-3,3,4,4-tetrahydrobenzo[c]phenanthrene (**3.18**),[353] (**3.19**),[354] and heterocyclic polycyclic compounds, one example of which is **3.20**.[355]

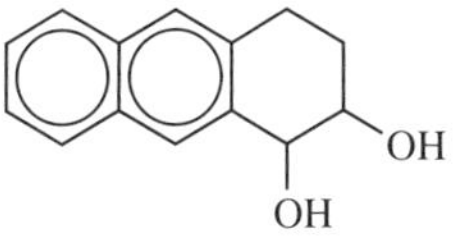

3.16 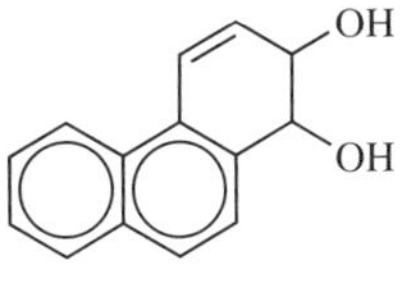 **3.17** 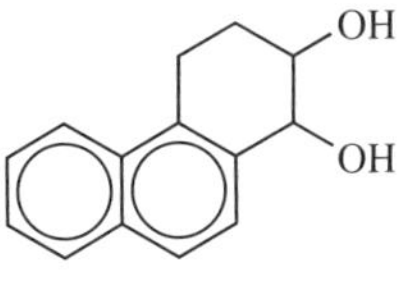 **3.18** 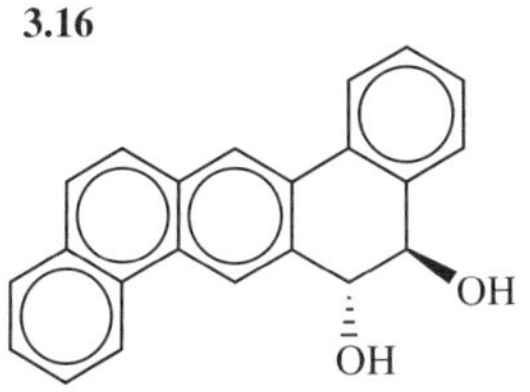 **3.19** 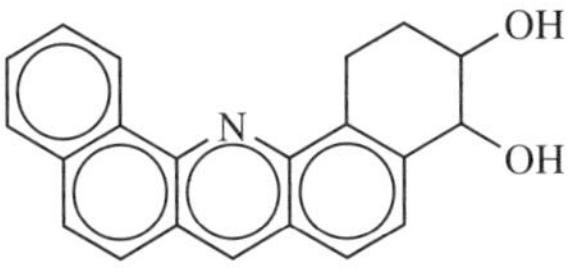 **3.20**

For diol **3.21**, reaction with α-methoxy-α-trifluoromethylphenylacetic acid (MTPA) or α-methoxyphenylacetic acid (MPA) caused a complicated mixture that was not suitable for NMR analysis.[356] Reaction with MAA in triethylamine as the solvent formed the bis-MAA ester in acceptable yield and the derivative was suitable for the analysis of absolute configuration. The assignment was confirmed through the use of circular dichroism.

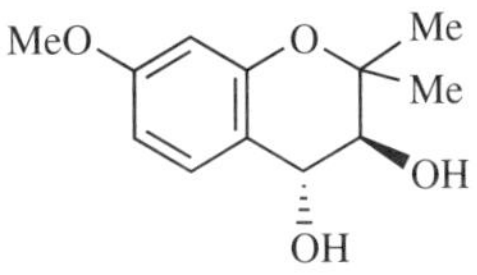

3.21

MAA has also been used to analyze monoalcohols. These include *trans*-2-*o*-tolyl cyclohexanol,[357] tetrahydro bromohydrin derivatives of chrysene (**3.22**),[352] naphthalene and anthracene,[358] and chloroquinols such as **3.23**.[359]

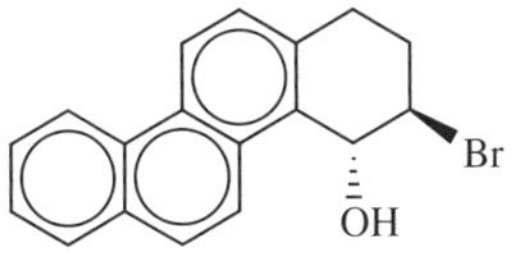

3.22 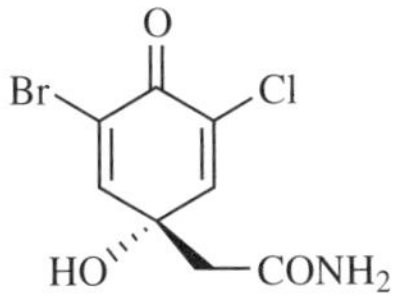 **3.23**

3.3. 2-(2,3-ANTHRACENEDICARBOXIMIDO)CYCLOHEXANE CARBOXYLIC ACID (3.24)

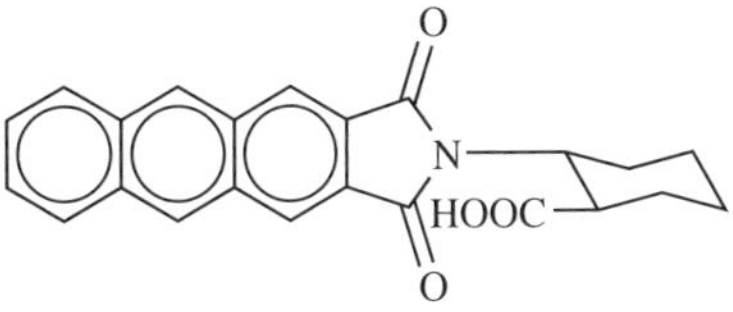

3.24

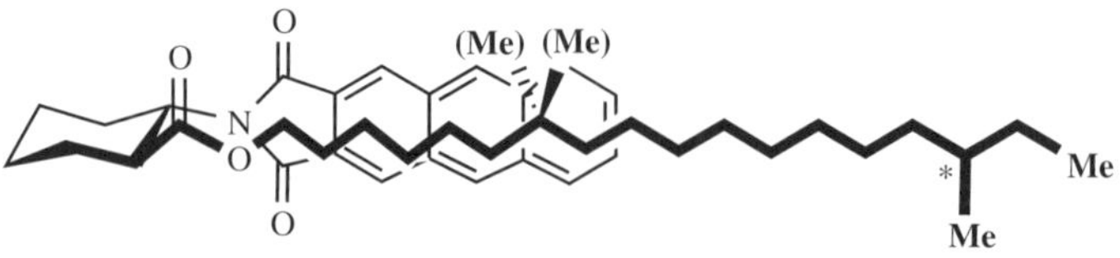

Figure 3.1. Shielding of the anthryl ring in derivatives of primary alcohols with 2-(2,3-anthracenedicarboximido)cyclohexane carboxylic acid.

3.24 is remarkable in its ability to distinguish chiral alcohols with remotely disposed chiral centers.[360–363] The ester derivatives of primary and secondary alkyl alcohols with 2-(2,3-anthracenedicarboximido)cyclohexane carboxylic acid position the alkyl chain directly over the shielding anthryl ring as seen in Figures 3.1 and 3.2. For primary alcohols, nonequivalent resonances in the ^{1}H NMR spectrum occurred up to the C9 group.[361,362] For secondary alcohols, it was possible to discriminate compounds with two alkyl groups differing by only one carbon and the chiral center at C16.[364]

Furthermore, the $\Delta\delta^{RS}$ values showed a characteristic positive and negative trend for methyl resonances at a branched position that alternated for odd and even carbons.[361] These trends were used to determine the absolute configuration of the alcohol. Similarly, a consistent trend of the terminal methyl resonances of secondary alcohols could be used to empirically assign the absolute configurations.[363]

3.4. 3β-ACETOXY-Δ5-ETIOCHOLENIC ACID (3.25)

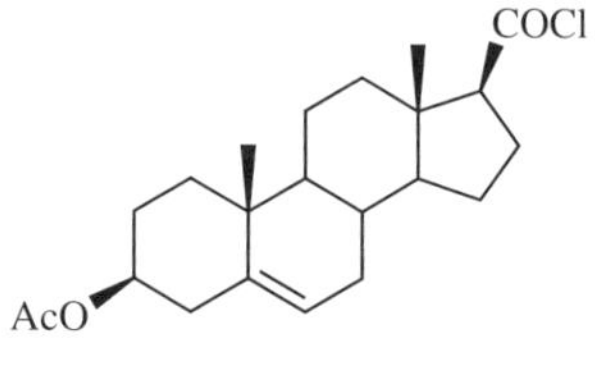

3.25

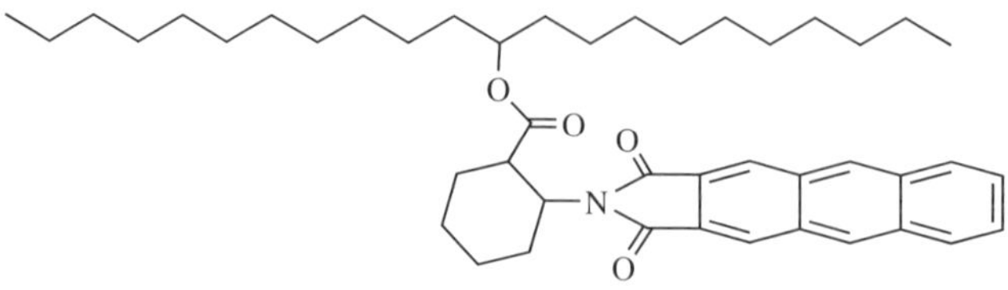

Figure 3.2. Shielding of the anthryl ring in derivatives of secondary alcohols with 2-(2,3-anthracenedicarboximido)cyclohexane carboxylic acid.

The compound 3β-acetoxy-Δ5-eticholenic acid can be used to assign the absolute configuration of *trans*-2-arylcyclohexanols,[365,366] 1-aryl-1-alkyl alcohols,[366,367] 1-aryl-1-alkyl amines,[366,368] and 1-arylethanethiols.[368] Shifts of the C18 methyl group of the steroid were diagnostic as the aryl substituent of the alcohol, amine, or thiol caused specific shielding depending on its absolute configuration. For example, the aryl group shielded the C18 methyl group in the (1S,2R)-derivative of *trans*-2-aryl cyclohexanol and deshielded it in the (1R,2S)-derivative.[365] The C18 methyl resonances typically appeared at shifts of about 0.5 and 0.7 ppm for the two diastereomers, a relatively unobstructed portion of most ^{1}H NMR spectra.[367] Discrimination was observed in methanol-d_4, benzene-d_6, and chloroform-d, although the greatest distinction was typically observed in chloroform-d.[366] Similar discrimination was observed for compounds with heteroaromatic substituent groups.

For the 1-arylethanethiol derivatives, the C18 resonance for the (S)-thiol derivative was consistently between 0.54 and 0.58 ppm, whereas that for the (R)-thiol derivative was between 0.69 and 0.71 ppm, enabling a reliable assignment of absolute configuration.[368]

3.5. NAPROXEN (3.26)

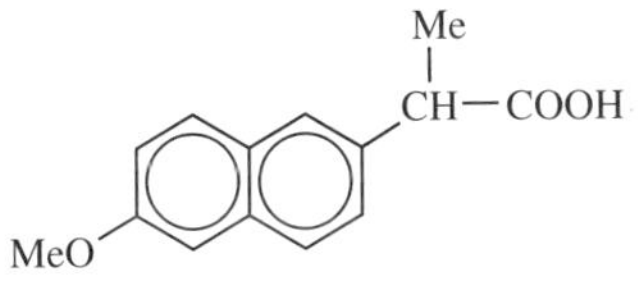

3.26

The utility of (S)-naproxen as a chiral solvating agent for sulfoxides was demonstrated.[369] Nonequivalence was observed in both the ^{1}H (0.015–0.031 ppm) and ^{13}C NMR spectra in substrates such as **3.27**. The extent of enantiomeric discrimination with naproxen was larger than that with MPA or ibuprofen, although (2-naphthyl)-methoxyacetic acid (2-NMA) was more effective for some of the substrates.

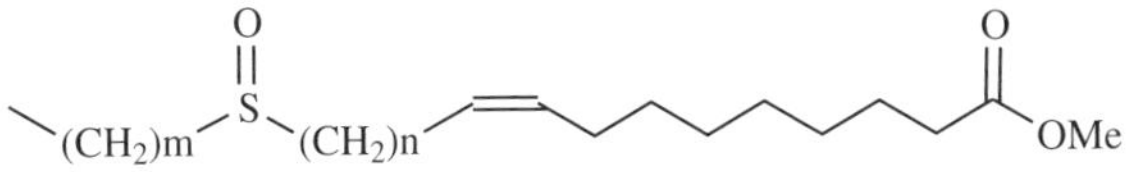

3.27

(S)-Naproxen has been used as a chiral derivatizing agent for the determination of optical purity of diethyl-1-hydroxy-, 2-hydroxyalkyl-, and aminoalkyl phosphonates (**3.28**).[370] The ^{31}P NMR spectrum was useful for determining optical purity, and with the 1-hydroxyalkyl phosphonates, the ^{31}P resonance of the (S)-isomer of the substrate was always at higher frequency than that of the (R)-isomer. The nonequivalence in the spectrum of the naproxen derivatives of hydroxyl compounds was comparable to those with MTPA and MPA, but larger with amines.

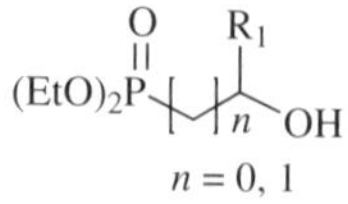

$$(EtO)_2P(O)-CH_2-[CH(R_1)]_n-OH$$
$$n = 0, 1$$

3.28

3.6. 2-*tert*-BUTYL-2-METHYL-1,3-BENZODIOXOLE-4-CARBOXYLIC ACID (3.29)

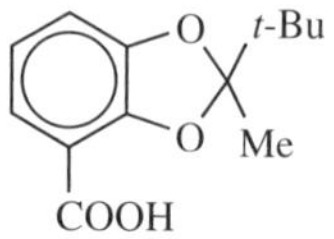

3.29

The absolute configurations of monosaccharides such as glucose, galactose, mannose, fructose, rhamnose, xylose, arabinose, and ribose can be determined by derivatization with 2-*tert*-butyl-2-methyl-1,3-benzodioxole-4-carboxylic acid.[371] **3.29** reacts with *per-O*-acetyl-D- and L-glycopyranoside-1-bromides at the 1-position. Most sugars formed only a single pyranoside anomer, although some did yield α-*axial* and β-*equatorial* pyranosides. Nevertheless, the shifts of the *tert*-butyl, methyl, and H1 signals in the ^{1}H NMR spectra correlated with the absolute stereochemistry about the anomeric position. Once this configuration is known, the stereochemistry of other sites can be obtained through an analysis of coupling constants. A set of relationships were provided that can be applied to other monosaccharides.

3.29 can also be applied to amino deoxy sugars.[372] After coupling to the reagent at position-1, the *per-O*-acetylated derivative was prepared. Analysis of the shifts of the *tert*-butyl and methyl resonances was used to assign absolute configurations.

3.7. *endo*-3-BENZAMIDONORBORNANE-2-CARBOXYLIC ACID (3.30)

3.30

3.30 and its corresponding 5-ene analog have been evaluated as chiral solvating agents for amines.[373] The reagents form salts with amines and the norbornane reagent generally provided slightly better chiral discrimination than the norbornene reagent, although the distinction was rather modest.

3.8. COUMARIN-BASED REAGENTS

Chiral derivatizing agents consisting of a coumarin unit attached to mandelic acid (**3.31**),[374] proline (**3.32**),[375] L-pyroglutamic acid (**3.33**),[376] and lactic acid[377] have been evaluated for determining the optical purity of amines and alcohols. Nonequivalence on the order of 0.01–0.18 ppm was observed for a wide variety of amines and alcohols, but none among these coumarin-based compounds was consistently the most effective discriminating agent. The resonance at C3 of the coumarin group provided a clear signal for analysis in the ^{1}H NMR spectrum. No racemization or kinetic resolution was observed during the derivatization.

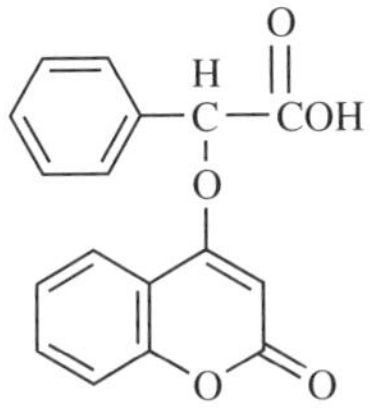

3.31

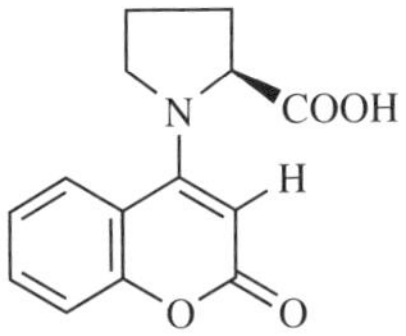

3.32

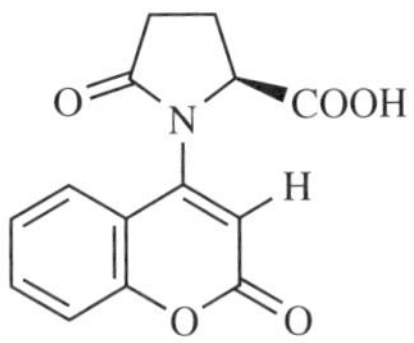

3.33

3.9. 2,2-DIPHENYLCYCLOPROPANE CARBOXYLIC ACID (3.34)

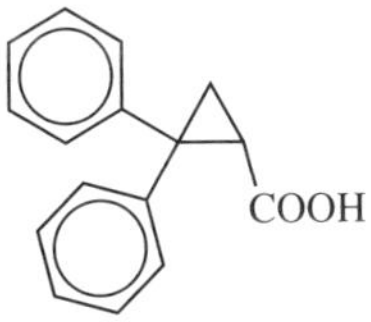

3.34

3.34 has been used to determine the enantiomeric purity of a variety of secondary alcohols.[378] Larger discrimination was observed than with the corresponding MTPA derivatives. No effort was made to see if the shifts could be used to assign absolute configuration.

3.10. AMIDE DERIVATIVES OF KEMP'S TRIACID

Kemp's triacid has served as a template for the preparation of chiral discriminating agents for amines.[379,380] The bis-amides with 1-phenylethylamine (**3.35**), 1-(1-naphthyl)ethylamine, or 1,2-diaminocyclohexane were evaluated. Mixing a compound such as **3.35**, which has one underivatized carboxylic acid group, with the amine substrate in chloroform-*d* formed the corresponding salt. The diaminocyclohexane reagent was ineffective. The naphthyl derivative generally provided the largest chiral recognition, although discrimination in the NMR spectrum was rather small.[379,380]

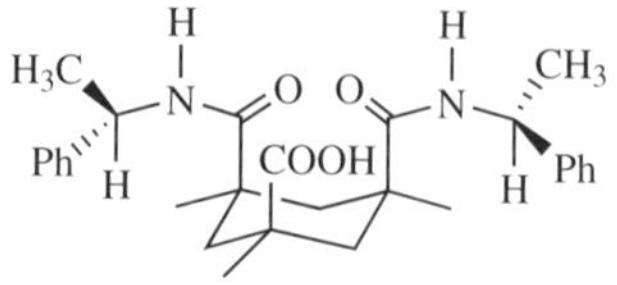

3.35

3.11. AMINO ACIDS

3.11.1. N-Boc-phenylalanine (3.36)

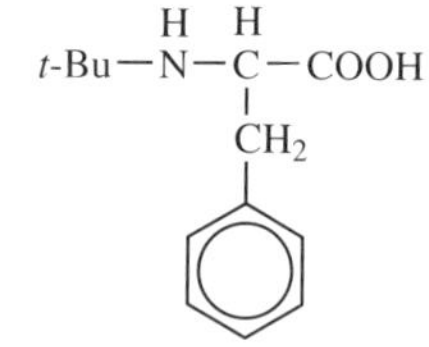

3.36

Esters of hydroxyl cyclopentenones were prepared using **3.36** and the optical purity was determined by ^{1}H NMR spectroscopy.[381] The optical purities of 1-hydroxyalkyl phosphonic acids (**3.37**) were determined using either N-Boc-L-phenylalanine or the corresponding L-valine.[382] Substrates of structure **3.37** were first converted to a dibenzyl derivative. Reaction of the alkyl hydroxyl group to form the corresponding ester provided diastereomers that exhibited nonequivalence of 0.06–0.60 ppm in the ^{31}P NMR spectrum.

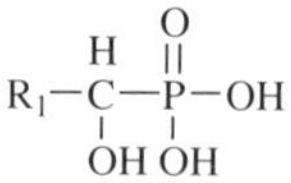

3.37

3.11.2. Isotopically Substituted Amino Acids

The L-*threo* and L-*erythro* [1-^{13}C-2,3-^{2}H] forms of phenylalanine, tyrosine, tryptophan, leucine, and histidine were prepared and incorporated into proteins.[383] The *pro*-(R) and *pro*-(S) positions of the isotopically labeled amino acids were distinguishable, and the vicinal coupling constants between the carbonyl carbon and prochiral β-protons were used to determine the conformation of side chains and fractional population of different rotamers.

3.12. (R)-(−)-(2,3,5,7-TETRANITRO-9-FLUORENYLOXIMINO)PROPANOIC ACID (3.38)

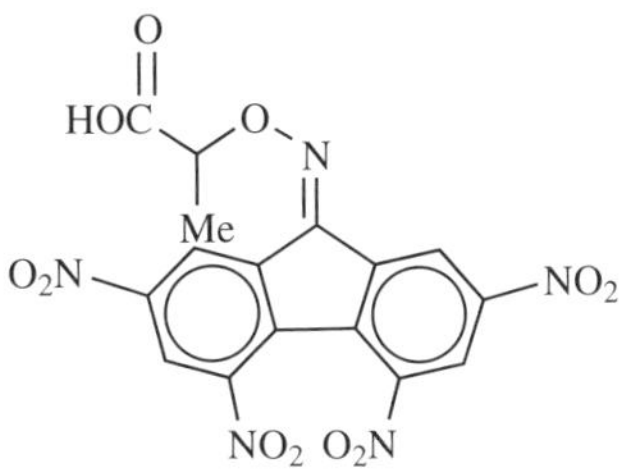

3.38

3.38 forms charge-transfer complexes with helicenes of structure **3.39**.[384] Certain resonances in the ^{1}H NMR spectrum of the helicene exhibit enantiomeric discrimination in the mixture.

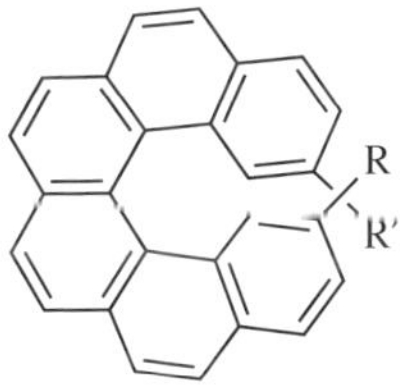

3.39

3.13. 2-CHLOROPROPANOIC ACID (2-Cl-PA)

2-Cl-PA has been used on a limited number of occasions for chiral NMR discrimination. One example was the determination of the optical purity of α-alkylated amino acids.[385] Neither 2-bromopropanoic acid nor MTPA was suitable for the analysis.

MTPA did not effectively react with α-substituted α-hydroxy acids or esters (**3.40**), whereas 2-Cl-PA did and provided diastereomeric ester derivatives in which the methyl doublet exhibited nonequivalent resonances in the ^{1}H NMR spectrum.[386]

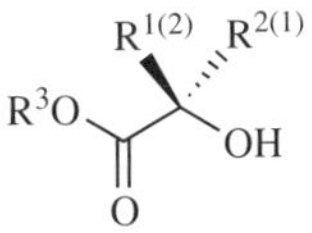

3.40

2-Cl-PA has also been used to assign the absolute configuration of hydroxyl groups in gluco-, galacto-, and mannopyranosides.[387] After protecting all of the hydroxyl groups except one, the 1-carboxyethyl ether derivative with 2-Cl-PA was prepared. The ^{13}C shifts varied depending on whether (R)- or (S)-2Cl-PA was used such that the absolute configuration of the monosaccharide could be assigned.

3.14. CHLOROFLUOROACETIC ACID

Chiral secondary alcohols react quickly with chlorofluoroacetic acid to form diastereomeric esters that exhibit distinct ^{1}H and ^{19}F NMR spectra.[388,389] No racemization occurred during the derivatization reaction. The shifts had empirical trends that correlated with the absolute configuration.

3.15. PERFLUOROPROPOXY PROPIONIC ACID/PERFLUORO ISOPROPOXY PROPIONIC ACID

Perfluoropropoxy propionic acid[390] and perfluoro isopropoxy propionic acid[391] have been used as chiral NMR derivatizing agents for amines, alcohols, and amino esters. The ^{19}F signal of the CF$_3$ group was conveniently monitored for the determination of optical purity, and the largest discrimination was observed in dimethylsulfoxide-d_6.[391] The perfluoro isopropoxy propionic acid was used to determine the optical purity of vinyl iodide **3.41** and vinyl alcohol **3.42** by monitoring the ^{19}F NMR spectrum.[392] The MTPA derivative of **3.42** could not be prepared, whereas the perfluoro reagent readily reacted.

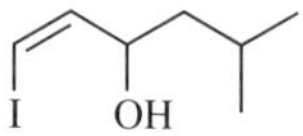

3.16. 2-METHYLBUTYRIC ACID

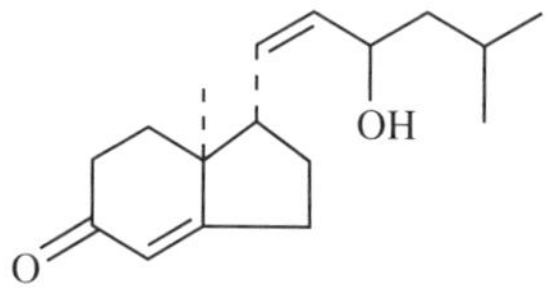

The absolute configurations of monosaccharides such as galactose, glucose, arabinose, and xylose were assigned on the basis of the ^{1}H NMR spectra of the *per-O-(S)*-2-methylbutyrate derivatives.[393] By examining D- and L-monosaccharides with known configurations, it was noted that there were small but significant shift differences that could be used as the basis of an empirical method. It was useful to have comparative standards of known configuration when applying the method.

3.17. LACTIC ACID AND DERIVATIVES

Metal complexes with ligands of structure **2.90** are chiral and form a pair of $\lambda\lambda\lambda\lambda$ and $\delta\delta\delta\delta$ enantiomers. Addition of lactate ion to the Yb(III) complex of **2.90** caused enantiomeric discrimination of certain resonances in the ^{1}H NMR spectrum, confirming the chirality of complexes of this type.[213]

3.17.1. TRIFLUOROLACTIC ACID (3.43)

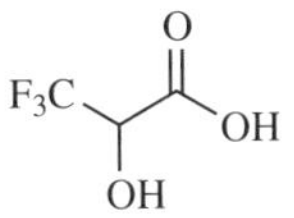

3.43

Trifluorolactic acid and its O-methyl and O-ethyl derivatives have been used as chiral derivatizing agents for alcohols.[394,395] The ^{19}F signal of the CF$_3$ group is conveniently monitored for the determination of optical purity. No racemization was observed during the derivatization procedure.

3.17.2. (S)-O-Acetyllactylchloride (3.44)

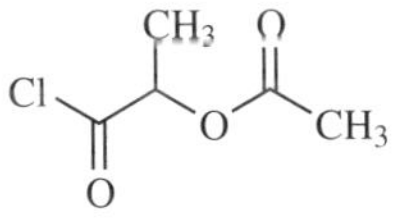

3.44

(S)-O-Acetyllactylchloride has been used as a chiral derivatizing agent to determine the optical purity of 2-furylcarbinols (**3.45**),[396] the primary-amine-containing compound **3.46**,[397] and 6-hydroxy-3-methylcyclohex-2-en-1-one (**3.47**).[398] The methyl doublet for the lactyl group split into two for the diastereomers and could be monitored for the analysis.

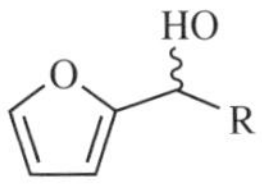

3.45

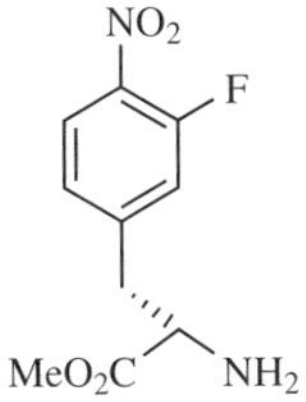

3.46

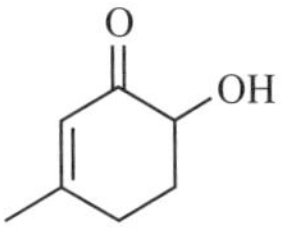

3.47

3.18. LASALOCID (3.48)

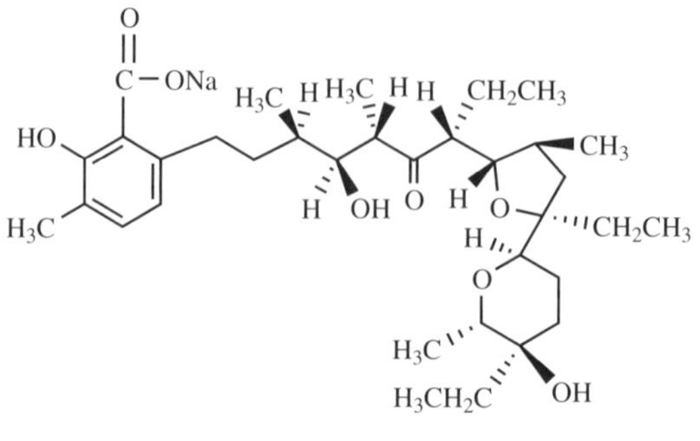

3.48

The compound 2-phenyl-1-(4-pyridyl)-3-(4-pyridyl-1-oxide)-2-propanol (**3.49**) was produced as a metabolite of a compound administered to rats, dogs, and a human subject.[399] The addition of lasalocid to solutions of **3.49** resulted in the formation of diastereomeric salts in which the resonances of the hydrogens α to the pyridine and pyridine oxide nitrogen atoms split.

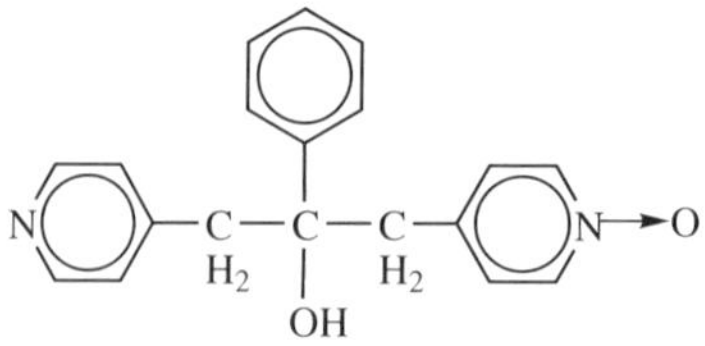

3.49

3.19. AXIAL CHIRAL CARBOXYLIC ACIDS

3.19.1. (+)-1-[2-Carboxy-6-(trifluoromethyl)phenyl]pyrrole-2-carboxylic Acid (3.50)

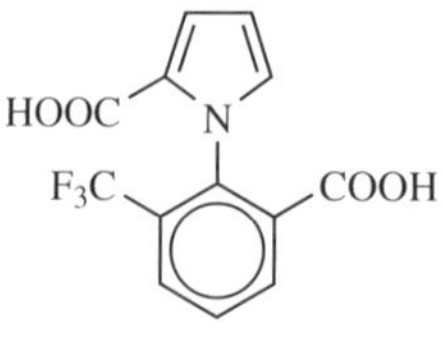

3.50

The axial chiral (+)-1-[2-carboxy-6-(trifluoromethyl)phenyl]pyrrole-2-carboxylic acid has been shown to be an effective chiral solvating agent for the determination of optical purity of oxetanes (**3.51**) and *cis*-but-2-ene-1,4-diols (**3.52**).[400] Hydrogen bonding between nitrogen and hydroxyl groups of the substrate and the carboxylic acid moieties of the discriminating agent accounted for the association in solution.

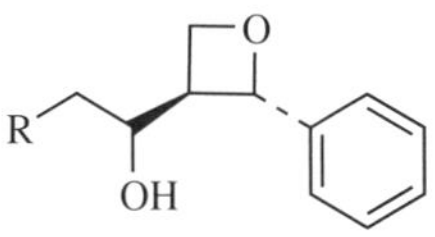

R = Et$_2$N
piperidyl

3.51

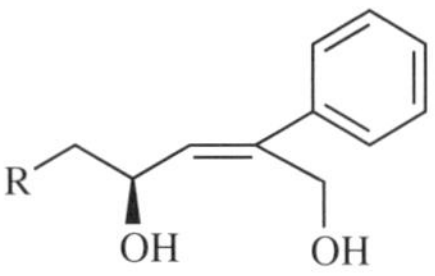

R = Et$_2$N
piperidyl

3.52

3.19.2. 2′-Methoxy-1,1′-binaphthyl-2-carboxylic Acid (MBNC) (3.53)

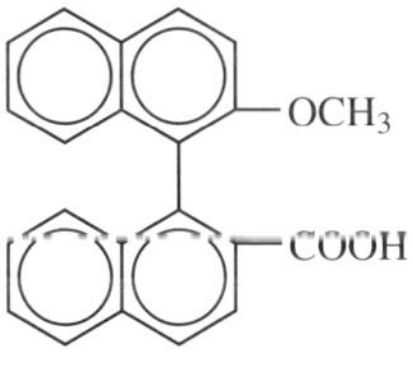

3.53

MBNC is an effective chiral derivatizing agent for α-chiral alkyl and alkyl aryl alcohols and amines.[401] Discrimination in the NMR spectrum of the diastereomers could also be enhanced by the addition of Eu(fod)$_3$ (fod = 6,6,7,7,8,8,8-heptafluoro-2,2-dimethyl-3,5-octanedione). In all cases, the methoxy resonance of the (aS,R)-derivative showed larger lanthanide-induced shifts than that of the (aS,S)-counterpart. Figure 3.3 illustrates how binding of the Eu(III) influenced the shift of the methoxy resonance, where L and M stand for the largest and medium-sized substituents. The (aS,R)-derivative exhibits less steric hindrance toward binding of europium.

The use of MBNC in assigning absolute configuration of secondary alcohols has been demonstrated.[402] Shielding by the naphthyl rings produced predictable shifts such that the $\Delta\delta^{RS}$ values correlated with the absolute stereochemistry. A more detailed NOE (nuclear Overhauser effect) study with derivatives of (−)-menthol and (−)-1-phenylethanol confirmed the conformational preferences and utility of the reagent for assigning the absolute configuration of secondary alcohols.

Molecular modeling studies of the MBNC derivatives indicated that two preferred conformations occurred, one with an *sp* (*syn-periplanar*) arrangement of the Ar$_1$–Ar$_2$

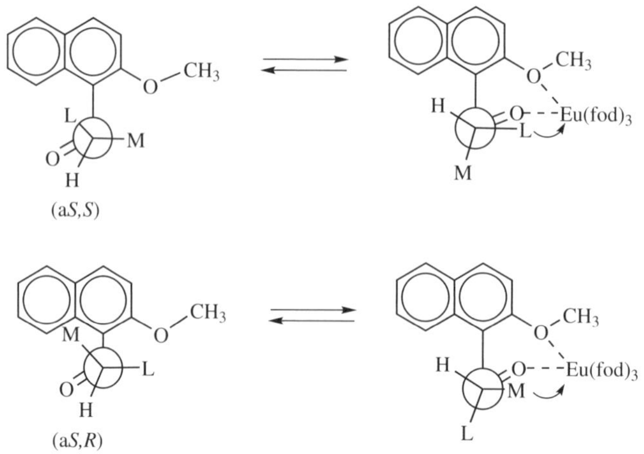

Figure 3.3. Binding of Eu(fod)₃ to MBNC derivatives of alcohols and amines. [Reproduced with permission.[401]]

and C=O bonds and the other with an *ap* (*anti-periplanar*) arrangement (Fig. 3.4). The *sp* conformer had the higher population of the two, but the shielding in the *ap* conformer was far more significant in producing the $\Delta\delta^{RS}$ values.[403] On the basis of these findings, the compound 1-(10-bromo-1-anthryl)-2-naphthoic acid (**3.54**) was prepared and evaluated as a chiral derivitizing agent for alcohols and amines. The greater shielding from the anthryl ring produced $\Delta\delta^{RS}$ values with the correct sign distribution to assign more reliably the stereochemistry of the alcohol or amine.[403]

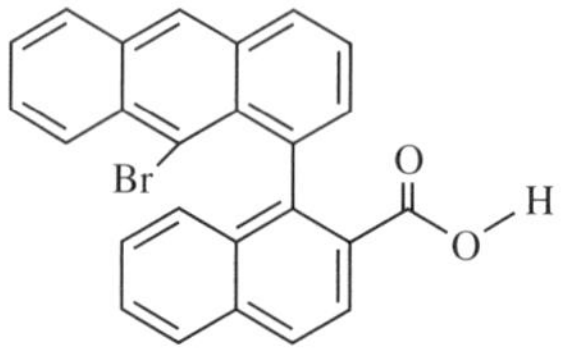

3.54

3.19.3. 2′-Methoxy-1,1′-binaphthalene-8-carboxylic Acid (3.55)

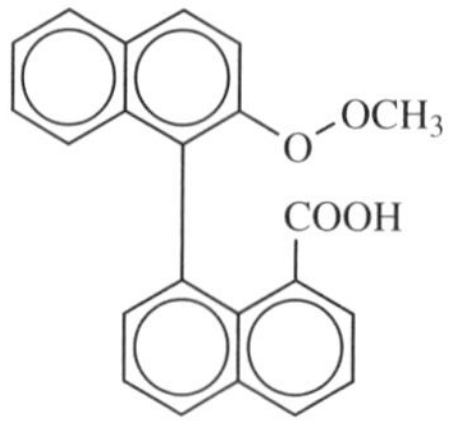

3.55

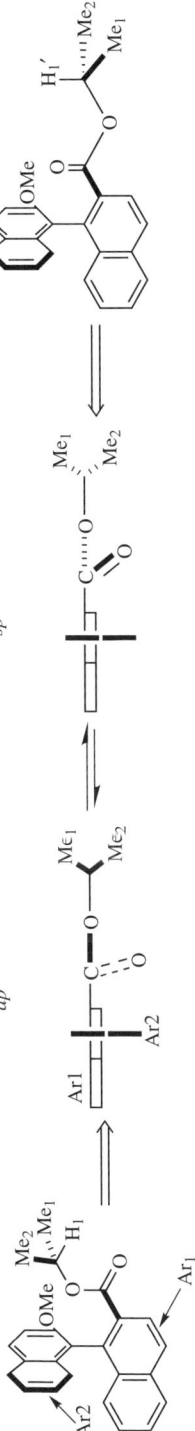

Figure 3.4. Conformational equilibrium of MBNC esters.

3.55 is an effective derivatizing reagent for assigning the absolute configuration of β-chiral primary alcohols.[404] A consideration of stabilizing CH–π interactions or destabilizing steric effects was used to predict the preferred rotamers. The coupling constants between H1a,b and H2 facilitate the determination of the major conformer. NOE correlations of the H1a,b signals of the primary alcohol with signals of the MBCA were then used to assign the absolute configuration. Relating the NOE of H1a,b to other protons of the alcohol provided the entire configuration of the alcohol. The analysis usually required only one enantiomer of **3.55**, although the reliability of the method will be enhanced by analyzing data with both enantiomers.[404]

3.19.4. 2′-Octylcarbamoyl-1,1′-binaphthyl-2-dicarboxylic Acid (3.56)

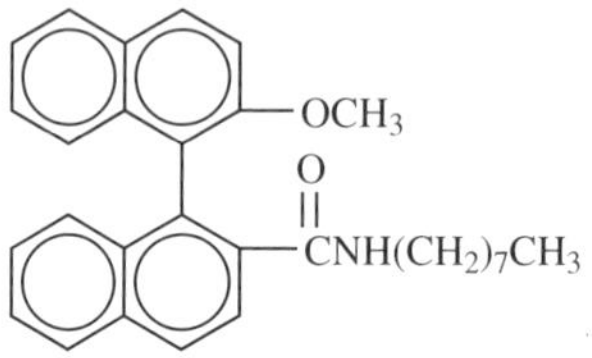

3.56

3.56 was examined as a soluble analog of a liquid chromatographic phase to determine aspects of the enantiomeric discrimination mechanism by NMR spectroscopy.[405] The 3,5-dinitrobenzoyl carbamate derivative of 1-phenylethanol (**3.57**) was studied and nonequivalence up to 0.38 ppm was observed for some resonances in its ^{1}H NMR spectrum with **3.56**.

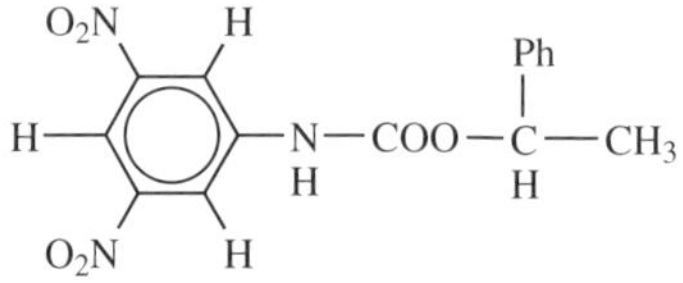

3.57

3.19.5. 2-(2′-Methoxy-1′-naphthyl)-3,5-dichlorobenzoic Acid (MNCB) (3.58)

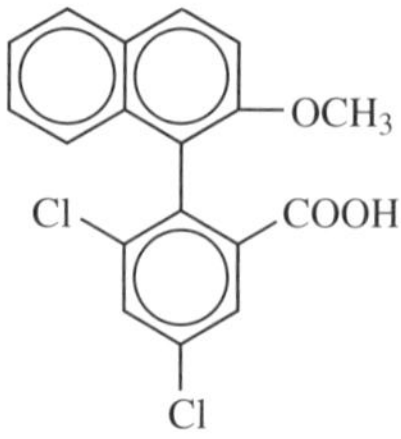

3.58

The absolute configuration of secondary alcohols can be determined after derivatization to their corresponding esters with MNCB.[402] The ester derivatives adopt a preferred conformation in which the naphthyl ring caused shielding that correlated with absolute stereochemistry. NOE experiments on the derivatives with (−)-menthol and (−)-1-phenylethanol[39] and molecular modeling[403] confirmed the conformational preferences used in predicting the signs of the $\Delta\delta^{RS}$ values.

The absolute configuration of planotriol monoacetate (**3.59**) was assigned by analyzing the $\Delta\delta^{RS}$ values from the ^{1}H NMR spectrum of the monoester of MNCB.[406] The absolute configurations of *ent*-bicyclogermacrene (**3.60**)[406] and the C8 hydroxy site of enokipodin C methylated at the C1 and C4 hydroxyl sites (**3.61**)[407] were also assigned on the basis of $\Delta\delta^{RS}$ values measured for the derivatives with MNCB. NOE data were used to further support the stereochemical assignments.

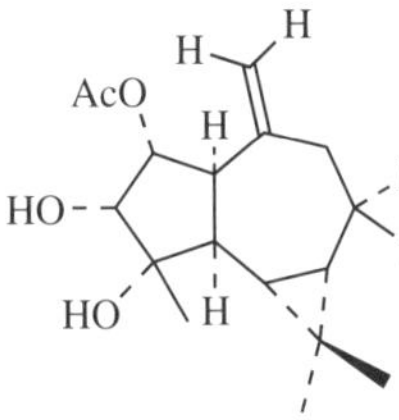

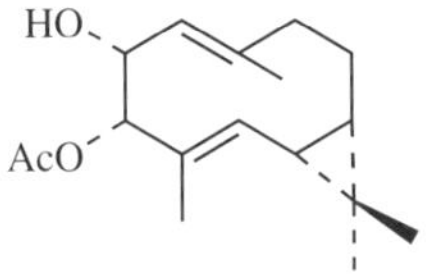

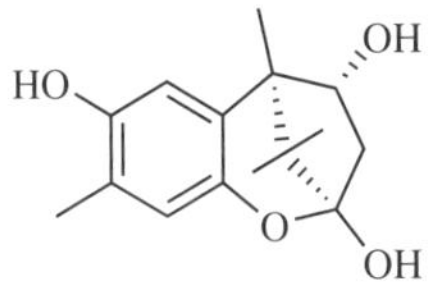

4

HYDROXYL- AND THIOL-CONTAINING REAGENTS

Chiral reagents with hydroxyl groups have been widely used for NMR analyses. Alcohols have been employed as chiral derivatizing agents, usually involving the conversion of carboxylic acids to esters. Certain diols have been used in the analysis of chiral ketones, whereas a lactol commonly known as Noe's reagent can be employed as a derivatizing agent to assign the absolute configurations of alcohols, amines, thiols, and carboxylic acids. Another strategy has been to use glycosidylation shifts in the spectra of aglycones prepared by the reaction of a secondary alcohol and glycoside for assigning absolute configurations of the alcohol. Analysis of the prochiral hydrogen atoms of methylene groups α to a carboxylic acid has been possible using certain derivatives prepared from chiral alcohols.

The compound 1-(9-anthryl)-2,2,2-trifluoroethanol (TFAE) is one of the most versatile chiral solvating agents ever developed. For certain classes of substrates, association with TFAE has a specific geometry, such that predictable shielding by the anthryl ring can be used to assign the absolute configuration. Many other compounds are capable of forming donor–acceptor complexes with TFAE through dipole–dipole and $\pi-\pi$ interactions in nonpolar solvents and are therefore candidates for chiral analysis with this reagent.

Axial chiral compounds with hydroxyl groups, the most notable example of which is 2,2'-dihydroxy-1,1'-binaphthalene, have also been applied as chiral NMR solvating and derivatizing agents.

Discrimination of Chiral Compounds Using NMR Spectroscopy, by Thomas J. Wenzel
Copyright © 2007 John Wiley & Sons, Inc.

4.1. 2,2,2-TRIFLUOROPHENYLETHANOL (TFPE) (4.1)

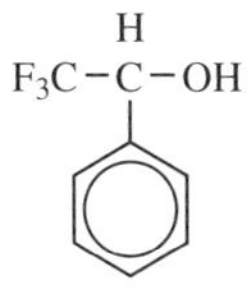

$$F_3C-\underset{\underset{\text{Ph}}{|}}{\overset{\overset{H}{|}}{C}}-OH$$

4.1

TFPE was first used as a chiral NMR solvating agent in 1968.[408] TFPE was used as the solvent in this early work and produced enantiomeric discrimination in the ^{1}H NMR spectra of nine amines including 1-(1-naphthyl)ethylamine (NEA). An especially interesting observation was that the shifts of enantiomeric amines with TFPE showed characteristic trends that correlated with the absolute configuration. For example, the structures in Figure 4.1 for the complex between TFPE and NEA show differential shielding caused by the phenyl ring for the two enantiomers. The π–π stacking interaction was important in stabilizing the complex. The corresponding 2,2,2-trifluoromethyl-1-cyclohexylethanol derivative was not nearly as effective in causing enantiomeric discrimination in the ^{1}H NMR spectra. Even in this first report, it was noted that the corresponding 2,2,2-trifluoromethyl-1-(1-naphthyl)ethanol derivative caused larger enantiomeric discrimination than the phenyl derivative because of greater shielding of the naphthyl ring.[408] The naphthyl derivative has not been used that much and instead the corresponding 2,2,2-trifluoro-1-(9-anthryl)ethanol (TFAE), often referred to as Pirkle's alcohol, was introduced a few years later and is now the reagent of choice for these analyses.

TFPE and TFAE, by virtue of their hydroxyl groups, have the potential to bind in nonpolar solvents to many compounds through dipole–dipole interactions and can be used to either determine the optical purity or in some cases assign the absolute configuration of substrates. The findings with TFPE that are described herein apply similarly to the use of TFAE. TFPE is effective for the analysis of optical purity and assignment of absolute configurations of sulfoxides.[409–411] The interaction involves hydrogen bonding to the sulfinyl group as shown in Figure 4.2.[411] The differential shielding caused by the phenyl group for the two sulfoxide enantiomers is apparent from the structures. The reagent has been used to distinguish sulfoxides that are chiral

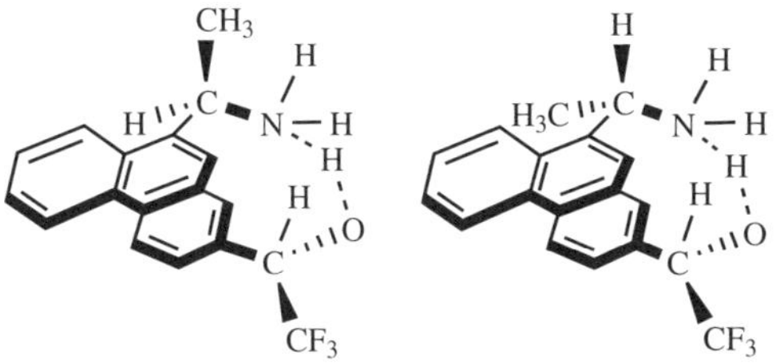

Figure 4.1. Complex between TFPE and the two configurations of NEA.

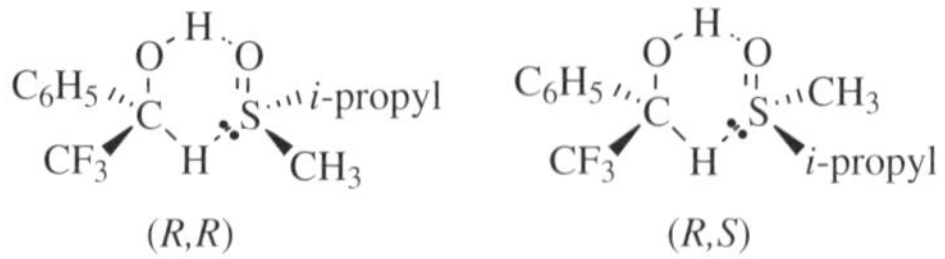

Figure 4.2. Association of TFPE with the two configurations of a sulfoxide.

only by virtue of ^{13}C substitution (e.g., dimethylsulfoxide in which one of the methyl groups is selectively ^{13}C labeled)[410] or ^{2}H substitution (dimethyl-1,1,1-d_3-sulfoxide).[409] Enantiomeric discrimination of sulfoxides was two- to fivefold larger using derivatives in which the phenyl ring of TFPE was replaced with a 9-(10-methylanthryl),[410] 1-naphthyl, 9-anthryl, or 3-pyrenyl group.[411]

The absolute configuration of the sulfoxides **4.2** and **4.3**, which were obtained through conversion of their corresponding sulfones, was assigned using NMR shift data with TFPE.[412] Chiral episulfides and episulfoxides (**4.4**) were enantiomerically discriminated in the presence of TFPE.[413] The extent of discrimination was larger for the sulfoxide. The resonances of the (S)-enantiomer of the sulfoxide consistently shifted to lower frequency in the presence of (−)-TFPE than those of the (R)-enantiomer.

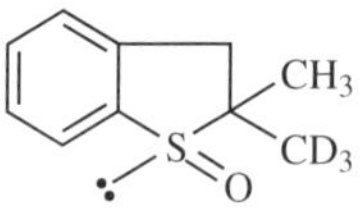

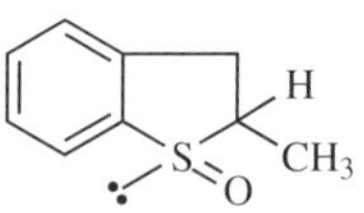

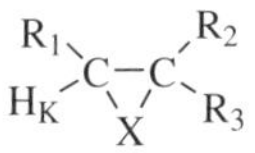

Compound **4.5** exists as four possible stereoisomers, and shifts in the NMR spectrum in the presence of TFPE were used to distinguish and assign the absolute configuration of all four isomers.[414] The analysis of the monosulfoxide was facilitated by analyzing corresponding disulfoxides of established configurations.

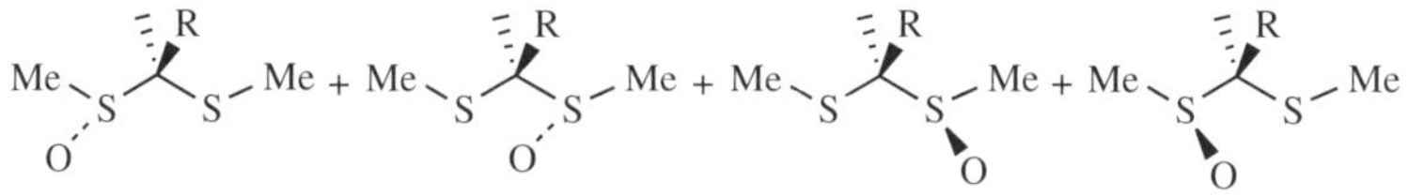

TFPE also caused enantiomeric discrimination in the ^{1}H NMR spectra of sulfinamides, sulfinates, sulfites, and thiosulfinates (**4.6**).[415] TFPE and the corresponding derivatives in which the phenyl ring was replaced with a 1-naphthyl or 10-methyl-9-anthryl group caused consistent trends in the shifts that correlated with the absolute configuration of alkyl and aryl sulfinates.[416] The association of sulfinate esters is more complex than that of sulfoxides because the alkoxy oxygen atom interacted with the reagent. In all cases, (S)-TFPE caused the resonances in the sulfinyl R group of the

(S)-enantiomer to have a lower frequency than that of the (R)-enantiomer, whereas the reverse was true for the hydrogen resonances of the alkoxy substituent group.

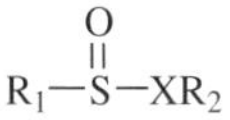

$$R_1-\underset{\underset{O}{\|}}{S}-XR_2$$
$$X = N, O, S$$

4.6

The NMR spectra of phosphine oxides (**4.7**) were enantiomerically discriminated in the presence of TFPE.[415,417] Similarly, TFPE was able to discriminate the ^{31}P resonances of the racemic pair [(S,S) from (R,R)] of the P,P'-diisopropyl-N,N,N',N'-tetramethylphosphoramide (**4.8**).[418] No evidence of the *meso*-isomer [(R,S)/(S,R)] was observed in the ^{31}P NMR spectrum.

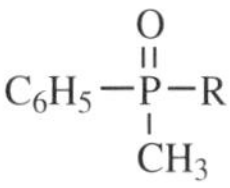

4.7

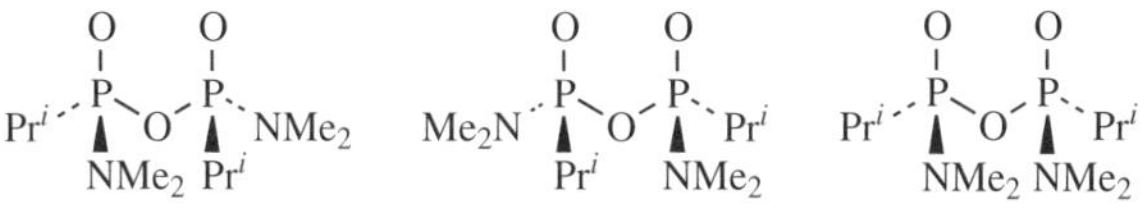

4.8 (S)-(S)   (R)-(R)   (R)-(S)

TFPE has been used to measure the optical purity and assign the absolute configuration of N,N'-dialkylaryl amine oxides such as N-methyl-N-ethyl-α-naphthylamine oxide (**4.9**) and N-methyl-N-ethyl aniline oxide.[415,419] 2,2,2-Trifluoro-1-(1-naphthyl)ethanol caused enantiomeric discrimination that was about twice as large as that with TFPE, whereas the corresponding 2,2,2-trifluoro-1-cyclohexylethanol derivative produced no enantiomeric discrimination. The trends in relative shifts with absolute configuration were explained by the association geometry shown in Figure 4.3. The structure has a hydrogen bond between the chiral alcohol and the amine oxide and a secondary attractive interaction between the acidic carbonyl hydrogen and the basic π-electron system of the oxide's aryl group.[419]

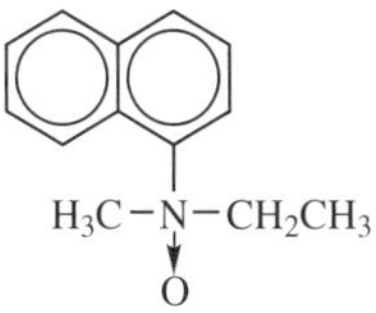

4.9

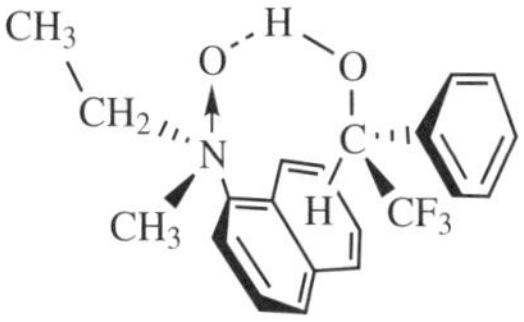

Figure 4.3. Association of TFPE with N-methyl-N-ethyl-α-naphthylamine oxide.

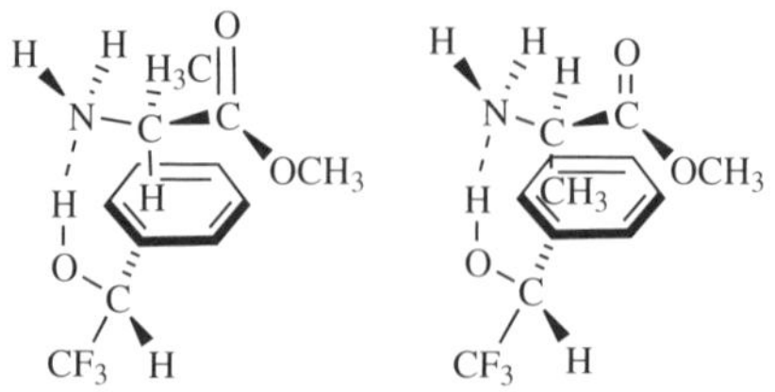

Figure 4.4. Association of TFPE with the two configurations of amino acids.

Shift data in the ^{1}H NMR spectra of α-amino acid methyl esters, disubstituted glycines, and β-amino acids with TFPE have been used to assign absolute stereochemistries.[420] A consistent observation with 15 amino acids was that (−)-TFPE caused the α-carbomethoxy and α-carbinyl resonances of the (S)-isomer to occur at higher frequency than those of the (R)-isomer, whereas the reverse order occurs for the resonances of the R group. The association of amino acids with TFPE is depicted in Figure 4.4. The steric bulk of the α-carbon substituents, the interaction of the carbonyl group with the π-electrons of the phenyl ring, and the strength of the hydrogen bond between the α-amino group and the hydroxyl group of the TFPE were the three factors involved in the association of the solvating agent and amino acids. The methoxy resonance was most conveniently monitored, but others showed discrimination as well.

The hydrogen resonance of chiral epoxides (**4.10**) was enantiomerically discriminated in the presence of TFPE.[421] Using (−)-TFPE, the resonance for the (S)-isomer was consistently at lower frequency than that of the (R)-isomer such that absolute configuration of the epoxide could be assigned.

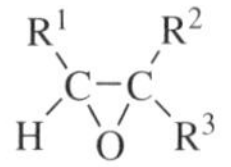

4.10

^{1}H NMR shift data with TFPE show consistent trends that can be used to assign the absolute configuration at the chiral nitrogen atom of oxaziridines (**4.11**).[422,423] The interaction of TFPE with oxaziridines is shown in Figure 4.5, and the structures show the differential shielding by the phenyl ring that can be used in assigning the stereochemistry. TFPE has been used to determine the optical purity of diaziridines of general structures **4.12** and **4.13**.[424]

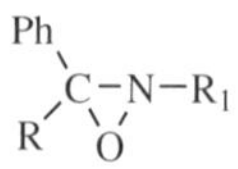

4.11 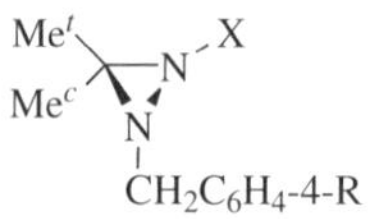   **4.12** 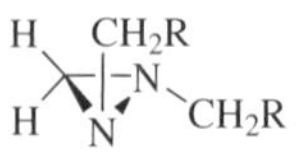   **4.13**

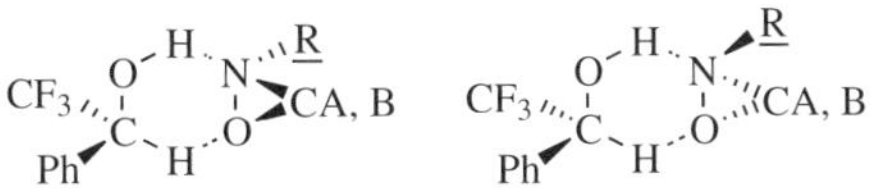

Figure 4.5. Association of TFPE with the two configurations of oxaziridines.

TFPE can be used in the analysis of the absolute configurations of imines such as **4.14**.[423] Figure 4.6 shows the interaction that can be used to rationalize the stereochemical assignment.

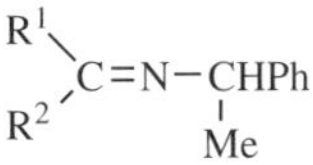

4.14

The ability of using a reagent like TFPE to distinguish the *meso*-isomer of a compound from the *d,l*-diastereomers was demonstrated for 2,3-diamino succinate (**4.15**).[425] The methine signals of the *meso*-isomer are anisochronous in the presence of TFPE and couple to each other, whereas the *d,l*-pair exhibits a singlet for each set of methine resonances with no coupling.

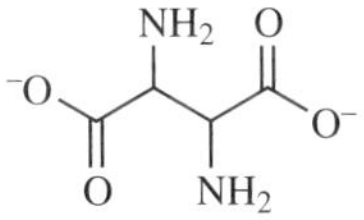

4.15

There are compounds for which the association with TFPE may not be known with enough certainty to assign absolute configurations, but the spectrum shows nonequivalence that is sufficient for determining optical purity. One example is the spirosulfurane **4.16**.[426] Four methyl singlets were observed in the ^{1}H NMR spectrum

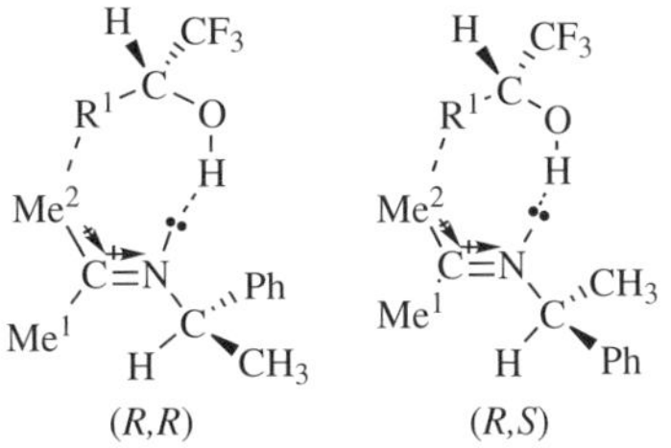

(R,R)   (R,S)

Figure 4.6. Association of TFPE with the two configurations of imines.

of **4.16** in the presence of TFPE, which was consistent with the trigonal bipyramid geometry about the sulfur atom.

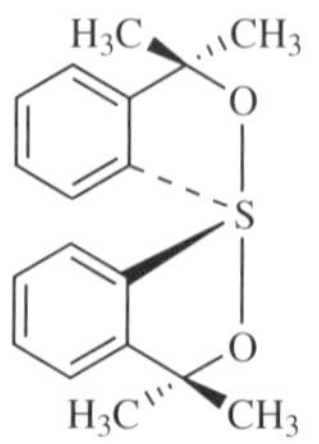

4.16

The enantiomeric purity of N-benzyl-3-isobutoxy-N-phenyl-2-(pyrrolidin-1-yl)propylamine (**4.17**) was determined by the addition of TFPE.[427] In this case, the compound was tritiated and the analysis was undertaken by ^{3}H NMR spectroscopy.

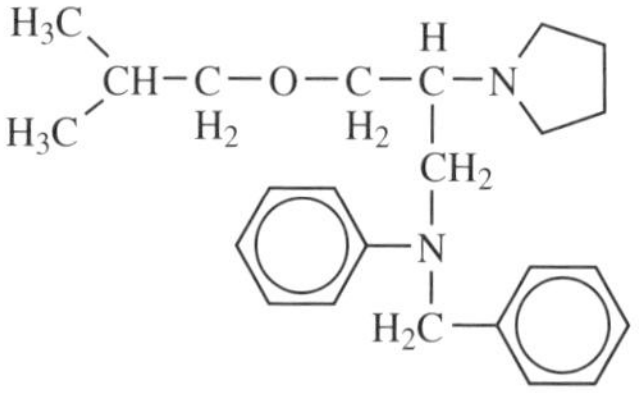

4.17

A study of the ligand substitution reaction of (R)- or (S)-1-phenylethylamine (PEA) for the valine-O-methyl ligand of trans-[PtCl$_2$(SR$_2$)(R,S-Val-O-Me)], in which the R group on the sulfur atom was methyl, benzyl, or tert-butyl, was facilitated through the addition of TFPE.[428] Enantiomeric discrimination of the resonances of the valine-O-methyl group in the presence of TFPE allowed a monitoring of the reaction.

The compound 2-bromo-2-methyl-1-(3-bromo-2-methoxy-4,5,6-trimethylphenyl)-1-propanone (**4.18**) exhibits chirality because of slow rotation.[429] The rate constant of enantiomerization was measured in the absence and presence of TFPE. The TFPE reportedly did not perturb the process, indicating its suitability for these types of measurements.

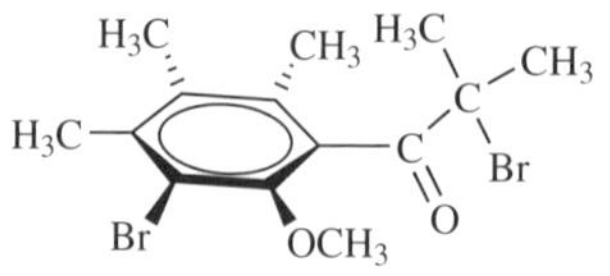

4.18

1,3-Alternate configurations of calixarenes (Fig. 4.7) such as **4.19** under conditions of slow exchange have aromatic protons that are enantiotopic by internal

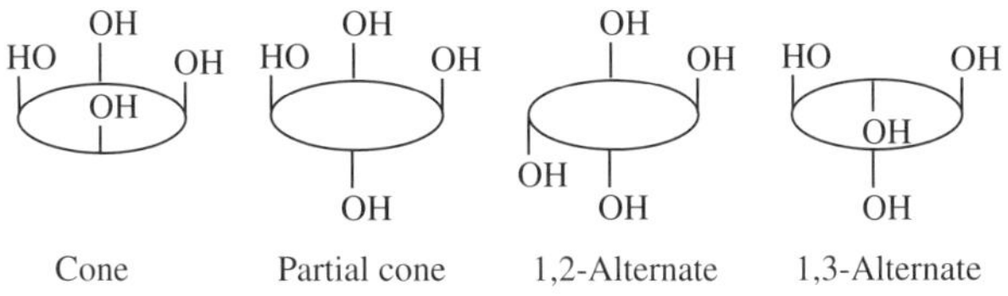

Figure 4.7. Four conformations of calix[4]arenes.

comparison.[430] Addition of TFPE caused discrimination in the ^{1}H NMR spectrum and two signals in the ^{13}C NMR spectrum for the carbonyl carbon. Warming the sample in the presence of TFPE facilitated a determination of the inversion barrier of **4.19**.

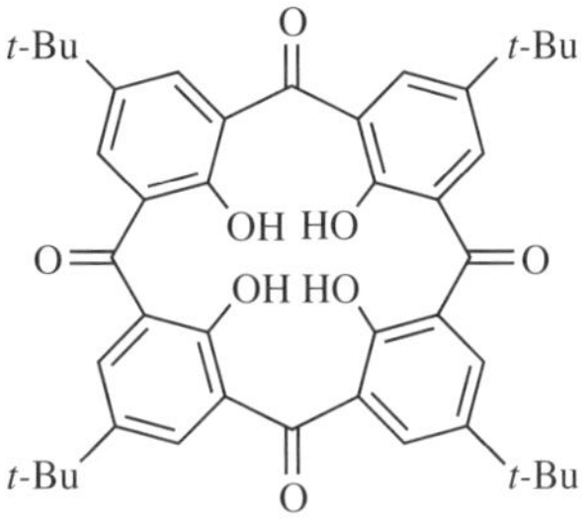

4.19

4.2. 2,2,2-TRIFLUORO-1-(9-ANTHRYL)ETHANOL (PIRKLE'S ALCOHOL (TFAE) (4.20)

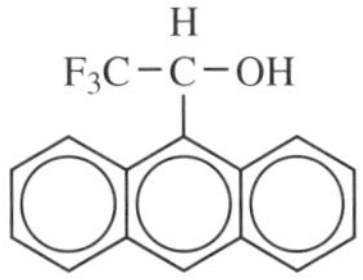

4.20

TFAE is one of the most widely used and versatile chiral NMR solvating agents. The following section will not be an exhaustive description of all of the uses of TFAE, but will provide examples of the types of analyses that it has facilitated. TFAE can be employed with certain families of substrates to assign absolute configurations. As will be shown, TFAE is also useful for measuring the optical purity of many types of compounds that have polar groups capable of association through dipole–dipole and π–π interactions.

4.2.1. Analysis of Sulfoxides

The enantiomeric discrimination observed in the NMR spectra of sulfoxides is much larger with TFAE than with TFPE because of greater shielding from the anthryl ring.[411] The binding mechanism between TFAE and sulfoxides is identical to that with TFPE (Fig. 4.2), and the larger enantiomeric discrimination provides for more reliable assignment of absolute configuration.

The utility of TFAE and α-methoxyphenylacetic acid (MPA) in causing enantiomeric discrimination in the ^{1}H and ^{13}C NMR spectra of homoallylic and allylic sulfoxides such as **4.21** and **4.22** was examined.[431] The nonequivalence was greater with TFAE than MPA and larger in benzene-d_6 than in chloroform-d. The ability to use the discrimination to assign the absolute configurations may be possible, although greater reliability would have been obtained if the analysis extended to more nuclei of the substrate.

4.21   **4.22**

4.2.2. Analysis of Lactones

TFAE is especially useful for assigning the absolute configuration of γ-lactones (**4.23**).[432,433] Lactones associate with TFAE as shown in Figure 4.8 and cause predictable differences in shielding that occur for the two configurations of the lactone. The discrimination with TFAE was superior to that with TFPE or the corresponding reagent with a naphthyl group, and nonequivalence of up to 0.1 ppm was common for some of the hydrogen resonances.[432] Enantiomeric discrimination was observed in both the ^{1}H and ^{13}C NMR spectra.[433] In lactones containing a dinitrophenyl substituent, there was some π–π stacking interactions that led to perturbations in the general trends of the shifts. It was possible to anticipate the different geometry with these, such that the shifts could still be used to assign absolute stereochemistries.

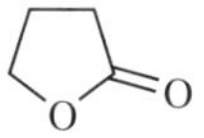

4.23

TFAE also caused enantiomeric discrimination in the ^{1}H NMR spectra of δ-lactones such as **4.24** and **4.25**.[434] Unlike the γ-lactones, analysis of the relative shifts of δ-lactones required a consideration of the conformation of the ring. Chiral recognition was generally larger if the lactone had substituents in the axial position.

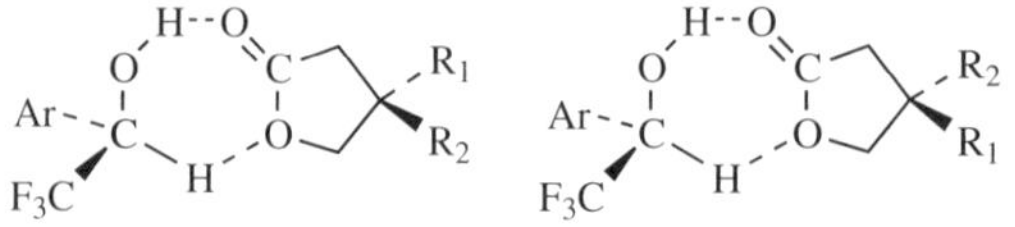

Figure 4.8. Association of TFPE and TFAE with the two configurations of lactones.

Trends in the shifts that allow the assignment of absolute configuration may be possible, but not enough examples were studied to know this with certainty.[434] Analyzing model δ-lactones with known configurations is likely warranted to enhance the reliability of assignments of absolute configuration.

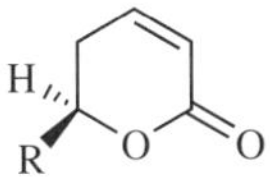

4.24

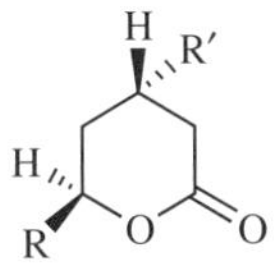

4.25

The nonequivalence in the NMR spectra of α,α-disubstituted and β-substituted β-propiolactones (**4.26**) with TFAE could be used to assign absolute configurations.[435] The structure of the TFAE–lactone complex provided reliable predictions of the differences in shielding for the two enantiomers.

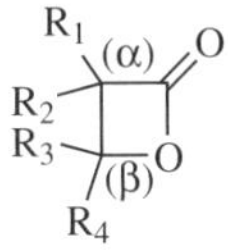

4.26

The absolute stereochemistry of the lactone moieties of annonaceous butenolides such as **4.27** was assigned on the basis of NMR data with TFAE.[436] Other chiral reagents including 1-phenylethylamine, 2,2′-dihydroxy-1,1′-binaphthalene, and 1-phenylethanol were tested but TFAE was the only one that provided satisfactory enantiomeric discrimination. Model lactones such as 5-methyl furan-2(5H)one were studied to confirm the stereochemical assignments. The other base sites in the annonaceous butenolides did not interfere with the binding of the lactone moiety to TFAE.

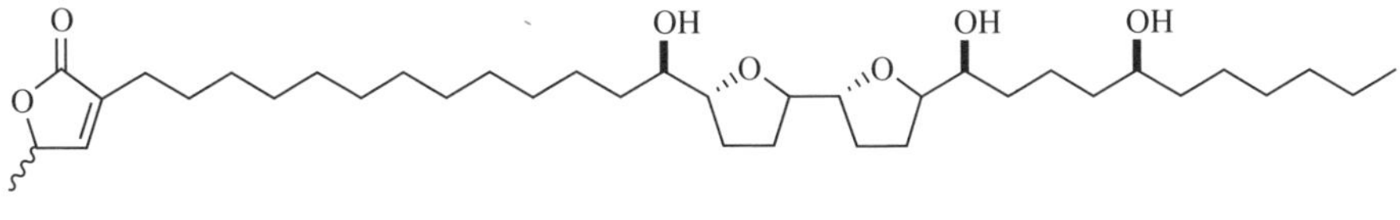

4.27

4.2.3. Analysis of Lactams

NMR shift data with TFAE have been used to determine the optical purity and assign the absolute configurations of γ-lactams such as 5-methyl-2-pyrrolidinone (**4.28**).[437] The lactam–TFAE complex involved a hydrogen bond between the hydroxyl group of TFAE and the carbonyl group of the lactam, and an interaction between the electron-deficient NH group and the π-system of the anthryl ring. The binding was specific enough that the sense of nonequivalence and absolute configuration could be assigned.[437]

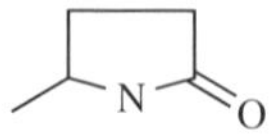

4.28

Shifts in the ^{1}H NMR spectrum of the four stereoisomers of (1-phenylethyl)-2-oxa-7-azabicyclo[3.2.0]heptan-6-one (**4.29**) with TFAE were used to assign the stereochemistry of the bridgehead positions.[438]

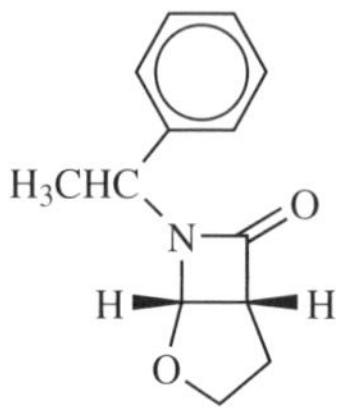

4.29

4.2.4. Analysis of Oxaziridines

Shifts in the ^{1}H NMR spectra of oxaziridines (**4.11**) with TFAE permitted the determination of optical purity.[439,440] It is also possible to assign the absolute configuration, although considerable care has to be taken.[440] For example, ozaziridines with an aryl substituent group had a secondary interaction with TFAE that caused additional shielding, complicating the relationship between the shifts and the absolute configuration.[439] In the absence of other basic sites that may associate with TFAE and alter the geometry, the structures in Figure 4.5 can be used to explain the relative shifts in the spectra of oxaziridine enantiomers in the presence of TFAE.[440]

The optical purities and absolute configurations of N-alkyloxaziridine-3,3-dicarboxylic esters (**4.30**) were determined on the basis of shift data with TFAE.[441] TFAE was also applied to the analysis of N-chloroaziridines.[442] The optical purity was readily determined but the trends in relative shifts for the two configurations were difficult to rationalize so that absolute configurations of the N-chloroaziridines could not be assigned with certainty.

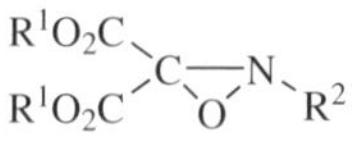

4.30

4.2.5. Analysis of Axial Chiral Compounds

Compounds of structure **4.31** exhibit axial chirality. This was unequivocally demonstrated by the doubling of certain resonances in the ^{1}H NMR spectrum of **4.31** in the presence of TFAE.[443] Similarly, xanthene-4,5-dicarboxamides (**4.32**) are axially chiral because the *anti*-conformation of the amide carbonyl groups causes them to point in different directions off the plane of the molecule.[444] A doubling of some of the signals in the ^{1}H NMR spectrum in the presence of TFAE confirmed the chirality of **4.32**.

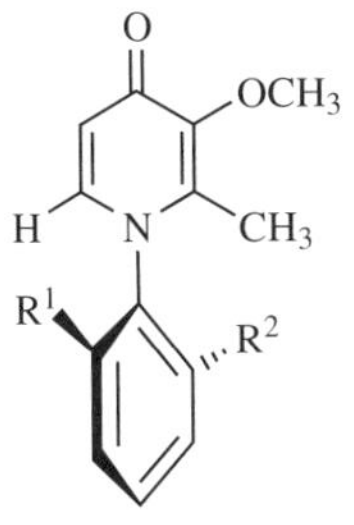

	1	2	3
R^1 =	Cl	CH$_3$	CH$_3$
R^2 =	H	H	CH$_3$

4.31

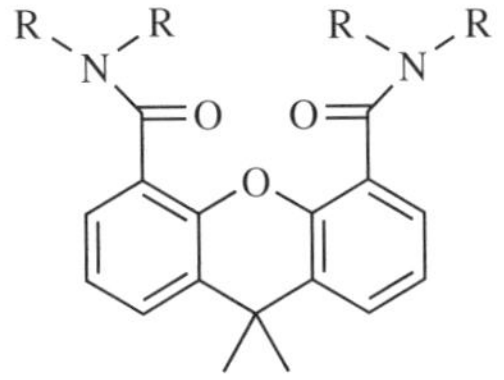

a (R = *i*-Pr)
b (R = Et)

4.32

4.2.6. Analysis of Compounds that are Chiral by Virtue of Slow Rotation

TFAE has been used on a number of occasions to study compounds that are chiral by virtue of slow rotation or have substituent groups that are enantiotopic by internal comparison. Addition of TFAE to *N*-aryl-tetrahydropyrimidines (**4.33**) caused a splitting of certain resonances because of slow rotation about the bond between the two rings.[445] The interconversion barrier was determined through a variable temperature study.

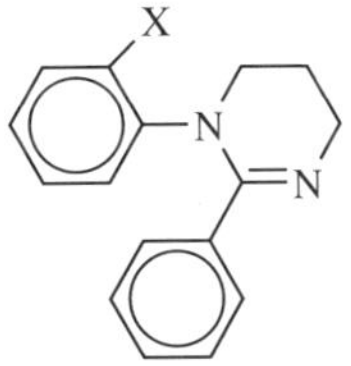

X = Cl
X = NO$_2$
X = CH$_3$

4.33

N-Alkyl-*N*-methyl-1-naphthyl amides (**4.34**) are chiral at low temperatures because of slow conformational averaging.[446,447] With an isopropyl R group, the methyl groups become diastereotopic at slow rotation and discrimination was observed in the presence of TFAE. It was possible to determine the barrier to rotation, although binding of the TFAE reportedly increased the rotational barrier.[447]

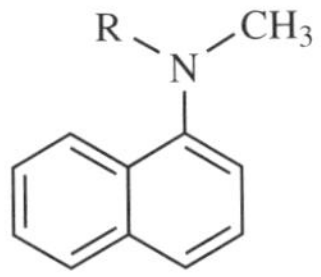

4.34

Hindered napthphyl ketones such as **4.35**[448] and phenyl ketones such as **4.36**[449] undergo slow rotation at low temperatures that causes the compounds to be chiral. For **4.35**, the 2-methyl resonance was readily monitored.[448] For **4.36**, enantiomeric discrimination in certain resonances in the ^{1}H or ^{13}C NMR spectra was observed in the presence of TFAE.[449] Similarly, the imine **4.37** has rotational enantiomers at low temperature.[450] If the substituent groups have prochiral nuclei, these show discrimination in the ^{1}H and ^{13}C NMR spectra. If there are no prochiral nuclei, addition of TFAE caused a splitting of certain signals for the enantiomers. The presence of the TFAE reportedly raised the rotational barrier by a small degree.[450]

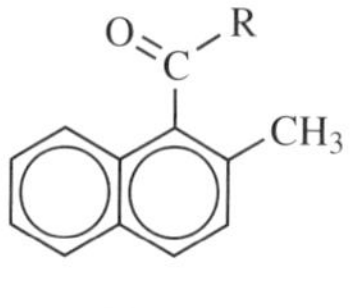

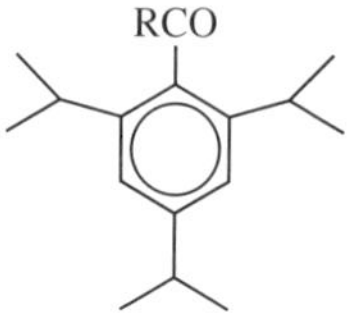

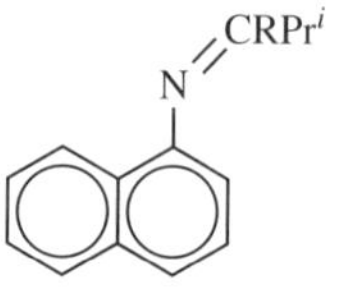

Hindered *O*-substituted oximes (**4.38**) are enantiomeric at low temperatures (−85°C) and addition of TFAE caused a discrimination of signals of the enantiomers.[451] The enantiomeric forms of thioamide **4.39** that result from slow rotation at 0°C exhibited distinct resonances in the ^{1}H NMR spectrum in the presence of TFAE.[452] Model compounds were used to aid in the analysis.

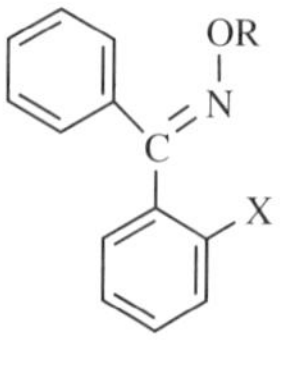

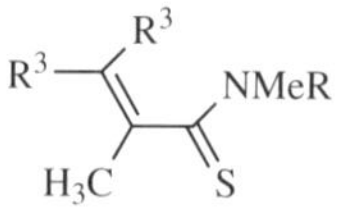

The sulfonamide bond of the *N*-(2,3-dinitrobenzenesulfenyl)acridone (**4.40**) is enantiotopic under conditions of slow rotation.[453] Addition of TFAE caused the enantiotopic groups of the acridone ring to become diastereotopic and anisochronous ($\Delta\delta$ of H5 of 18.7 Hz at 400 MHz). The enantiomeric interconversion of the methylene moieties of the cryptate species **4.41** is frozen out at a temperature of −50°C, which was confirmed by a doubling of certain resonances in the presence of TFAE.[454]

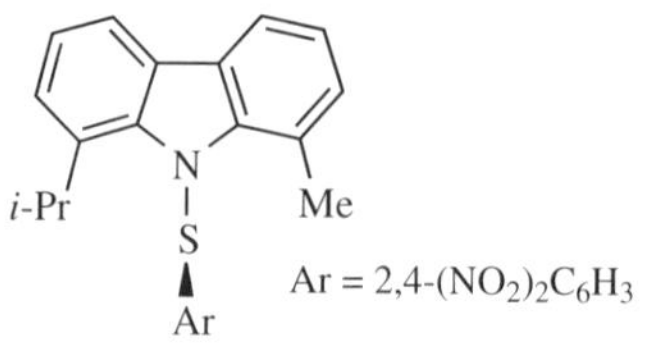

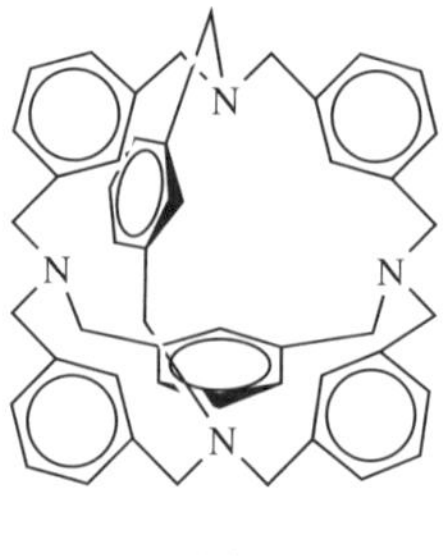

4.2.7. Analysis of Metal Complexes

TFAE has been used to determine the optical purity of metal complexes with a ligand that is also capable of binding to TFAE. Binding of TFAE to the sulfinyl group of niobium complexes **4.42** caused enantiomeric discrimination of the cyclopentadiene resonances.[455] The chirality of the Cu(I) bipyridyl polyether complex with **4.43** was confirmed by a doubling of certain resonances in the ^{1}H NMR spectrum at $-28°$C in the presence of TFAE.[456] The chirality of Cu(I)[2]catenates with two bipyridine-crown ether type ligands (**4.44**) was demonstrated by a doubling of some of the resonances in the ^{1}H NMR spectrum in the presence of TFAE.[457] Splitting of the bridging methylene resonances of the palladium complex of 2,6-lutidinyl biscarbene (**4.45**) in the presence of TFAE confirmed the chirality of the complex.[458]

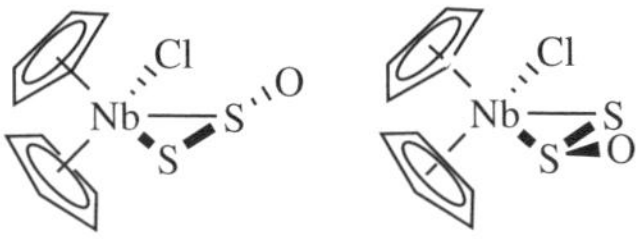

4.42

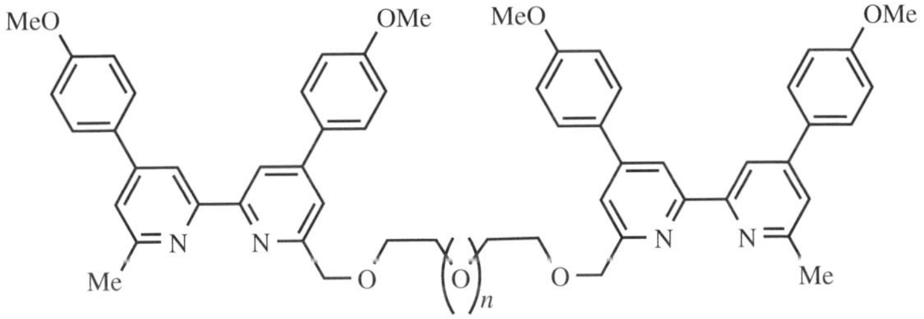

1a (*n*=3)
1b (*n*=4)

4.43

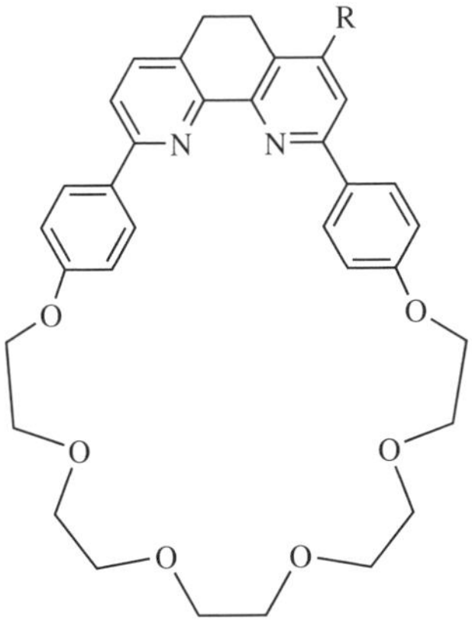

4.44

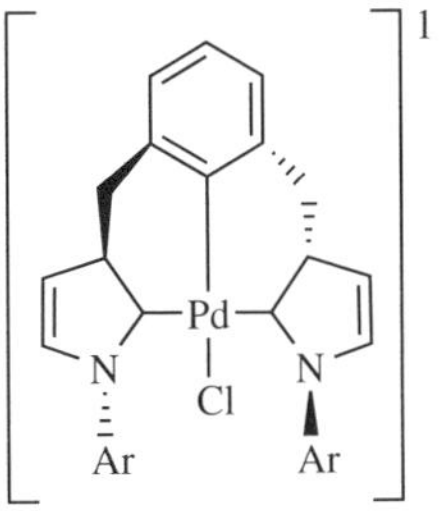

4.45

4.2.8. Analysis of Cyclophosphazenes

TFAE has been used to analyze the optical purities of cyclophosphazenes, an example of which is **4.46**.[459–463] A doubling of the ^{31}P NMR resonances of **4.46** in the presence of TFAE showed that the compound was racemic.[459] The chirality of cyclophosphazenes **4.47**, which have one stereogenic center, was confirmed by observing a doubling of ^{31}P resonances in the presence of TFAE.[461] Similarly, cyclotriphosphazene trispiranes, an example of which is **4.48**, were shown to be chiral based on the splitting of the ^{31}P resonances in the presence of TFAE.

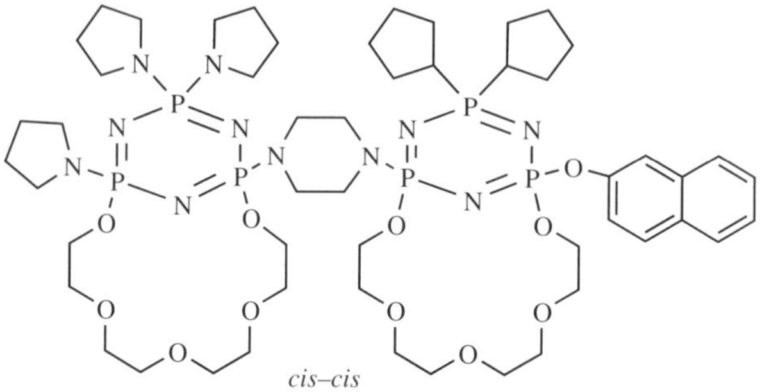

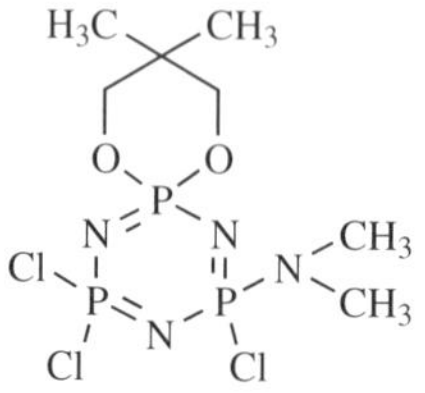

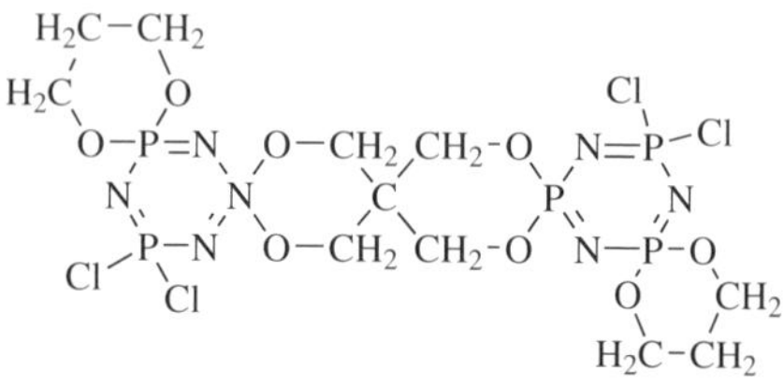

Caution was necessary in the analysis of certain cyclic phosphazenes with TFAE. The three spermine-bridged *gem*-disubstituted cyclotriphosphazatrienes (**4.49**) initially appeared to exhibit too many peaks in the ^{31}P NMR spectrum in the presence of TFAE.[460] These compounds have both a racemic pair [(*S,S*) and (*R,R*)] and a *meso* pair [(*S,R*) and (*R,S*)]. The racemic pair gives two signals in the presence of TFAE, which was the expected result. The *meso*-isomers in which a long carbon chain (C5 and C7) separated the two units also gave two ^{31}P signals, which was not originally expected. The extra signal was caused by having two TFAE units bound to the *meso* form of the cyclophosphazine (CP) at the same time, [(*R*)TFAE-CP(*R*)-CP(*S*)-(*R*)TFAE]. In effect, the two halves of the associated complex were essentially a diastereomeric pair with different ^{31}P NMR shifts.^{460}

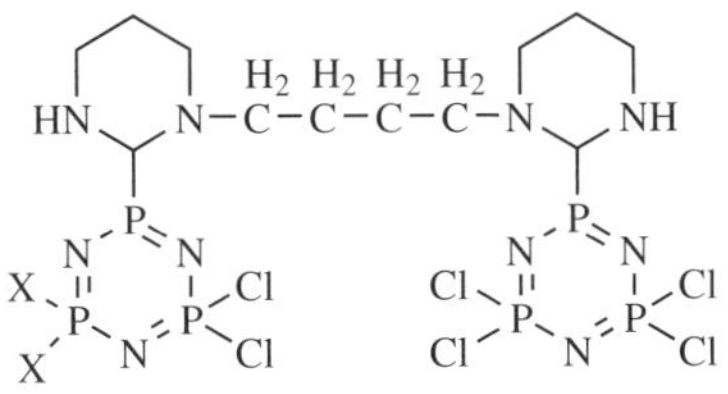

	X
a	phenyl
b	t-Bul-NH
c	O-(CH$_2$)$_3$-O

4.49

TFAE has been used to distinguish the *cis*-isomer, which is *meso*, and the *trans*-isomer, which is a racemic pair, of 1,3-disubstituted cyclotriphosphazenes (**4.50**).^{462} The ^{31}P NMR spectrum of the *trans*-isomer showed two signals in the presence of TFAE, one for each enantiomer, whereas the ^{31}P resonance of the *cis*-isomer did not split in the presence of TFAE. The shifts in the spectrum of the NH derivative of **4.50** in the presence of TFAE were larger than those in the corresponding *N*-methyl derivative, indicating that TFAE complexed more effectively with the less sterically hindered NH derivative.^{462}

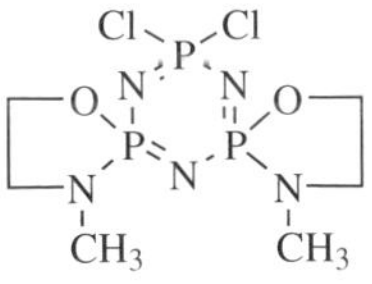

4.50

4.2.9. Analysis of Phosphine Oxides

TFAE has been used to determine the enantiomeric purity of the *cis*- and *trans*-phosphine oxides **4.51** on the basis of the ^{31}P NMR spectrum.^{464} The enantiomeric purity of the phosphine oxide **4.52** was also determined through the use of TFAE.^{92} The C3 methylene protons and diastereotopic methoxy signals of **4.52** gave distinct resonances for the two enantiomers.

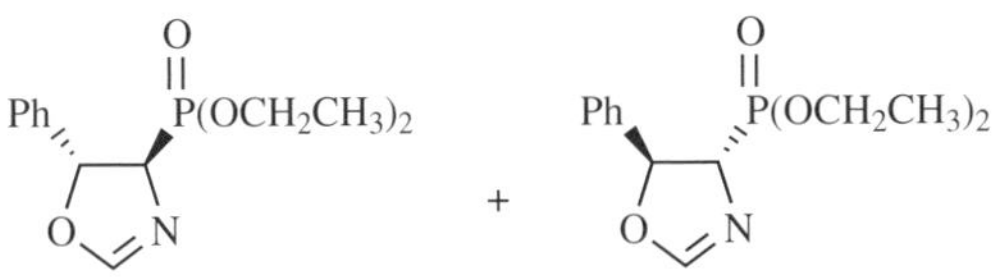

4.51 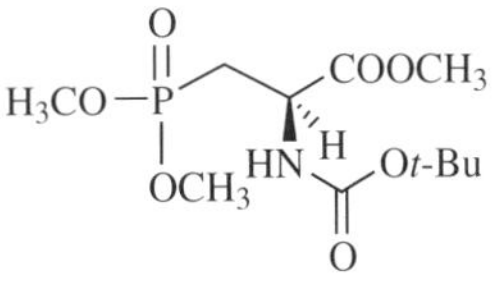   **4.52**

4.2.10. Analysis of Calixarenes

TFAE has been used on several occasions to demonstrate that calixarene systems are chiral. The chirality of the 1,4-2,5-calix[8]bis-crown-4 (**4.53**) was confirmed by a doubling of certain resonances in the ^{1}H NMR spectrum with TFAE.[465] The isomers of 5,11-di-*tert*-butylcalix[4]arene-25,26,27,28-tetrol (**4.54**) that have an AABB or CDCD substitution pattern are inherently chiral. This was unequivocally demonstrated through the observation of a doubling of some resonances in the ^{1}H NMR spectrum in the presence of TFAE.[466]

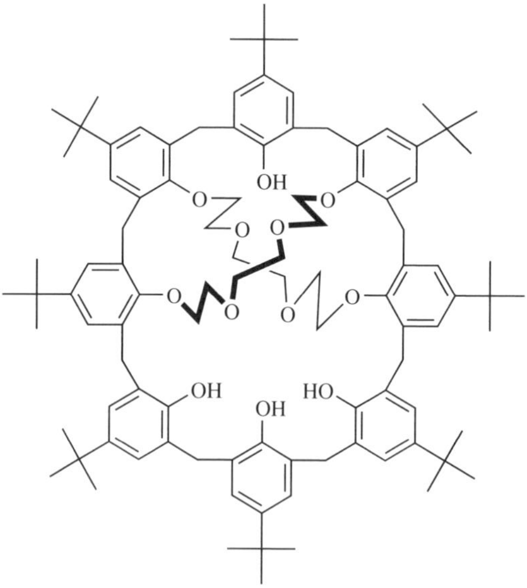

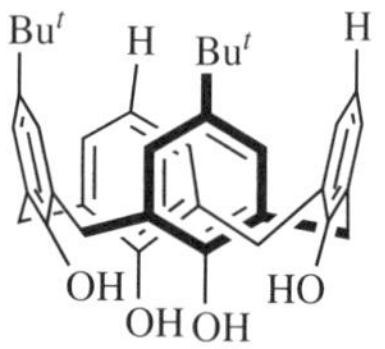

A calixarene in the alternate 1,2- and partial cone arrangements that bound to Cr(CO)$_3$ through an aromatic ring (**4.55**) showed a splitting of some resonances in the presence of TFAE, indicating that each isomer was a pair of racemates.[467] Similarly, most of the resonances in the ^{1}H NMR spectrum of a dixanthylene double calix[6]-arene (**4.56**) were split in the presence of TFAE.[468] Finally, a doubling of resonances in the spectra of asymmetrically substituted calix[4]arenes (**4.57**) in the presence of TFAE confirmed the chirality of these compounds.[469]

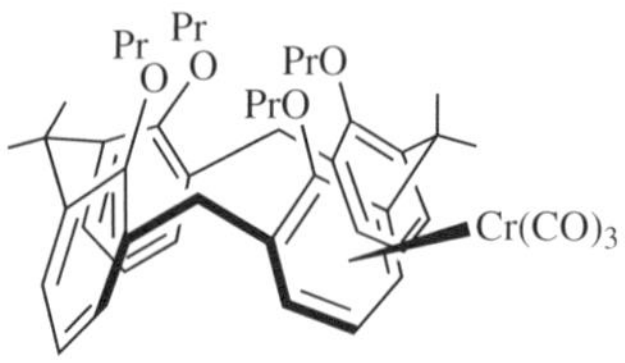

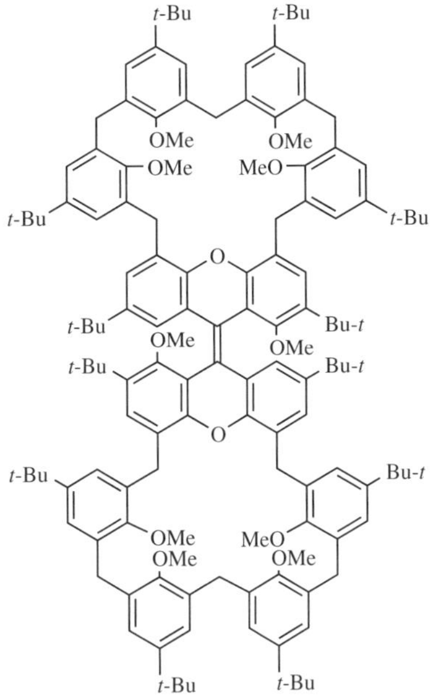

4.56

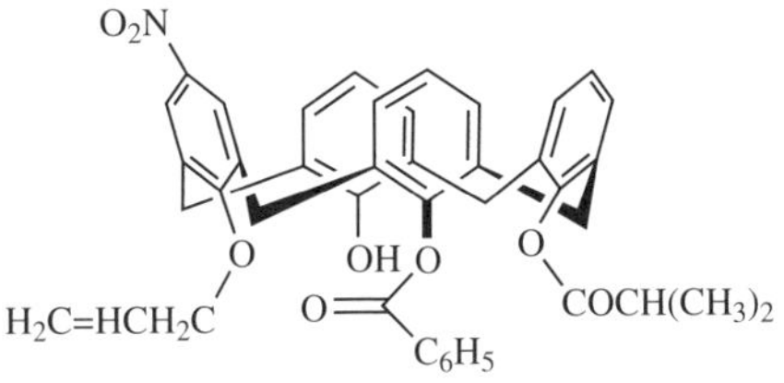

4.2.11. Analysis of Other Substrates

By evaluating the effect of TFAE on model compounds such as the isobutylester of mandelic acid, it was possible by analogy to assign the absolute configuration of pisolithin B (**4.58**).[470]

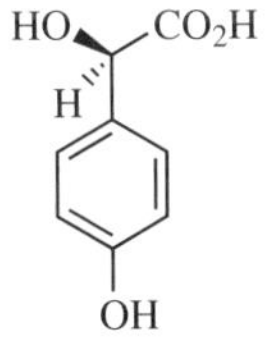

4.58

TFAE was used to determine the optical purity and assign the absolute configuration of the (2R,2'R)-(+)- and (2S,2'S)-(−)-*threo*-phenidates (**4.59**).[471] Shielding of the methoxy group by the anthryl ring was a key in making the assignment, and an analysis of each of the pure drug enantiomers with (R)-(+)-TFAE and (S)-(−)-TFAE confirmed the stereochemistry.

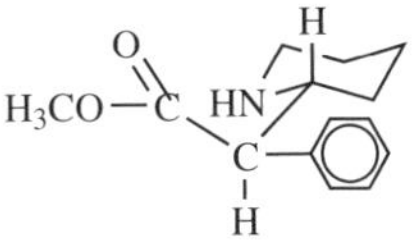

4.59

NMR shift data of cyclic and acyclic α-acyloxyketones (**4.60** and **4.61**) with TFAE were used to determine the optical purity.[472] Nonequivalence of 0.01–0.136 ppm was observed in the ¹H NMR spectra. It was also possible to esterify the α-hydroxyketones to the corresponding acetate and analyze the product with TFAE.[472] The use of TFAE with the corresponding acetates was reportedly more reliable than the results obtained with derivatives of MTPA.

4.60 **4.61**

The absolute stereochemistry of compound **4.62** was established using Eu(hfc)₃ (hfc = 3-heptafluorobutyryl-*d*-camphor) and comparing the results with model compounds. The assignment was confirmed by analyzing the corresponding methyl ester with TFAE.[473] Enantiomeric discrimination was observed in the ¹H NMR spectrum of lacinilene C (**4.63**) in the presence of TFAE.[474] Both the hydroxyl and carbonyl groups of **4.63** were involved in association with the hydroxyl and methine groups of TFAE. The shift order of the enantiomers could be rationalized on the basis of the binding interaction so that absolute configurations were assigned as well.

	R¹	R²	R³	R⁴
(1)	menthyl	OH	Me	Me
(2)	menthyl	Me	OH	Me

4.62

4.63

A procedure for determining the enantiomeric purity and absolute configuration of chiral allenes using TFAE has been described.[475] The allene was converted to a chiral methoxy ester ether via a methoxy mercuration. The resulting methoxy ether (Fig. 4.9)

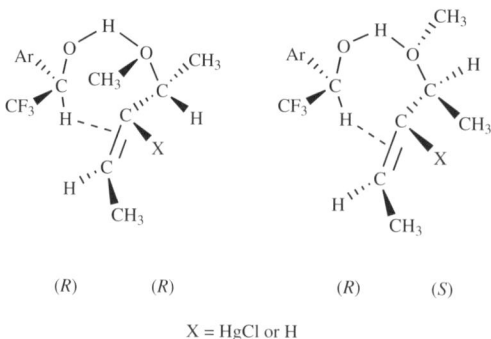

X = HgCl or H

Figure 4.9. Association of TFAE with the two configurations of methoxy ether of allenes.

was analyzed with TFAE. When the subsituents on each side of the allene were identical, only one methoxy ether was formed, whereas two ether derivatives resulted if the substituent groups were different.

A procedure for assigning the absolute configurations of N,N-dimethyl amino acids using TFAE has been reported.[476] The amino acid was first amidated with 3,5-dinitroaniline. The ^{1}H NMR spectrum of the derivative in the presence of TFAE showed enantiomeric discrimination in the resonances for the α-, C-methyl, and N-methyl hydrogen atoms, and the relative shifts exhibited a systematic trend that correlated with the absolute configuration.[476]

The optical purities of cyclic amines **4.64**[477] and **4.65**[478] and acyclic amine (**4.66**)[479] were determined in the presence of TFAE. The optical purity of N^6-alkyl derivatives of (R)-3-(adenin-9-yl)-2-hydroxypropanoic acid, one example of which is **4.67**, was determined by recording the ^{1}H NMR spectrum in the presence of TFAE.[480]

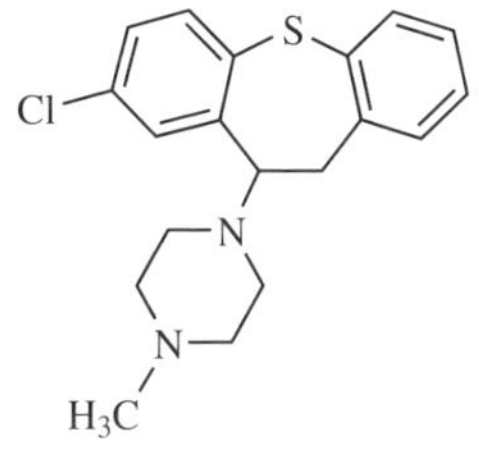

4.64

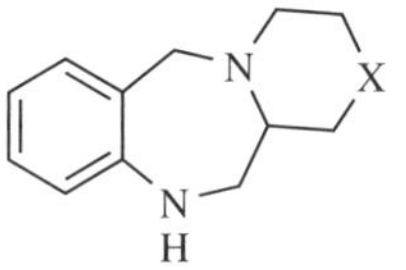

4.65
a: X = NMe
b: X = S
c: X = O

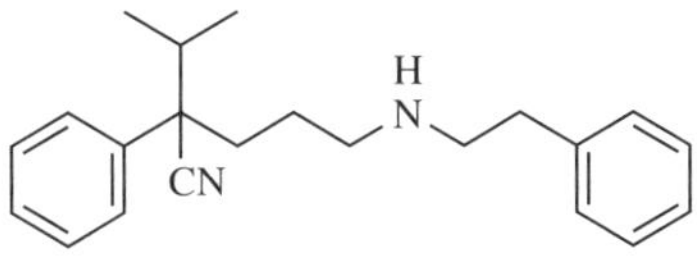

4.66

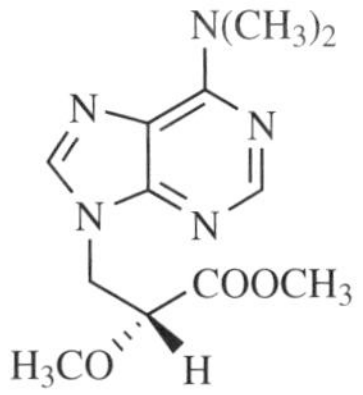

4.67

The optical purity of Δ^2-oxazolines-1,3 (**4.68**) was determined on the basis of ^{1}H and ^{19}F NMR spectra with TFAE.[481] ^{19}F spectra were recorded with substrates in which one of the substitutent groups had a fluorine atom. Binding involved the hydroxyl group of TFAE with the nitrogen atom of the oxazoline. Also, π–π interactions between the anthryl ring and substituent groups with an aryl ring were observed. The effectiveness of TFAE and TFPE was compared and the TFAE produced larger nonequivalence in the NMR spectrum.

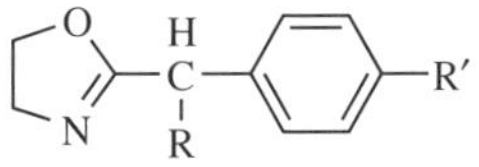

4.68

TFAE is an effective chiral NMR solvating agent for compounds with oxygen functionalities. The methoxy singlet of **4.69** split into two signals in the presence of TFAE.[482] The optical purity of the silicon and germanium complexes (**4.70**) with a hydroxyl group was determined by the use of TFAE.[483]

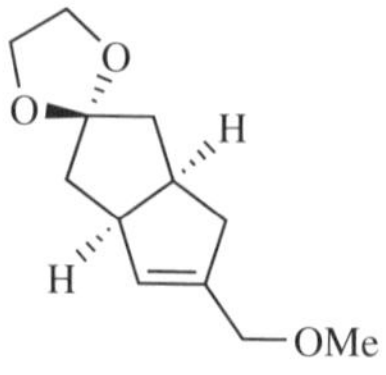

4.69

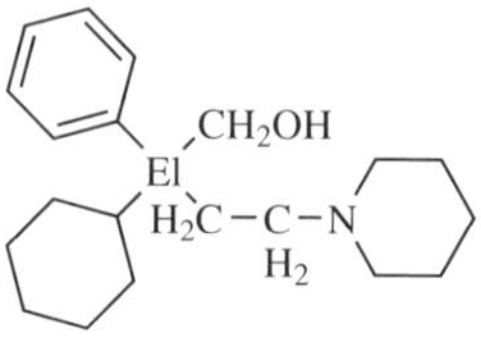

4.70
1: El = Si, 2: El = Ge

The methylene hydrogens in citrate (**4.71**) are prochiral. The NMR spectrum of trimethyl citrate in carbon tetrachloride-benzene-d_6 in the presence of TFAE showed four different CH signals.[484] An enzymatically deuterated derivative with selective incorporation of deuterium was used to confirm the assignments of the individual methylene resonances.

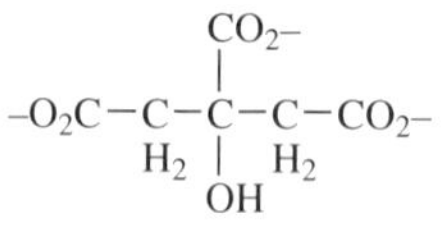

4.71

4.2.12. Use of Lanthanide Chelates with TFPE and TFAE

The potential of coupling organic-soluble achiral lanthanide species such as Eu(fod)$_3$, in which fod is 6,6,7,7,8,8,8-heptafluoro-2,2-dimethyl-3,5-octanedione, to TFPE– and TFAE–substrate systems was demonstrated with sulfoxides[485] and lactones.[433] The hydroxyl group of TFPE and TFAE did not associate that well with the lanthanide ion because of the electron-withdrawing trifluoromethyl group and phenyl ring. As a result, the non-TFPE- or TFAE-bound form of the sulfoxide or lactone associated preferentially with the europium ion. Provided the two enantiomers had different association constants with TFPE or TFAE, the enantiomer with the lower association constant with the solvating agent was more available to bind with the lanthanide and its spectrum exhibited larger lanthanide-induced shifts. At low concentrations of lanthanide relative to TFPE or TFAE and substrate, the enantiomeric discrimination was enhanced. Raising the concentration of Eu(fod)$_3$ too high eventually stripped the substrate from the chiral solvating agent, such that the enantiomeric discrimination began to diminish. In addition to facilitating the analysis of enantiomers because of the larger discrimination, the magnitude of the lanthanide-induced shifts could be used to provide insights into whether the enantiomeric pairs have similar association constants with the chiral solvating agent.

Enantiomeric discrimination in the ^{1}H NMR spectrum of oxadiazine **4.72** was probed by the addition of TFPE.[486] Addition of Eu(fod)$_3$ caused an enhancement in the enantiomeric discrimination, facilitating the determination of the optical purity. Addition of Eu(fod)$_3$ to a spirosulfurane (**4.16**) caused no enhancement in enantiomeric discrimination, showing that both enantiomers had similar association constants with the chiral solvating agent.[426]

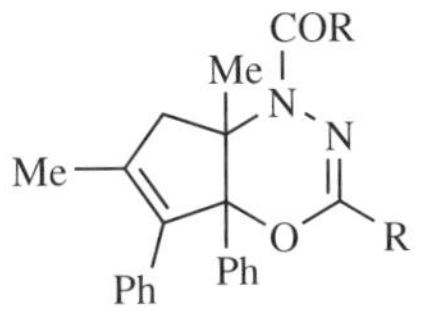

4.72

4.2.13. Use as a Chiral Derivatizing Agent

TFAE has been used as a chiral derivatizing agent for carboxylic acids. For example, the enantiomeric purity of **4.73** was determined by converting the carboxylic acid group to its corresponding TFAE ester.[487] The dithioacetal methine proton was monitored in the ^{1}H NMR spectrum.

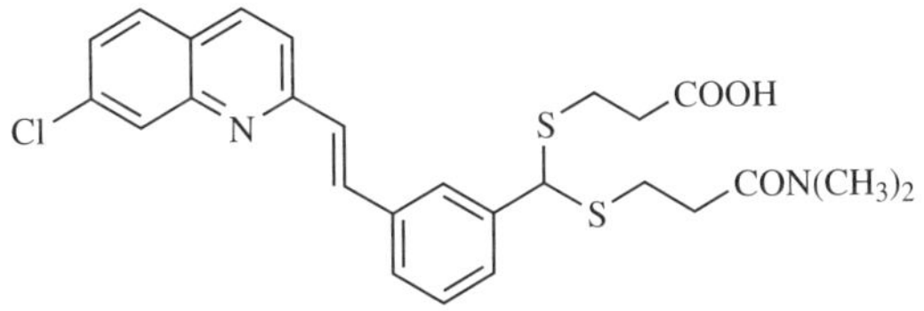

4.73

A series of chiral α-chloro carboxylic acids such as 2-chloro-3-methylbutanoic acid, 2-chloro-3-methylpentanoic acid, and 2-chloro-4-methylpentanoic acid were prepared from α-amino acids.[488] The optical purities were determined by analyzing the ^{1}H NMR spectra of the corresponding esters with TFAE.

4.3. OTHER ANTHRYL-BASED REAGENTS

4.3.1. Analogs of TFAE

Several other anthryl-containing reagents that are analogs of TFAE have been explored for their utility as chiral NMR solvating agents. The effectiveness of

2-(9-anthryl)-, 2-(1-naphthyl)-, 2-(2-naphthyl)-, and 2-phenyl-2-ethanol in causing enantiomeric discrimination in the spectra of methyl *p*-tolyl sulfoxide and 2-deuterio-2-methyl-2,3-dihydrobenzo[b]thiophene-1-oxide in chloroform-*d* and carbon tetrachloride was compared.[489] The largest enantiomeric discrimination was observed with the anthryl derivative. The enantiomeric discrimination in the ^{1}H NMR spectra of 1-phenylethylamine (PEA) and 1-phenylethylmethyl ether with TFAE was compared to that of analogous derivatives with $-CF_2CF_3$, $-CCl_2CF_3$, and $-CF(CF_3)_2$ groups.[490] TFAE was most effective for PEA, whereas the derivative with the $-CF_2CF_3$ group caused the largest discrimination in the spectrum of 1-phenylethylmethyl ether.

Reagents in which the CF_3 group of TFAE was replaced with a *tert*-butyl,[491] 1-adamantyl,[492] or pentafluorophenyl[493] group have been examined as chiral NMR solvating agents. The *tert*-butyl derivative was tested on menthyl-*p*-toluene sulfonate, 9-(1-amino-2,2-dimethylpropyl)-9,19-dihydroanthracene (**4.74**), α-methoxyphenylacetic acid (MPA), and 1-phenyl-1,2-ethanediol, although the enantiomeric discrimination was usually larger with TFAE.[491] The adamantyl derivative caused nonequivalence in the ^{1}H NMR spectra of MPA and 1-phenyl-1,2-ethanediol, but the effectiveness was not compared to TFAE.[492] The effectiveness of the pentafluorophenyl derivative was evaluated with 1-(1-naphthyl)ethylamine and 1-aminoindane but was not compared with TFAE.[493]

4.74

The compound α,α'-bis(trifluoromethyl)-1,8-anthracene dimethanol (**4.75**) was shown to be an effective chiral solvating agent for five substrates with hydroxyl and/or amine groups.[494] α,α'-Bis(trifluoromethyl)-9,10-anthracene dimethanol (**4.76**) was also evaluated for its utility as a chiral NMR discriminating agent.[495,496] The synthesis of **4.76** produced the (*R,R*)-, (*S,S*)-, and (*S,R*)-(*meso*)-derivatives that needed to be purified. **4.76** reportedly produced larger enantiomeric discrimination in the NMR spectra of 2-phenylethane-1,2-diol, NEA, *trans*-stilbene oxide (**4.77**), and fluoxetine (**4.78**) than TFAE.[495] **4.76** may be preferred for substrates with two or more remote binding sites that can complex in a bidentate manner. For example, α-substituted β-dicarbonyl compounds such as **4.79**, which are the product of a Michael addition, likely formed two hydrogen bonds with **4.76**, one involving a carbonyl group of the β-dicarbonyl moiety and the other a carbonyl group of the amide.[496] Compounds with carbonyl groups that were separated by four bonds or that had a C=O or cyano group at 1,5-relative positions formed a bidentate complex with **4.76** as well. The ability to

OTHER ANTHRYL-BASED REAGENTS

assign absolute configurations of the compounds based on the shifts with **4.76** was not discussed.[496]

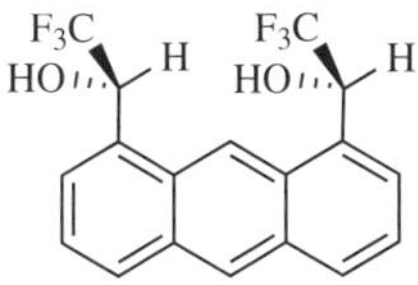

4.75

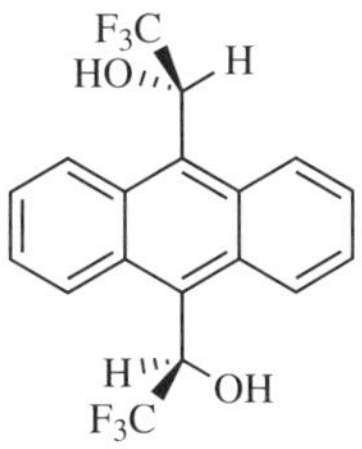

4.76

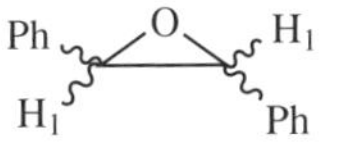

4.77

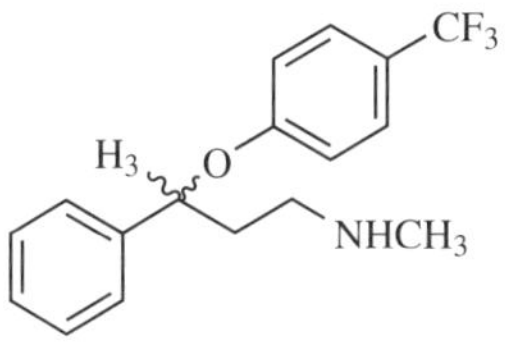

4.78

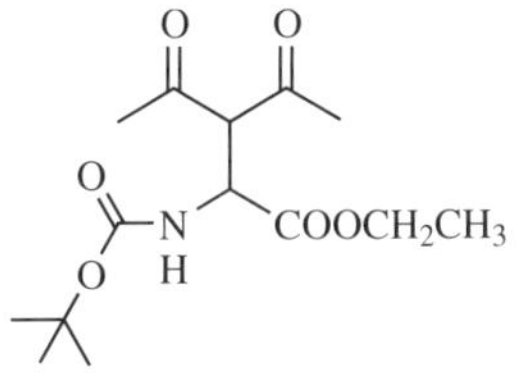

4.79

A bis-anthryl analog of TFAE (α,α'-bis(trifluoromethyl)-10,10'(9,9'-bianthryl) dimethanol—**4.80**) was evaluated for its effectiveness as a chiral NMR solvating agent.[497] Enantiomeric discrimination similar to TFAE was observed in the NMR spectra of PEA, 1-(1-pyrenyl)ethylamine, cis-1-amino-2-indanol, and fluoxetin (**4.78**). Finally, the perdeutero TFAE was prepared and recommended for the analysis of substrates with aromatic hydrogen resonances that may otherwise overlap with those of TFAE.[498]

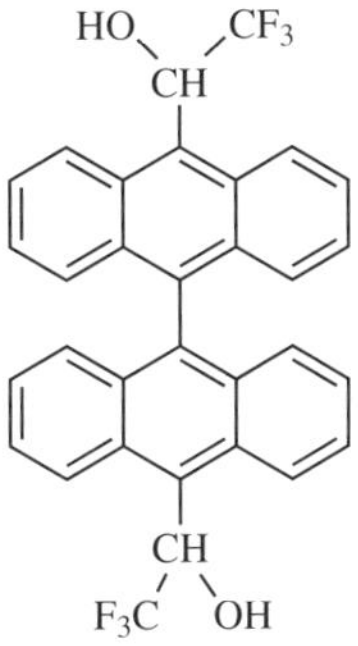

4.80

It is not clear that any of these other reagents offer enough advantages over TFAE to recommend their use. As shown, TFAE is an especially effective and versatile chiral NMR solvating agent. For certain classes of compounds, the association with TFAE is known with enough certainty to assign the absolute configuration. In many other cases, TFAE associates with substrates through dipole–dipole or π–π interactions, often causing enantiomeric discrimination in the NMR spectrum that can be used for the determination of optical purity.

4.3.2. Ethyl-2-(9-anthryl)-2-hydroxyacetate (4.81)

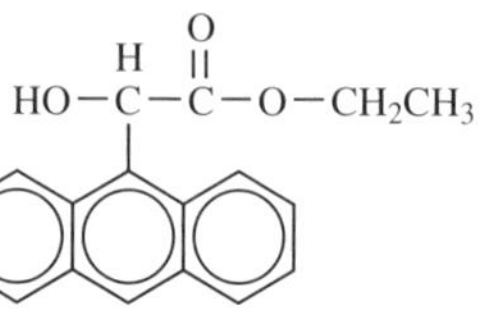

4.81

4.81 is an especially effective reagent for the analysis of absolute configurations of α-chiral carboxylic acids.[499,500] Molecular mechanics calculations of the corresponding esters indicated a predominance of the *anti-periplanar* (*ap*) over the *syn-periplanar* (*sp*) conformer. Low-temperature NMR studies of the derivatives showed a further population of the *ap* conformer and confirmed the conformational preference. The conformer preference of ester derivatives of carboxylic acids was found to be better than that of amides, which enhanced the differences in shielding caused by the anthryl ring.[500] Testing a variety of model α-chiral carboxylic acids, derivatives with **4.81** produced $\Delta\delta^{RS}$ values as high as 0.4 ppm. The $\Delta\delta^{RS}$ values for the **4.81** derivatives were compared to those with the alcohols shown in Figure 4.10. Except for *trans*-2-phenyl-1-cyclohexanol, which was comparable to 9-AHA, 9-AHA was superior to all others.[500]

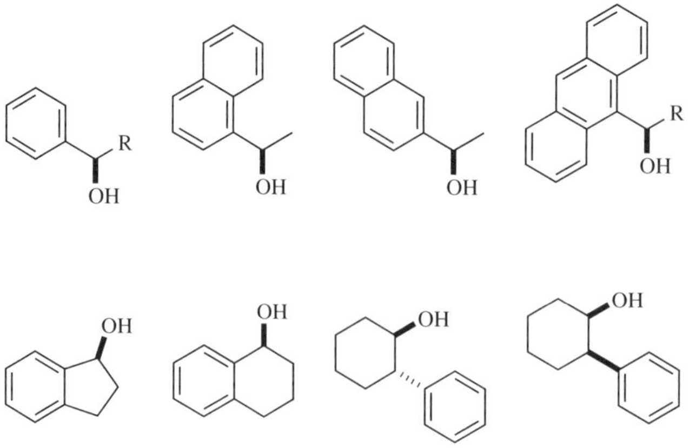

Figure 4.10. Alcohols compared to 9-AHA for the analysis of α-chiral carboxylic acids.

4.3.3. 2-(2,3-Anthracenedicarboximido)-1-cyclohexanol (4.82)

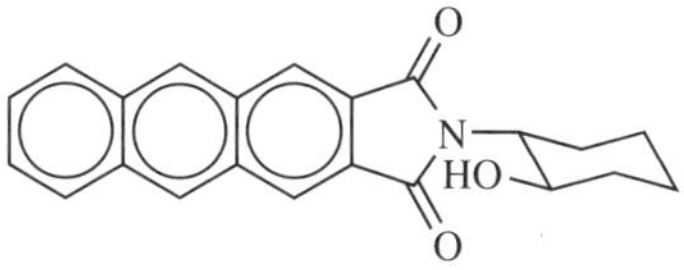

4.82

4.82 is an effective reagent for the analysis of carboxylic acids, especially those in which the stereogenic center is remote from the acid moiety. For example, the methyl resonance in racemic 12-methyl pentadecanoic acid was split into two signals in the ^{1}H NMR spectrum of its diastereomeric esters with **4.82**.[501] The configuration of the cyclohexane unit relative to the anthryl ring positions the alkyl chain of the carboxylic acid over the ring, thereby optimizing differences in the shielding over such long distances.

4.4. 2,2,2-TRIFLUORO-1-(1-PYRENYL)ETHANOL (4.83)

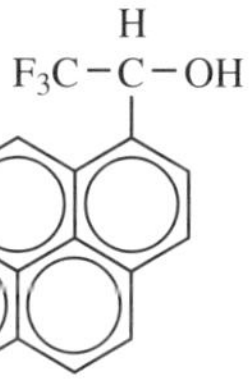

4.83

Early on in the development of aryl trifluoromethyl ethanol derivatives as chiral NMR solvating agents, 2,2,2-trifluoro-1-(1-pyrenyl)ethanol was examined with sulfoxides and observed to produce greater discrimination than TFPE.[411] 2,2,2-Trifluoro-1-(1-pyrenyl)ethanol was also evaluated for its ability to cause enantiomeric discrimination in the ^{1}H NMR spectra of PEA, NEA, 1-phenylethane-1,2-diol, and 1-(1-naphthyl)ethanol.[502,503] The pyrene reagent was more effective than TFAE for 1-phenylethane-1,2-diol, whereas the reverse trend was noted for the other three substrates.

4.5. 2-(TRIFLUOROMETHYL)BENZHYDROL (4.84)

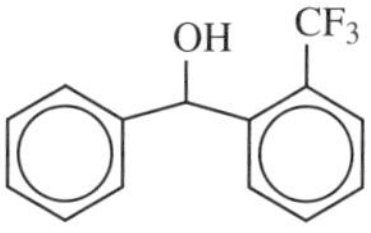

4.84

4.84 has been shown to be an effective chiral derivatizing agent for the analysis of chiral carboxylic acids.[504] Derivatives of fifteen different carboxylic acids with a stereogenic center at either the α- or β-position exhibited nonequivalence in the ^{19}F NMR spectrum.

4.6. 1-PHENYLETHANOL (4.85)

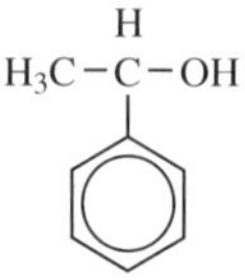

4.85

4.85 has been used on occasions as a chiral solvating or derivatizing agent, but is generally not as effective as TFPE and TFAE. The ^{1}H NMR spectra of thiazoline-2-thione derivative (**4.86**), camphor, and cocaine (**4.87**) exhibited enantiomeric discrimination in the presence of **4.85**.[505]

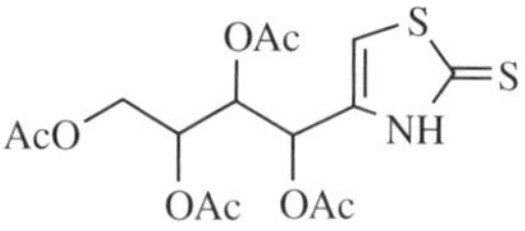

4.86

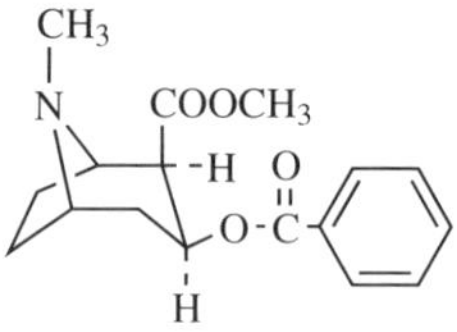

4.87

The optical purity of [2.2] metacyclophane-4-carboxylic acid (**4.88**) was determined by analyzing the methyl resonances of the diastereomeric ester derivatives with **4.85**.[506] The ^{19}F NMR spectrum of the **4.85** esters was used for determining the optical purity of compounds such as **4.89**.[507]

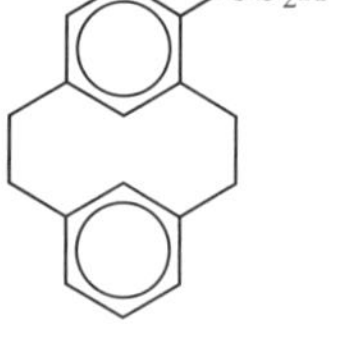

4.88

4.89

Diastereomeric esters of 2-(5,6-dichloro-3-indolyl)propionic acid (**4.90**) with **4.85** exhibited distinct resonances in the NMR spectra.[508] The shielding caused by the phenyl ring for a set of 2-(3-indolyl)propionic acids with known configurations was used to rationalize a method to assign the absolute configuration of unknowns.

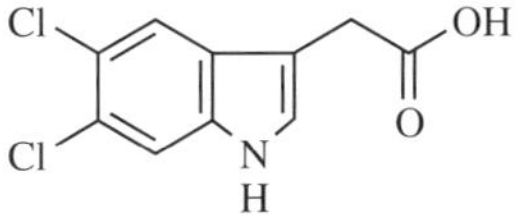

4.90

The compound 1-perfluorohexyl-1-phenylethanol, as well as the corresponding octyl and decyl derivatives, was prepared by reacting the perfluoroalkyl iodide with acetophenone.[509] The ^{19}F NMR spectra recorded in toluene-d_8 showed nonequivalence of diastereotopic CF_2 groups at temperatures below 0°C. Five of the CF_2 groups showed diastereotopic resolution when the samples were cooled to −80°C.

Compounds of structure **4.91** are potentially chiral if there is slow rotation about the disulfide bond.[510] The NMR spectrum of the derivative of **4.91** in which the R groups were (R)-**4.85** had two methyl doublets, confirming the chirality of the structure. If a racemic mixture of **4.85** was used to prepare the disulfide, four methyl doublets were observed in the ^{1}H NMR spectrum.

R−O−S−S−O−R

4.91

4.7. METHYL MANDELATE (MM) (4.92)

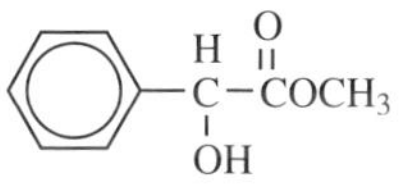

4.92

The methyl ester of mandelic acid has been utilized as a chiral derivatizing agent for carboxylic acids.[341,511] The hydroxyl group of MM reacts with the substrate's carboxylic acid group to form an ester. Derivatives of N,N-dimethyl amino acids with methyl-(S)-(+)-mandelate provided diastereomers with distinct resonances in the ^{1}H NMR spectrum.[476]

MM has been used to determine the optical purity of α-deuterated carboxylic acids such as [2,3-^{2}H$_2$]propanoic acid,[332] 2-fluoropropionic acid, 2-chloropropionic acid, and monodeuterofluoroacetic acid[512] by analyzing the ^{1}H and ^{2}H NMR spectra in benzene-d_6.

The pro-(R) and pro-(S) hydrogen atoms in decanoic acid that has a single deuterium at C2 exhibited distinct resonances in the ester derivative with MM.[511,513,514] The pro-(R) signal in the derivative was at a lower field than that of the pro-(S) such that the exact position of the deuterium substitution could be determined. The decanoic acid was obtained by the activity of dehydrase on **4.93** and facilitated an analysis of the mechanism of β-hydroxydecanoylthioester dehydrase. The ethyl ester of mandelic acid has also been used to determine the site of deuterium substitution at the prochiral methylene group α to a carboxylic acid.[341] By comparing the shifts to compounds with known configurations, it was possible to assign the site of deuterium substitution to the *pro-(S)* or *pro-(R)* site. The optical purity was readily assessed using ^{2}H NMR spectroscopy.

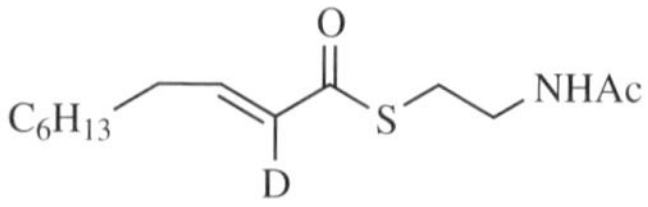

4.93

An attempt was made to study the site of substitution of a single deuterium α to the carboxylic acid group of an oleic acid compound (**4.94**) by analysis of its (S)-(+)-MM ester. However, a significant loss of deuterium occurred in the derivatization reaction.[515] The corresponding octanoic acid obtained through conversion of **4.94** could be analyzed as its (S)-MM ester. The *pro-(R)* hydrogen resonance was at a higher frequency than that of the *pro-(S)* in the mandelate derivative.

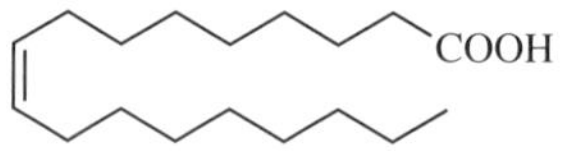

4.94

The potential of using (S)-MM to assign the absolute configuration of a range of α-chiral carboxylic acids and one β-chiral carboxylic acid was explored.[516] Characteristic shielding from the phenyl ring could be used in assigning the stereochemistry. The (S)-MM derivatives had a preferred conformation that caused the methyl or hydrogen resonances at the α-position of the (S)-isomer of the substrate to resonate at a lower frequency than those of the (R)-isomer. Conversely, the methyl and hydrogen resonances at the β-position of the (S)-isomer resonated at a higher frequency than those of the (R)-isomer. The distinction was far less pronounced for the substrate chiral at the β-position.

4.8. OCTAHYDRO-8,9,9-TRIMETHYL-5,8-METHANO-2*H*-1-BENZOPURAN-2-OL (NOE'S REAGENT) (4.95)

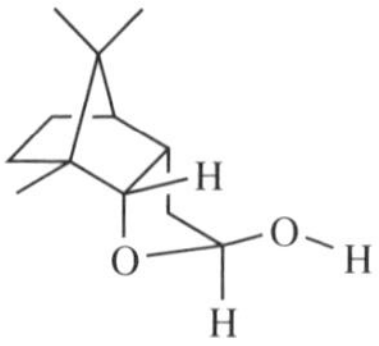

4.95

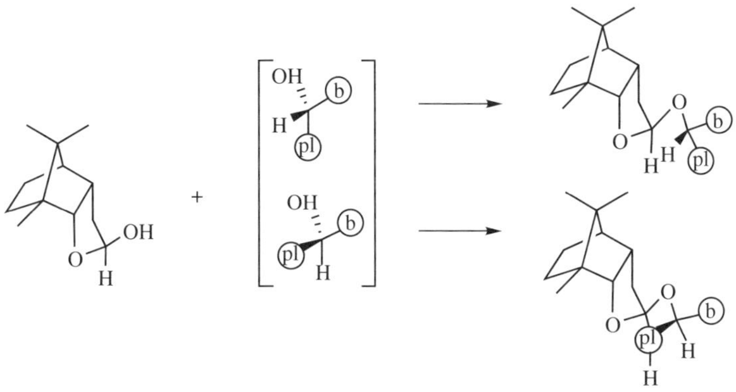

Figure 4.11. Acetal derivatives formed by the reaction of the two configurations of a secondary alcohol with Noe's reagent. [Reproduced with permission.[518]]

The lactol octahydro-8,9,9-trimethyl-5,8-methano-2H-1-benzopuran-2-ol, often referred to as Noe's reagent, is a versatile compound for the analysis of optical purity and assignment of absolute configuration of alcohols, amines, thiols, and carboxylic acids. Reaction of Noe's reagent with an alcohol leads to the formation of an acetal derivative as shown in Figure 4.11.[517,518] The acetal adopted a preferred conformation that led to predictable shifts for certain resonances, thereby enabling the assignment of absolute stereochemistries. In particular, a bonding interaction occurs between the electron lone pair of the oxygen atom and the σ* orbital of the Csp^3–Csp^2 bond. Alcohols, amines, or thiols with a bulky group (b) and a planar or linear group (pl) produced a stabilizing n–σ* interaction and were amenable to study with this reagent.[518] The shifts of the ^{13}C NMR resonances around the glycosidic linkage of secondary alcohols were also used to assign the absolute configuration.[519]

Noe's reagent was used to assign the absolute configurations of a series of α-hydroxy substituted nitriles, aldehydes, and alkynes.[520,521] X-ray crystallographic data confirmed the idealized conformations that were assumed in assigning the absolute configurations.[521] The *exo* form of Noe's reagent has been used to assess the optical purity of a series of α-hydroxy carboxanilides (**4.96**) by ^{1}H NMR spectroscopy.[522]

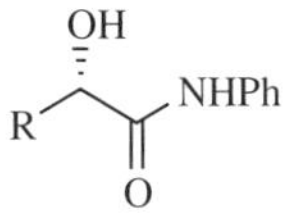

4.96

Coupling of **4.97** to the *exo* and *endo* forms of Noe's reagent provided diastereomers in which the original rules put forward for determining the preferred conformation and effects in the NMR spectrum could be used to assign the absolute stereochemistry of the carbinol.[523,524]

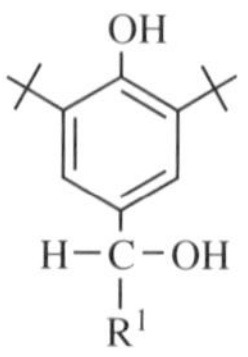

4.97

Noe's reagent was effective in assigning the absolute configuration of lactic acid derivatives (**4.98**).[525] The acetal linkage was formed from the hydroxyl group of the lactic acid with the *exo*-(*S*) form of Noe's reagent. The reaction was run directly in the NMR tube. The H2 signal for the derivative with the (*S*)-enantiomer of the lactic acid was at higher frequency than that of the (*R*)-enantiomer. Nonequivalence of up to 0.4 ppm was observed for the resonances for the hydrogen atom at the chiral center of the lactic acid and up to 4.0 ppm for the resonances for the acetal carbon atom. Similar trends were noted in the ^{1}H NMR spectrum of lactamide derivatives attached through an acetal linkage of the hydroxyl group of the lactamide. There was no evidence to suggest that amide derivatives adopted a different conformation than that of the esters.[525]

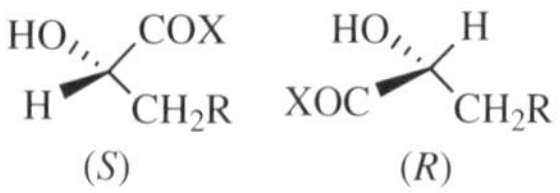

4.98

4.9. MENTHOL (4.99)

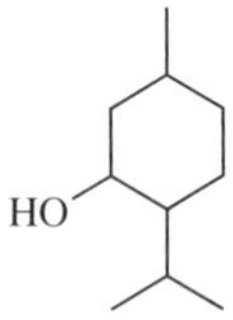

4.99

The use of menthol as a chiral NMR derivatizing agent dates back to some of the earliest reports in this field. The first application involved the analysis of *n*-alkyl phenyl phosphinates (**4.100**).[526,527] Shielding by the phenyl ring of the substrate caused characteristic shifts of the methyl resonances of the menthyl moiety that correlated with the absolute configuration at the phosphorus atom. The analysis of partially deuterated analogs showed that the *pro*-(*S*) methyl group of the (−)-menthol derivative was more shielded.[527] Replacing the phenyl ring of the substrate with a cyclohexyl group eliminated the shielding of the methyl group. The use of menthol was also suitable for the analysis of the configuration of diarylphosphinates such as β-naphthylphenylphosphinates.

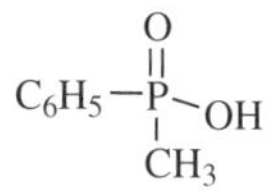

$$C_6H_5-\underset{\underset{CH_3}{|}}{\overset{\overset{O}{\|}}{P}}-OH$$

4.100

The optical purity of 2-(phosphonomethyl)pentanedioic acid (**4.101**) was determined by the analysis of its derivative with menthol.[528] The carboxylic acid groups were first converted to their corresponding benzyl esters. One of the phosphorus hydroxyl groups was then derivatized with menthol. Incorporation of the menthol creates a new stereogenic center at the phosphorus so that four peaks were observed in the ^{31}P NMR spectrum of a racemic mixture. Only two peaks were observed in the spectrum of a pure enantiomer.

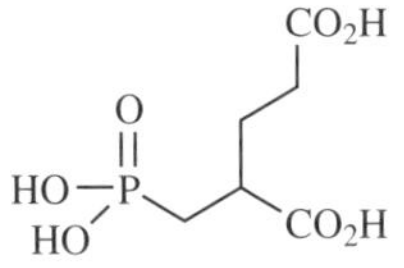

4.101

Menthol has been used as a reagent for the analysis of chiral silanes.[529-531] Reaction of methylphenyl-1-naphthyl silane with menthol yields the corresponding menthoxy silane **4.102**.[529] Discrimination of the silyl methyl group for the two diastereomers was observed in the ^{1}H NMR spectrum. In certain cases, reaction of (−)-menthol with silane did cause some loss of configuration.[530]

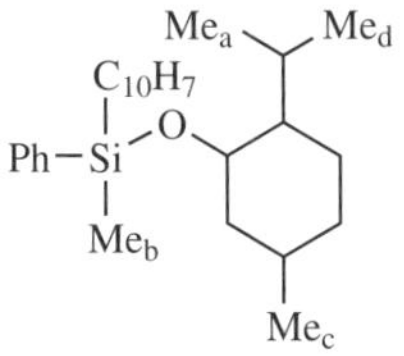

4.102

Menthol has been most widely used for the analysis of carboxylic acids. The (−)-menthol ester of 2-bromo-2,3,3,3-tetrafluoropropionic acid produced diastereomers, the ^{19}F NMR spectra of which showed distinct resonances for both the CF and CF$_3$ groups.[532] The optical purity of α-bromomercuryphenyl acetic acid (**4.103**) was determined by analyzing the ^{1}H NMR spectrum of its menthyl ester.[533] The optical purities of organotin compounds (**4.104**) were determined on the basis of adding Eu(fod)$_3$ to their menthyl esters.[534]

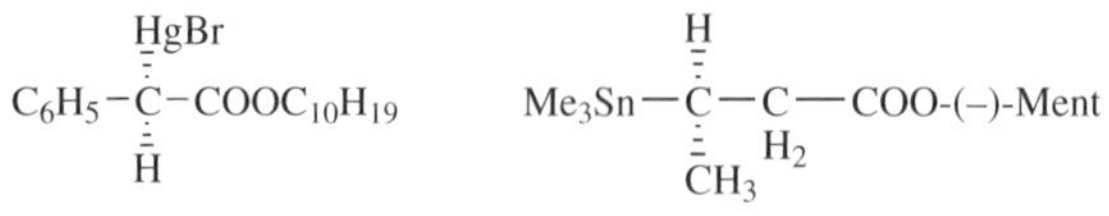

$C_6H_5-\overset{HgBr}{\underset{H}{C}}-COOC_{10}H_{19}$

4.103

$Me_3Sn-\overset{H}{\underset{CH_3}{C}}-\underset{H_2}{C}-COO\text{-}(-)\text{-Ment}$

4.104

The absolute configuration of 2-(5,7-dichloro-3-indolyl)propionic acid (**4.105**) was assigned on the basis of NMR shift data of its menthyl ester.[535] Analysis of model 2-(3-indolyl)propionic acids of known configurations revealed trends in the shifts of the menthyl methyl resonances that could be used to reliably assign the absolute configuration of the unknown.

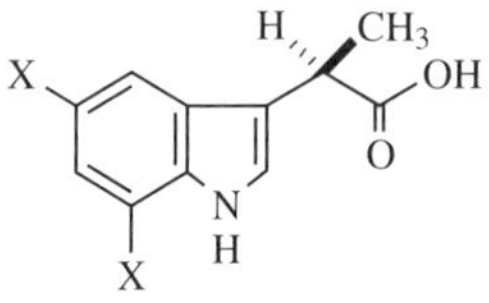

4.105

The absolute configurations of 2-hydroxy-2-(1-naphthyl)propionic acid,[536] 2-methoxy-2-(2-naphthyl)propionic acid, and 2-hydroxy-2-(2-naphthyl)propionic acid[537] were unequivocally assigned on the basis of $\Delta\delta^{RS}$ values of the menthyl resonances in the corresponding (−)-menthol esters.

The chirality of a menthyl derivatized calix[4]resorcarene (**4.106**) was demonstrated by a doubling of certain resonances in the ^{1}H NMR spectrum in chloroform-d.[538] The optical purity of spiro[3.3]heptane-2,6-dicarboxylic acid (**4.107**) was determined on the basis of the ^{13}C NMR spectrum of the corresponding bis(−)-menthyl ester.[539]

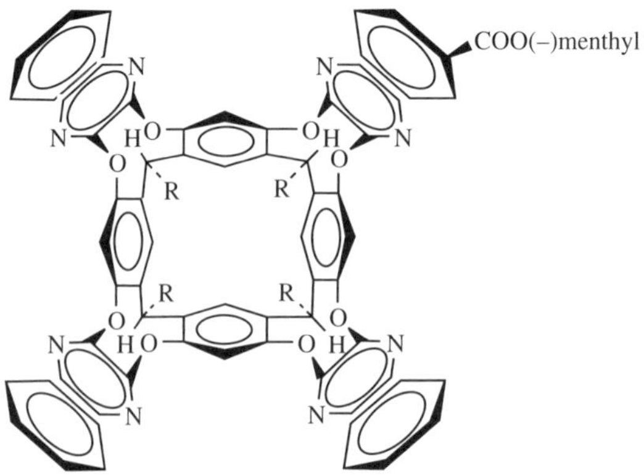

4.106

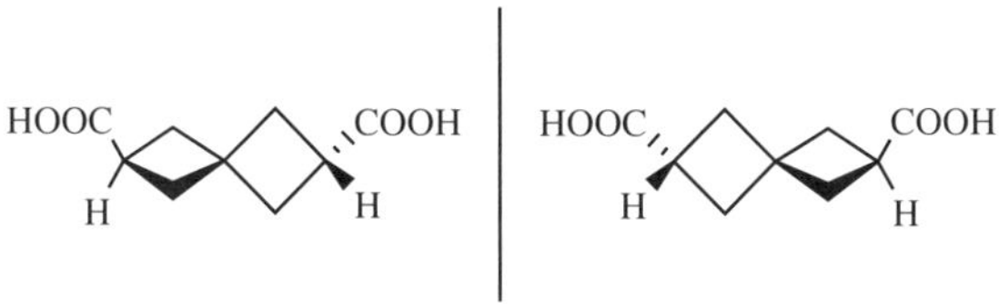

4.107

4.10. BORNEOL (4.108)

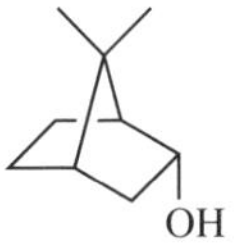

4.108

The enantiomeric purity of chlorofluoroacetic acid was determined by an analysis of the ^{19}F NMR spectrum of its ester with (−)-borneol.[540] The enantiomeric discrimination in the ^{19}F NMR spectrum was larger than that observed with esters prepared from 2-octanol.

4.11. (−)-10-MERCAPTOISOBORNEOL (4.109)

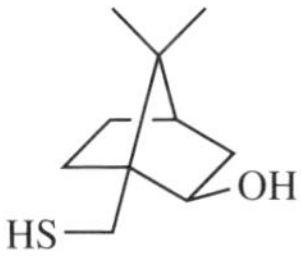

4.109

The compound (−)-10-mercaptoisoborneol was reacted with enones such as **4.110** to prepare the corresponding sulfide (**4.111**).[541] The 9-methyl resonances of the isoborneol unit were monitored for the determination of optical purity of the original enone.

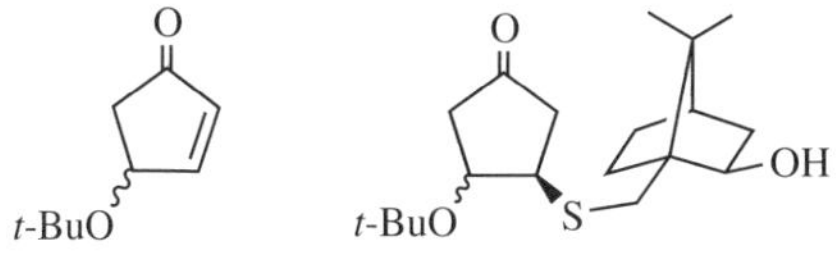

4.110 **4.111**

4.12. (R)-(−)-PANTOLACTONE (4.112)

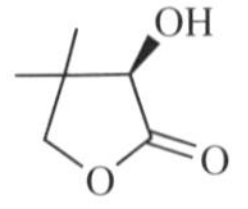

4.112

Derivatives of 2-methylbutanoic, 2-methylpentanoic, and 2-methylhexanoic acid with **4.112** provided diastereomeric esters that exhibited nonequivalent resonances in either the ^{1}H or ^{13}C NMR spectrum.[542] The method was then applied to determine the optical purity of 2-methyl butanoic acid found in strawberries.

4.13. *trans*-BIS(HYDROXYDIPHENYLMETHYL)-2,2-DIMETHYL-1,3-DIOXACYCLOPENTANE (4.113)

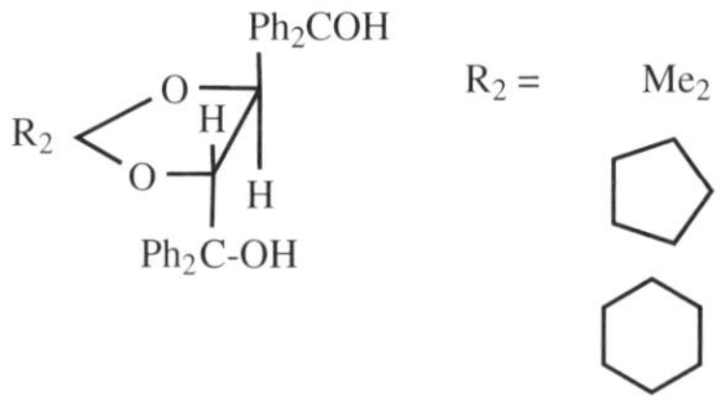

4.113

Reagents of structure **4.113** with two methyl groups, a cyclopentyl or cyclohexyl group, have been evaluated as chiral NMR solvating agents.[543–546] Shifts in the ^{1}H NMR spectra of a variety of primary and secondary amines, cyanohydrins, and amino acid esters generally exhibited characteristic trends that correlated with the absolute stereochemistry.[544,545] For example, with acyclic amines the shift of the α-methine proton correlated with the absolute configuration, whereas the β-methyl proton of cyclic amines was diagnostic in assigning the stereochemistry.[544]

A solid-state X-ray study of the complex of 1-phenylethylamine and 2-methylpiperidine with **4.113** showed that a hydrogen bond between the hydroxyl group of the reagent and the nitrogen atom of the substrate was an important interaction.[543] The solid-state structures correlated with the trends observed in the NMR spectra.

The optical purity of *N*-benzyl- and *N*-ethyl-3-hydroxypyrrolidine (**4.114**) was determined by analyzing the ^{13}C NMR spectrum in the presence of **4.113**.[546]

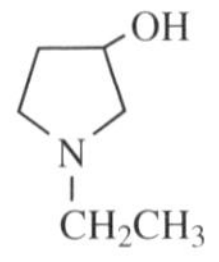

4.114

4.14. 2-HYDROXYMETHYL-4,6-DIMETHYL-2-PHENYL-1,3-DIOXANE (4.115)

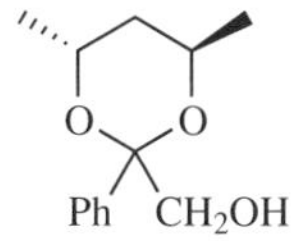

4.115

Reaction of **4.115** with carboxylic acids chiral at the α-position provided diastereomeric esters with distinct resonances in the ^{1}H NMR spectrum.[547]

4.15. 2-BUTANOL

4.116 was reacted with (R)- and (S)-2-butanol to generate the corresponding esters.[548] It was possible to assign the absolute configuration using comparative spectra of the two diastereomers.

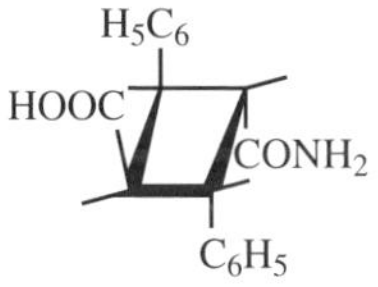

4.116

Calix[4]resorcarene **4.117** is inherently chiral. Alkylation of a racemic mixture of **4.117** with (S)-(−)-2-methyl butyl tosylate to generate the corresponding 2-methyl butoxy derivatives gave two sets of signals for the diastereomers in methylene chloride-d_2.[549] Using a starting resorcarene of known configuration, it was possible to assign the signals in the discriminated spectrum.

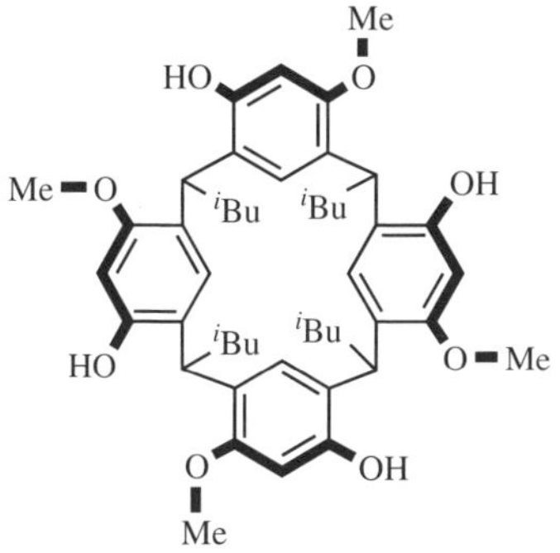

4.117

4.16. ASSIGNMENT OF ABSOLUTE CONFIGURATION USING GLYCOSIDATION SHIFTS

The derivatization of an enantiomeric pair of secondary alcohols with D-glucose or D-mannose leads to diastereomeric aglycones that have pronounced and systematic differences in their ^{13}C NMR spectra.[550–552] These differences are observed in the C1' resonance of the glucone, as well as the α- and β-carbons of the alcohol unit. For alcohols with two adjacent methylene groups to the hydroxyl carbon, the *pro*-(S) carbon of the alcohol was more shielded than the *pro*-(R) carbon for the aglycone.[550] Shifts of the γ-carbon of the alcohol were generally too small to be useful in the analysis.

The absolute configurations of secondary alcohols could be assigned by comparing the ^{13}C shifts in pyridine-d_5 of sugar and alcohol resonances in the aglycone derivative to those of the corresponding methylglucoside ($\Delta\delta_S$) and underivatized alcohol, respectively ($\Delta\delta_A$).[552] Characteristic trends were observed for the $\Delta\delta_S$ values of C1' and $\Delta\delta_A$ values of Cα and Cβ that correlated with the absolute stereochemistry. A set of rules were presented and the effectiveness of the method was demonstrated for aliphatic and cyclic alcohols.

The method was subsequently extended to the assignment of absolute configurations of secondary allylic and benzylic alcohols.[553] Shifts of the ^{13}C resonances of 2–4 ppm for the C1' position and 5–8 ppm for the Cα position were observed. The β-D-glucopyranoside or corresponding tetraacetate β-D-glucopyranoside derivatives could be used in the analysis.

With a high-field NMR spectrometer, it was possible to extend the method to the analysis of ^{1}H NMR spectra.[554] Derivatives of secondary alcohols with tetra-*O*-acetylglucose showed specific shifts for the β-protons of the alcohol that correlated with absolute configuration. The method was applicable to alcohols with methylene groups at both β-positions, one or two substituents at the *syn*-β-position, one or two substituents at the *anti*-β-position, and one or two substituents at both the *syn*- and *anti*-positions.

The derivatization of D-glucose, D-mannose, or L-rhamnose with racemic 2-butanol and subsequent conversion of each to the tetra-*O*-acetyl derivative provided two diastereomers in which a combination of molecular modeling, long-range coupling constants, and NOESY data could be used to assign the stereochemistry of the alcohol.[555]

The rules for assigning absolute configurations of such aglycones can be applied either to secondary alcohols that are specifically converted to their corresponding glucopyranoside derivative or to naturally occurring glycosides. The absolute configuration of the naturally occurring 5,6,7,8-tetrahydro-6-β-D-xylopyranosyloxy-2-naphthalene carboxylic acid (**4.118**) was assigned on the basis of the ^{13}C NMR shifts.[556] Xylose derivatives of alcohols with known configurations were also examined and showed that the trends were comparable with those observed for the glucose derivatives.

4.118

ASSIGNMENT OF ABSOLUTE CONFIGURATION USING GLYCOSIDATION SHIFTS 139

Shifts in the ^{13}C NMR spectra have been used to assign the absolute configuration of particular sites in naturally occurring ionol glycosides such as **4.119**[557–560] and **4.120**.[561]

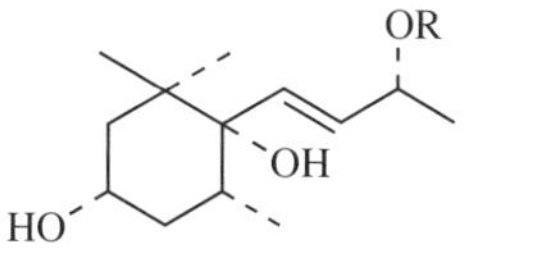

R = β-d-Glucopyranosyl

4.119

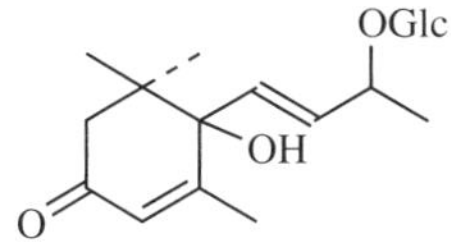

4.120

The stereochemistry of the C14 site in **4.121** was assigned on the basis of ^{13}C NMR shifts of its glucopyranoside derivative.[562] Shifts in the ^{13}C resonances at C3 and C5 of the 4-*O*-α-D-glucopyranosyl derivative of **4.122** were used to assign the configuration at C4.[563]

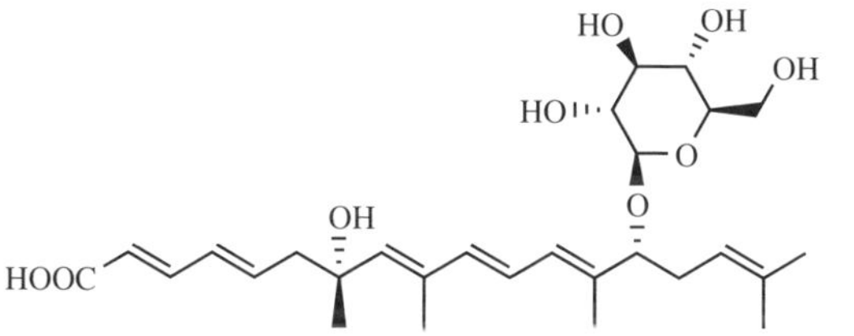

4.121

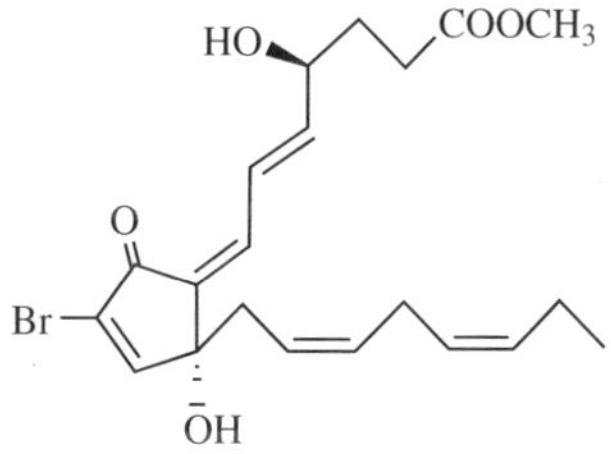

4.122

An analysis of ROESY data for the peracetylated β-D-glucopyranoside derivative of a negastignane 3,9-diol (**4.123**) was performed and confirmed the rationale for assigning the absolute configuration of compounds based only on shifts in the ^{13}C NMR spectra.[564]

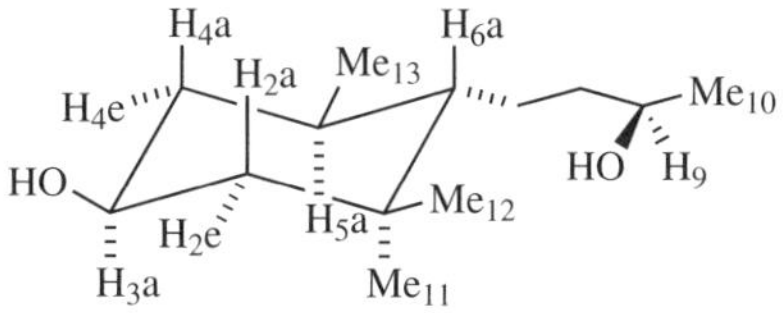

4.123

Glucopyranoside derivatives of secondary alcohols prepared by using commercially available tetra-*O*-benzoyl-β-glucosepyranosyl bromide exhibited shifts in the ^{1}H NMR spectra that correlated with the absolute stereochemistry.[565,566] Shifts of the aglycone with D-glucose compared to the underivatized alcohol or of the D- and

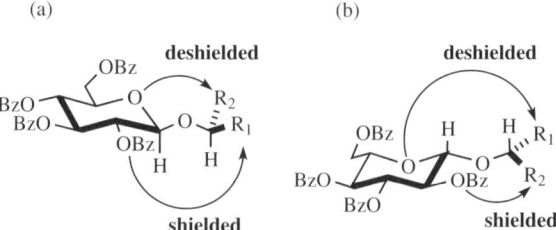

Figure 4.12. Configurational correlation models for secondary alcoholic (a) β-D-glucopyranosides and (b) β-L-glucopyranosides. [Reproduced with permission from American Chemical Society.[565]]

L-glucone derivatives were used in assigning the absolute configuration. As seen from the structures in Figure 4.12, protons *anti* to the endocyclic glucopyranoside oxygen (*O*-5) were shielded from the benzoyl group ($\Delta\delta < 0$), whereas those *syn* were deshielded ($\Delta\delta > 0$). Fifteen alcohols were examined and the shift differences between the D- and L-glucopyranoside derivatives were larger than those between the D-glucopyranoside derivative and the underivatized alcohol.[565] The distinction was reportedly greater than that with α-methoxy-α-trifluoromethylphenylacetic acid (MTPA), but was not compared with the prior work using glucose or tetra-*O*-acetylglucose.

A conformational analysis of the 2,3,4,6-tetra-*O*-benzoyl-β-glucopyranoside derivatives of secondary alcohols indicated that the C5–C6 bond exhibited three conformers (*gauche–gauche* (*gg*), *gauche–trans* (*gt*), and *trans–gauche* (*tg*)) (Fig. 4.13), the population of which depended on the structure of the aglycone.[566] The rotamer populations correlated with the absolute configuration of the alcohol, which related to the stereoelectronic *exo*-anomeric effect. The C1' proton of the D-glucopyranosyl of the (*R*)-alcohol was consistently at lower frequency than that of the (*S*)-alcohol. Furthermore, the $J_{H5,H6}$ coupling constant for the (*S*)-alcohol was often larger by 0.2–0.9 Hz than that observed for the (*R*)-alcohol, although this trend was reversed with cholesterol, cholestanol, and dimethylmaleate.

The optical purity of 1-(4-benzyloxy-3-methoxyphenyl)-3-hydroxy-2-(2-methoxyphenoxy)propanone (**4.124**) was determined by analyzing the ^{1}H or ^{13}C NMR spectrum of the tetra-*O*-benzoylglucopyranoside derivative of the 3-hydroxy position.[567]

4.124

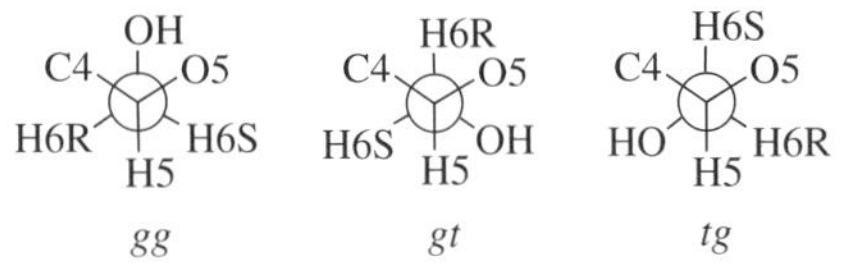

ASSIGNMENT OF ABSOLUTE CONFIGURATION USING GLYCOSIDATION SHIFTS 141

Figure 4.13. Three conformations of the 2,3,4,6-tetra-*O*-benzoyl-β-glucopyranoside derivatives of secondary alcohols.

The absolute configuration of the C10–C19 fragment of nystatin A (**4.125**), which contains a D-mycosamine unit, was assigned by utilizing a series of two-dimensional NMR methods including NOESY and ROESY.[568] Through-space contacts between the D-mycosamine unit and the macrocycle facilitated the assignment.

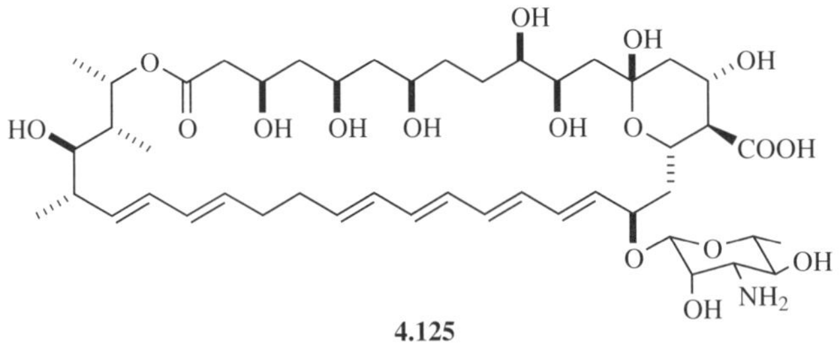

The β-D-galactopyranosyl derivatives of L- and D-threonine and L- and D-allothreonine had nonequivalence in the ^{13}C shifts of the anomeric C1 and threonine methyl signals of up to 8.5 and 6 ppm that correlated with the absolute configuration.[569] The ^{13}C NMR spectrum could then be monitored to see if any racemization occurred during the synthesis of the glycopeptides. Being able to use two signals, one from the amino acid and the other from the glycoside, was an advantage of this system.

Reaction of the glycal epoxide **4.126** with a substrate such as 2-hexanol produced the corresponding glycosides (**4.127**). ^{13}C NMR shifts of the corresponding *O*-methylated species were used to determine the absolute configuration of the 2-hexanol.[570]

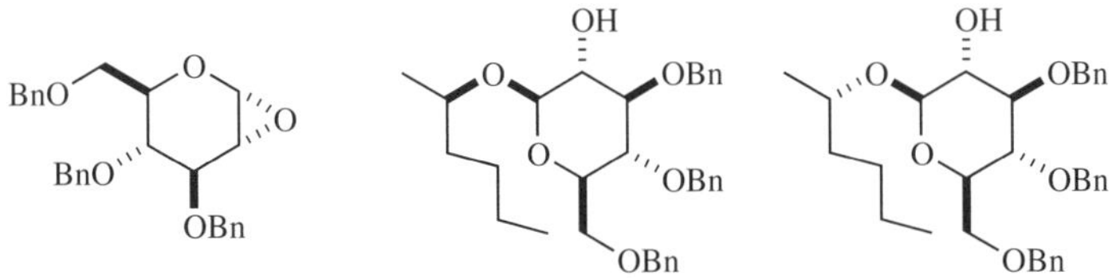

4.16.1. Analysis of Di- and Polysaccharides

If the anomeric configuration and absolute configuration of one monosaccharide of a disaccharide are known, shifts in the ^{13}C NMR spectrum can be used to assign the absolute configuration of the second monosaccharide unit.[571] The shifts in the ^{13}C NMR spectra of 19 different 1,1′-disaccharides were examined, and the shifts of the anomeric carbon of the (R,R)- and (S,S)-derivatives were consistently smaller than those of the (S,R)-derivative.[572]

The *pro*-(R) and *pro*-(S) positions of disaccharides linked through the 1–6 positions were distinct in the ^{1}H NMR spectrum.[573] Stereoselective monodeuteration of such a derivative (**4.128**) facilitated an analysis of the populations of the *gg*, *gt*, and *tg* conformers, such that differences in the coupling constants could be used to assign the absolute stereochemistries.

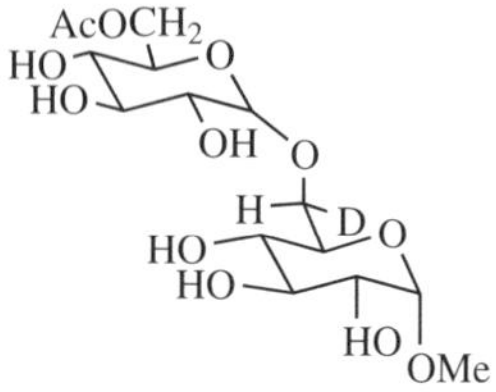

4.128

The ^{13}C chemical shifts of diastereomeric trisaccharides of the form LLL/DDD, LDL/DLD, DLL/LDD, and DDL/LLD were used to develop a computer model to predict the structure of branched polysaccharides at the sites where the branches occur.[574]

4.16.2. β-D- and β-L-Fucofuranoside/Arabinofuranoside (4.129)

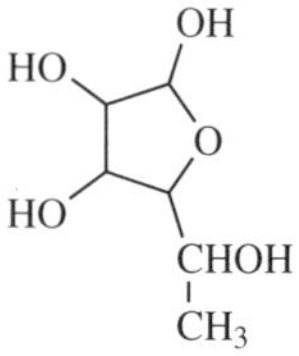

4.129

The use of β-D- and β-L-fucofuranosides to determine the absolute configurations of secondary[575,576] and tertiary alcohols[577] as well as 1,2-glycols[578] has been described. The furanoside was converted to its tetraacetate derivative, which was then reacted with the alcohol. Alkaline hydrolysis of the acetate groups provided the necessary derivative for NMR analysis. Although most reports on this system have used fucofuranosides, arabinofuranoside was equally effective for the analysis of alcohols.[579] An advantage of the use of arabinofuranoside was that its tetraacetate was more readily prepared than that of fucofuranoside. Both the D- and L-arabinofuranose are commercially available.

Either the ^{1}H or ^{13}C NMR spectrum in pyridine-d_5 showed distinct differences that correlated with the absolute configuration of the alcohol. The differences were caused by different solvent-induced shifts of the pyridine on the aglycon. The glycosidation shifts in the ^{13}C NMR spectra also showed specific trends for symmetrical and unsymmetrical substituent groups that correlate with the stereochemistry of the alcohol. The same patterns occurred with glucopyranoside derivatives, although the more rigid linkage in the five-membered fucofuranoside derivative led to greater discrimination.[575] The utility of the method was demonstrated on a wide range of secondary alcohols. A set of four rules for determining the absolute configuration using ^{1}H and ^{13}C NMR data were provided.[575]

The method was successfully applied to determine the absolute configuration of the C3 hydroxy groups in betulinic acid (**4.130**) and 3-epibetulinic acid (**4.131**).[576] The acids were first converted to the methyl esters prior to conversion to the fucofuranoside derivative. The analysis was done using ^{13}C NMR in pyridine-d_5.

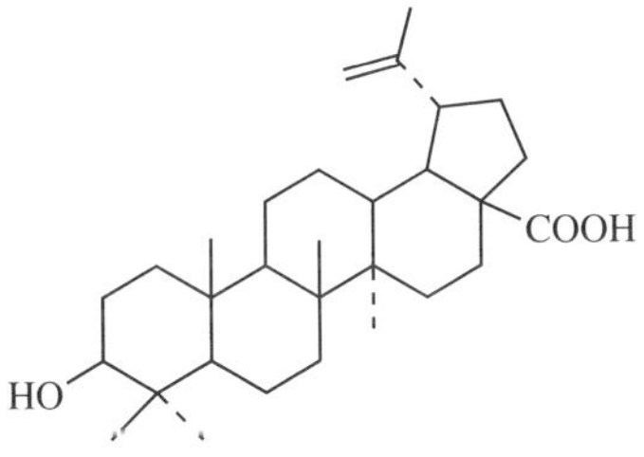

4.130 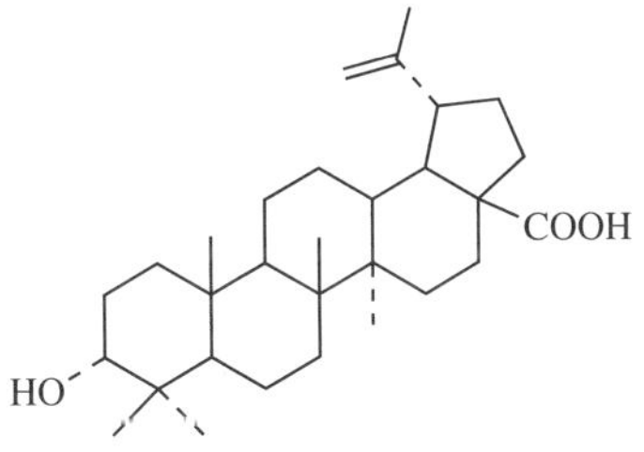   **4.131**

The utility of the method for tertiary alcohols was examined by studying a series of steroids with tertiary alcohols on six- or five-membered rings.[577] Systematic patterns to the shifts that correlated with the absolute configuration were noted in the ^{1}H and ^{13}C NMR spectra in pyridine-d_5 of five of the six steroids studied. The compound that gave somewhat ambiguous results had a hydroxyl group on a five-membered ring.

Five steroids with a 1,2-glycol unit comprising a secondary and a tertiary alcohol, one example of which is **4.132**, were analyzed by the fucofuranoside method.[578] The compounds included derivatives with secondary hydroxyl groups in the axial or equatorial position as well as *cis*- and *trans*-glycols. Reaction occurred at the secondary hydroxyl group and gave mostly the β-anomer. The shift patterns in the ^{1}H and ^{13}C NMR spectra in pyridine-d_5 agreed with the earlier studies of secondary alcohols. The ^{13}C NMR data were deemed more useful for the analysis, although the ^{1}H NMR spectra could also be used for additional confirmation of the assignment. Shifts in the anomeric, α-, and two β-carbons showed systematic shielding and deshielding effects that could be rationalized based on the absolute configuration,[578] and the effects were great enough that solvents other than pyridine-d_5 could be used.

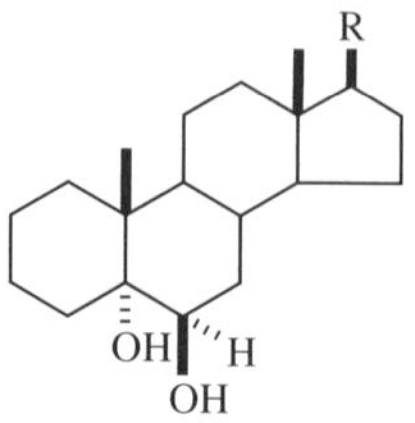

4.132

4.17. LINEAR DEXTRINS

The permethyl oligosaccharide species MeGly$^{1\alpha}$Fru$_4$ caused enantiomeric discrimination in the ^{1}H NMR spectra of 2-propylester salts of amino acids.[580] The cationic ammonium ion of the amino acid associated with the oligosaccharide.

The ^{1}H NMR spectrum of tetrahelicene 1,12-dimethylbenzo[c]phenanthrene-5,8-dicarboxylic acid (**4.133**) exhibited chiral discrimination in the presence of linear dextrins comprising of glucose units.[581] The series from glucose (G1) to maltoheptaose (G7) was evaluated. Enantiomeric recognition was observed in the presence of G4 through G7 and was largest with G7.

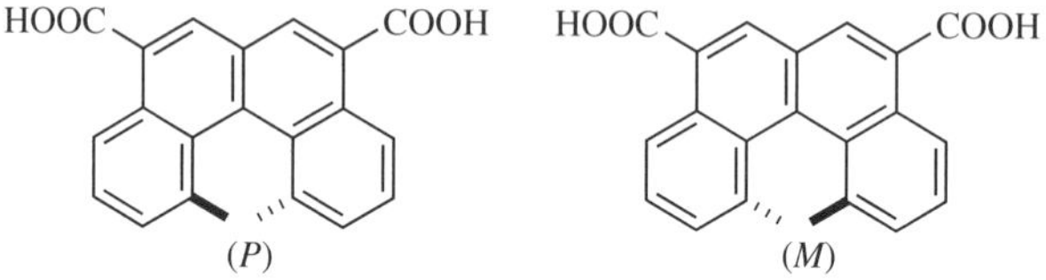

(P)   (M)
4.133

4.18 AXIAL CHIRAL (ATROPOISOMERIC) ALCOHOLS

Certain compounds exhibit a phenomenon known as axial chirality. Compounds that are axial chiral do not have a stereogenic center such as that formed by four different constituents on a carbon atom, but instead have steric constraints that force the molecule to adopt a spatial geometry that is chiral. The classic example is the compound 2,2′-dihydroxy-1,1′-binaphthyl, commonly known as BINOL. Steric constraints force the two naphthyl rings to lie in different planes, such that the molecule exists as a pair of enantiomers. Axial chiral compounds have been commonly exploited in the development of strategies for chiral NMR discrimination.

4.18.1. 2,2′-Dihydroxy-1,1′-binaphthalene (BINOL) (4.134)

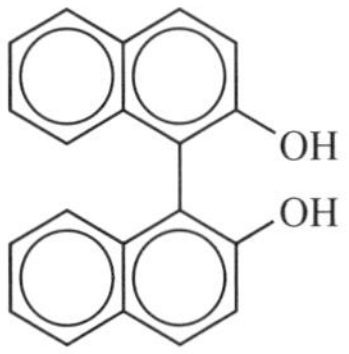

4.134

Because of the potential for hydrogen bonding of the hydroxyl groups and shielding from the aromatic rings, BINOL has been used on a number of occasions to affect chiral discrimination in NMR spectroscopy. For example, BINOL has been shown to be an effective chiral NMR solvating agent for alkyl and aryl alkyl alcohols, sulfoxides, selenoxides, and amines.[582] Nonequivalence on the order of 0.02–0.05 ppm was observed in the ^{1}H NMR spectrum. It was also shown that Eu(fod)$_3$ (fod = 6,6,7,7,8,8,8-heptafluoro-2,2-dimethyl-3,5-octanedione) could be added and often enhanced the discrimination by preferentially binding to and shifting the resonances of the enantiomer that bound less strongly to BINOL.

It was shown in a subsequent study that the shielding effects of BINOL on sulfoxides, amines, and amino alcohols consistently correlated with the absolute configuration of the substrate as illustrated in Figure 4.14.[583] Another report found consistent trends in the shifts of resonances in the ^{1}H NMR spectra or aryl, alkyl, and dialkyl sulfoxides in the presence of BINOL that correlated with absolute stereochemistry.[584] The chiral recognition in the spectra of sulfoxides with BINOL was usually larger than that with N-(3,5-dinitrobenzoyl)-1-phenylethylamine, another well-known reagent for the analysis of sulfoxides described in Section 6.6.1.

The effectiveness of BINOL and TFAE in causing enantiomeric discrimination in the NMR spectra of 1-phenylethylamine, 1-phenylethanol, camphor, phenyl vinyl sulfoxide, and other alcohols and amines has been compared.[585] In most cases, the enantiomeric discrimination was larger with BINOL.

BINOL has been used to determine the optical purity of phenylalkylamine alkaloids such as cathinone (**4.135**), norepinephrine (**4.136**), norpseudoephedrine,

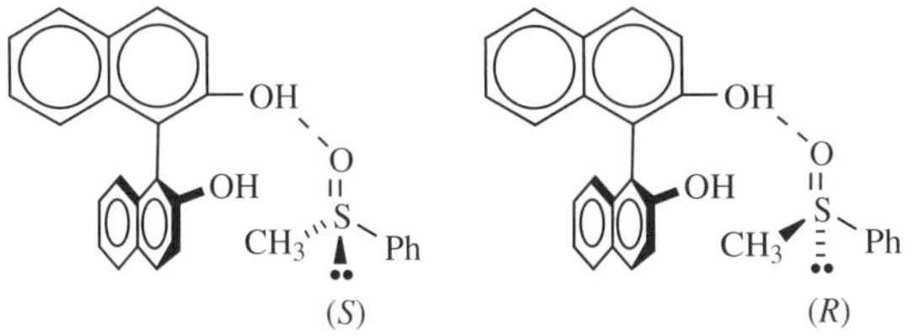

Figure 4.14. Association of BINOL with the two configurations of methyl phenyl sulfoxide.

methamphetamine (**4.137**), and methcathinone (**4.138**).[586,587] The optical purity of **4.139** was determined by analyzing the ^{1}H NMR spectrum in the presence of BINOL.[588]

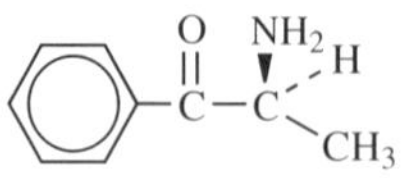

4.135

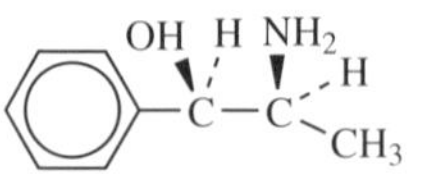

4.136

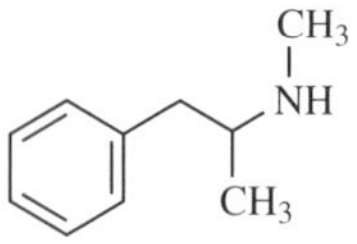

4.137

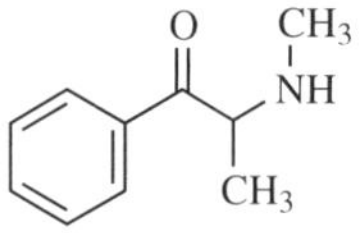

4.138

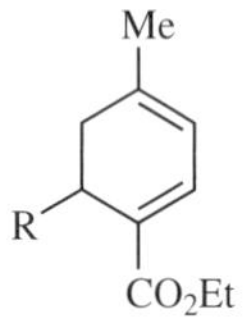

R = —CH=C(CH$_3$)$_2$
—C$_6$H$_5$

4.139

The compound 1,2-bis(*N*-benzoyl-*N*-methylamino)benzene (**4.140**) is chiral by virtue of rotational isomers.[589] At low temperatures, only one enantiomer was observed. Raising the temperature leads to racemization, as evidenced by the splitting of the *N*-methyl resonances in the presence of BINOL.

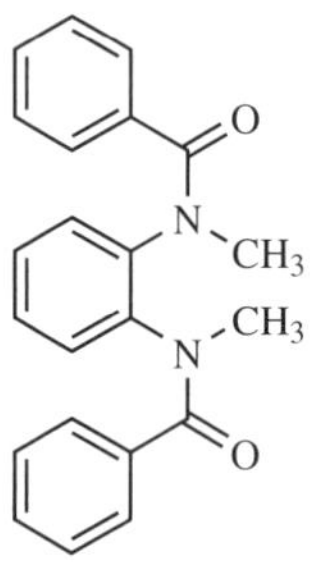

4.140

The relative effectiveness of BINOL, its corresponding 2-hydroxy-2′methoxy derivative, and a 2-hydroxy-2′-dimethoxyphosphate derivative in causing enantiomeric discrimination in the *N*-methyl signals of the *trans*, *exo–cis*, and *endo–trans* isomers of **4.141** was examined.[590] The largest chiral discrimination was generally observed with BINOL. The *trans*-isomer had particularly large shifts and discrimination with BINOL, likely reflecting a geometry that enabled a simultaneous association of the hydroxyl and amine groups of the substrate with the hydroxyl groups of

BINOL. BINOL and its 2-hydroxy-2′-methoxy derivative caused enantiomeric discrimination in the ^{1}H NMR spectra of other N-methyl amino alcohols as well.[591]

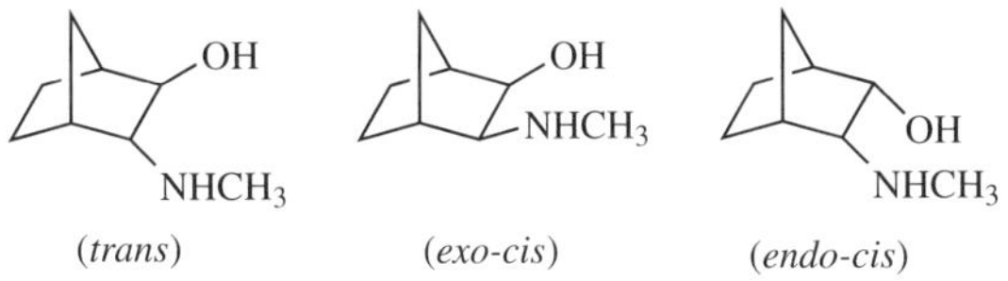

(trans) (exo-cis) (endo-cis)

4.141

4.18.2. 8,8′-Dihydroxy-1,1′-binaphthalene (4.142)

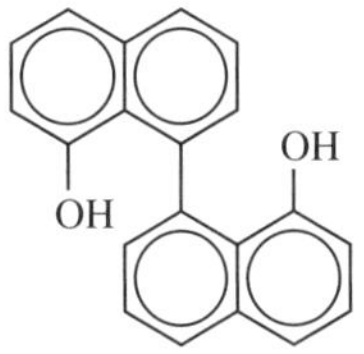

4.142

The utility of **4.142** as a derivatizing agent for α-chiral carboxylic acids such as isobutyric acid and its derivatives has been demonstrated.[592] A monoester was formed, and specific and predictable shielding by the aromatic ring produced $\Delta\delta^{RS}$ values that correlated with the absolute configuration. Nonequivalence in both the ^{1}H and ^{13}C NMR spectra was observed.

4.18.3. Alkyl-Substituted Binaphthols

A variety of alkyl-substituted BINOL derivatives (**4.143**) were examined for their ability to cause chiral recognition in the NMR spectra of quinine and quinidine.[593] The BINOL derivatives have a minor groove by the 2,2′-substituents and a major groove by the 7,7′-substituents. Substrates associated at the minor groove. One or more hydroxyl groups were needed to discriminate the alkaloids, although there was some evidence for π–π stacking in the interaction between the substrates and the BINOL derivatives.

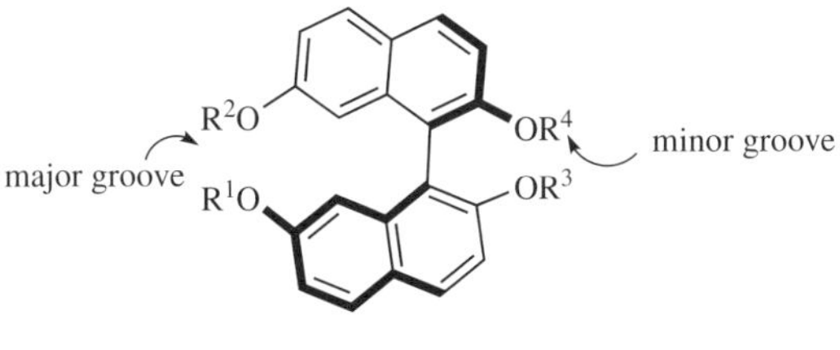

4.143

4.18.4. 2′-Hydroxy-1,1′-binaphthyl-2-yl[3,5-dinitrobenzoyl)amino](phenyl)acetate (4.144)

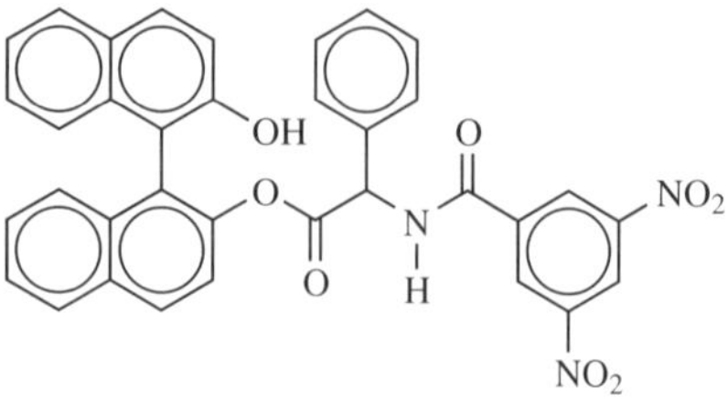

4.144

A derivative of (R)-BINOL with a single (R)- or (S)-N-(3,5-dinitrobenzoyl)phenylglycine unit was evaluated as a chiral solvating agent for benzamines, aryl alkyl amines, and 3,5-dinitrobenzoyl derivatives of amino alcohols.[594] Nonequivalence as large as 0.13 ppm was observed in the spectra of substrates with a protonated nitrogen and aryl group. Tertiary alkyl aryl amines were not successfully discriminated. **4.144** was the most effective for many of the substrates.

4.18.5. 1,6-Di(o-chlorophenyl)-1,6-diphenylhexa-2,4-diyne-1,6-diol (4.145)

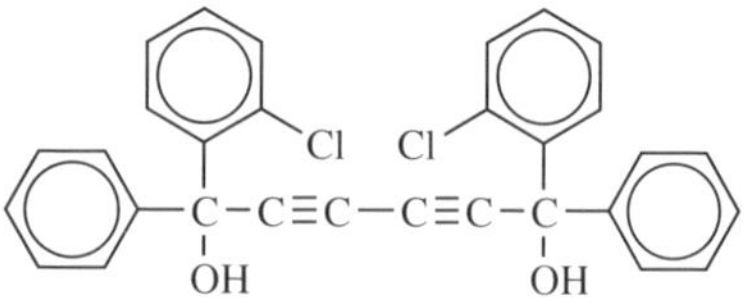

4.145

4.145 has been shown to cause enantiomeric discrimination in the ^{1}H NMR spectra of alkyl and aryl alkyl amines, phosphinates, arsenoxides,[582] sulfoxides, and amino alcohols.[583] In some cases, the phenyl ring of the reagent caused specific shielding that correlated with the absolute configuration of the substrate.

4.18.6. 4,4′6,6′-Tetrachloro-2,2′-bis(hydroxydiphenylmethyl)biphenyl (4.146)

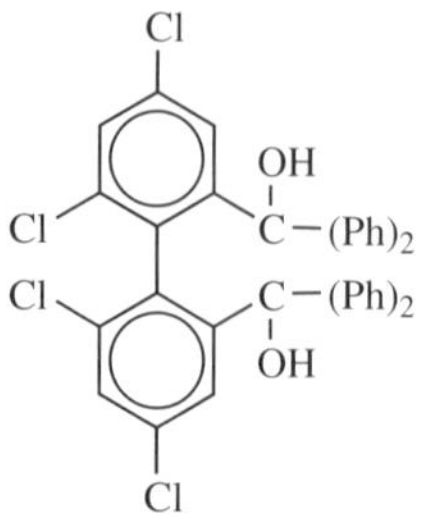

4.146

DIOL AND DITHIOL REAGENTS

4.146 was evaluated as a chiral solvating agent and found to cause discrimination in the NMR spectra of amines, alcohols, amine N-oxides, phosphinates, phosphine oxides, arsine oxides, sulfoxides, sulfoximine, and selenoxides.[595] The reagent was effective for a wider variety of substrates than BINOL. The shifts in the spectra of sulfoxides correlated with the absolute configuration.

4.19. DIOL AND DITHIOL REAGENTS

4.19.1. Butane-2,3-diol/Butane-2,3-thiol

Butane-2,3-diol and butane-2,3-thiol are useful chiral derivatizing agents for the analysis of chiral ketones. The procedure involves the conversion of the ketone such as **4.147** into a pair of diastereomeric ketals (**4.148**) using (R)-(−)-butane-2,3-diol.[596] The initial report on these reagents demonstrated their use in determining the optical purity of 3-substituted cyclohexanones by analyzing the ^{13}C NMR spectra of the ketal. No kinetic resolution or racemization was observed in the derivatization reaction and larger discrimination was observed for cyclic than for acyclic ketones.[597] The enantiomeric discrimination was usually greater with butane-2,3-thiol than with butane-2,3-diol, although most follow-up reports have used butane-2,3-diol.

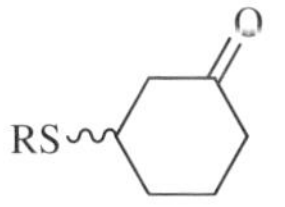

4.147

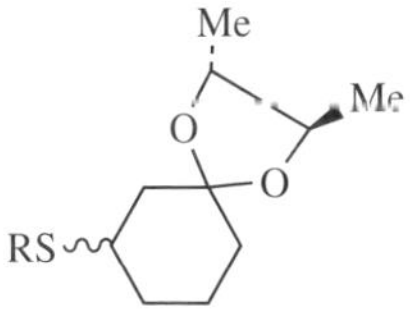

4.148

The optical purity of the *trans*-isomer of **4.149** was determined by analyzing the ^{1}H or ^{13}C NMR spectra of the cyclic ketone formed with butane-2,3-thiol.[598]

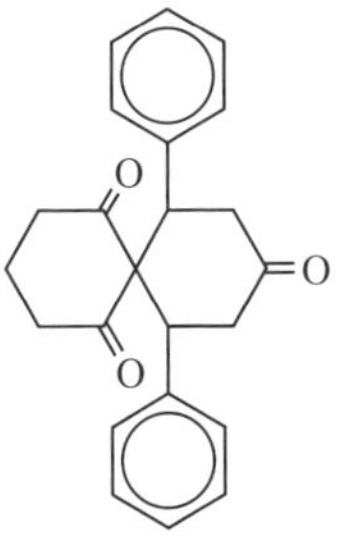

4.149

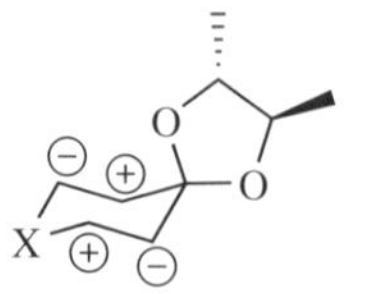

Figure 4.15. Shift directions of the ^{13}C resonances of the ketal derivatives of cyclohexanones.

2-Alkylcyclohexanones such as the 2-ethyl and 2-benzyl derivatives can also be analyzed with this method.[599] The ^{13}C NMR spectrum was used and no kinetic resolution or epimerization occurred in the derivatization reaction. The method was generally ineffective for acyclic and cyclic ketones of other ring sizes.

A detailed study of the utility of this method for 39 ketones including 2- and 3-substituted cyclohexanones, 2-alkyltetrahydropyran-4-ones, and 2- and 3-alkyltetrahydrothiopyran-4-ones was reported.[600] For substrates with a six-membered ring and preference for the chair conformation, the ^{13}C NMR data correlated with the absolute stereochemistry. Figure 4.15 shows a model of the shift direction as it correlates with absolute configuration. The ^{13}C shifts could not be used to reliably assign the absolute configuration of acyclic ketones, cyclopentanones, and cycloheptanones.

The optical purity of **4.150** was determined by analyzing the ^{13}C NMR spectrum of its corresponding acetal.[601] The ^{31}P NMR spectrum of the acetal of 4-(diphenylphosphinyl)-3-phenylcycloheptanone (**4.151**) showed distinct peaks for the two diastereomers that could be used to determine the optical purity.[602]

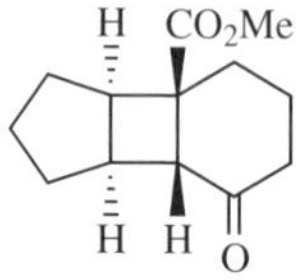

4.150 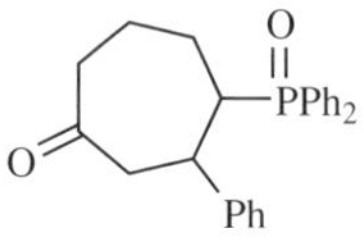   **4.151**

4.19.2. 2,2-Dimethyl-1,3-propanediol

Reaction of 2,2-dimethyl-1,3-propanediol with a chiral carboxylic acid leads to the formation of a diester that has diastereotopic methyl and methylene groups.[304] The (R,R)- and (S,S)-derivatives are isochronous with only one methyl resonance, whereas the *meso*-(R,S)-isomer has different methyl groups and shows two signals. The relative area of these peaks can be related to the enantiomeric purity of the carboxylic acid. The utility of the method was demonstrated for a series of 2-substituted alkanoic acids.[304]

4.19.3. 1,1,2-Triphenyl-1,2-ethanediol

The compound 1,1,2-triphenyl-1,2-ethanediol forms associated adducts with carboxylic acids such as 2-chloropropanoic acid, mandelic acid, and α-hydroxyl phosphonate (**4.152**). The resulting diastereomers had distinct resonances in the ^{1}H NMR spectrum.[603]

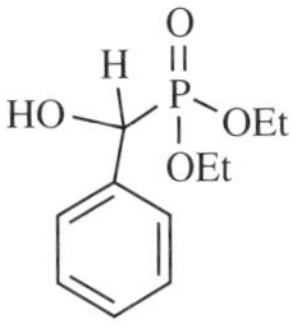

4.152

4.19.4. Ethylene Dithiol

Reaction of ethylene dithiol (HSCH$_2$CH$_2$SH) with 4,10-dimethyldecal-3-one derivatives such as **4.153** produces the corresponding ethylenedithioacetal derivative (**4.154**).[604] The relative configuration of the methyl group α to the dithioacetal linkage could be determined by analyzing its shift in benzene-d_6, as the resonance of the methyl group in the equatorial position was deshielded relative to that in the axial position.

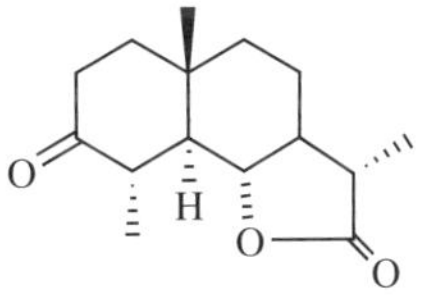

4.153 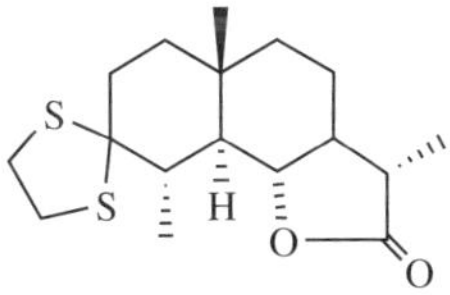   **4.154**

4.19.5. α,α,α′,α′-Tetraphenyl-1,3-dioxolane-4,5-dimethanols (TADDOL) (4.155)

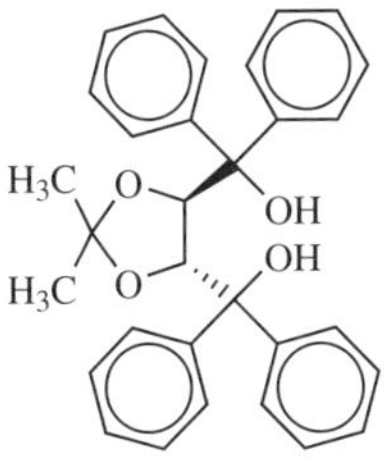

4.155

TADDOL has been shown to be an effective chiral NMR solvating agent for determining the optical purity of cyanohydrins (**4.156**)[605,606] and a variety of other aliphatic alcohols.[605] Nonequivalence of up to 0.12 ppm was observed in the ^{1}H NMR spectrum. Discrimination was also observed in the ^{13}C NMR spectrum.

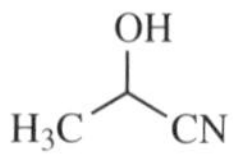

$$H_3C \overset{OH}{\underset{}{\diagup\!\!\diagdown}} CN$$

4.156

5

AMINE-BASED REAGENTS

Primary, secondary, and tertiary amines have been employed as chiral NMR derivatizing and solvating agents. The compounds 1-phenylethylamine (PEA) and 1-(1-naphthyl)ethylamine (NEA) have an extensive record of use for the analysis of carboxylic acids either as salts formed in solution or as amide derivatives. PEA and NEA are effective with other classes of compounds as well. Secondary amines such as ephedrine have found utility in chiral NMR discrimination studies. Tertiary amines, the most important example of which is quinine, have been exploited primarily as chiral NMR solvating agents. Because quinine can associate with substrates through dipole–dipole and π–π stacking interactions involving multiple functionalities, it has been exploited in the analysis of many classes of substrate compounds.

Amino acids have also been used to effect chiral discrimination in NMR spectroscopy. This includes the analysis of peptides. Phenyl glycine methyl ester is an especially effective reagent for assigning the absolute configurations of carboxylic acids. In a few cases, suitable amine compounds have been used as chiral derivatizing agents for the analysis of aldehydes or ketones. The method involves a reaction to form the corresponding imine. Diamine reagents have also been exploited as chiral derivatizing or solvating agents. In some cases, these are more effective in causing chiral recognition than common monoamine reagents.

Finally, shift data with two chiral amine solvating agents, N,α-dimethylbenzylamine and bis-1,3-methylbenzylamine-2-methylpropane, have been incorporated into ^{13}C and ^{1}H NMR databases that can be used to assign the stereochemistries of certain structural motifs in natural products.

Discrimination of Chiral Compounds Using NMR Spectroscopy, by Thomas J. Wenzel
Copyright © 2007 John Wiley & Sons, Inc.

5.1. PRIMARY AMINES

5.1.1. 1-Phenylethylamine (PEA) (5.1)/1-(1-Naphthyl)ethylamine (NEA) (5.2)/ 1-(9-Anthryl)ethylamine (AEA) (5.3)

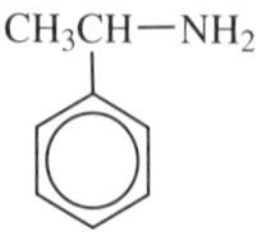

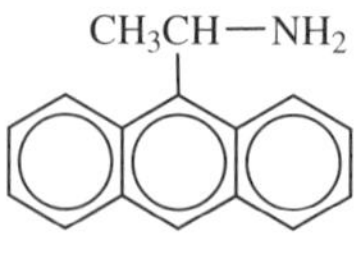

The first report of a chiral NMR solvating agent was by Pirkle in 1966 and involved the use of PEA as a solvent to discriminate the ^{19}F NMR signals of 2,2,2-trifluoro-1-phenylethanol.[607] Soon after it was shown that NEA could be used to produce enantiomeric discrimination in the ^{1}H NMR spectra of aryl alkyl carbinols[608] and in the ^{19}F NMR spectra of **5.4** in which R was either an aryl or an alkyl substituent group.[609] These reagents also caused discrimination in the ^{13}C NMR spectra of carbinols.[610]

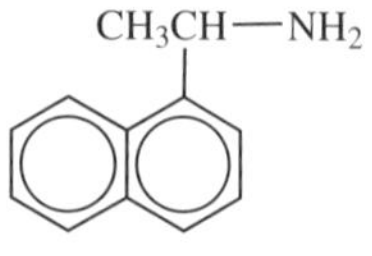

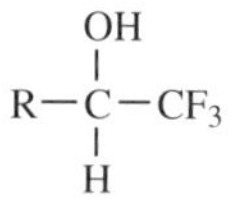

PEA and NEA associate with carbinols through the formation of hydrogen bonds. The geometry of the associated complex of NEA with aryl alkyl carbinols was specific enough such that the shifts caused by shielding of the aromatic ring of NEA correlated with the absolute configuration of the substrate.[611] The expected geometry of the associated complex was used to rationalize the shifts in the (R)- and (S)-enantiomers of 3-O-acetyl-(R)-2,3-dihydroxy-2-methylpropanoate (**5.5**)[612] and lacinilene C methyl ether (**5.6**).[474]

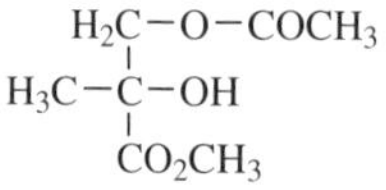

There are two strategies by which primary amines such as PEA, NEA, and AEA can be used for the analysis of carboxylic acids. One involves mixing the amine with the acid to form diastereomeric salts. The second is to form the corresponding amide through a derivatization reaction.

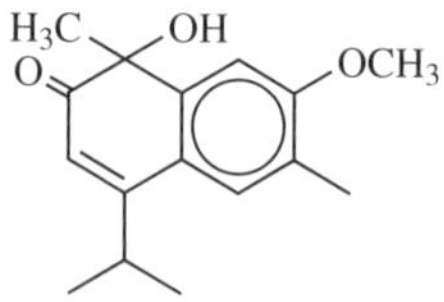

The occurrence of chiral discrimination in the ^{1}H NMR spectra of carboxylic acids in the presence of PEA[613,614] and NEA[615,616] was noted soon after the first reports on the use of NMR spectroscopy for chiral discrimination. These reagents were effective for analyzing carboxylic acids in solvents such as methanol-d_4,[613] carbon tetrachloride, chloroform-d, benzene-d_6,[614] dimethylsulfoxide-d_6, and pyridine-d_5.[616]

PEA has been used to determine the optical purity of carboxylic acids with a selenide functionality at the α-position (**5.7**) through the formation of a salt in chloroform-d.[617] The ^{1}H NMR spectrum did not provide sufficient enantiomeric discrimination, whereas the ^{77}Se NMR spectrum did. The analog of **5.7** with the selenide functionality β to the carboxylic acid was not enantiomerically discriminated in the presence of PEA.

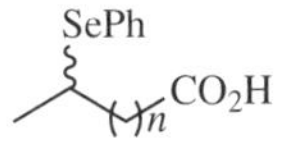

$n = 0, 1, 2$

5.7

Chiral recognition in the ^{1}H NMR spectrum of 2(R),3(S)-[(dimethylmethylene)dioxy]-5(R)-hydroxy-1(S)-carboxy-4(R)cyclopentane carboxylate (**5.8**) was found to be much larger with NEA than with PEA.[618] A comparison of the enantiomeric discrimination in the NMR spectra of carboxylic acids with PEA, AEA, and 1,2 diphenyl-1,2-diaminoethane showed that the best results were obtained with the AEA, presumably because of larger shielding from the anthryl ring.[100] Similarly, AEA caused larger enantiomeric discrimination in the ^{1}H NMR spectra of carboxylic acids in chloroform-d than PEA or NEA.[619] The corresponding 9-(1-aminoethyl)phenanthrene reagent was unsuitable for use as precipitates are formed in chloroform-d.[619]

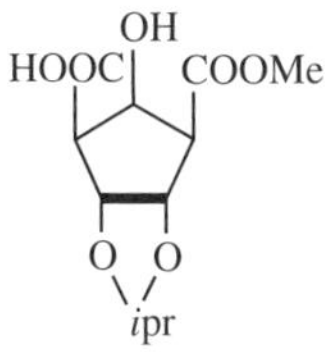

5.8

Amide derivatives with PEA have been used to determine the optical purity of carboxylic acids such as (methyl)ferrocene-α-carboxylic acid (**5.9**),[620] 2-bromo-2,3,3,3-tetrafluoropropionic acid,[532] and 2-(4-dimethylvinylphenyl)propionic acid (**5.10**).[621] In the latter example, comparison of the shifts to those of ibuprofen with a known stereochemistry allowed the assignment of the absolute configuration of **5.10**. The optical purity of **5.11** was determined by measuring the ^{1}H NMR spectrum of the amide derivative with PEA.[622] The methoxy resonance exhibited the largest discrimination. A folded conformation that caused shielding of the methoxy resonance for one of the enantiomers was the likely cause of the large degree of recognition in the NMR spectrum.

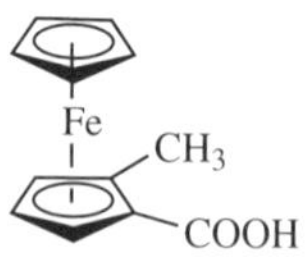

5.9 (ferrocene derivative with CH₃ and COOH)
5.10 (CH₃)₂C=CH–C₆H₄–CH(CH₃)–COOH
5.11 (tetrahydronaphthalene derivative with CH₃, H₃C, H₃CO, CO₂H)

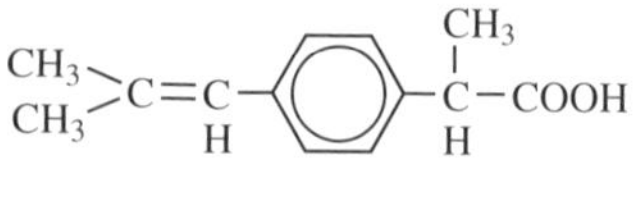

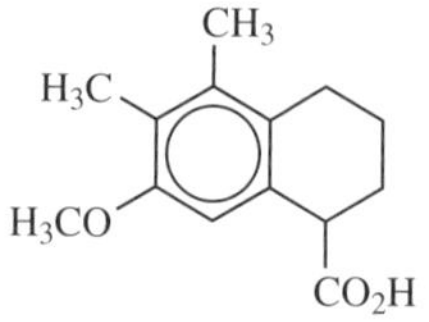

Amide derivatives of carboxylic acids prepared with PEA and NEA adopt a preferred conformation and shielding by the aromatic ring can often be used to assign the absolute configuration.[623] A detailed analysis on the use of PEA, NEA, and AEA amides to assign the absolute stereochemistry of 3-methylcarboxylic acids has been undertaken.[624] The extent of discrimination for the methyl resonances of the diastereomers occurred in the order AEA > NEA > PEA, although NEA and PEA were considered preferable because of their commercial availability. The amide derivatives had major and minor conformations as illustrated in Figure 5.1, and differential shielding from the aryl group correlated with the absolute configuration of the β-position. The β-methyl resonance of the *syn*-diastereomer was deshielded relative to that of the *anti*-diastereomer.[624] All discernible resonances of the carboxylic acid that showed nonequivalence were used to increase the reliability of the assignment.

The use of NEA and PEA was subsequently extended to carboxylic acids with a variety of substituent groups at the β-position.[625] Larger discrimination was observed in the spectra of the derivative with NEA. Conformational preferences allowed differential shielding in the *syn*- and *anti*-rotamers to be rationalized in a way that could be used to reliably determine the absolute configuration.[625]

Amide derivatives with PEA have been used to assign the absolute configurations of ketoprofen (**5.12**),[626] N-phthaloyl derivatives of amino acids,[627] 3,3′4,4′-tetramethyl-1,1′-diphosphaferrocene-2-carboxylic acid (**5.13**),[628] and 2-fluoro- and 2-methyl hexanoic acids.[629] It is especially noteworthy that the terminal methyl resonances of the hexanoic acids were distinct in the ¹H NMR spectra of the

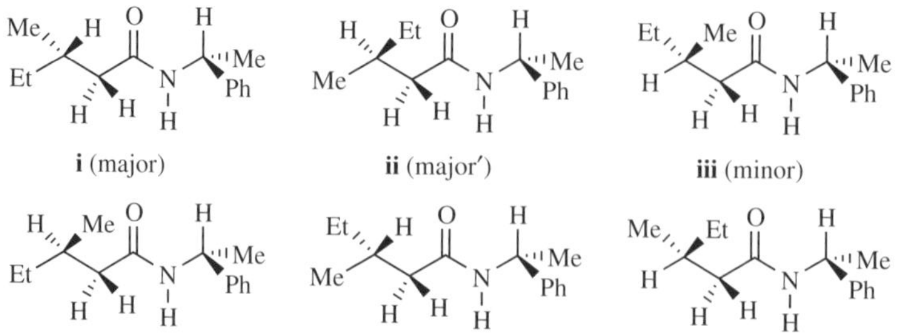

i (major) **ii** (major′) **iii** (minor)

Figure 5.1. Major and minor conformations of the amide derivatives of carboxylic acids with PEA or NEA.

diastereomeric amide derivatives.[629] It was possible to assign the absolute configuration of 3-(trichloromethyl)butanoic acid by comparing NMR data of its PEA amide with those observed for similar compounds with known configurations.[630]

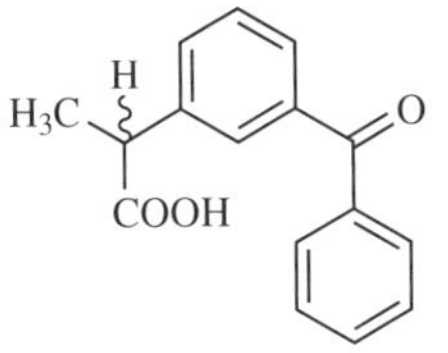

5.12

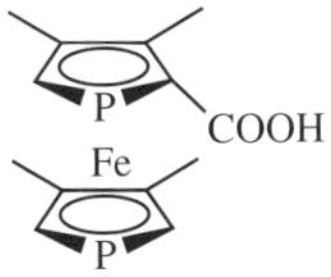

5.13

The shifts in the PEA amide derivatives of aryl propionic acids such as **5.12**, suprofen (**5.14**), indoprofen (**5.15**), and naproxen generally correlated with absolute configuration.[631] Furthermore, the shifts caused in the spectra by adding Eu(fod)$_3$ (fod = 6,6,7,7,8,8,8-heptafluoro-2,3-dimethyl-3,5-octanedione) showed consistent trends that further confirmed the stereochemical assignment.[631,632]

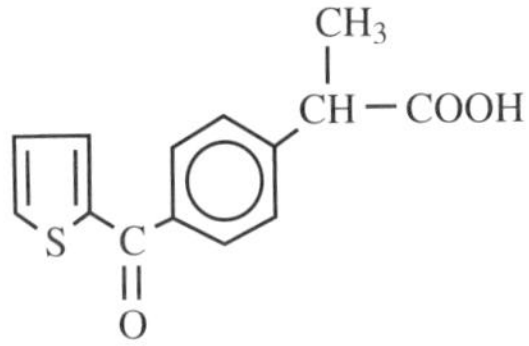

5.14

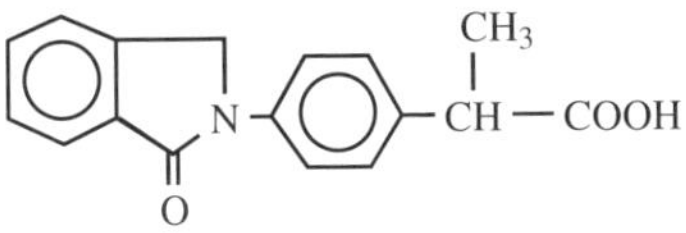

5.15

Solvent-induced shifts with NEA have been used to assign the absolute configuration of the methyl esters of α-hydroxy-α-trifluoromethylphenyl acetic acid[633] and other substituted mandelic acid derivatives.[634] A consistent trend in the shifts of either the methoxy or trifluoro resonance with configuration was observed.

PEA and NEA have been used to generate diastereomeric salts of phosphorus thioacids of structure **5.16**.[635,636] Chiral discrimination was observed in the ^{1}H, ^{13}C, and ^{31}P NMR spectra. The enantiomeric distinction with PEA, NEA, N-benzyl-1-phenylethylamine, and N,N-dimethyl-1-phenylethylamine was compared and was largest with NEA.[636] Aliphatic amines did not produce enantiomeric discrimination in the NMR spectra of **5.16**. It was possible to discriminate compounds such as methyl methyl-d_3 substituted phosphinothioic acid and O-methyl-O-methyl-d_3-phosphorthioic acid that are chiral by virtue of deuterium substitution. Differential shielding caused by the aromatic group of the amine could be used to assign the stereochemistry. Using (−)-PEA, the P-methyl resonance of the (S)-enantiomer was consistently at lower frequency than that of the (R)-enantiomer.[636]

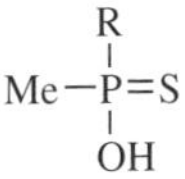

5.16

The effectiveness of PEA, NEA, and ephedrine in causing enantiomeric discrimination in the ^{31}P NMR spectra of 1-hydroxyalkyl phosphonic acids (**5.17**) and their benzyl esters (**5.18**) was compared.[637] The largest nonequivalence (up to 0.36 ppm) was observed with NEA and of those solvents that were tested, chloroform-d provided the best results.

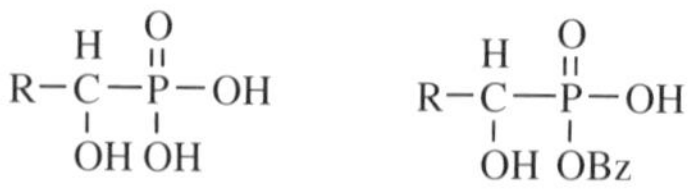

5.17   **5.18**

PEA has been used to successfully analyze the optical purity and assign the absolute configuration of compounds with a P–Cl bond such as methyl p-nitrophenyl phosphorochloridothionate (**5.19**)[638] and O-methyl ethylphosphonochloridothionate (**5.20**).[639] PEA displaces the chloro group to form the corresponding phosphoramide. Shielding by the phenyl ring of the PEA facilitated the assignment of absolute configuration of these compounds.

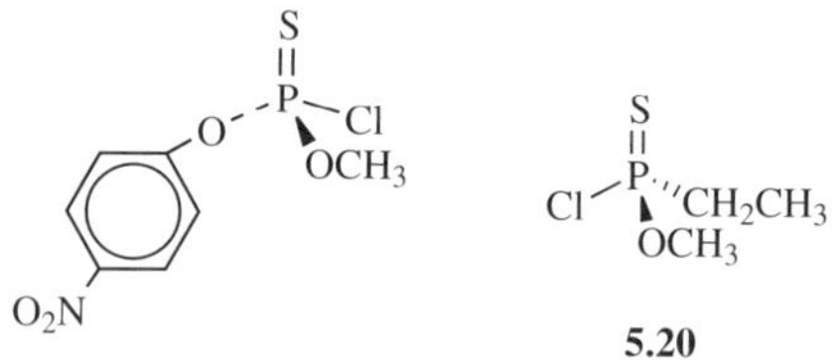

5.19   **5.20**

PEA derivatives of sulfonyl chlorides (**5.21**) caused discrimination in the methylene and methyl resonances of the resulting diastereomers.[640] The shielding caused by the aromatic group was different for the two enantiomers as seen in Figure 5.2 and could be used to assign the stereochemistry of these compounds.

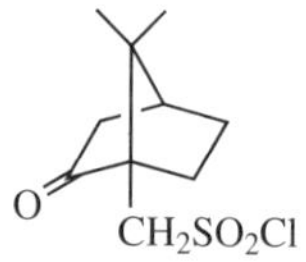

5.21

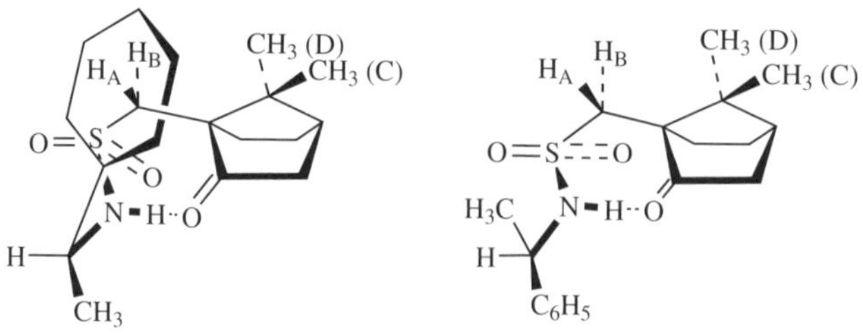

Figure 5.2. Shielding by the phenyl ring in PEA derivatives of sulfonyl chlorides.

PRIMARY AMINES

Two procedures for using PEA[641] or NEA[642] to determine the enantiomeric purity of ketones have been described. One involved the conversion of the ketone to an acid oxime by reaction with NH_2OCH_2COOH.[642] Addition of NEA to the acid oxime in chloroform-d formed a salt and enantiomeric discrimination was observed in the NMR spectrum. The second involved a reductive amination of an enone with PEA perchlorate.[641]

The optical purities of α-amino aldehydes were determined through the use of a reagent that contained optically pure PEA or NEA was used as the chiral moiety.[643] An optically pure semicarbazide with a PEA or NEA unit (**5.22**— R = phenyl or naphthyl) was prepared and reacted with the α-Boc-aminoaldehyde to provide the corresponding semicarbazone derivative (**5.23**). The ^{1}H NMR spectrum had distinctive Boc peaks that could be used to determine optical purity.

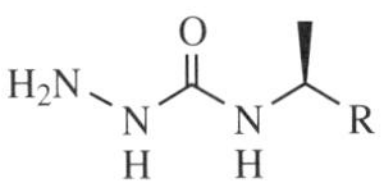

5.22

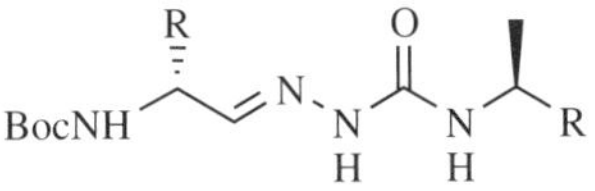

5.23

PEA has been used as a derivatizing agent to determine the optical purity of chiral isocyanates (**5.24**).[644] The methoxy resonance of the corresponding urea derivative was conveniently monitored. The optical purity and absolute configurations of 3-substituted 1-[1'-(S)-phenylethyl]pyrrolidinones (**5.25**) could be determined by NMR spectroscopy.[645] Shielding by the phenyl ring of the PEA group provided information about the preferred conformation and served to differentiate the two configurations.

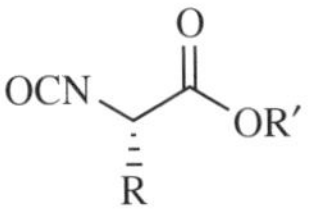

5.24

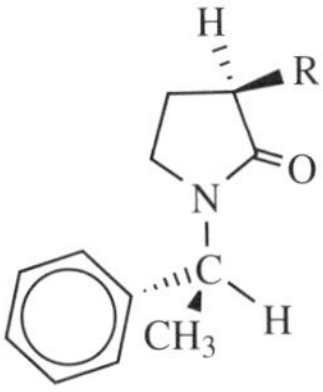

5.25

Cp*Ir–amino acid complexes (**5.26**) form an adduct with PEA (**5.27**) that exhibits two signals in the ^{1}H NMR spectrum in chloroform-d, thereby indicating the presence of two enantiomers. The resolved signals enabled the determination of the optical purity of **5.26**.[646] The chirality of zinc complexes with 3,4-dihydro- and 7,8-dihydro-5-(o-pivaloylaminophenyl)-10,15,20-triphenylporphyrin (**5.28**) was demonstrated by adding optically pure PEA or ephedrine.[647] Certain of the porphyrin resonances split into two peaks in the ^{1}H NMR spectrum recorded in the presence of the optically pure amine ligand.

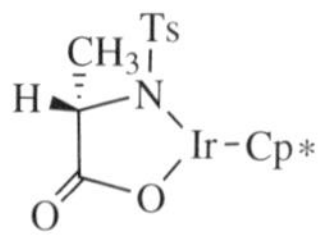

5.26

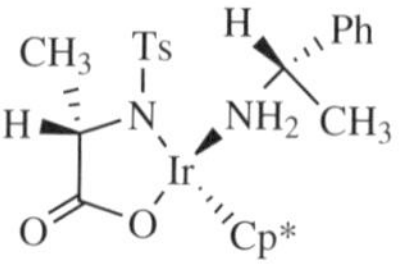

5.27

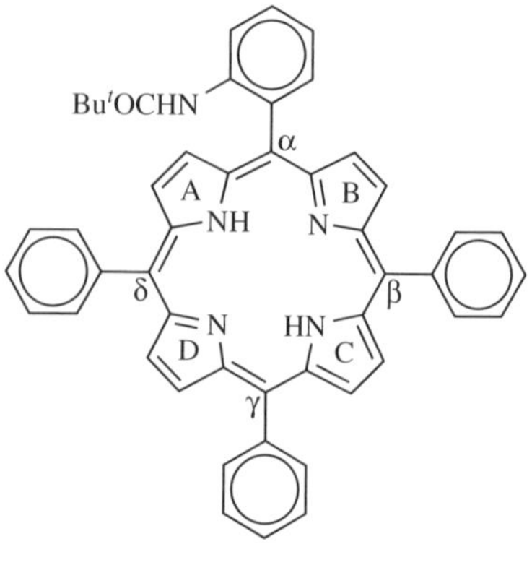

5.28

The optical purity of quinone methides was determined with PEA (**5.29**).[648] The PEA reacts to form **5.30** and the ^{1}H NMR spectra of the resulting diastereomers had distinct resonances for certain of the hydrogen atoms.

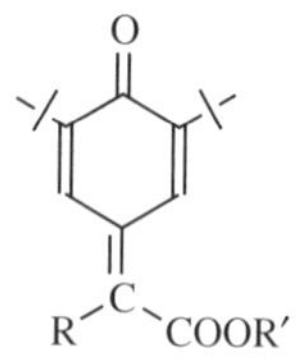

5.29

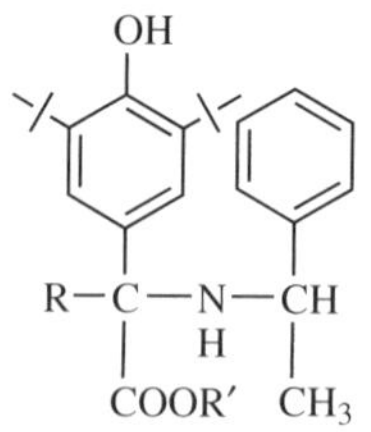

5.30

PEA has been used as a chiral solvating agent to determine enantiomeric purities or demonstrate the chirality of certain chemical compounds. For example, the spectra of a series of spirocyclic oxaphosphates (**5.31**) were examined in the presence of PEA, NEA, and (S)-(−)-1-amino-2-(methoxymethyl)pyrrolidine (AMP) in chloroform-d.[649] The ^{31}P NMR spectrum showed a slight nonequivalence (0.013 ppm) in the presence of PEA or AMP.

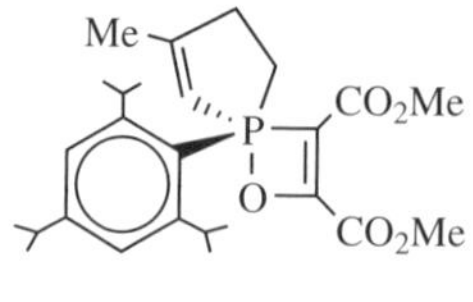

5.31

A *tert*-butyl calix[4]arene derivative with different substituent groups attached through the phenolic oxygen atoms (**5.32**) was shown to be chiral through a doubling of certain resonances in the ^{1}H NMR spectrum in the presence of PEA.[650]

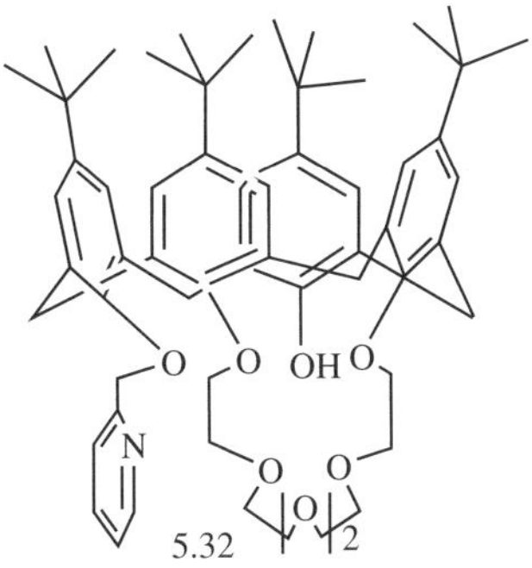

5.1.2. Fluorinated Aryl Amines

Other reagents that have somewhat analogous structures to PEA and NEA have been explored for their utility as chiral NMR discriminating agents. These include several fluorinated analogs such as 2,2,2-trifluoro-1-phenylethylamine (**5.33**),[651] β-trifluoromethyl-β-methoxy-β-phenylethylamine (**5.34**),[652-654] α-amino-α-trifluoromethylphenyl acetonitrile (**5.35**),[655] and 2-fluoro-2-phenyl-1-amino ethane (**5.36**).[656] The ^{19}F NMR spectrum was conveniently monitored with derivatives of these reagents. Most of these studies involved the analysis of carboxylic acids either as salts or as amide derivatives. **5.34** was also used for the analysis of ketones.[653,654] Reaction of the ketone with **5.34** produced four diastereomeric ketimines and either of the ^{1}H, ^{19}F,[653] or ^{13}C[654] NMR spectra could be used to determine the enantiomeric purity of the starting ketone.

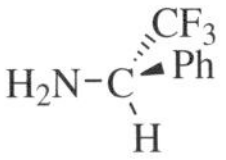

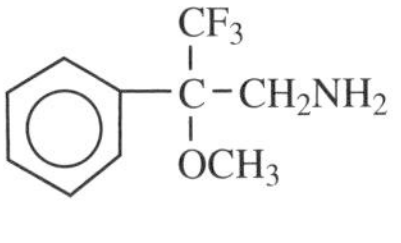

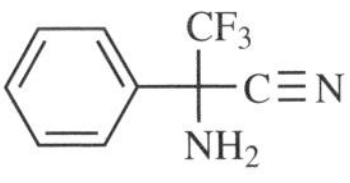

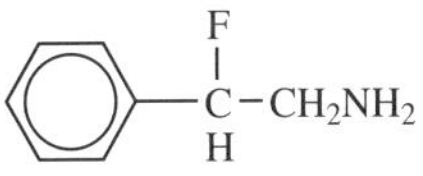

5.1.3. 9-(1-Amino-2,2-dimethylpropyl)-9,10-dihydroanthracene (5.37)

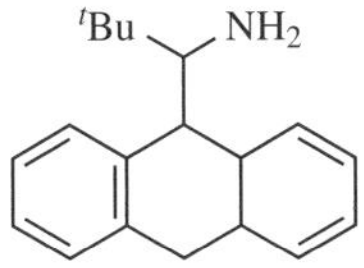

The compound 9-(1-amino-2,2-dimethylpropyl)-9,10-dihydroanthracene (**5.37**), which is quite similar to 9-AEA, functioned as a chiral solvating agent for carboxylic acids and alcohols and produced enantiomeric discrimination in the ¹H NMR spectra.[657]

5.1.4. (1S,2S)-1-Phenyl-2-amino-3-methoxy-1-propanol (5.38)

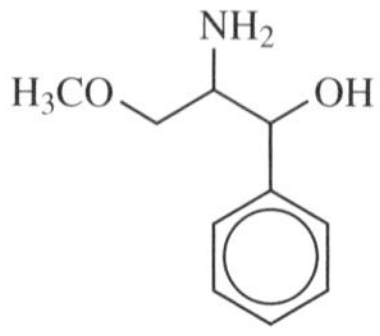

5.38

5.38 and (R)-phenylglycinol (**5.39**) have been used to analyze racemic phthalides.[658] The phthalides were first converted to the corresponding thionophthalide and then reacted with **5.38** or **5.39** as shown in Figure 5.3. Certain resonances of the diastereomers were distinct in the ¹H NMR spectrum. Shielding caused by the phenyl ring could possibly be used to assign the absolute configuration of the phthalide, although more reliable results were obtained by using **5.38** instead of **5.39**.[658] Shielding of the methoxy group of **5.38** by the aryl ring of the phthalide enabled the determination of absolute configuration.

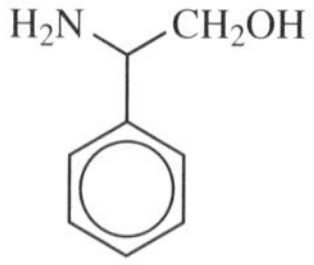

5.39

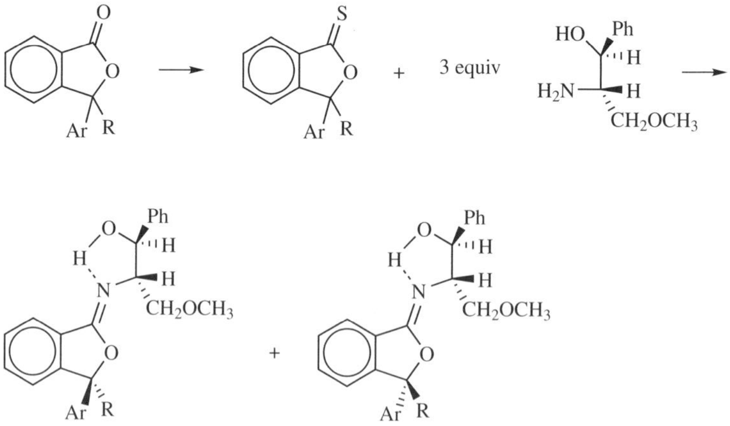

Figure 5.3. Reaction sequence of racemic phthalides with phenylglycinol.

5.1.5. Phenylglycinol (5.39)

Derivatives of 2-, 3-, and 4-alkyl branched carboxylic acids with **5.39** were suitable to assign the absolute stereochemistry based on the conformational preference of the amide derivative and differential shielding by the phenyl ring.[659]

The configuration of the stereogenic center β to the aldehyde in (*E,Z*)-2,3-dihydrofarnesal (**5.40**) was assigned on the basis of ^{1}H NMR data of a phenylglycinol derivative.[660] The compound was oxidized to the corresponding carboxylic acid and shielding by the phenyl ring in the amide derivative caused specific shifts that could be used to assign the configuration.

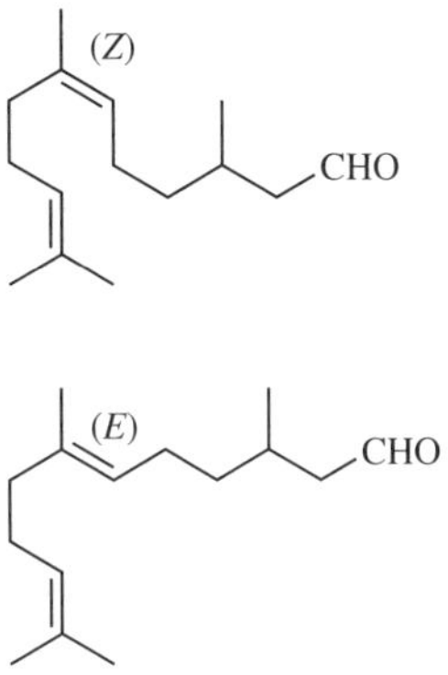

5.40

5.1.6. 1-(3-Aminopropyl)-(5*R*,8*S*,10*R*)terguride (5.41)

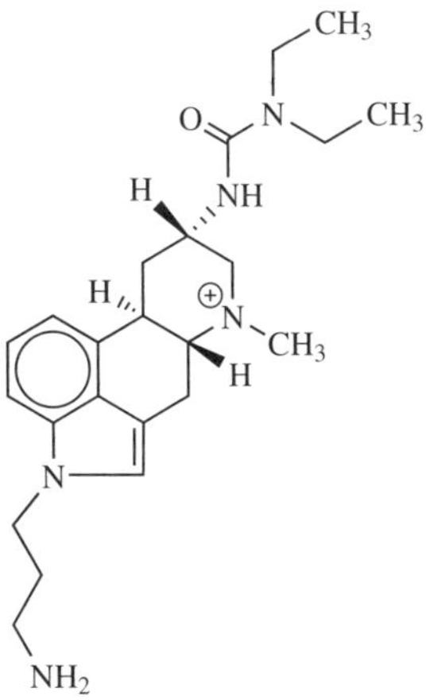

5.41

5.41 has been used as a liquid chromatographic phase to separate chiral carboxylic acids.[661] It was also noted that **5.41** caused chiral recognition in the ^{1}H NMR spectrum of naproxen.

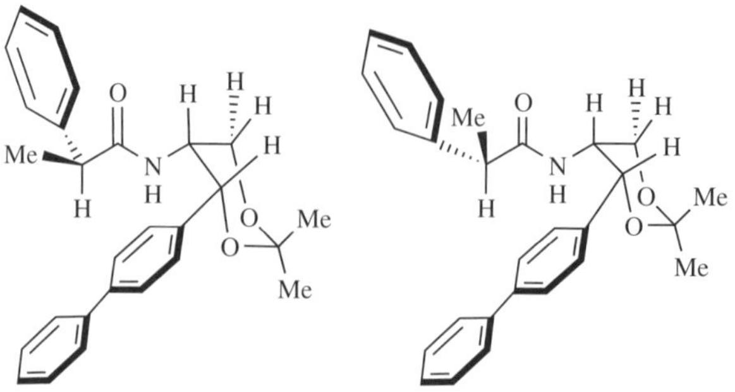

Figure 5.4. Conformational preference of amide derivatives of carboxylic acids with 5-amino-4-biphenyl-2,2-dimethyl-1,3-dioxan.

5.1.7. 5-Amino-4-aryl-2,2-dimethyl-1,3-dioxans (5.42)

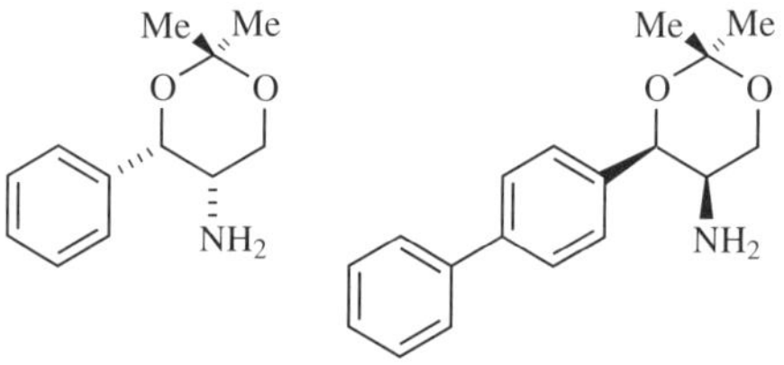

5.42

The phenyl and 4'-biphenyl derivatives of 5-amino-4-aryl-2,2-dimethyl-1,3-dioxans (**5.42**) have been used to assign the absolute configurations of carboxylic acids.[662] The biphenyl analog gave the largest chiral recognition in the NMR spectrum. The biphenyl unit was preferentially oriented in an *sp (syn-periplanar)* conformation as shown in Figure 5.4. The conformational preference was confirmed by NOE (nuclear Overhauser effects) data and produced a *syn*-orientation of the carbonyl group and 1,3-dioxan residue on the nitrogen. Shielding from the aromatic rings could then be rationalized and used to assign the stereochemistry of the carboxylic acid. The utility of this reagent was also demonstrated on one example of a secondary sulfonamide (**5.43**).[662] The NH proton was acidic enough to serve as a site for the formation of diastereomeric complexes.

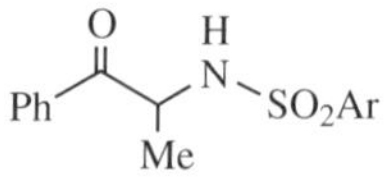

5.43

5.1.8. Amino Acids

5.1.8.1. Peptides One of the earlier applications on the use of NMR spectroscopy for chiral discrimination involved the realization that diastereomeric *N*-substituted di-and tripeptides of alanine and phenylalanine had different chemical shifts in the ^{1}H NMR spectrum.[663,664] The methyl resonance of the alanine group was conveniently monitored. The spectra of the dipeptides with other amino acids showed enantiomeric discrimination as well. The discrimination was larger if an amino acid with an aromatic substituent group as in tyrosine or phenylalanine was in the dipeptide.[663] The discrimination observed in the spectra of diastereomeric dipeptides was subsequently applied to determine the degree of racemization that occurred with various peptide coupling systems.[665–669] In these studies, either the methyl resonance of an alanine residue or the methoxy resonance of an amino acid methyl ester was monitored.

The optical purity of 5,5,5,5′,5′,5′-hexafluoroleucine was determined by reacting its N-Boc derivative with a *tert*-butyl-protected L-serine methyl ester (**5.44**).[670] Nonequivalence of the *t*-Boc, *t*-butyl, and methoxy resonances was observed in the ^{1}H NMR spectrum of the two diastereomers. Similarly, the enantiomeric purity of homopentafluorophenylalanine was determined by coupling it to **5.44** and recording the ^{19}F NMR spectrum.[671] The reaction of N-thiocarboxyanhydrides (**5.45**) with L-phenylalanine produced dipeptides.[672] The optical purity was readily determined by measuring the area of the two doublets for the methyl resonances.

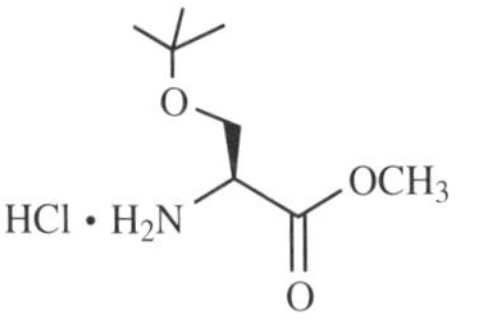

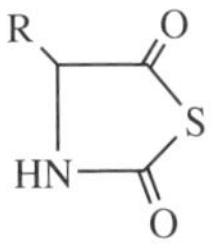

5.1.8.2. Phenylglycine methyl ester (PGME) (5.46)/Phenylglycine dimethylamide PGME and phenylglycine dimethylamide are effective chiral reagents for the analysis of the absolute configuration of carboxylic acids.[36,673–675] The amide derivatives adopt a preferred conformation as shown in Figure 5.5. Shielding by the phenyl group causes $\Delta\delta^{RS}$ values that can be used to assign the absolute configuration.[673] Both X-ray and NOE data confirmed the preferred conformation of the PGME derivatives.

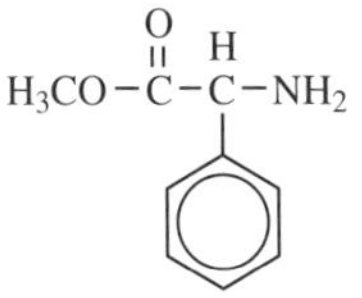

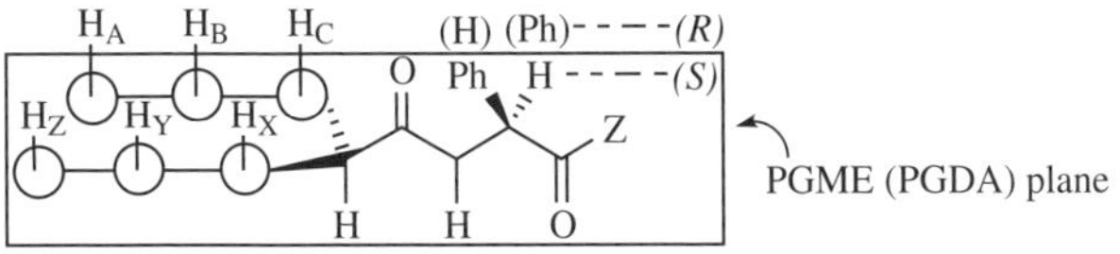

Figure 5.5. Conformational preference of PGME amide derivatives of carboxylic acids.

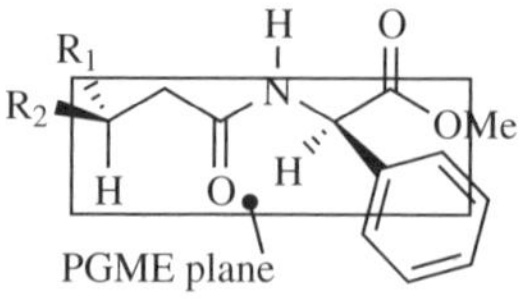

Figure 5.6. Conformational preference of PGME amide derivatives of β,β-substituted propionic acids.

The stereochemistry of carboxylic acids with phenyl substituents such as 2-methyl-4-phenyl butanoic acid and 2-methyl-3-phenylpropionic acid has been assigned by analyzing the NMR spectra of the PGME amides.[674] The presence of the phenyl ring in the substrates did not complicate the assignment. The absolute configurations of 2-methoxy-2-(1-naphthyl)propionic acid and 2-methoxy-2-(2-naphthyl)propionic acid were assigned on the basis of NMR data of the PGME amides.[676] The absolute stereochemistry of a wide range of α-hydroxy, α-alkoxy, and α-acyloxy-α,α-disubstituted acetic acids was also assigned on the basis of shifts of resonances of the PGME amides.[675] Predictable shielding by the phenyl group in PGME was used to assign the absolute configuration of β,β-substituted propionic acids.[675] Figure 5.6 illustrates the conformational preference of PGME amides of β,β-substituted propionic acids that was used to justify the assignment of absolute configuration based on the measured $\Delta\delta^{RS}$ values.[675] PGME was also applicable to compounds that are easily converted to carboxylic acids. This includes the oxidative cleavage of a methyl vinyl carbinol to give an α-hydroxy carboxylic acid, the ozonolysis of an olefin, and the oxidative cleavage of glycols.[675]

In a study of furoic acid derivatives (**5.47**), the shifts observed in the (R)-PGME derivative were all positive except for the resonance of the α-hydrogen.[36] The reverse trends were observed for the (S)-PGME amide. These PGME amides adopted an unexpected conformation because of a hydrogen bond between the amide hydrogen and the ether oxygen, such that the conventional expectations for the signs of the $\Delta\delta^{RS}$ values did not apply. The pattern of shifts was studied for (R)- and (S)-tetrahydro-2-furoic acids of known configuration and the trends were used to assign the stereochemistry of compounds with unknown configurations through its PGME amide.[36]

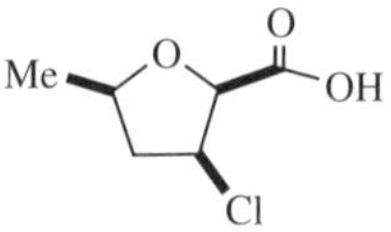

5.47

PGME derivatives have been used to assign the absolute configuration of chiral sites in a number of natural products.[677–686] These include the assignment of the C-18 carbon of **5.48**,[677] the peroxide α-methyl acetate moiety of compounds such as **5.49**,[678] the C2 carbon of **5.50**,[681] and the C8 carbon of **5.51**.[683]

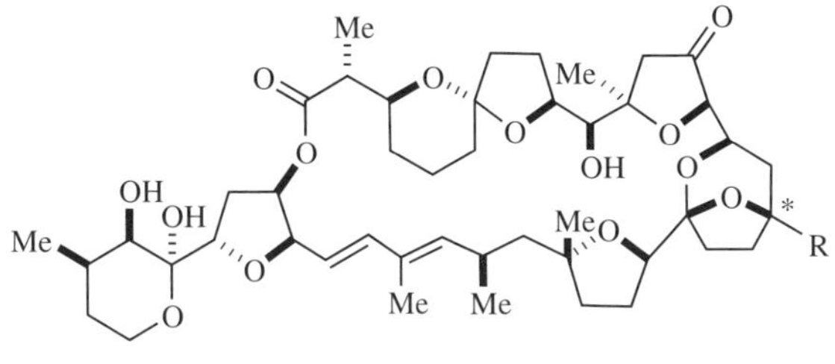

5.48

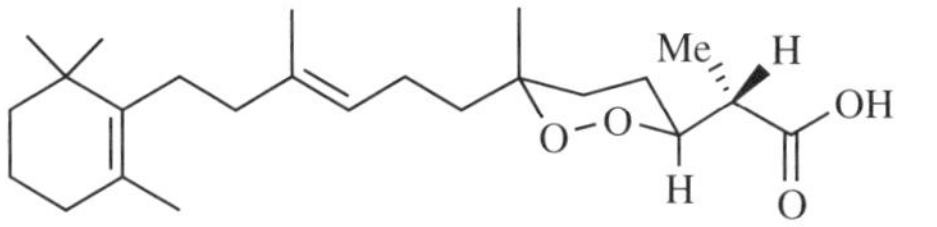

5.49

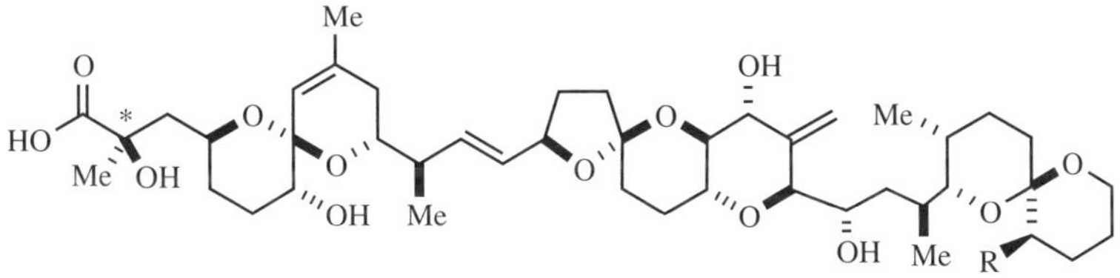

5.50

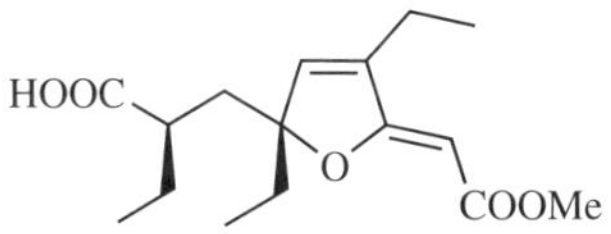

5.51

For **5.52**, the absolute configuration at C5 was assigned by first oxidizing the hydroxyl group to the corresponding carboxylic acid and then analyzing the $\Delta\delta^{RS}$ values of the amide derivatives with (*R*)- and (*S*)-PGME.[679] Compound **5.53** was converted to **5.54**, and the stereochemistry was assigned on the basis of $\Delta\delta^{RS}$ values of the PGME amides.[680]

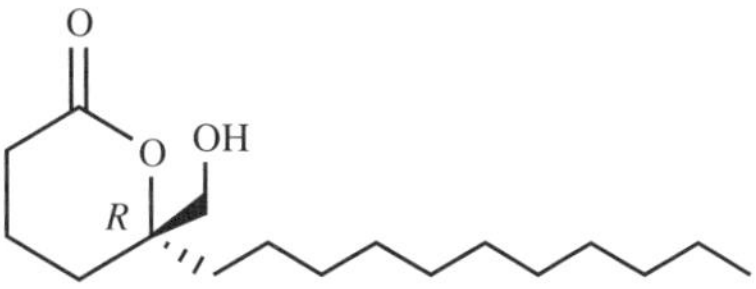

5.52 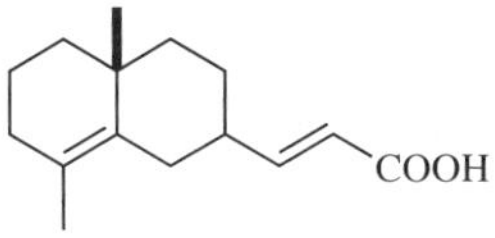   **5.53** 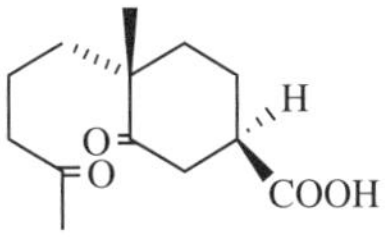   **5.54**

The absolute configuration of the position β to the C1 carboxylic acid group in gambieric acid (**5.55**) was assigned on the basis of $\Delta\delta^{RS}$ values of its amides with PGME.[687] The stereochemistry at C9 was similarly assigned on the basis of esters with α-methoxy-α-trifluoromethylphenylacetic acid (MTPA). Combining the known stereochemical information with NOE data and coupling constants, it was then possible to assign the configurations of the other sites in **5.55**.

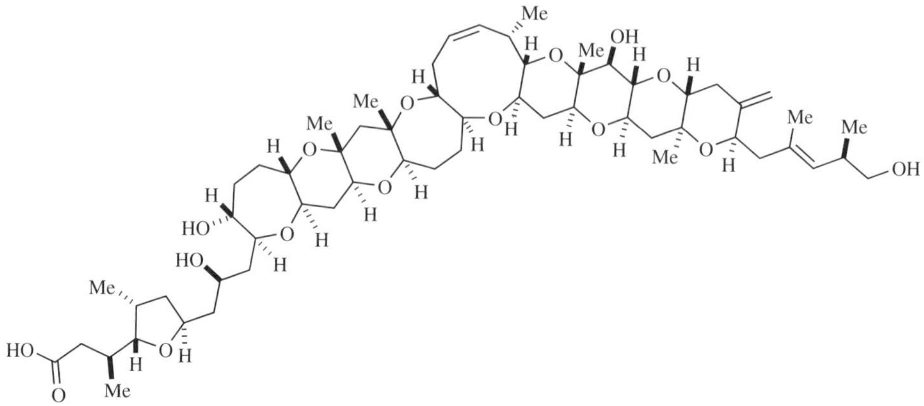

5.55

The stereochemistry of the acyclic secondary hydroxyl group in **5.56** could not be reliably assigned on the basis of shifts of its derivatives with MTPA.[685] Conversion of the β,β-substituted carboxylic acid group to the PGME amide provided $\Delta\delta^{RS}$ values that could be used to assign the absolute configuration of the acidic secondary carbinol site.

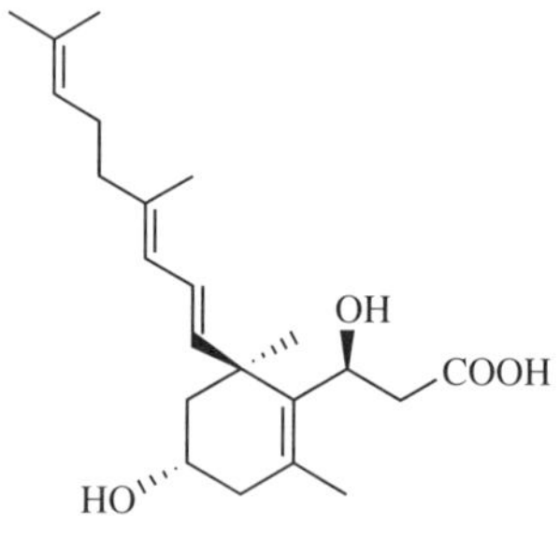

5.56

The utilization of PGME in assigning the absolute configurations of carboxylic acids has been reviewed.[19]

5.1.8.3. L-Cysteine (5.57) The absolute configuration of aldehyde **5.58** was assigned on the basis of its 2-substituted thiazolidine derivative **5.59** with **5.57**.[688] The absolute configuration of the stereocenters in the thiazolidine was assigned on the basis of NOE data.

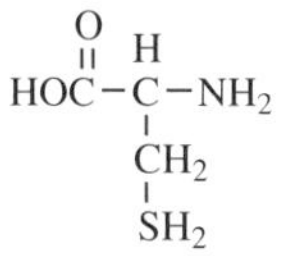

PRIMARY AMINES   169

5.57 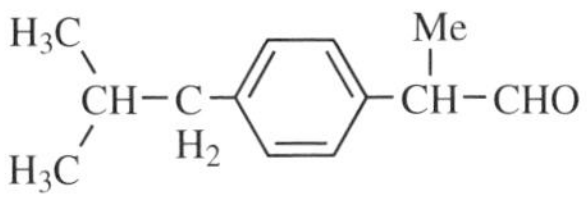 , **5.58** 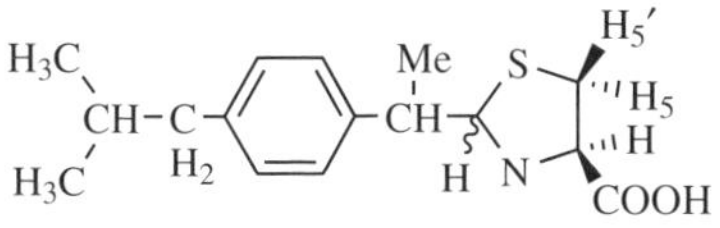 , **5.59**

5.1.8.4. L-Proline (5.60)

The optical purity of thiophosphorane **5.61** was determined on the basis of its derivative with L-proline methyl ester.[639] Nonequivalence was observed in the ^{1}H and ^{31}P NMR spectra of the diastereomers.

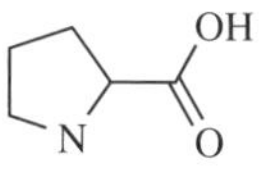

5.60 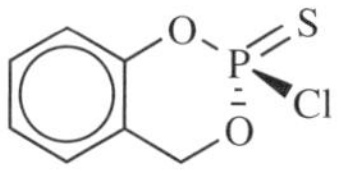 , **5.61**

5.1.9. (1R,2R)-1-(1',8'-Naphthalimide)-2-aminocyclohexane Derivatives

A series of (1R,2R)-1-(1',8'-naphthalimide)-2-aminocyclohexane derivatives (**5.62**) have been used as chiral solvating agents for determining the optical purity of α-chiral carboxylic acids.[689] The mixtures form salts in chloroform-d and discrimination was observed in the ^{1}H NMR spectrum. Amino acids and aryl alkyl compounds were among those that were studied. Shielding from the naphthyl ring of **5.62** caused larger discrimination in the spectra than that observed using PEA as the chiral solvating agent.

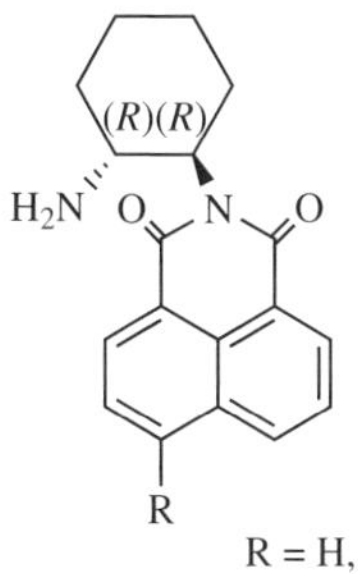

R = H, Br, N(CH$_3$)$_2$

5.62

5.1.10. 1-(1-Naphthyl)-2,2-dimethylpropylamine (5.63)

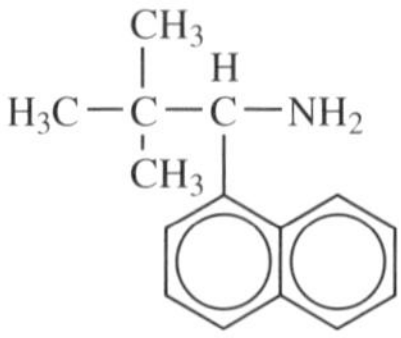

5.63

5.63 has been shown to be an effective chiral solvating agent for carboxylic acids and alcohols.[690] Salts are formed with carboxylic acids in chloroform-d and chiral discrimination is observed in the ^{1}H NMR spectrum. Nonequivalence in the range of 0.01–0.10 ppm was found.

5.1.11. 1-Methoxy-2-aminopropane

The utility of eight commercially available amines for analyzing the optical purity of α-substituted aldehydes was compared.[691] The amines reacted with the aldehydes to form the corresponding imine as shown in Figure 5.7. Of the amines tested, 1-methoxy-2-aminopropane was particularly effective. The reaction occurred immediately and only the E-configuration of the imine was formed initially. However, epimerization did occur slowly on standing, such that the NMR spectrum should be measured within 5 min of the derivatization step. No kinetic resolution was observed. The CH resonance of the imine exhibits excellent splitting in the diastereomers. In the case of hydrocinnamaldehyde (**5.64**), which has two chiral centers, four distinct imine resonances were observed in the NMR spectrum.

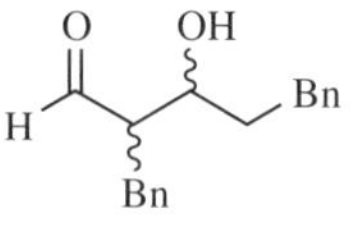

5.64

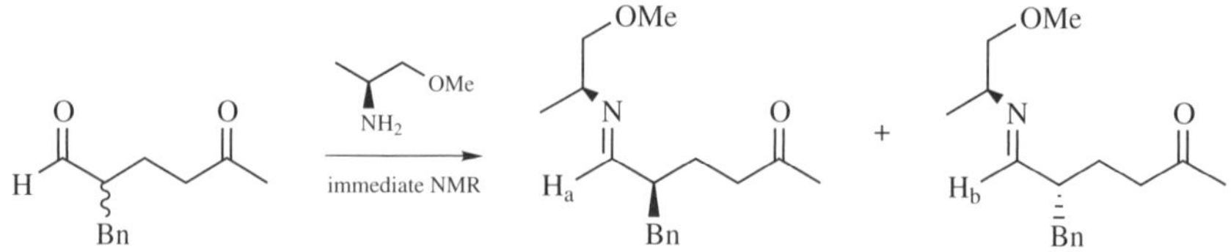

Figure 5.7. Reaction of 1-methoxy-2-aminopropane with aldehydes to form the corresponding imine.

5.1.12. (+)-2-Amino-1-methoxymenth-8-ene (5.65)

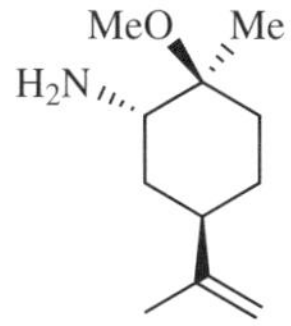

5.65

A procedure for determining the optical purity of 2-substituted aldehydes using **5.65** has been described.[692] Reaction of the aldehyde with the reagent forms the corresponding aldimine (**5.66**). The CH resonance of the imine functionality was conveniently monitored. No racemization was observed in the procedure.

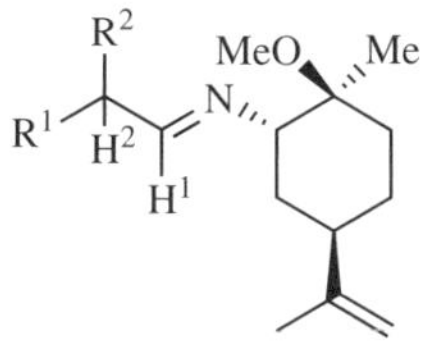

5.66

5.2. SECONDARY AMINES

5.2.1. Ephedrine (5.67)

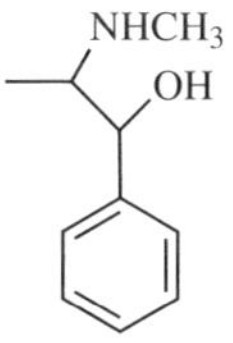

5.67

Ephedrine has been used on several occasions as a chiral NMR reagent for the analysis of phosphorus-containing compounds.[693–697] For example, the enantiomeric discrimination in the ^{1}H NMR spectrum of o-methylisopropylphosphonothioic acid (**5.68**) was larger in the presence of ephedrine than that in the presence of PEA and could be used to determine the optical purity of the substrate.[693]

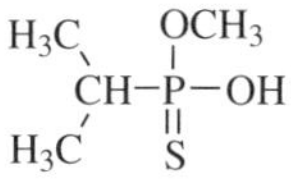

5.68

The effectiveness of ephedrine, NEA and PEA in causing enantiomeric discrimination in the ^{31}P NMR spectra of N-protected 1-amino alkylphosphonates (**5.69**) was compared.[694] Nonequivalence up to 0.8 ppm was observed with ephedrine, which was greater than those with PEA and NEA. Association of ephedrine with the phosphonate did cause some of the *cis*- and *trans*-configuration for certain of the Z,N-protected species and monoesters. This resulted in more ^{31}P signals in the spectrum than originally expected. Care is therefore required when interpreting the results when applying ephedrine to this class of compounds.

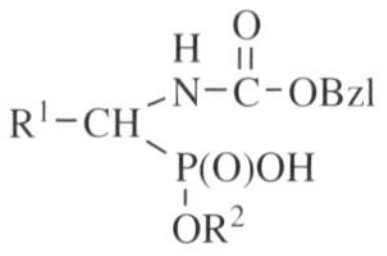

5.69

The effectiveness of NEA and ephedrine in causing chiral recognition in the ^{1}H and ^{31}P NMR spectra of N-phthaloyl-1-aminoalkylphosphonate monoesters (**5.70**) was compared.[695] Greater nonequivalence was observed in the ^{31}P NMR spectrum of the salts formed with ephedrine, whereas NEA produced larger discrimination in the ^{1}H spectra. Ephedrine has also been used to analyze the optical purity of α-amino phosphonic acids.[696] Conversion of the α-aminophosphonic acid to the corresponding N-phthaloyl protected compound and formation of a salt with (−)-ephedrine caused enantiomeric discrimination in the ^{31}P NMR spectrum.

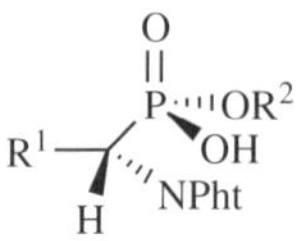

5.70

The reaction of ephedrine with an alkyl phosphorodichloridate causes a mixture of 1,3,2-oxazaphospholidin-2-ones that are epimeric at phosphorus (Fig. 5.8).[697] The diastereomers could be separated by chromatographic methods, and NOESY data were used to assign the absolute configuration about the phosphorus atom.

Figure 5.8. Reaction of ephedrine with an alkyl phosphorodichloridate to form the 1,3,2-oxazaphospholidin-2-one.

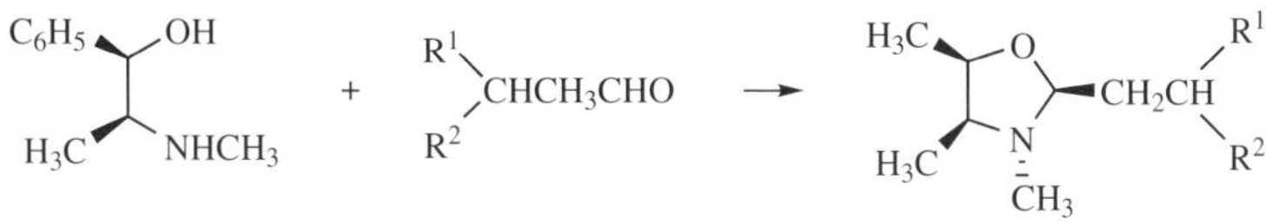

SECONDARY AMINES

Figure 5.9. Reaction of 3-substituted aldehydes with (1R,2S)-(−)-ephedrine to produce the corresponding oxazolidine.

The addition of ephedrine to **5.71** in chloroform-*d* formed a salt and the enantiomeric discrimination of the methoxy resonance in the ^{1}H NMR spectrum was much larger than that with PEA.[698]

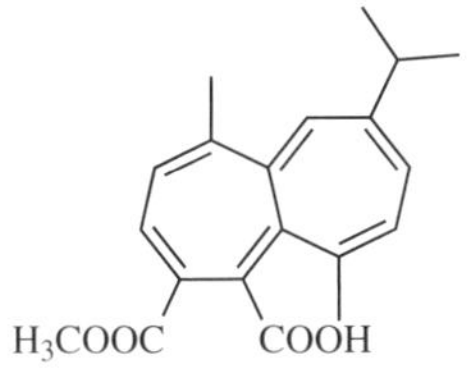

5.71

Finally, ephedrine has been used to assign the absolute configuration of 3-alkyl and 3-aryl substituted aldehydes.[699] Reaction with (1R,2S)-(−)-ephedrine produced the corresponding oxazolidine as shown in Figure 5.9. The reaction produced only the (2S)-configuration of the product when performed under usual conditions. The C2 signal in the ^{13}C NMR spectrum exhibited distinct resonances for the two diastereomers and the shifts were found to correlate with the absolute configuration. Only the ^{1}H NMR spectra of the aryl derivatives showed pronounced discrimination.[699]

5.2.2. 2-Methyl Piperidine

The *pro-(R)* and *pro-(S)* hydrogen resonances of the CH$_2$D group in the monodeuterated 2-methyl piperidine derivative (**5.72**) were nonequivalent by 0.014 ppm in the ^{1}H NMR spectrum.[700] Theoretical calculations of the preferred conformations of the compound were used to assign the resonances. In a subsequent report, (R)-CH^{2}H^{3}HCO$_2$H (CHDTCOOH) was converted to the corresponding ditosylamide compound [(R)-CHDTN(Tos)$_2$]. Reaction of the tosylated species with 2-methyl piperidine produced **5.73**, and the optical purity of the final compound was determined by ^{3}H NMR spectroscopy.[701]

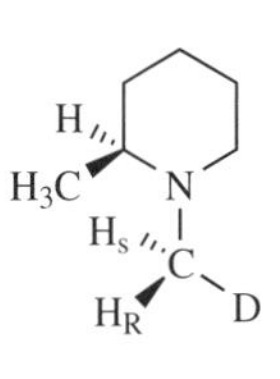

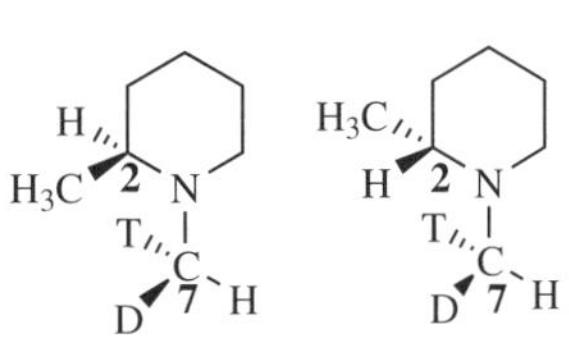

5.72   **5.73** (2S,7S)   (2R,7R)

5.2.3. (*N*-Methyl)-α-isosparteinium cation (5.74)

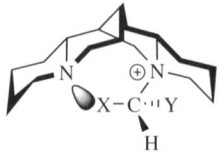

5.74

Subsequent reports showed that **5.74** produced even greater discrimination in the prochiral hydrogens of a CH₂D group and of the enantiomeric CHDT group than that achieved with 2-methylpiperidine.[702,703] Nonequivalence of the *pro-(R)* and *pro-(S)* hydrogens with **5.74** was 0.043 ppm compared to 0.014 ppm with 2-methylpiperidine.[702] Formation of a hydrogen bond between a hydrogen of the methyl group and a lone pair of the second nitrogen as shown in **5.74** was responsible for the discrimination. For the CHDT derivatives, the ^{3}H signal for the (S)-derivative was −0.049 ppm downfield of that for the (R)-derivative.[703] Reaction of the CHDT-N(tos)₂ with the reagent did lead to some loss of configuration such that the optical purity of the CHDTN(tos)₂ could not be determined. Milder methylation conditions using methyl-*p*-toluenesulfonate or methyl triflate did work, such that the analysis of the optical purity of the corresponding CHDT species using **5.74** would be possible.[703]

5.2.4. *N*-Boc-1-(1-naphthyl)ethylamine (5.75)

CH₃CH—N—*t*-Butyl
         H

5.75

The catenane **5.76** exhibits helical chirality because of slow exchange at sufficiently low temperatures. This was unequivocally demonstrated by a doubling of resonances that occurred at 197 K in the presence of **5.75**.[704]

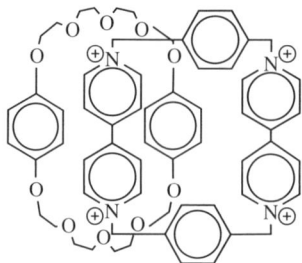

5.76

TERTIARY AMINES

5.2.5. *N*-Methyl-D-(−)-glucamine (5.77)

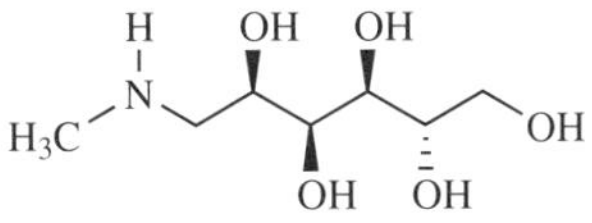

5.77

Lanthanide complexes of the ligand 1,4,7,10-tetraazacyclodecane-*N,N′,N″,N‴*-tetramethylene phosphonate (DOTP) (**5.78**) are racemic.[705] Addition of **5.77** to a sample of the Eu(III) chelate with DOTP produced splitting of certain resonances in the ^{1}H, ^{13}C, and ^{31}P NMR spectra. Rather than binding to the metal ion, the glucamine instead was believed to bind in a chelate manner to two oxygen atoms of the phosphonate groups.

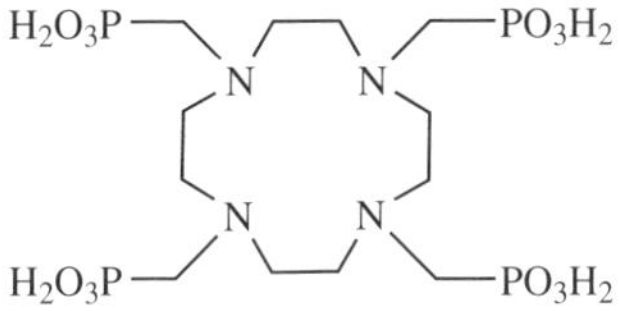

5.78

5.2.6. (*S*)-2-(Diphenylmethyl)pyrrolidine (5.79)

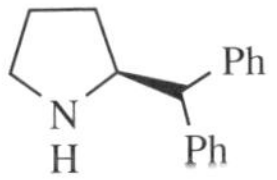

5.79

5.79 has been used as a chiral solvating agent for carboxylic acids and a limited number of secondary alcohols.[706] The reagent formed salts with several alkyl and alkyl aryl carboxylic acids and the resulting ^{1}H NMR spectra exhibited enantiomeric discrimination.

5.3. TERTIARY AMINES

5.3.1. Quinine (5.80)/Cinchonidine (5.81)/Quinidine (5.82)

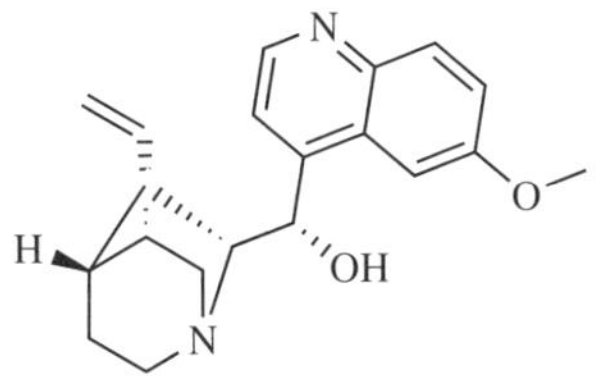

5.80 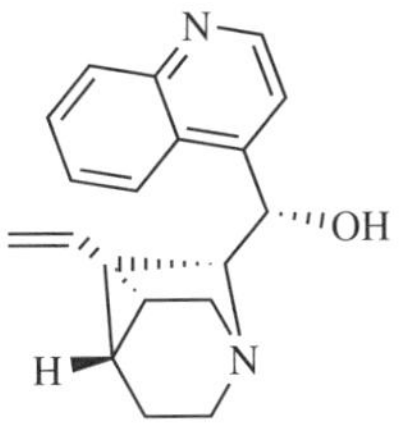 **5.81** 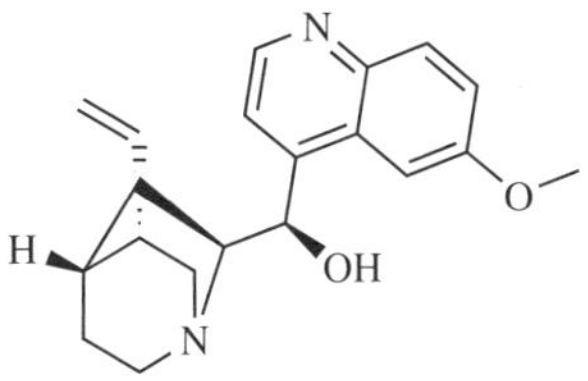 **5.82**

Quinine and derivatives of quinine have been applied on a number of occasions as chiral NMR discriminating agents. Most of these studies involve its use as a chiral solvating agent. The variety of functional groups and donor–acceptor sites in quinine render it a suitable discriminating agent for a number of classes of substrates. Quinine formed salts with alkyl and aryl substituted carboxylic acids and caused enantiomeric discrimination in certain ^{13}C resonances of the substrate.[707]

Quinine has been shown to be an effective chiral NMR solvating agent for a variety of alkyl aryl carbinols and binaphthyl derivatives (**5.83**).[708] Enantiomeric discrimination was observed in the ^{1}H and ^{19}F (when applicable) NMR spectra in chloroform-d. Detailed studies of the site of interactions of substrates with quinine using NOE data generally showed that the hydroxyl group was more important in interactions than the nitrogen atom.[709]

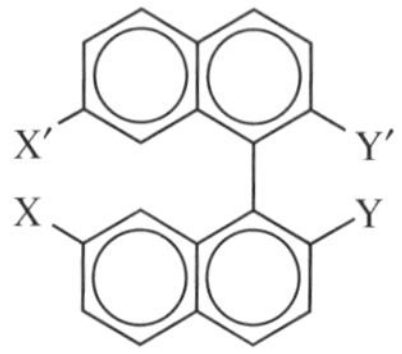

5.83

However, a detailed NMR study of the association of quinine with 7,7′-bis(1-propen-3-oxy)-2,2′-dihydroxy-1,1′-binaphthyl (**5.84**)[710] and 2′-(2-propoxy)-1,1′-binaphthyl-2-ol (**5.85**)[711] showed that the quinine associated with substrates through multiple binding sites. Chiral discrimination of **5.84** was observed in both the alkoxy and aromatic portions of the ^{1}H NMR spectrum. Enantiomeric discrimination was observed for the alkoxy resonances of **5.85**.[711]

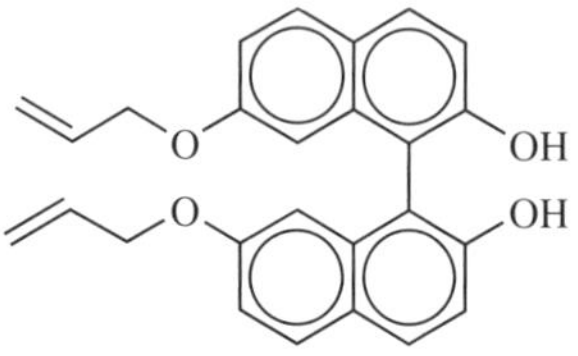

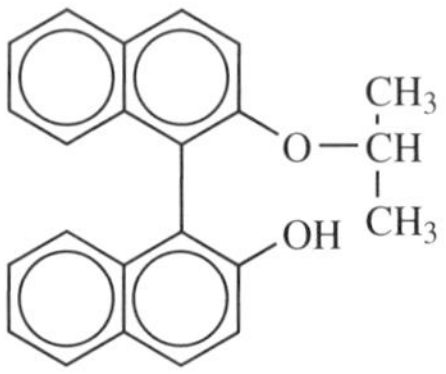

5.84   **5.85**

Quinine was used in the first reported enantiomeric NMR discrimination of cyclic hemiacetals (**5.86**) and methyl acetals (**5.87**).[712] Nonequivalence was observed in methylene chloride-d_2 and tetrahydrofuran-d_8, and quinine was more effective than PEA.

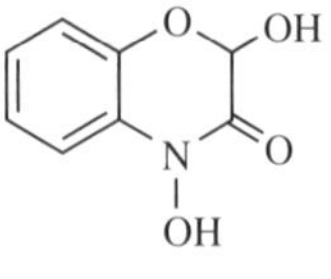

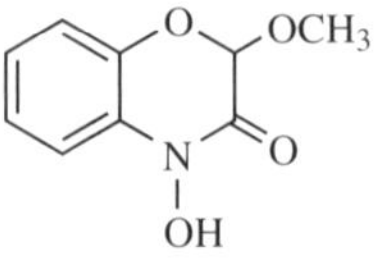

5.86   **5.87**

TERTIARY AMINES 177

Quinine has been used to determine the optical purity of β-hydroxy esters (**5.88**).[713] Methyl, ethyl, and *t*-butyl esters were examined, and the *t*-butyl resonance was an especially suitable one to monitor. Most of the β-hydroxy esters had aryl substituent groups, although enantiomeric discrimination was also observed in the compound with a cyclohexyl group.

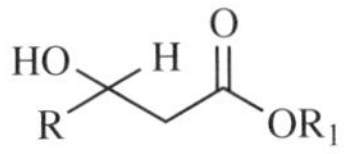

5.88

The optical purities of α-trifluoromethylated hydroxyl compounds of structures **5.89** were determined by monitoring the ^{19}F NMR spectrum in the presence of quinine.[714] Similarly, the optical purity of **5.90** could be determined by recording the ^{1}H NMR spectrum with quinine.[715] The absolute configuration of each of the pure enantiomers of **5.90** was also assigned on the basis of NOE studies of the associated complex.

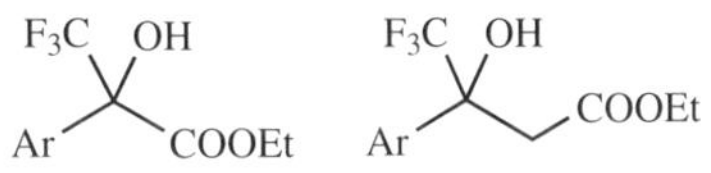

5.89

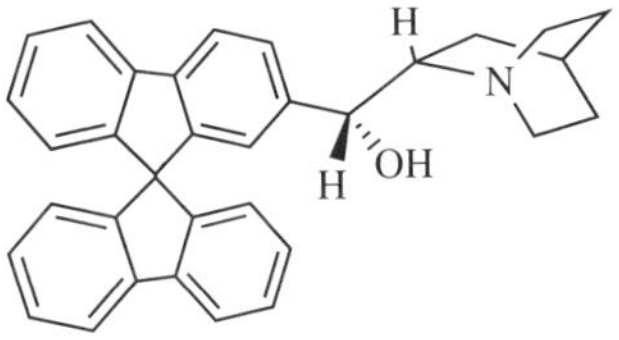

5.90

A strategy for using quinine to determine the optical purity of alcohols and carboxylic acids,[716,717] *N*-acetyl amino acids, and *N*-diacetyl dipeptides[718] has been reported. The substrate is reacted with bis(triphenyltin)oxide as shown in Figure 5.10. The resulting tin complex was coordinatively unsaturated and associated with an amine. Either quinine or PEA was effective in causing enantiomeric discrimination in the NMR spectrum of the tin complex, although quinine generally produced larger nonequivalence of the signals.[718]

Quinine or similar alkaloid reagents have been used to analyze the optical purity of phosphorus-containing compounds. The ^{31}P NMR spectra of diethyl-1- (**5.91**) and diethyl-2-hydroxy alkyl phosphonates (**5.92**) exhibited enantiomeric discrimination

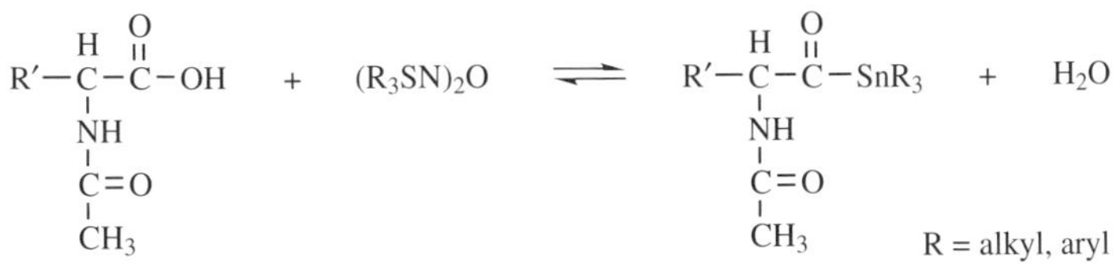

Figure 5.10. Reaction of *N*-acetyl amino acids and dipeptides with bis(triphenyltin)oxide.

in the presence of quinine.[719] Addition of PEA, brucine, cinchonine, or ephedrine to these substrates did not cause enantiomeric discrimination in the spectrum. Quinine also caused nonequivalence in the ^{31}P NMR spectra of 1-hydroxyphosphinothioic acids (P=S instead of P=O) but not the corresponding 2-hydroxy compounds.[719]

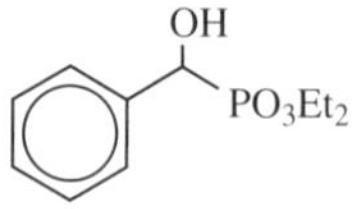

5.91

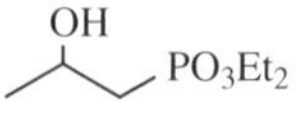

5.92

Quinine has been used to enantiomerically discriminate the ^{31}P NMR spectra of dihydroxyethane phosphonates (**5.93**).[720] In a mixture of the (1*S*,2*S*)-, (1*R*,2*R*)-, (1*R*,2*S*)-, and (1*S*,2*R*)-isomers, the two ^{31}P signals were split into four in the presence of quinine. By analyzing similar compounds with known configurations, it was possible to use the NMR shifts to assign the stereochemistry of the unknowns.

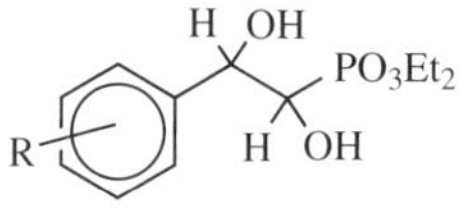

5.93

The optical purity of α-monofluoroalkylphosphonic acids such as **5.94** was determined by recording the ^{19}F NMR spectrum in chloroform-*d* in the presence of quinidine.[721] The ^{31}P NMR spectrum of the phosphonium salt **5.95** exhibited enantiomeric discrimination in the presence of cinchonidine.[722]

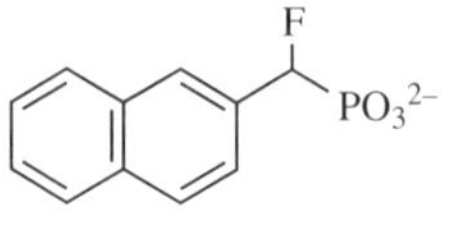

5.94

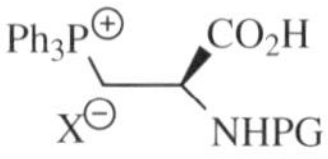

5.95

Quinine has been used as a chiral solvating agent to discriminate the NMR spectrum of 2-deuterio-2,3-dihydro-2-methyl-6-nitrobenzothiophene-1-oxide (**5.96**).[723] The association reportedly involved hydrogen bonding of the hydroxyl group of quinine to the sulfoxide group of **5.96**.

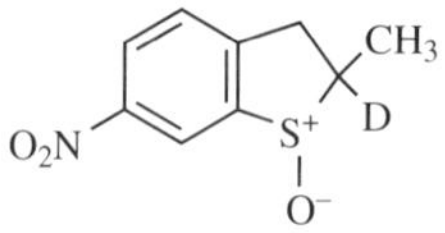

5.96

Quinine derivatives with a carbamate group at either the C9[724,725] or C11[726] position have been examined for their utility as chiral NMR solvating agents. Derivatives with a variety of carbamate substituent groups were tested and one with a 1-naphthyl group (**5.97**) tended to give the largest chiral recognition in the

TERTIARY AMINES

spectra of many of the substrates examined.[725] Quinine by itself was not effective in discriminating the aryl-substituted amines and amino acids examined in this study. Substrates derivatized with a dinitrophenyl (DNP) moiety showed particularly large discrimination, presumably because of a π–π interaction between the DNP ring and the aryl ring of the carbamate groups. A quinine derivative with substitution of the naphthyl carbamoyl group at the C11 position (**5.98**) produced enantiomeric discrimination that was similar to the corresponding C9-derivatized quinine for the DNP-derivatized substrates.[726] The C11 carbamate was generally more effective as a chiral solvating agent for the underivatized substrates examined in this study.

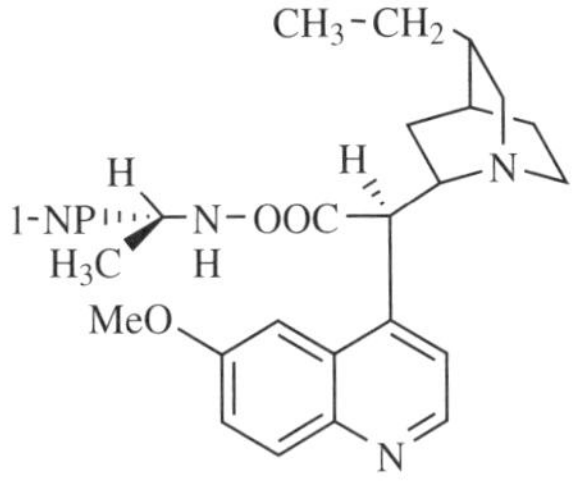

5.97

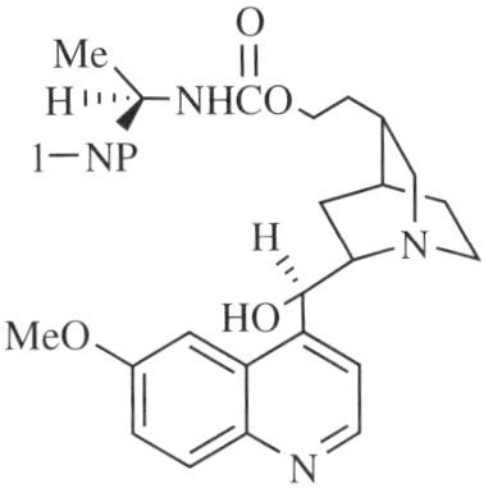

5.98

Quinine has been used as a template to prepare a set of chiral liquid chromatographic stationary phases that exhibit remarkable selectively toward enantiomeric substrates. Soluble analogs of these phases have been used to study the mechanism of association by NMR spectroscopy. This included the enantiomeric discrimination that occurred for *N*-(3,5-dinitrobenzoyl)-leucine (DNB-Leu) in methanol-d_4 with the cinchona alkaloid **5.99**.[727] Alternatively, solid-phase NMR spectroscopy has been used to study the association mechanism of DNB-Leu with a similar cinchona alkaloid stationary phase.[728] Differences in cross-peak intensity in trNOESY spectra were noted between the more strongly and weakly bound enantiomers.

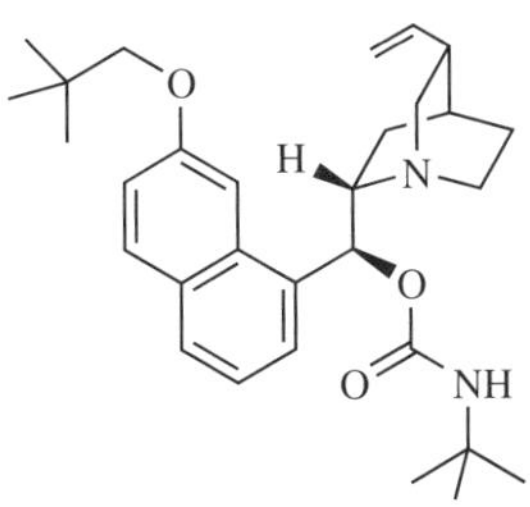

5.99

The possibility of using the heteronuclear multiple quantum correlation (HMQC) method to remove overlap of proton resonances was demonstrated using quinine as a chiral solvating agent with DL-isocitric lactone (**5.100**), ibuprofen, *O*-acetylmandelic acid, and *trans*-3-(4-methoxyphenyl)glycidic acid methyl ester.[729] The overlap in proton resonances was completely resolved by differences in the carbon chemical shifts, demonstrating the utility of HMQC spectra for chiral analysis.

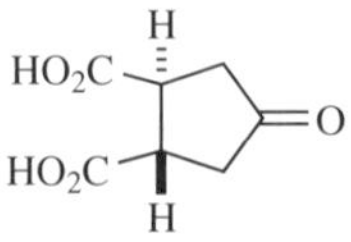

5.100

5.3.2. 2,8-Dimethyl-6H,12H-5,11-methanodibenzo[b,f][1,5]diazocine (Troger's Base) (5.101)

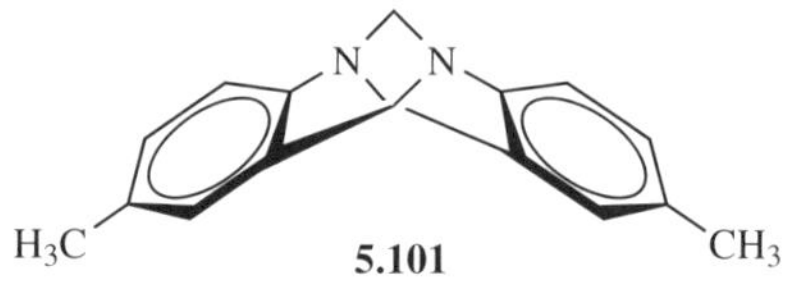

5.101

The compound 2,8-dimethyl-6H,12H-5,11-methanodibenzo[b,f][1,5]diazocine, better known as Troger's base (**5.101**), exists with a concave structure. The utility of **5.101** as a chiral solvating agent for aryl alkyl secondary and tertiary alcohols in chloroform-d has been demonstrated.[730]

5.3.3. Brucine (5.102)

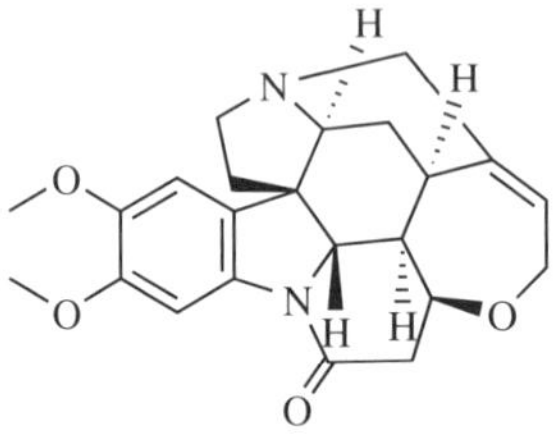

5.102

Brucine was used to selectively precipitate enantiomers of α- and β-haloacetylenic alcohols (**5.103**) that have two chiral sites at the 1,2- or 1,3-positions, and α,β-dichloroacetylenic alcohols (**5.104**) that have stereogenic carbons at the 1,2,3-positions.[731] Enantiomeric discrimination was also noted in the ^{1}H NMR spectrum of the mixtures.

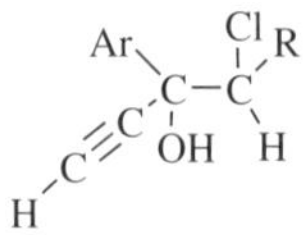

5.103 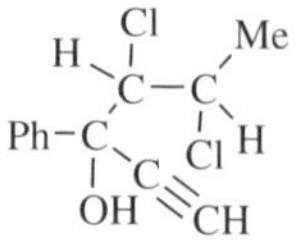   **5.104**

Resonances in the ^{13}C NMR spectrum of the dimenthyl ester of spiro[3.3]heptane-2,6-dicarboxylic acid (**5.105**) exhibited enantiomeric discrimination in the presence of brucine.[539] This was used to confirm the optical purity of **5.105**.

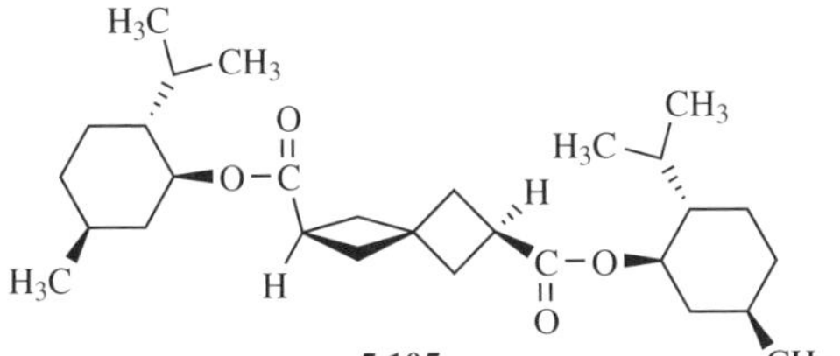

5.105

5.3.4. Strychnine (5.106)

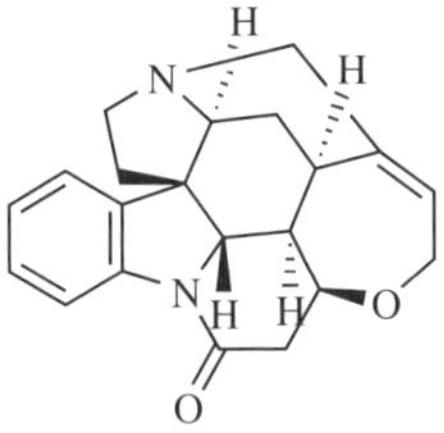

5.106

The optical purity of fluorochloroiodoacetic acid was determined by measuring the ^{19}F NMR spectrum of its salt with (−)-strychnine.[732] Different chemical shifts for the ^{19}F resonances were observed depending on whether the molecule had the ^{35}Cl or ^{37}Cl isotope.

5.3.5. (*R*)- and (*S*)-Tetrahydroisoquinoline

The potential of using HPLC–NMR to analyze mixtures of chiral compounds has been demonstrated using atracurium besylate (**5.107**), a compound with four chiral centers that is formulated as a mixture of *cis*- and *trans*-isomers with (*R*)- and (*S*)-tetrahydroisoquinoline.[733] A total of 10 isomers were separated, which included four racemic pairs and two *meso* compounds. Certain ^1H resonances were diagnostic in the analysis and assigned using a combination of one- and two-dimensional methods. The NMR spectra enabled the assignment of the two *meso* compounds and allowed the identification of each of the enantiomeric pairs. HPLC with circular dichroism detection was then used to assign the absolute configuration of each enantiomer.

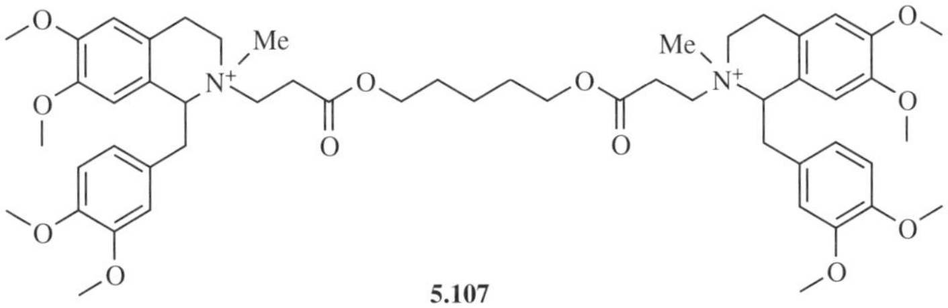

5.107

5.4. DIAMINE REAGENTS

5.4.1. 1,2-Diphenyl-1,2-diaminoethane (5.108)

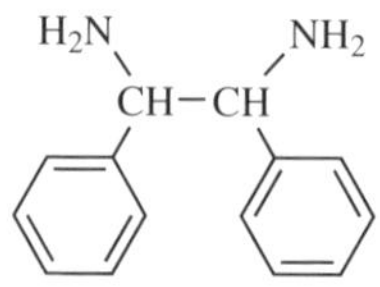

5.108

The utility of **5.108** for determining the optical purity of mono- and di-α-aryl propionic as well as α-deuterio- and α-halocarboxylic acids has been demonstrated.[734,735] Nonequivalence of 0.03–0.27 ppm was observed in the ^{1}H NMR spectra of the diastereomeric salts in chloroform-d and benzene-d_6. **5.108** was added to carboxylic acids at a 2:1 acid–amine ratio.[735] Enantiomeric discrimination in the spectra of carboxylic acids with **5.108** was larger than that with PEA, which was attributed to the presence of the second aromatic ring.[735] The relative shifts in the spectra of α-phenylpropionic acids exhibited a similar sense of nonequivalence that correlated with the absolute configuration. **5.108** was reportedly suitable for the analysis of α-halo acids that were prone to racemization.[734] The addition of **5.108** to primary carboxylic acids of general structure RCH$_2$COOH formed salts and caused distinct resonances for the *pro-(R)* and *pro-(S)* hydrogen atoms.[735]

5.108 has been used to determine the optical purity of 2-bromohexadecanoic acid,[736] 2-(2,4,5-trichlorophenoxy)propanoic acid,[737] nicotinic acid (**5.109**), and quinolinic acids such as **5.110**.[738] In the latter case, enantiomeric discrimination in pyridine-d_5 was typically better than that observed in chloroform-d.

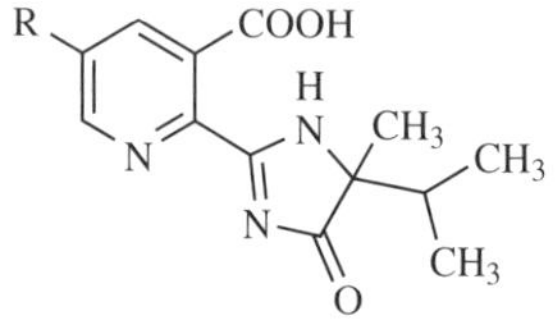

5.109 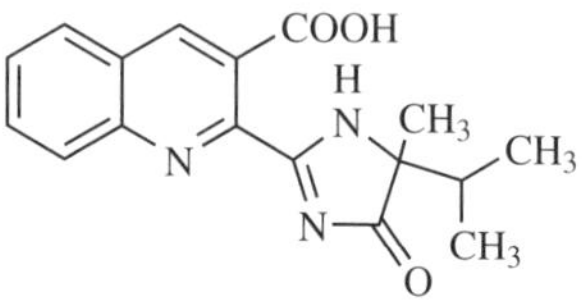   **5.110**

5.108 has been used as a chiral solvating agent for chiral alcohols of structure **5.111**.[739] The methoxy signals of **5.111** were nonequivalent and the results with **5.108** were better than those observed with PEA and NEA.

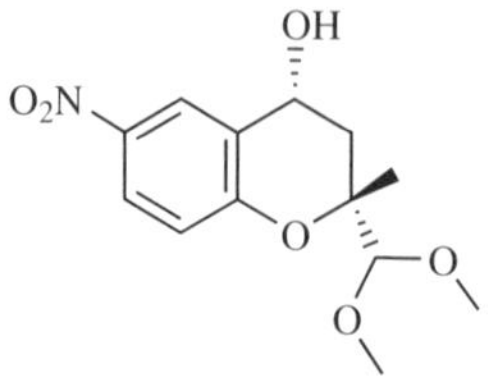

5.111

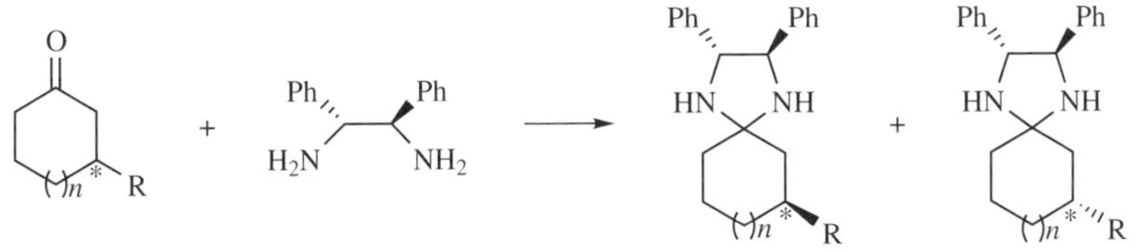

Figure 5.11. Reaction of 1,2-diphenyl-1,2-diaminoethane with 3-substituted cyclohexanones to form the corresponding aminal.

5.108 reacts immediately with 3-substituted cyclohexanones to form the corresponding aminals as shown in Figure 5.11.[740] The resulting diastereomers had distinct resonances in the ^{13}C NMR spectrum in benzene-d_6 or chloroform-d. The reaction with 3-substituted cyclopentanone was slower, but the procedure was still effective for analyzing the optical purity of these compounds. The method did not work with acyclic ketones and enones.

5.4.2. N,N′-Substituted 1,2-diphenyl-1,2-diaminoethane

Aldehydes react with N,N′-dimethyl-1,2-diphenyl ethylenediamine to form the corresponding imidazolidines as shown in Figure 5.12.[741] The diastereomeric products had distinct resonances in the ^1H and ^{13}C NMR spectra. For β-substituted aldehydes, the phenyl ring of the amine group caused differential shielding that depended on the bulkiness of the substituents of the aldehyde. Tests on a variety of aldehydes confirmed that the shielding effects could be used to reliably assign the absolute configurations of β-substitutes aldehydes.[741]

Similarly, N,N′-dimethyl-1,2-bis[m-trifluoromethyl)phenyl]-1,2-ethanediamine (**5.112**) has been shown to react with aldehydes to form the corresponding imidazolidines.[742] In this case, ^{19}F NMR spectroscopy could be used as well as ^1H and ^{13}C NMR to determine optical purity. A similar reaction did not occur with ketones.

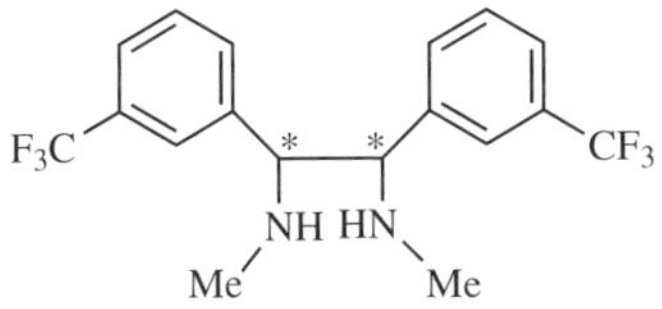

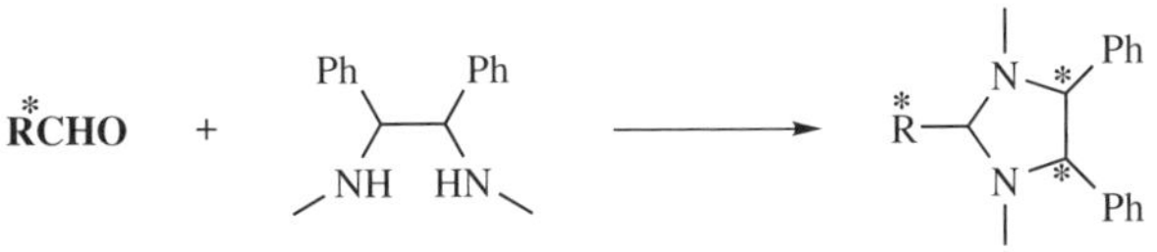

Figure 5.12. Reaction of N,N′-dimethyl-1,2-diphenyl ethylenediamine with aldehydes to form the corresponding imidazolidines.

The Corey chiral controller (R,R)-1,2-di-(2,4,6-trimethylbenzylamino)-1,2-diphenylethane (**5.113**) was evaluated as a chiral NMR solvating agent for a variety of carboxylic acids, sulfonic acids, and cyclic β-diketones.[743] **5.108** was ineffective with keto acid **5.114** because it formed a Schiff base, whereas the Corey controller was suitable for the analysis. The corresponding dibenzyl derivative of **5.113** was not nearly as effective as the dimesityl derivative. The resonances of hydrogen atoms up to five bonds away from the acid group showed enantiomeric discrimination in the spectra of some substrates.

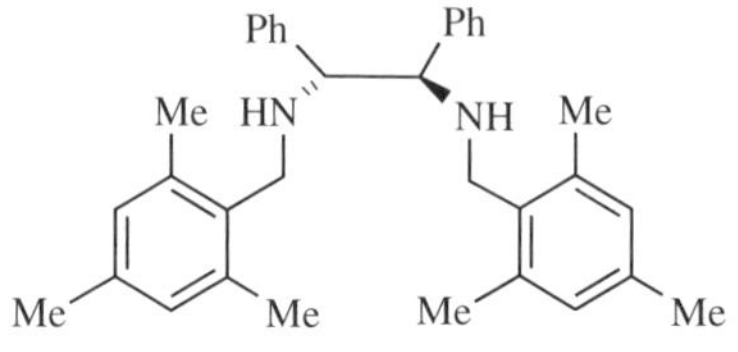

5.113

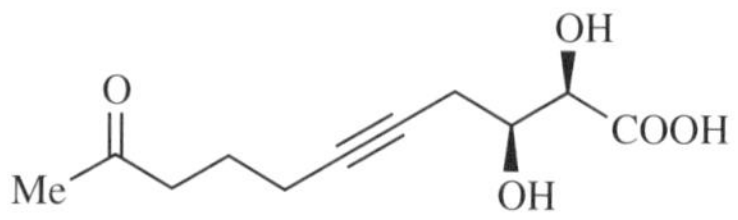

5.114

5.5. DATABASES USING AMINE CHIRAL SOLVATING AGENTS

5.5.1. *N*,α-Dimethylbenzylamine (DMBA)

A strategy for constructing a database using ^{13}C NMR shift data was described in Section 2.2.12. A second strategy for constructing NMR databases is to record the ^{13}C chemical shifts for each different diastereomer of a particular structural motif in a chiral solvent and to calculate the difference between the actual chemical shift for a particular diastereomer and the average chemical shift for all the diastereomers. Profiles similar to those described in the ^{13}C method are obtained, and the known profile that best matches the pattern of the unknown profile provides the configuration. Several chiral solvents were evaluated and perdeuterated DMBA (**5.115**) was an especially suitable one for this method.[744] Chemical shifts in dimethylsulfoxide-d_6 or methanol-d_4 were used as the reference point. If the spectrum of the compound dissolved in perdeuterated DMBA exhibited broadened resonances, as was observed with oasymcin A, a small amount of dimethylsulfoxide (10%) was added to sharpen the resonances without compromising the method.[744] An additional variant was to calculate $\Delta\delta^{RS}$ values with (*R*)- and (*S*)-DMBA for each of the diastereomers to provide a different set of profiles such as that shown in Figure 5.13 that could be compared with the profile obtained for an unknown.

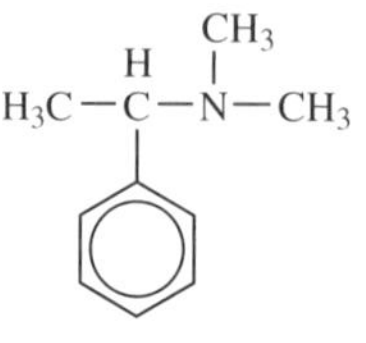

5.115

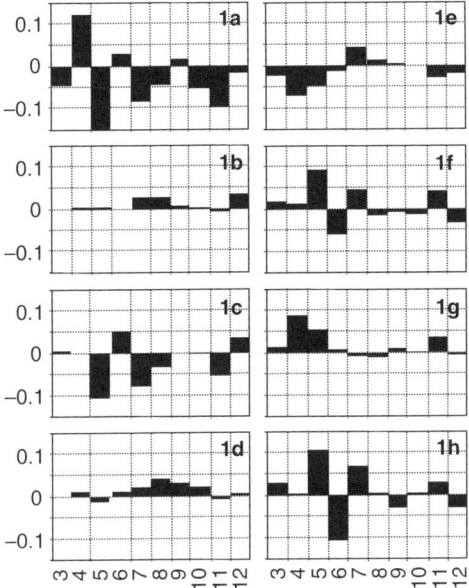

Figure 5.13. Difference in carbon chemical shifts of 1a–h (100 MHz) between (R)- and (S)-DMBA. The x- and y-axes represent carbon number and $\delta_R - \delta_S$, respectively. [Reproduced with permission from American Chemical Society.[744]]

Overall, DMBA databases for the contiguous propionate (Fig. 5.14),[744] 1,3-diol-, 1,3,5-triol- (Fig. 5.15), and 1,3-diol-2-methyl motifs (Fig. 5.16)[745] as well as the motif shown in Figure 5.17[746] have been developed. Several of these were then used to assign the stereochemistry of sites in oasomycin A (**2.31**).[745] Once the relative stereochemistries were known for a compound, assigning the absolute configuration of one site using established NMR methodologies allowed all the other absolute stereochemistries to be assigned.

During the construction of these databases, it was observed that certain ^{13}C resonances of *meso* compounds were desymmetrized in the presence of DMBA.[747] For example, the C2 and C3 carbons in the *meso syn*-1,3-diol (**5.116**) are desymmetrized from the C7 and C8 carbons in the presence of DMBA. It was also noted that the 1,3,5-triol relied on a 1,3-diol substructure, such that

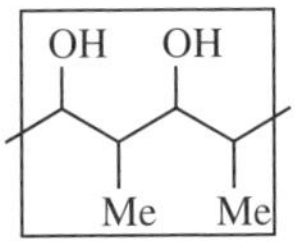

Figure 5.14. Structural motif for two contiguous propionate units.

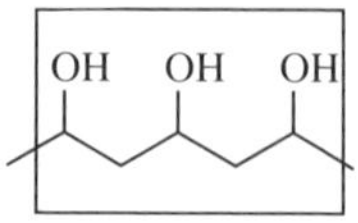

Figure 5.15. 1,3,5-Triol motif.

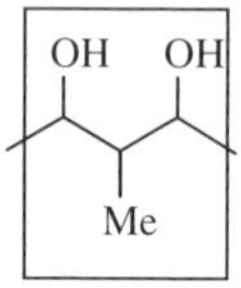

Figure 5.16. 2-Methyl-1,3-diol motif.

appropriate 1,3-diol profiles could be added together to predict the profiles for the corresponding 1,3,5-triols.[747]

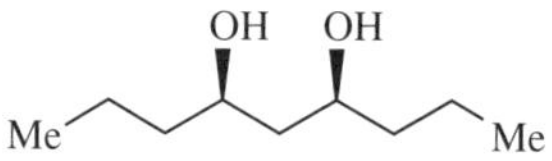

5.116

The ^{13}C NMR database for the motif in Figure 5.17 was then used to assign relative stereochemistries for several mycolactones (**5.117**).[746] It was also noted in this report that the method could be extended to ^{1}H profiles as well, and that ^{1}H $\Delta\delta^{RS}$ data with (R)- and (S)-DMBA could be used to assign the absolute configuration.

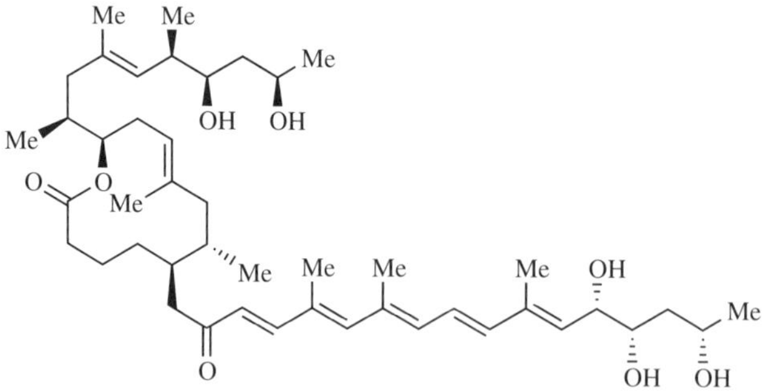

5.117

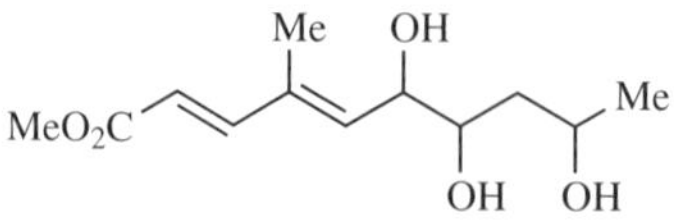

Figure 5.17. Structural motif used for constructing DMBA database.

5.5.2. Bis-1,3-methylbenzylamine-2-methylpropane (BMBA-pMe)

The absolute configuration of acyclic secondary alcohols can be assigned by considering the chemical shift behaviors of the two adjacent carbons in the bidentate solvent BMBA-pMe (**5.118**).[748] Several bidentate solvents were screened and BMBA-pMe was the most effective. Figure 5.18 illustrates the binding of BMBA-pMe to a secondary alcohol. The discrimination of the enantiotopic carbons in 2-propanol and other compounds on binding to BMBA-pMe indicated the likelihood that this method would be suitable for assigning absolute stereochemistries. The shifts of the two carbons adjacent to the hydroxyl groups are measured with (R,R)- and (S,S)-BMBA-pMe. The signs of $\Delta\delta[\delta(R,R)-\delta(S,S)]$ were then calculated for the two adjacent carbon atoms and are characteristic of the absolute configuration of the secondary alcohol. The two hydroxyl groups in 1,4- and 1,5-diols functioned independently in the presence of DMBA-pMe and the absolute configuration of each could be assigned using the method. The hydroxyl groups of 1,2- and 1,3-diols could not be treated independently, but instead had to be analyzed as a structural cluster.[748] A detailed analysis of the effect of DMBA-pMe on 1,2-diols led to a set of rules using $\Delta\delta[\delta(R,R)-\delta(S,S)]$ values that could be used to assign the relative and absolute configurations of 1,2-*syn*- and 1,2-*anti*-diols.[749]

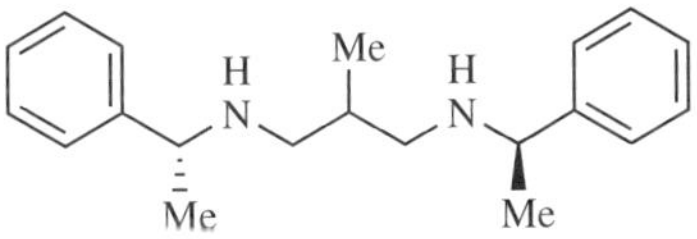

5.118

The applicability of the BMBA-pMe method for assigning the stereochemistry of cyclic and biaryl secondary alcohols and acyclic *tert*-alcohols was evaluated.[750] A set of empirical rules for assigning the absolute configuration of the hydroxyl site based on ^{13}C NMR shift data with (R,R)- and (S,S)-BMBA-pMe was devised and is shown in Figure 5.19. The utility of the method was demonstrated with menthol, cholesterol, borneol, isoborneol, and a variety of other substrates.

The compound tetrafibricin (**5.119**) has five independent stereoclusters. Using a combination of the differential ^{13}C database method and data in perdeuterated DMBA

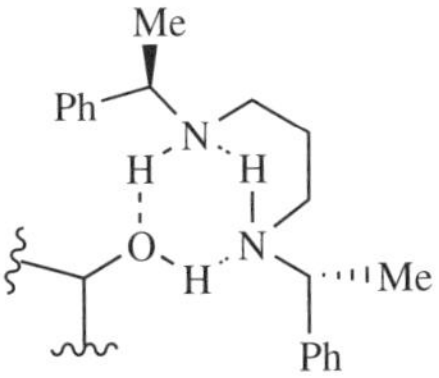

Figure 5.18. Binding of BMBA-pMe to a secondary alcohol.

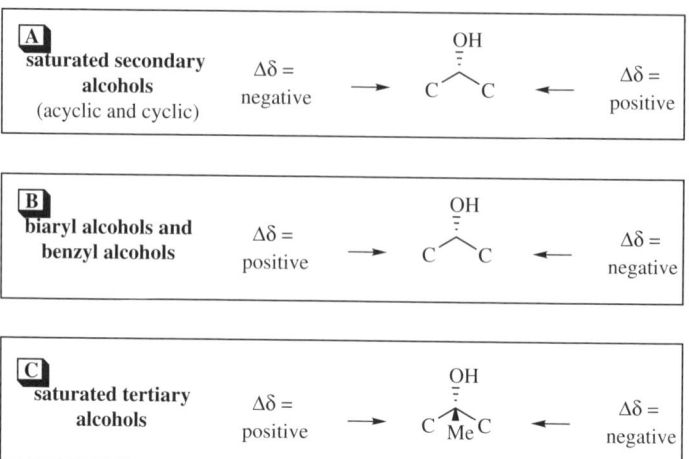

Figure 5.19. Relationship between the absolute configuration of alcohols and the sign of $\Delta\delta$ in the chiral bidentate NMR solvent BMBA-pMe: $\Delta\delta = \delta(R,R) - \delta(S,S)$.

and BMBA-pMe, it was possible to assign the complete stereochemistry of **5.119**.[751] The validity of the method was also confirmed by performing the assignment based on the established method of MTPA esters.

5.119

A general advantage of the database method is that there is no need to derivatize the compound or carry out degradation reactions to smaller functional units. Also, it is possible to simultaneously assign multiple stereogenic sites within the compound.

The database approach has subsequently been applied to the α-substituted α-hydroxy-β-aminoester moiety of taxanes **5.120** and **5.121**.[752] Epimeric 2′-metaxanes **5.120** and **5.121** were used to construct the database. The conclusions reached using the database approach were validated by using the more complex and time consuming J-value assignment method. The database approach was noted as being especially applicable to a large series of diastereomeric compounds as might result from a combichem process.[752]

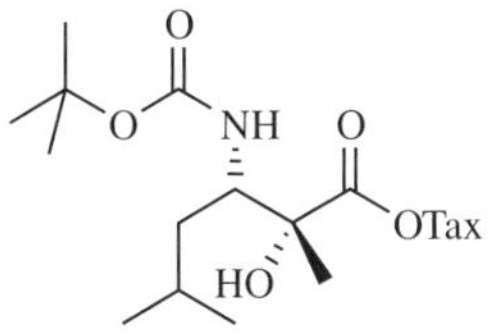

5.120

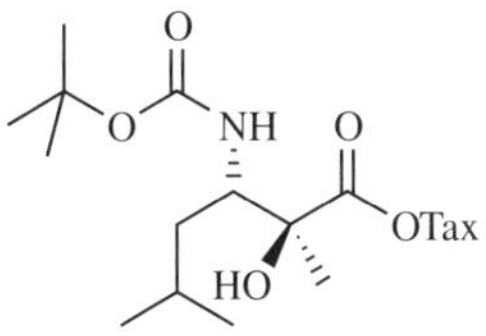

5.121

5.5.3. (*R*-α-Methoxy-α-trifluoromethylphenylacetic acid (MTPA)/BMBA-pMe

A database constructed of ^{1}H NMR shift data from the (*R*)-MTPA derivatives of 80 chiral secondary alcohols and ^{13}C NMR shifts of 94 compounds measured in (*R,R*)- and (*S,S*)-BMBA-pMe was tested on 20 new compounds to predict (1) the sign of the difference in the chemical shift of opposite stereoisomers and (2) the difference in chemical shift between the two chiral solvents.[753] The predictions were correct in each of the 20 test compounds, and the strategy for performing the predictions was described.

6

MISCELLANEOUS ORGANIC-BASED CHIRAL DERIVATIZING AND SOLVATING AGENTS

A number of reagents that encompass other classes of compounds have been evaluated either as chiral solvating or derivatizing agents. These include amide- or lactam-containing compounds that have the ability to associate with substrates through dipole–dipole interactions. In many examples of amides or other classes of reagents, an aromatic functionality is incorporated into the reagent to provide the potential for π–π interactions. Aldehyde- or ketone-containing reagents have been used to analyze amines or diols through the formation of the corresponding imine or acetal derivatives. Reagents with an isocyanate moiety react with amines or alcohols to form the corresponding urea or carbamate derivatives, which are then often distinguishable by NMR spectroscopy. A variety of heterocyclic ring compounds have been shown to cause chiral recognition in the NMR spectra of selected substrates, either through solvation or derivatization methods. The reagents described in this chapter have typically been used in a small number of applications, although several are commercially available and are therefore candidates for more widespread usage.

The phenomenon of self-discrimination has been observed for solutions of a number of compounds in nonracemic mixtures. This occurrence can be used to determine optical purity of such systems or, as has been described in other places in the text, has led to the development of general reagents for chiral recognition.

Discrimination of Chiral Compounds Using NMR Spectroscopy, by Thomas J. Wenzel
Copyright © 2007 John Wiley & Sons, Inc.

AMIDES

High-throughput methods for measuring optical purity have been developed. These methods are likely to have particular significance in the analysis of optical purities of combinatorial synthetic libraries.

6.1. AMIDES

A variety of reagents with amide units have been utilized as chiral NMR solvating agents. The positive and negative sites in the amide moiety provide points for dipole–dipole interactions. These reagents often incorporate an aromatic ring for π–π stacking and substituent groups that provide potential steric effects. Several of these were initially developed for use in liquid chromatographic applications and then soluble analogs were exploited for their utility in NMR spectroscopy.

6.1.1. N-(3,5-Dinitrobenzoyl)-1-phenylethylamine (DNB-PEA) (6.1)

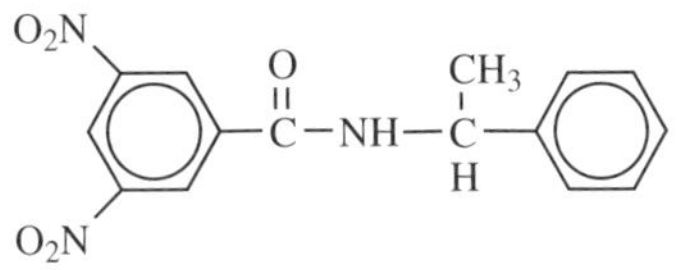

6.1

DNB-PEA was shown to be an effective chiral solvating agent for the determination of optical purities of sulfoxides.[754] Twenty-five different substrates were examined in this initial report. The enantiomeric discrimination was larger in carbon tetrachloride than chloroform-d, and the corresponding reagent with a naphthyl ring replacing the 3,5-DNB ring was not as effective.

The optical purity of **6.2** was determined using DNB-PEA.[755] Nonequivalence of one of the benzylic hydrogen resonances was noted. The optical purity of a comparable sulfoxide derivative to **6.2** with a naphthyl ring replacing the phenyl ring was also determined with DNB-PEA.[194]

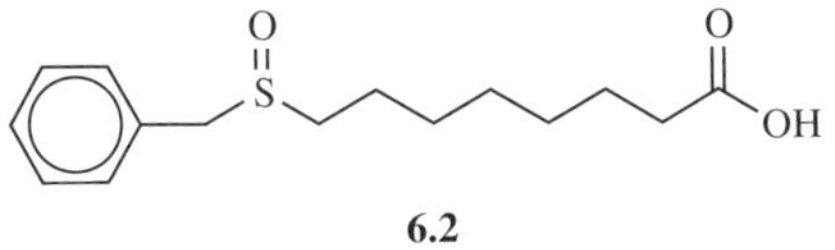

6.2

The optical purity of 3-butylthiolane-1-oxide (**6.3**) was determined by measuring the ^{1}H NMR spectrum in the presence of DNB-PEA.[756] Coupling constants were also used to assign the configuration of the compound.

6.3

The methyl singlet and methylene resonances of **6.4** and the corresponding phenyl, 2-furyl, and 2-thienyl derivatives were enantiomerically discriminated in the presence of DNB-PEA.[757] The enantiomeric discrimination with DNB-PEA was larger than that with Eu(hfc)$_3$ (hfc = 3-heptafluorobutyryl-d-camphor).

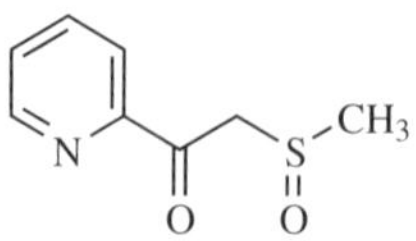

6.4

The compound mesoridazine (**6.5**) has two chiral centers and the ^{1}H NMR spectrum shows two sulfoxide methyl singlets, one for each of the diastereomeric isomers.[758] The spectrum of **6.5** in the presence of DNB-PEA at 500 MHz has four singlets for the methyl group.

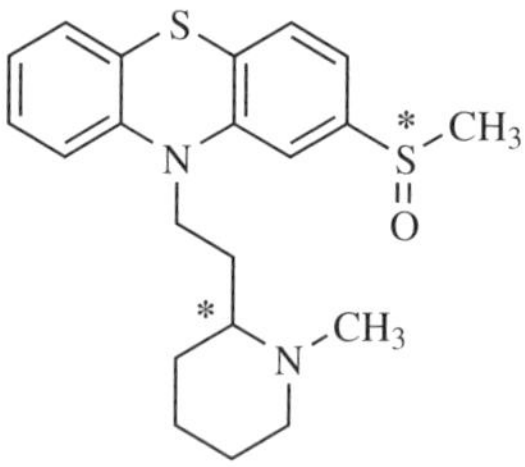

6.5

The optical purity of phosphine oxides can be determined using DNB-PEA.[759] Substrates that were asymmetric either at the phosphorus atom or at carbon α to the phosphorus were enantiomerically discriminated in this first report.

The optical purity at the C7 position of compound **6.6** was determined through the use of DNB-PEA.[760] The ^{31}P NMR spectrum in chloroform-d was used for the analysis. The optical purities of chiral phosphine oxide **6.7** and similar derivatives with different substituents on the naphthyl ring were determined with DNB-PEA.[761]

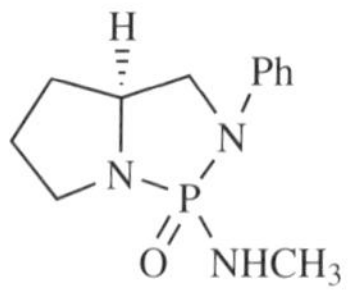

6.6 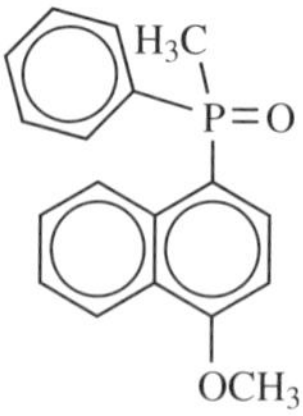   **6.7**

The effectiveness of DNB-PEA and the corresponding DNB-(1-(1-naphthyl)ethylamine (DNB-NEA) in causing enantiomeric discrimination in the NMR spectra of

AMIDES

6.8, **6.9**, and **6.10**, as well as some derivatives of these was compared.[762] Nonequivalence was observed in the ^{1}H and ^{31}P NMR spectra for most substrates, and the splitting was larger with DNB-NEA than with DNB-PEA. For 2-phospholene-1-oxide derivatives such as **6.11**, the H2 resonances of the R_P enantiomers were always at lower frequency than those of the S_P isomer. Similarly, the ^{31}P signal for the R_P enantiomer was consistently at higher frequency than that of the S_P isomer.

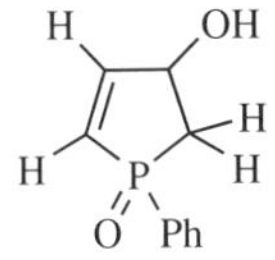

6.8 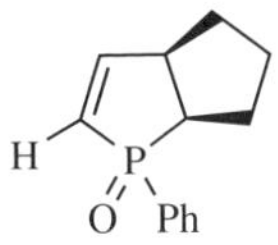   **6.9** 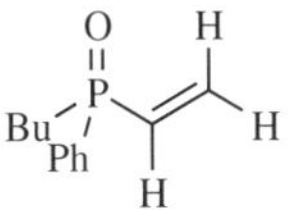   **6.10** 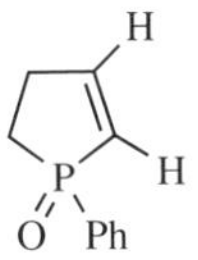   **6.11**

DNB-PEA has also been shown to cause enantiomeric discrimination in the NMR spectra of certain amides, esters, and alcohols with electron-rich substituent groups.[763–765] Chiral recognition was observed in the ^{1}H and ^{19}F NMR spectra of fluorine-containing compounds such as **6.12**, but not in the ^{13}C NMR spectra.[763] Enantiomeric discrimination in the NMR spectra of a variety of aromatic esters and alcohols was examined with DNB-PEA and several other chiral solvating agents that contained electron-deficient aromatic rings, to better understand the recognition mechanism. The best discrimination was obtained when the chiral center was rigidly bound to the aromatic ring. Nonpolar solvents enhanced the extent of distinction.[764] A comparative study of the effect of DNB-PEA and analogs with more electron-rich aromatic rings in producing enantiomeric discrimination in the NMR spectra of 2-phenyl-2-(2-piperdyl)acetamides (**6.13**) was reported. DNB-PEA was the most effective among those studied in causing chiral discrimination in the ^{1}H NMR spectrum in chloroform-d.[765]

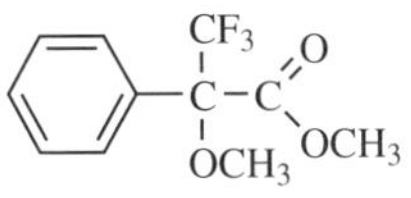

6.12 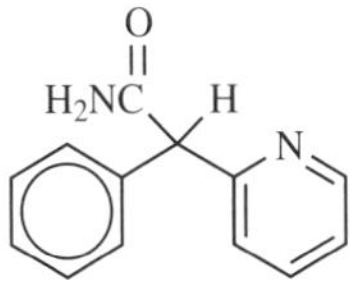   **6.13**

6.1.2. N-(3,5-Dinitrobenzoyl)-L-leucine (DNB-Leu) (6.14)

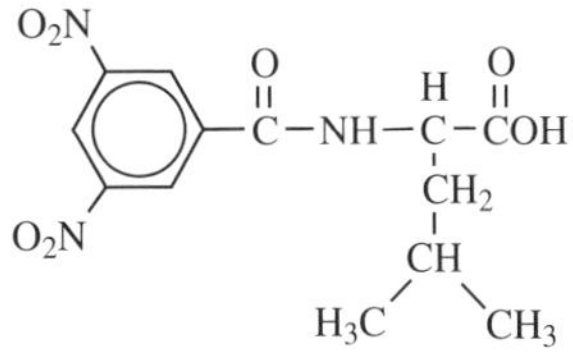

6.14

At the time it was first introduced, a *N*-(3,5-dinitrobenzoyl)-L-leucine (DNB-Leu) unit bonded to silica gel provided one of the most versatile liquid chromatographic stationary phases for the separation of enantiomers.[766] Organic-soluble analogs of DNB-Leu prepared as either *N*-propylamide or ester derivatives were subsequently exploited as chiral NMR solvating agents.

DNB-Leu has been shown to cause chiral recognition in the ^{1}H and ^{13}C NMR spectra of benzodiazepinones (**6.15**), lactones, amides, and sulfoxides.[767] Nonequivalence of 0.05–0.10 ppm was typically observed in the ^{1}H NMR spectrum, although resonances of the NH proton of some substrates had chiral discrimination as large as 0.5 ppm. The enantiomer that exhibited the greater liquid chromatographic retention consistently exhibited the larger shifts in the NMR spectrum, meaning that the large amount of data available from liquid chromatographic studies could be applied to the NMR analyses. NOE studies of the mechanism of interaction between *N*-(2-naphthyl)alanine methyl ester and DNB-Leu confirmed the importance of π stacking of the electron-rich and electron-deficient aromatic rings.[768,769]

6.15

The utility of adding Eu(fod)$_3$ (fod = 6,6,7,7,8,8,8-heptafluoro-2,2-dimethyl-3,5-octanedione) to mixtures of the ethyl ester of DNB-Leu and substrates such as sulfoxides and benzodiazepinones to enhance the discrimination in the ^{1}H and ^{13}C NMR spectra has been demonstrated.[770] The europium chelate preferentially binds to the unbound form of the substrate rather than DNB-Leu, causing larger lanthanide-induced shifts for the enantiomer with the lower association with DNB-Leu. Provided the concentration of europium was not so high as to strip significant amounts of the substrate off the DNB-Leu, substantial improvements in enantiomeric discrimination were realized. The chiral discrimination was so significant that spectra could be obtained at elevated temperatures (typically 50°C) to reduce exchange broadening with the lanthanide chelate, while still providing discrimination much larger than DNB-Leu–substrate mixtures at ambient probe temperature.[770]

The underivatized acid form of DNB-Leu can be solubilized in chloroform-*d* through the addition of triethylamine.[771,772] This soluble salt is also an effective chiral NMR solvating agent for sulfoxides, amines, and alcohols. A very different effect was observed when Eu(fod)$_3$ or Pr(fod)$_3$ was added to the acid form of DNB-Leu as compared to the ester of DNB-Leu. With the acid form, the DNB-Leu anion bonded to the lanthanide chelate to create an anionic species of the form [Ln(fod)$_3$(DNB-Leu)]$^-$, and the substrate enantiomer with the higher association constant with DNB-Leu exhibited larger lanthanide-induced shifts. This suggested that a complex of the form [Ln(fod)$_3$(DNB-Leu-substrate)]$^-$ is formed in which the substrate is bonded to the DNB-Leu unit and not directly to the lanthanide. In some cases, the enantiomeric discrimination with the [Ln(fod)$_3$(DNB-Leu)]$^-$ was much larger than observed with only DNB-Leu or with the combination of the ester of DNB-Leu with Eu(fod)$_3$. One notable example involved a significant nonequivalence of the *p*-tolyl methyl resonance of methyl *p*-tolyl sulfoxide (**6.16**) in the presence of [Ln(fod)$_3$(DNB-Leu)]$^-$.[771,772] No other DNB-Leu system caused chiral recognition of this resonance, which is rather remote from the chiral center.

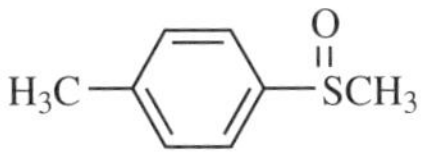

6.16

6.1.3. *N*-(3,5-Dinitrobenzoyl)phenylglycine (6.17)

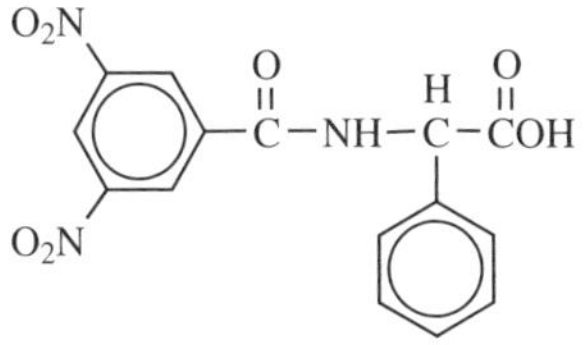

6.17

The utility of **6.17** as a chiral solvating agent for mono and diamines was compared to α-methoxy-α-trifluoromethylphenylacetic acid (MTPA) and mandelic acid.[104] These mixtures form diastereomeric salts and none of the three reagents was consistently the most effective for the range of substrates that were studied.

6.1.4. *N*-(3,5-Dinitrobenzoyl)-4-amino-3-methyl-1,2,3,4-tetrahydrophenanthrene (Whelk-O-1) (6.18)

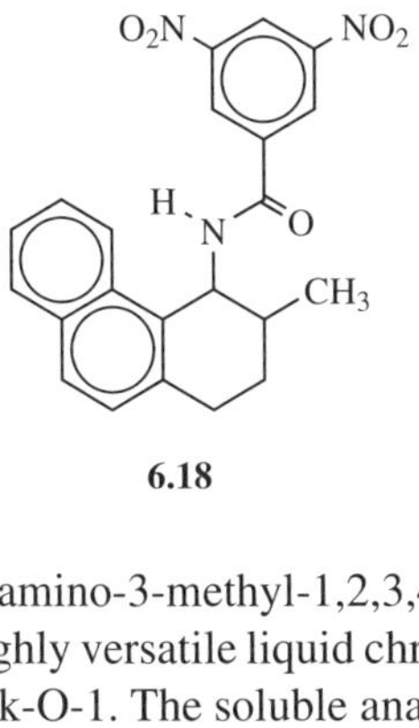

6.18

The *N*-(3,5-dinitrobenzoyl)-4-amino-3-methyl-1,2,3,4-tetrahydrophenanthrene unit has been incorporated into a highly versatile liquid chromatographic stationary phase commonly referred to as Whelk-O-1. The soluble analog is an effective chiral NMR solvating agent for a wide range of substrates.[773–777] Furthermore, the extensive amount of available liquid chromatographic data can be correlated with shifts in the NMR spectrum to assign absolute configurations.

Whelk-O-1 has a cleft between the DNB and phenanthrene units. Early NMR spectroscopic studies focused on confirming the mode of interaction of model substrates such as naproxen derivatives[773] and other naphthyl containing compounds[774] by ^{1}H NMR shifts and NOE data.

The ^{1}H NMR spectrum of **6.19**, which is chiral by virtue of deuterium substitution on one of the aromatic rings, exhibited enantiomeric discrimination of the aromatic signals in the presence of Whelk-O-1.[775] The extent of discrimination was enhanced by cooling the sample to $-20°$C. Ester derivatives of several chiral diaryl methanols (**6.20**) with *o*-, *m*-, and *p*-substituents could be enantiomerically discriminated in the presence of Whelk-O-1.[776]

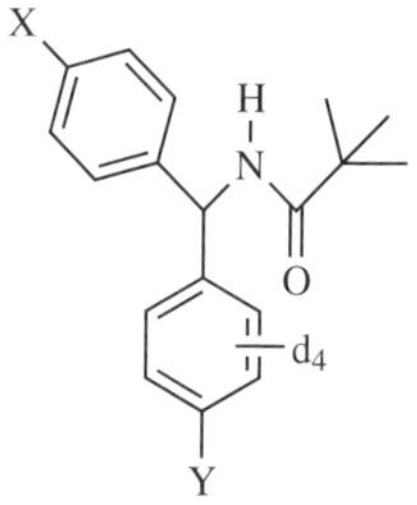

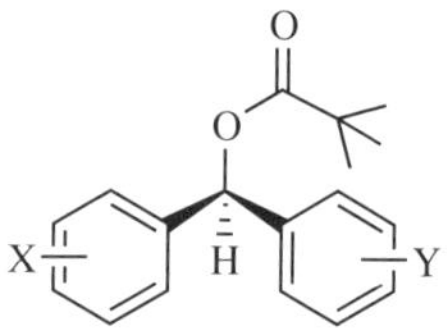

A more recent report explored the utility of Whelk-O-1 as a chiral NMR solvating agent on 20 different substrates that included epoxides, amides, lactones, lactams, alcohols, sulfoxides, and primary amines.[777] By raising the concentration of the chiral solvating agent or lowering the temperature, it was possible to produce enantiomeric discrimination in the spectra of 18 of the 20 substrates. The only two that were not successfully discriminated had only aromatic rings and no nitrogen or oxygen atoms. Some compounds that could not be separated by the Whelk-O-1 liquid chromatographic stationary phase were discriminated in the NMR spectrum with the soluble analog.

6.1.5. *N*-1-(1-Naphthyl)ethyltrifluoroacetamide (6.21)

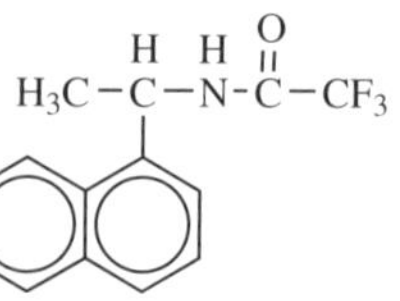

(**6.21**) is an effective chiral solvating agent for sulfoxides, with enantiomeric discrimination typically comparable to that observed with DNB-PEA.[754] In a later study,

AMIDES

it was shown that Eu(fod)$_3$ could be added to mixtures of **6.21** and sulfoxides to significantly enhance the enantiomeric discrimination in the NMR spectrum.[770] The europium bounded preferentially with the substrate enantiomer that had the lower association with **6.21**.

6.1.6. 1-(1-Naphthyl)ethyl Urea Derivatives of Amino Acids (NEU-AA) (6.22)

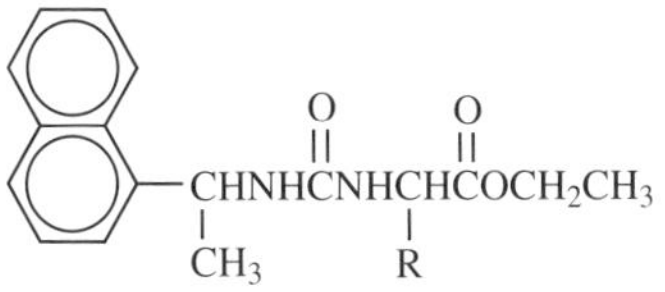

6.22

Reaction of (R)-(−)-1-(1-naphthyl)ethyl isocyanate with the amino acids valine, leucine, *tert*-leucine, and proline produced the corresponding urea derivatives. Ethyl esters of these were examined as organic-soluble chiral solvating agents for amines, sulfoxides, alcohols, and carboxylic acids.[772,778] The *tert*-leucine (R = *tert*-butyl) and valine (R = *iso*-propyl) derivatives were the most effective of those studied, and the valine derivative (NEU-Val) was identified as the reagent of choice. It was further shown that Eu(fod)$_3$ and Pr(fod)$_3$ could be added to enhance the enantiomeric discrimination. Evidence indicated that the lanthanide preferentially bounded to the substrate rather than the chiral solvating agent, such that the enantiomer with the lower association constant with the NEU-Val had larger lanthanide-induced shifts in its NMR spectrum.

The carboxylic acid forms of NEU-AA could be solubilized in chloroform-*d* by the addition of triethylamine and were also effective chiral solvating agents for sulfoxides, amines, and alcohols[771] and compounds such as 1-benzoyl-2-*tert*-butyl-3-methyl-4-imidazolidinone (**6.23**) and 4-benzyl-2-oxazolidinone (**6.24**).[772] Addition of lanthanide chelates such as Eu(fod)$_3$ to these mixtures resulted in the formation of a tetrakis chelate anion of the form [Eu(fod)$_3$(NEU-AA)]$^-$. In a reversal of what was observed for the ester forms of NEU-AA, the substrate enantiomer that exhibited the higher association with the NEU-AA exhibited larger lanthanide-induced shifts in the presence of [Eu(fod)$_3$(NEU-AA)]$^-$.[771,772] In many cases, addition of the lanthanide ion to form the [Eu(fod)$_3$(NEU-AA)]$^-$ species enhanced the enantiomeric discrimination in the resulting NMR spectrum.

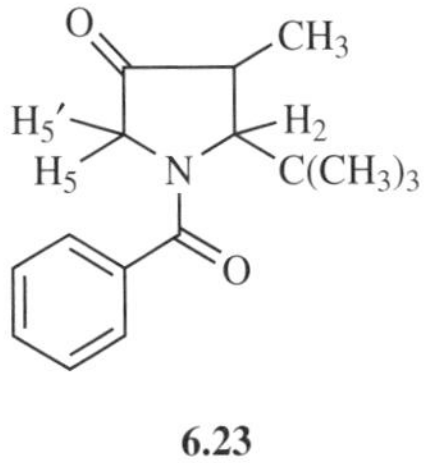

6.23

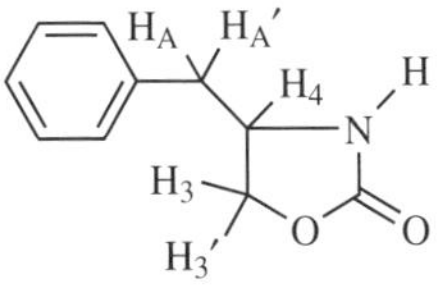

6.24

6.1.7. Bis Allyl Amide Derivatives

6.1.7.1. N,N'-Diallyl-L-tartardiamide bis(4-tert-butylbenzoate) (6.25) **6.25** was primarily developed for polymerization into a stationary phase for use in liquid chromatography. Studying either the polymer by solid state NMR spectroscopy, or solution phase work using a soluble analog, discrimination of the methine proton of O,O'-dibenzoyl tartaric acid (**6.26**) was noted.[779–781]

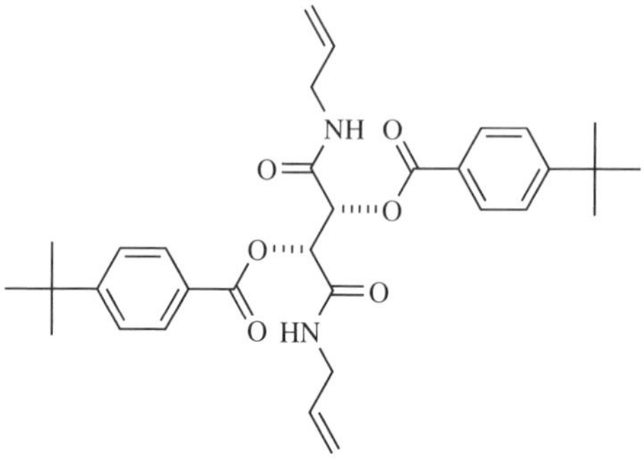

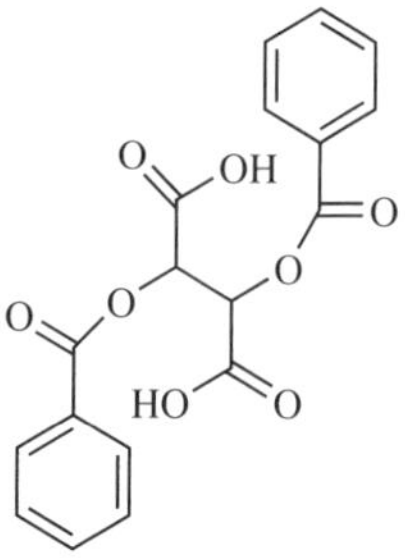

6.1.7.2. (S,S)-trans-Acenaphthene-1,2-dicarboxylic acid bis allylamide (6.27) **6.27** produced enantiomeric discrimination in the methine resonance of O,O'-dibenzoyl tartaric acid (**6.26**).[781–783] The enantiomeric discrimination ranked in the order: carbon tetrachloride > chloroform-d > methylene chloride-d_2. The chiral discrimination with **6.25** was larger than that with **6.27**.

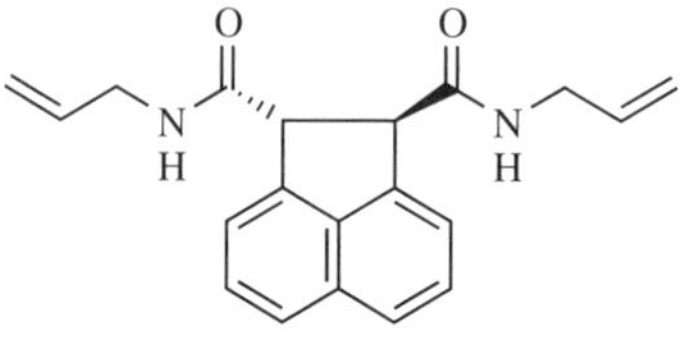

6.1.8. (R)-Phenylglycinol-N-3,5-dinitrobenzoyl-O-triethoxysilylpropylcarbamate (6.28)

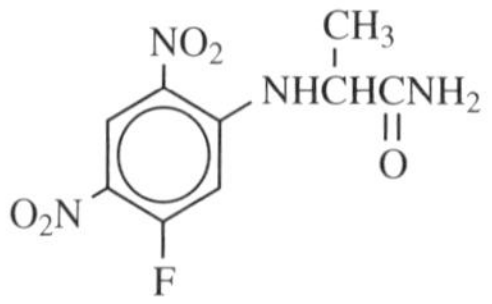

6.28 was developed for use as a stationary liquid chromatographic phase. The soluble analog **6.29** produced nonequivalence as high as 0.16 ppm in the H8 aromatic signal in the ^{1}H NMR spectra of a series of N-acylnaphthyl alkyl amines (**6.30**).[784]

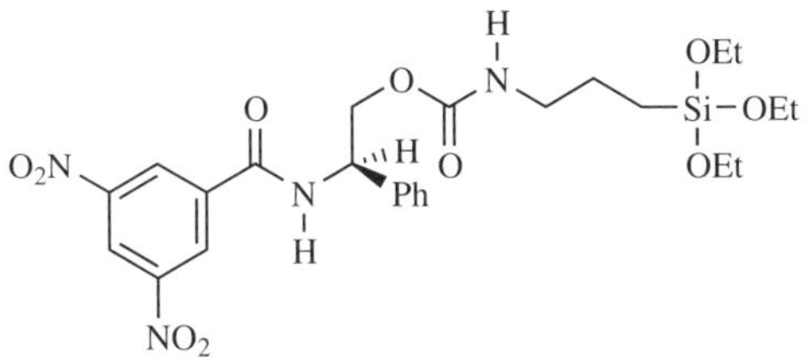

6.29

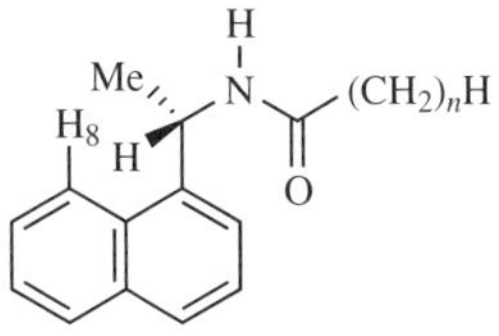

6.30

$n = 1, 2, 3, 5, 11$

6.1.9. N-Methyl Amide of (R)-N-Acetyl-4-methoxy Phenyl Glycine (6.31)

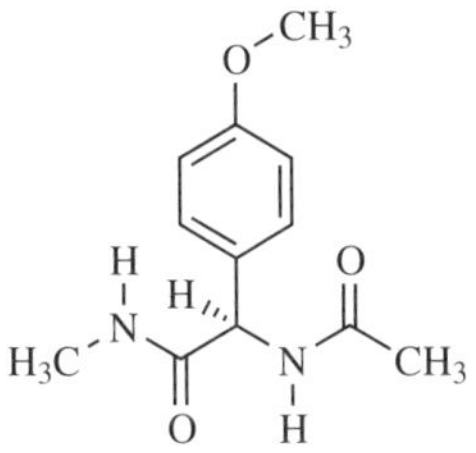

6.31

6.31 is an effective chiral NMR solvating agent for N-(3,5-dinitrobenzoyl) (DNB) derivatives of α-amino acid amides and esters.[785] Nonequivalence of up to 0.78 ppm was observed for the NH resonance, and on the order of 0.1–0.2 ppm for the aromatic hydrogen resonances of the DNB ring. For the four amino acids examined in this work, the NH resonance of the (S)-enantiomer consistently shifted further, whereas the opposite trend was observed for the DNB signals.

6.1.10. N-(n-Butylamide) of (S)-2-Phenylcarbamoyloxypropionic Acid (6.32)

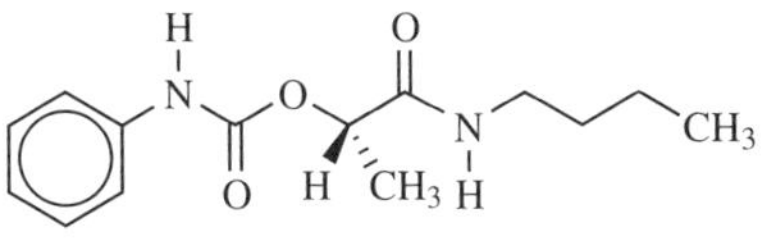

6.32

6.32 is an effective chiral solvating agent for the analysis of N-(3,5-DNB) derivatives of amino acid methyl esters.[786,787] Chiral discrimination of the methoxy signal and the two hydrogens of the DNB ring was observed in the ^{1}H NMR spectrum. The NH proton exhibited nonequivalence as high as 0.54 ppm.[787]

6.1.11. 2,2′-Oxybis[N-(1-phenylethyl)acetamide] (6.33)

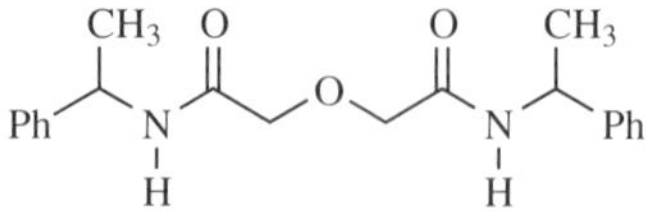

6.33

6.33 has been shown to be an effective chiral solvating agent for amides.[788] Primary or secondary amines can be analyzed by conversion to their corresponding amide with acetic or pivalic acid. Alternatively, chiral carboxylic acids can be analyzed after conversion to an amide.

6.1.12. Tetrapeptide Species

Tetrapeptide species that contain an amino isobutyric acid (Aib) unit (X-Trp-Aib-Gly-Leu-NH-Ar; X=H, Ar = phenyl or 3,5-dimethylphenyl) have been shown to cause chiral recognition in the NMR spectra of quaternary ammonium ions such as 2-hydroxymethyl-1,1-dimethylpyrrolidinium iodide (**6.34**) and 2-butylcarbamoyl-1,1-dimethylpyrrolidinium iodide (**6.35**), although the largest nonequivalence was only 0.015 ppm in the ^{1}H NMR spectrum.[789]

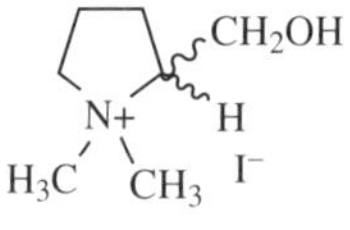

6.34 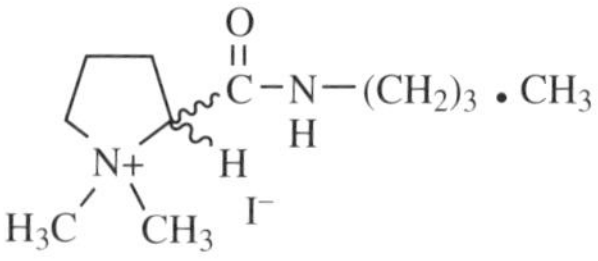   **6.35**

6.1.13. Tetraamidic Selector

A compound prepared by attaching two (S)-phenylalanine groups to a 3,6,9-oxadecanoyl bridge (**6.36**) was used as a chiral chromatographic phase.[790] A ^{1}H NMR study of the chiral recognition mechanism with N-trifluoroacetyl-phenylalaninate methyl and n-butyl esters noted nonequivalence discrimination of up to 0.21 ppm for certain resonances in the ^{1}H NMR spectrum.

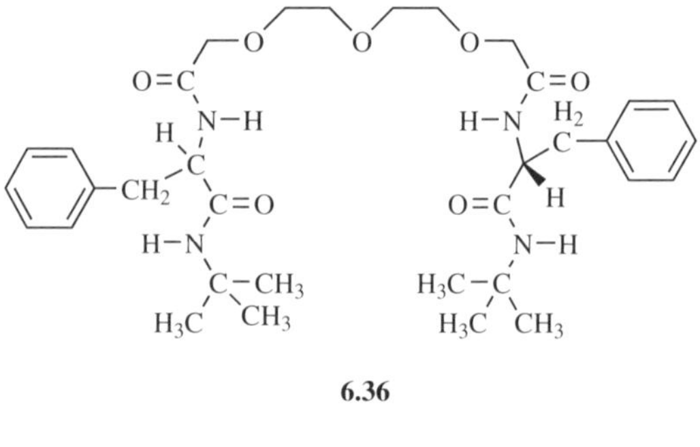

6.36

6.2. LACTAMS

The chiral lactam **6.37** has the ability to associate with and cause enantiomeric discrimination in the ^{1}H NMR spectra of substrates with amide and amide-like functionalities.[791,792] For example, 7-substituted 3-azabicyclo[3.3.1]nonan-2-ones associate with **6.37** through hydrogen bonding as shown in Figure 6.1.[791] Only the enantiomer with a complementary structure favorably associated with and showed sizeable shifts in its NMR spectrum. The NH resonance exhibited a particularly large shift, provided an apolar solvent such as benzene-d_6 or toluene-d_8 was used. Because of the requirement of a complementary structure, it was also possible to assign the absolute configuration of the substrate.

6.37

The effectiveness of **6.37** was examined for a range of lactams of varying ring sizes, quinolones such as **6.38**, and oxazolidinones such as **6.39**.[792,793] The enantiomer that had its bulky substituent groups pointing away from the shielding substituent of **6.37** exhibited greater association and larger shifts. Larger discrimination was observed in benzene-d_6 than chloroform-d, but some substrates could even be analyzed in acetone-d_6. The spectra of six-membered lactams showed less enantiomeric discrimination than five-membered lactams.

6.38   **6.39**

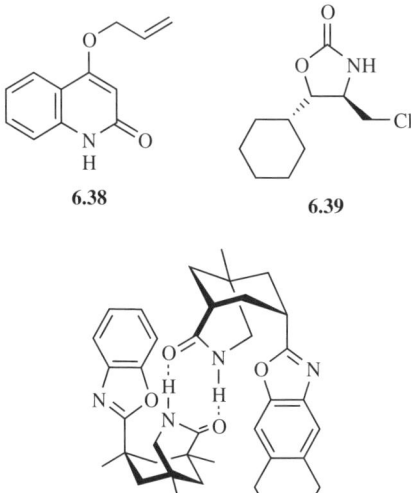

Figure 6.1. Binding interaction of 7-substituted 3-azabicyclo[3.3.1]nonan-2-ones.

6.2.1. Dihydropyrimidone (6.40)

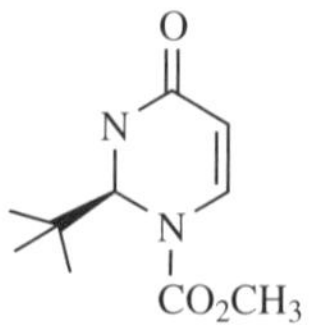

6.40

Dihydropyrimidone **6.40** has been used to assign the absolute stereochemistry of α-chiral carboxylic acids.[794] Carboxylic acids react at the lactam nitrogen to create an amide of structure **6.41**. Certain signals in the ^{1}H NMR spectrum of the amide derivative of the dihydropyrimidone showed characteristic $\Delta\delta^{RS}$ values that correlated with absolute configuration. The method could be reliably applied to compounds of similar structure; however, with dialkylcarboxylic acids, the $\Delta\delta^{RS}$ values were much smaller requiring more caution when performing the analysis.[794]

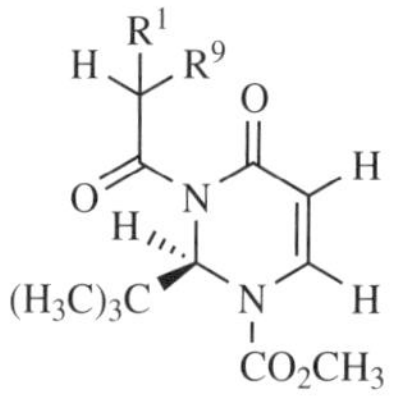

6.41

6.2.2. Amino Acid Isocyanurate Derivatives

Five amino acid isocyanurate derivatives of general structure **6.42** were evaluated for their ability to cause chiral recognition in the ^{1}H NMR spectrum of 2,2'-dihydroxy-1,1'-binaphthalene in chloroform-d.[795] Discrimination occurred only with derivatives in which all three of the nitrogen atoms were substituted. The largest nonequivalence (0.11 ppm) was observed for the hydroxyl proton of the substrate with the leucine derivative.

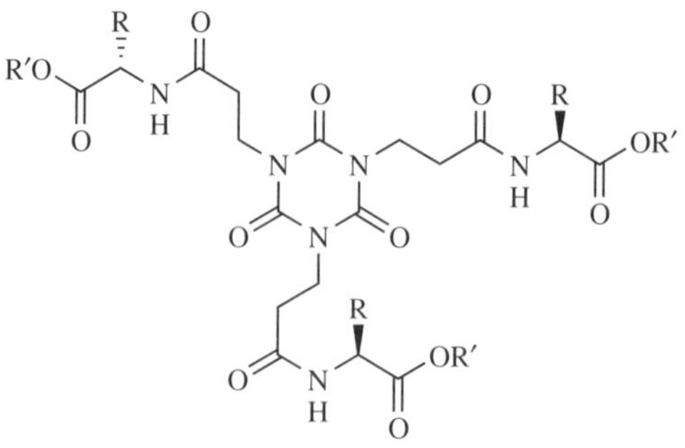

6.42

6.3. ALDEHYDES

6.3.1. 15-Formyl-14-hydroxy-2,8-dithia[9](2,5)pyridinophane (6.43)

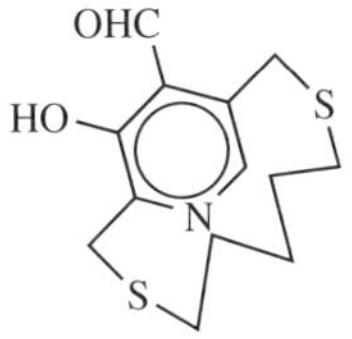

6.43

6.43 reacted with the amine group of α-deuterated α-amino acids to produce the corresponding imine (**6.44**).[796] The azomethine resonance of the diastereomeric products was resolved in the ^{1}H NMR spectrum in benzene-d_6, facilitating the determination of optical purity.

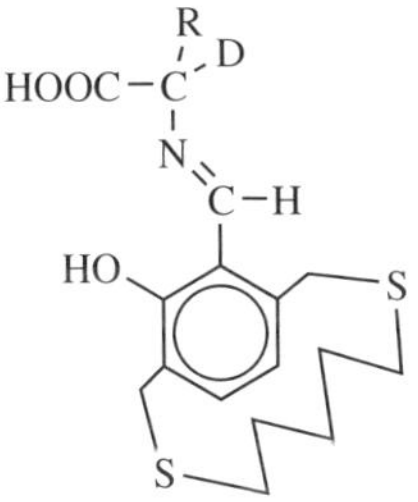

6.44

6.3.2. 2′-Methoxy-1,1′-binaphthalene-8-carbaldehyde (6.45)

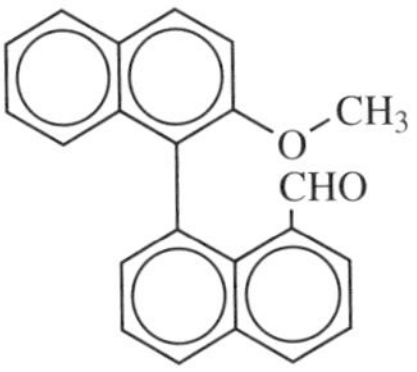

6.45

The absolute configuration of 1,2- and 1,3-diols were assigned by preparing the corresponding acetal (**6.46**) of **6.45**.[797] A single diastereomeric acetal was obtained and each diol configuration had specific and predictable NOE connectivities. The reliability of the method was confirmed by assigning the absolute configuration of eight diols with known configurations.

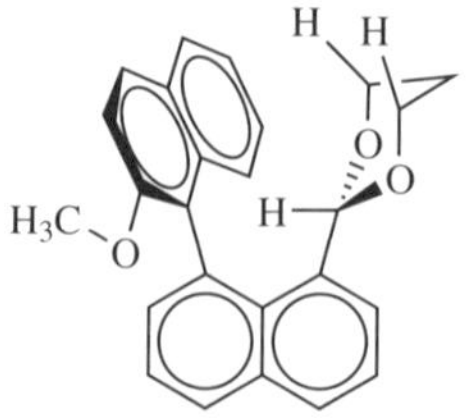

6.46

6.3.3. 2-Hydroxy-2'-nitrobenzoate-3-aldehyde-1,1'-binaphthalene (6.47)

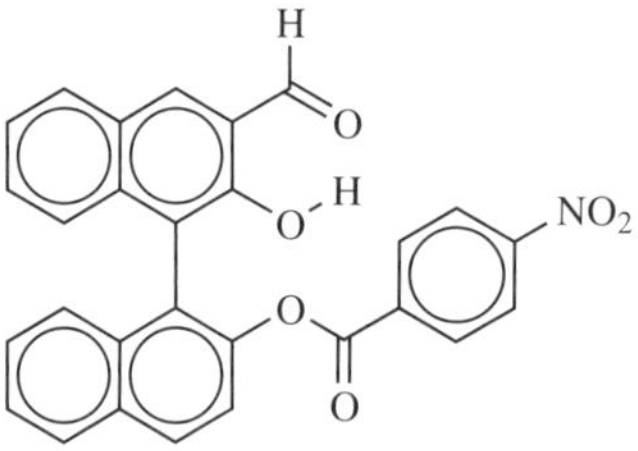

6.47

Amines and amino acids react with **6.47** to form an intramolecular resonance-assisted hydrogen-bonded imine as shown in Figure 6.2.[798] The resonance of the hydrogen atom that associated with the nitrogen atom shifted substantially (as high as 13 ppm) to higher frequencies. The derivative with L-alanine has a shorter hydrogen bond than with D-alanine, and therefore the hydrogen resonance occurred at higher frequency. The trend was consistent and could be used to assign the absolute configuration of amino acids. No epimerization was observed in the derivatization reaction.

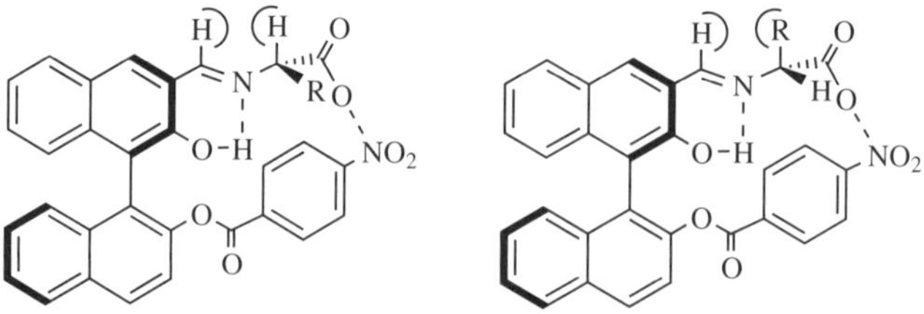

Figure 6.2. Intramolecular resonance-assisted hydrogen-bonded imine derivative formed by the reaction of an amine or an amino acid with 2-hydroxy-2'-nitrobenzoate-3-aldehyde-1,1'-binaphthalene.

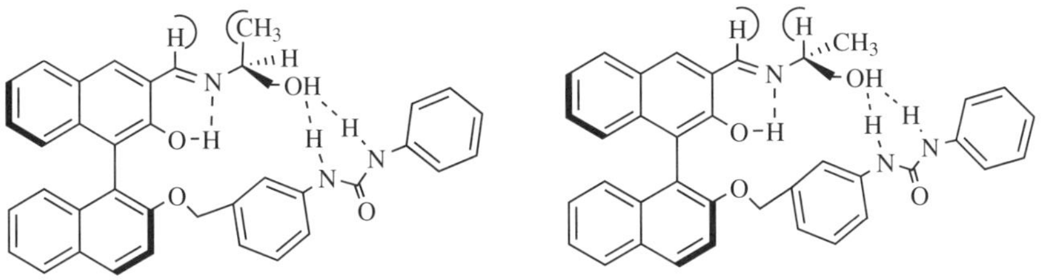

KETONES 205

Figure 6.3. Intramolecular interaction in the product of 1,2-amino alcohols with 2-hydroxy-2'-substituted-3-aldehyde-1,1'-binaphthalene.

6.3.4. 2-Hydroxy-2'-substituted-3-aldehyde-1,1'-binaphthalene (6.48)

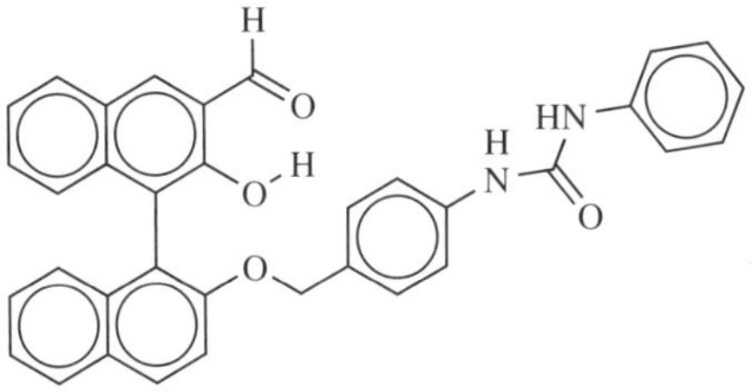

6.48

6.48 was especially suited to the analysis of chiral 1,2-amino alcohols.[799] The amino alcohol formed an imine by reacting at the aldehyde group. The substrate was effectively locked in place via resonance-assisted hydrogen bonds as shown in Figure 6.3. The binding always occurred with the same sense of stereoselectivity such that the (R)-enantiomer of the substrate was more stabilized than the (S)-enantiomer.

6.4. KETONES

6.4.1. l-Menthone (6.49)

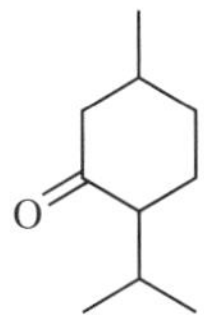

6.49

As described in Section 4.19.1, 2,3-butanediol reacts with ketones to produce the corresponding acetal, which can then be used to determine the optical purity and in some cases to assign the configuration of the ketone. The situation can be reversed,

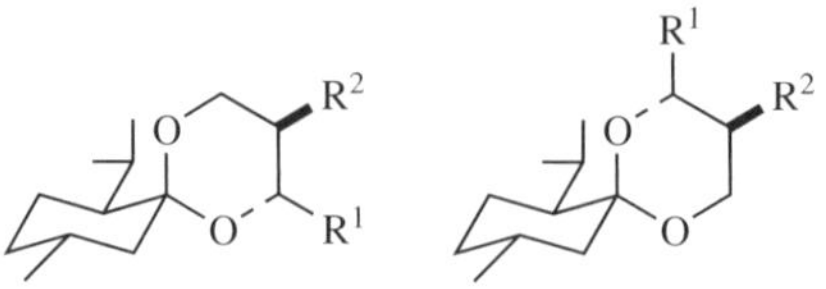

Figure 6.4. Diastereomeric spiroacetals produced by the reaction of menthone with a 1,3-diol.

such that the utility of l-methone for assigning the configuration of 1,3-alkanediols has been demonstrated.[800] Reaction of l-menthone and the diol produced diastereomeric spiroacetals as shown in Figure 6.4. Of the four possible products, only the two stereochemically rigid ones shown in Figure 6.4 were obtained. These were separated by column chromatography and long-range anisotropy effects of the menthone ring caused specific trends in the ^{1}H shifts that correlated with the absolute stereochemistry of the diol.[800] To assign the absolute configuration of diols that were enantiomerically pure, derivatives with d- and l-menthones were prepared and the ^{1}H NMR spectra analyzed.

6.4.2. (S)-(+)-2-Propylcyclohexanone (6.50)

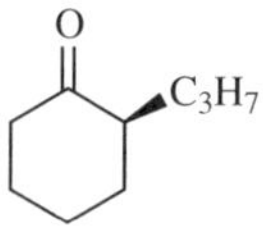

6.50

(S)-(+)-2-propylcyclohexanone has been used to measure the optical purity of 1,2-glycols.[801] The acetal carbon formed in the derivatization reaction added a new chiral center so that up to four diastereomers were produced. The ^{13}C signal of the C2 or C6 carbon was monitored. The method was sutiable for analyzing tertiary glycols such as 1-methyl-1-phenyl-1,2-ethanediol.

6.5. ISOCYANATES

Isocyanates readily react with amine and alcohol functionalities to form the corresponding urea and carbamate derivatives. Chiral isocyanate compounds have been used on a number of occasions as derivatizing agents for enantiomeric analysis via NMR spectroscopy.

6.5.1. 1-Phenylethyl Isocyanate (PE-I) (6.51)

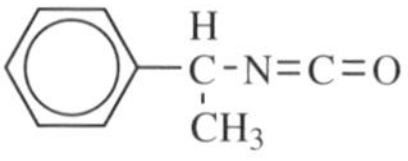

6.51

ISOCYANATES

Compounds **6.52**,[802] **6.53**,[803] **6.54**,[804] and **6.55**[805] are examples of substrates whose optical purity has been determined by analyzing the NMR spectra of their derivatives with PE-I. Compounds **6.52**[802] and **6.54**,[804] which have aryl hydroxyl groups, preferentially reacted at the amine group to form a urea derivative. Often the methyl doublet of the PE-I functionality was useful for determining the optical purity. In the case of **6.56**, the bis-PE-I derivative was formed.[805]

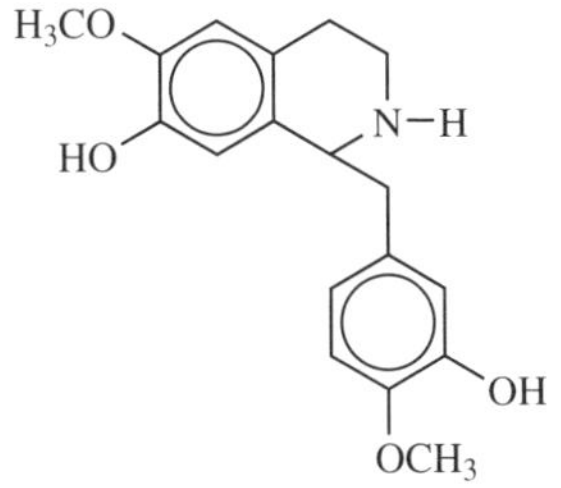

6.52

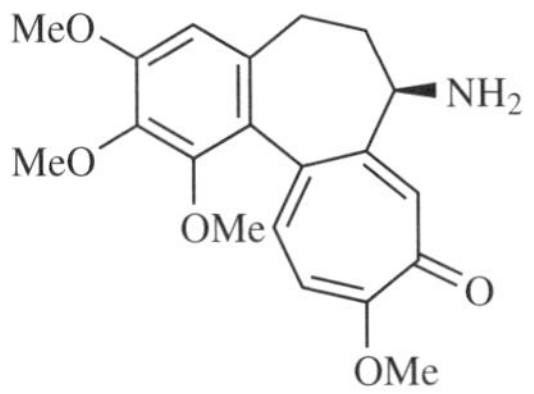

6.53

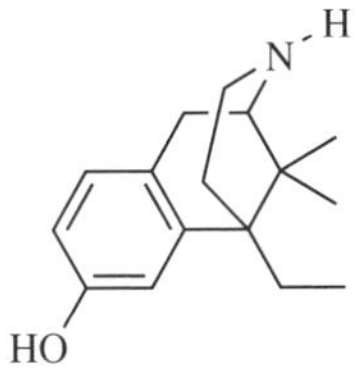

6.54

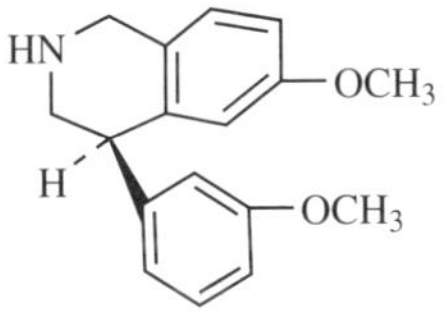

6.55

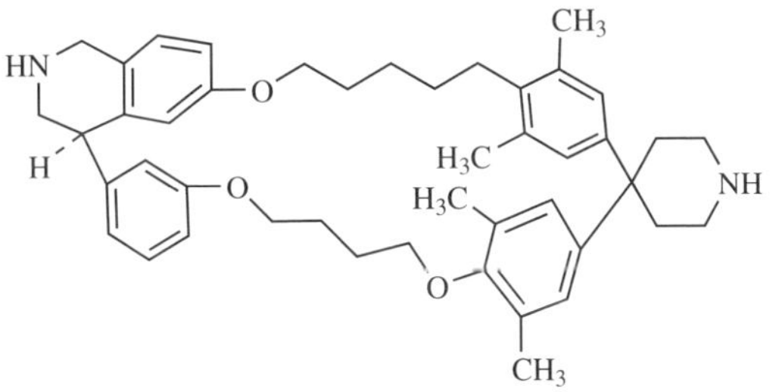

6.56

Carbamate derivatives with PE-I have been used for the analysis of optical purity of 1-(1-isoquinolinyl)-1-(2-pyridyl)ethanol (**6.57**)[806] and terbutaline (**6.58**).[807] For the PE-I derivative of 2,2-dimethyl-1-(1-naphthyl)propanol,[808] slow rotation about the amide bond resulted in both Z- and E- rotamers, each of which could be assigned.[808] NOE data could then be used to determine the absolute configuration of the alcohol.

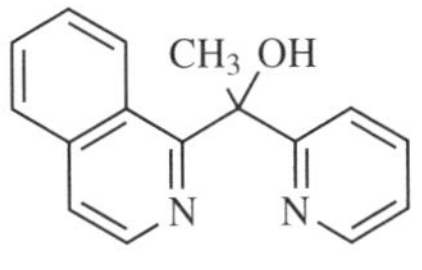

6.57

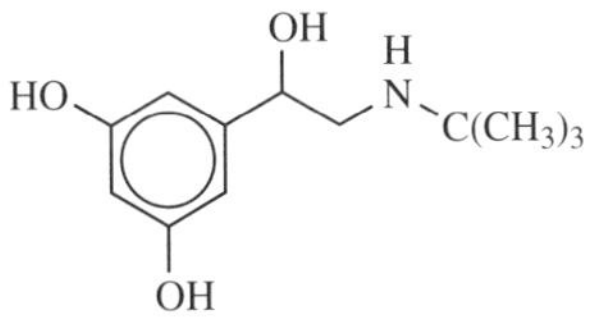

6.58

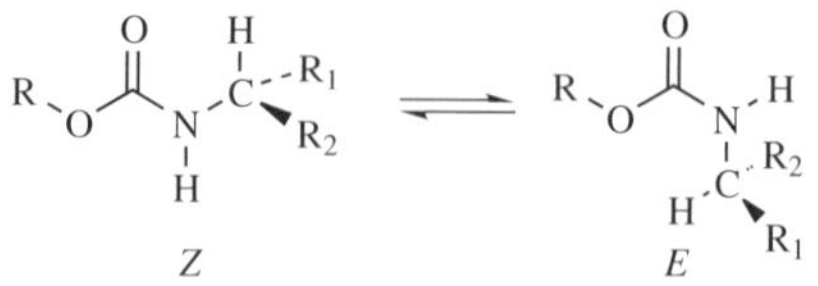

Figure 6.5. The Z- and E-rotamers of the carbamates of alkyl aryl carbinols.

6.5.2. 1-(1-Naphthyl)ethyl Isocyanate (NE-I) (6.59)

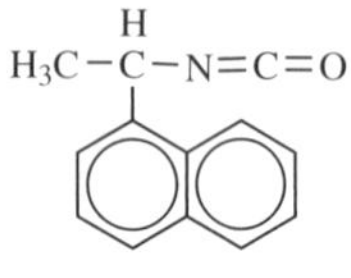

6.59

NE-I has been used as a chiral derivatizing agent for alcohols. The carbamate product with alkyl aryl carbinols existed in both the Z- and E-rotamers at a ratio of 90:10 (Fig. 6.5).[809] The diastereomeric conformers formed by the reaction of NE-I with secondary carbinols exhibited discrimination in the ^{1}H and ^{13}C NMR spectra. Furthermore, shielding in the more dominant Z-rotamer with the naphthyl ring could be used to assign the absolute configuration.[810] The corresponding derivatives with PE-I showed a similar trend but the phenyl ring caused considerably less shielding than the naphthyl ring.

The NE-I carbamate derivatives of the methyl esters of 12-hydroxy-(Z)-9-octadecenoic acid and 9-hydroxy-(Z)-12-octadeceneoic acid were used to determine optical purity and assign absolute configurations.[811] Shielding from the naphthyl ring was sufficiently large such that nonequivalence of the methoxy resonance of 0.01–0.04 ppm was observed, even though it was nine bonds away from the chiral center. A knowledge of the structure of the carbamate, as shown in Figure 6.6, enabled an assignment of the absolute configuration on the basis of the shielding of the naphthyl ring.

The optical purity of fluoxetine (**6.60**) was determined by analyzing the NMR spectrum of the corresponding urea derivative of the secondary amine with NE-I.[812]

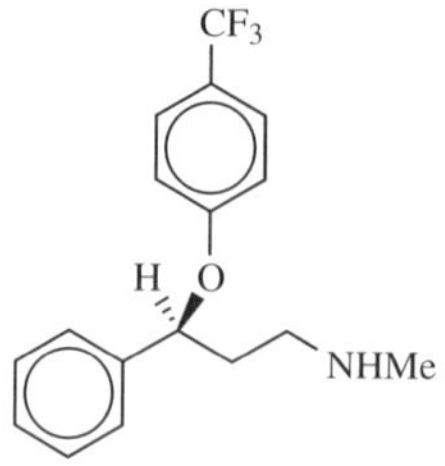

6.60

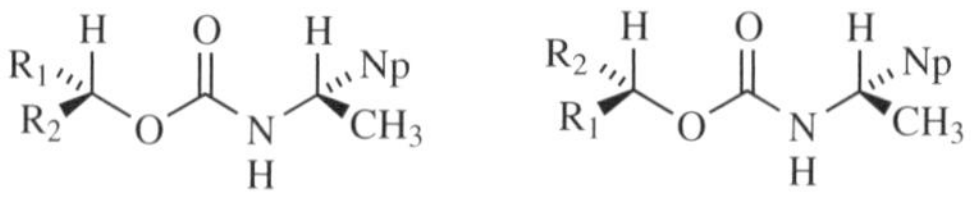

Figure 6.6. Conformation of diastereomeric carbamates produced by the reaction of NE-I with hydroxy carboxylic acids.

Both NE-I and PE-I have been used to analyze the optical purity and absolute configurations of vinyl ethers of general structure **6.61**.[813] Under high pressure, the isocyanates underwent a [2+2] cycloaddition with the vinyl ether to form the corresponding bicyclic or monocyclic azetidinone (**6.62**). The aryl ether group adopted a preferred conformation, such that specific shielding by the aromatic ring could be used to assign the absolute stereochemistry. Lanthanide-induced shifts with Eu(tfc)$_3$ (tfc = 3-trifluoroacetyl-d-camphor) were used in conjunction with these derivatives to confirm the conformational preference of the azetidinone derivative.

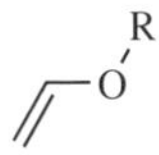

6.61

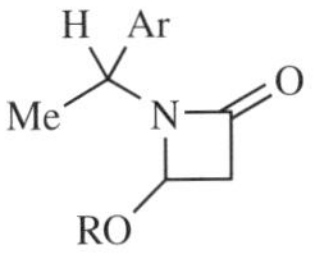

6.62

6.5.3. Phenylethyl Isothiocyanate (PE-IS) (6.63a)/Naphthylethyl Isothiocyanate (NE-IS) (6.63b)

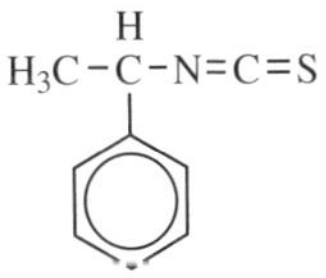

6.63a

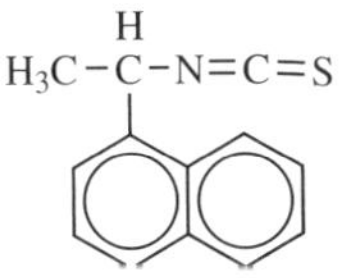

6.63b

The enantiomeric purity of a variety of chiral primary and secondary amines has been determined by analyzing their derivatives with PE-IS and NE-IS.[814] An advantage of the isothiocyanate reagents relative to the corresponding isocyanates is their enhanced stability in water. The reagents reacted rapidly with amines in n-hexane, chloroform-d, methanol-d_4, dimethylsulfoxide-d_6, and deuterium oxide, and discrimination was observed in the ^{1}H and ^{13}C NMR spectra. The methyl signal was an especially useful one to monitor. The optical purity of **6.64** was determined by analyzing its derivative with PE-IS.[210] The analysis of alcohols with this reagent was complicated by the presence of two tautomeric forms of the product.[814]

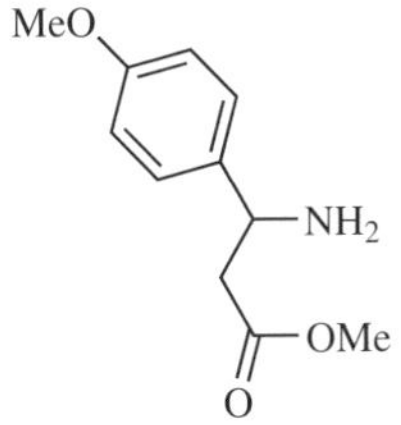

6.64

6.5.4. α-Methoxy-α-(trifluoromethyl)benzyl Isocyanate (6.65)

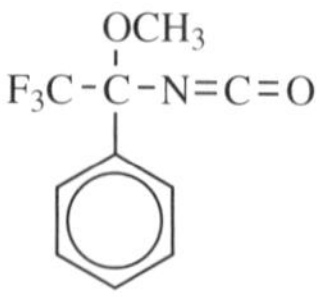

6.65 is an effective chiral NMR derivatizing agent for the analysis of primary and secondary amines and secondary alcohols.[815] The reaction with aliphatic amines, cyclic amines, and aryl alkyl amines occurred rapidly enough that it could be done directly in the NMR tube using chloroform-d as the solvent. Either the ^{1}H or ^{19}F NMR spectrum was used for the determination of optical purity. The products adopted a Z-conformation as shown in Figure 6.7. The preferred conformation caused shielding of the methyl group by the aromatic ring in the (S,R)-derivative but not the (S,S)-derivative, such that the absolute configurations of amines could be assigned. The reaction with alcohols occurred more slowly and the results were less reliable, such that this reagent is not recommended for the study of alcohols.[815]

6.5.5. 2,3,4,6-Tetra-O-acetyl-β-D-glucopyranosyl Isothiocyanate (6.66)

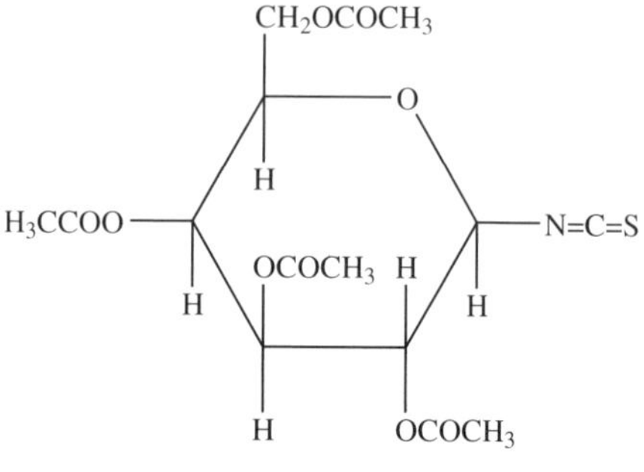

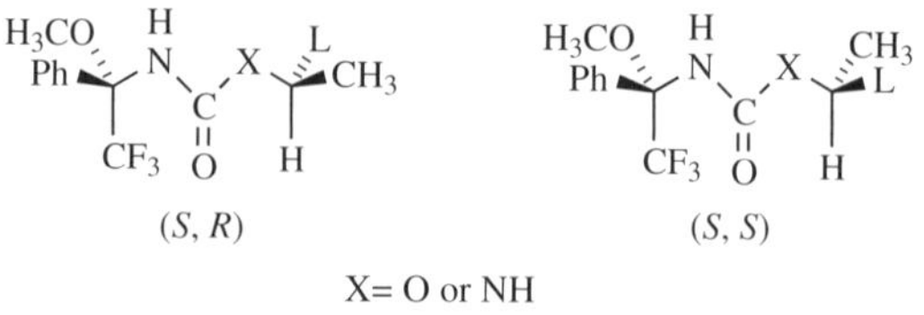

Figure 6.7. Z-Conformation of diastereomeric carbamates produced by the reaction of secondary amines or alcohols with **6.65**.

The optical purity of methamphetamine (**6.67**) was determined by analyzing the ^{1}H NMR spectrum of its derivative with **6.66**.[816]

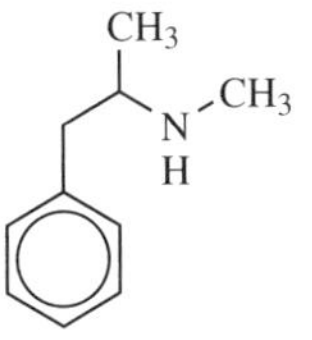

$$\underset{\textbf{6.67}}{\text{PhCH}_2\text{-CH(CH}_3\text{)-N(H)-CH}_3}$$

6.5.6. (S)-2-Chloro-2-fluoroethanoyl Isocyanate (6.68)

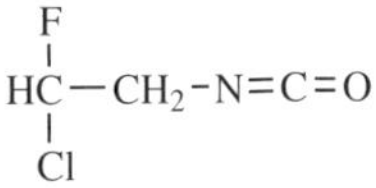

$$\underset{\textbf{6.68}}{\text{HC(F)(Cl)}-\text{CH}_2-\text{N}=\text{C}=\text{O}}$$

Acyclic, cyclic, and alkyl aryl secondary alcohols undergo a rapid reaction with **6.68** to produce diastereomers with distinct resonances in the ^{1}H NMR spectra.[817] Except for those alcohols with simple alkyl groups, the carbamate derivative adopted a preferred conformation that resulted in $\Delta\delta^{RS}$ values that correlated with the stereochemistry of the alcohol. No racemization was observed during the derivatization step.

6.6. MISCELLANEOUS REAGENTS

6.6.1. 2,2-Dimethoxypropane

The relative configurations of 1,2-diols can be determined on the basis of the acetonides or 4,5-disubstituted-2,2-dimethyl-1,3-dioxolane (**6.69**) derivatives prepared from 2,2-dimethoxypropane.[818] Additivity rules based on the *cis-* and *trans-*groups of the product could be used to distinguish the relative configuration through analysis of the ^{13}C NMR spectrum.

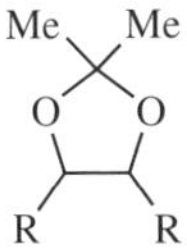

$$\underset{\textbf{6.69}}{\text{2,2-dimethyl-4,5-di-R-1,3-dioxolane}}$$

The method was subsequently applied to the analysis of the relative configuration of the C2 and C3 positions in **6.70**.[819] Smaller nonequivalence between the ^{13}C signals of the *gem*-dimethyl groups for the *syn*-diol (0.8 ppm) in contrast to that for the *anti*-diol (about 3 ppm) enabled the assignment. The relative configuration of the diol units of **6.71**,[820] **6.72**,[821] and **6.73**[822] was assigned on the basis of this method.

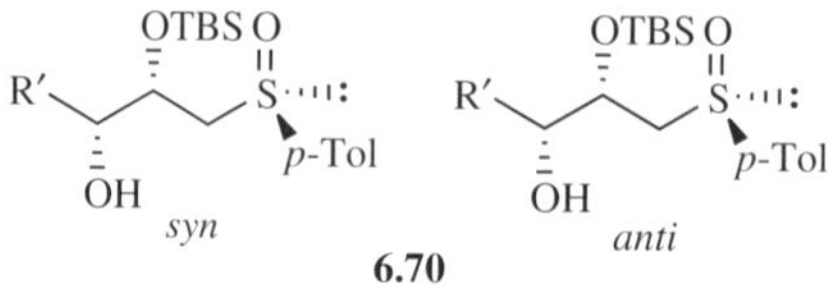

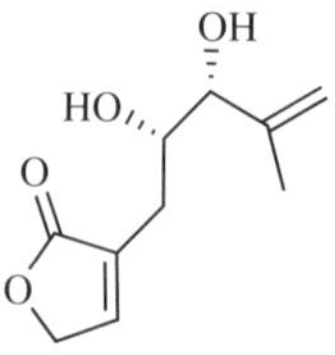

This reagent has been used to assign the configuration of 1,3-diols such as (2R,3R,7Z)-2-aminotetradec-7-ene-1,3-diol (**6.74**)[823] and about C1 in 2-(*N*-benzoylamino)-1-(2-furyl)-1,3-butanediols (**6.75**).[824] For **6.75**, an analysis of the coupling constants for the *cis*-form (chair conformer) versus *trans*-form (twist-boat conformer) was used in assigning the configuration.[824]

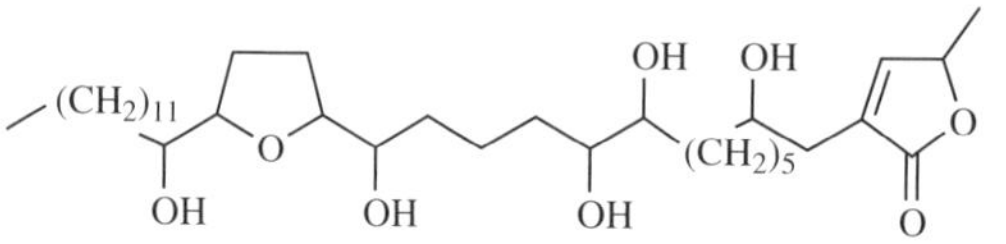

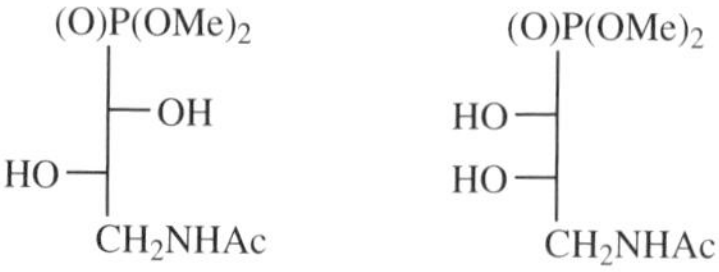

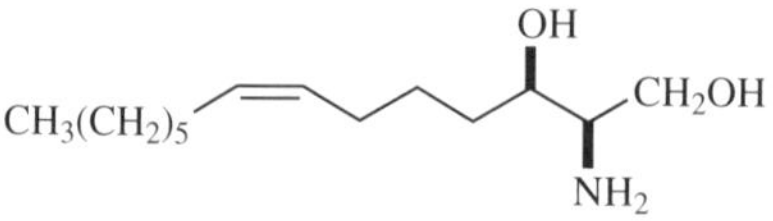

6.6.2. (−)-Chloromethylmenthyl Ether (6.76)

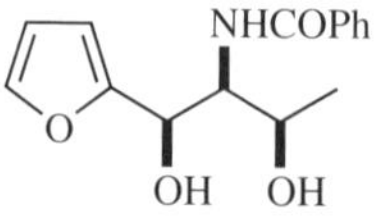

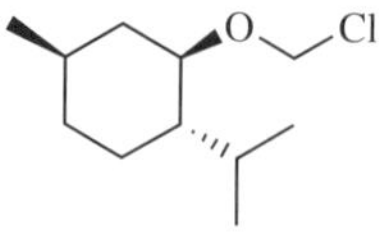

The enantiomeric purity of the dihydrodiol of benzo[*a*]pyrene (**6.77**) was determined using (−)-chloromethylmenthyl ether as a chiral derivatizing agent.[825] The reaction produces a bis formaldehyde acetal derivative (**6.78**) and the methylene signals are distinct in chloroform-d and benzene-d_6.

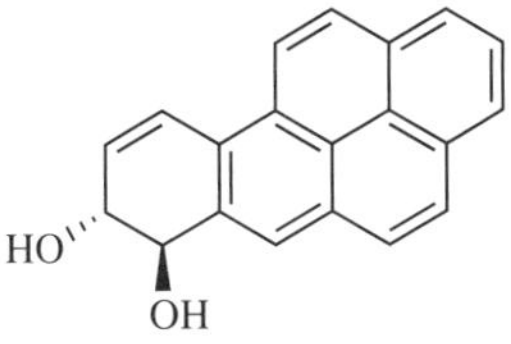

6.6.3. 2-Oxazolidones

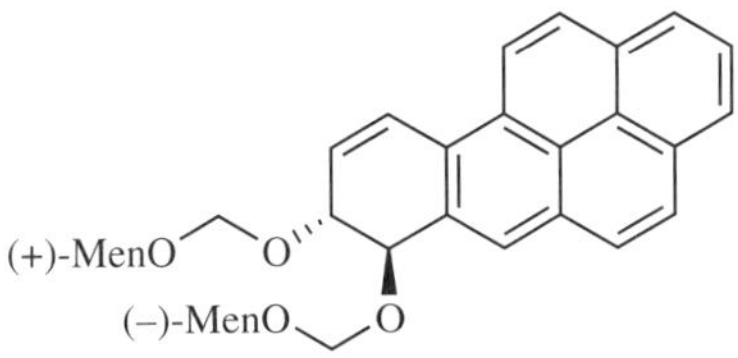

cis-4,5-Diphenyl-2-oxazolidone (**6.79**) and similar compounds react with isocyanates of primary amines to give the corresponding allophanate derivative (**6.80**).[826] **6.80** formed an intramolecular hydrogen bond between the NH and the carbonyl oxygen atom that locked in the conformation. The ^{1}H NMR spectrum was used to determine the optical purity and assign the absolute configuration of the isocyanate, since the aryl rings of the oxazolidone had a consistent sense of nonequivalence and caused specific shielding of the groups of the amine substrate.

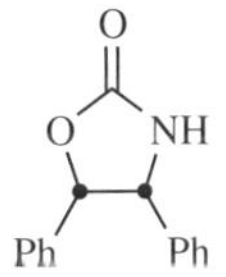

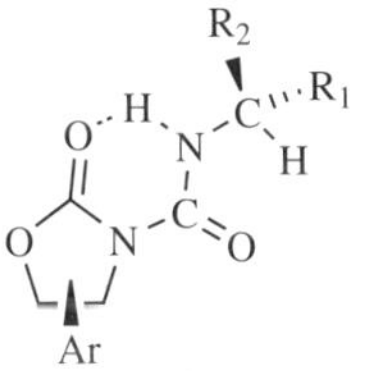

Conversion of **6.79** to its corresponding carbamoyl chloride (**6.81**) provided a compound that reacted directly with primary amines to form diastereomeric allophanates (**6.82**).[826] The derivatives were analogous in structure to those with the isocyanates and the resulting ^{1}H NMR spectra were used to determine optical purity and assign absolute configuration.

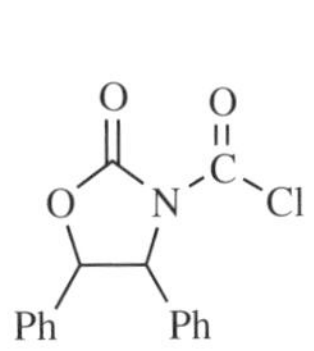

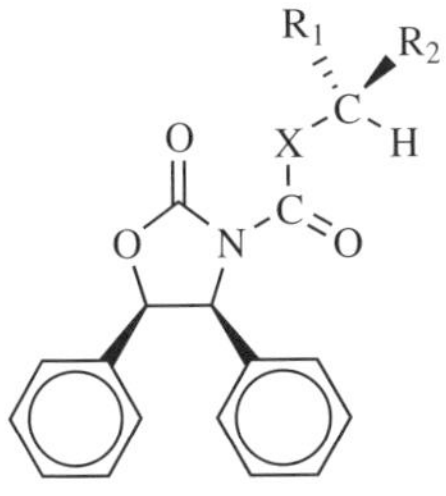

4-Methyl-5-phenyl-2-oxazolidinone (**6.83**) has been used as a chiral derivatizing agent for α-halo carboxylic acids and *O*-substituted α-hydroxycarboxylic acids.[827]

The carboxylic acids reacted at the NH to produce the corresponding amide derivative, and chiral recognition was observed in the resulting ^{1}H NMR spectra.

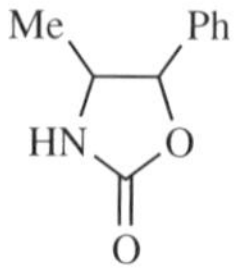

6.83

6.6.4. 5(R)-Methyl-1-(chloromethyl)-2-pyrrolidinone (6.84)

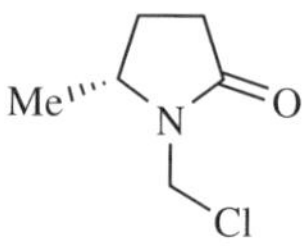

6.84

6.84 reacts with chiral alcohols to produce the corresponding N-(alkoxymethyl)-2-pyrrolidinones (**6.85**).[828] Resolution of the two diastereotopic N-methylene resonances of the derivatives with alcohols, such as menthol (**6.85**), 2-butanol, and 1-phenylethanol, was used to determine optical purity. These resonances appeared between 4.3 and 5.2 ppm and were generally in a part of the spectrum that was free of other signals. No kinetic resolution was observed in the derivatization reaction.

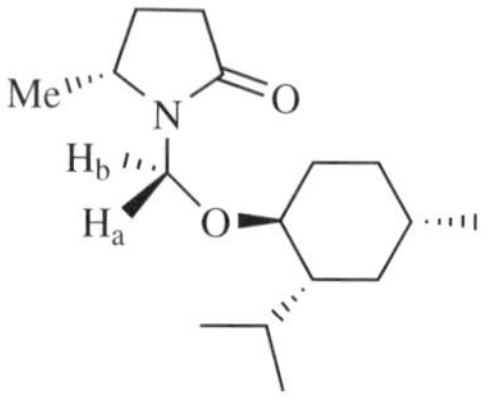

6.85

Amines were also analyzed using **6.84** by an analogous procedure to that for alcohols.[829] The methylene signals between the two nitrogen atoms were especially diagnostic. The method was demonstrated with 1-phenylethylamine and valine ethyl ester and no kinetic resolution was observed during the derivatization.

A reagent incorporating a methyl-5-carboxylate group in place of the 5-methyl group (**6.86**) was used to analyze the optical purity and assign the absolute configuration of alcohols by formation of **6.87**.[830] The methoxy signal was ideal for determining optical purity. NOE interactions between the diastereotopic N-methylene resonances were distinct for the two configurations of the alcohols and could be used to assign the absolute configuration.

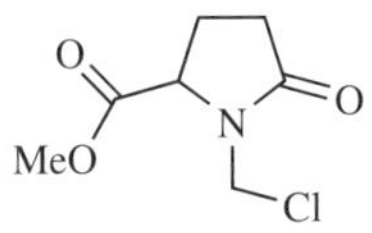

6.86

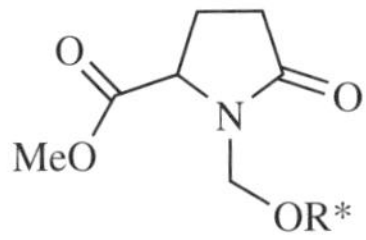

6.87

6.6.5. 5-Methyl-5-phenylpyrroline *N*-oxide (6.88)

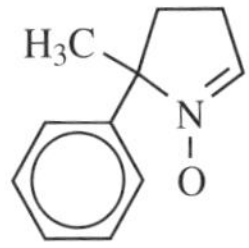

6.88

6.88 undergoes a 1,3-dipolar addition to allenes as illustrated in Figure 6.8.[831] Specific and predictable shieldings by the phenyl ring caused distinct resonances in the spectra of the diastereomers. In addition, the coupling pattern for the H5 resonance varied with the dihedral angles in a way that correlated with the absolute stereochemistry of the allene.

6.6.6. (*S*)-2[(*R*)-Fluoro(phenyl)methyl]oxirane (6.89)

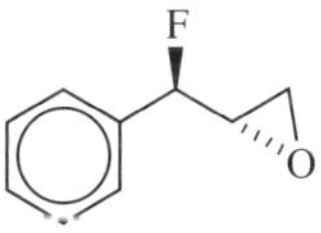

6.89

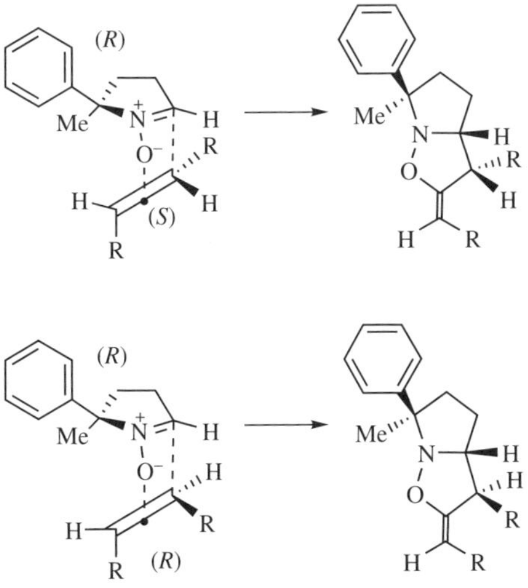

Figure 6.8. Product of the 1,3-dipolar addition of 5-methyl-5-phenylpyrroline-*N*-oxide with allenes.

6.89 reacted with α-chiral primary and secondary amines through a regioselective ring opening to produce **6.90**.[832] The ^{1}H, ^{13}C, and ^{19}F NMR spectra of the derivatives of alkyl amines, aryl alkyl amines, amino alcohols, and amino esters were used to determine optical purity. No kinetic resolution was observed and the ^{19}F spectrum was the easiest to monitor, as only two signals occurred for the pair of diastereomers.

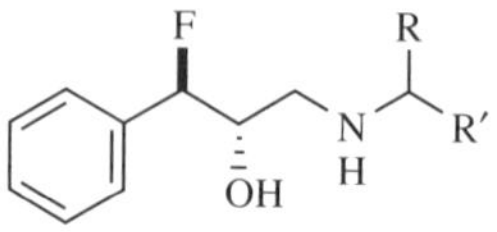

6.90

6.6.7. (S)-Triazine Selector

A series of (S)-triazine receptors (**6.91**) substituted with 1-(1-naphthyl)ethylamine,[833,834] 1-phenylethylamine,[835] and 1-(9-anthryl)ethylamine groups were evaluated as chiral NMR solvating agents.[836] In most cases, all three substituent groups of the triazine were not the same. A wide range of substrates that encompassed compounds such as N-3,5-dinitrobenzoyl (DNB) derivatives of amines, amino alcohols and amino acids, and 3,5-dinitrophenyl (DNP) derivatives of carboxylic acids were analyzed.[833,834] In these cases, the aromatic resonances of the DNB or DNP ring were especially useful to monitor for chiral recognition. Certain underivatized compounds such as methyl esters of amino acids,[833] carbinols, diols such as 2.2′-dihydroxy-1,1′-binaphthalene, propargyl alcohols (**6.92**), and some underivatized carboxylic acids such as ibuprofen were analyzed with these triazine derivatives as well. The triazine derivative with one anthryl and two naphthyl units (**6.91**) was generally the most effective.[836] The greater conformational rigidity of the larger anthryl- and naphthyl-containing groups, as well as the greater diamagnetic anisotropy shielding of the rings, likely explained the effectiveness of **6.91** over the others.

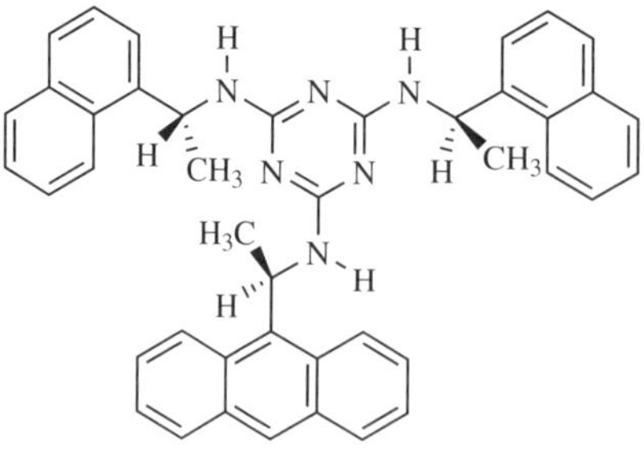

6.91

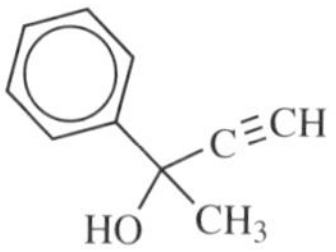

6.92

6.6.8. 2-Deuterio-2,3-dihydro-2-methyl-6-nitrobenzothiophene-1-oxide

The compound 2-deuterio-2,3-dihydro-2-methyl-6-nitrobenzothiophene-1-oxide (**5.96**) underwent a π–π interaction with aryl methyl carbinols of the form ArCH(CH$_3$)OH

(Ar = phenyl, naphthyl, and 9-anthryl).[723] Enantiomeric discrimination of the methyl, methine, and hydroxyl signals were observed in the ^{1}H NMR spectrum of the substrate. **5.96** was more effective than the analogous reagent without a nitro group.

6.6.9. 3a-Benzhydryl-3,3a,4,5-tetrahydro-2H-cyclopenta[b]furan (6.93)

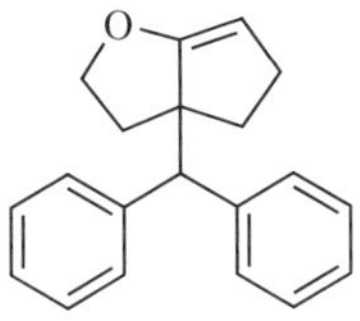

6.93

The cyclic alkene **6.93** reacted with 2-alkanols to form alkoxide derivatives of structure **6.94**.[837] The two diastereomers exhibited distinct methine hydrogen and carbon resonances. The shifts of these resonances also showed trends that were consistent with the absolute configuration of the alkanol.

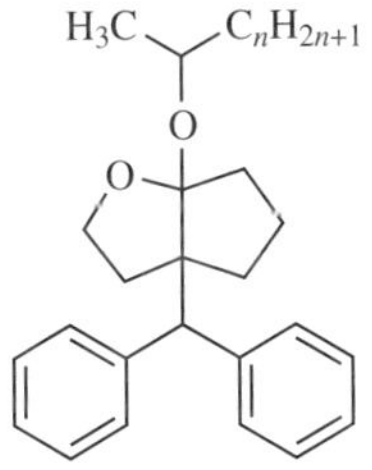

6.94

6.6.10. Camphor-10-sulfonic acid (CSA)(6.95)

6.95

Camphor-10-sulfonic acid (CSA) has been used to determine the optical purity of a variety of amines and alcohols. Conversion of aryl akyl amines into their corresponding camphor sulfonates provided diastereomeric derivatives in which the methyl and methylene (CH$_2$SO$_2$ unit) resonances of the camphor were resolved.[838] Reaction of *trans*-1-amino-2-dimethylaminocyclohexane (**6.96**)[839] or 3-aminothiolane (**6.97**)[840] with CSA formed the corresponding amide derivatives, and the ^{1}H NMR spectra were used to determine optical purity.

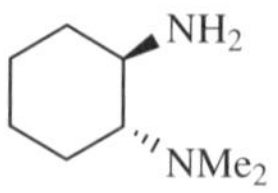

The compound 1,8-di(3′-pyridyl)naphthalene (**6.98**) undergoes restricted rotation that leads to *syn*-[(*meso*– (*R,S*)] and *anti*-[(racemic- (*R,R*) and (*S,S*)] forms.[841] Addition of (*R*)-CSA in acetone-d_6 formed diastereomeric salts that enabled the *syn*- and *anti*-forms to be distinguished. Association of two (*R*)-CSAs with **6.98** formed diastereomeric pairs [(*R-RR-R*) and (*R-SS-R*)] that had nonequivalent resonances in the NMR spectrum. The *meso*-form (*R-RS-R*) has diastereotopic faces. Warming up the sample causes a rapid interconversion between *syn*- and *anti*-forms as evidenced by a simplification of the NMR spectrum.

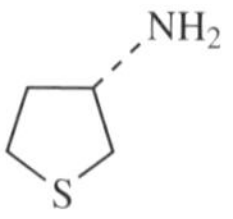

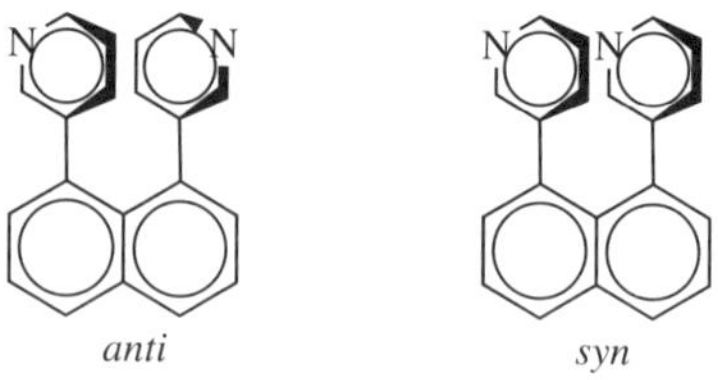

CSA esters of isoquinuclidines (**6.99**)[842] and bis-CSA derivatives of substituted 2,2′-dihydroxy-1,1′0-binaphthalene (**6.100**)[843] have been used to determine optical purities. The diastereotopic methylene signals of the isoquinuclidine derivatives were most conveniently monitored in the analysis.

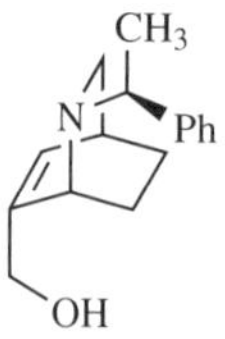

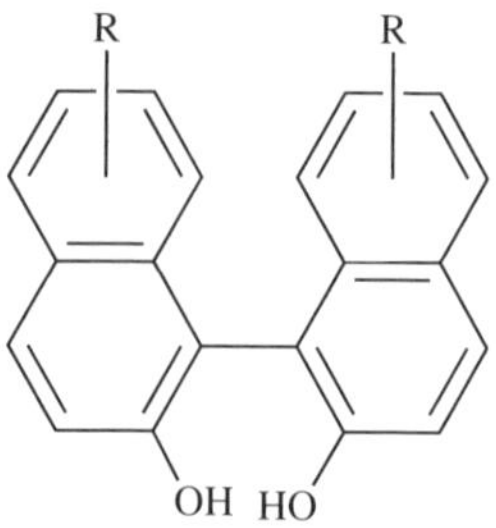

MISCELLANEOUS REAGENTS

A doubling of certain resonances in the ^{1}H NMR spectrum of a fullerene-dibenzo[18]crown-6 conjugate (**6.101**) in the presence of CSA was used to demonstrate the chirality of **6.101**.[844]

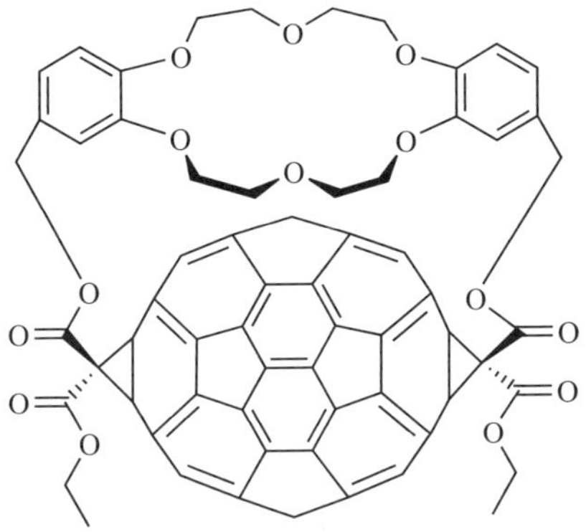

6.101

Cobalticinium complexes of structure **6.102** have planar chirality.[845] Addition of tetrabutylammonium camphor-10-sulfonate caused enantiomeric discrimination of the NH signal of the cobalt complex. The methyl resonance, when R_1 was a methyl group, also split into two signals in the ^{1}H NMR spectrum.

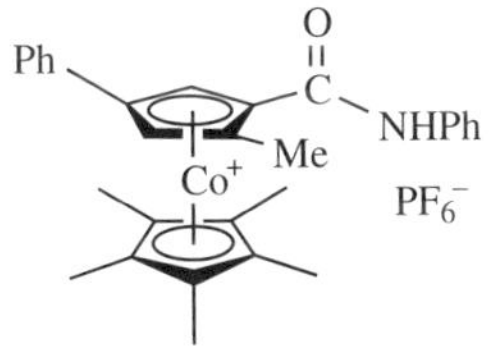

6.102

6.6.11. Menthyl Chloroformate (6.103)

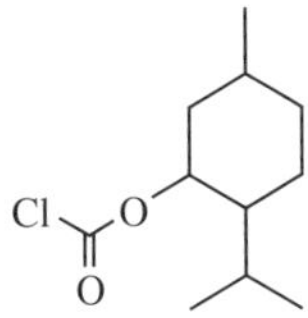

6.103

6.103 has been used to determine the optical purity of an amine and hydroxyl containing compounds. For example, **6.103** reacted with 6-tetrahydrofolate as shown in Figure 6.9 to form diastereomers that exhibited nonequivalent resonances in the

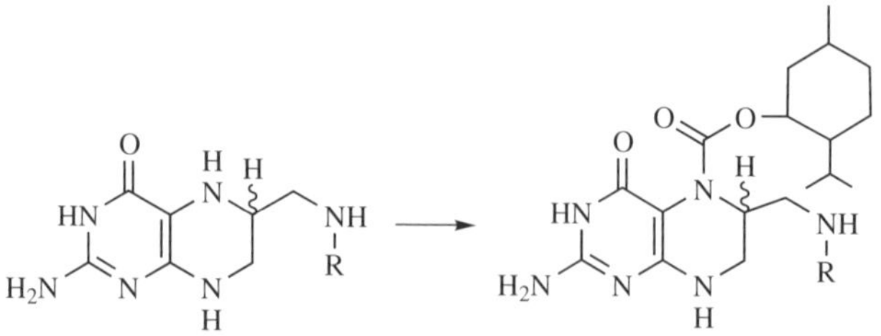

Figure 6.9. Product of the reaction of menthyl chloroformate with 6-tetrahydrofolate.

^{1}H NMR spectrum in dimethylsulfoxide-d_6.[846] The optical purity of furobenzofuran **6.104** was determined by analyzing the ^{1}H NMR spectrum of the resulting menthylcarbonate derivative prepared by the reaction of **6.103** at the hydroxyl group.[847]

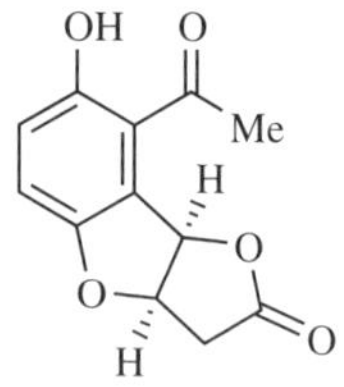

6.104

6.6.12. (N-Methylphenylsulfoximidoyl)methyl Lithium

(N-Methylphenylsulfoximidoyl)methyl lithium reacted with cyclic ketones as shown in Figure 6.10.[848] Shielding from the phenyl ring caused specific trends in the ^{1}H and ^{13}C NMR spectra that were used to assign the stereochemistry of the ketone. The shielding of the β'-hydrogens and β'-carbons were especially diagnostic. The utility of the method was demonstrated on eight different ketones.

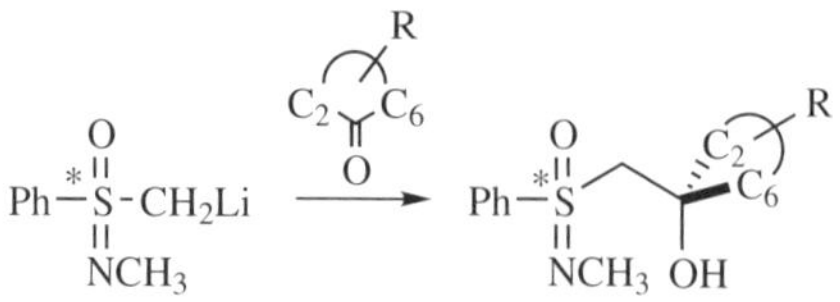

Figure 6.10. Reaction of (N-methylphenylsulfoximidoyl)methyl lithium with cyclic ketones.

6.6.13. Coumarin Dimer

The *anti* head-to-head coumarin dimer (**6.105**) is an effective chiral derivatizing agent for analyzing the optical purity of amines and alcohols.[849] Reaction with an amine or alcohol produced the corresponding diamide (**6.106**) or diester. No condensing agent was needed for amines. Alcohols were reacted as their lithium alcoholate. Three diastereomeric diamides or diesters were formed [(R,R), (R,S)/(S,R), and (S,S)] with areas 1:2:1 for a racemic mixture. Peak areas were used to determine optical purity. The utility of the method was demonstrated for several alkyl and alkyl aryl amines and alcohols.

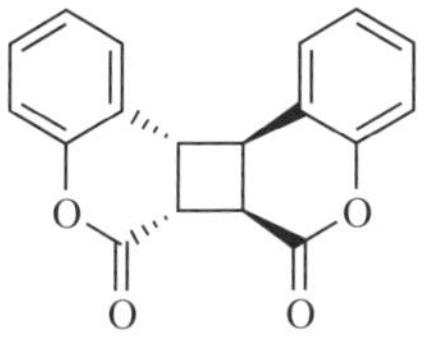

6.105 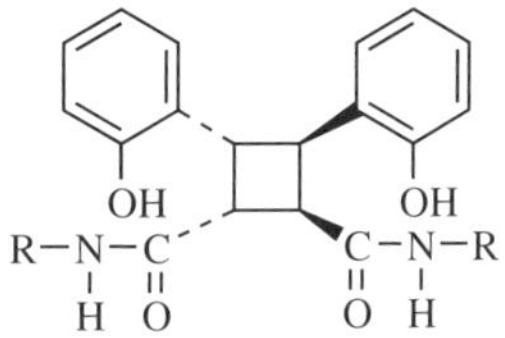   **6.106**

A similar monolactone monoamide derivative (**6.107**) was used to measure the optical purity of α- and β-chiral amines and α-amino alcohols.[850] In contrast to the product with **6.105**, only two diastereomeric amides were formed with **6.107** and the optical purity was determined by direct integration of the peaks. The enantiomeric discrimination in the spectra of amines was larger than that with MTPA. The shifts for resonances of amine substrates chiral at the α-carbon exhibited a consistent trend that correlated with absolute configuration.

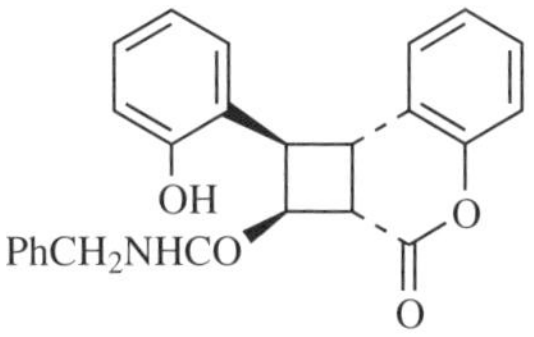

6.107

6.6.14. Di-*O*-benzoyl Tartrate/Di-*O*-*p*-toluoyl Tartrate

Tartrate species have primarily been used to selectively crystallize enantiomeric compounds, although there are a few examples of their utility in chiral NMR applications. The di-*O*-benzoyl tartrate anion (**6.108**) caused enantiomeric discrimination in the ^{1}H and ^{31}P NMR spectra of **6.109**.[851] The ^{1}H NMR spectra of several phenothiazine-5-oxides, such as **6.110**, in chloroform-*d* exhibited nonequivalent resonances in the presence of di-*O*-benzoyl tartaric acid.[852] A charge-transfer complex formed between **6.108** and **6.110**. Certain resonances in the ^{1}H NMR spectrum of cizolirtine (**6.111**), a pyrazole-containing compound, exhibited chiral discrimination in the presence of di-*O*-*p*-toluoyl tartaric acid.[853]

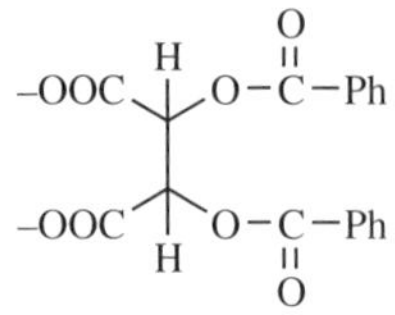

6.108, 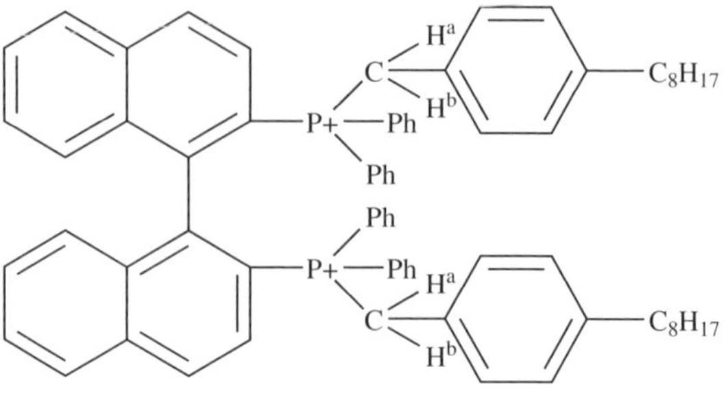 **6.109**, 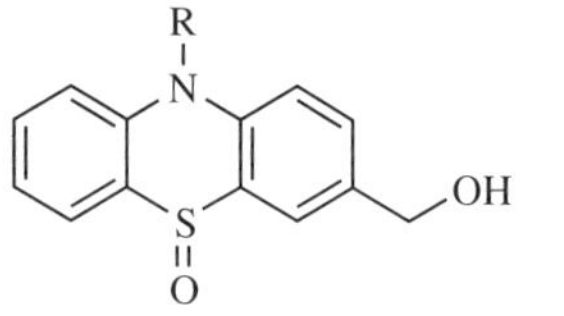 **6.110**, 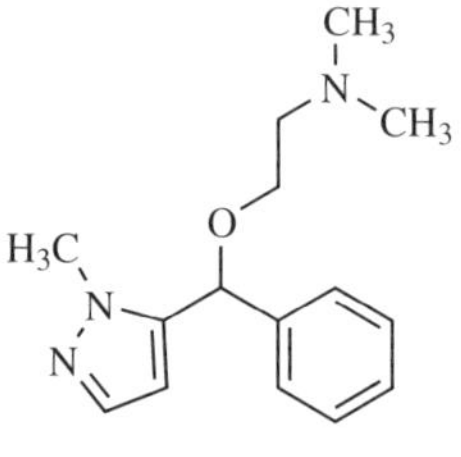 **6.111**

6.6.15. Dibenzoyl-L-tartaric Acid Anhydride (6.112)

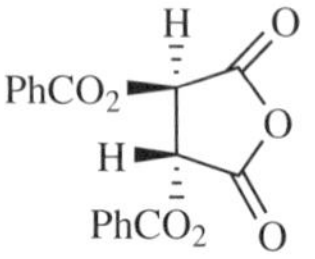

6.112

Amino acids, alkyl amines, and aryl amines react with **6.112** to form the corresponding tartarimide (**6.113**).[854] The methine resonances of the tartaric acid unit in the diastereomeric products were resolved by 0.01–0.09 ppm, and the resonance of the derivative of the (*S*)-enantiomer was always at higher frequency than that of the (*R*)-enantiomer. The methine resonance was an excellent one to monitor since its chemical shift (about 6 ppm) was usually unobstructed by other resonances in the spectrum.

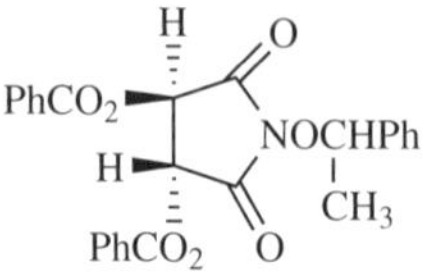

6.113

6.6.16. DNA

The binding of nucleotide fragments CAAGG (L and D configuration) and GUUCC (L and D configuration) were explored by ^{1}H NMR spectroscopy.[855] The chemical shifts were different for the diastereomeric pairs and the NMR data could be used to study the binding strengths of the associated species.

6.6.17. 5,8-Bis(aminomethyl)-1,12-dimethylbenzo[c]phenanthrene (6.114)

The helicene **6.114** binds to calf thymus DNA.[856] It was observed that the right-handed (*P*)-helicene bound more strongly than the left-handed (*M*)-helicene, and splitting of some resonances of a racemic mixture of **6.114** occurred in the resulting ^{1}H NMR spectrum.

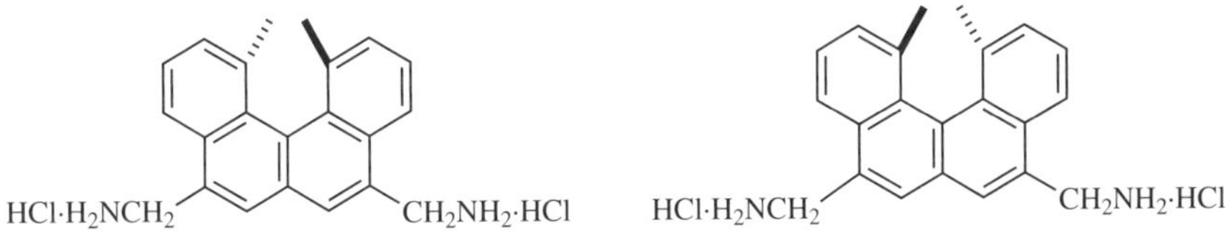

(*P*)
right-handed helix

(*M*)
left-handed helix

6.114

6.6.18. 2′-Methoxy-1,1′-binaphthalene-2-carbohydroxymoyl Chloride (MBCC) (6.115)

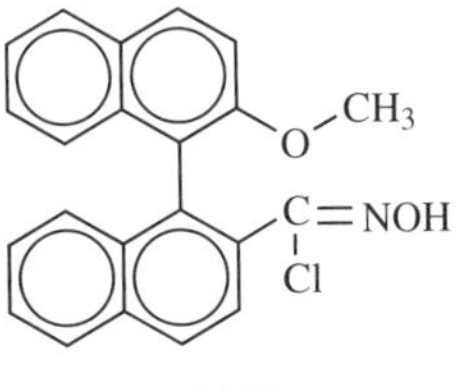

6.115

6.115 has been employed as a chiral derivatizing agent to assign the absolute configuration of substrates with an alkene group. For example, the sesquiterpene kelsoene (**6.116**) reacted with MBCC at the isopropenyl group to create a 4,5-dihydroisoxazole derivative (**6.117**—Ar = binaphthyl group).[857] The bulkiness of the binaphthyl group constrained the molecule to a preferred rotamer, such that chemical shifts and NOE data could be used to assign the absolute configuration of the alkene. The method was also applied to the analysis of prespatane (**6.118**).

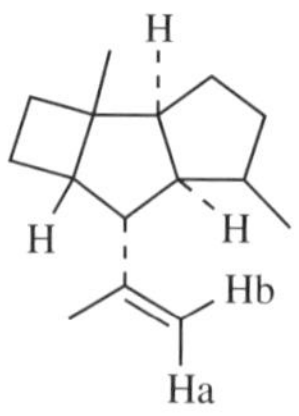

6.116 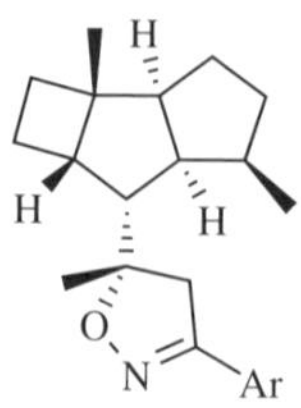   **6.117** 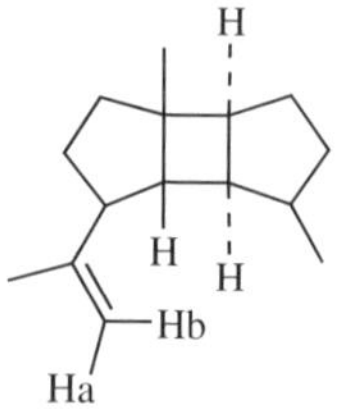   **6.118**

With α- and β-pinenes (**6.119**), MBCC added stereoselectively at the less-hindered side of the double bond to form 4,5-dihydroisoxazoles with two stable conformers.[858] NOE data of the alkene hydrogen atoms that were incorporated into the isoxazole ring of the product were dependent on the absolute configuration of the alkene.

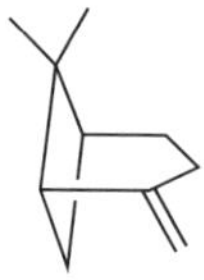

6.119

MBCC has been used to assign the absolute configuration of the macrophomate synthase inhibitor 2-carboxymethyl-1-methoxy-bicyclo[2.2.2]oct-5-ene-2-carboxylic acid (**6.120**).[859] Reaction occurred at the cyclic alkene group to create the corresponding 4,5-dihydroisoxazole. Reaction of (a*R*)- and (a*S*)-MBCC at the isopropenyl group of decipinone (**6.121**) enabled the assignment of absolute configuration at C7 and C13 by a consideration of shifts and NOE data.[860]

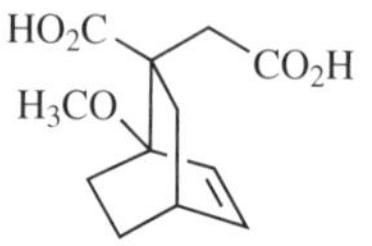

6.120 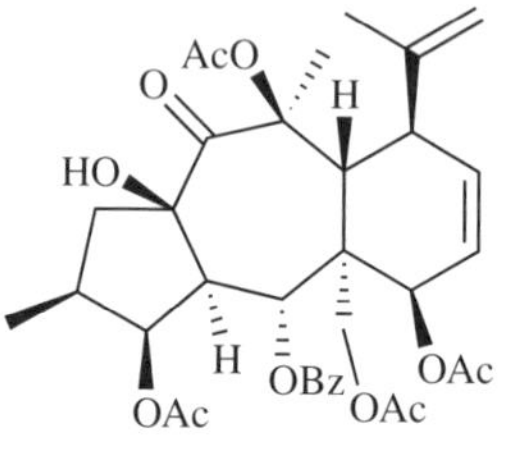   **6.121**

6.7. SELF-DISCRIMINATION OF CHIRAL COMPOUNDS

There are a number of chiral compounds in which self-discrimination has been observed in the NMR spectra of nonracemic mixtures. Cases in which self-discrimination of particular compounds led to the development of reagents generally useful for chiral discrimination are described elsewhere in the text. For self-discrimination to occur, the enantiomers must associate with each other in solution. In nonracemic mixtures, the proportions of each enantiomer that are self-associated will differ, creating the potential for different chemical shifts for each enantiomer in the ^{1}H NMR spectrum.

The more enriched the mixture the larger is the difference in chemical shifts. The areas of the peaks can potentially be used to determine optical purity. A theoretical treatment of the self-discrimination in such systems has been reported.[861] This report urges caution in using simple integration of the peaks as a means of determining optical purity, because of the potential for homoenantioselectivity in solution. Enantiomeric self-discrimination was first noted in 1969 for solutions of dihydroquinine (**6.122**) in chloroform-d.[862]

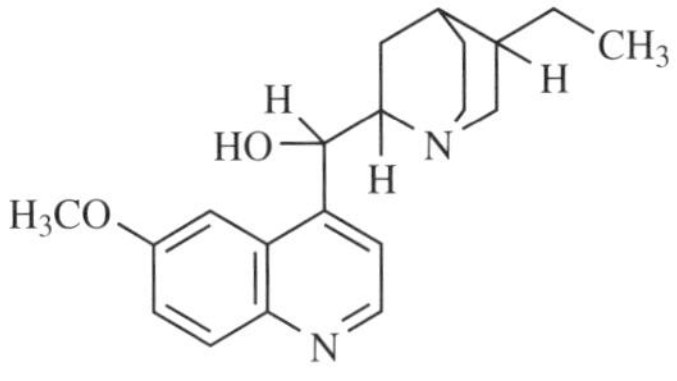

6.122

Amino acid derivatives such as **6.123** formed hydrogen bonded dimers in carbon tetrachloride, such that different shifts were observed for the LL and LD dimers for nonracemic mixtures.[863] No doubling of the resonances was observed in a polar solvent such as dimethylsulfoxide-d_6, presumably because it minimized any self-association.

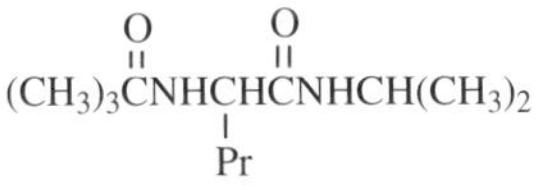

$$(CH_3)_3\overset{O}{\overset{\|}{C}}NHCHCNHCH(CH_3)_2$$
$$\overset{|}{Pr}$$

6.123

Protected dipeptides that contained either 1-aminocyclopropane carboxylic acid (**6.124**) or α-aminoisobutyric acid unit (**6.125**) at the C-terminus exhibited chiral recognition in the ^{1}H NMR spectra of nonracemic mixtures due to self-discrimination.[864] Enantiomeric discrimination was not observed if a glycine unit replaced **6.124** or **6.125**, or if the measurements were made in dimethylsulfoxide-d_6. The NH resonance exhibited the largest nonequivalence.

6.124    **6.125**

Carboxamides such as **6.126** formed hydrogen-bonded arrays in chloroform-d that exhibited enantiomeric discrimination in the ^{1}H NMR spectra of nonracemic mixtures.[865] No discrimination was observed in methanol-d_4 or dimethylsulfoxide-d_6.

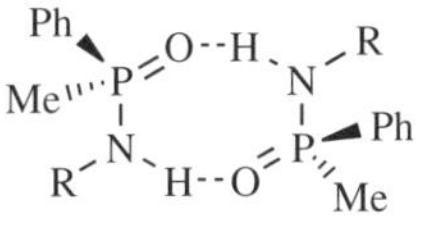

6.126

The methyl esters of 2-[(2H-1,2,4-benzothiadiazine-1,1-dioxide-3-yl)thio]propanoic acids (**6.127**) associated in nonpolar solvents such as chloroform-d.[866] Chiral discrimination of the methoxy, NH, and several aromatic proton resonances was observed in the ^{1}H NMR spectra for nonracemic mixtures.

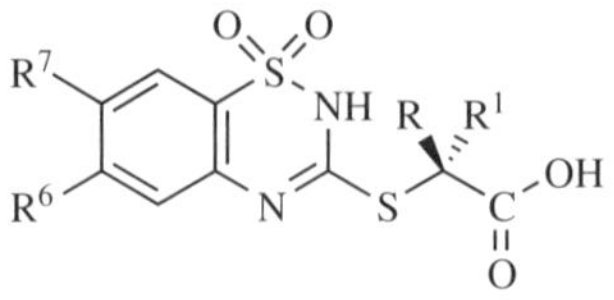

6.127

The compound 4-(4-fluorophenyl)-3-hydroxymethyl-1-methylpiperidine (**6.128**) underwent a self-aggregation in chloroform-d, and the ^{1}H and ^{13}C NMR spectra were used to determine the optical purity of nonracemic mixtures.[867]

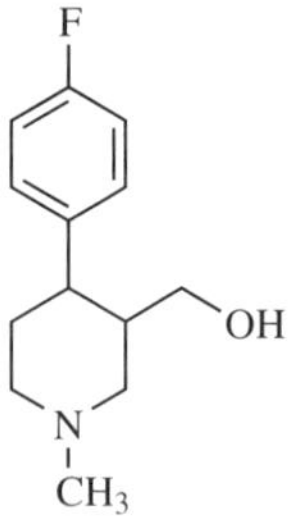

6.128

Subphthalocyanines of structure **6.129** are 14 π-electron aromatic macrocycles with three nitrogen-fused diiminoisoindole units.[868] The molecules have a curved structure and are inherently chiral. In solution, these formed dimers between opposite enantiomers that led to a self-discrimination observed in the NMR spectrum.

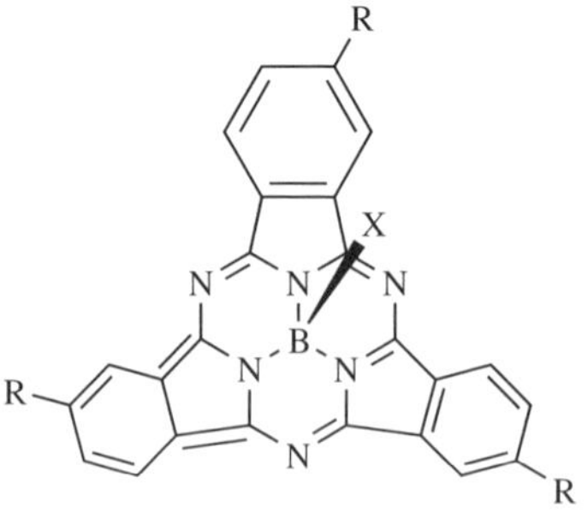

6.129

Ruthenium trischelate complexes with two bipyridine ligands and a third ligand that is eilatin (**6.130**), isoelatin (**6.131**) or tpphz (**6.132**) self-associate through π-stacking of the ligands and self-discrimination was observed in the NMR spectra of nonracemic mixtures.[869] Of particular interest was that the discrimination even occurred in a polar solvent such as acetonitrile-d_3.

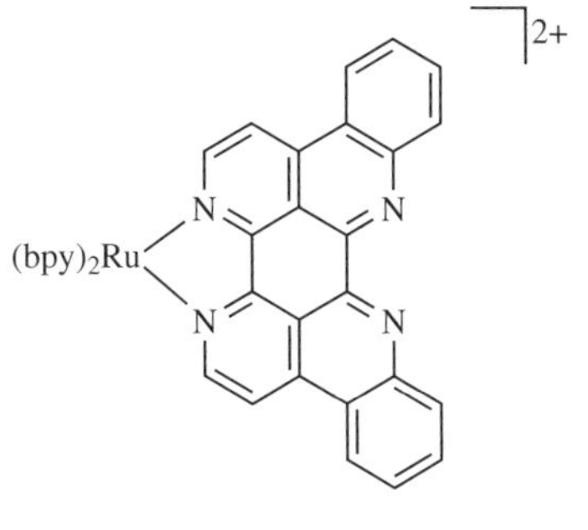

6.130

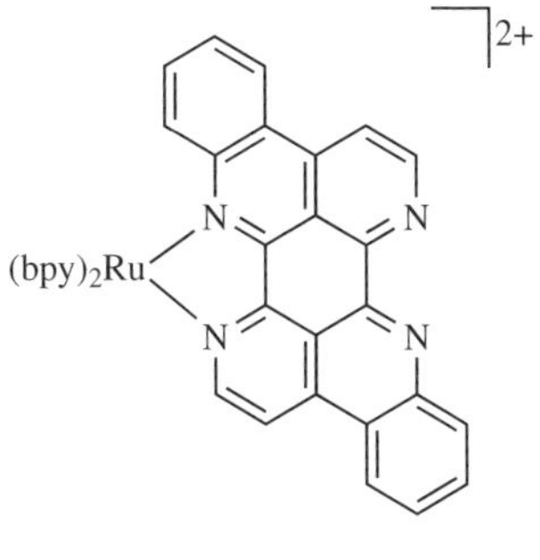

6.131

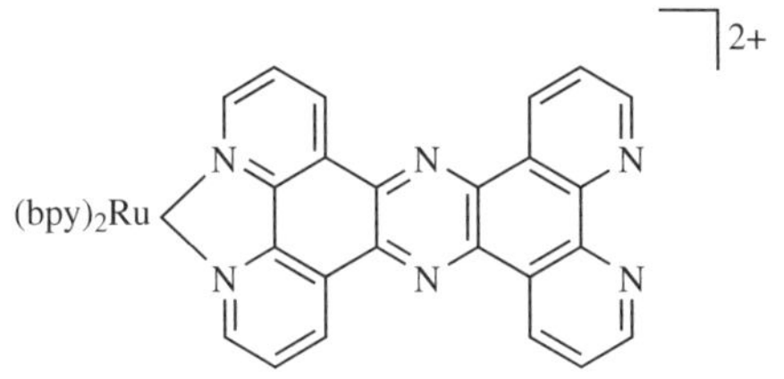

6.132

6.8. HIGH-THROUGHPUT OPTICAL PURITY MEASUREMENTS

Strategies for determining optical purities on a high-throughput basis, as would be desirable for a combinatorial library of compounds, have been described. One of these strategies involved the use of enantiotopic groups that are rendered diastereotopic when a neighboring prochiral center undergoes a chirality-generating reaction.[870] The method depends on the use of suitably designed isotopically chiral materials. An example involved the ruthenium-catalyzed reduction of a ketone to its corresponding alcohol (Fig. 6.11).[870] A ^{13}C-labeled methyl group was used to generate the chiral center.

A high-throughput NMR assay using a flow cell and autosampler to perform up to 1400 ee determinations per day has been described.[871] In a subsequent report, chemical shift imaging and a 19 capillary system enabled an increase to

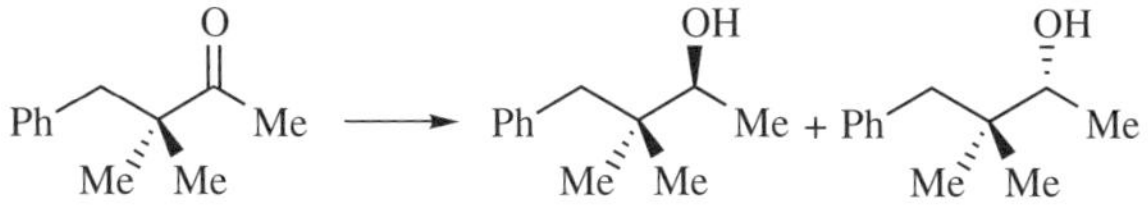

Figure 6.11. ^{13}C-Labeled alcohol products analyzed using a high-throughput NMR system.

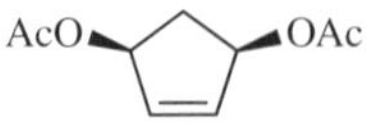

AcO⬩⟨⟩⬩OAc

6.133

5600 determinations per day.[872] The utility of these methods was demonstrated on several systems. One involved the use of MTPA as a chiral derivatizing agent. Another involved the incorporation of ^{13}C isotopic labeling into compounds to generate pseudo enantiomers. The use of the method for analyzing kinetic resolution of racemic 1-phenyl ethyl acetate from racemic 1-phenylethanol, the kinetic resolution of racemic 2-phenylpropionic acid methyl ester, and the desymmetrization of *meso*-1,4-diacetoxy cyclopent-2-ene (**6.133**) was also demonstrated.[871] The utility of the faster method was shown for the lipase-catalyzed hydrolytic kinetic resolution of *rac*-1-phenylethylacetate from *rac*-1-phenylethanol. In this example, the (S)-enantiomer was labeled with ^{13}C and the (R)-enantiomer was not. As the reaction proceeded in the racemic mixture and kinetic resolution occurred, the ratio of (R)- to (S)-enantiomer changed and the ratio of the ^1H intensities of the (R)-resonance (a singlet) changed relative to that of the (S)-resonance (a doublet because of ^{13}C coupling).[872]

7

REAGENTS INCORPORATING PHOSPHORUS, SELENIUM, BORON, AND SILICON ATOMS

Reagents incorporating phosphorus, selenium, boron, and silicon atoms have been exploited for chiral NMR recognition. An advantage of the phosphorus- and selenium-containing reagents has been the utility of the ^{31}P and ^{77}Se NMR spectra in facilitating the analysis of optical purity. The simplicity of the ^{31}P and ^{77}Se NMR spectra relative to the ^{1}H spectra and the larger shift dispersion make these especially suitable for analysis. Analyses with the boron- and silicon-containing reagents have usually involved the ^{1}H and/or ^{13}C NMR spectra.

Most of the heteroatom systems are used as chiral derivatizing agents. In the case of phosphorus, boron, and silicon, the reaction of the substrate usually occurs at the heteroatom. Reaction of the substrate with the selenium reagents usually involves another moiety than the selenium atom, and the selenium is incorporated for the purpose of serving as an NMR probe nucleus. The selenium reagents have been used as chiral derivatizing agents for carboxylic acids, alcohols, amines, and alkyl halides.

Phosphorus reagents have been extensively studied and a wide variety of them have been reported in the literature. These include P(III) and P(V) reagents. The P(V) reagents are generally more stable but the P(III) reagents usually provide larger discrimination of the enantiomeric substrate. An important set of chiral NMR discrimination reagents based on phosphorus involve propeller-like ionic systems that associate through ion pairing with oppositely charged substrates. In the aggregate, phosphorus-containing compounds used as either solvating or derivatizing agents have been applied to a wide range of substrate classes.

Discrimination of Chiral Compounds Using NMR Spectroscopy, by Thomas J. Wenzel
Copyright © 2007 John Wiley & Sons, Inc.

Boron-containing agents have been used to analyze monofunctional alcohols and diols. Silyl-containing reagents have been employed in the analysis of amines, alcohols, and carboxylic acids. A number of applications of boron and silicon reagents involve the analysis of substrates for which more common reagents (e.g., α-methoxy-α-trifluoromethylphenylacetic acid) are ineffective.

7.1. PHOSPHORUS-CONTAINING REAGENTS

A wide variety of phosphorus-containing reagents have been developed for chiral NMR discrimination. The primary use of these reagents has been to determine optical purity rather than to assign absolute configurations. A virtue of phosphorus-containing reagents is the relative simplicity of the ^{31}P NMR spectrum, which typically consists of a singlet that splits into two resonances for the two diastereomeric complexes. Integration of the resonances allows the determination of enantiomeric purity. Reagents with P(V) and P(III) functionalities have been employed. The P(V) reagents, which have either a phosphine oxide (P=O) or a phosphine sulfide (P=S) group, tend to be more stable than the P(III) reagents. However, the extent of enantiomeric discrimination in the ^{31}P NMR spectrum of P(V) reagents is usually considerably smaller than with P(III) reagents. Many of these reagents are not commercially available but are prepared from commercially available phosphorus compounds and optically pure substituents.

The P(III) reagents are especially sensitive to oxidation. Many P(III) and P(V) reagents must be kept free of water and are prepared and stored in a suitable solvent under an inert atmosphere. The reagent is then removed via syringe and added to a solution of the substrate. A majority of phosphorus reagents are used as chiral derivatizing agents with suitably reactive substrates. A few of the phosphorus reagents are employed as chiral solvating agents.

7.1.1. Phosphorus(V) Reagents

7.1.1.1. Phosphinic Amides (Phos1) In 1976 it was first reported that the nonracemic mixtures of (N-phenyl)methylphenyl phosphinic amide (**7.1**) and (N-p-nitrophenyl)methyl phosphinic amide (**7.2**) exhibited a doubling of resonances in the ^{1}H NMR spectrum in chloroform-d.[873,874] Racemic mixtures contained only one resonance. The splitting of resonances occurred because of the formation of a dimeric complex as shown in Figure 7.1 that caused a chiral self-recognition. Shielding by the N-phenyl group accounted for the differential shifts in the spectra and a doubling of the P-methyl resonances could be used to determine optical purity.

Ph\\P⫽O
Me/ \\NHPh
7.1

Ph\\P⫽O
Me/ \\NHC$_6$H$_4$NO$_2$ -p
7.2

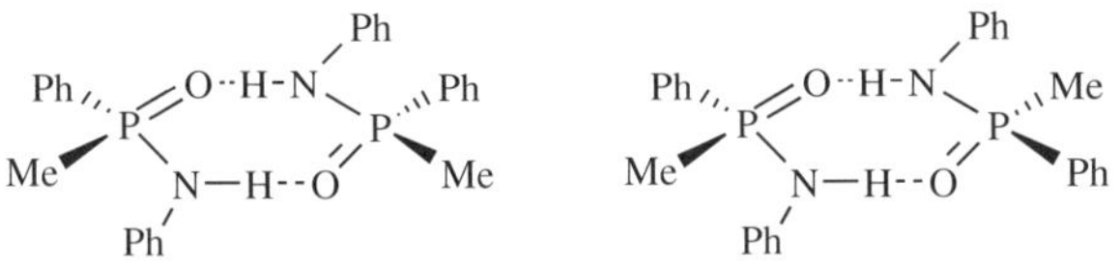

Figure 7.1 Dimer formed by the association of (N-phenyl)methylphenyl phosphinic amide.

It was possible to use optically pure **7.1** to discriminate the enantiomers of **7.2** and vice versa. The spectra of phenyl-*t*-butylphosphinic amides or (N-phenyl)-*t*-butyl phosphinic amides either in nonracemic mixtures or with **7.1** or **7.2** were enantiomerically discriminated as well.[874] The resonances of the enantiotopic methyl groups in (N-phenyl)dimethyl phosphinic amide were distinct in the presence of optically pure **7.1** because of differential shielding of the N-phenyl group.[874] The structure of these dimeric complexes is well understood, such that the differential shielding caused by the N-phenyl group can be used to assign the absolute configuration of the substrate.

The self-discrimination that occurred in nonracemic mixtures of phosphinic amides has been used to determine the optical purity of methanephosphonic acid N,N'-bis(1-phenylethyl)diamides (**7.3**).[875,876]

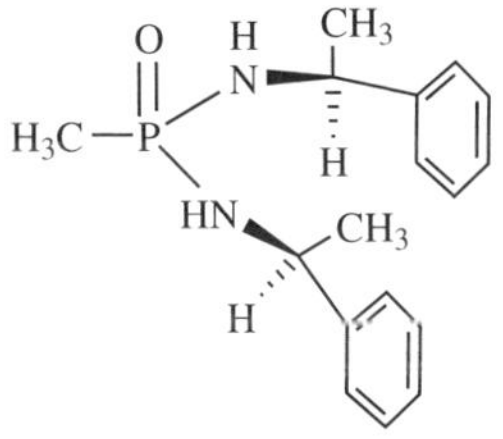

7.3

7.1.1.2. Phosphinothioic Acids (Phos2) The more versatile and commonly used reagents than the phosphinic amides are methylphenylphosphinothioic acid (**7.4**) and *t*-butylphenylphosphinothioic acid (**7.5**) (Phos2).[877] These reagents function similarly to the phosphinic amides, forming an association complex as shown in Figure 7.2. **7.5** is generally more effective than **7.4**. The utility of Phos2 for the enantiomeric discrimination of several phosphinic amides and O-methyl-O-hydrogen phenylphosphonothioate (**7.6**) was demonstrated in this initial report.[877]

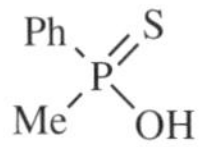

7.4 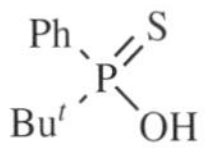   7.5 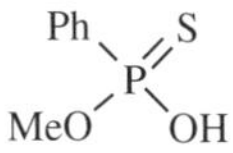   7.6

The self-association of thiophosphinic acids is a general phenomenon, such that the optical purity of 4-methoxyphenylmethylthiophosphinic acid (**7.7**) was conveniently monitored during a purification process by recording the ^{1}H NMR spectrum.[878]

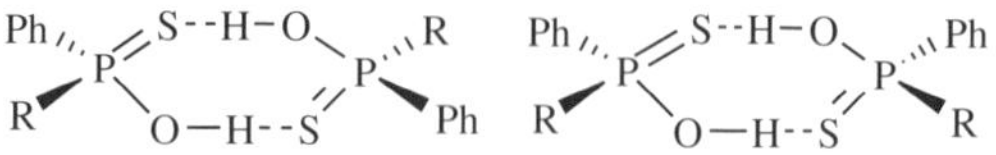

Figure 7.2 Dimer formed by the association of phosphinothioic acids.

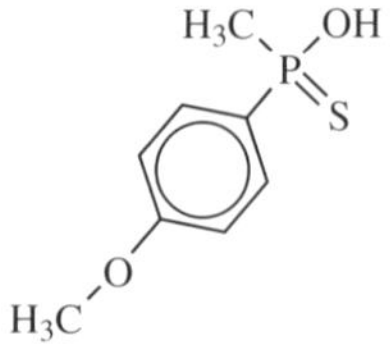

7.7

Phos2 has been used in several studies to discriminate compounds with a phosphine oxide group.[879–887] This includes phosphinate esters [RPhP(O)OR'] and thiophosphinates [RPhP(O)SR'].[879] The enantiotopic methyl and methoxy groups of Me$_2$P(O)OMe and PhP(O)(OMe)$_2$ were discriminated in the presence of Phos2. The shifts in the ^{1}H NMR spectra of the phosphinate esters caused by shielding from the phenyl group of Phos2 exhibited a general trend that correlated with absolute configuration. However, unlike phosphinic amides or acids, the phosphinate esters only have a hydrogen bond acceptor group and not a hydrogen bond donor group, so the assignment of absolute configuration requires more caution.

Phos2 has been used to determine the optical purity of allyl (**7.8**) or butenyl phosphine oxides (**7.9**).[880,886] The vinyl or *t*-butyl signals exhibited enantiomeric discrimination in the ^{1}H NMR spectrum. The ^{1}H NMR spectra of phosphine oxides (R^{1}R^{2}R^{3}P=O) with a variety of functional R groups exhibited enantiomeric discrimination in the presence of Phos2.[883,884]

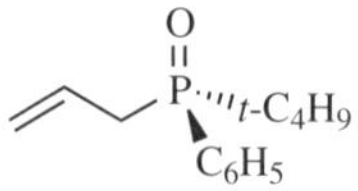

7.8 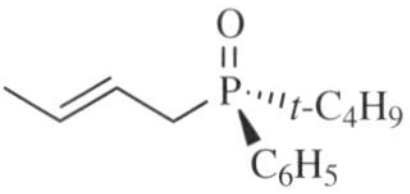   **7.9**

The ^{1}H NMR spectrum of the corresponding phosphine oxide of phosphite **7.10** exhibited chiral discrimination in the presence of Phos2.[885] The optical purities of *tert*-α-hydroxyphosphonates of structure **7.11** were determined using Phos2.[887] The ^{31}P and ^{1}H NMR spectra were suitable for the analysis.

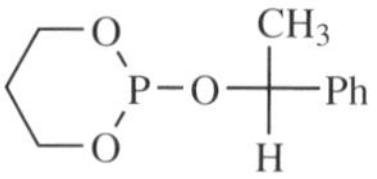

7.10 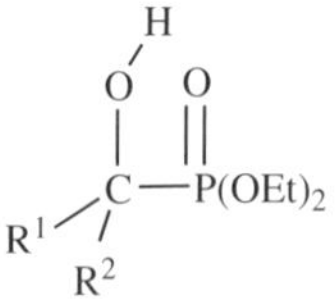   **7.11**

The optical purities of a variety of α-hydroxy, α-acetoxy, α-chloroacetoxy, α-azido, α-phthalimidooxy, and α-aminooxyphosphonates were determined in mixtures with

Phos2.[881] Nonequivalence in the ^{1}H and ^{31}P NMR spectra was observed in chloroform-d and benzene-d_6 although the distinction was larger in benzene. For α-hydroxyphosphonates, the ^{31}P resonance of the (R)-isomer was consistently at lower frequencies than that of the (S)-isomer in the presence of (S)-Phos2. The reverse order of shifts was consistently observed for the ^{1}H resonance of the α-hydrogen.[881] No consistent trends were observed for the other α-substituted classes of substrates.

The effectiveness of Phos2 in causing chiral recognition in the ^{31}P NMR spectrum of diethyl-1-hydroxyalkylphosphonates and diethyl-2-hydroxyalkylphosphonates (**7.12**) was compared with that of quinine.[719] Larger discrimination was observed in the spectra of the 1-hydroxy compounds with Phos2, whereas only quinine was effective in causing nonequivalence in the spectra of the 2-hydroxy derivatives. Discrimination in the t-butyl hydrogen resonance of the (R)-Phos2 and (S)-Phos2 adducts of secondary phosphane oxides (**7.13**) was used to measure the optical purity.[882]

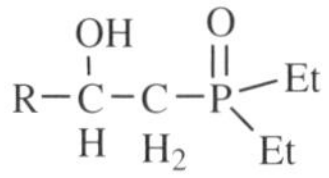

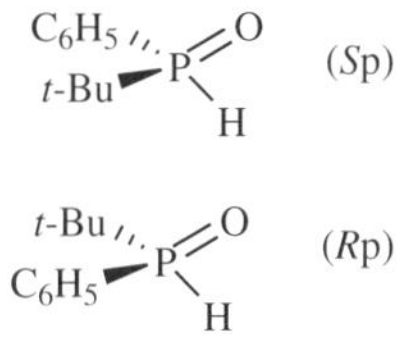

Phos2 is effective in causing enantiomeric discrimination in the ^{1}H NMR spectra of sulfoxides.[888] Nonequivalence of a few hundredths of a ppm was observed out to the γ-proton positions. The magnitude of the discrimination was larger than that observed with 2,2,2-trifluoro-1-(9-anthryl)ethanol, 2,2′-dihydroxy-1,1′-binaphthalene, and N-(3,5-dinitrobenzoyl)-α-methylbenzylamine.

Phos2 is an effective chiral solvating agent for amine oxides, including tertiary amine oxides of general structure **7.14**.[889] With amine oxides, discrimination of enantiotopic methyl and methylene resonances of the alkyl substituents has been observed. Large chiral recognition was observed for the resonances of the N-methyl groups in the substrate, and also for the methyl resonance of a pentyl substituent group. The chiral discrimination in the spectra of N-oxides with Phos2 was larger than that observed with 2,2,2-trifluoro-1-(9-anthryl)ethanol.[889] The enantiomeric discrimination in the ^{1}H NMR spectrum of **7.15** and related compounds was evaluated with Phos2, 2,2′-dihydroxy-1,1′-binaphthalene, and 2,2-dimethyl-4,5-bis(diphenylhydroxymethyl)-1,3-dioxolane (TADDOL).[890] The largest discrimination was obtained with Phos2 and the association likely occurred through the N-oxide functionality rather than the hydroxyl groups.

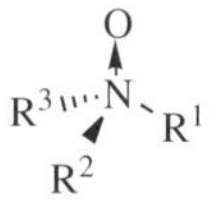

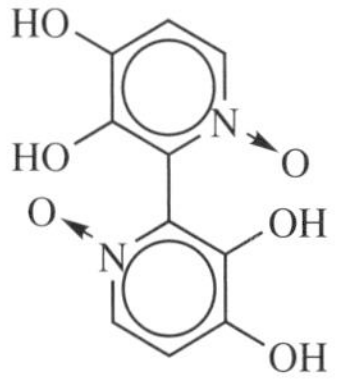

The utility of Phos2 as a chiral NMR solvating agent for alcohols, diols, thiols, mercaptoalcohols, amines, amino alcohols, and hydroxyl acids has been demonstrated.[891] Of this group of substrates, amines had the largest discrimination, which was likely because of stronger association with Phos2. Diols and amino alcohols are likely associated with Phos2 in a bidentate manner.

NMR data with Phos2 could be used to distinguish the *erythro* and *threo* configurations of a wide variety of 1,2-amino alcohols.[892] The complex of these substrates with Phos2 had an internal hydrogen bond between the hydroxyl proton and ammonium group. This led to a preferred conformation in which a larger coupling constant was observed between the methine resonances of the *threo* configuration than in the *erythro* configuration.

Phos2 has been used to determine the optical purity of 2α-hydroxypinan-3-one (**7.16**) and its corresponding oxime (**7.17**).[893] Certain resonances in the ^{1}H NMR spectra exhibited nonequivalence in chloroform-d and benzene-d_6. Enantiomeric discrimination of one of the aromatic and methoxy signals of the isoquinoline alkaloid crispine A (**7.18**) was observed in the ^{1}H NMR spectrum in the presence of Phos2.[894]

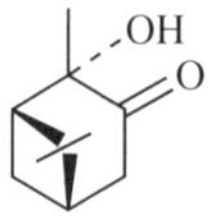

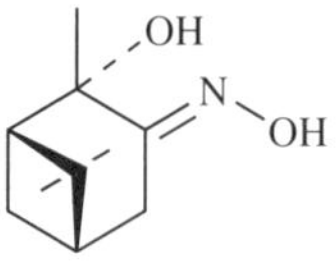

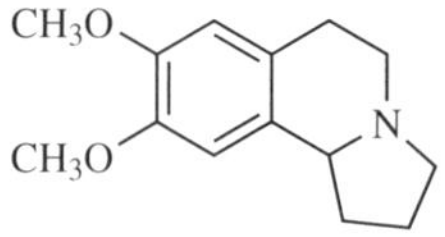

7.1.1.3. cis-2-Chloro-3,4-dimethyl-5-phenyl-1,3,2-oxazophospholidin-2-one (Phos3)

cis-2-Chloro-3,4-dimethyl-5-phenyl-1,3,2-oxazophospholidin-2-one and its corresponding sulfide (**7.19**–Phos3) contain an ephedrine moiety and have been used as chiral NMR derivatizing agents. Chiral amines and alcohols displace the chlorine atom of Phos3 to produce the corresponding amide or ester derivative.[895] Larger discrimination was generally observed in the ^{31}P NMR spectrum of the thio analog of Phos3. The effectiveness of the reagent was demonstrated for several primary amines and primary and secondary alcohols.

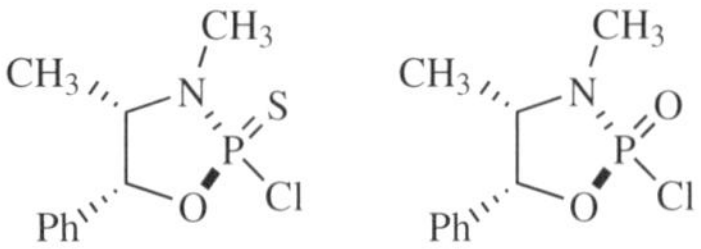

An especially interesting application of Phos3 was in the analysis of a thiophosphate monoester (**7.20**) that is chiral only by virtue of having two different oxygen isotopes. Reaction of **7.20** with Phos3 displaced the chlorine atom to give four

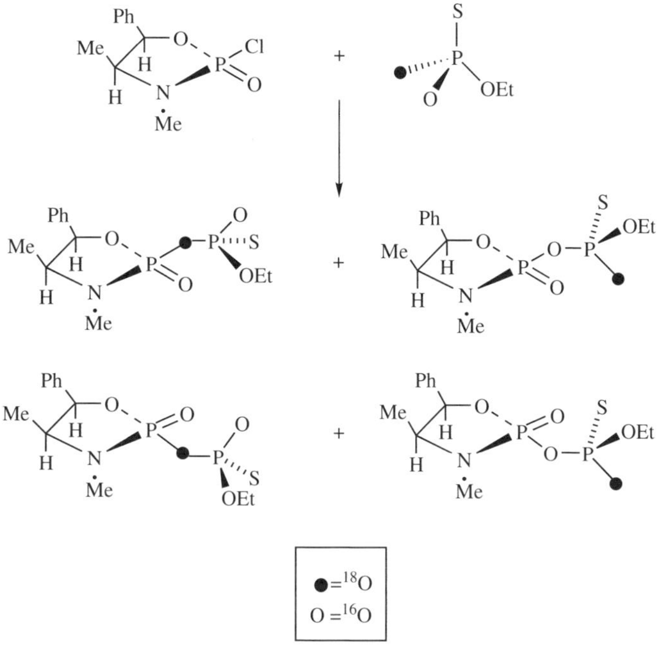

Figure 7.3 Stereochemical analysis of ethyl (S_P)-[^{16}O,^{18}O]thiophosphate by ^{31}P NMR spectroscopy of the product following reaction with cis-3-chloro-3,4(S)-dimethyl-5(S)-phenyl-1,3,2-oxazaphospholidin-2-one. [Reproduced with permission from American Chemical Society.[896]]

products as shown in Figure 7.3.[896] The different ^{31}P signals could be assigned on the basis of whether the ^{18}O or ^{16}O atom was involved in the bridge because the two oxygen isotopes have different effects on the chemical shifts of the phosphorus resonance. The ^{31}P spectra was used to perform a configurational analysis of the thiophosphate.[896] A later study found that some compounds epimerized during the derivatization reaction, especially when pyridine, N-methylimidazole, or N,N-dimethylaminopyridine were used as the nucleophilic catalyst.[897]

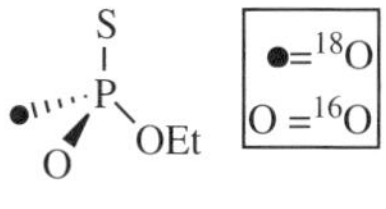

7.20

A reagent analogous to Phos3 except that a hydrazine group replaces the chlorine atom (**7.21**) has been used to analyze the optical purity of ketones.[898] The ketone reacts with **7.21** to form the corresponding hydrazone (**7.22**). The analysis was complicated because of the formation of both syn- and anti-diastereomeric hydrazones.

236 REAGENTS INCORPORATING PHOSPHORUS, SELENIUM, BORON, AND SILICON

Also, a number of ketones did not fully react as evidenced by fewer peaks than expected in the NMR spectrum.

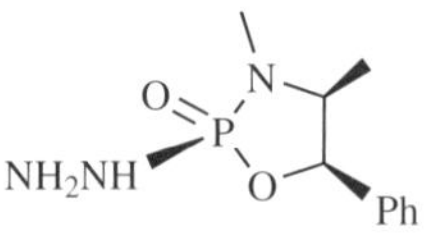

7.21

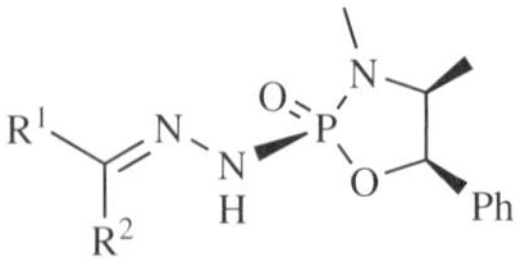

7.22

7.1.1.4. (2-Chloro-4(R),5(R)-dimethyl-2-oxo-1,3,2-dioxaphospholane (Phos4) (7.23) Phos4, which incorporates a 2,3-dimethyl-1,4-butanediol moiety, was shown to be effective for determining the optical purity of primary and secondary alcohols.[899] The reagent has been used in several other reports to analyze the enantiomeric purity of alcohols by monitoring the ^{31}P NMR spectrum of the diastereomeric products.[900–902]

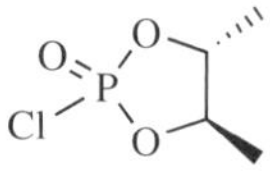

7.23

7.1.1.5. L-Menthylphenylchlorophosphine Oxide (Phos5) (7.24) The enantiomeric purity of mono-1-deuteroethanol was analyzed using Phos5 as a chiral derivatizing agent.[903] Distinct resonances were observed in the ^{31}P, ^{1}H, ^{2}H, and ^{13}C NMR spectra of the diastereomeric products, although the ^{31}P and ^{2}H spectra were judged the best to use for the analysis. The discrimination with Phos5 was greater than that observed with Phos3. Phos5 was used to study the extent to which enzymes selectively deuterated ethanol. By using enantiomerically enriched samples of known configuration, the signal that corresponded to the *pro-(R)* site of the deuterated ethanol could be assigned.

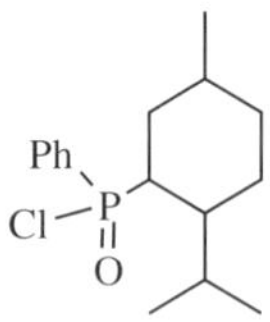

7.24

7.1.1.6. 2-Chloro-5,5-dimethyl-4-phenyl-1,3,2-dioxaphosphorinane-2-oxide (Phos6) (7.25) Reaction of a chiral primary or secondary amine with Phos6 yields diastereomeric products that exhibit nonequivalence in the ^{1}H and ^{31}P NMR spectra.[904] A later study extended the use of Phos6 to other primary and secondary alcohols and amino esters.[905] The derivatization was carried out directly in the NMR tube in benzene-d_6, with no loss of configuration and no kinetic resolution. The ^{31}P NMR spectrum was used for the analysis.

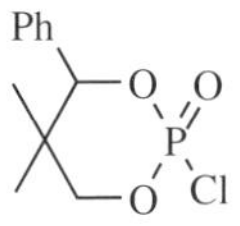

7.25

An analogous reagent in which the chlorine atom was replaced by a hydrogen atom ((S)-2H-2-oxo-5,5-dimethyl-4(R)-phenyl-1,3,2-dioxaphosphorinane–**7.26**) was effective for the analysis of the optical purity of unprotected amino acids in aqueous solution.[906] The use of **7.26** was recommended over **7.25** because the hydrogen analog is not susceptible to hydrolysis. **7.26** was effective in analyzing the enantiomeric purity of alcohols, amines, amino alcohols, thiols, and esters of amino acids as well. The hydrogen atom is displaced and the corresponding amide or ester derivative results. Nonequivalence of 0.25–1.2 ppm was observed in the ^{31}P NMR spectrum of the diastereomeric derivatives with amino acids.[906] The only side product involved the formation of a small quantity of a pyrophosphate, but the ^{31}P shift of the by-product was so different that it was easily identified and did not interfere with the analysis. The use of **7.26** was recommended over a prior phosphorus-containing water-soluble reagent that contained 2-butanol groups (**7.27**) that had been developed in the same research lab.[907] Analysis of nonracemic mixtures indicated that the configuration about the phosphorus atom in Phos6 was retained on reaction with amines but inverted on reaction with alcohols.[906] Nevertheless, the ^{31}P signals were separated for the diastereomeric derivatives and the optical purity could be determined irrespective of whether the original configuration of the phosphorus was retained or inverted.

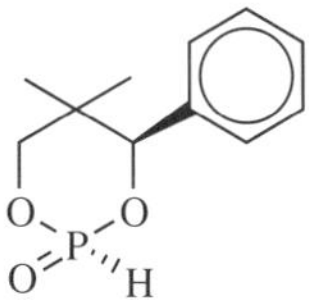

7.26 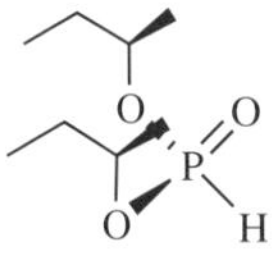 **7.27**

7.1.1.7. O,O-Di-(2-(S)-(N,N-diethyl-2-hydroxypropyl)phosphonate (Phos7) (7.28) Phos7 has been used as a chiral derivatizing agent for the analysis of unprotected amino acids in aqueous solution.[908] Phos7 was not as effective as Phos6 for the analysis of alcohols and amines, but with amino acids, the discrimination in the ^{31}P NMR spectra of the diastereomeric derivatives with Phos7 showed a pronounced dependence on pH. This pH dependency indicated that the NH$^+$ moiety of Phos7 associated with a carboxylate group of the amino acid as shown in Figure 7.4. The

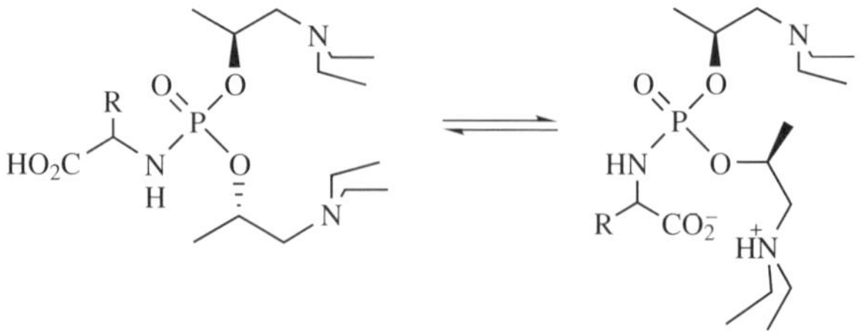

Figure 7.4 Intramolecular interaction in the product of Phos7 with amino acids.

conformational locking led to large nonequivalence in the spectra of certain of the amino acids.[908]

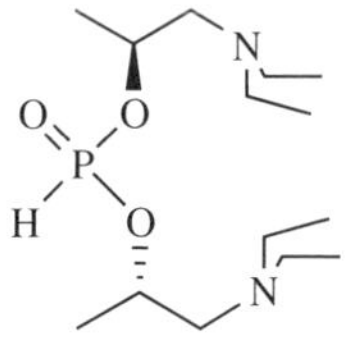

7.28

7.1.1.8. 2-Chloro-3-phenyl-1,3,2-diazaphosphabicyclo[3.3.0]octane-2-oxide (Phos8) (7.29) The diazacyclophosphamide Phos8 has interesting properties as a chiral derivatizing agent for amines and alcohols.[909] Reaction of Phos8 with amines or alcohols led to the corresponding phosphamide or ester without any loss of stereochemistry at the phosphorus. The ^{31}P and ^{1}H NMR spectra were used for the analysis. The phenyl ring caused specific shielding that could be used to assign the absolute configuration of the substrate. Figure 7.5 illustrates the differential shielding that the phenyl group exerts on L_1 and L_2 of the substrate that can be used in assigning the stereochemistry.[909]

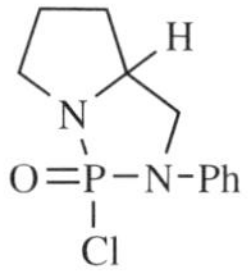

7.29

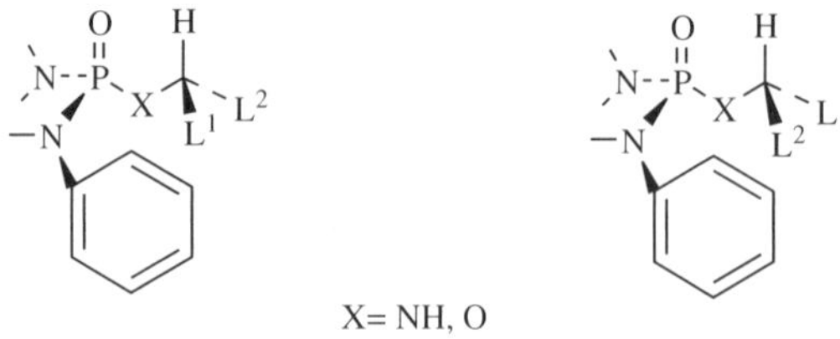

X= NH, O

Figure 7.5 Configurations of diazacyclophosphamides that can be used in assigning the absolute stereochemistry.

7.1.1.9. Methyl Phosphonic Dichloride (Phos9) (7.30)/Methyl Phosphonothioic Dichloride (Phos10) (7.31) Another strategy for determining enantiomeric purity with phosphorus reagents is to use dichloride compounds such as Phos9 and Phos10. Reaction of an alcohol or thiol with Phos9 or Phos10 leads to the formation of diesters or thioesters of the form CH_3-P(O)-(XR$_2$) (X = O or S).[910] The resulting ^{31}P NMR spectrum of an enantiomeric pair has three signals, one of which corresponds to the chiral pair [(S,S) and (R,R)] and the other two are for the *meso* derivatives [(S,R) and (R,S)]. For a racemic mixture of a substrate, the integrated intensity is 2:1:1 for the chiral, *meso*-(S,R), and *meso*-(R,S) isomers. For nonracemic mixtures, the areas of the three peaks can be related to the optical purity of the substrate using the method originally described by Horeau.[911]

7.30   7.31

In the initial report of Phos9, the chiral recognition was compared with that of the corresponding ethyl, propyl, butyl, and benzyl derivatives.[910] The methyl derivative generally provided the largest nonequivalence in the ^{31}P NMR spectrum.[910,912]

Phos10 was similarly effective for amines and amino esters.[913,914] A comparison of the methyl and phenyl phosphonothioic dichloride indicated that larger nonequivalence (values of 0.25–1.62 ppm in the ^{31}P NMR spectrum) was observed with the methyl derivative.[913] Phos10 was subsequently applied to the analysis of the methyl ester of 2-amino-3,3-dimethylbutanoic acid.[915] The substrate was dissolved in chloroform-*d* and treated with an excess of Phos10 to prepare the derivative.

Chiral thiols and alcohols can also be reacted with phosphorus trichloride, a reagent that will be described later, in a scheme similar to that with Phos9. A comparison of the effectiveness of Phos9 to PCl$_3$ indicated that PCl$_3$ was generally recommended for alcohols, whereas Phos9 was preferable for thiols.[910]

Further studies on the utility of Phos9 as a chiral derivatizing agent for the analysis of thiols have been undertaken.[912,916,918] In one study, the effects of electronnegativity, steric constraints, and temperature on the analysis were explored.[912] It was noted that Phos9 was more effective than Phos10 for thiols. Phos9 was effective for the analysis of α-thiol carboxylic acids, β-thiolcarboxylic acids,[916] aliphatic thiols, secondary benzylic thiols, and α-thiolamides.[912] Reaction occurred at the thiol group in the bifunctional substrates. Lowering the temperature of the sample caused a much larger distinction among the resonances of the diastereomers.[912]

A series of chiral secondary (R^1CH(SH)CO$_2$H) and tertiary-α-mercaptocarboxylic acids (R^1R^2CH(SH)CO$_2$H) were analyzed using Phos9.[917] The acids were esterified before conversion to the diastereomeric phosphono dithioates. The discrimination in the ^{31}P NMR spectrum of the tertiary thiols was smaller than with the secondary thiols, although baseline separation was obtained for all of the substrates studied.

7.1.1.10. O,O-Diphenyldithioic Acid (Phos11) (7.32) An alternative to the use of Phos9 or Phos10 is to employ Phos11 as the derivatizing agent.[918] Chiral alcohols displace the phenoxy groups to give products of structure $(RO)_2P(=S)SH$. The hydrogen atom is in fast exchange between the two sulfur atoms rendering them equivalent. As a result, only two signals are observed in the ^{31}P spectrum, one for the (R,R)/(S,S) forms and the other for the (R,S)/(S,R) forms. Separation of the peaks was not as large as with Phos9 and Phos10, but was still sufficient to determine enantiomeric purity. Addition of an optically pure chiral tertiary amine, such as (−)-cinchonidine (**7.33**), to the alkoxy derivative formed diastereomeric salts in which the ^{31}P signal of the (R,R)- and (S,S)-species was split into two signals, while the ^{31}P signals of the (R,S)- and (S,R)-complexes were not affected.[918]

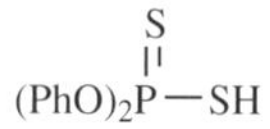

$(PhO)_2P-SH$
$\overset{S}{\|}$

7.32

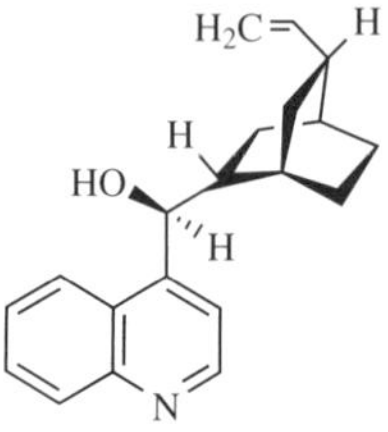

7.33

An analogous application of the Horeau method has been demonstrated with a series of chiral bis(alkylalkoxythiophosphoryloxy)phosphonates (**7.34**).[919] Integration of the three signals for the (S,S)/(R,R)-, (S,R)-, and (R,S)-isomers enabled the determination of optical purity.

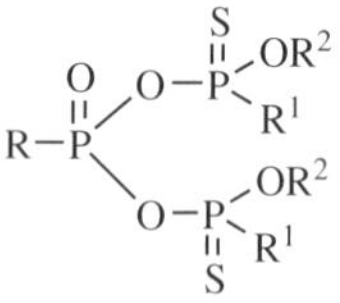

7.34

7.1.1.11. 1,1′-Binaphthyl-2,2′-diylphosphoric Acid (Phos12) (7.35) The addition of Phos12 to secondary and tertiary amines formed chloroform-soluble salts that exhibited chiral discrimination in the ^1H and ^{13}C NMR spectra.[920] The recognition was larger than that observed with α-methoxy-α-trifluoromethylphenylacetic acid (MTPA). No discrimination was observed in acetonitrile-d_3. The reagent was also an effective chiral solvating agent for alcohols, but Phos12 had to be solubilized by addition of an equivalent of pyridine.

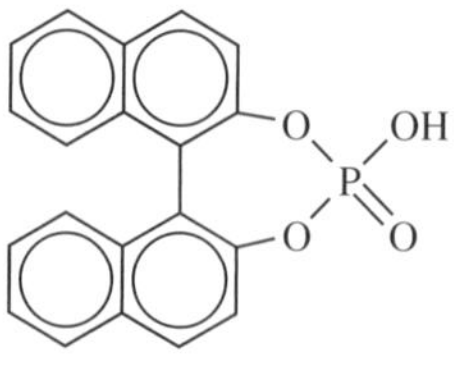

7.35

The methoxy resonance of 5-methoxy-2-methyl-2-dipropylaminotetralin (**7.36**) was enantiomerically discriminated in the presence of Phos12 in acetone-d_6.[921] In addition to integrating the two methoxy signals to determine the optical purity, another procedure for samples highly enriched in one of the enantiomers was described. This involved comparing the area of the ^{13}C satellite signal of the methoxy signal of the major enantiomer to the area of the methoxy signal of the minor enantiomer. Knowing that ^{13}C is 1.108% abundant, quantification of the minor enantiomer was more accurate as two peaks of more similar area were being compared.

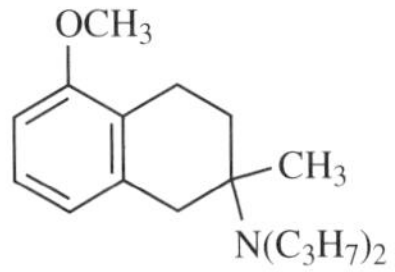

7.36

The optical purities of nicotine (**7.37**) and nicotine-like compounds were determined by analyzing the ^{1}H NMR spectrum in the presence of Phos12.[922] Phos12 was used to purify 7-chloro-4-{4-[ethyl(2-hydroxyethyl)amino]-1-methyl-butylamino}-quinoline (hydroxychloroquine—**7.38**) by fractional crystallization. The purification achieved through the process was conveniently monitored by examining the ^{1}H NMR spectrum of the mixture prior to the crystallization.[923]

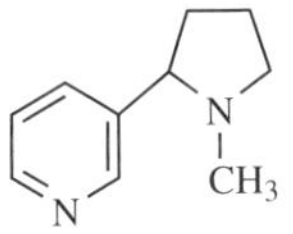

7.37 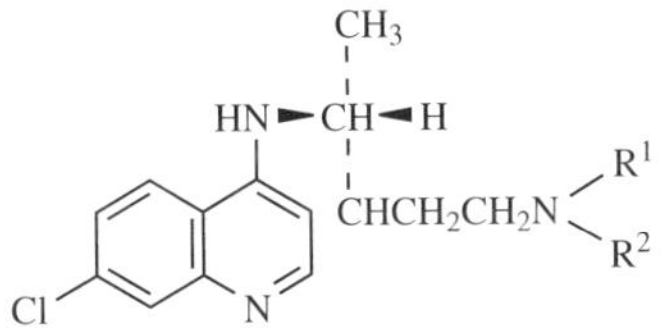   **7.38**

Phos12 has been used to confirm the chirality of certain complex molecules. One example involved a dicopper(I) trefoil knot with *m*-phenylene bridges (**7.39**).[924] Addition of Phos12 to **7.39** led to a doubling of some of the resonances in the ^{1}H NMR spectrum. Another example involved a fullerene-dibenzo[18]-crown-6 conjugate (**6.101**) that had planar chirality.[844] Addition of Phos12 in chloroform-*d* caused a doubling of several resonances in the ^{1}H NMR spectrum.

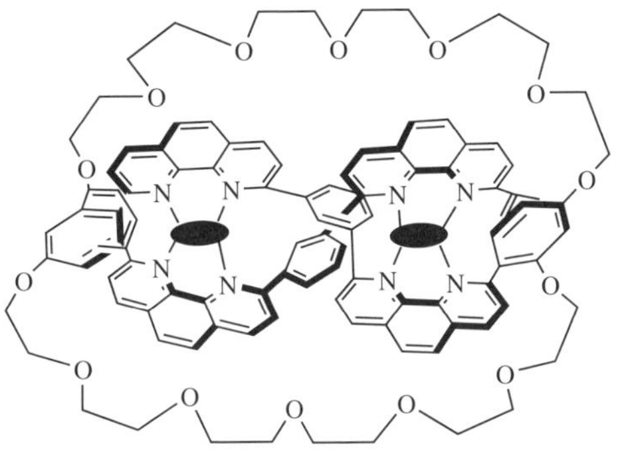

7.39

7.1.2. Phosphorus(III) Reagents

7.1.2.1. Diazaphospholidines (Phos13–Phos20) One general family of P(III)-containing chiral derivatizing agents involves compounds with an optically pure diamine group and leaving group such as a chlorine atom (**7.40**) or dimethylamine moiety (**7.41**). Figure 7.6 shows many of the diamine ligands that have been used in

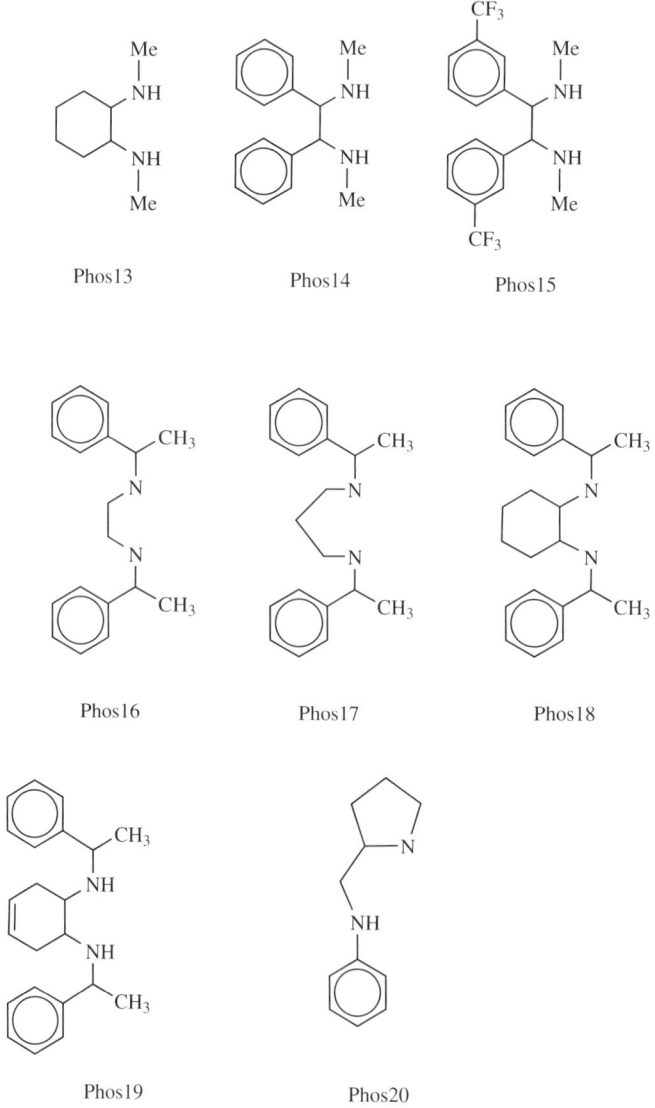

Figure 7.6 Diamine ligands used with diazaphospholidine reagents.

preparing reagents of this type. The reagents were prepared by reacting the diamine with phorphorus trichloride.

7.40 **7.41**

In the first report with compounds of this general type, significantly larger recognition was observed in the ^{31}P NMR spectrum of the P(III) derivatives of primary, secondary, or tertiary alcohols compared to the P(V) analogs (P = O and P = S) of Phos13 and Phos14.[925] Nonequivalence of up to 6 ppm was observed in the ^{31}P NMR spectrum with Phos13. The derivatization reaction was run directly in the NMR tube at room temperature in benzene-d_6. No kinetic resolution was observed.[925]

In a subsequent study, the utility of Phos13 as a chiral NMR derivatization reagent was extended to other primary, secondary, and tertiary alcohols and thiols.[926] Discrimination was observed in the ^{31}P, ^{1}H, and ^{13}C NMR spectra, although the ^{31}P spectrum, which has only two peaks for the two diastereomeric complexes, was often the most convenient for determining optical purity. Phos13 is stable for months under an inert atmosphere and was stored as a 0.2 M solution in toluene. An appropriate aliquot of the reagent was then added to the sample in an NMR tube with benzene-d_6. Phos13 was effective with hindered secondary alcohols such as menthol, neomenthol, and borneol as well as hindered *tert*-alcohols. Compounds in which the hydroxyl group was α- or β- to the chiral center could be analyzed. The reagent was mild toward sensitive functionalities such as epoxides, halides, amines, esters, and ketones. No epimerization of enolizable positions or β-eliminations was observed, although propargyl alcohols or 1,2- and 1,3-diols underwent a cyclization reaction that complicated the analysis.[926]

Addition of sulfur to the P(III) product converted it to the corresponding phosphorus sulfide derivative.[926] The ^{31}P NMR spectrum of the P(V) derivative still showed two signals, although the enantiomeric discrimination was less than that of the P(III) derivative. More important was that the P(V) derivative was enough stable to permit chromatographic separation of the two diastereomers.

Unlike Phos13, Phos14 was found to be suitable for use with propargyl alcohols such as **7.42**, polyalcohols such as 1,2- and 1,3-diols, and amino alcohols.[927] Phos15 offers the possibility of using the ^{19}F NMR spectrum in addition to the ^{31}P, ^{1}H, and ^{13}C spectra to monitor optical purity. Phos15 was effective for determining the optical purity of primary, secondary, and tertiary alcohols; primary and secondary amines; phenols; and thiols. A procedure for adding PCl$_3$, the optically pure diamine, substrate, chloroform-*d*, and a tertiary amine directly in an NMR tube to carry out the derivatization was described. The ^{31}P spectrum could be run within 5 min with no apparent kinetic resolution of the substrates.[927]

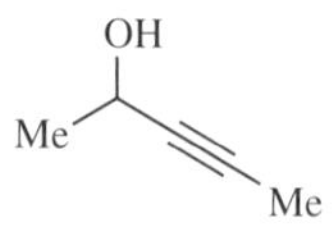

7.42

Alkoxycarbonyl β-ketoesters (**7.43**) reacted with Phos15 through the hydroxyl group.[928] The resulting ^{31}P NMR spectra were used to determine the optical purities of the alkoxy β-ketoesters.

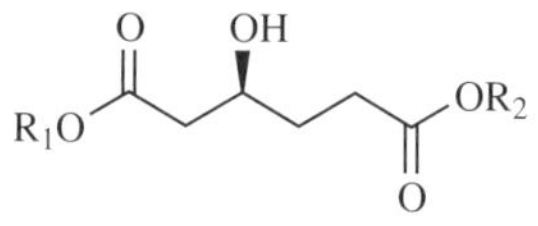

7.43

An analysis of the optical purity of several hydroxybiaryls such as **7.44** and **7.45** was accomplished with Phos14.[929] A reaction mixture of the diamine, PCl$_3$, the hydroxyl biaryl, and sulfur produced the corresponding bis-diaza phosphorus sulfide (**7.46**). The ^{31}P NMR or ^{1}H NMR spectra in a mixture of benzene-d_6/methylene chloride-d_2 exhibited distinct resonances for the diastereomers that could be used to determine the optical purity of the hydroxyl biaryl.

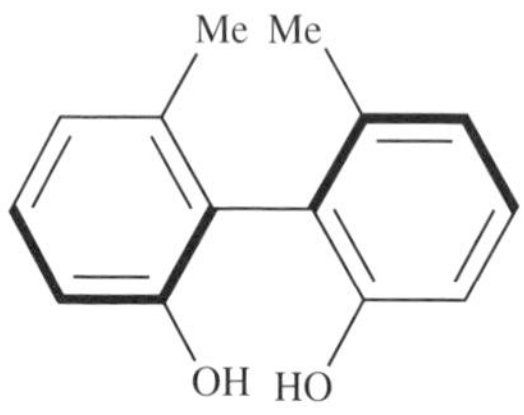

7.44 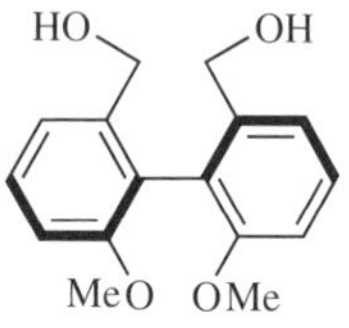   **7.45** 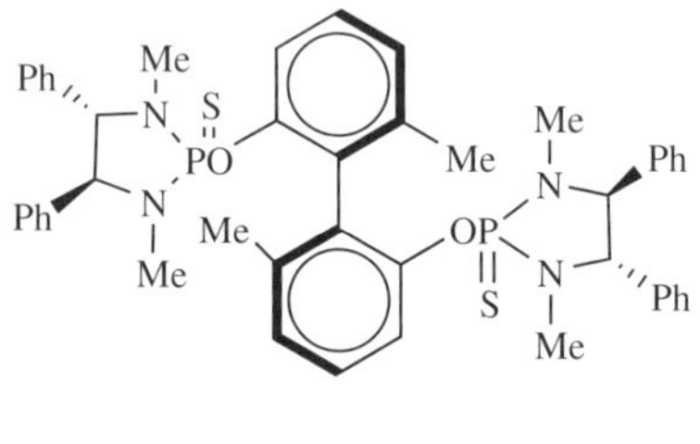   **7.46**

The utility of Phos13 and Phos14 as chiral reagents for the analysis of carboxylic acids has been investigated.[930] Nine carboxylic acids were examined. Phos13 usually provided a greater degree of chiral recognition than Phos14. Substrates with the stereogenic center α to the carboxylic acid were successfully discriminated, whereas two substrates with the chiral center β to the carboxylic acid were not. Analysis of a substrate with a carboxylic acid and hydroxyl group was complicated since some reaction occurred at both sites. The products with ibuprofen and naproxen underwent a rapid decomposition. Addition of sulfur to the mixture produced the corresponding P(V) sulfide derivative. The discrimination in the ^{31}P NMR spectrum was not as large as with the P(III) derivative, but it was still sufficient to permit the determination of optical purity.[930]

Several diamine-phosphorus reagents contain a bridging linkage between two N-1-phenylethylamine (PEA) units (Phos16–19). The reagents with an ethylene (Phos16) or propylene (Phos17) link between the N-PEA units were effective for the analysis of alcohols, amines, thiols, α-thiol acids, and α-thiol esters.[931] The reagents were stable toward racemization and no kinetic resolution was observed. An advantage of Phos16 over Phos13 is its improved stability toward hydrolysis or oxidation. Phos16 is reportedly stable toward moisture and was stored in solution or in the solid state for 6 months without any noticeable decomposition.

In a comparative study, the nonequivalence in the ^{31}P NMR spectra of alcohol derivatives of Phos18 was larger for most substrates than that observed for Phos13.[932] Another study of the effectiveness of several reagents of this general category of diazaphospholidines for the enantiomeric discrimination of secondary alcohols was undertaken.[933] Derivatives with the phenyl substituents (Phos14) or cyclohexene group (Phos19) generally provided the largest distinction in the ^{31}P NMR spectrum.

Phos16 was an effective reagent for the enantiomeric analysis of α-hydroxyphosphonate esters (**7.47**) by ^{31}P NMR spectroscopy.[934,935] The chloride form of Phos16 was used because of its enhanced reactivity. The derivatization reaction occurred in only few seconds at room temperature. The ^{31}P signals for the P(III) and P(V) atoms were split in each of the diastereomeric products. The P(III) signal generally had a larger extent of nonequivalence (1.7–5.8 ppm) than the P(V) signal (less than 0.2 ppm).[935]

7.47

The products of Phos16 with α-hydroxyphosphonates had P–P and P–H coupling constants that correlated with the conformation of the substrate.[935] Furthermore, the frequency of the ^{31}P signals of the P(III) atom showed a consistent trend that correlated with the absolute configuration of the α-hydroxyphosphonate. Phos16 was then applied in a study of the effectiveness of asymmetric catalysts.

The enantiomeric discrimination in the ^{31}P NMR spectra of phenyl carbinols with Phos13 showed a consistent trend that correlated with absolute configuration.[936] Using (R,R)-cyclohexylamine as the diamine, the signal for the diastereomer of the (R)-enantiomer of the alcohol was at a lower frequency than that of the (S)-enantiomer for 16 different substrates. In cases where only one enantiomer of the substrate was available, its absolute configuration was assigned by analysis of the derivatives with the (R,R)- and (S,S)-cyclohexylamine reagents.[936]

Phos20 was shown to be a useful reagent for determining the optical purity of aliphatic and cyclic halohydrins (**7.48**).[937] The reagent was stable for few weeks under an inert atmosphere, and nonequivalence of 0.2–12.9 ppm was observed in the ^{31}P resonances of the diastereomeric products.

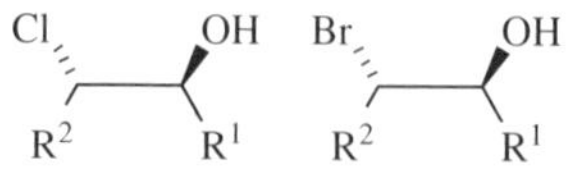

7.48

A caution when using Phos16 and other of phosphorus reagents is that the phosphorus atoms in the diastereomeric products can have very different spin-lattice relaxation times.[935] It is essential to incorporate a sufficiently long delay time between pulses to insure that complete relaxation occurs so that the peaks can be reliably integrated.

7.1.2.2. 1,3,2-Dioxaphospholanes (Phos21–Phos27) P(III)-containing reagents of structure **7.49** that incorporate bidentate oxygen-containing groups have been explored for their utility as chiral NMR derivatizing reagents. Figure 7.7 shows the structure of many dihydroxy compounds used to prepare derivatizing agents of this type. Phos21 incorporates a diethyl or diisopropyl tartrate unit.[938] Reaction of Phos21 with primary, secondary, or tertiary alcohols provided diastereomeric complexes that exhibited distinct resonances that were separated by 0.2–1.5 ppm in the ^{31}P NMR spectrum. No kinetic resolution was observed with Phos21.

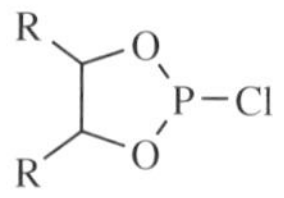

7.49

The diethyl ester form of Phos21 has been used to determine the enantiomeric purity of carboxylic acids.[939] The effectiveness of Phos21 for analyzing carboxylic acids was compared to the P(III) and P(V) analogs of Phos13, and Phos21 generally produced the best results. No kinetic resolution was observed with the derivatization of carboxylic acid substrates.

Primary and secondary alcohols have been analyzed with a P(III) reagent incorporating 2,2'-dihydroxy-1,1'-binaphthalene (Phos22)[940,941] or 1,4:3,6-dianhydro-D-mannitol unit (Phos23).[940] The distinction of the ^{31}P resonances was larger with Phos22 than with Phos23, and also larger for the P(III) than for P(V) (P = O) analog.[940]

Phos22 has been used to analyze the optical purity of primary amines.[942] In a study on the use of Phos22 for alcohol substrates, no kinetic resolution occurred for aromatic alcohols and baseline discrimination was observed in the ^{31}P NMR spectrum.[943] Aliphatic alcohols that were examined did undergo kinetic resolution on reaction with Phos22 and the ^{31}P resonances were not baseline resolved.

Phos24, which has a butane-2,3-diol unit, and Phos25, which has a 1,2-diphenylethane-1,2-diol unit, have been evaluated as chiral NMR derivatizing agents with several primary and secondary alcohols.[944] Similar to the findings in other studies, the nonequivalence in the spectra with the P(III) reagent was larger than with the P(V)

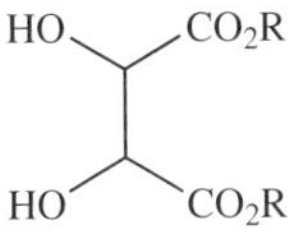

R = Et or *i*Pr

Phos21

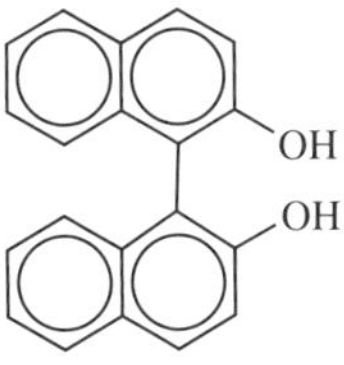

Phos22

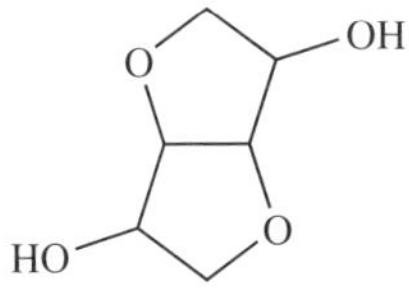

Phos23

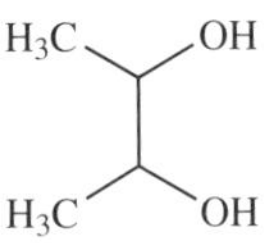

Phos24

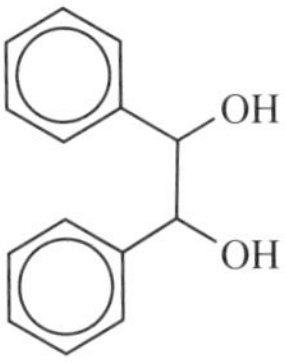

Phos25

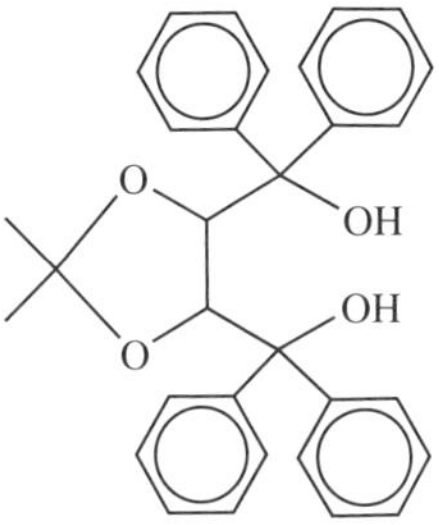

Phos26

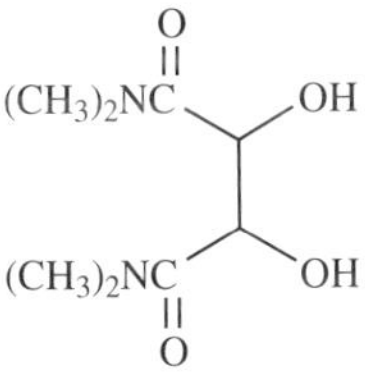

Phos27

Figure 7.7 Dihydroxy ligands used with 1,3,2-dioxaphospholane reagents.

analog. Also, the distinction was larger by a factor of about 2 for Phos24 than for Phos25 and compared favorably to Phos21.

P(III) and P(V) reagents with a TADDOL unit (Phos26) have been evaluated for their ability to cause discrimination in the ^{1}H and ^{31}P NMR spectra of diastereomeric derivatives with alcohols or carboxylic acids.[945] Unlike most other phosphorus-containing reagents, some substrates exhibited larger enantiomeric discrimination with the P(V) analog of Phos26. Success with carboxylic acids was observed for substrates only with a stereogenic center α to the carboxylic acid. The optical purity of primary or secondary alcohols with the stereogenic center α or β to the hydroxy group could be analyzed with the P(III) form of Phos26. The ^{31}P shift differences in the spectrum of diastereomeric menthol complexes with Phos 12, Phos24, Phos 25, and Phos26 were in the order Phos24 > Phos25 > Phos26 > Phos12.

Phos27 incorporates an *N,N*-dimethyl derivative of tartaric acid. When tested on a wide range of alcohols, Phos27 was judged better than Phos21 because there were not as many by-products in the derivatization reaction.[946] Phos21, Phos22, and Phos27 have been applied to the analysis of chiral epoxides (**7.50**).[947,948] With epoxides, the ring opened up resulting in an -OR functionality bonded to the phosphorus atom. The analysis was complicated with unsymmetrical epoxides because two regioisomers can form in the derivatization.[947] The regioisomer preference depended on common regularities of electrophillic additions, and also varied with the particular phosphorus reagent employed.[947] None of the three reagents was preferable for all of the epoxide substrates studied.

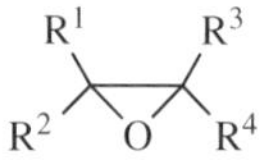

7.50

7.1.2.3. Dimenthyl Chlorophosphite (Phos28) (7.51)

Phos28 has been applied to the analysis of amino acids and peptides.[948] The amine functionality of the substrate reacted with Phos28. The reagent was stable for weeks in an inert atmosphere, and because of the two menthyl groups, retention or inversion of configuration about the phosphorus yields the same diastereomers. Nonequivalence of 0.1–1.48 ppm was observed in the ^{31}P NMR spectrum. The diastereomeric products of **7.52**, which has a chiral center 11 atoms away from the site of derivatization, had distinct ^{31}P resonances.[949]

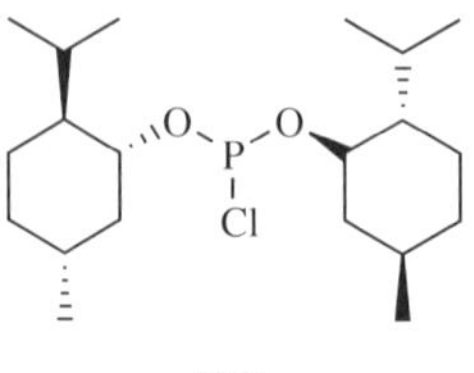

7.51 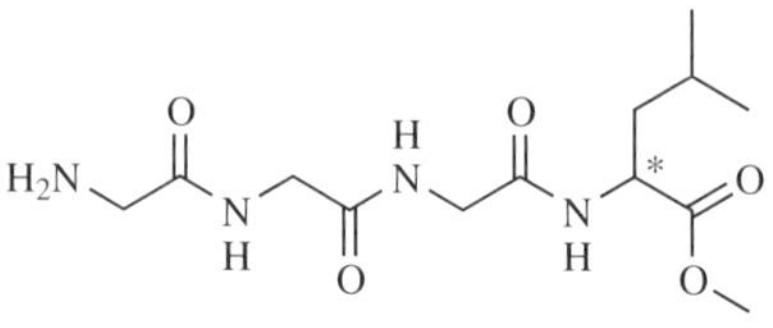   **7.52**

7.1.2.4. Dichloromenthylphosphine (Phos29) (7.53)

Phos29 has been evaluated as a chiral discriminating agent for 1,2-diols, 1,3-diols, and secondary diamines.[950,951] Substrates reacted in a matter of minutes in a bidentate manner with the phosphorus atom. Discrimination in the ^{31}P NMR spectrum was observed. No kinetic resolution was observed.[950]

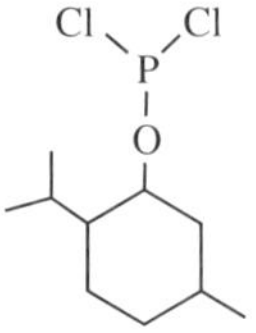

7.53

7.1.2.5. Phosphorus Trichloride (Phos30) A procedure for analyzing the enantiomeric purity of chiral alcohols using PCl₃ has been described.[952] The reaction conditions used in the analysis yield a phosphonate derivative as shown in Eq 7.1.

$$3ROH + PCl_3 = (RO)_2P(=O)H + RCl \quad (7.1)$$

As with other reagents in which two substrate molecules bind to each phosphorus atom, a racemic mixture of the alcohol yields one ^{31}P signal of area two for the *dl*-pair [(*R,R*)/(*S,S*)], and two signals of area one for each of the *meso* [(*R,S*) and (*S,R*)] compounds. The area of the signals for nonracemic mixtures can be used to determine the enantiomeric purity. The method was effective for the analysis of primary and secondary benzylic and allylic alcohols, α-hydroxyamides, and α-hydroxyesters.[952] The method was subsequently applied to aryl alcohols as well.[953]

Under different conditions, PCl₃ reacts with alcohols to produce trialkyl phosphites of structure P(OR)₃.[954] The ^{31}P spectrum showed signals for the (*R,R,R*)/(*S,S,S*)-, (*R,R,S*)-, and (*S,S,R*)-derivatives and their relative intensities could be used to determine the optical purity of the alcohol.

7.1.2.6. Hexaalkylphosphorus Triamides (Phos31) Chiral alcohols react with hexaalkylphosphorus triamides such as (Et₂N)₃P (Phos31) in a 2:1 ratio as shown in Eq 7.2.[955] The resulting ^{31}P NMR spectrum has three signals corresponding to the chiral pair [(*R,R*)/(*S,S*)] and two different *meso*-isomers [(*S,R*) and (*R,S*)]. The area of the signals can be used to determine optical purity.

$$(Et_2N)_3P + 2R^*OH = (R^*O)_2PNEt_2 \quad (7.2)$$

7.1.2.7. α-1-Phenylethylamino Trisaminoamine (Phos32) (7.54) Phos32, in which R is an α-1-phenylethylamino unit, is effective for the chiral discrimination of azides.[956] Azides react with Phos32 to form compounds of structure **7.55**. The diastereomeric products of neomenthyl (**7.56**) and other azides exhibited distinct resonances in the ^{31}P NMR spectrum. Phos13 was evaluated as well but was not an effective reagent for the analysis of azides.

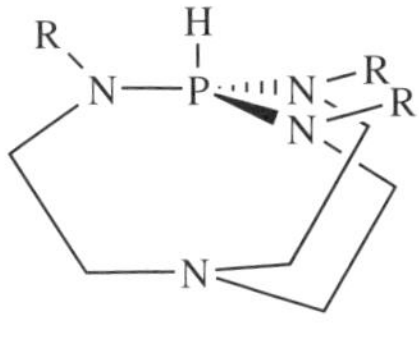

7.54 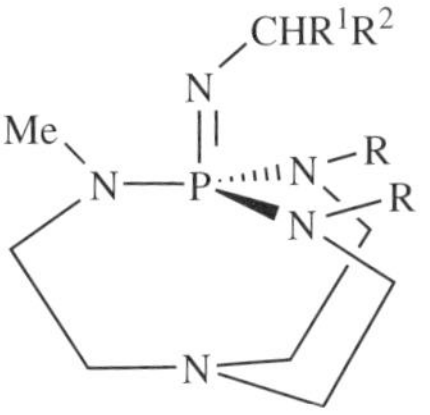   **7.55** 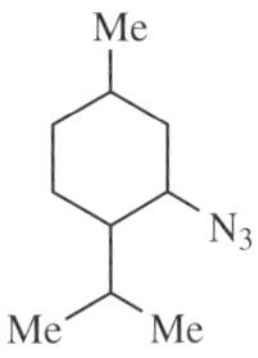   **7.56**

7.1.2.8. [5] HELOL Phosphite (Phos33) The compound [5] HELOL phosphite is a helically grooved sensor that can be used to analyze the optical purity of primary and secondary alcohols, phenols, and amines.[957] Carboxylic acids were also analyzed after being coupled to 2-aminophenol. Since this initial report, a similar derivative that is easier to prepare but just as effective as a chiral derivatizing agent has been reported (**7.57**).[958]

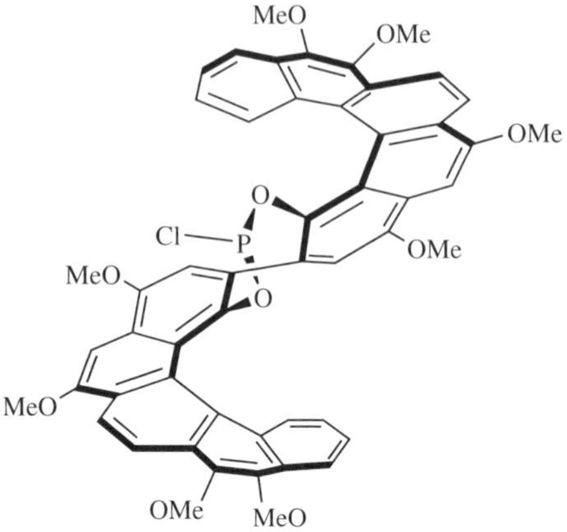

7.57

The substrates reacted with the PCl group to form the corresponding phosphite or phosphoramide derivative. The diastereomers have nonequivalent ^{31}P resonances in chloroform-d or acetonitrile-d_3. No kinetic resolution was observed with these reagents. It is of special significance that these reagents worked for substrates with remote chiral centers, since the substrates bound into and likely altered the chiral groove. For phenylmethyl alcohols of structure **7.58**, nonequivalence of 0.027 ppm was observed in the ^{31}P NMR spectrum for the diastereomeric derivatives in which $n = 7$.[957] For carboxylic acids, distinct ^{31}P resonances were observed for substrates in which the stereocenter was up to six bonds removed from the carboxylic acid group.

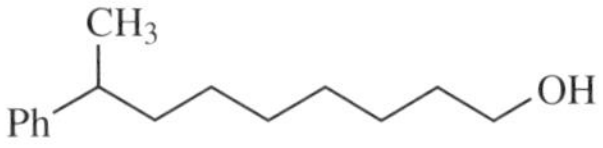

7.58

7.1.3. Ionic Reagents—TRISPHAT, BINPHAT, and BINTROP

Three ionic phosphorus-containing reagents, (tris[tetrachlorobenzene-1,2-bis(olato)-phosphate(V)) (**7.59**—TRISPHAT), (bis[tetrachlorobenzene-1,2-bis(olato)mono ([1,1′]-binaphthalenyl-2,2-diolato)phosphate(V)) (**7.60**—BINPHAT), and (bis(tropyliumdiolato)mono ([1,1′]-binaphthalenyl-2,2-diolato)phosphate(V)) (**7.61**– BINTROP) have been used as chiral NMR solvating agents. TRISPHAT has D_3 symmetry. BINPHAT and BINTROP have C_2 symmetry. The anionic TRISPHAT and BINPHAT reagents are useful with certain cationic substrates. The cationic BINTROP is useful with certain anionic substrates. Matching the symmetry of the discriminating

PHOSPHORUS-CONTAINING REAGENTS 251

agent to that of the substrate often causes greater chiral recognition.

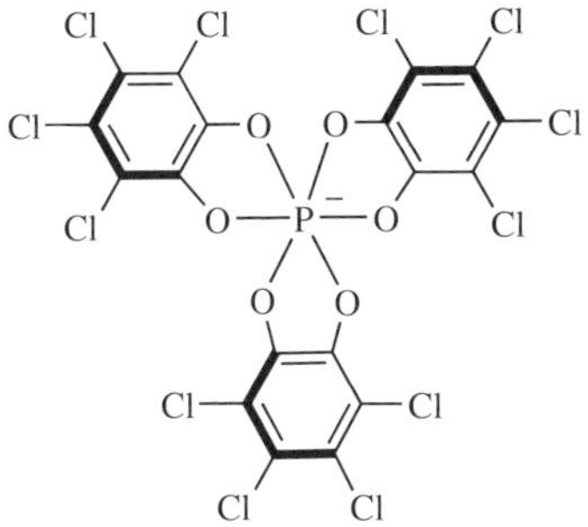

7.59

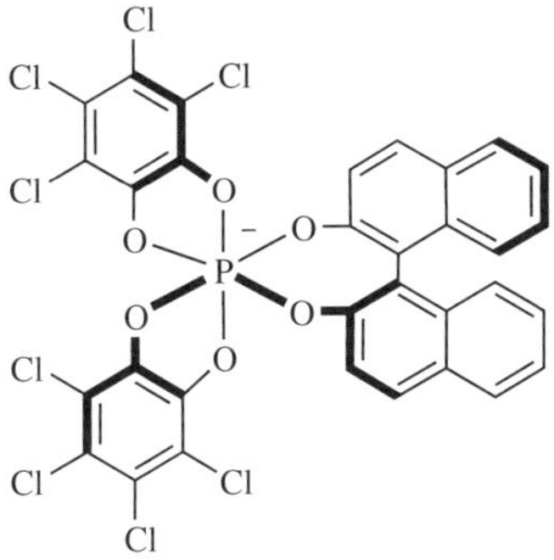

7.60

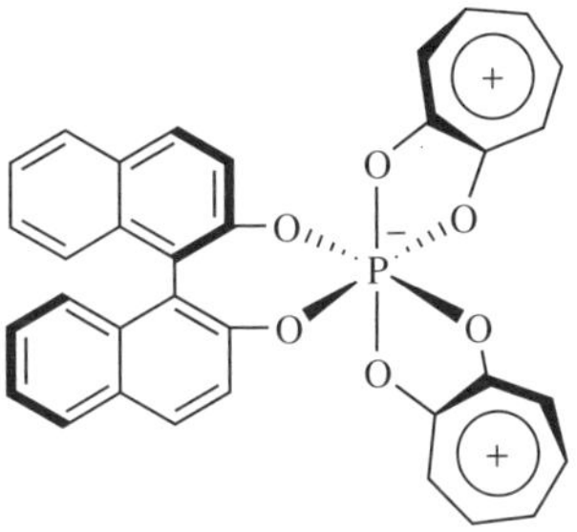

7.61

The first application of TRISPHAT, which is chiral because of the propeller-like arrangement of the three tetrachlorophenyl units, was in discriminating chiral cationic metal chelates such as [Ru(bpy)$_3$]$^{2+}$ (bpy = 2,2′-bipyridine) and [Ru(phen)$_3$]$^{2+}$ (phen = 1,10-phenanthroline) that have a similar propeller-like structure.[959] Nonequivalence in the ^{1}H NMR spectrum of up to 0.177 ppm and 0.135 ppm was observed for certain resonances of the two complexes. Distinct resonances were even observed in dimethylsulfoxide-d_6, a relatively polar solvent that might have been expected to minimize the formation of an ion pair between the metal cations and TRISPHAT. Resonances in the ^{1}H NMR spectrum of the iron chelate [Fe(eilatin)$_3$]$^{2+}$ (**7.62**) also exhibited enantiomeric discrimination in a polar solvent consisting of 90:10 acetone-d_6/chloroform-d.[960]

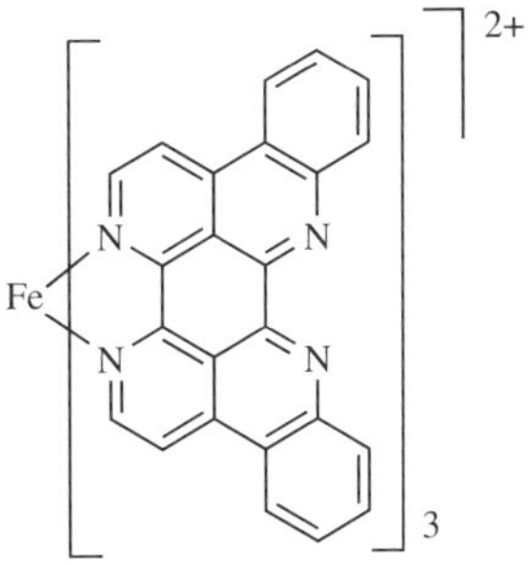

7.62

Monofluorinated analogs of TRISPHAT with fluorine atoms at the C3 or C4 position of one of the aromatic rings (**7.63**) were studied with [Ru(4,4′-Me$_2$bpy)$_3$]$^{2+}$ to investigate the mode by which TRISPHAT associated with metal complexes to effect chiral discrimination.[961] The anions all favored homochiral association along the C_3 axis of the cations. Data from ^{1}H-^{19}F-HOESY experiments showed contact between the anion and metal cation complex that was rationalized by the alignment of the propellers along their C_3 axes.

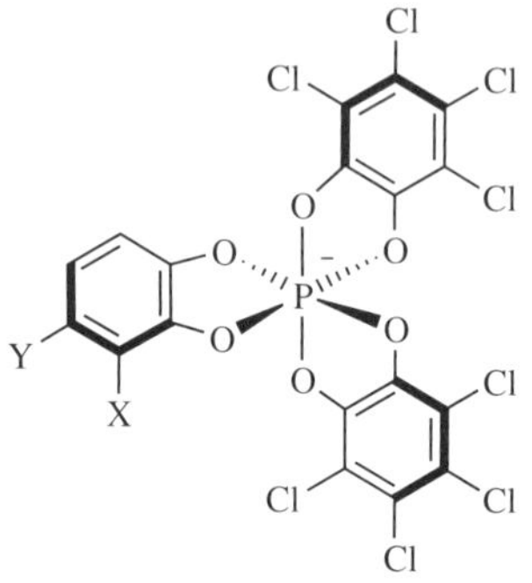

7.63

TRISPHAT has since been used to determine the enantiomeric purity of a variety of other chiral complexes similar to [Ru(phen)$_3$]$^{2+}$ and [Ru(bpy)$_3$]$^{2+}$ that differ in the identity of one of the ligands. This included the analysis of [Ru(bpy)$_2$(ppy)]$^+$ (ppy = -phenylpyridine) in acetone-d_6,[962–964] **7.64** in chloroform-d,[965] [Ru(bpy)$_2$ (quo)]$^+$ (quo = 8-quinolate)[963,964] and [Ru(bpy)$_2$(dcbpy)]$^+$ (dcbpy = 2,2 ′dipyridine-4-carboxylic acid) in 95:5 methylene chloride-d_2/acetonitrile-d_3.[966] Nonequivalence as large as 0.45 ppm was observed for resonances in the ^{1}H NMR spectra of these compounds.

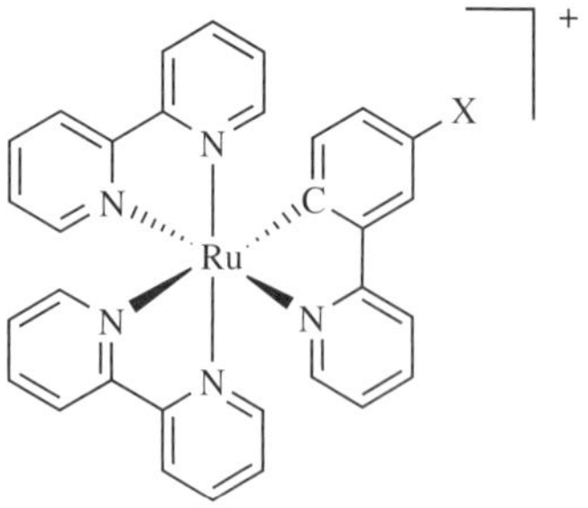

7.64

The effectiveness of TRISPHAT and Eu(tfc)$_3$ (tfc = 3-trifluoroacetyl-d-camphorate) in causing enantiomeric discrimination in the ^{1}H NMR spectrum of [Ru(phen)$_2$(py)$_2$]$^{2+}$, which has C_2 symmetry, was compared.[967] The europium reagent

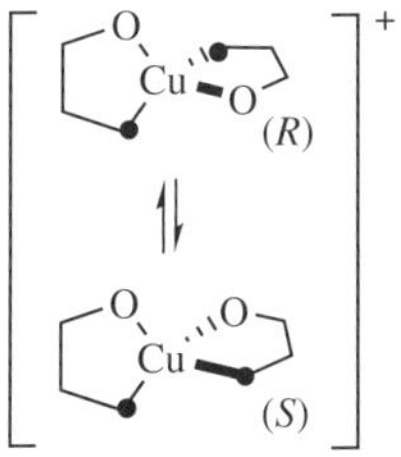

Figure 7.8 Chirality of bis(diimine)copper(I) complexes.

produced significant line broadening at 600 MHz that eliminated the possibility of observing chiral discrimination in the spectrum. Line broadening with TRISPHAT was not a problem and clear distinction of enantiotopic groups was observed.

The ruthenium complex with two 2,9-dimethyl-1,10-phenanthroline (dmp) ligands [cis[Ru(dmp)$_2$(CH$_3$CN)$_2$][PF$_6$]$_2$ is chiral because of steric hindrance of the methyl groups on neighboring ligands. Addition of TRISPHAT to the complex in acetone-d_6 caused enantiomeric discrimination in the ^{1}H NMR spectrum that could be used to determine its optical purity.[968]

TRISPHAT has also been used to discriminate cationic metal complexes with tetrahedral coordination geometries. Bis(diimine)copper(I) complexes of the form [CuL$_2$]$^+$ in which L is 2-alkylphenanthroline, 6-alkylbipyridine, and 2-iminopyridine (**7.65**) are chiral by virtue of the arrangement of the substituent groups on the two ligands as shown in Figure 7.8.[969] The ^{1}H NMR spectrum of these complexes in chloroform-d in the presence of TRISPHAT showed enantiomeric discrimination for certain resonances. At elevated temperatures, the ligands underwent fast exchange such that the system was no longer chiral. The racemization barrier was determined in the presence of TRISPHAT, which did not alter the measured value.[969]

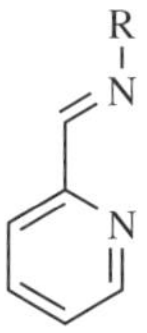

7.65

The optical purities of chiral metal complexes with cyclopentadienyl ligands have been analyzed using TRISPHAT. The ^{1}H NMR spectra of the Rh and Ir complexes of the form [Cp*M(2-alkyl-phenoxo)][BF$_4$] (**7.66**) in chloroform-d exhibited enantiomeric discrimination in the presence of TRISPHAT.[970] The optical purities of the iodide salts of compounds such as **7.67**, which have a cationic ammonium functionality in the ligand, were determined by recording the ^{1}H NMR spectrum in the presence of TRISPHAT.[971]

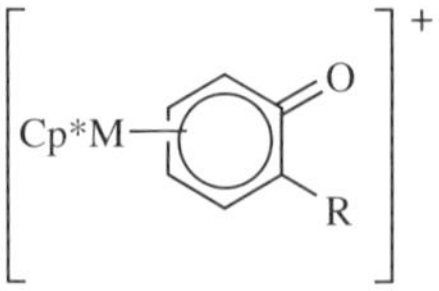

7.66 [Cp*M–(arene-C(=O)R)]+

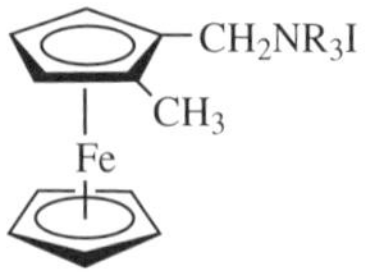

7.67 Ferrocene with CH$_2$NR$_3$I and CH$_3$ substituents on one Cp ring

TRISPHAT is an effective chiral NMR solvating agent for metal carbonyl complexes as well. The ^{1}H NMR spectra of cationic manganese tricarbonyl complexes of substituted anisoles and toluenes such as **7.68** and **7.69** showed chiral discrimination in the presence of TRISPHAT.[972] A mixture of benzene-d_6 containing at least 20% acetone-d_6 was needed to solubilize the tetrafluoroborate or hexafluorophosphate salt of the metal complex, although the presence of the TRISPHAT did enhance the solubility. The lower the proportion of acetone in the solvent. The larger the nonequivalence that was observed.

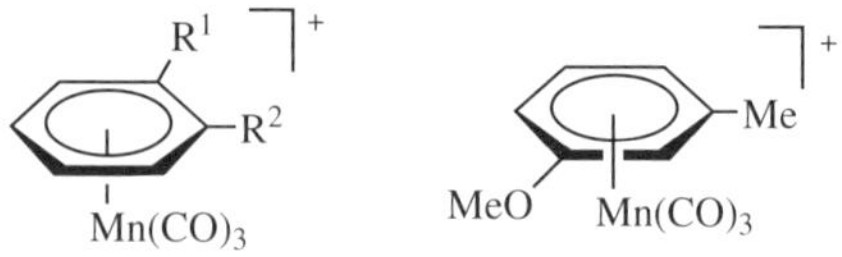

7.68 [(R^1,R^2-arene)Mn(CO)$_3$]+

7.69 [(MeO, Me-arene)Mn(CO)$_3$]+

Chromium tricarbonyl complexes of substituted benzaldehydes and nitrones, one example of which is **7.70**, are neutral but have electron-deficient groups by virtue of the aldehyde and nitrone moieties. Addition of TRISPHAT caused chiral recognition of certain resonances in the ^{1}H NMR spectrum.[973] Larger discrimination was observed for the nitrones, which were prepared by conversion of the aldehyde using appropriate conditions.

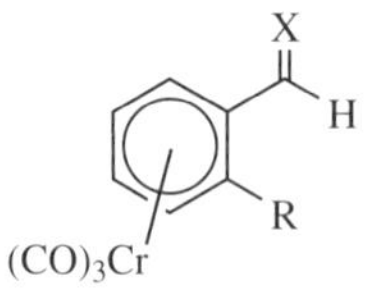

7.70 (CO)$_3$Cr complex of R-substituted benzaldehyde/nitrone (X=H)

Chiral bimetallic systems have been analyzed with TRISPHAT. The ^{31}P and ^{1}H NMR spectra of the bimetallic ruthenium–chromium system (**7.71**) (depe = 1,2-bis(diethylphosphino)ethane) in methylene chloride-d_2 exhibited nonequivalent resonances in the presence of TRISPHAT.[974] The optical purity of the bimetallic *ortho*-chloropalladated 2-[tricarbonyl(η^6-phenyl)chromium]pyridine complex (**7.72**) was determined by recording the ^{1}H NMR spectrum in the presence of TRISPHAT.^{975}

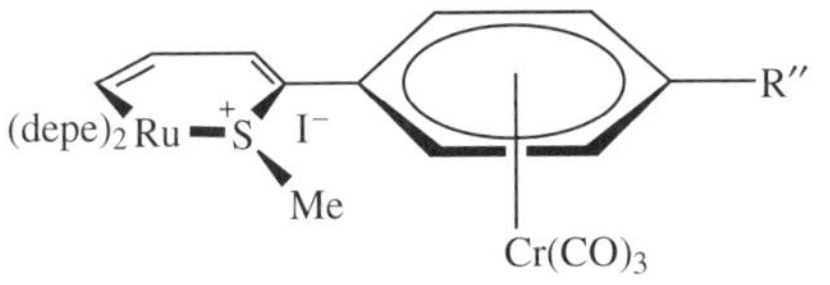

7.71

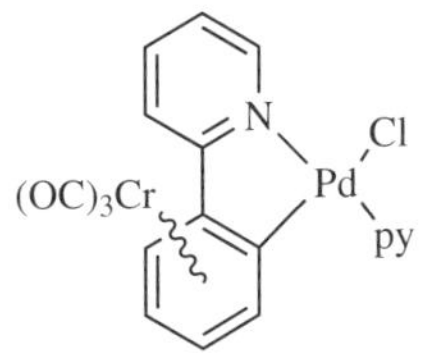

7.72

The ^{1}H NMR spectrum of the racemic dinuclear ruthenium complex *trans*-[bis(-Cp*Ru)carbazolyl] (**7.73**) exhibited some split signals in the presence of TRISPHAT in 95:5 benzene-d_6/acetone-d_6.^{976} Attempts to crystallize the TRISPHAT complex of **7.73** resulted in isolation of only the ($S_P S_P$) ruthenium complex. The X-ray structure showed π stacking of aromatic groups of the anion with the C_2-planar cation.

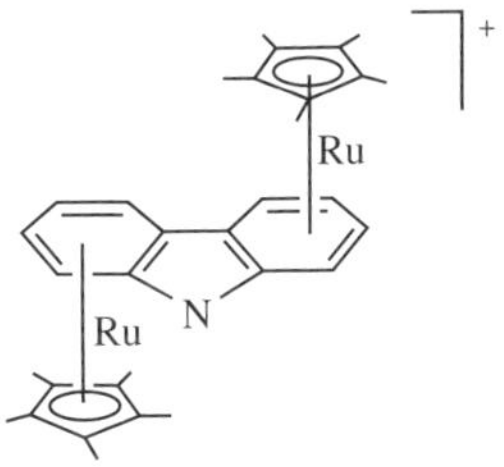

7.73

The dinuclear cobalt complex [(Co$_2$(CO)$_4$-μ,η^2,η^2-(-H$_2$CC=CH$_2$-) (dppm)$_2$][BF$_4$]$_2$ (**7.74**) has a two-bladed propeller geometry. Addition of TRISPHAT caused chiral discrimination in the ^{1}H and ^{31}P NMR spectra in methylene chloride-d_2.^{977} This was the first example in which TRISPHAT was used to enantiomerically distinguish a C_2-symmetrical metal complex. All prior reports involved metal centers with D_3 symmetry.

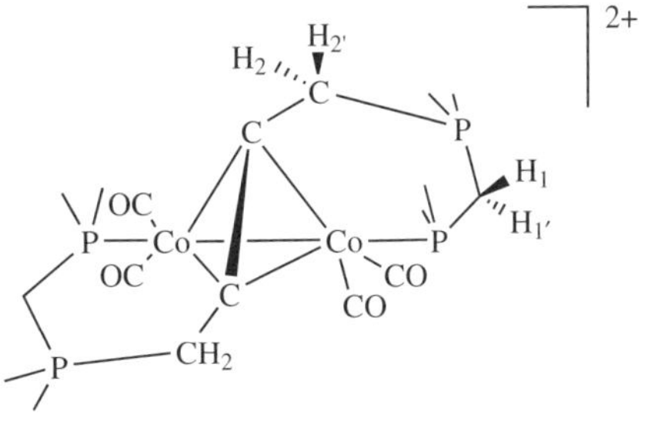

7.74

The diiron and dicobalt complexes with 1,2-bis[4-(4′-methyl)-2,2′-bipyridine] ligands (**7.75**) form triple helicate species that exhibited enantiomeric discrimination in the ^{1}H NMR spectrum in the presence of TRISPHAT.^{978,979} The nonequivalence in the spectrum of the iron species was up to 0.3 ppm in a mixture of 80–20 acetonitrile-d_3/dimethylsulfoxide-d_6.^{978} Increasing the proportion of dimethylsulfoxide-d_6 reduced the enantiomeric discrimination, although the spectra still had separate resonances when recorded in pure dimethylsulfoxide-d_6. As typical for many systems analyzed with TRISPHAT, only one enantiomer of the complex showed appreciable shifts in the NMR spectrum. The TRISPHAT anion has a high degree of selectivity in its association with chiral metal complexes.^{978}

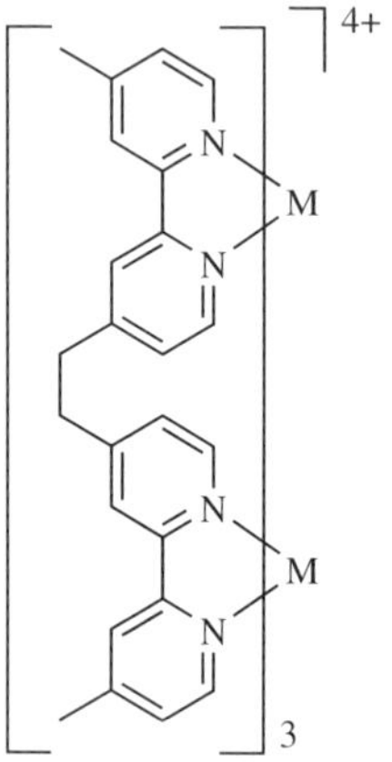

7.75

The optical purity of the trirhodium supramolecular cluster compound with Cp* units and 5-chloro-2,3-dioxopyridine ligands (**7.76**) was determined through the use of TRISPHAT.^{980} Iron complexation with 2,5-bis([2,2'] bipyridin-6-yl)pyrazine (**7.77**) results in a self-assembled species of the form [Fe$_4$L$_4$](PF$_6$)$_8$.^{981} The enantiomeric purity of the assembled species was determined in the presence of TRISPHAT in acetonitrile-d_3.

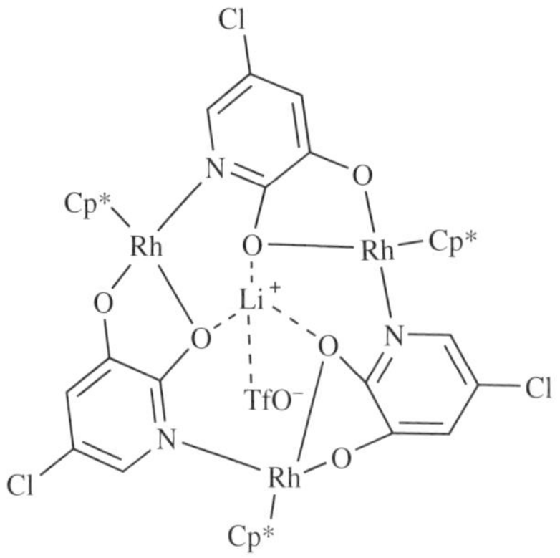

7.76

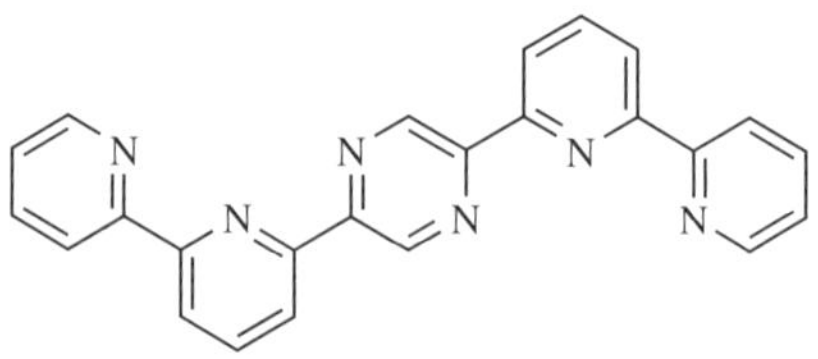

7.77

PHOSPHORUS-CONTAINING REAGENTS 257

TRISPHAT and BINPHAT have been applied to cationic species other than metal complexes. Quaternary phosphonium salts such as **7.78** and **7.79** exhibited enantiomeric discrimination in the ^{1}H and ^{31}P NMR spectra in the presence of TRISPHAT.[982] The impressive chiral recognition ability of the TRISPHAT ion was demonstrated with **7.79**, which is chiral by virtue of the deuterated methyl group. The analysis was successful in chloroform-d or benzene-d_6.

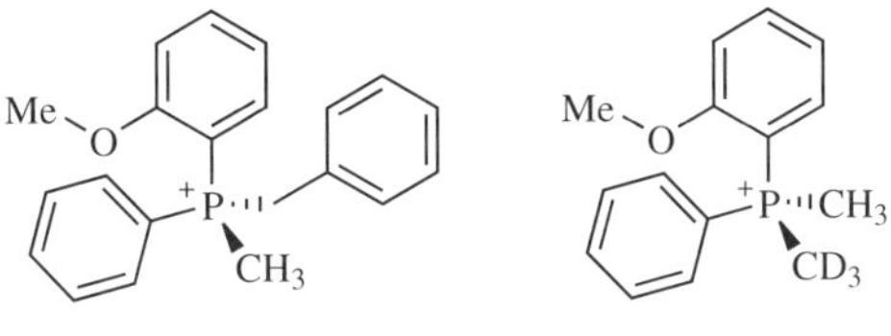

7.78   **7.79**

The nonequivalence in the ^{1}H and ^{31}P NMR spectra of triphenylphosphonium salts such as **7.80** was much better using BINPHAT than TRISPHAT.[983] Addition of dimethylsulfoxide-d_6 (5%) to chloroform-d was necessary to solubilize some of the salts.

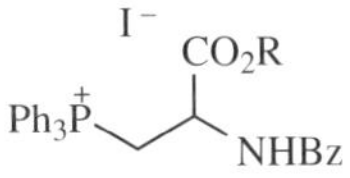

7.80

BINPHAT is also preferred over TRISPHAT for the analysis of quaternary amines. Spirobi[dibenzazepinium] cations undergo a dynamic process that leads to *meso* (S_4 symmetry) and chiral (D_2 symmetry) forms as shown in Figure 7.9.[984] BINPHAT was far more effective than TRISPHAT at distinguishing these species and facilitated a variable temperature NMR study of this system. The ^{15}N NMR spectrum was the best to monitor as the two sides of the *meso*-isomer bound individually to BINPHAT compounds, which complicated the analysis of the ^{1}H NMR spectrum.[984]

The effectiveness of TRISPHAT, BINPHAT, and HYPHAT (**7.81**) in causing enantiomeric discrimination in the NMR spectra of chiral diquaternary amines such as **7.82** was compared.[985] BINPHAT was the best of the three and provided

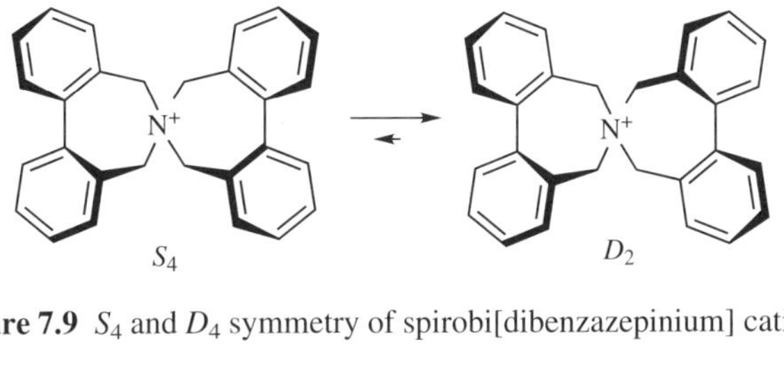

Figure 7.9 S_4 and D_4 symmetry of spirobi[dibenzazepinium] cations.

nonequivalence of up to 0.56 ppm in the ^{1}H NMR spectrum. Mixtures of dimethyl-sulfoxide-d_6-chloroform-d were needed to solubilize some of the ammonium salts.

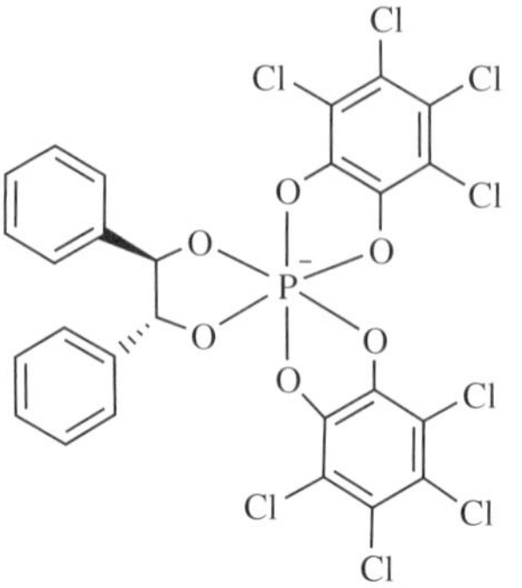

7.81

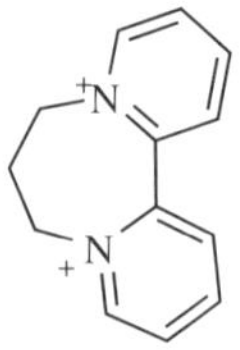

7.82

BINPHAT was used to measure the optical purity of C_2-symmetric monomethinium dyes (**7.83**).[986] Nonequivalence of up to 0.18 ppm was observed in the spectra. TRISPHAT was not an effective chiral solvating agent for these compounds.

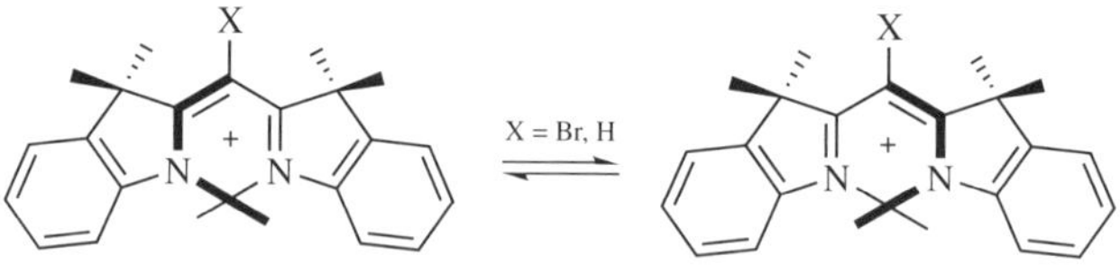

X = Br, H

7.83

The ^{1}H NMR spectra of a variety of chiral quaternary amines with aliphatic, cyclic, and bicyclic units exhibited enantiomeric discrimination in the presence of BINPHAT.[987] Substrates with chiral centers α or β to the ammonium group were successfully analyzed. Nonequivalence up to 0.29 ppm was observed and the presence of aromatic rings in the substrates did not inhibit the analysis.

The helicene compound **7.84** is chiral because of steric hindrance of the methoxy groups. The optical purity of **7.84** could be monitored in the ^{1}H or ^{13}C NMR spectra in the presence of BINPHAT.[988]

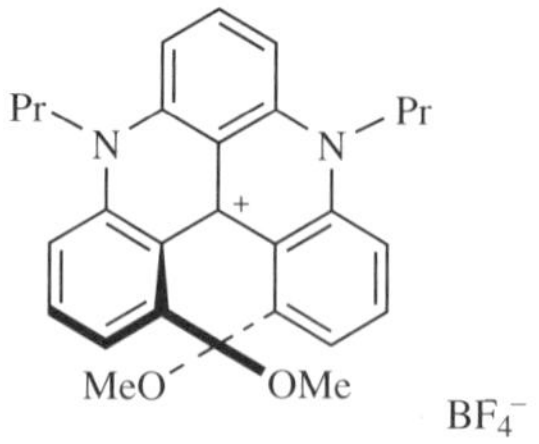

7.84

BINPHAT is an effective chiral shift reagent for thiiranium ions such as **7.85**.[989] The enantiomeric discrimination with BINPHAT was much greater than that with

TRISPHAT, which was related in part to the strength of the respective ion pairs formed in solution. In chloroform-d, nonequivalence of up to 0.25 ppm was observed in the ^{1}H NMR spectrum with BINPHAT.

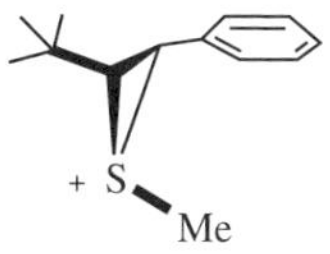

7.85

Even though TRISPHAT is generally recommended over BINPHAT for the analysis of chiral cationic metal complexes, the ^{1}H NMR spectrum of a binuclear chromium–ruthenium complex (**7.86**) showed enantiomeric discrimination in the presence of BINPHAT but not in TRISPHAT.[990]

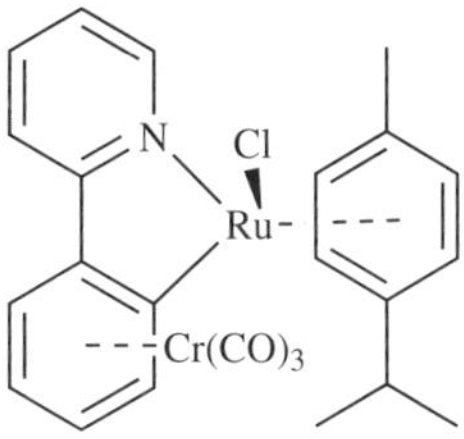

7.86

The potential of using different diffusion rates to study chiral recognition has been demonstrated using ion pairs of chiral cations with TRISPHAT and BINPHAT.[991] Diffusion measurements were performed using ^{1}H, ^{19}F, and ^{31}P pulsed field gradient spin-echo techniques. The diffusion constant was found to depend on the diastereomeric structure of the ion pair, with the less abundant stereoisomer having a faster diffusion rate. The difference in diffusion rates exhibited a pronounced solvent and concentration dependence. While differences occurred in methylene chloride-d_2, no distinction occurred in chloroform-d, reportedly since all of the ions were paired.[991] As the use of NMR spectroscopy to measure rates of diffusion becomes more common, this technique will likely find applications with other chiral NMR discriminating agents as well.

BINTROP (**7.61**), which has been the focus of only limited study to date, is a C_2-symmetric cationic phosphorus compound that offers the potential to enantiomerically discriminate chiral anions.[992] BINTROP was found to cause enantiomeric discrimination in the ^{31}P NMR spectrum of TRISPHAT and the ^{1}H NMR spectrum of bis(3,5-di-*tert*-butylbenzenediolato)borate (**7.87**) in chloroform-d.[992]

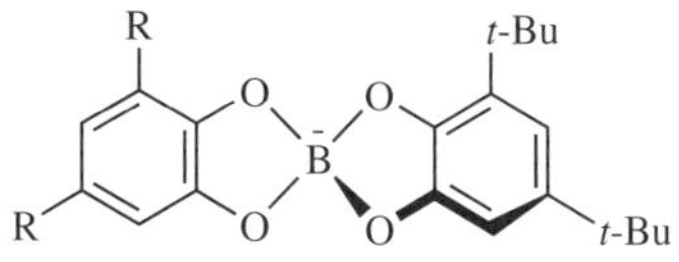

7.87

A catenane compound (**5.76**) under conditions of slow exchange exhibits helical chirality.[704] Addition of BINTROP to samples of the cationic **5.76** at 197 K in acetone-d_6 showed a splitting of certain resonances, confirming the chiral nature of the compound.

The utilization of TRISPHAT and BINPHAT in asymmetric reactions and as chiral NMR shift reagents was recently reviewed.[993] The use of TRISPHAT is recommended for organometallic and metalloorganic species, whereas BINPHAT is recommended for use with organic cations.

7.1.4. Configurational Analysis of Phosphates

7.1.4.1. Oxygen Isotope Methods   Selective incorporation of oxygen isotopes into phosphate compounds has been used to study aspects of their absolute configuration. The presence of an ^{18}O atom in cyclic[^{18}O]-2'deoxy adenosine monophosphate (AMP) caused a perturbation of the ^{31}P signal that was 3.1 Hz if the oxygen was in an equatorial position and 1.2 Hz if in an axial orientation.[994] By monitoring a starting phosphate compound of known configuration, it is then possible to see if a subsequent reaction occurs with retention or inversion of configuration.

Several procedures using a [$^{16}O,^{17}O,^{18}O$]thiophosphate or phosphate species to assign the absolute configuration of phosphorus atoms in phosphate products have been developed.[995–997] One caution is that incorporation of the ^{18}O does not usually occur to 100%, so extra peaks will be observed in the ^{31}P NMR spectrum. One method involved incorporating the labeled thiophosphate into adenosine 5'-O-[(2S)-thiotriphosphate] (SP-ATPBS).[995] The (R)- and (S)-isomers of the product had different oxygen isotopes in the bridging position, such that the isotopic effect on the ^{31}P signal could be related to the original configuration of the phosphate. This procedure was used to show that the 3-phosphyglycerate-kinase-catalyzed reaction proceeded by inversion at phosphorus.[995]

A second procedure involved the reaction of [$^{16}O,^{17}O,^{18}O$]phosphate monoester with propane-1,2-diol to form cyclized products as shown in Figure 7.10.[996] The

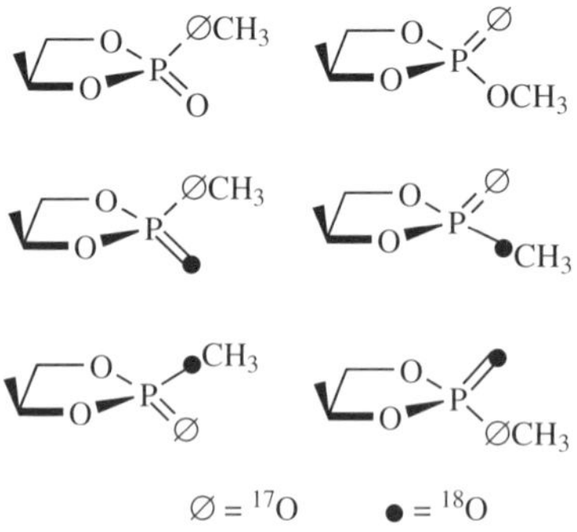

$\emptyset = {}^{17}O$ ● $= {}^{18}O$

Figure 7.10 The three cyclic diesters derived by the cyclization reaction of propane-1,2-diol with [$^{16}O,^{17}O,^{18}O$]phosphate monoester.

methyl and methoxy groups have *syn-* and *anti-*arrangements for the two configurations. Since the ^{18}O in the methoxy had a different effect on the ^{31}P signal than the ^{18}O in P = O, it was possible to assign the absolute configuration. This procedure was used to assign the configuration of 1,2-dipalmitoyl-*sn*-glycero-3-phosphoethanolamine (**7.88**) (DPPE).[998] The two diastereomers of DPPE were converted into the corresponding (R_P)- and (S_P)-1-[^{16}O,^{17}O,^{18}O]phosphor-(R)-propane-1,2-diol and assigned configurations based on the differential ^{18}O perturbation effects on the ^{31}P signals.

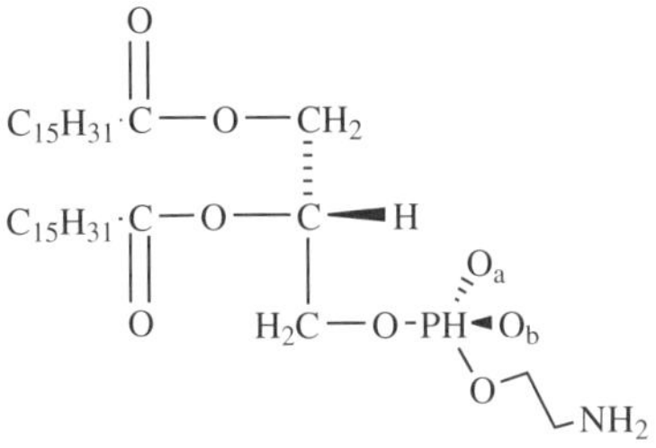

7.88

This procedure has been applied more generally to the stereospecific cyclization of a chiral phosphate monoester of a chiral diol into a cyclic phosphate diester.[999] The diester was then esterified to give a diastereomeric cyclic phosphate triester. The validity of the method was demonstrated with isotopically labeled D-glucose-6-phosphate and adenosine-5′-phosphate. Figure 7.11 shows the potential outcome for D-glucose depending on whether the phosphorus configuration was retained or inverted. The effect of the ^{18}O atom shift on the ^{31}P signals was used to determine whether the ^{18}O atom is singly or doubly bonded to the phosphorus atom, as the higher the bond order the larger the isotope shift. With this known, it was possible to assign the configuration of the phosphate. The method was useful in studying whether retention or inversion of configuration occurred in phosphoryl transfer reactions.[999]

The third, and somewhat analogous procedure, involved the reaction of [^{16}O,^{17}O,^{18}O]thiophosphate with (*S*)-2-iodo-1-phenylethanol (**7.89**) to form the corresponding thiophosphate (*S*)-ester.[997] After cyclization and esterification as shown in Figure 7.12 for one of the configurations of the phosphorus atom, the stereochemistry of the *cis-* and *trans-*methyl esters could be assigned on the basis of the ^{18}O perturbations of the ^{31}P resonances. This allowed a determination of whether the configuration was retained or inverted. The method was reportedly more reliable than the first one described above, which reportedly led to a substantial loss of the labeled oxygen and some racemization because of the instability of the intermediate in the reaction.[995]

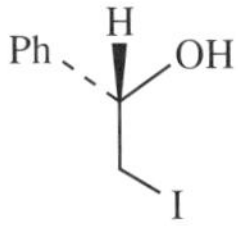

7.89

262 REAGENTS INCORPORATING PHOSPHORUS, SELENIUM, BORON, AND SILICON

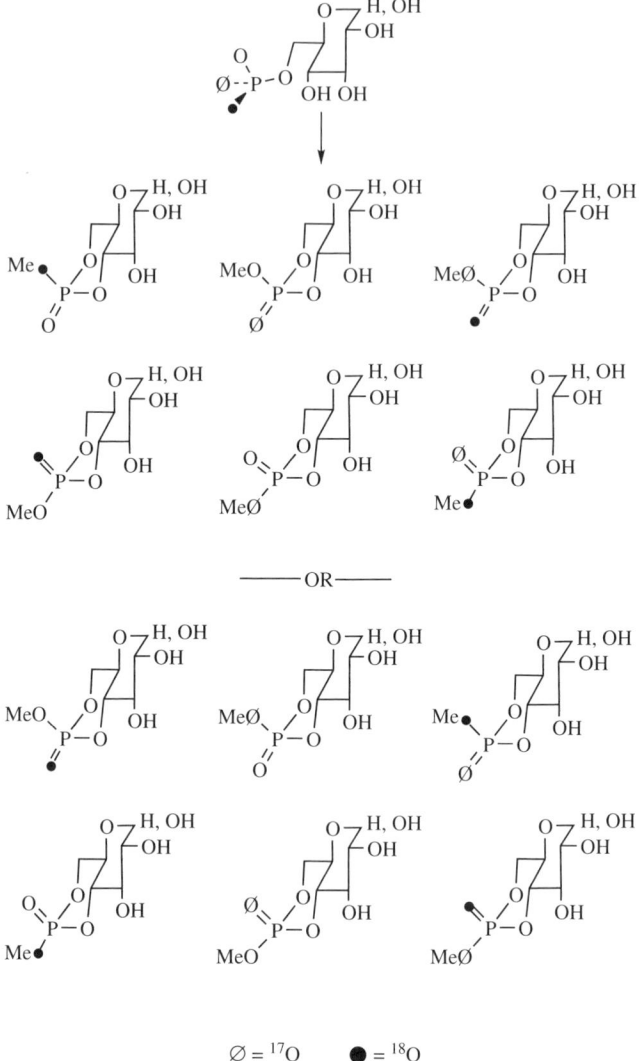

Figure 7.11 Potential products after the cyclization and methylation of D-glucose 6-[(S)-^{16}O,^{17}O,^{18}O]phosphate. Top group occurs if there is retention of configuration at phosphorus. Bottom group occurs if there is inversion of configuration at phosphorus. [Reproduced with permission from American Chemical Society.[999]]

The effect of ^{18}O on ^{31}P chemical shifts has been used to assign the absolute configuration of phosphate groups in triaminecobalt(III)adenosine triphosphate and to follow the phosphorus configuration through the course of the reaction.[1000]

^{17}O NMR has also been used to study the configuration about the phosphorus atom in phosphate species. The configuration of chiral phosphodiesters (**7.90**) was assigned

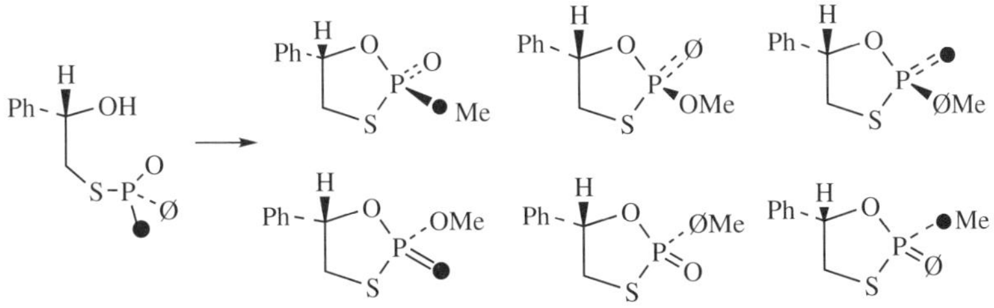

PHOSPHORUS-CONTAINING REAGENTS

Figure 7.12 Cyclic thiophosphate esters derived by the cyclization reaction of (S)-2-iodo-1-phenylethanol with [^{16}O,^{17}O,^{18}O]thiophosphate.

on the basis of ^{17}O shifts, since the ^{17}O atoms in the axial or equatorial positions have different chemical shifts.[1001] Similarly, ^{17}O NMR spectroscopy has been used to assign the absolute configuration about the phosphorus atom in thymidine 3′,5′-monophosphate.[1002] The labeled thymidine 3′,5′-monophosphate was cyclized to a diester as described for earlier diol functionalities and then converted to a methyl triester (**7.91**). The signals for ^{17}O atoms in the equatorial and axial positions of the cyclic thymidine methyl 3′,5′-monophosphate were distinct from each other, which facilitated the analysis of the configuration about the phosphorus.

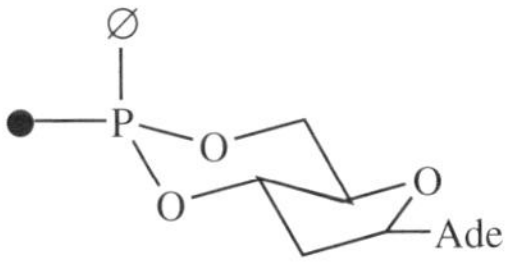

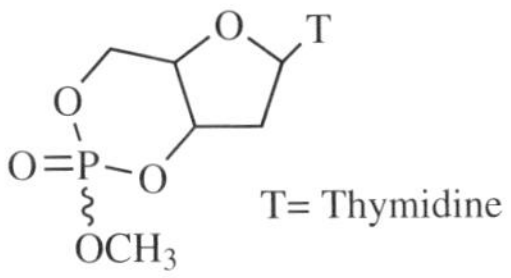

7.1.4.2. ROESY Studies Through-space connectivities that are obtained with a two-dimensional technique, such as ROESY, have been used to assign absolute configurations of the phosphorus atom in phosphate groups. One example involves assigning the configuration of the phosphate group in 5′-O-monomethoxytrityl-2′-O-deoxyribonucleoside 3′-O-[O-(4-nitrophenyl]methanephosphonates (**7.92**).[1003] NOE connectivities of the P-methyl group were used in assigning the configuration. The same technique has been used to assign the stereochemistry of the phosphate in deoxyribonucleoside methanephosphonamidates (**7.93**).[1004] NOE connectivities of the P-methyl group were used in distinguishing the two configurations.

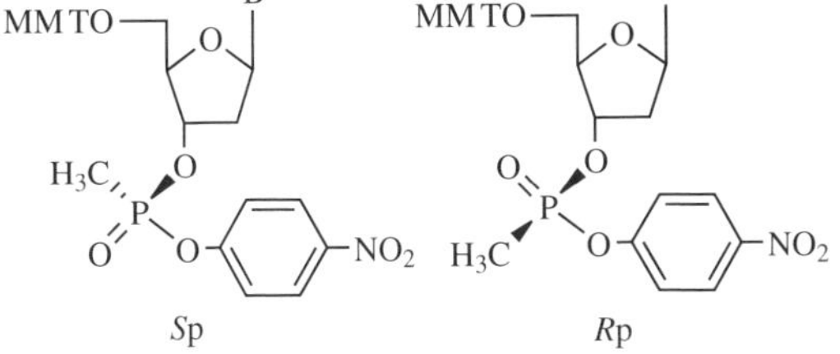

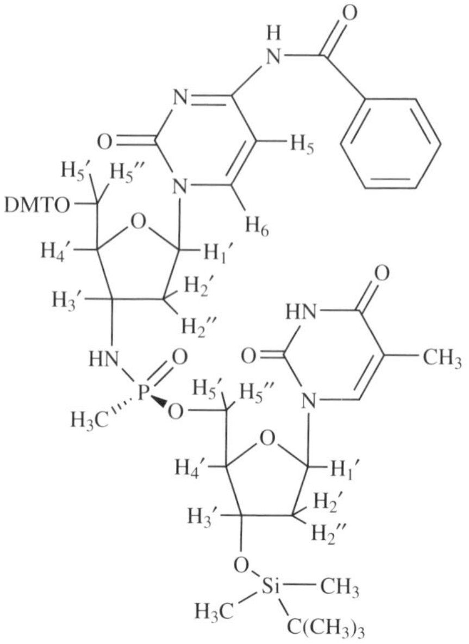

7.93

7.1.5. Summary

A comprehensive review of the use of phosphorus-based reagents for chiral NMR discrimination was published in 1995.[1005] The advantages and disadvantages of the different P(V) and P(III) systems developed at that time were described in that report. A few of the phosphorus reagents have been used in a relatively wide variety of applications. Many of the others have only been evaluated in exploratory studies by the investigators who developed the reagents. In the aggregate, a diverse range of compounds are amenable to study with phosphorus-based reagents.

7.2. SELENIUM-CONTAINING REAGENTS

7.2.1. 4-Methyl-5-phenyloxazolidine-2-selone (7.94)

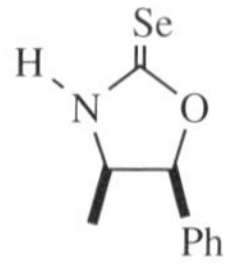

7.94

7.94 and derivatives with substituents other than the methyl and phenyl groups are effective chiral NMR derivatizing agents for several classes of substrates. The analysis of optical purity was best performed using ^{77}Se NMR spectroscopy. One advantage is that the single ^{77}Se resonance splits into two singlets for the diastereomeric products. Another advantage of ^{77}Se NMR spectroscopy is the enormous shift range (about 3400 ppm), such that minor changes in chirality were often detected as differences in the ^{77}Se NMR spectrum.[1006]

Carboxylic acids reacted at the NH group of **7.94** to produce the corresponding amide.[1006] 5-Methylheptanoic acid is chiral at the δ-position and has a stereogenic center eight atoms removed from the selenium in its derivative with **7.94**. The resulting diastereomers of racemic 5-methylheptanoic acid had different chemical shifts for the ^{77}Se signals. The diastereomers produced by the reaction of lipoic acid (**7.95**) with **7.94** have ^{77}Se signals that differed in chemical shift by 0.119 ppm.[1007] The enantiomers of 1-[^{2}H$_1$]-1-phenylethanol (**7.96**), which are chiral by virtue of deuterium substitution, were distinguished in the ^{77}Se NMR spectrum of their amide derivatives with **7.94**.[1006] The ^{77}Se NMR spectrum actually exhibited four signals, two from the diastereomeric pairs of monodeuterated compounds, one from the diproteo compound, and the other from the dideuterio compound.

7.95

7.96

Tests with 2-phenylbutanoyl and propanoyl chloride indicated that coupling to **7.94** proceeded with no racemization or epimerization. If the selenium compound was the limiting reactant, kinetic resolution did occur. Therefore, an excess of the selenium reagent must be used in the derivatization. In a study of seven *N*-boc amino acids, no discernible epimerization or racemization occurred on coupling to **7.94**.[1008] A consistent trend was that the chemical shift of the ^{77}Se resonance of the D-amino acid derivative was deshielded relative to that of the L-derivative.

Conditions for optimizing the preparation and analysis of carboxylic acid and acid chloride derivatives with **7.94** have been studied.[1009] Eleven solvents were tested and the largest enantiomeric discrimination was observed in the most nonpolar solvents such as cyclohexane-d_{12} and benzene-d_6. An interesting observation was that the separation of the ^{77}Se signals was greater at higher temperatures. This is contrary to what is typically observed with many chiral discriminating agents.

Attempts to determine the optical purity of oxazole **7.97**, a fragment of a larger natural product, were made by attaching chiral derivatizing agents at the corresponding C26-carboxylic acid, C26-alcohol, and C32-amine. Neither the chlorodioxaphospholane reagent nor α-methoxy-α-trifluoromethylphenylacetic acid (MTPA) was successful for the analysis of the substrate. Nonequivalence of approximately 2 ppm was observed in the ^{77}Se NMR spectrum of the diastereomeric amide derivatives of **7.97** with **7.94**.[1010]

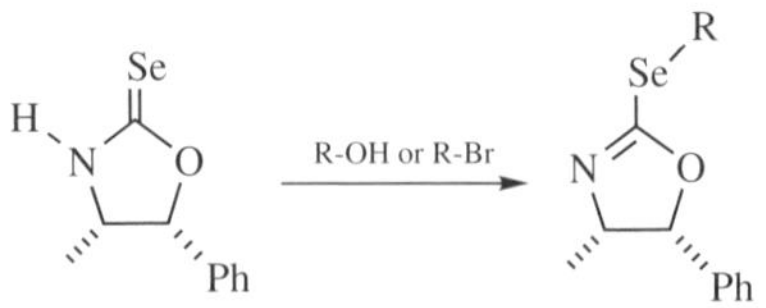

Figure 7.13 Reaction of alcohols or alkyl halides with 5-methyl-5-phenyloxazolidine-2-selone.

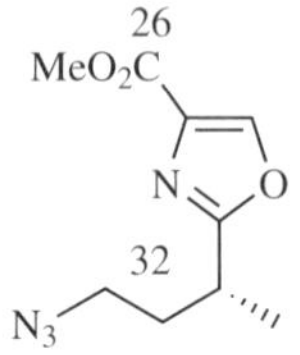

7.97

Primary and secondary alcohols react with **7.94** in the presence of triphosgene to produce the corresponding selenide derivative as shown in Figure 7.13.[1011,1012] Tertiary alcohols could not be analyzed with this method. The diastereomeric derivatives with 4-phenyl-1-pentanol, which has a remotely disposed chiral center, exhibited nonequivalence of 0.3 ppm in the ^{77}Se NMR spectrum.[1011]

Amines react with **7.94** in the presence of triphosgene to form the corresponding carbamoyl derivative (**7.98**).[1013] The amine adduct formed an intramolecular hydrogen bond that stabilized the *anti*-carbonyl relationship to the selenium as shown in the structure. Discrimination was observed in the ^{77}Se NMR spectrum of the derivative with 4-phenyl-1-amino pentane. The analysis of secondary amines was complicated by the observation of both *Z*- and *E*-conformers of the carbamoyl as there is no intramolecular hydrogen bond to stabilize a single conformation.

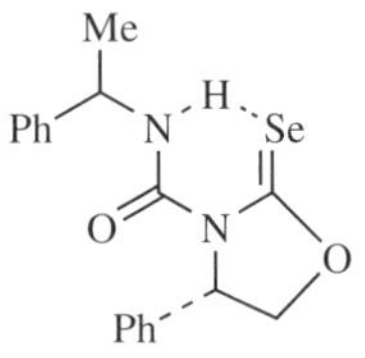

7.98

Alkyl halides such as 1-bromo-3-phenylbutane reacted with **7.94** to produce the corresponding selenide (Fig. 7.13).[1012] The diastereomers were readily distinguishable in the ^{77}Se NMR spectrum.

7.2.2. (*R*,*R*)-*N*,*N'*-Dimethylcyclohexyl-1,2-diazaselenophospholine

The utility of Phos10 for the analysis of chiral alcohols was previously described. Conversion with sulfur to the corresponding pentavalent P(V) sulfur derivative improved the stability of the product but the ^{31}P NMR spectrum showed less discrimination between the diastereomers. An alternative is to prepare the corresponding (R,R)-N,N'-dimethylcyclohexyl-1,2-diazaselenophospholine derivative (**7.99**) in which a selenium atom replaces the sulfur atom.^{1014} The selenium reagent was found to be more stable than the sulfur analog. Measurement of the ^{77}Se NMR spectrum was recommended for chiral alcohols with remotely disposed chiral centers (2, 3, and 4 bonds removed from the hydroxyl group), whereas the ^{31}P NMR spectrum was recommended for analysis of α-chiral alcohols.^{1014}

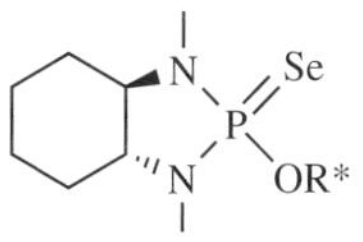

7.99

The use of **7.94** and **7.99** as chiral NMR discriminating agents have been reviewed.^{1015,1016}

7.2.3. 2-Phenylselenopropionic Acid (7.100)

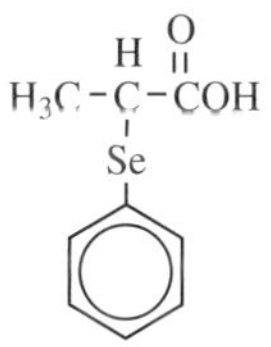

7.100

7.100 is an effective reagent for the analysis of chiral alcohols.^{1017} Reaction to produce the corresponding diastereomeric esters resulted in shift differences of 2.1 ppm in the ^{77}Se NMR spectrum for 2-octanol and 0.41 ppm for 1,2-dilaurylglycerol. No racemization about the chiral center of the selenium reagent was observed in the derivatization reaction.

7.2.4. 1,2,3-Selenadiazoles (7.101)

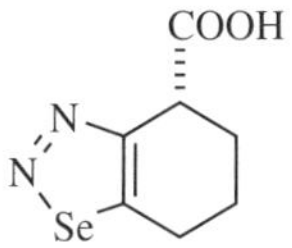

7.101

Reaction of **7.101** with menthol produced diastereomers that showed distinct resonances in the ^{77}Se NMR spectrum.[1018] The cyclohexyl reagent (**7.101**) was more effective than the corresponding compound with a cyclopentyl ring.

7.3. BORON-CONTAINING REAGENTS

Several chiral NMR derivatizing agents based on boric acid derivatives have been described. In most cases, an optically pure chiral group that is often a diol is bound to the boron of the boric acid species. The boron-diol reagent is then reacted with the enantiomeric substrate to prepare a pair of diastereomers that are analyzed by NMR spectroscopy. Alternatively, boric acid species with a single substituent group have been used to analyze chiral diols or other substrates that can bind to boron in a bidentate manner.

7.3.1. 2-(1-Methoxyethyl)phenylboronic Acid (7.102)

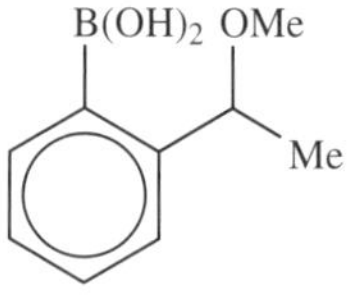

7.102

7.102 is an effective chiral NMR derivatizing agent for determining the optical purity of 1,2-diols, 1,3-diols, 2-hydroxyacids, and 2-aminoalcohols.[1019] The reaction of a 1,2-diol with **7.102** is illustrated in Figure 7.14. The derivation was carried out by just mixing the compounds in the NMR solvent with some ground up molecular sieves. No kinetic resolution was observed as there was complete conversion to the ester. The methoxy signal was conveniently monitored and the largest enantiomeric discrimination was observed in benzene-d_6.

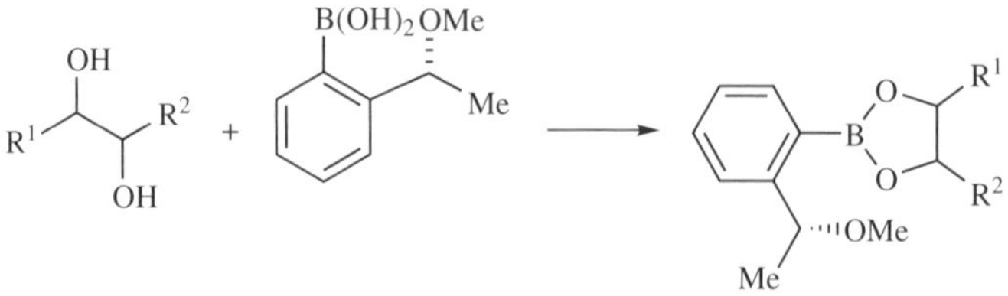

Figure 7.14 Reaction of a 1,2-diol with MPBA.

The utility of **7.102** in assigning the absolute configuration of *cis*-diols was subsequently demonstrated.[1020] For *cis*-arene diols such as **7.103**, the methoxy signal of the derivative with (−)-**7.102** was always at lower frequency than the derivative with (+)-**7.102**, whereas the reverse trend was observed for the methyl resonance. Diols with a different structural motif, such as that in **7.104**, showed a different trend. By comparing the trend in known *cis*-diols with similar structures to an unknown, it was possible to use **7.102** to assign the absolute stereochemistry.

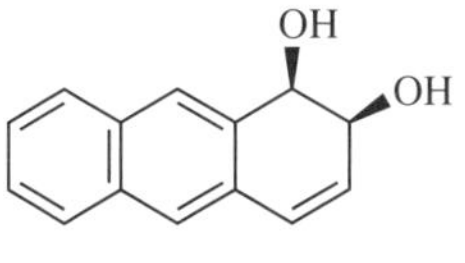

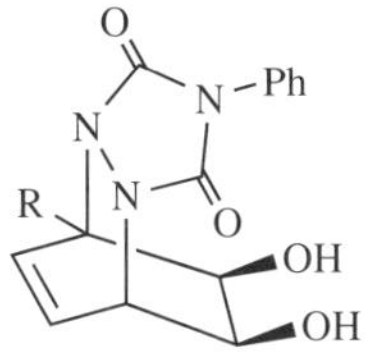

The absolute configurations of *cis*-diols produced through bacterial action (**7.105**, as one example)[1021] or by a biphenyl dioxygenase catalyzed sequential asymmetric *cis*-dihydroxylation of polycyclic arenes and heteroarenes (**7.106** as one example)[1022] have been determined by analyzing the ^{1}H NMR spectrum of their derivatives with **7.102**. The optical purity of **7.107**, produced by a bacterial dioxygenase-catalyzed dihydroxylation, was determined using **7.102** and agreed with the results obtained using α-methoxy-α-trifluoromethylphenylacetic acid as a derivatizing agent.[1023]

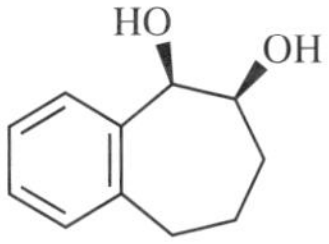

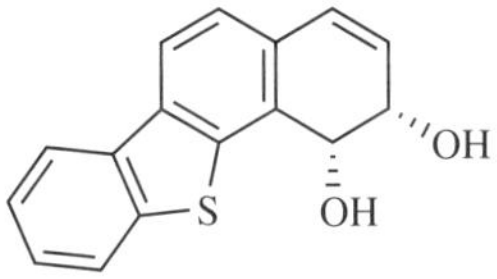

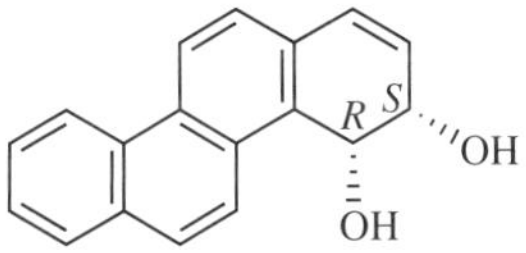

7.3.2. Phenylboronic Acid (7.108)

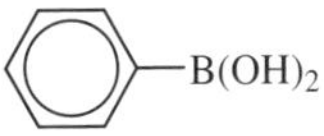

Reaction of the *erythro*- and *threo*-forms of 2-(1-naphthyl)-1-phenylpropane-1,2-diol with **7.108** produced the two cyclic phenyl boronate species **7.109**.[1024] The chemical shift of the methyl resonance varied considerably depending on whether the phenyl ring was *cis* (shielded) or *trans* to it. The different shielding enabled the assignment of the absolute configuration of the diol.

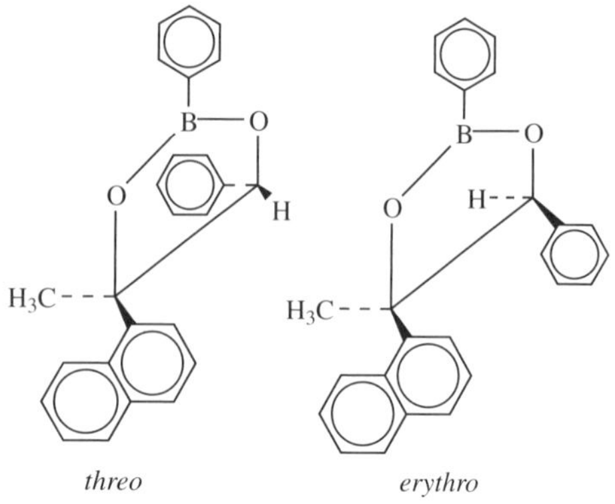

threo   *erythro*

7.109

7.3.3. Boric Acid [B(OH)₃]

The reaction of triols such as **7.110** with boric acid formed a cyclic borate species involving the two hydroxyl groups at the C12 and C14 positions.[1025] A comparison of the shifts of the carbon resonances to data available in the literature on closely related compounds facilitated the assignment of the stereochemistry at C14.

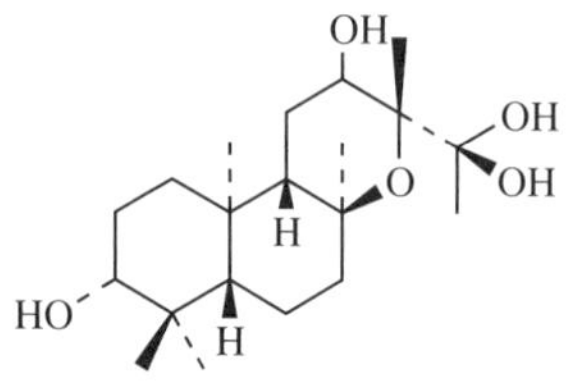

7.110

7.3.4. Camphanylboronic Acid (7.111)

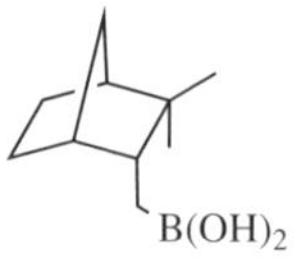

7.111

The determination of the optical purity of 1,2-diols was performed by analyzing the ^{13}C NMR spectra of the boronate derivatives with camphanylboronic acid.[1026] Nonequivalence of ^{13}C signals from the camphanyl unit and substrate were observed in benzene-d_6. The reaction occurred at room temperature and was done directly in the NMR tube.

7.3.5. 4S,5R-4-Methyl-5-phenyl-1,3,2-oxazaborolidine (Ephedrine Borane) (7.112)

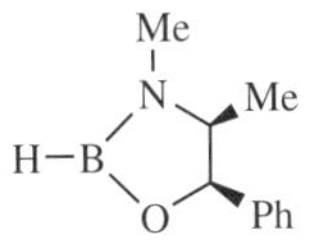

7.112

The borane species prepared from ephedrine has been shown to be an effective reagent for determining the optical purity of secondary alcohols with aliphatic and alkyl aryl substituents, one example of which is **7.113**.[1027] The borate (**7.114**) formed rapidly in toluene-d_8 at 20°C. No kinetic resolution was observed and the N-methyl resonances as well as others showed discrimination in the ^{1}H NMR spectrum.

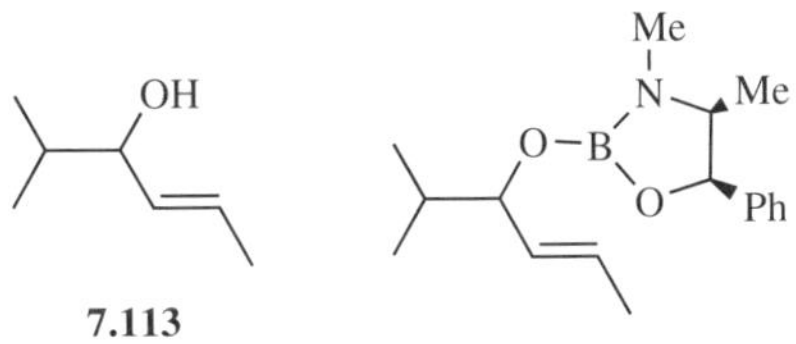

7.113

7.114

7.3.6. (S)-(+)-N-Acetylphenylglycine Boronic Acid (7.115)

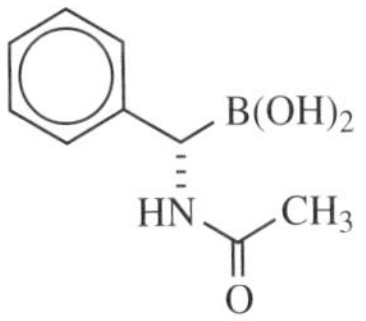

7.115

7.115 was applied to the analysis of the optical purity of 1,2-diols.[1028] Quantitative derivatization occurred in 1 h in tetrahydrofuran. Aliphatic and aromatic substrates with primary, secondary, and tertiary hydroxyl groups were analyzed and nonequivalence of 0.070–0.220 ppm in the ^{1}H and 0.12–0.89 ppm in the ^{13}C NMR spectrum of the diastereomeric products was observed. The discrimination was larger than that observed with 2-(1-methoxyethyl)phenylboronic acid, camphanylboronic acid, and menthyldichlorophosphate.

7.3.7. 1-Benzamido-1-(2-methylphenyl)methaneboronic Acid (7.116)

In an effort to further improve the enantiomeric analysis of 1,2-diols, seven boronic acid reagents with optically pure aryl-containing groups were examined.[1029] Of those studied, **7.116** was the most effective and the chiral recognition was larger than that observed with the acetylphenylglycine derivative.

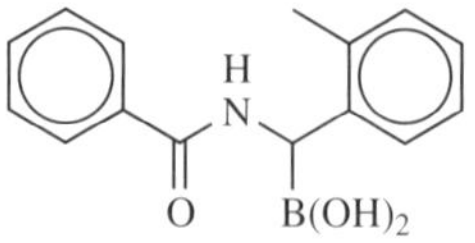

7.116

7.3.8. 2-Formylphenylboronic Acid

2-Formylphenylboronic acid can be coupled with other optically pure reagents to analyze chiral amines and diols. For example, a mixture of a chiral amine, 2,2′-dihydroxy-1,1′-dinaphthalene, and 2-formylphenylboronic acid in chloroform-d with some molecular sieves reacted within 5 min to form the corresponding borate–imine complex as shown in Figure 7.15.[1030] Discrimination in the ^{1}H NMR spectra of the diastereomeric products was used to determine the optical purity of α-arylethyl amines, α-methylalkylamines, β-aminoethers, α-aminoesters, and β-aminoesters. For some compound categories, the relative shifts of the diastereomers correlated with the absolute configuration. Discrimination was observed in the spectra of amines in which the stereogenic center was up to five bonds removed from the amine group.

A procedure for analyzing the optical purity of 1,2-, 1,3-, and 1,4-diols using a 2-formylphenylboronic acid derivative has been described.[1031] Reaction of the boronic acid, diol, and optically pure 1-phenylethylamine produced diastereomeric iminoboronate esters as shown in Figure 7.16. The imine signal was especially diagnostic as it showed large splitting and was in a clear portion of the NMR spectrum. Spectra were run in acetone-d_6.

7.3.9. 2,2′-(1,2-Phenylene)bis[(4R,5R)-4,5-diphenyl-1,3,2-dioxaborolane (7.117)

Association of 1-phenylethylamine (PEA) with **7.117** produced an adduct in which the lone pair of the nitrogen atom associated with one boron atom and the NH formed a hydrogen bond with a borate oxygen.[1032] The ^{1}H NMR spectrum in chloroform-d

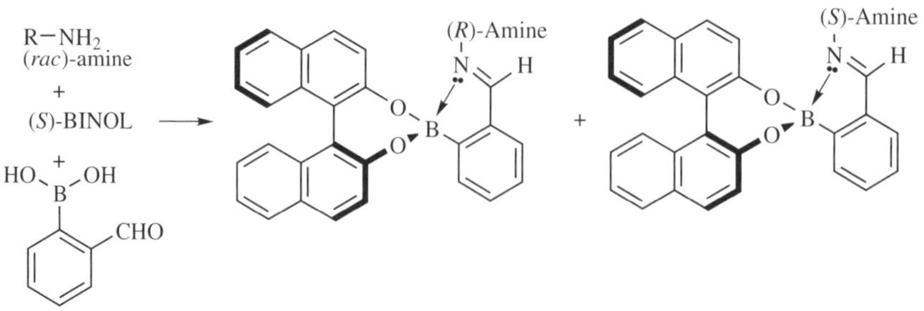

Figure 7.15 Borate–imine complex formed by the reaction of 2,2′-dihydroxy-1,1′-dinaphthalene, 2-formylphenylboronic acid, and a chiral amine.

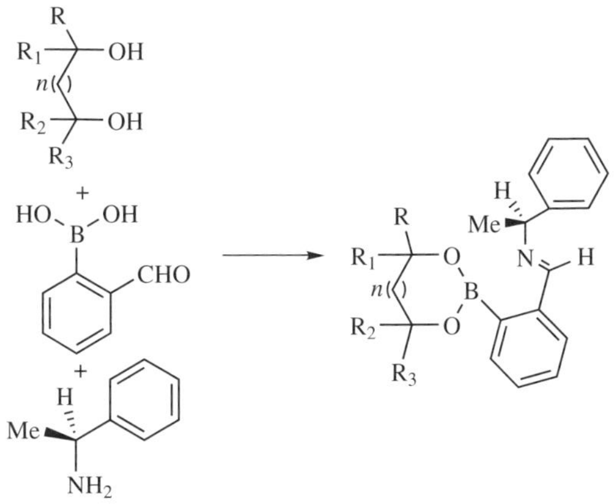

BORON-CONTAINING REAGENTS 273

Figure 7.16 Iminoboronate esters produced by the reaction of boronic acid, 1-phenylethylamine, and a diol.

showed nonequivalence of the methine (0.10 ppm) and methyl (0.07 ppm) resonances of the PEA.

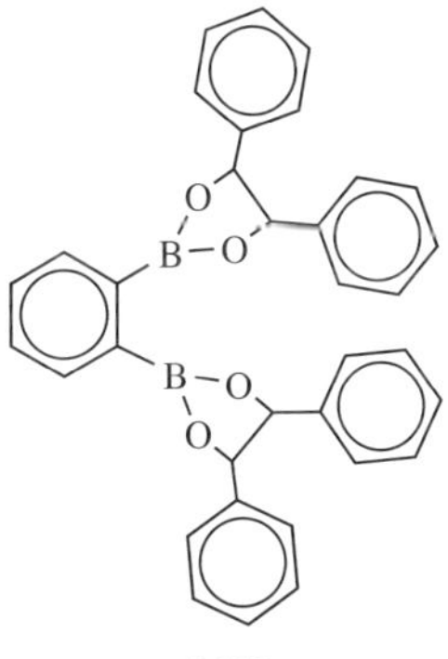

7.117

7.3.10. (1-Chloroalkyl)boronates

Diol derivatives of (1-chloroalkyl)boronates (**7.118**) were further reacted with sodium 2-pyridyl thionate as shown in Figure 7.17 for 2,3-pinanediol.[1033] The

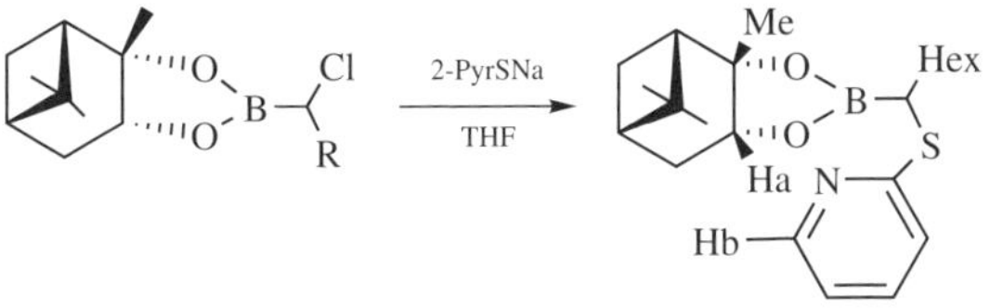

Figure 7.17 Reaction of diol derivatives of (1-chloroalkyl)boronates with sodium 2-pyridyl thionate.

resonances of Ha in the pinane unit or Hb of the pyridyl ring showed significantly different shifts in the spectra of the two diastereomers. No kinetic resolution occurred and the boron was configurationally stable during the reaction. Furthermore, the shifts of the resonances showed trends that correlated with absolute configuration. The corresponding boron compounds with 1,2-dicyclohexyl-1,2-ethanediol could be analyzed in a similar manner, but the discrimination was not as large as with 2,3-pinanediol.

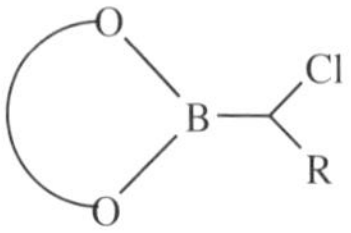

7.118

7.4. SILYL-CONTAINING REAGENTS

A number of silicon-containing compounds have been evaluated as chiral NMR discriminating agents.

7.4.1. Menthylphenylsilylacetic Acid (7.119)

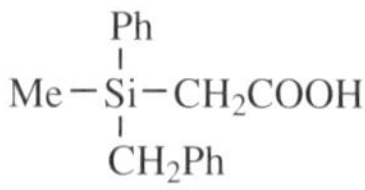

7.119

7.119 has been used as a chiral derivatizing agent for amines and alcohols.[1034,1035] Formation of the corresponding amide or ester caused nonequivalence of the silyl methyl resonance of up to 0.3 ppm, although the distinction with alcohols was smaller than that with amines. Because of the high degree of shielding, the silyl methyl resonance was in an especially clean part of the ^{1}H NMR spectrum.

7.4.2. (Benzylmethylphenylsilyl)methylamine (7.120)

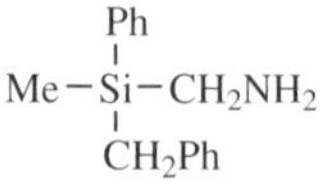

7.120

Using **7.120**, it was possible to analyze the enantiomeric purity of carboxylic acids after preparation of the corresponding diastereomeric amide derivatives.[1035]

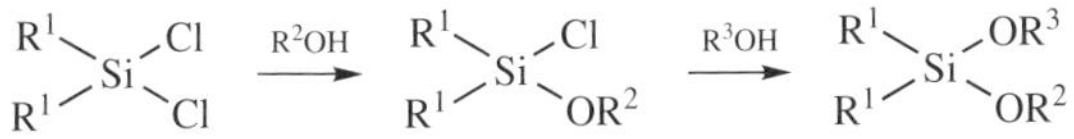

Figure 7.18

Reaction scheme used to prepare silyl acetals from dialkyldichlorosilanes.

7.4.3. Monochlorosilanes

A more common procedure with silicon-containing reagents is to bond the chiral compound directly to the silicon atom. In the case of alcohols, reaction with a dialkyldichlorosilane leads to the formation of a silyl acetal. For example, a dichlorodimethyl or dichlorodiphenyl silane was first reacted with an alcohol of 100% known optical purity (OR^2 in Figure 7.18) such as menthol, borneol, or methyl mandelate to create a monochlorosilane intermediate. The intermediate was then reacted with the unknown alcohol (OR^3) and the product analyzed by ^1H or ^{13}C NMR spectroscopy.[1036,1037] The methyl signal of the optically pure alcohol group was especially convenient to monitor. Larger discrimination was observed for alcohols with an aryl substituent group.

Chloromenthoxydiphenylsilane (**7.121**) was subsequently evaluated for its utility in assigning absolute configurations of secondary alcohols.[1038–1040] Reaction of the (+) and (−) forms of **7.121** with optically pure chiral allylic alcohols such as **7.122** provided diastereomers that could be separated by chromatography. Subsequent analysis of each diastereomer by NMR spectroscopy enabled an assignment of the absolute configuration of the alcohol. The absolute configuration of pinoresinol (**7.123**) was assigned on the basis of an analysis of the bis-silyl derivative with **7.121**.[1039] The splitting of the methyl doublet of the menthoxy group was largest in benzene-d_6. Even though the chiral site was removed from the point of attachment of the silyl group, nonequivalence of 0.004 ppm was still observed for the diastereomers. The utility of **7.121** was also demonstrated with citronellol (**7.124**), N-acetyl tryrosine methyl ester, and 5,9-dimethyl-8-decen-1-ol. In some cases, derivatization with MTPA could not be used because of an elimination process during the reaction. The milder conditions required to prepare derivatives with **7.121** enabled the analysis to be performed.[1038,1040] The compound 11-hydroxyoctadeca-9(Z),12(Z),15(Z)-trieneoic acid methyl ester was one example where MTPA did not work but analysis with **7.121** was successful.

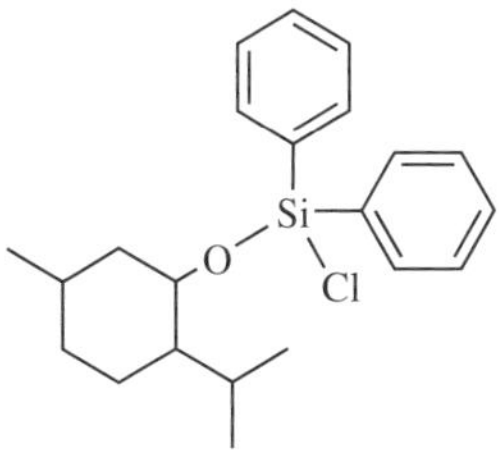

7.121

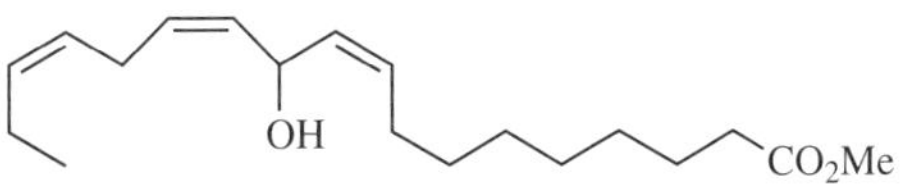

7.122

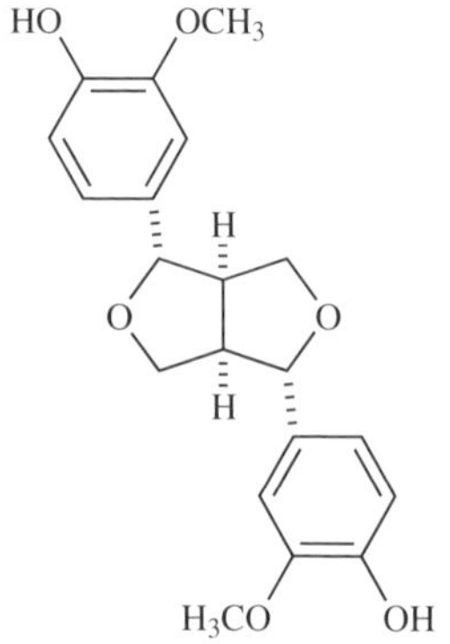

7.123 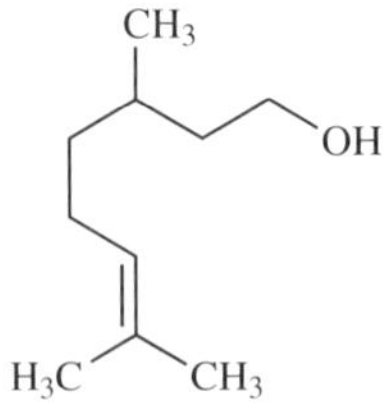 / **7.124**

The corresponding chlorodiphenylsilane reagent with a 2-bromo-α-methylbenzyl group (**7.125**) in place of the menthoxy group was also effective for the enantiomeric analysis of secondary alcohols and has less spectral overlap than the menthoxy group.[1038]

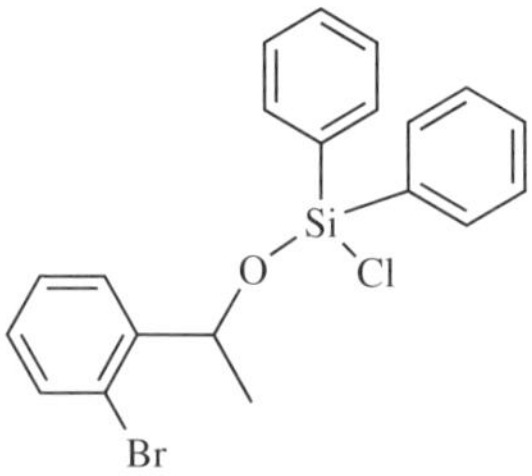

7.125

Chlorodimethyl-(S)-α-(trifluoromethyl)benzyl alcohol silane (**7.126**) reacts with chiral secondary alcohols to form a silyl acetal with a preferred conformation (Fig. 7.19) in which the phenyl ring exhibits predictable shielding of the substituent groups of the alcohol.[1041] The $\Delta\delta^{RS}$ values for L_1 and L_2 were used to assign the absolute configuration of a variety of alkyl, cyclic, and aryl alkyl alcohols. The

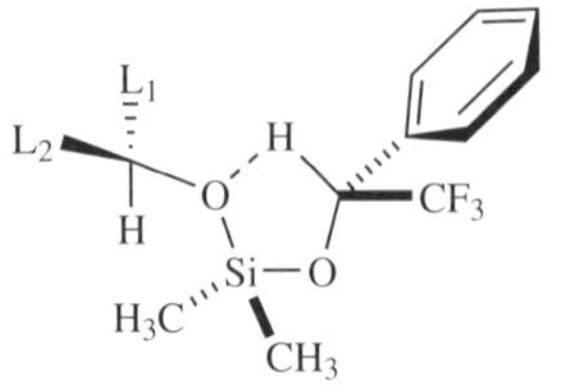

Figure 7.19 Conformational model of the chlorodimethyl-(S)-α-(trifluoromethyl)benzyl alcohol silane derivative of a secondary alcohol.

derivatization reaction occurred under mild conditions and sterically hindered secondary alcohols resistant to standard esterification conditions readily reacted.

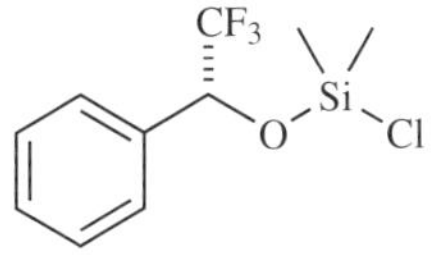

7.126

Dichlorodiphenylsilane was used for the analysis of optical purity of alcohols.[1042] Preparation of a dialkoxy derivative leads to the formation of *meso* [(S,R) and (R,S)] and *dl* [(S,S) and (R,R)] diastereomeric acetals. The ^{13}C resonance of the chiral carbon in substrates, such as 2-butanol, 2-octanol, and menthol, was resolved and the ratio of the peaks was used to determine optical purity.

# 8

HOST COMPOUNDS AS CHIRAL NMR DISCRIMINATING AGENTS

Host-guest strategies have been exploited in the development of chiral NMR discriminating agents. Most host-guest systems are used for the purposes of determining optical purity, although in some cases the interaction is specific enough to allow the assignment of absolute configuration. Most noteworthy among the host-guest complexes are the cyclodextrins. Through appropriate derivatization, cyclodextrin systems spanning the range of every common NMR solvent have been developed. The cyclodextrins form host-guest complexes with a diverse group of substrates making them widely applicable for the analysis of chiral compounds.

By contrast, crown ethers, another important class of host compound, are generally applicable only to the analysis of chiral primary amines, although the utility of a commercially available chiral crown ether for the NMR analysis of secondary amines was recently reported.

Calixarenes and calixresorcarenes have been the focus of more limited chiral NMR studies, although the extent to which these systems are subject to chemical modification that can alter their solubility and chiral selectivity makes it likely that further exploration of these systems will occur.

There are many reports describing the application of specialized receptor compounds. Some of these have been specifically designed to bind carboxylic acid groups in a receptor cavity. Others bind diol functionalities. Finally, a number of other macrocycles with peptide and amide groups, glucan units, and other functional units that can be used for chiral NMR discrimination have been reported.

Discrimination of Chiral Compounds Using NMR Spectroscopy, by Thomas J. Wenzel
Copyright © 2007 John Wiley & Sons, Inc.

8.1. CYCLODEXTRINS

8.1.1. Introduction

Cyclodextrins (CDs) are cyclic oligosaccharides formed from D-glucose units; the most common CDs have six (α), seven (β), or eight (γ) glucose units. The CDs have a basket-shaped geometry in which the narrower opening is ringed with the primary hydroxyl groups of the glucose units and the wider opening is ringed with the secondary hydroxyl groups (Fig. 8.1). The α-, β-, and γ-CDs have different sizes, which can then be tailored to the size of potential substrate compounds. The more complementary the fit, generally the better the enantiomeric discrimination. The underivatized or native CDs are soluble in water, although β-CD is considerably less soluble (maximum concentration of 20–25-mM) than α- and γ-CDs. As such, they represent one of only a few water-soluble systems available for chiral NMR discrimination.

The ability to derivatize CDs through either the primary or secondary hydroxyl groups provides a virtually unlimited set of potential chiral discriminating agents. The nature of the attached substituent groups alters the solubility of the CDs, such that organic-soluble derivatives are available as well. Another important factor with certain CD derivatives involves the degree of substitution (DS) of substituent groups on the CD. Some CD derivatives are prepared with highly specific sites of attachment and DS, whereas others are indiscriminately substituted with variable DS.

An extensive body of literature exists on the application of CD derivatives in the gas and liquid chromatographic separation of enantiomers. More recently, CDs have been applied to chiral separations in capillary electrophoresis. In the aggregate, these studies serve to demonstrate the wide array of chiral compounds that can be analyzed using CDs.

There are two recent reviews on the use of NMR spectroscopy to study the chiral recognition mechanism of CDs.[1043,1044] These authors were primarily interested in using CDs to effect separation in capillary electrophoresis and discussed the use of NMR spectroscopy to study CD-substrate geometry. Procedures for the use of two-dimensional techniques such as NOESY and ROESY measurements, the measurement of binding constants, and utilization of relaxation rate measurements were presented.[1043] These articles do not focus specifically on the use of CDs as chiral NMR discriminating agents. The native CDs are commercially available. Because of their utility in separation techniques and other applications, a number of neutral and ionic CD derivatives are commercially available.

CDs have been widely used to determine the optical purity of compounds by NMR spectroscopy. The geometry of association of a particular substrate or family of

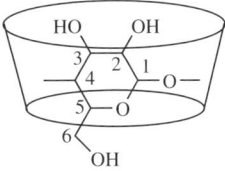

Figure 8.1. Cone-shaped representation of a cyclodextrin superimposed with one D-glucose subunit.

substrates with a specific CD is usually not understood well enough to use NMR shift data to assign absolute configurations, unless it involves empirical observations about trends in shifts for similar compounds.

The discussion herein does not describe every example of the use of a CD for chiral NMR discrimination, but is meant to provide an overview of the range of CD derivatives that have been employed in NMR applications and the types of compounds that have been examined.

8.1.2. Native Cyclodextrins

The use of native CDs as NMR shift reagents was first recognized in 1975 with the report that α-CD could solubilize achiral hydrocarbons such as *p*-cymene and *p*-isopropyltoluene in deuterium oxide at 50°C.[1045] The first use of CDs for chiral NMR discrimination was in 1977 with the report that β-CD caused enantiomeric discrimination of the ^{19}F resonance of 1-phenyl-2,2,2-trifluoroethanol.[1046] In the same report, it was noted that addition of β-CD distinguished the ^{19}F signals of the prochiral trifluoromethyl groups in **8.1** and **8.2**.[1046] β-CD was subsequently shown to cause a small amount of enantiomeric discrimination in the ^{1}H NMR spectrum of pirprofen (**8.3**).[1047]

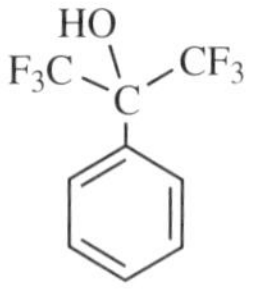

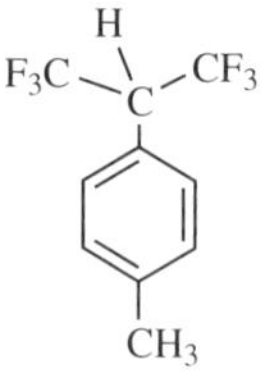

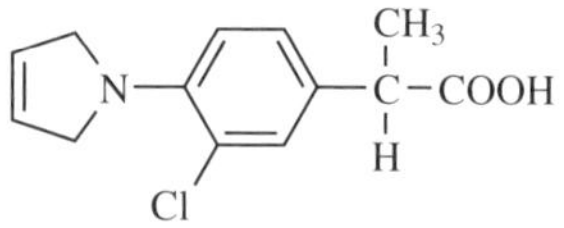

The first comparison of the ability of α-, β- and γ-cyclodextrins to cause chiral discrimination in ^{1}H NMR spectroscopy involved a study of propranolol (**8.4**).[1048] The influence that the different cavity sizes had on the resulting enantiomeric discrimination was demonstrated. No shifts were observed in the spectrum of **8.4** with α-CD, presumably because the naphthyl ring was too large to fit into the cavity. The spectrum of **8.4** with β-CD showed small shifts and the ^{1}H resonances of the methylene group adjacent to the NH group exhibited some enantiomeric discrimination. The shifts in the ^{1}H NMR spectrum of **8.4** were largest with γ-CD. The chiral recognition was also greatest with γ-CD and some of the aromatic signals showed nonequivalence as well.[1048]

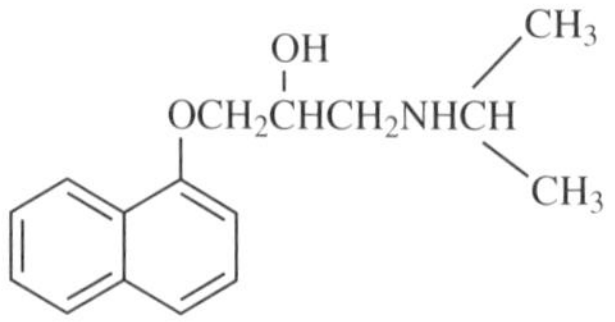

CYCLODEXTRINS

The versatility of native CDs as water-soluble chiral NMR discriminating agents was subsequently demonstrated in a study of a wide range of substrate compounds.[1049] Enantiomeric discrimination was observed in the ^{1}H NMR spectra of antihistamines such as dimethindene maleate (**8.5**), carbinoxamine maleate, doxylamine succinate, neobenodine hydrochloride, and pheniramine hydrochloride, as well as for analgesics such as alphaprodine (**8.6**), methadone hydrochloride, propoxyphene hydrochloride, and sodium ibuprofen (**8.7**).

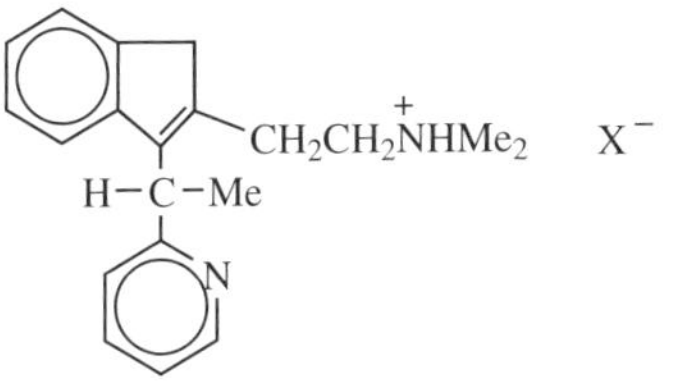

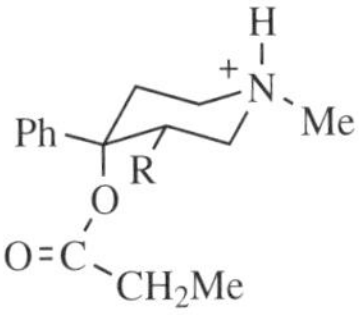

The majority of compounds for which native CDs have been used to measure enantiopurity contain an aromatic ring. The sizes of the CD cavities are complementary to phenyl (α and β) and bicyclic (β and γ) aromatic ring systems. Insertion of the hydrophobic aromatic ring into the relatively hydrophobic CD cavity provides a driving force for host-guest association in water.

For the 1-*O*-allyl- and 1-*O*-benzyl myoinositol derivatives (**8.8**), α- and β-CDs provided different levels of chiral discrimination for different derivatives, hence it is worth testing both on any analogous compound.[1050] For fenoldopam (**8.9**), the chiral discrimination with β-CD was greater than that with α- or γ-CD.[1051] A similar result was observed with 9-[(phenylmethyl)amino]-1,2,3,4-tetrahydroacridine-1-ol maleate (**8.10**).[1052] The benzylic resonances of **8.10** were nonequivalent in the ^{1}H NMR spectrum with β-CD but not with α- or γ-CD. For the dihydrochloride salt of telenzepine (**8.11**), the hydrogen and methyl resonances of the thienyl ring were enantiomercially discriminated in the presence of β- and γ-CDs but not α-CD.[1053] **8.11** has a major and a minor conformer caused by slow rotation about the amide bond that leads to two sets of resonances in the ^{1}H NMR spectrum of the substrate. Only the minor conformer complexed with β-CD and showed a splitting of peaks in the NMR spectrum. With γ-CD, both conformers associated and both sets of peaks split in the NMR spectrum.

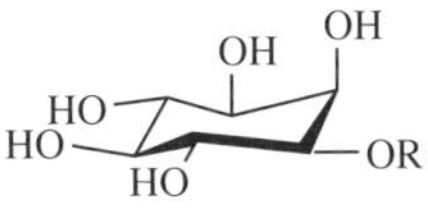

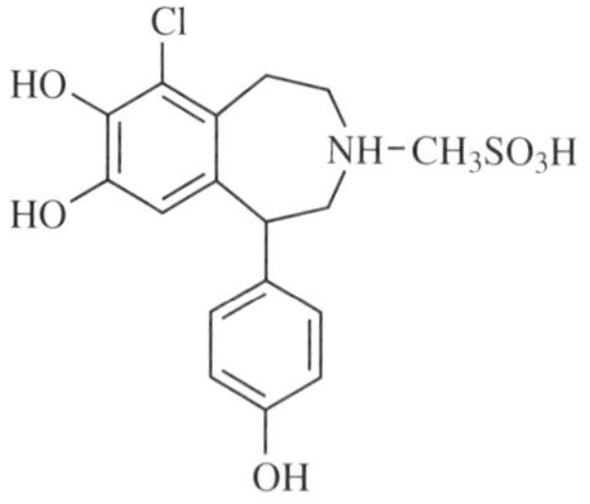

The ^{1}H NMR spectrum of the cationic compound amlodipine maleate (**8.12**) exhibited enantiomeric discrimination in the presence of γ-CD that was comparable in magnitude to the distinction caused by carboxymethyl and sulfobutylether derivatives of β-CD.[1054] Both the ^{1}H and ^{13}C NMR spectra of *cis*-ketoconazole tartrate (**8.13**) showed chiral recognition in the presence of β-CD.[1055]

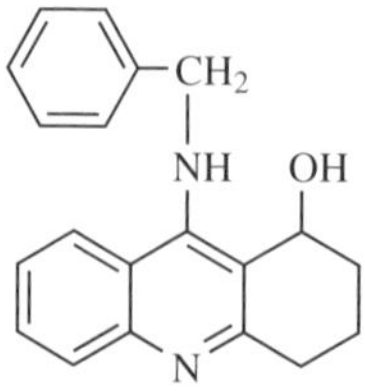

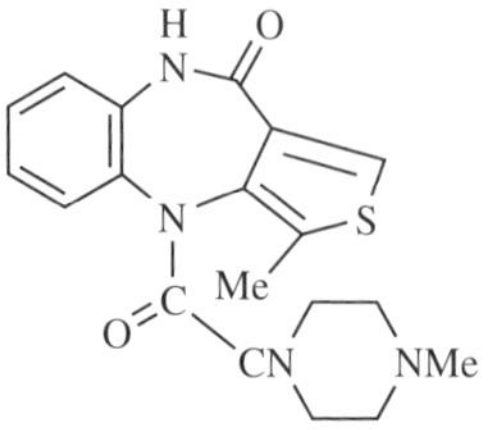

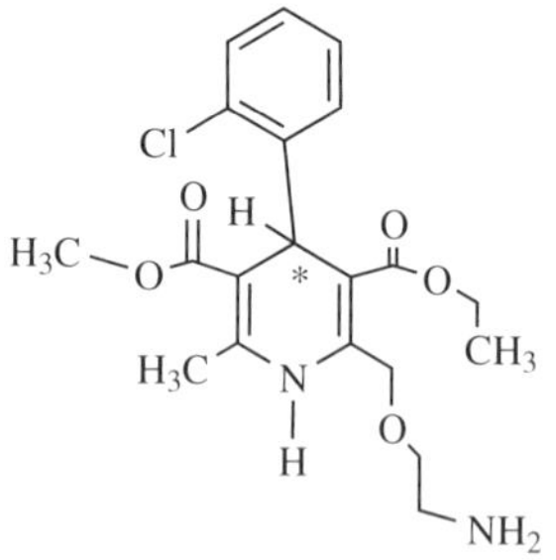

A comparison of the effectiveness of α-, β-, and γ-CDs and heptakis(2,6-di-*O*-methyl)-β-CD at causing enantiomeric discrimination in the ^{1}H NMR spectrum of cizolirtine (**8.14**) and its corresponding carbinol derivative was examined.[1056] The best results were observed with β-CD and nonequivalence of 0.007–0.083 ppm was observed for certain resonances.

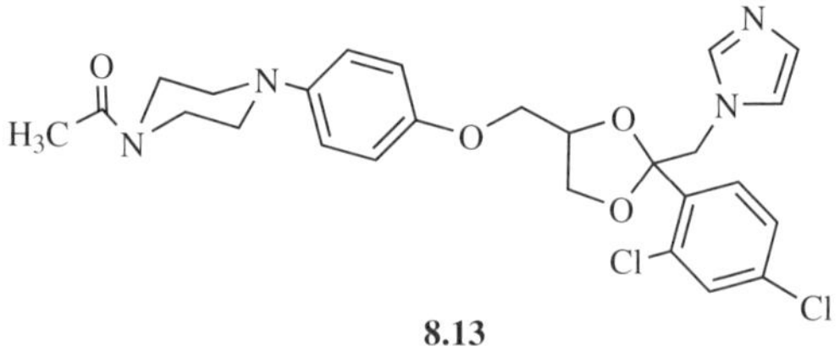

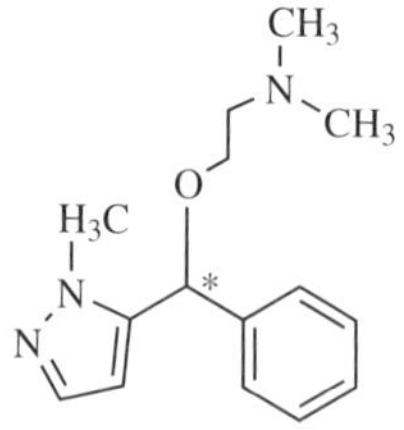

An excellent example of the utility of CDs for chiral analysis involves the discrimination in the ^{1}H NMR spectrum of metolachlor ESA (**8.15**) with γ-CD. The methoxy resonance appears as eight singlets in the presence of γ-CD.[1057] These represent the resonances for each of the four stereoisomers (aS,5S; aR,5S; aR,5R; aS,5R) of the *s-trans* and *s-cis* compounds.

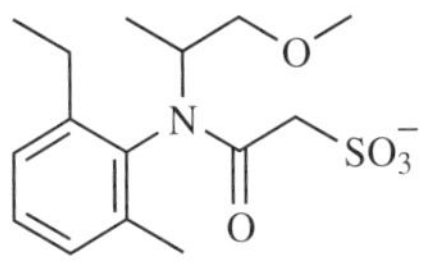

8.15

The optical purity of 1-[2-carboxy-6-(trifluoromethyl)phenyl]pyrrole-2-carboxylic acid (**8.16**) was determined by analysis of the ^{19}F NMR spectrum in the presence of β-CD.[1058] Both β- and γ-CDs caused appreciable chiral discrimination in certain of the ^{1}H resonances of the helical compounds **8.17**.[1059,1060] Other CDs, including α-CD and heptakis(2,6-di-*O*-methyl)-β-CD and heptakis(tri-*O*-methyl)-β-CD were ineffective at discriminating **8.17**.[1060]

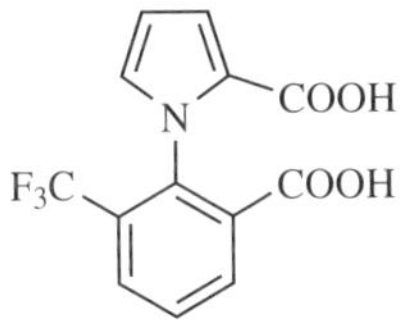

8.16 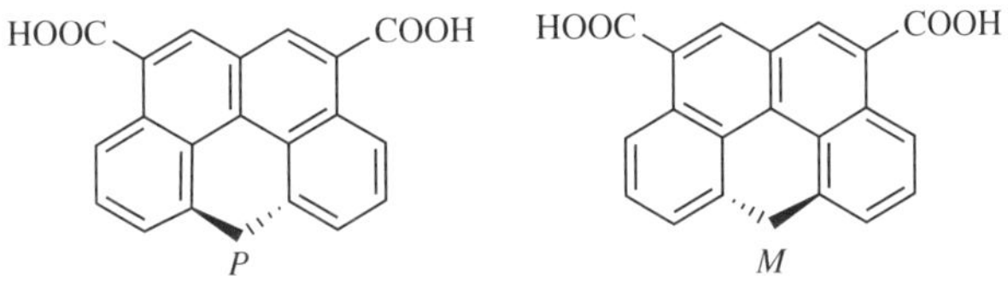   **8.17**

The trifluoromethyl signals of α-methoxy-α-trifluoromethylphenylacetic acid (MTPA) were enantiomerically discriminated in the presence of β-CD.[1061] A procedure for analyzing the optical purity of MTPA in diluted samples of urine by measuring the NMR spectrum in the presence of β-CD was described. This facilitated the study of the in vivo inversion of the stereocenter of MTPA in rats.

The optical purities of amino acids with aromatic rings either present in the substituent group or added through a derivatization reaction have been determined using native CDs. The ^{19}F signal of amino acid derivatives such as *N*-(*p*-fluorophenyl)glycine and *N*-(*p*-fluorobenzoyl)valine was enantiomerically discriminated in the presence of α-CD.[1062] An interesting observation was that the trends of the relative shifts of the (*R*)- and (*S*)-enantiomers were reversed for the protonated and deprotonated substrates.

The addition of β-CD caused a small nonequivalence for some of the alkyl ^{1}H resonances of the 3,5-dinitrophenyl derivatives of valine, leucine, and methionine.[1063] Resonances of the β-naphthylamide derivative of alanine exhibited chiral discrimination in the presence of β-CD, but no shifts in the presence of α-CD.[1064] The ^{1}H NMR spectra of dipeptides with an aromatic residue such as Ala-Phe and Ala-Tyr exhibited chiral discrimination in the presence of β-CD.[1065] The distinction in the spectra of dipeptides was comparable to that with heptakis(6-*O*-sulfo)-β-CD and better than that with heptakis(2,3-di-*O*-acetyl)-β-CD.

The ^{1}H NMR spectra of benzo[*lmn*][3,8]phenanthroline derivatives of amino acids such as tryptophan (**8.18**) showed enantiomeric discrimination in the presence of β-CD.[1066] No distinction was observed with γ-CD. On the basis of mass spectral evidence, it was proposed that the substrate formed a ternary complex with γ-CD that likely reduced the chiral recognition.

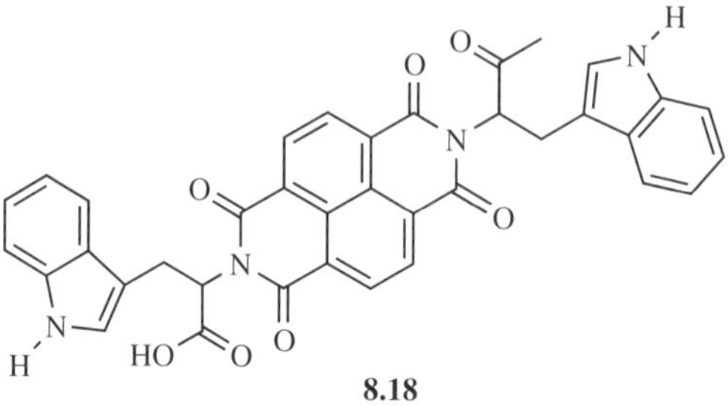

8.18

In some cases, the formation of a ternary complex with γ-CD may have a beneficial effect on the degree of chiral recognition. For example, the ^{1}H NMR spectrum of the benzo[de]isoquinoline-1,3-dione derivative of tryptophan (**8.19**) exhibited chiral discrimination in the presence of γ-CD, but not with α- or β-CD.[1067] Evidence suggested that **8.19** formed a 2:1 complex with the γ-CD that was not possible with the smaller α- or β-CD. It also appeared likely that one of the two guests had the benzo[de]isoquinoline ring in the cavity, whereas the other had the tryptophan ring inserted.

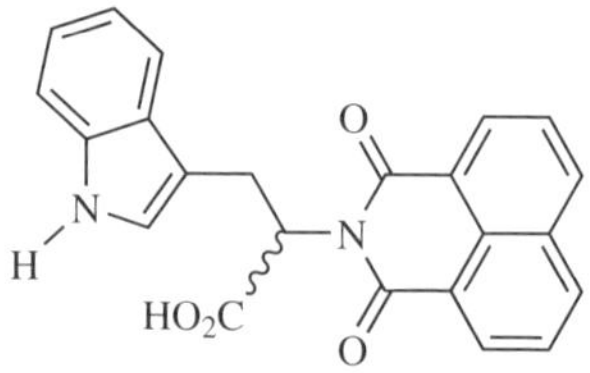

8.19

Native CDs have been shown to cause enantiomeric discrimination in the ^{31}P NMR spectra of aminophosphonic (**8.20**) and aminophosphinic (**8.21**) acids[1068] and their N-benzyloxycarbonyl protected analogs.[1069] It was desirable to use as high a cyclodextrin-to-substrate ratio as possible, typically 10:1. In most cases, the best results were observed with α-CD and nonequivalence as large as 0.6 ppm was observed in the ^{31}P NMR spectrum.[1068] In addition, the chiral recognition was greater at lower pH values.

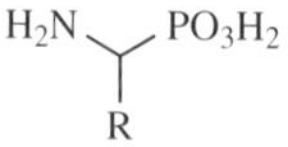

8.20 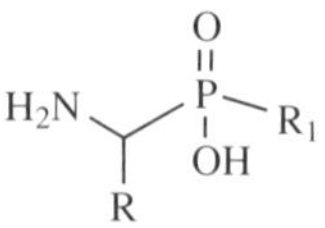   **8.21**

Native CDs are also candidates for the chiral NMR analysis of compounds without an aromatic ring. cis-Decalin (**8.22**) has two invertomers and three of the signals in the ^{13}C NMR spectrum split in the presence of β-CD.[1070]

8.22

Several reports on the chiral recognition of α-pinene (**8.23**) with CDs have been published.[1071–1074] Among the native CDs, the best enantiomeric discrimination was observed with α-CD.[1073] Permethyl α-, β-, and γ-CDs were also evaluated but were not as effective as α-CD.[1073] Peaks in the ^{13}C NMR spectrum of α-pinene split in the presence of α-CD into dimethylsulfoxide-d_6 and dimethylformamide-d_7.[1074]

8.23

The ^{1}H NMR spectrum of camphor (**8.24**) exhibited nonequivalence of 0.01–0.12 ppm in the presence of α-CD.[1075] The sodium salt of abscisic acid (**8.25**) and other compounds with similar structures showed a splitting of several resonances in the ^{1}H NMR spectrum in the presence of γ-CD.[1076] Discrimination was not observed in the NMR spectrum with α- and β-CDs.

8.24   **8.25**

The Yb(III) complexes of tetraaza macrocyclic ligands containing three phosphinate groups and one carboxamide and in which one of the R groups is a phenyl ring (**8.26**) are enantiomeric. The ^{31}P resonance of the Yb(III) complex splits into two signals in the presence of β-CD.[1077] The Yb(III) complexes associated with β-CD by insertion of the phenyl ring into the cavity.

8.26

8.1.2.1. Lanthanide Coupling to Native Cyclodextrins

The coupling of paramagnetic lanthanide ions to CDs is a means of enhancing the enantiomeric discrimination in the NMR spectrum.[1078–1080] The lanthanide ion was bound to the CD through a diethylenetriaminepentaacetic acid (DTPA)-like ligand. The DTPA moiety was attached by reacting DTPA-dianhydride with a monoethylenediamine (EN) (Fig. 8.2)[1078] or monoamine (NH)[1079] CD derivative. By varying the conditions used to prepare the tosyl intermediate, it was possible to attach the DTPA ligand at either the 6- (primary) or 2- (secondary) position. Dy(III) was then added as its nitrate salt in small aliquots to solutions of the CD-DTPA. Most samples were run at 50°C to reduce the exchange broadening. The warmer temperature did reduce the magnitude of the dysprosium-induced shifts and hence the chiral discrimination. But considerable enhancements in enantiomeric distinction were still observed when compared to samples of the native CD and substrate at ambient temperature.

Enhancements in enantiomeric discrimination in the ^{1}H NMR spectra of carbinoxamine (**8.27**), pheniramine, doxylamine, propranolol (**8.4**), and the sodium salt of tryptophan were observed. These often went from no observable discrimination to several hundredths of a ppm. For the EN-linked derivatives, addition of Dy(III) to the 2-substituted CD caused far greater improvements in chiral discrimination than those observed with the 6-substituted CD.[1078] Chiral recognition with CDs usually involves differences in association at the secondary face. Presumably, having the Dy(III) closer to the site of chiral differentiation accounted for the effectiveness of the Dy(III) complex of the 2-CD-EN-DTPA over the 6-CD-EN-DTPA derivative. The Dy(III) complexes of the DTPA derivative with an NH link at the 6-position were more effective than the corresponding 6-EN-linked compound.[1079] Presumably,

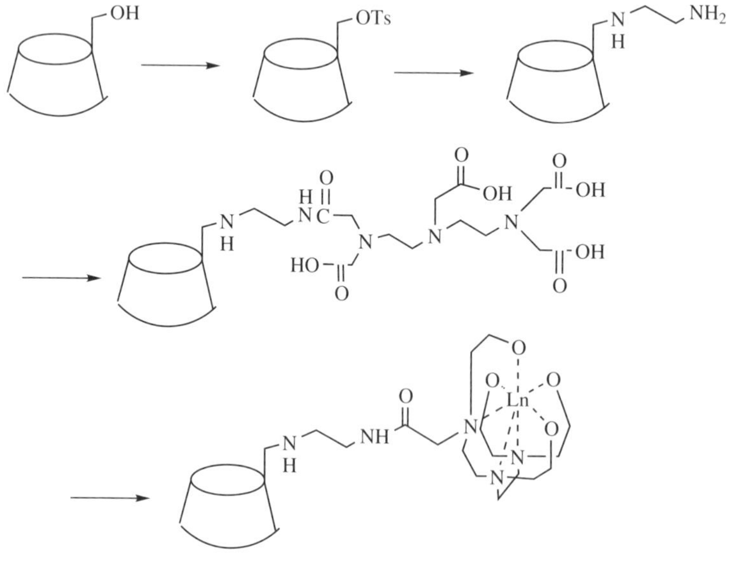

Figure 8.2. Scheme used to couple lanthanide ions to cyclodextrins through a DTPA amide unit.

the shorter amine tether positioned the Dy(III) closer to the CD cavity and increased the magnitude of the dysprosium-induced shifts and enantiomeric discrimination. The Dy(III) complex of the DTPA derivative with an NH link at the 2-position was not as effective as the corresponding 2-EN-linked compound.[1079] In this case, evidence indicated that the shorter tether caused the Dy(III)-DTPA moiety to restrict access of the substrate into the cavity.

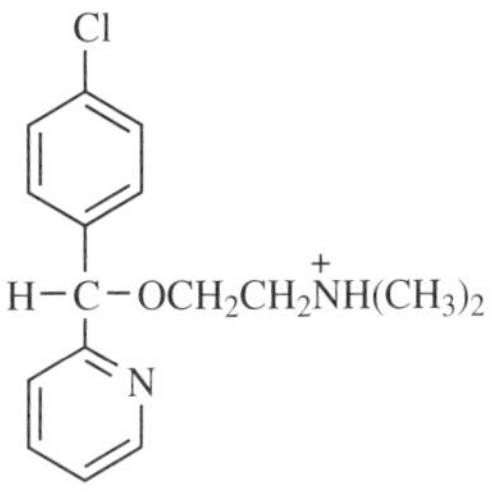

8.27

An alternative means of binding a lanthanide to CD involved coupling a 1,4,7-tri(carboxylatomethyl)-1,4,7,10-tatraazacyclododecan-10-yl group at the 6-position (**8.28**).[1080] The effect of the Eu(III) complex of **8.28** on histidinate (**8.29**), tyrosinate, phenylalaninate, and tryptophate was examined. Addition of Eu(III) shifted the ^{1}H resonances of all of the substrates, but the only improvement in enantiomeric discrimination was observed in the spectrum of histidinate.

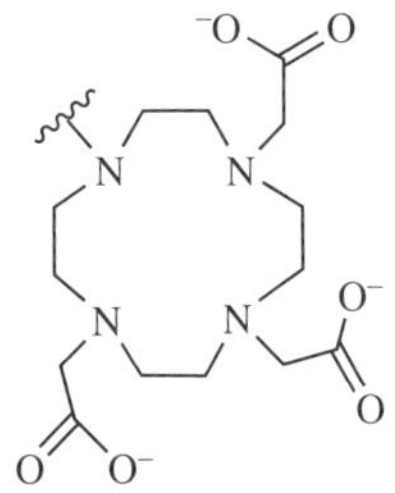

8.28 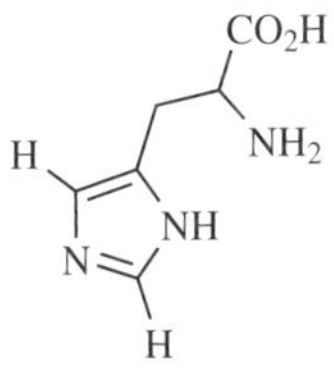   **8.29**

8.1.3. Neutral Cyclodextrin Derivatives

8.1.3.1. Hexakis(2,3,6-tri-O-methyl)-α-cyclodextrin (TM-α-CD)/Heptakis(2,3,6-tri-O-methyl)-β-cyclodextrin (TM-β-CD) – (Permethyl CDs) A somewhat unexpected observation is that TM-β-CD is more soluble in water than native β-CD. This enhanced solubility is most likely the reason why TM-β-CD was more effective in causing enantiomeric discrimination in the ^{19}F NMR spectrum of (5(Z)-7-([2RS,4RS,5SR]-4-o-hydroxyphenyl-2-trifluoromethyl-1,3-dioxan-5-yl)heptenoic acid (**8.30**) than β-CD.[1081] Although α-CD was more effective than TM-α-CD in causing enantiomeric discrimination in the ^{1}H NMR spectrum of α-pinene (**8.23**),

TM-β-CD was the only reagent of α- and β-CDs and their corresponding permethyl derivatives that caused enantiomeric discrimination in the ^{13}C NMR spectrum.[1071]

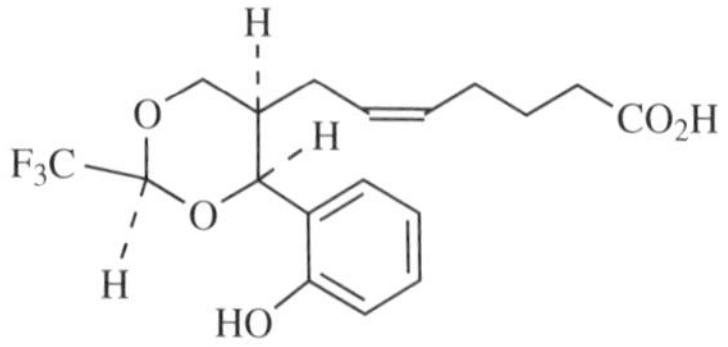

8.30

The enantiomeric discrimination in the ^{1}H NMR spectra of 1,1'-binaphthyl-2,2'-diyl hydrogen phosphate (**8.31**) and 1,1'-binaphthyl-2,2'-dicarboxylic acid (**8.32**) in deuterium oxide was much greater with TM-β-CD than with β-CD.[1082] TM-β-CD also caused chiral recognition in the ^{31}P NMR spectrum of **8.31**.

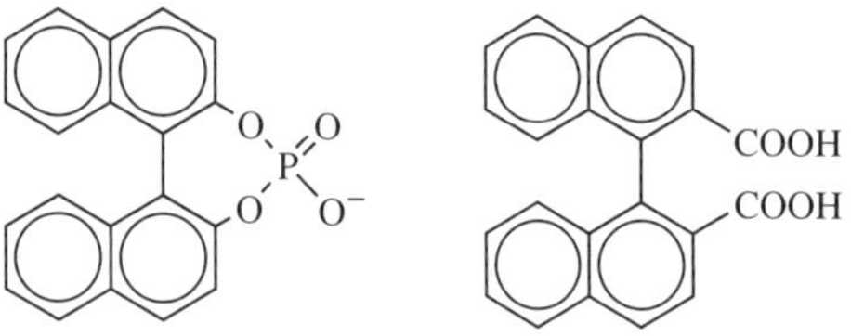

8.31   **8.32**

The effectiveness of TM-β-CD, β-CD, and a biphenyl-capped CD in producing chiral discrimination in the spectra of mandelic acid, 1-phenylethylamine, and 2-phenylpropionic acid in deuterium oxide was compared.[1083] TM-β-CD was the only reagent that caused discrimination in the spectra of all three substrates. Nonequivalence on the order of 0.02–0.09 ppm was observed in the ^{1}H NMR spectra.

Chiral discrimination in the ^{13}C NMR spectrum of 2-isopropyl-2,8-bis(3,4-dimethoxyphenyl)-6-methyl-6-azaoctannitrile (verapamil – **8.33**) in the presence of β-CD and TM-β-CD in deuterium oxide was examined.[1084] Even though the shifts were larger with β-CD, the enantiomeric distinction was greater with TM-β-CD.

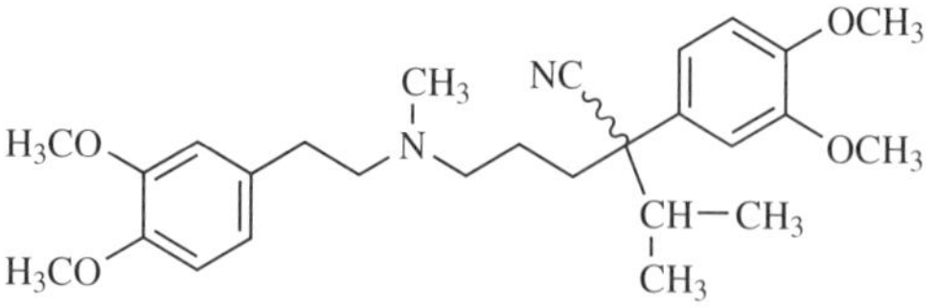

8.33

Chiral recognition was observed in the ^{1}H NMR spectra of metal complexes such as Ru(phen)$_3^{2+}$ and Ru(bipy)$_3^{2+}$ with TM-α-CD but not with TM-β-CD.[1085] The association of these cationic metal complexes with TM-α-CD was much less than with an anionic thiocarboxymethylated CD, although the enantiomeric discrimination with these two CD derivatives was comparable.[1085]

TM-β-CD is especially effective in causing enantiomeric discrimination in the ^{1}H NMR spectra of chiral trisubstituted allenes (**8.34**).[1086–1089] In comparison with TM-α-CD,[1089] heptakis(2,6-di-O-methyl)-β-CD,[1086,1088] and heptakis(2,3,6-tri-O-acetyl)-β-CD,[1086] TM-β-CD was consistently the best reagent for the analysis of allenes. Nonequivalence of 10-20 Hz at 300 MHz was observed for the allene hydrogen resonance. The resonances of the alkyl substituent groups showed the same trend. Comparable levels of discrimination were observed in methanol-d_4 and dimethylsulfoxide-d_6, but no distinction occurred in acetone-d_6.[1086,1088] It was of particular significance that TM-α-CD caused deshielding of the allene hydrogen that was consistently larger for the (S)-enantiomer.[1087,1088]

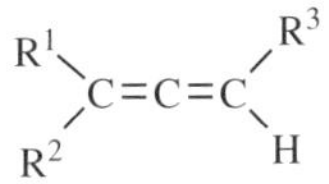

$$R^1\diagdown_{R^2}C=C=C\diagup^{R^3}_{\diagdown H}$$

8.34

The utility of TM-β-CD and TM-α-CD for the analysis of aromatic hydrocarbon substrates containing phenyl, 1-naphthyl, and 2-naphthyl rings was examined.[1089,1090] ^{1}H NMR spectra recorded methanol-d_4. The CDs were effective for all of the substrates with a phenyl and a 2-naphthyl substituent, but only for some with a 1-naphthyl substituent. The nonequivalence was generally quite small (1–4 Hz), but in most cases was larger with TM-α-CD than with TM-β-CD.[1089]

The use of TM-β-CD in methanol-d_4 to determine the optical purity of 1-phenylethanol was evaluated.[1091] The methine resonance exhibited negligible nonequivalence (0.005 ppm) at a probe temperature of 15°C. Reducing the temperature to $-50°$C caused an enhancement in the distinction to 0.016 ppm, although the enantiomers were not fully resolved. Complete enantiomeric discrimination of greater than 0.1 ppm was observed in the ^{13}C NMR spectrum.[1091]

8.1.3.2. Heptakis(2,6-di-O-methyl)-β-cyclodextrin (2,6-DM-β-CD) 2,6-DM-β-CD caused a small extent of nonequivalence (0.01 ppm) in some of the resonances in the ^{1}H NMR spectrum of alanine β-naphthylamide.[1064] The enantiomers of N-methacryloyl-DL-phenylalanine methyl ester (**8.35**) exhibited distinct resonances in the spectrum with 2,6-DM-β-CD.[1092] In polymerization reactions of **8.35**, the presence of 2,6-DM-β-CD influenced which isomer was preferentially incorporated into the polymer by its preferential complexation of one of the enantiomers.

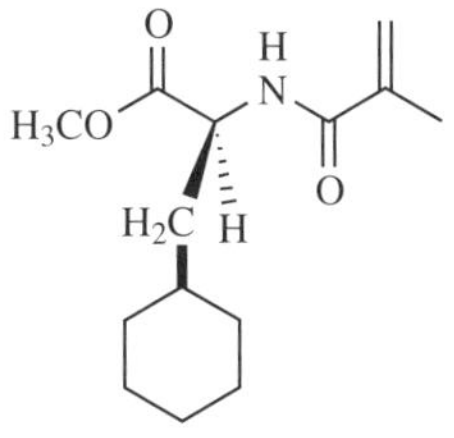

8.35

8.1.3.3. Benzylated and Benzoylated Cyclodextrins The utility of benzoyl ($-OC(O)C_6H_5$) and benzyl ($-O-CH_2C_6H_5$) substituted CDs as chloroform-soluble chiral discriminating agents for 3,5-dinitrophenyl derivatives of chiral amines, amino alcohols, alcohols, carboxylic acids, and amino acids has been explored.[1093,1094] These include hexakis(2,3-di-*O*-benzoyl)-α-CD,[1093] hexakis(2,3,6-tri-*O*-benzoyl)-α-CD,[1093] hexakis(6-*O*-benzoyl)-α-CD,[1094] heptakis(2,3,6-tri-*O*-benzoyl)-β-CD,[1094] hexakis(2,3-di-*O*-benzyl)-α-CD,[1094] and hexakis(6-*O*-benzoyl-2,3-di-O-benzyl)-α-CD.[1094] The resonances of the internal hydrogen atoms did not shift on addition of the substrate, an indication that association occurred external to the cavity and presumably involved π–π interactions of the electron-deficient dinitrophenyl ring with the benzoyl and benzyl rings. The CDs that had substituent groups at primary or secondary sites only were generally more effective than those that were exhaustively substituted.[1094] Nonequivalence in the spectra was modest and typically in the range of 2–7 Hz at 300 MHz.

8.1.3.4. Carbamoylated Cyclodextrins Carbamate derivatives of cellulose represent the most widely used class of chiral liquid chromatographic phases today. Carbamoylated CDs ($-OC(O)NH-R$) have been explored for their utility as chiral NMR discriminating agents.[1095,1096] The ^{1}H NMR spectrum of atenolol (**8.36**) with the perphenylcarbamate derivative of β-CD showed enantiomeric discrimination, but the interactions were quite weak and distinction was only minimal.[1095]

$$\text{OCH}_2\text{CHCH}_2\text{NHCH}$$

with OH and CH₃ substituents, *chiral center, and para-CH₂CONH₂ on the benzene ring.

8.36

A series of α- and β-CDs with 3,5-dimethylphenylcarbamate groups at the 2-, 3-, and 6-positions, mixed carbamoylated-silylated (3,5-dimethylphenylcarbamate at the 2- and 3-positions, *tert*-butyldimethylsilane at the 6-position), and partially carbamoylated (3,5-dimethylphenylcarbamate at the 2- and 3-positions) were evaluated as chloroform-soluble chiral NMR discriminating agents.[1096] Of the various reagents, the *per*-carbamoyl derivative was the most effective, although the mixed 2,3-carbamoyl-6-silylated derivative was quite comparable. The carbamoyl derivatives were more effective than a series of CDs with benzoyl and benzyl groups studied in the same laboratory.[1093,1094] Nonequivalence of 3–40 Hz (at 300 MHz) was observed in the ^{1}H NMR spectra of *N*-(3,5-dinitrobenzoyl) amino acid methyl esters. The carbamoylated CDs caused chiral recognition in the spectra of carboxylic acids and alcohols as well. The spectra of amines did not show enantiomeric discrimination; however, the corresponding amide derivatives with phenyl, 3,5-dimethoxyphenyl, *p*-nitrophenyl, or 3,5-dinitrophenyl groups did.[1096] There was no evidence that the substrates entered the cavity of the CD. Instead, the substrates associated through polar groups and π–π interactions with the external carbamoyl groups.

The effectiveness of mixed methylated/carbamoylated β-CDs [heptakis(2,3-di-*O*-methyl-6-*O*-(3,5-dimethylphenylcarbamoyl))-β-CD and heptakis(6-*O*-methyl-2,3-di-*O*-(3,5-dimethylphenylcarbamoyl))-β-CD] as chiral NMR solvating agents was compared to TM-β-CD and other partially methylated CDs for apolar trisubstituted allenes and polar compounds modified by the addition of a 3,5-dinitrophenyl ring.[1097] The TM-β-CD derivative was the most effective for the trisubstituted allenes. Using *N*-(3,5-dinitrobenzoyl)alanine methyl esters as a test substrate, no enantiomeric distinction occurred with TM-β-CD whereas the carbamoylated derivatives did cause chiral discrimination in the NMR spectrum.

8.1.3.5. Octakis(3-O-butanoyl-2,6-di-O-pentyl)-γ-cyclodextrin (BP-CD) BP-CD was developed for use in gas chromatographic separations. The compound is soluble in organic solvents such as chloroform and cyclohexane. The ^{1}H NMR spectrum of enflurane (**8.37**), isoflurane (**8.38**), and desflurane (**8.39**) showed only marginal enantiomeric discrimination when mixed with BP-CD in chloroform-*d*.[1098,1099] In the more nonpolar cyclohexane-d_{12}, both the ^{1}H and ^{19}F NMR spectra of the substrates exhibited chiral discrimination. For example, the ^{1}H resonances of **8.37** were separated by 0.13 ppm in the presence of BP-CD.

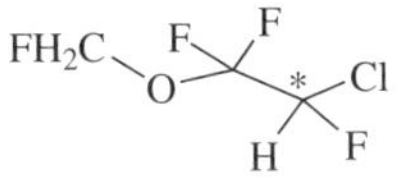

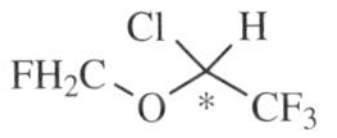

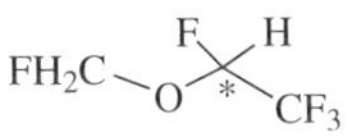

8.1.3.6. Hydroxyethyl and Hydroxypropyl Cyclodextrins (HE-CD, HP-CD) A few reports have examined the ability of HE- and HP-CDs to function as chiral NMR discriminating agents.[1051,1069,1081,1100,1101] In each case, the enantiomeric discrimination was quite small, and in those instances in which other CDs were examined, the HE-CD and HP-CD derivatives were not as effective.[1069,1081]

8.1.3.7. Heptakis(2,3-di-O-acetyl)-β-CD (2,3-DA-β-CD) The utility of 2,3-DA-β-CD for measuring the optical purity of a variety of phenethylamines has been examined.[1093,1102–1104] The discrimination compared favorably to that obtained with heptakis(6-*O*-acetyl)-β-CD[1103,1105] and was either comparable to[1105] or more effective than native β-CD.[1102,1103] Nonequivalence in the ^{1}H NMR spectrum was typically on the order of a few hundreds of a ppm. The reagent has been shown to cause a small degree of chiral recognition in the ^{1}H NMR spectra of alanine β-naphthylamide (**8.40**)[1064] and the dipeptides Ala-Tyr and Ala-Phe.[1106]

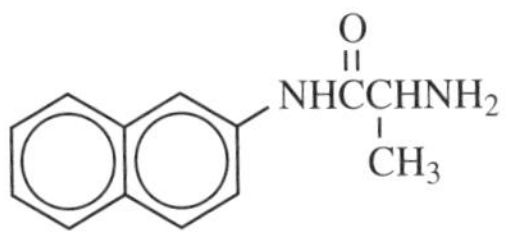

8.1.3.8. Mono(2-acetylamino-2-deoxy)-, Mono(3-acetylamino-3-deoxy)-, and Mono(6-acetylamino-6-deoxy)-β-cyclodextrin (AA-β-CD)
The extent to which 2-AA-β-CD, 3-AA-β-CD, and 6-AA-β-CD (-NHC(O)CH$_3$) caused chiral recognition in the ^{1}H NMR spectra of mandelic acid, methyl mandelate, and N-acetyl-α-phenylglycine was compared.[1107,1108] The 2- and 3-AA-β-CD derivatives were the most effective.

8.1.3.9. Mono-3,6-anhydro-β-cyclodextrin
A comparison of the chiral discrimination in the ^{1}H NMR spectrum of dothiepine (**8.41**) with mono-3,6-anhydro-β-CD (**8.42**), mono-3,6-anhydro-heptakis(2-O-methyl)hexakis(6-O-methyl)-β-CD, and mono-3,6-anhydro-heptakis(2,3-di-O-methyl)hexakis(6-O-methyl)-β-CD was performed.[1109,1110] The best results were observed with the nonmethylated and dimethylated derivatives.

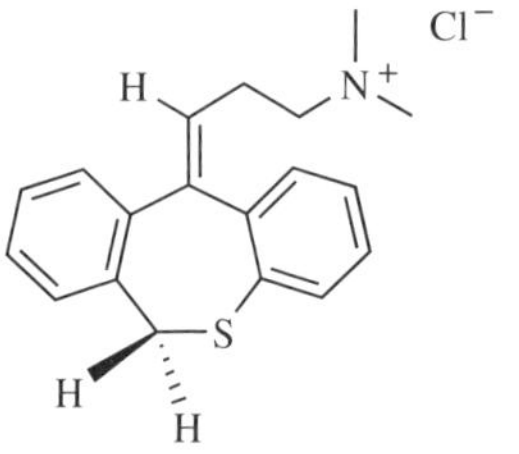

8.41

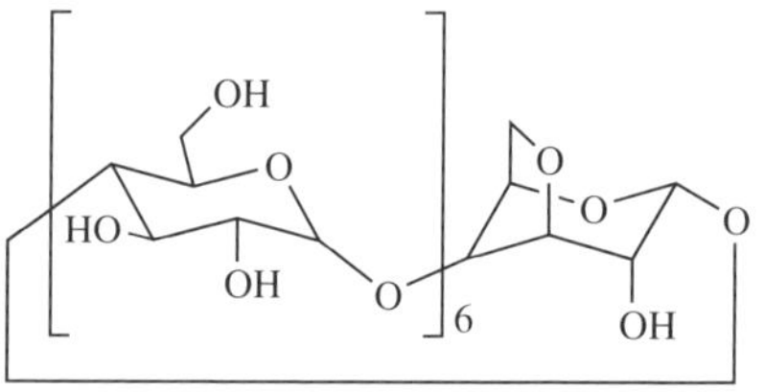

8.42

8.1.3.10. Heptakis(3-O-acetyl-2,6-di-O-pentyl)-β-cyclodextrin
Enantiomeric discrimination was observed in the ^{1}H NMR spectrum of methyl-2-chloropropionate, methyl-2-bromopropionate, methyl lactate, and methyl mandelate in cyclohexane-d_{12} in the presence of heptakis(3-O-acetyl-2,6-di-O-pentyl)-β-cyclodextrin.[1111]

8.1.3.11. Octakis(2,6-di-O-pentyl-3-O-trifluoroacetyl)-γ-cyclodextrin
The ^{1}H and ^{19}F NMR spectra of fluoroether anesthetics enflurane (**8.37**), isoflurane (**8.38**), desflurane (**8.39**), and other analogs exhibited enantiomeric discrimination in the presence of octakis(2,6-di-O-pentyl-3-O-trifluoroacetyl)-γ-cyclodextrin in cyclohexane-d_{12}.[1112] The resolution in the ^{1}H spectra was typically a few hundredths of a ppm.

8.1.3.12. Cyanoethylated-β-cyclodextrin
A cyanoethylated β-CD with a DS of 3.8 was shown to produce small nonequivalence (0.005-0.005 ppm) in the ^{1}H NMR spectrum of naproxen (**8.43**) in deuterium oxide.[1113]

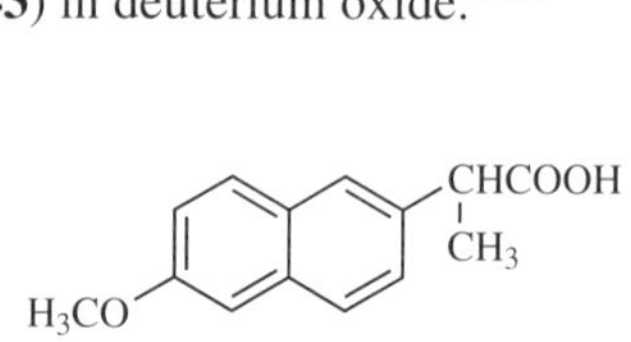

8.43

8.1.3.13. Per-O-octyl-α-cyclodextrin

The per-*O*-octyl derivative of α-CD caused a slight degree of enantiomeric discrimination (0.02 ppm) for the resonances of the *ortho*- and *meta*-hydrogen atoms of the protonated form of ephedrine (**8.44**) in chloroform-*d*.[1114]

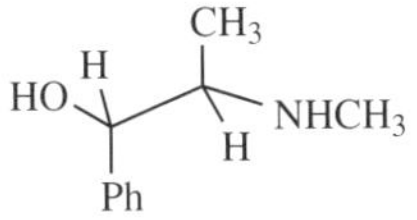

8.44

8.1.3.14. Biphenyl-capped-β-cyclodextrin

A biphenyl-capped β-CD derivative with the substituent attached through two primary hydroxyl groups (**8.45**) was shown to cause chiral recognition in the ^{1}H NMR spectrum of mandelic acid.[1083] Ring current effects of the biphenyl group were believed to be responsible for the nonequivalence of the *o*-, *m*-, and *p*-hydrogen resonances.

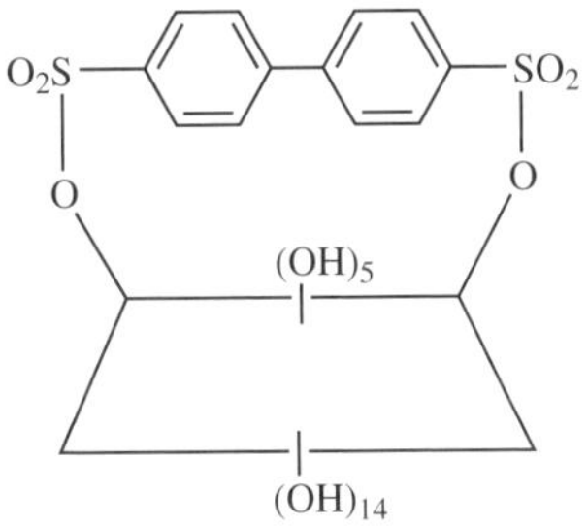

8.45

8.1.3.15. Heptakis(2,3,6-tri-O-acetyl)-β-CD (TA-β-CD)

The compound TA-β-CD is soluble in water and caused chiral discrimination in the ^{1}H NMR spectrum of α-pinene (**8.23**).[1071] The distinction with TA-β-CD was greater than that with TM-β-CD and comparable to that with α-CD.

The nonequivalence in the ^{1}H NMR spectrum of mandelonitrile (**8.46**) was measured with succinyl-β-CD, HP-α-CD, HP-β-CD, 2-hydroxyhexyl-β-CD, 2-HP-β-CD, TA-α-CD, TA-β-CD, TA-γ-CD, TM-β-CD and TM-γ-CD.[1115] The TA-β-CD produced the largest discrimination of all of the CDs examined in this report.

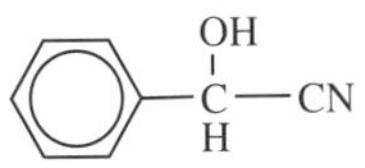

8.46

8.1.4. Anionic Cyclodextrin Derivatives

8.1.4.1. Carboxymethylated Cyclodextrins (CM-CD)

Anionic carboxymethylated CDs with $-OCH_2CO_2^-$ groups have been used to effect chiral separations in capillary electrophoresis. The preparation of CM-CD derivatives usually involves an indiscriminate substitution where CM groups may be at the 2-, 3- and 6-positions. An analysis of the synthesis showed that substitution at the 2-position predominated, although some substitution at the 3- and 6-positions occurred.[1116] A number of studies have noted how indiscriminately substituted CM-β-CD is an effective chiral NMR discriminating agent for cationic substrates.

The nonequivalence produced in the ^{1}H NMR spectra of three cationic substrates pheniramine, chlorpheniramine, and brompheniramine (**8.47**) was compared to those with native CDs, indiscriminately derivatized CM-CDs, and carboxymethylated α-, β-, and γ-CDs selectively derivatized at the 6-position or 2-position.[1117] The indiscriminately substituted CM-CDs in this study had high DS (6-9 CM units). The best results were obtained with the indiscriminately substituted CM-CDs, followed by the 2-substituted CM-CDs. Among the indiscriminately substituted α-, β-, and γ-derivatives, no one was consistently the most effective. The α- and β-CM-CDs generally produced the largest nonequivalence for the proton resonances of the pyridyl ring, whereas the γ-CM-CD produced the largest for those of the phenyl ring. Evidence indicated that the substrates formed 2:1 complexes with γ-CD that likely accounted for the enhanced discrimination of the phenyl resonances.

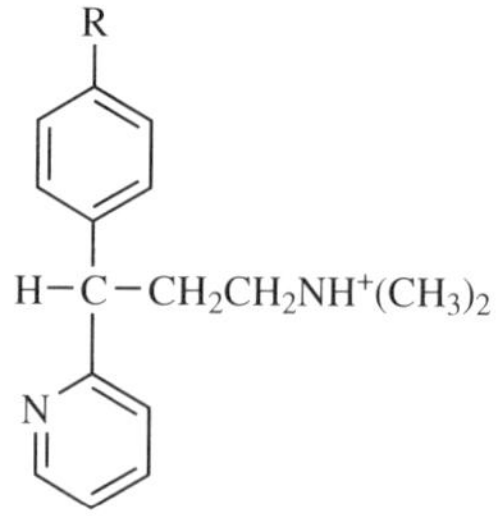

R= H, Cl, Br

8.47

CM-β-CD was more effective in causing enantiomeric discrimination in the ^{1}H and ^{13}C NMR spectra of metomidate (**8.48**) than native β-CD.[1118] Enantiomeric discrimination was eliminated at very acidic and basic pH, with pH 6 being the optimal value.

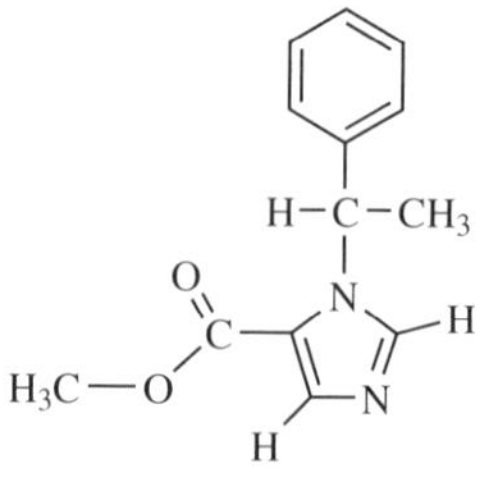

8.48

The ^{1}H NMR spectrum of cationic oxamniquine (**8.49**) exhibited no chiral discrimination in the presence of neutral CDs such as α-, β- and γ-CDs, hydroxyethyl-β-CD, and hydroxypropyl-β-CD.[1119] The use of CM-β-CD and sulfobutylether-β-CD, another anionic CD derivative, did cause nonequivalence in the ^{1}H NMR spectrum. The H1' and H4' signals had the greatest distinction and CD-β-CM was the more effective of the two anionic reagents.

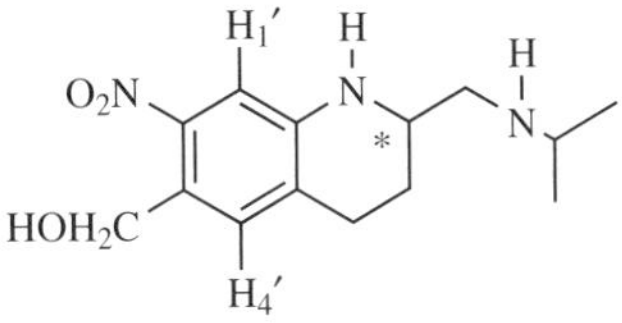

8.49

The enantiomeric discrimination in the ^{1}H NMR spectrum of amlodipine maleate (**8.12**) with γ-CD, CM-β-CD, and sulfobutylether-β-CD was compared.[1054] While the values were rather comparable, CM-β-CD did produce the largest nonequivalence (0.032 ppm for H10) of any of the reagents.

Other cationic substrates for which CM-β-CD was more effective than β-CD include dimethindene (**8.50**),[1120] chlorpheniramine (**8.51**),[1121] metoprolol (**8.52**),[1122] and noradrenaline (**8.53**).[1123] The ^{1}H NMR spectrum of **8.51** showed no enantiomeric discrimination in the presence of TM-β-CD (P1254). CM-β-CD produced nonequivalence in the ^{13}C NMR spectrum of chlorpheniramine[1121] and noradrenaline.[1123]

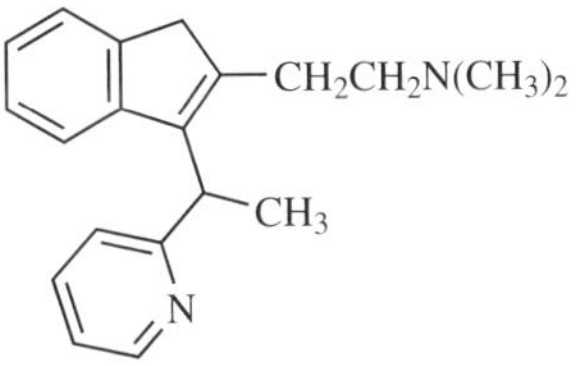

8.50 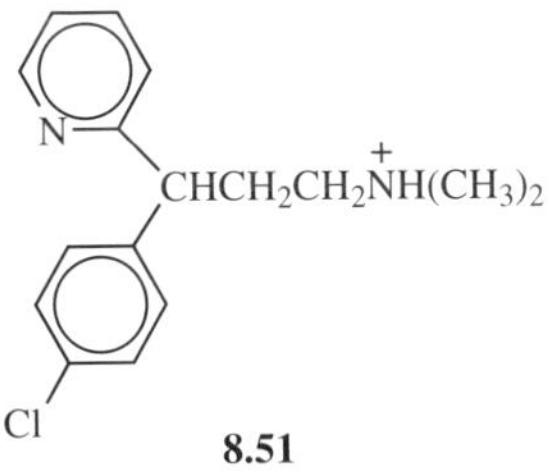   **8.51**

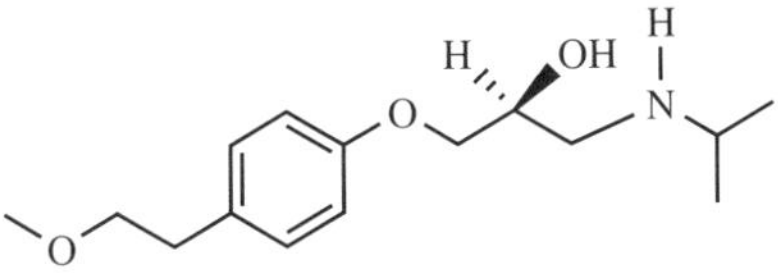

8.52 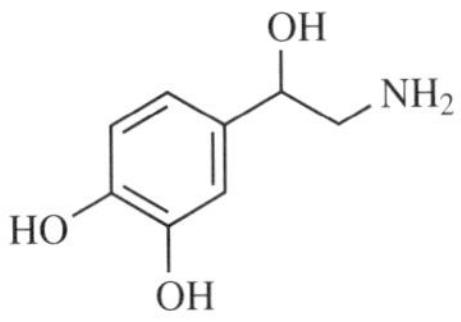   **8.53**

CM-β-CD does not have to be used exclusively with cationic substrates. The spectrum of 1,1'-binaphthyl-2,2'-diylhydrogenphosphate (**8.31**), a neutral compound, with CM-β-CD exhibited nonequivalence of the H8 resonance that was more than twice as large with CM-β-CD than with β-CD.[1124]

The anionic groups of CM-β-CD provide sites where paramagnetic lanthanide ions can bind to the CD. Addition of lanthanides such as Yb(III) and Dy(III) have been shown to cause significant enhancements in the enantiomeric discrimination.[1125,1126] For example, the methine resonance of carbinoxamine exhibited nonequivalence on the order of 0.5 ppm in the presence of indiscriminately substituted CM-β-CD and Dy(III).[1125,1126] Peak separation of about 0.4 ppm was observed in the methyl resonance of doxylamine in the presence of indiscriminately substituted CM-β-CD and Dy(III). Enhancements with Yb(III) were less than those with Dy(III) by a factor of about three, but the nonequivalence was still appreciably larger than that observed in the absence of Yb(III). The broadening with Yb(III) was also considerably less than that with Dy(III).

The enhancements in chiral discrimination with CM-β-CD-lanthanide mixtures were so large that considerably lower concentrations of CM-β-CD could be used. For example, when adding Dy(III) to indiscriminately substituted CM-β-CD, only 1:10 CD-substrate ratios were needed to obtain baseline resolution for the resonances of some substrates.[1125]

8.1.4.2. Heptakis(6-carboxymethylthio-6-deoxy)-β-cyclodextrin (6-CMT-β-CD)
The utility of anionic CD derivatives with a thiocarboxymethyl group ($-S-CH_2-CO_2^-$) at the 6-position as chiral NMR discriminating agents has been explored.[1085,1127–1129] With amino acid methyl esters at pH 7, the host and guest have opposite charges and the heptakis-6-CMT-β-CD produced greater nonequivalence of the methoxy resonance than either a mono-6-CMT-β-CD derivative or native α- and β-CDs.[1127,1129]

The heptakis-CMT-β-CD derivative has been used to analyze the optical purities of complexes of the form $M(phen)_3^{n+}$.[1085,1128] Complexes with Ru(III), Rh(III), Fe(II), Co(II), and Zn(II) were examined.[1085] Neutral CDs such as native α-, β-, and γ-CDs as well as TM-β-CD did not cause discrimination in the ^1H NMR spectrum. In the case of $Ru(phen)_3^{2+}$, addition of 6-CMT-β-CD caused nonequivalence up to about 0.2 ppm for some resonances. Association seemed to require not only some inclusion of the metal complex in the CD cavity but also ion pairing between the cationic complex and anionic thiocarboxymethyl groups. The octakis(CMT-6)-γ-CD derivative was not as effective for these metal complexes and it was proposed that the cavity was too large for a complementary fit.[1085]

8.1.4.3. Sulfobutylether-β-cyclodextrin (SBE-β-CD)
Sulfobutylether derivatives of β-CD are anionic species that have found application in capillary electrophoresis separations and have been explored in a limited manner for their application as chiral NMR reagents for cationic substrates. These are usually prepared as indiscriminately substituted derivatives with various DS.

Addition of SBE-β-CD to structural analogs of 1,2,3,4,10,14b-hexahydro-2-methyldibenzo[c,f]pyrazino[1,2-a]azepine (mianserine - **8.54**) caused nonequivalence in the ^{13}C NMR spectra.[1130] The order of discrimination for **8.54** was SBE-β-CD > CM-β-CD > β-CD, although comparisons of the relative effectiveness of the anionic SBE-β-CD and CM-β-CD varied from substrate to substrate. For example, an SBE-β-CD with a DS of four caused greater nonequivalence in the spectrum of metomidate (**8.48**) than CM-β-CD.[1118] For amlodipine maleate (**8.12**), the enantiomeric discrimination caused in the ^{1}H NMR spectrum by SBE-β-CDE and CM-β-CD was comparable.[1054] Both SBE-β-CD and CM-β-CD caused chiral recognition in the ^{1}H NMR spectrum of oxamniquine (**8.49**), but the distinction was larger with CM-β-CD.[1119] A consistent observation among these reports was that the anionic CDs were more effective in causing chiral recognition in the spectra of cationic substrates than neutral CDs.

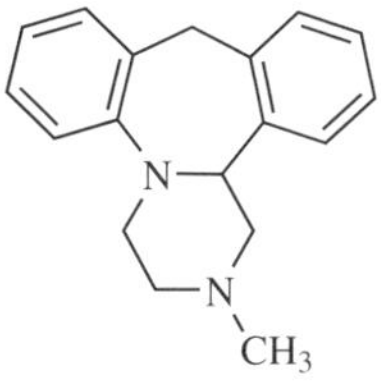

8.54

For vericonazole (**8.55**), resonances of the pyrimidine and triazole ring exhibited enantiomeric discrimination in the presence of SBE-β-CD.[1131] A comparison of derivatives with DS of 6.5, 4.5, and 1.0 showed that the larger shifts and enantiomeric distinction occurred for the highest DS. Nonequivalence of 40–106 Hz at 400 MHz was observed for some of the proton resonances. Native α-, β-, and γ-CDs produced essentially no shifts in the spectrum. Hydroxypropyl- and hydroxyethyl-β-CD derivatives provided enantiomeric discrimination about two-thirds the size of that with SBE-β-CD, whereas the distinction with CM-β-CD was about one-third the size.[1131] The need to possibly evaluate a range of anionic CDs with cationic substrates is evident from these findings.

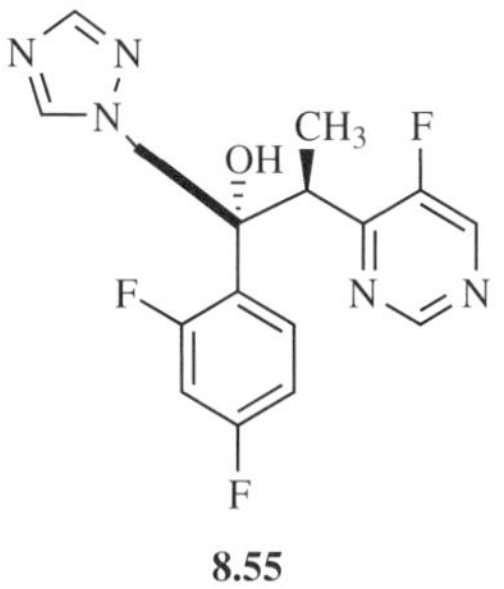

8.55

8.1.4.4. Heptakis(2,3-di-O-acetyl-6-sulfo)-β-CD (HDAS-β-CD) Enantiomeric discrimination in the ^{1}H NMR spectrum of clenbuterol (**8.56**) in deuterium oxide was compared with HDAS-β-CD (-OAc, -OSO$_2$-Na$^+$) and native β-CD.[1132] The β-CD caused distinct aromatic resonances for the two enantiomers, whereas HDAS-β-CD caused nonequivalence of the *t*-butyl resonance. Chiral discrimination in the

^{1}H NMR spectrum of dimethindene (**8.5**) was observed in the presence of HDAS-β-CD.[1133]

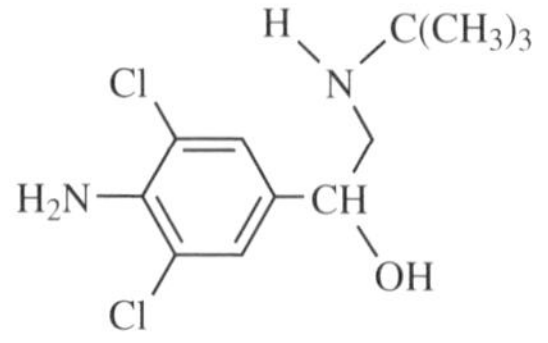

8.56

The corresponding octakis(2,3-di-O-acetyl-6-sulfo)-γ-CD (ODAS-γ-CD) was tested for its ability to cause chiral recognition in the ^{1}H, ^{13}C, and ^{19}F NMR spectra of **8.57**.[1134] The greatest nonequivalence was observed in the ^{19}F spectrum, although the corresponding octakis(6-sulfo)-γ-CD (OS-γ-CD) was more effective than ODAS-γ-CD since the substrate entered more deeply into the cavity of OS-γ-CD.[1134]

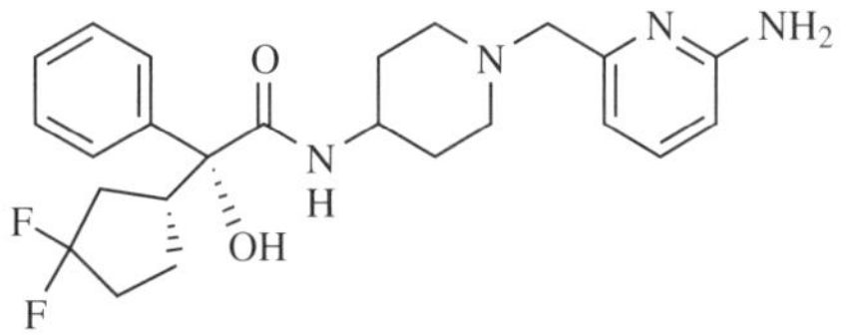

8.57

8.1.4.5. Sulfated Cyclodextrins (S-CD) Sulfated cyclodextrins (—O—SO$_3^-$) have been examined in a limited number of applications as chiral NMR solvating agents. For example, heptakis(6-sulfo)-β-CD was shown to cause nonequivalence in the ^{1}H NMR spectra of dimethindene (**8.5**)[1133] and the dipeptides Ala-Phe and Ala-Tyr.[1065] The distinction in the spectra of the dipeptides with 6-S-β-CD was comparable to that with native β-CD and larger than that with 2,3-DA-β-CD and HDAS-β-CD. The octakis(6-sulfo)-γ-CD was more effective in causing chiral discrimination in the ^{1}H, ^{13}C, and ^{19}F NMR spectra of **8.57** than the corresponding ODAS-β-CD reagent.[1134]

An indiscriminately sulfated β-CD with a DS of nine is commercially available and the enantiomeric discrimination in the spectra of cationic substrates such as pheniramine, propranolol (**8.4**), carbinoxamine (**8.27**), and doxylamine was compared to native β-CD and a commercially available CM-β-CD.[1125,1126] Nonequivalence in the spectra with the highly anionic sulfated derivative was larger than that observed with CM-β-CD or β-CD. It was also possible to add a paramagnetic lanthanide ion such as Dy(III) to S-β-CD to enhance the discrimination. Nonequivalence of up to 0.15 ppm was realized on addition of Dy(III) to these mixtures.

8.1.5. Cationic Cyclodextrin Derivatives

8.1.5.1. Amino-substituted Cyclodextrins Amino-substituted CDs, with appropriate adjustment of pH, provide cationic derivatives that can be more effective for anionic substrates than native CDs. The amino-substituted CDs used in NMR studies have been derivatized at the 6-position.[1127,1129,1135-1137]

The methyl resonance of 2-phenylpropionic acid was enantiomerically discriminated in the presence of a monoamine derivative, 6^A-amino-6^A-deoxy-β-CD, at pH 1 but not at pH 6.4.[1135] Chiral discrimination has been observed in the ^{1}H NMR spectra of 1,1′-binaphthyl-2,2′-diylphosphate (**8.31**) and 1,1′-binaphthyl-2,2′-dicarboxylate (**8.32**) with 6^A-amino-6^A-deoxy-β-CD.[1136]

The effectiveness of 6^A-amino-6^A-deoxy-β-CD and $6^A,6^D$-diamino-$6^A,6^D$-dideoxy-β-CD in discriminating several chiral carboxylic acids was compared to several neutral CDs.[1137] The cationic CDs were more effective at causing chiral recognition than the neutral CDs.

Mono- and heptakis-6-amino-substituted β-CDs were examined for their ability to cause nonequivalence in the ^{1}H NMR spectra of N-acetylated amino acids of tryptophan, phenylalanine, leucine, and valine.[1127] Both of the amino-substituted CDs caused enantiomeric discrimination, whereas native α- and β-CDs did not. A pH of 6 was recommended as the optimal value and in most cases the heptakis-substituted CD caused greater discrimination than the mono-substituted CD.

8.1.5.2. Bis(6-trimethylammonio)-β-cyclodextrin (TMA-β-CD) Enantiomeric discrimination was observed in the ^{1}H NMR spectrum of benzyloxycarbonyl-glutaric acid in a phosphate buffer at pH 6.9 in the presence of TMA-β-CD.[1138] Addition of native β-CD did not cause any observable chiral recognition.

8.1.5.3. Mono(6-xylylenediamine)-β-cyclodextrin (X-β-CD) The enantiomeric discriminating ability of X-β-CD (-NH-C_6H_4-CH_2NH_2) was examined with 2-phenylpropionic acid and methoxy phenyl acetic acid.[1139] A small nonequivalence of the methyl resonances was observed. Native β-CD was not found to discriminate the spectra of these substrates.

8.1.6. Summary

CDs are effective chiral solvating agents for an exceptionally wide array of substrates as they form guest complexes with many classes of compounds. Water- and organic-soluble CDs are available, further enhancing their utility for chiral analysis. A number of commercially available CDs are useful candidates for chiral NMR discrimination. The native α-, β-, and γ-CDs are useful reagents in water. Similarly, TM-α-CD and TM-β-CD function in water and sometimes are more effective than the native CDs. TM-CDs are also soluble in organic solvents such as methanol-d_4.

For cationic substrates, the use of anionic CDs is often warranted and leads to enhanced enantiomeric discrimination when compared to neutral CDs. The indiscriminately substituted carboxymethyl, sulfobutylether, and sulfated derivatives are potential candidates for these applications. Another possibility with these anionic CDs is to add a paramagnetic lanthanide ion. Binding of the lanthanide ion to the anionic CD often enhances the enantiomeric discrimination.

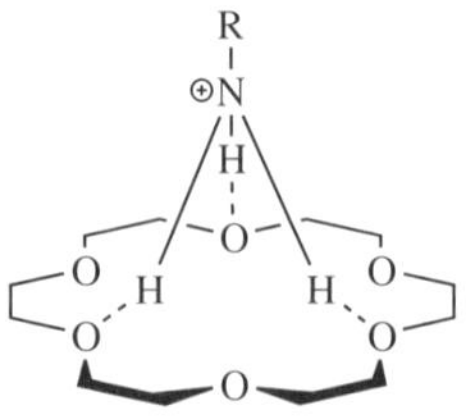

Figure 8.3. Interaction of a protonated primary amine with an 18-crown-6 ether.

For anionic substrates, the use of cationic amine-substituted CDs has not been extensively studied, but those studies that have been done suggest that the use of cationic CDs with anionic substrates is warranted over the use of neutral CDs.

8.2. CROWN ETHERS

8.2.1. Introduction

The most common chiral crown ethers incorporate the 18-crown-6 structure (**8.58**) as this is ideally suited to associate with protonated primary amines through the formation of three hydrogen bonds as illustrated in Figure 8.3. The use of chiral crown ethers for recognition of chiral protonated amines was first exploited by Cram and co-workers. Using a crown ether that incorporated two 1,1'-binaphthylene-2,2'-dihydroxy groups (**8.59**), enantiomeric discrimination was observed in the ^{1}H NMR spectrum of 1-phenylethylamine (PEA).[1140] Since the initial work by Cram, a diverse variety of chiral crown ether and crown ether-like compounds have been evaluated as chiral recognition agents. The study of these reagents has often focused on their potential application for liquid chromatographic applications, although in many cases NMR spectroscopy has been used to study aspects of the geometry of the associated complex. Chiral discrimination has often been noted in the NMR spectra in these studies.

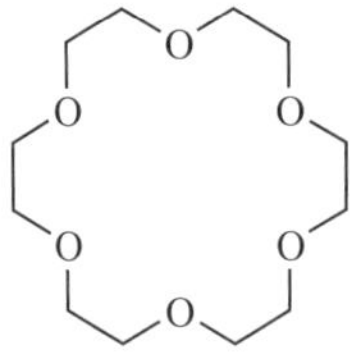

8.58
8.59 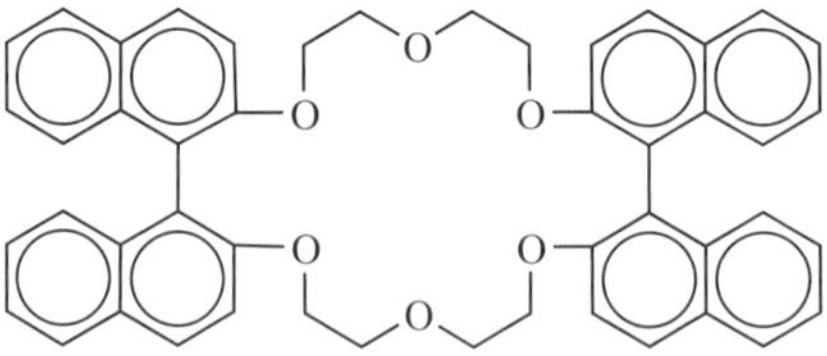

Many of the chiral crown ethers and crown ether-like compounds described in the literature require rather involved procedures for their synthesis and purification, meaning that commercial availability is necessary to truly promote their use in NMR applications. Several crown ethers are marketed as chiral liquid chromatographic phases. Only one crown ether, the (+) and (−) isomers of (18-crown-6)-2,3,11,12-tetracarboxylic acid (**8.60**), is available commercially for use in NMR applications.

Comparative NMR studies of **8.60** with some other crown ethers that show high recognition properties in liquid chromatographic applications indicate that it typically provides the highest degree of enantiomeric discrimination in NMR spectra. Furthermore, while crown ethers are usually only effective for the recognition of primary amines, a recent report has shown that **8.60** is also effective for the discrimination of secondary amines.[1141] **8.60** is clearly the crown ether of choice for NMR applications. Even so, a brief discussion of the range of crown ether and crown ether-like compounds that have been used in NMR applications will be provided.

8.60

8.2.2. (18-Crown-6)-2,3,11,12-tetracarboxylic Acid (18-C-6-TCA)

The first application of 18-C-6-TCA for chiral NMR discrimination was in 1998.[1142] Although primarily interested in the application of 18-C-6-TCA for liquid chromatographic separations, the authors noted that 18-C-6-TCA caused enantiomeric discrimination in the ^{1}H NMR spectrum of 1-(1-naphthyl)ethylamine (NEA) and alanine β-naphthylamide in methanol-d_4.[1142] The X-ray structure of NEA with 18-C-6-TCA was later published.[1143] The protonated NEA was generated directly in methanol-d_4 by the addition of perchloric acid.[1143]

The broad applicability of 18-C-6-TCA as a chiral NMR discriminating agent for primary amines including amino acid methyl esters such as valine, leucine, alanine, phenylglycine, tryptophan and lysine,[1144] underivatized amino acids, α-methyl amino acids,[1145] amines and amino alcohols[1144] was demonstrated soon thereafter. Analysis of underivatized amino acids with 18-C-6-TCA in methanol-d_4 was facilitated by adding a stoichiometric equivalent of deuterium chloride.[1145] Nonequivalence of 10-30 Hz at 300 MHz was common in the spectra of the underivatized amino acids and α-methyl amino acids.

For primary amines, it was possible to mix the neutral amine with 18-C-6-TCA in methanol-d_4.[1144,1146] A neutralization reaction between 18-C-6-TCA and the amine occurred, thereby forming the corresponding ammonium ion and monocarboxylate of 18-C-6-TCA. Shifts obtained by mixing a neutral amine with 18-C-6-TCA were essentially identical to those obtained by adding the ammonium salt to 18-C-6-TCA.[1144,1146]

Further studies showed the versatility of 18-C-6-TCA in causing chiral recognition in the ^{1}H NMR spectra of primary amines.[1146,1147] One study examined the chiral discrimination in the ^{1}H NMR spectra of sixteen amino acids, five alkyl amines, and seven amino alcohols.[1147] For the amino acids, the (+)-isomer of 18-C-6-TCA consistently caused the methine resonance of the D-enantiomer to shift to lower frequency than that of the L-enantiomer in deuterium oxide.[1147] Aliphatic alkyl amines and amino alcohols tended to exhibit relatively low enantiomeric discrimination in the presence of 18-C-6-TCA,[1147] although nonequivalence was still observed in the ^{1}H NMR spectrum of compounds such as 2-amino-1-butanol and 2-amino-octane.[1146,1147] Enantiomeric recognition was much greater when the amine or amino alcohol had an aromatic substituent group.[1147] 18-C-6-TCA also caused chiral discrimination in the ^{1}H NMR spectra of compounds such as baclofen (**8.61**),[1147] primaquine (**8.62**),[1147] aminoglutethimide (**8.63**),[1146] 1-amino-2-indanol (**8.64**),[1146] and 1-aminoindane.[1146] In some cases, nonequivalence of up to 0.25 ppm was observed in the ^{1}H NMR spectra of primary amines in the presence of 18-C-6-TCA.

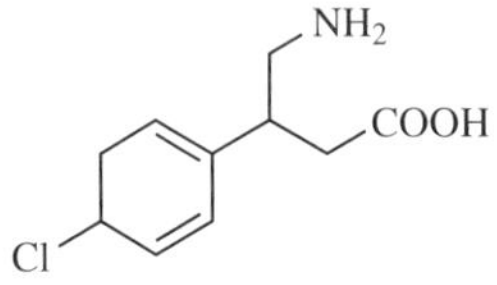

8.61

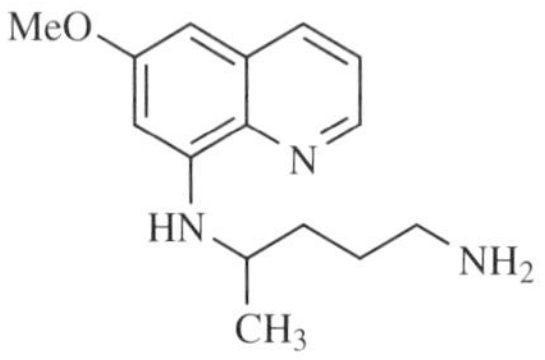

8.62

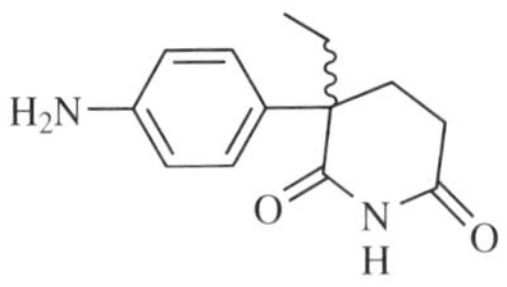

8.63

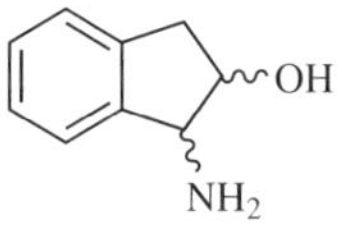

8.64

Analysis of chiral amines with 18-C-6-TCA can be performed in deuterium oxide, methanol-d_4, and acetonitrile-d_3.[1144,1147] The discrimination in methanol-d_4 and acetonitrile-d_3 was larger than that in deuterium oxide. An advantage of methanol-d_4 was the wide range of ammonium salts that dissolve without needing particular anions to enhance the solubility.[1144]

Although 18-C-6-TCA has usually been applied in ^{1}H NMR spectroscopy, chiral discrimination in the ^{13}C NMR spectra of alanine and alanine methyl ester in methanol-d_4 was noted as well.[1148]

An interesting observation with 18-C-6-TCA was that the addition of paramagnetic Yb(III) often enhanced the nonequivalence in the resulting NMR spectrum.[1144,1146] The Yb(III) was added as aliquots of a concentrated stock solution of Yb(III)nitrate in methanol-d_4. The Yb(III) bound to the carboxylate groups of 18-C-6-TCA. It was best to add the Yb(III) to a mixture of neutral 18-C-6-TCA and a protonated amine rather that the monocarboxylate form of 18-C-6-TCA.[1144] In the case of the monocarboxylate species, binding of the Yb(III) was too strong and unacceptable broadening occurred in the resulting NMR spectrum.

18-C-6-TCA has recently been shown to be an effective reagent for the analysis of secondary amines by NMR spectroscopy.[1141] This observation was unexpected as 18-crown-6 ethers usually only bind to protonated primary amines. Protonated secondary

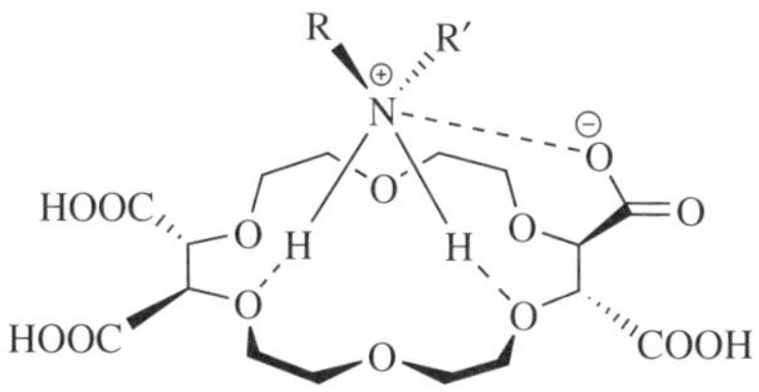

CROWN ETHERS 303

Figure 8.4. Interaction of a protonated secondary amine with the carboxylate ion of 18-C-6-TCA.

amines do not have the ability to form three hydrogen bonds to oxygen atoms of the crown unit. Furthermore, steric hindrance in secondary amines inhibits binding to crown ethers. With 18-C-6-TCA, the addition of a neutral secondary amine resulted in the formation of its protonated analog and the monocarboxylate form of 18-C-6-TCA. The substrate likely associates via two hydrogen bonds and an ion pairing interaction as illustrated in Figure 8.4. Baseline discrimination was observed in the ^{1}H NMR spectrum of a variety of secondary amines including N-methylamino acids, aryl alkyl amines, and proline.

8.2.2.1. Amide Derivatives of (18-Crown-6)-2,3,11,12-tetracarboxylic Acid Amide derivatives prepared by the reaction of tryptophan, PEA, NEA, 1-cyclohexylethylamine, and methylamine with the tetraacid chloride of 18-C-6-TCA were examined for their effectiveness as chiral NMR discriminating agents.[1146] Only the methylamine derivative was comparable to 18-C-6-TCA in its effectiveness, and, the use of the amide derivatives is not warranted for the analysis of primary amines.

8.2.3. (18-Crown-6)-2,3,11,12-derivatives

The enantiomeric discrimination in the ^{1}H and ^{13}C NMR spectra of PEA with crown ethers of structure **8.65** was examined in chloroform-d and methanol-d_4.[1149,1150] The compounds with R^1, R^2, R^4, and R^6 provided the best chiral recognition of the methyl doublet, but this was smaller than what has been observed with 18-C-6-TCA.

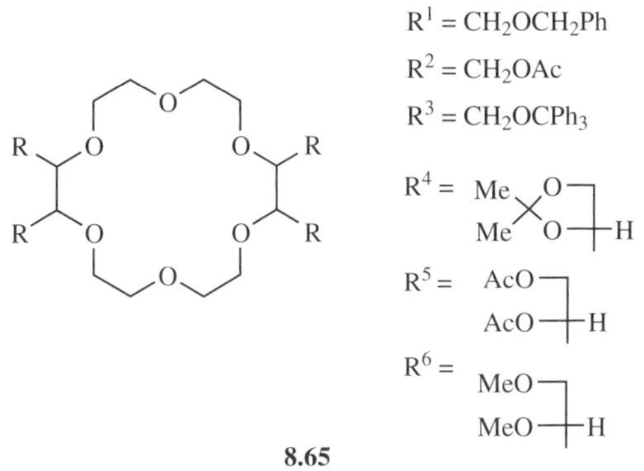

$R^1 = CH_2OCH_2Ph$

$R^2 = CH_2OAc$

$R^3 = CH_2OCPh_3$

$R^4 = $ Me, O / Me, O—H

$R^5 = $ AcO / AcO—H

$R^6 = $ MeO / MeO—H

8.65

The derivative of **8.65** in which R is a phenyl group was examined for its effectiveness at causing discrimination in the ^{1}H NMR spectra of the methyl ester hydrochloride salts of phenylglycine, phenylalanine, and tryptophan.[1151] The methoxy resonance exhibited nonequivalence on the order of 0.02–0.07 ppm.

Compound **8.66** was evaluated as a chiral NMR solvating agent with PEA, NEA, the methyl ester hydrochloride salts of phenylglycine, alanine, and phenyl alanine.[1152] Spectra were obtained in 9:1 chloroform-d/methanol-d_4. The crown produced nonequivalence in the ^{1}H NMR spectra from 0.01 to 0.06 ppm.

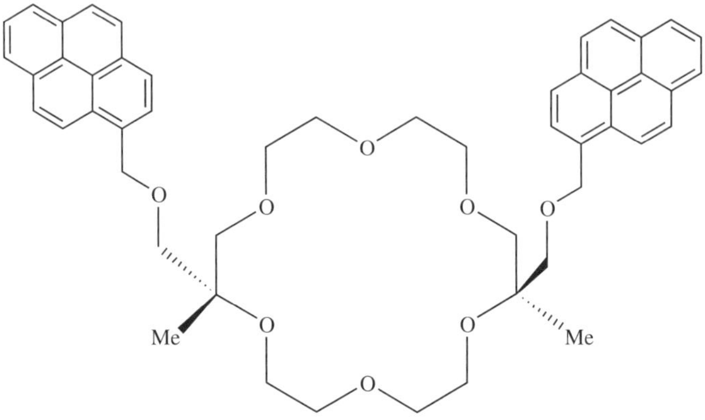

8.66

Crown **8.67**, which was developed for liquid chromatographic applications, was evaluated as a chiral NMR discriminating agent in chloroform-d, acetotrile-d_3, and methanol-d_4.[1153] The nonequivalence in ^{1}H NMR spectra with **8.67** was routinely larger than that observed with the binaphthyl crown (**8.68**) described by Cram and co-workers.[1154–1156] The enantiomeric discrimination with **8.67** was generally better in methanol-d_4 and acetonitrile-d_3 than in chloroform-d. Chiral distinction in the ^{1}H NMR spectra of amino acid esters, PEA, NEA, and α-(1-aminoethyl)-4-hydroxybenzylalcohol was observed. Subsequent studies have shown that **8.67** was not as effective as 18-C-6-TCA at causing enantiomeric discrimination.[1144] Furthermore, **8.67** is not commercially available and its synthesis is rather elaborate.

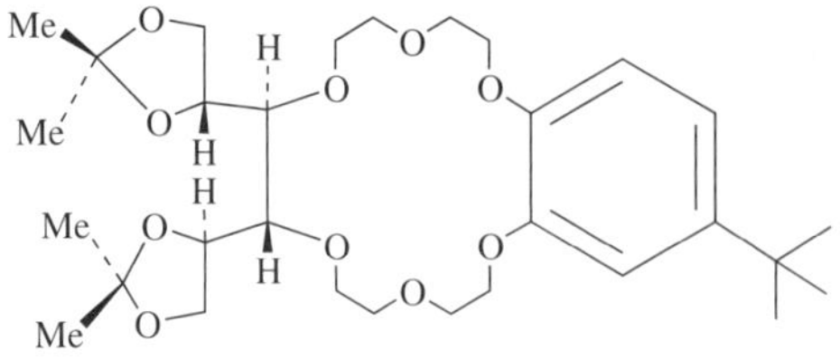

8.67

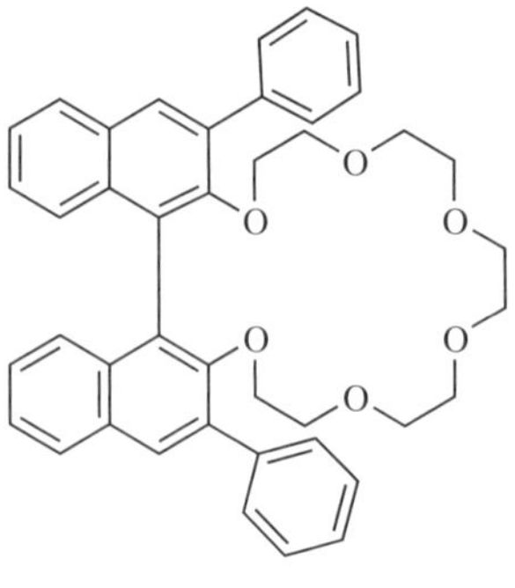

8.68

Chiral recognition in the spectra of substrates in the presence of **8.67** in chloroform-d and acetonitrile-d_3 could be improved by the addition of anionic tetrakis-lanthanide β-diketonate species of the form Ln(fod)$_4^-$, in which fod is 6,6,7,7,8,8,8-heptafluoro-2,2-dimethyl-3,5-octanedione.[1153] Addition of Eu(fod)$_3$/Ag(fod) to chloroform solutions of **8.67** with ammonium chloride salts resulted in the precipitation of silver chloride. The Eu(fod)$_4^-$ species associated with the ammonium cation in the bulk solution. Therefore, the enantiomer with the lower association constant with **8.67** exhibited larger europium-induced shifts, leading to enhancements in the enantiomeric discrimination. Improvements in the distinction of certain resonances of alanine ethyl ester hydrochloride, valine methyl ester hydrochloride, and leucine methyl ester hydrochloride were noted. In acetonitrile-d_3, the use of Pr(fod)$_4^-$, which produces larger shifts and Eu(III), formed in situ from Pr(fod)$_3$ and Ag(fod), caused similar improvements in chiral recognition.

8.2.4. Crown Ethers Incorporating a 2,2′-Dihydroxy-1,1′-binaphthalene Unit

Chiral crown ethers derived from 2,2′-dihydroxy-1,1′-binaphthalene were the first ones used for chiral NMR analysis.[1140] The enantiomeric recognition properties of a wide variety of crown ethers with 1,1′-binaphthalene units were examined in these early reports.[1154-1156] The crown ether with phenyl substituents at the 3- and 3′-positions (**8.68**) was generally the most effective and was developed into a commercially available liquid chromatographic column. **8.68** is not commercially available as a pure compound for use in NMR studies. A comparison of the chiral discrimination in the ^{1}H NMR spectra of a variety of substrates in the presence of **8.67** and **8.68** showed that **8.67** usually produced larger nonequivalence.[1153]

8.2.5. Glycoside-derived Crown Ethers

A large family of crown ethers incorporates glycoside units such as β-D-glucopyranoside and β-D-galactopyranoside.[1157] Many of the studies cited in this review article did not specifically evaluate the effectiveness of the glycoside-derived crown ethers for their effectiveness in chiral recognition in NMR spectroscopy. Those that have often only reported NMR data on PEA. In an evaluation of the relative chiral discrimination of 11 glycoside-derived crown ethers and some of their alkyl derivatives in chloroform-d, compound **8.69** caused nonequivalence of 0.23 ppm for the methine resonance of phenylglycine methyl ester hydrochloride.[1158]

8.69

Compound **8.70** which incorporates a β-D-galactopyranoside unit is soluble in chloroform and its effectiveness as a chiral NMR discriminating agent was examined.[1159] The ^{1}H NMR spectra of the hydrochloride salts of valine ethyl ester, alanine ethyl ester, and leucine methyl ester in the presence of **8.70** exhibited a small amount of nonequivalence. Conversion of **8.70** to the corresponding diol (**8.71**) provided a methanol- and acetonitrile-soluble crown that was far more effective as a chiral NMR discriminating agent. Nonequivalence on the order of 2–9 Hz (at 300 MHz) in the ^{1}H NMR spectra of the hydrochloride salts of amino acid methyl esters, NEA, 1-cyclohexylethylamine, and 1-(4-hydroxyphenyl)-2-aminopropanol in methanol-d_4 and valine ethyl ester, alanine ethyl ester, and PEA in acetonitrile-d_3 was observed in the presence of **8.71**.[1159] The diol functionality of **8.71** also provided a moiety that complexed with paramagnetic lanthanide ions. Addition of Yb(III) to solutions of **8.71** with substrates in methanol had no effect. Presumably, the polar methanol associated with the Yb(III) and eliminated association at the diol functionality. In acetonitrile-d_3, the Yb(III) did bind to **8.71** and caused enhancements in the chiral discrimination.

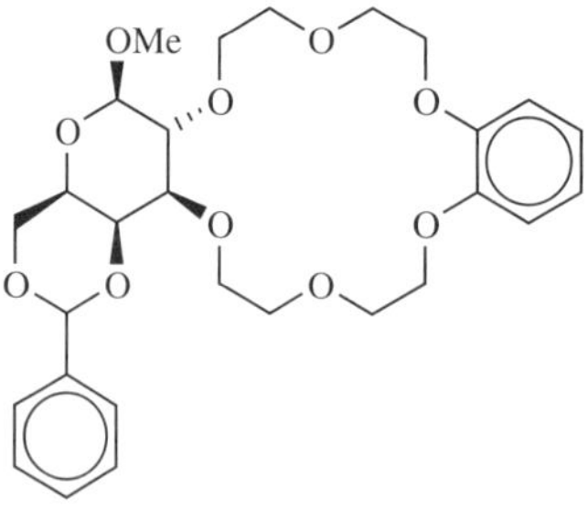

8.70

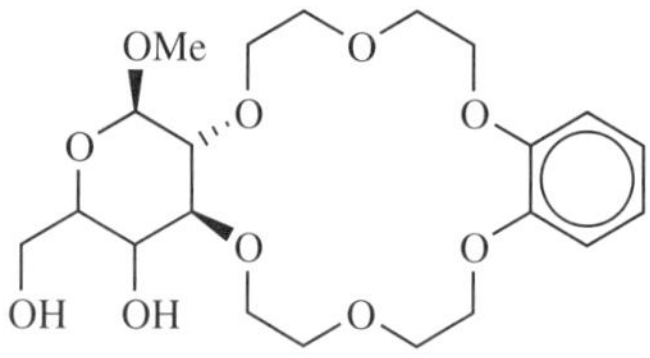

8.71

8.2.6. Crown Ether-like Compounds with Nitrogen-heterocyclic Units

A variety of chiral crown-like compounds that incorporate one or more pyridine (**8.72**),[1158,1160–1169] pyrimidine (**8.73**),[1170] or acridine (**8.74**)[1171] units into the macrocycle have been prepared and examined for their chiral recognition properties. The nature of the R substituent group has a significant effect on the chiral recognition[1163–1165] and compounds with methyl, phenyl, or *t*-butyl groups were most

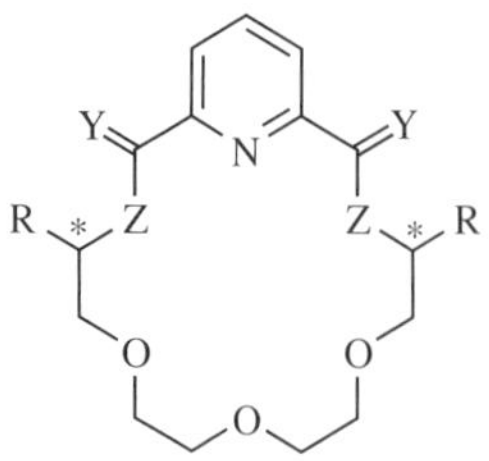

Y= O; Z= O, NH, NCH$_3$
Y= S; Z= O, NH, NCH$_3$
Y= H$_2$; Z= O, NH
R= various alkyl or phenyl

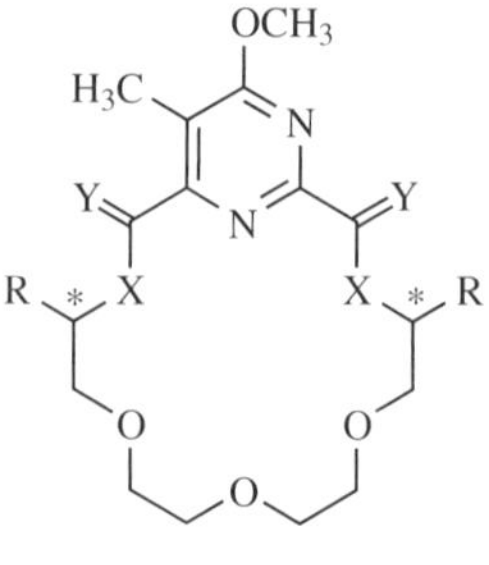

8.72

8.73

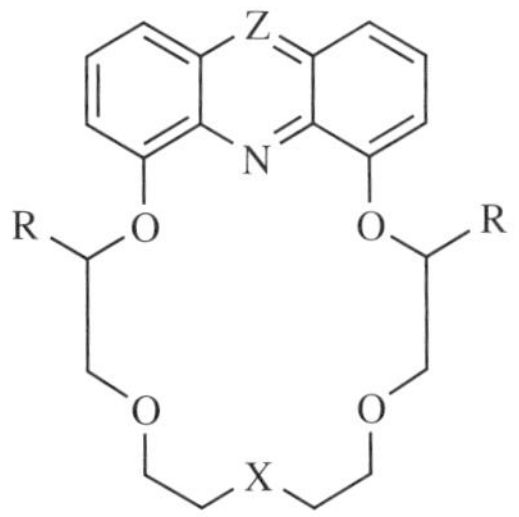

CROWN ETHERS

8.74

effective. Compounds of this type were soluble in chloroform, methylene chloride, methanol, and acetonitrile.[1161,1164,1168] NMR experiments were typically used to measure relative association constants for pairs of enantiomers rather than the extent of chiral discrimination in the ^{1}H NMR spectrum. NMR experiments usually involved only a few substrates, most notably NEA. This makes it difficult to compare the effectiveness of these pyridyl-based compounds to other chiral crown ethers.

Analogous crowns prepared using a pyrimidine or pyrimidone group instead of the pyridine unit showed better enantiomeric discrimination for NEA in a 50:50 mixture of chloroform-d/methanol-d_4.[1170]

The derivative of **8.72** with X = O and R = phenyl was effective in distinguishing hydrogen atoms that are enantiotopic by internal comparison in the *meso*-configured bis-ammonium salt **8.75**. At −83°C, the spectrum indicated that the crown selectively bound one of the enantiotopic groups of **8.75**.[1167] The corresponding cyclohexyl derivative of **8.72** also showed an enantioselective molecular recognition for the pair of enantiomers that resulted from rapid interconversion of the chair conformers.

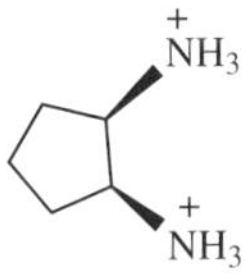

8.75

A crown ether-like compound containing a bipyridine unit (**8.76**) was shown to produce a small amount of separation of the methine resonance of the hydrochloride salt of phenyl glycine methyl ester.[1172]

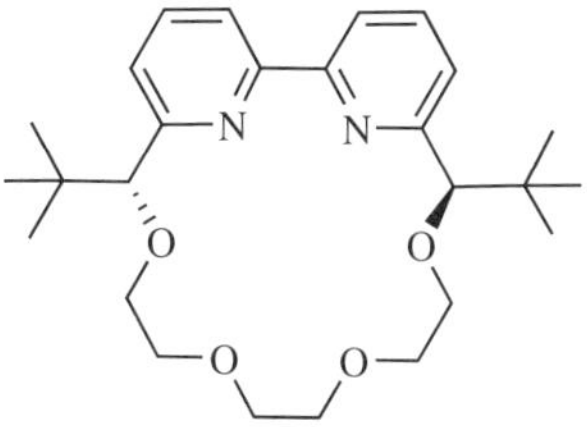

8.76

8.2.7. Triazole-18-crown-6 Ether

The chiral recognition properties of several 18-crown-6 derivatives with a triazole ring as part of the crown (**8.77**) toward PEA[1162] and NEA[1162,1173] have been examined. The derivative of **8.77** with the *n*-dodecyl substituent was the most effective in causing enantiomeric discrimination in the ^{1}H spectrum of NEA.

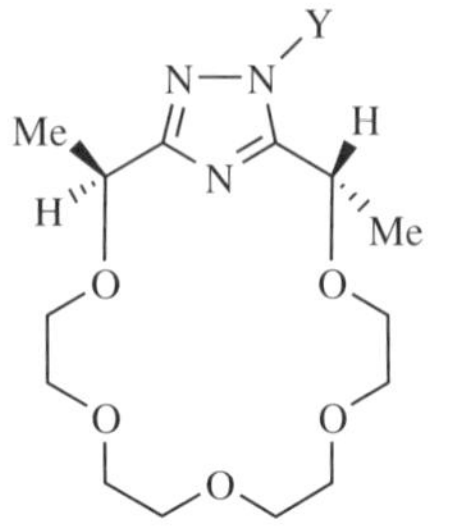

Y= CH$_2$COO cholesteryl
Y= *n*-dodecyl

8.77

8.2.8. Crown Ether-like Compounds with Phenol Units

The recognition properties of a series of chiral crown ether compounds with a phenol or methoxy aromatic ring (**8.78** and **8.79** as examples) have been explored.[1146,1174–1180] These crown ethers were primarily developed for chiral analyses based on visual color changes, chromatographic separations, and mass spectrometric measurements. Most of the NMR experiments involved studies to measure association constants and learn details of the crown-substrate geometries rather than to explore their utility as chiral NMR shift reagents.[1176–1178] The phenol-containing crowns can be used with neutral amines, since the phenol hydrogen will protonate the amine and facilitate formation of the crown-substrate complex.[1179] A few studies do provide examples of chiral discrimination in the ^{1}H NMR spectra. For example, **8.80** caused enantiomeric discrimination in the ^{1}H resonances of **8.81** in chloroform-*d*.[1180] **8.79** with an adamantyl R group caused pronounced nonequivalence of the methine resonance of methionine methyl ester hydrochloride (**8.82**) in chloroform-*d*.[1175,1181]

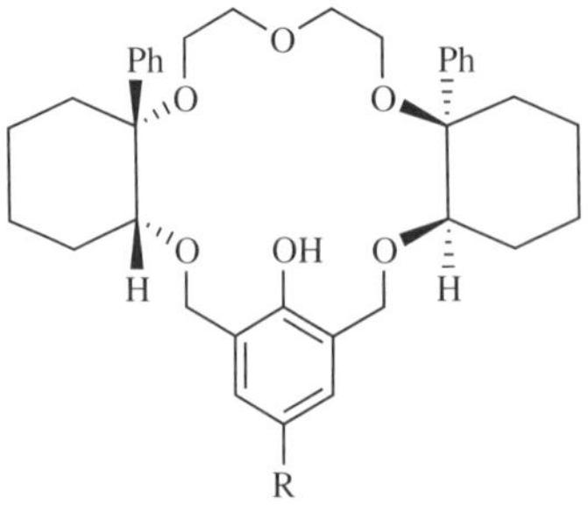

8.78

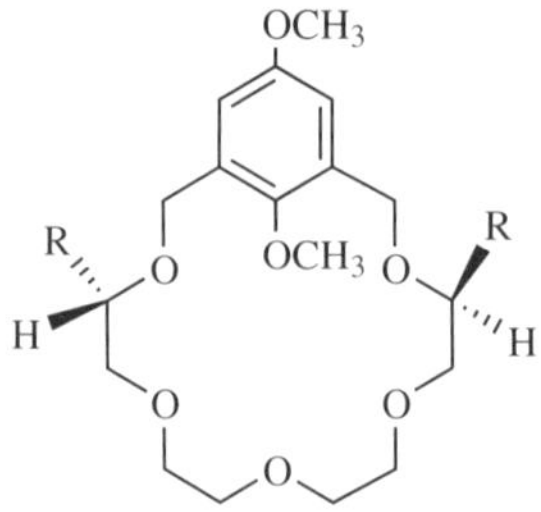

8.79

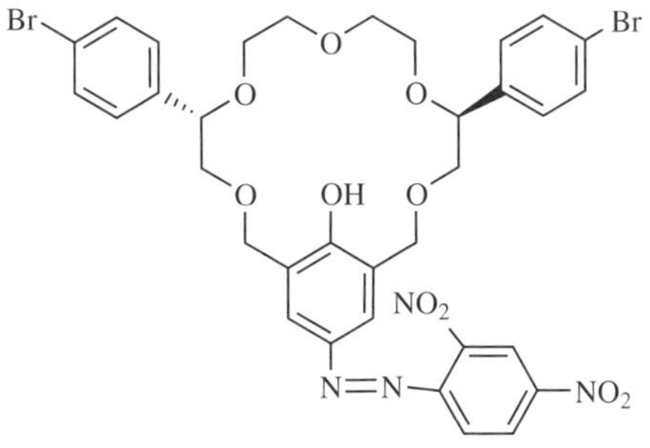

CROWN ETHERS 309

8.80

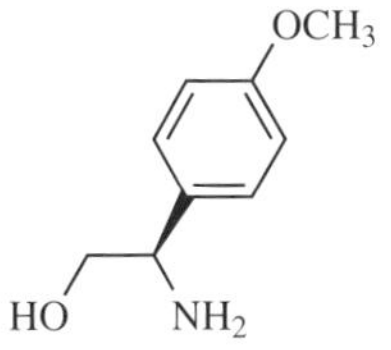

8.81

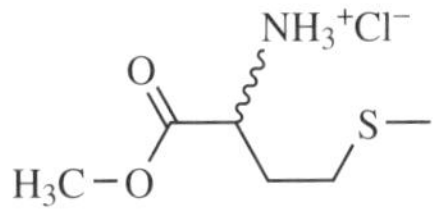

8.82

Similarly, crown **8.83** caused excellent chiral recognition in the ^{1}H NMR spectrum (400 MHz) of methionine methyl ester hydrochloride in chloroform-d.[1146] The carboxylate group of **8.83** provided a site for the binding of lanthanide ions; however, addition of Yb(III) as Yb(fod)$_3$ to chloroform-d and Yb(III)nitrate to methanol-d_4 solutions of **8.83** had minimal effects on the spectra of the substrates and essentially did not enhance the enantiomeric discrimination.

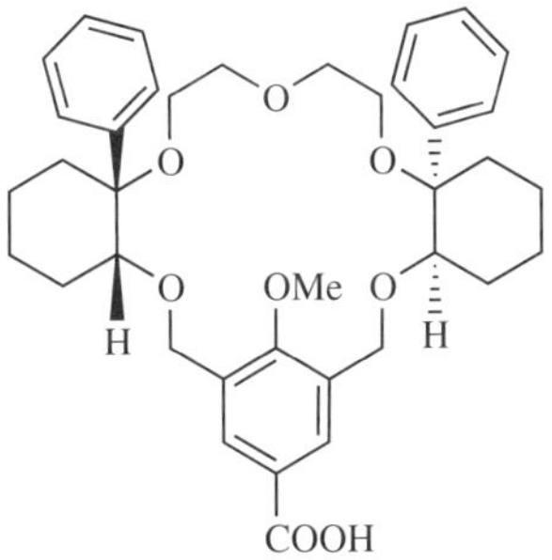

8.83

8.2.9. Aza Crown Ethers

Chiral crown ether-like compounds in which one or more of the oxygen atoms is replaced with a nitrogen atom have been prepared and their chiral recognition properties studied. The enantiomeric discrimination in the ^{1}H NMR spectra of PEA in nonpolar solvents in the presence of diaza-18-crown-6 (**8.84**) and diaza-15-crown-5 (**8.85**) derivatives has been measured.[1182,1183] The discrimination was better at lower temperatures and the derivative of **8.84** with X = H$_2$, R^1 = H, R^2 = Pr^i, and $R^3 = CH_2Ph$ caused nonequivalence of about 0.03 ppm and 0.3 ppm for the methyl and methine resonances of PEA, respectively.[1183]

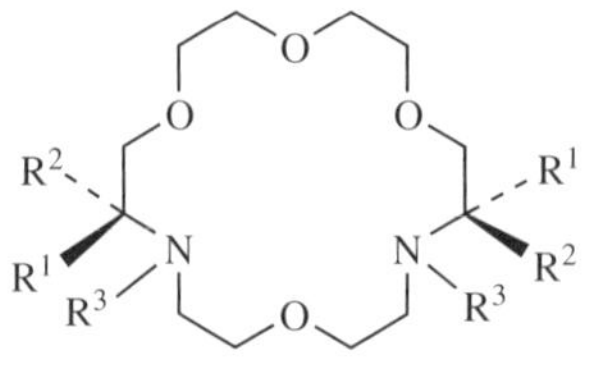

8.84

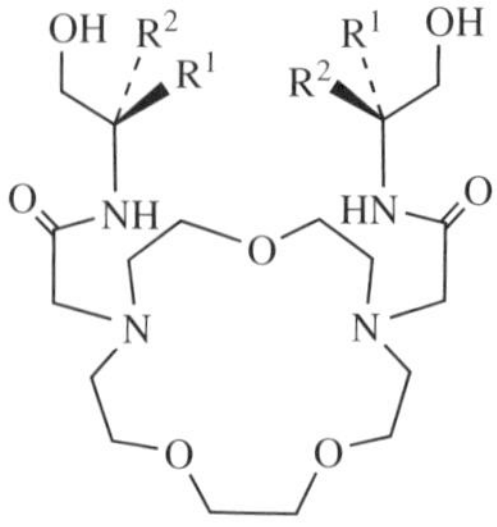

8.85

Specific selectors for zwitterionic amino acids that incorporate an aza-18-crown-6 and a guanidinium moiety have been developed (**8.86**). The NH_3^+ ion associates with the crown and the guanidinium moiety associates with the carboxylate ion as shown in Figure 8.5.[1184,1185] The selective nature was studied by extraction, and the system has not been extensively evaluated for chiral NMR discrimination purposes.

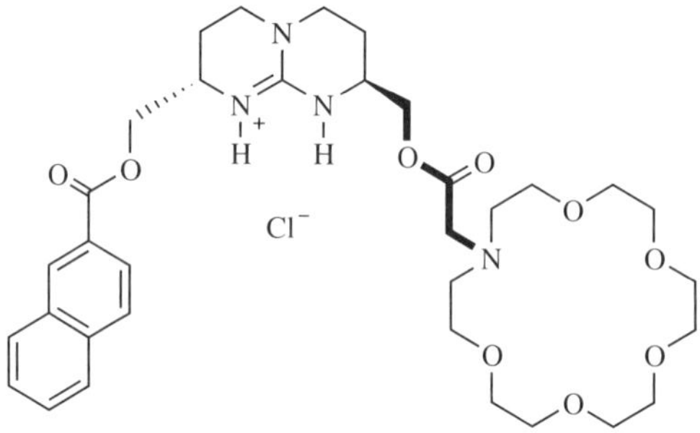

8.86

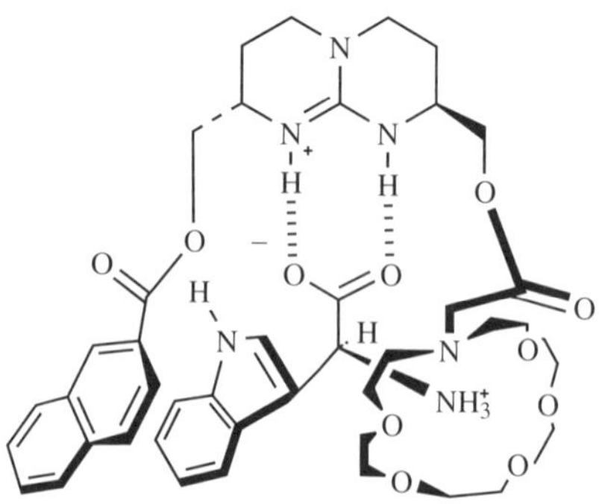

Figure 8.5. Interaction between L-tryptophan and **84.8**.

8.3. CALIXARENES AND CALIXRESORCARENES

8.3.1. Introduction

Calix[4]arenes (**8.87**) and calix[4]resorcarenes (**8.88**) are basket-shaped compounds formed from the condensation of appropriate aldehydes with phenol and resorcinol respectively. The classic calixarenes are prepared exclusively with formaldehyde. Calixarene baskets with four units are the most common, although varying the reaction conditions allows the preparation of compounds with larger cavity sizes. Calixresorcarenes can be prepared with virtually any aldehyde, although these are formed exclusively with four units. Varying the nature of the bridging unit in calix[4]resorcarenes allows the preparation of compounds with a wide range of solubilities.

Chiral derivatives that are potentially useful in NMR studies are typically obtained by attaching optically pure moieties to the calixarene or resorcarene platform. A potential complication with calixarenes and calix[4]resorcarenes is that the aromatic rings can invert leading to the formation of a series of conformational isomers (Fig. 8.6). The cone conformation is the desired form. For calix[4]arenes, conversion of the phenol moieties to propoxy groups is frequently done to prevent ring inversion and to lock in the cone conformation. Restricting calix[4]resorcarenes to the cone conformation is less readily accomplished, but many calix[4]resorcarenes naturally favor the cone conformation.

Calixarenes and calix[4]resorcarenes have found only limited application as chiral NMR discriminating agents to date. One problem is that the formation of host-guest

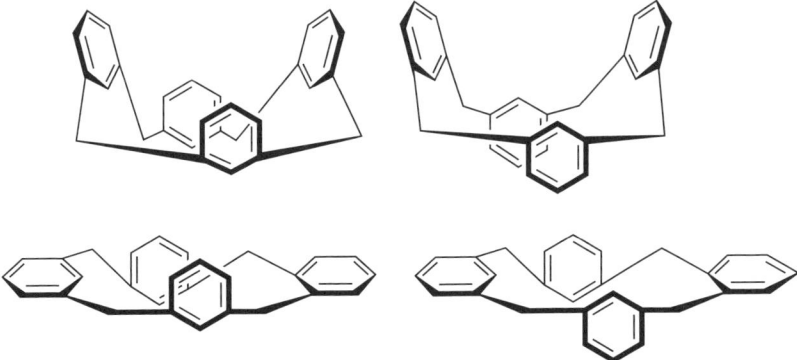

Figure 8.6. Conformational isomers of calix[4]resorcarenes.

complexes of organic substrates with organic-soluble calixarenes and calix[4] resorcarenes often does not occur to that significant extent. A water-soluble calix[4] resorcarene system is one exception, as this has excellent recognition properties in NMR spectroscopy.[1186–1188]

8.3.2. Calix[4]resorcarenes

An L-prolinylmethyl calix[4]resorcarene (**8.89**) that incorporates anionic sulfonate groups in the bridges is an especially effective water-soluble chiral NMR discriminating agent for aromatic compounds.[1186–1188] **8.89** adopts a cone conformation in water. Substantial upfield shifts in the aromatic resonances of the substrate (up to 2 ppm) indicate that host-guest association involved insertion of the hydrophobic aromatic ring into the cavity as shown in Figure 8.7. Chiral discrimination involved interactions between the proline residues and the chiral alkyl group of the substrate.

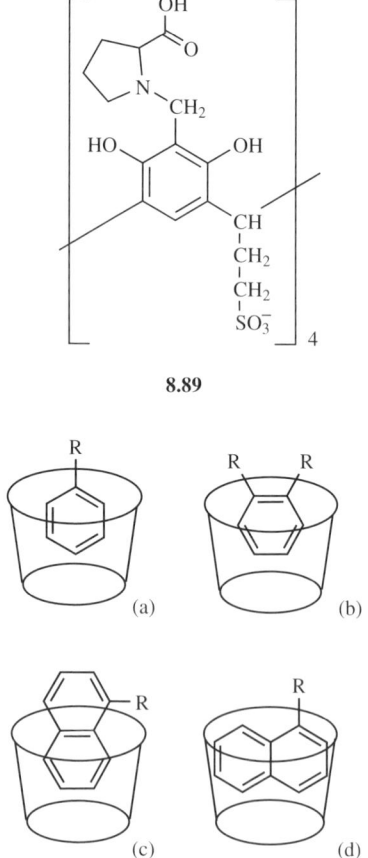

Figure 8.7. Geometries of association of substrates with **87.8**.

The initial report on **8.89** demonstrated its effectiveness for phenyl-containing compounds such as 1-phenylethanol, phenethane-1,2-diol, mandelic acid, phenylalanine methyl ester hydrochloride, and the benzyloxycarbonyl derivatives of serine and valine.[1186] The enantiomeric discrimination of aromatic resonances was generally largest for the *ortho*-position to the substituent group and least for the *para*, which was consistent with the *ortho*-hydrogen being closer to the L-prolinyl units. Chiral discrimination in the ^{1}H NMR spectrum of PEA,[1187] phenylglycine methyl ester hydrochloride, and 1-amino-2-indanol (**8.64**)[1188] with **8.89** has also been observed.

The nonequivalence in the ^{1}H spectrum of phenylalanine methyl ester hydrochloride with **8.89** was larger than that of the closely related phenylglycine methyl ester hydrochloride, likely reflecting more selective chiral recognition of the larger alkyl substituent group.[1188] Also, for phenylalanine methyl ester hydrochloride, the methoxy resonance of the D-enantiomer shifted further with **8.89**, whereas the reverse shift order was observed for the aromatic resonances. The reversal in shift order indicated that the diastereomeric nature of the associated complexes was responsible for the discrimination. The NMR spectra of aliphatic compounds such as 2-butanol, 2-pentanol, and 3-methyl-2-pentanone showed no enantiomeric discrimination in the presence of **8.89**, although the methyl doublet of 2-methyl pentanoic acid was discriminated.[1186]

In a subsequent report, it was shown that **8.89** is also an effective chiral NMR discriminating agent for water-soluble bicyclic aromatic compounds such as 1-(1-naphthyl)ethylamine (NEA), propranolol (**8.4**), and tryptophan methyl ester hydrochloride.[1187] The magnitude of the shifts in the spectra of these substrates suggested that the bicyclic compounds had different geometries of association with **8.89**. NEA and tryptophan formed a complex with one of the two rings in the cavity (Fig. 8.7c), whereas the shifts in the ^{1}H NMR spectrum of propranolol (**8.4**) with **8.89** indicated that both rings were inserted into the cavity (Fig. 8.7d). Presumably, interactions of the alkyl substituent group with the L-prolinylmethyl moieties accounted for these geometric differences. It was proposed that **8.89** likely adopts a flattened cone conformation (Fig. 8.6) to accommodate a substrate such as propranolol.[1187]

Association constants for several substrates with **8.89** were measured and are considerably larger for bicyclic compounds than for phenyl compounds, an observation in agreement with the greater hydrophobicity of the bicyclic substrates. The reduced association of the anionic sodium salt of tryptophan compared to that of the cationic tryptophan methyl ester hydrochloride suggested that some degree of ion pairing between anionic carboxylate groups on the proline and the cationic amine functionality occurred. An interesting observation was that the relative magnitudes of the association constants of the D- and L-tryptophan with **8.89** were reversed for sodium tryptophan and tryptophan methyl ester hydrochloride.[1187]

Ortho-substituted pyridine rings also formed insertion complexes with **8.89** as seen in a study of the chiral discrimination in the spectra of carbinoxamine (**8.27**), pheniramine, chlorpheniramine (**8.51**), brompheniramine, and doxylamine.[1188] With chlorpheniramine, brompheniramine, and carbinoxamine, the presence of the halogen atom *para* to the alkyl group inhibited binding of the aromatic ring within the cavity of **8.89** and complexation occurred through insertion of the pyridine ring.

For pheniramine and doxylamine, the shifts indicated that both the aromatic and pyridine rings associated to some degree with **8.89**. The discrimination in the spectra of several amine substrates with **8.89** was larger than that observed with (18-crown-6)-2,3,11,12-tetracarboxylic acid.

Calix[4]resorcarenes analogous to **8.89** except with an N-methyl alanine, N-methyl valine, and N-methyl leucine in place of the proline units were examined for their utility as chiral NMR shift reagents.[1188] In all three cases, the ^{1}H NMR spectrum of the calix[4]resorcarenes was broadened indicating that several conformations were present in solution. The valine and leucine reagents were ineffective in chiral NMR studies. The N-methyl alanine was the best of these three, but not recommended over **8.89**.

An elaborate hemicarcerand compound consisting of two calix[4]resorcarene units coupled together through a 1,1′-binaphthalene unit and $-O(CH_2)_4-O-$ linkages formed a cavity that bound aryl alkyl alcohols, sulfoxides, alkyl halides, and alkyl alcohols.[1189] Nonequivalence in the ^{1}H NMR spectrum up to 0.36 ppm was observed.

A phthalazine bridged calix[4]resorcarene compound (**8.90**) was prepared with optically pure chiral substituent groups.[1190] *Trans*-1,2-cyclohexanediol formed complexes with **8.90** that underwent slow exchange on the NMR time scale. Furthermore, while both enantiomers of the cyclohexanediol bound in the cavity, one associated in excess over the other indicating that chiral recognition occurred.

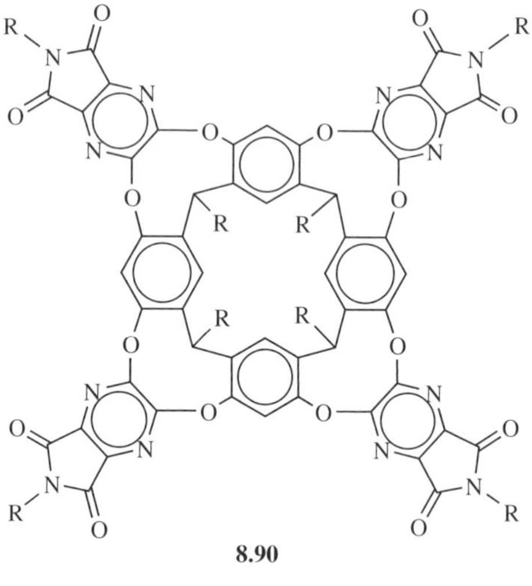

8.90

A phthalazine-bridged calix[4]resorcarene similar to **8.90** has been shown to aggregate in solution in a manner that created a cylindrical cavity with a guest compound such as (R)-propylene sulfide (**8.91**). Each individual capsule was chiral and these aggregates were referred to as constellational diastereomers.[1191] Three molecules fit into the cavity and they did not exchange positions. Two sets of signals were observed, one for the molecule in the center and two for the molecules at the end of the cavity. The resonances were assigned by adding non-optically pure samples of the guest. The optical purity of guests could be determined by analysis of the NMR spectrum.

8.91

Other phthalazine-bridged calix[4]resorcarenes have been shown to form self-assembling capsules around chiral carboxylic acids such as mandelic and α-bromobutyric acids.[1192] Multiple substrate compounds fit into the cavity of the dimer, and some compounds exhibited a diastereomeric selectivity as to which enantiomer preferentially entered the cavity. Mixtures of the resorcarene with isopropyl chloride and (R)-styrene oxide (**8.92**) formed an encapsulated mixture in which both substrates were in the cavity and the enantiotopic methyl groups of the isopropyl chloride exhibited distinct resonances.

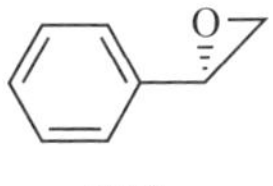

8.92

8.3.3. Calixarenes

Homazacalix[4]arenes (**8.93**)[1193,1194] and bis-homodiazacalix[4]arenes (**8.94**)[1195,1196] with optically pure amino acids incorporated into one or two of the bridging units caused enantiomeric discrimination in the ^{1}H NMR spectrum of α-methylbenzyl trimethylammonium iodide. Mono-amino acid derivatives with valine and tryptophan were studied.[1193,1194] The corresponding derivative with a t-butyl group instead of a methyl group on the phenol ring did not bind to α-methylbenzyl trimethyl ammonium iodide because of steric effects.[1193] Di-amino acid derivatives with valine, tryptophan,[1195] and tyrosine[1196] groups were studied.

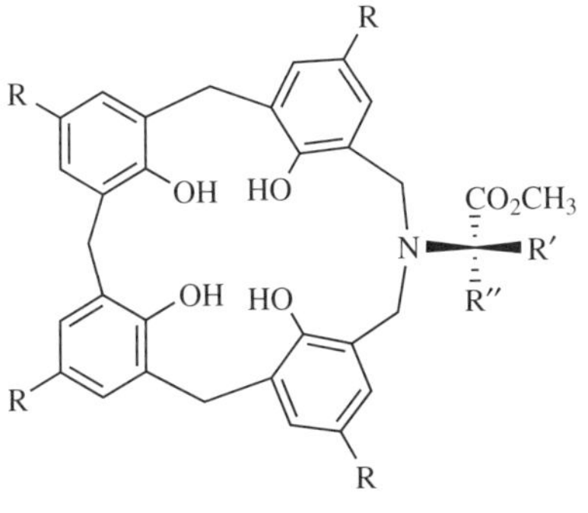

8.93 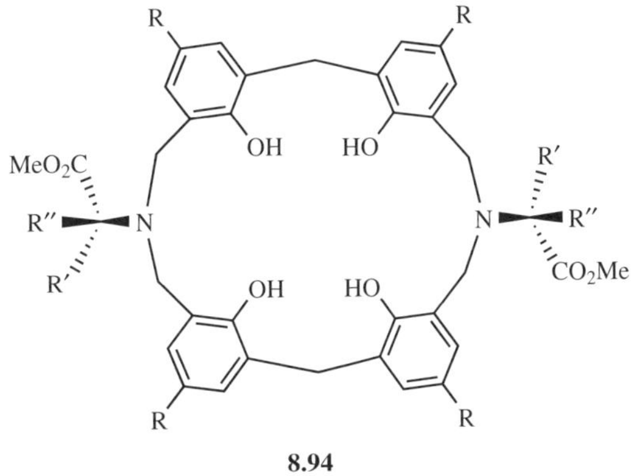   **8.94**

Calix[4]arene systems that included an azacrown unit (**8.95**) have been evaluated as chiral NMR discriminating reagents. The ^{1}H and ^{13}C NMR spectra of the dibenzoate of tartaric and mandelic acid exhibited nonequivalence in the presence of **8.95** in chloroform-d.[1197]

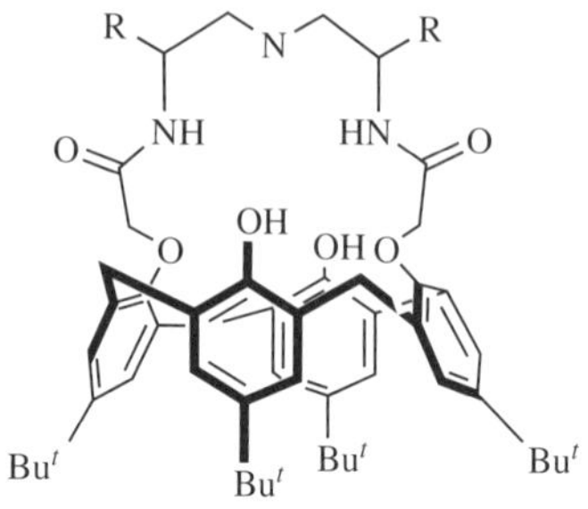

8.95

The ^{1}H NMR spectra of several primary aliphatic and aromatic amines and amino ester hydrochloride salts exhibited chiral discrimination in the presence of the thiacalix[4]crown **8.96**.[1198] Nonequivalence as large as 0.143 ppm was observed in the spectrum of 1-(1-naphthyl)ethylamine (NEA). The enantiomeric distinction was small for secondary and tertiary amines. Binding of these substrates with the crown moiety was likely the mode of association.

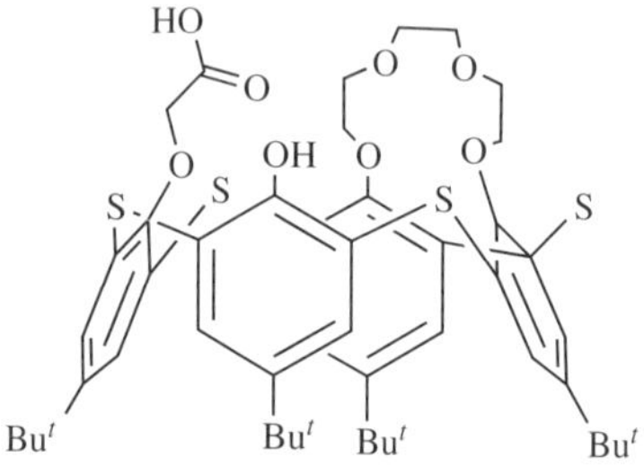

8.96

Cycloenantiomeric hexahomooxacalix[3]arenes (**8.97**) caused a slight amount of enantiomeric discrimination in the ^{1}H NMR spectrum of phenylalanine ethyl ester hydrochloride in 9:1 chloroform-d/methanol-d_4.[1199] Calix[4]arenes with optically pure amino acid groups attached through the phenol oxygen were evaluated as chiral NMR shift reagents.[1125] The t-butyl ester of L-alanine (**8.98**) was the most effective of those studied, but only a small amount of chiral recognition was observed in the ^{1}H NMR spectrum of the N-(3,5-dinitrobenzoyl) derivative of NEA.

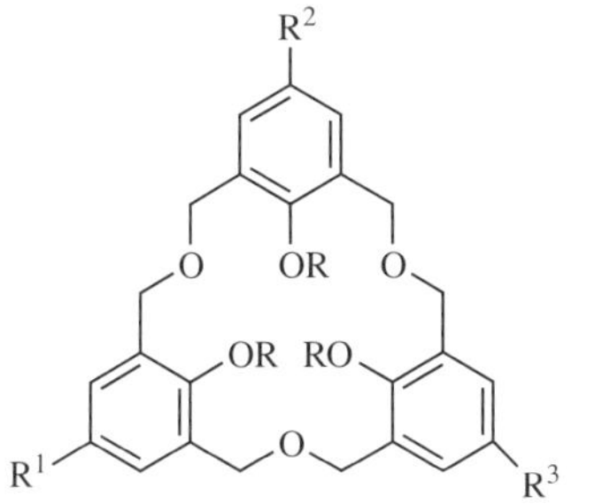

8.97

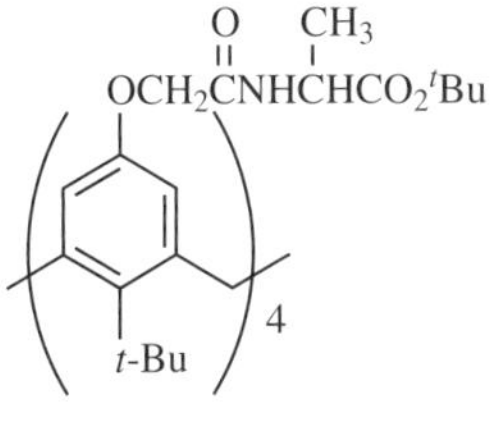

8.98

Calix[4]arenes with two dansylated alanine or phenylalanine substituent groups (**8.99**) caused a very slight distinction in the ^{1}H NMR spectrum of anionic alanine or phenylalanine in chloroform-d.[1200] A similar calix[4]arene with two optically pure substituent groups incorporating alanine or phenylalanine units (**8.100**) was developed primarily for its potential as a chromogenic chiral receptor.[1201] The ^{1}H NMR spectrum of the phenylglycinate ion showed a splitting of the methine resonance that was larger for the alanine dertivative.

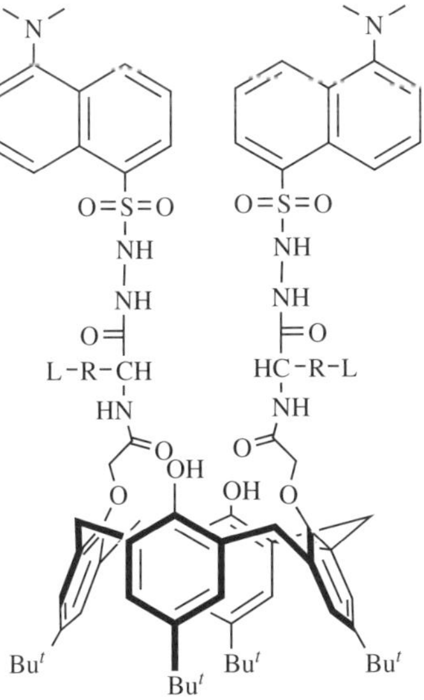

8.99

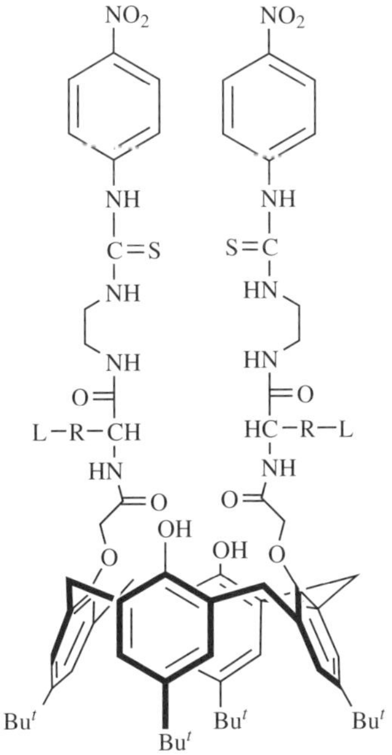

8.100

An enantiopure calix[6]arene prepared from an optically pure *tris*(aminoethyl) amine (tren) ligand (**8.101**) has been examined for its potential utility as a chiral solvating agent.[1202] The tren moiety is bound to alternate phenol sites of the calix[6]arene and locks the compound into the cone conformation. This enhances the potential for specific host–guest association relative to what occurs in flexible and rapidly isomerizing calix[6]arene systems. Nonequivalence was observed in the ^{1}H NMR spectrum of propane-1,2-diol and 4-methylimidazolidin-2-one (**8.102**) in chloroform-d.

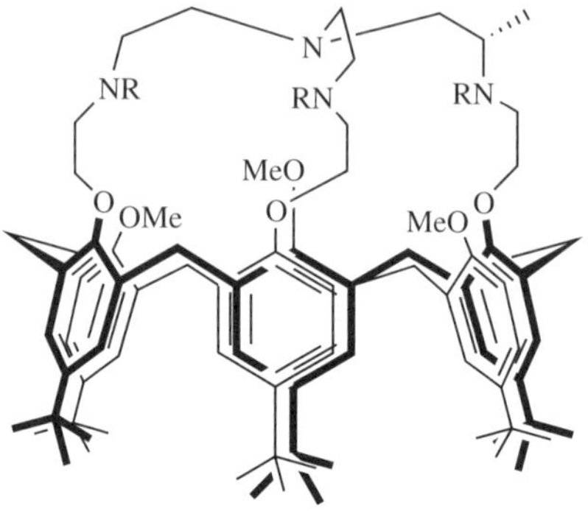

8.101

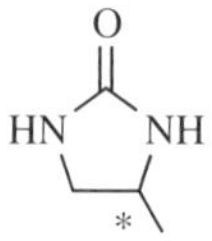

8.102

Metallacalix[n]arenes ($n = 4,6$)(**8.103**) can be formed in which a palladium species with either (R,R)- or (S,S)-1,2-diaminocyclohexane is complexed with 4,6-dimethyl-2-hydroxypyrimidine (**8.104**) bridging units.[1203] Addition of adenosine-5′-monophosphate to a racemic mixture of the metallacalixarene caused a splitting of some of the calixarene resonances in the ^{1}H NMR spectrum.

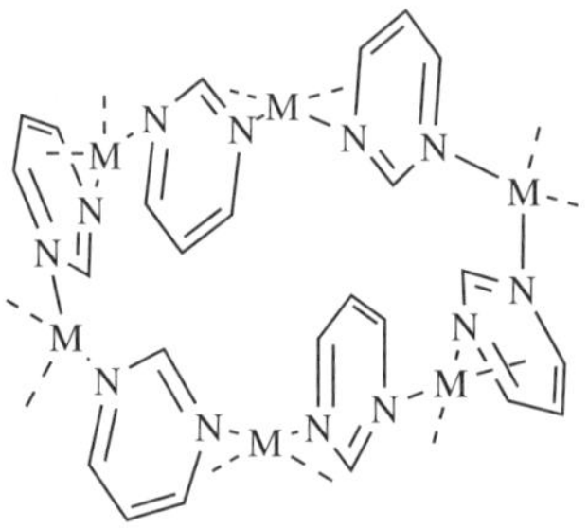

8.103

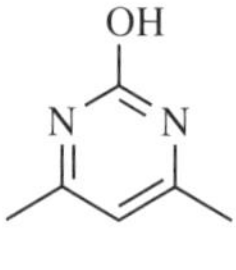

8.104

Calix[4]arene dimelamines (**8.105**) self-assembled into enantioselective receptors when mixed with compounds like 5,5-diethyl barbituric acids (**8.106**)[1204] or cyanuric acid derivatives.[1205] Mixing these assemblies with the (R)-enantiomer of amines such as PEA, alanine, and naphthyl dimelamine caused formation of an aggregate with only M-helicity.[1204] Although the (S)-enantiomer caused the aggregate of P-helicity, the self-assembled receptors caused enantiomeric discrimination in the NMR spectra of aryl alkyl, alkyl, and cyclic carboxylic acids.[1205]

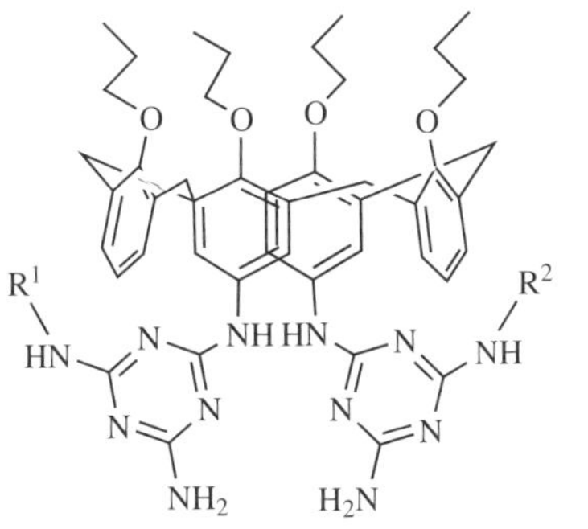

CALIXARENES AND CALIXRESORCARENES

8.105

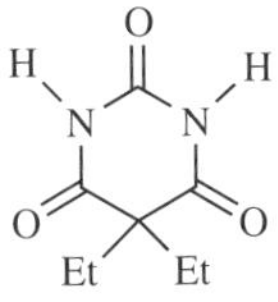

8.106

There are two strategies by which achiral calixarenes have been used in enantiomeric recognition. One involved the use of an achiral calix[4]arene (**8.107**) that formed a chiral dimer in a solvent such as p-xylene-d_{10}.[1206] The dimer encapsulated compounds such as (+)-nopinone (**8.108**), (−)-myrtenal (**8.109**), (+)-camphor (**8.110**), and tricyclene (**8.111**), and certain calixarene resonances in the ^{1}H NMR spectrum were split.

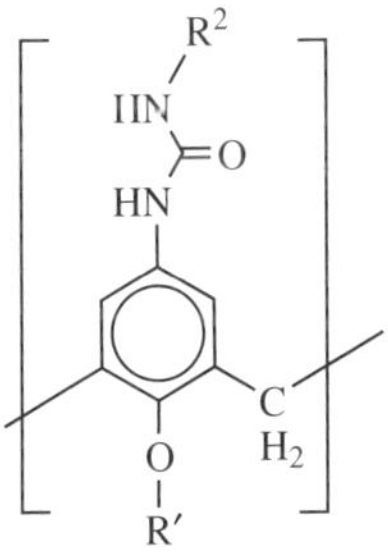

8.107 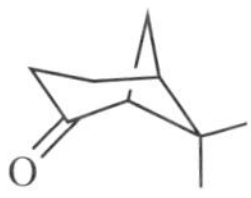   **8.108** 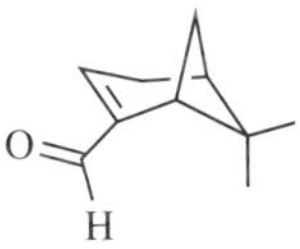   **8.109**    **8.110**    **8.111**

A second approach involved adding optically pure PEA or NEA to calix[6]arene to form a chiral host-guest complex.[1207] When chiral sulfoxides such as p-tolyl ethyl sulfoxide, phenyl methyl sulfoxides, and other aryl alkyl sulfoxides were added to the amine-calix[6]arene mixture, enantiomeric discrimination in the ^{1}H NMR spectrum of the sulfoxide was observed, presumably because the PEA and sulfoxide were simultaneously within the calixarene cavity. Enantiomeric discrimination was not observed in the absence of the calixarene.

8.3.4. Summary

The calixarene or resorcarene systems described in this section are not commercially available. Of those that have been studied as chiral NMR shift reagents, only the water-soluble **8.89** merits particular attention. Water-soluble chiral shift reagents are not that common, and this particular reagent is especially effective for compounds with an aromatic ring. It can be prepared in two steps from commercially available reagents.

8.4. RECEPTOR COMPOUNDS FOR CARBOXYLIC ACIDS

Guanidinium-based compounds such as (**8.112**),[1208] (**8.113**),[1209] and (**8.114**)[1210] selectively bind carboxylate species or carboxylic acids as shown in Figure 8.8. With compound **8.112**, enantiomeric discrimination was observed in the spectra of the anions of N-acetyl-DL-alanine, 2-methylbutyric acid, lactic acid, 2-bromo butyric acid, and DL-phenylalanine in acetonitrile-d_3.[1208] Receptor **8.113**, which binds in a similar manner, caused chiral discrimination in the spectra of naproxenate and mandelate.[1209] Compound **8.114** caused nonequivalence of up to 0.1 ppm in the NH resonance of DL-phenylalanine in chloroform-d.[1210]

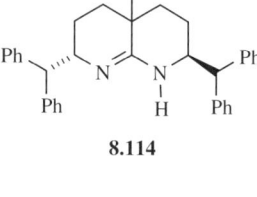

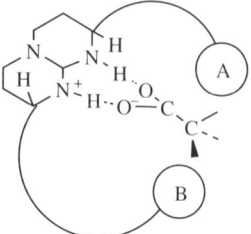

Figure 8.8. Interaction of carboxylate species with guanidinium-based reagents.

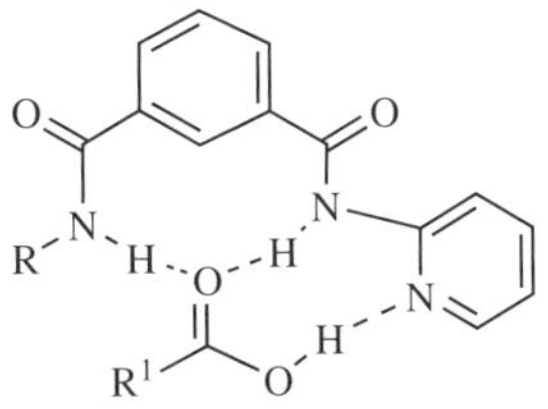

RECEPTOR COMPOUNDS FOR CARBOXYLIC ACIDS 321

Figure 8.9. Interactions of carboxylic acids with **113.8**.

Receptors **8.115**, in which R = phenyl, naphthyl, or cyclohexyl, have been evaluated for their ability to distinguish enantiomers of carboxylic acids.[1211,1212] The receptor forms three hydrogen bonds with carboxylic acids as shown in Figure 8.9. The chiral discrimination was largest with the naphthyl derivative and smallest with the cyclohexyl group.[1212] Nonequivalence of up to 0.283 ppm was observed for certain resonances in the spectrum of naproxen. A variety of other carboxylic acids were discriminated as well, and the enantiotopic resonances of prochiral acids such as isobutyric acid and benzyl carboxylic acid were resolved.[1211]

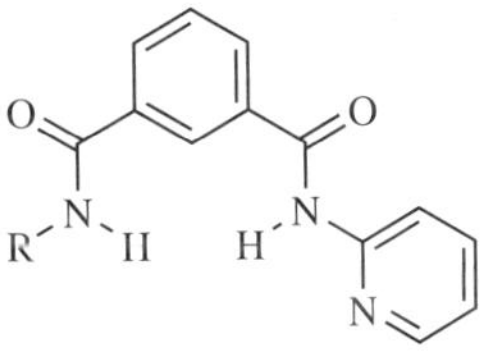

8.115

Certain resonances in the NMR spectra of tetrabutyl ammonium salts of carboxylic acids with **8.116** in chloroform-d exhibited nonequivalence of 0.03–0.11 ppm.[1213] Alkyl- and aryl-containing substrates were examined. Most had the chiral site α to the carboxylic acid, but discrimination was also observed in the spectrum of **8.117**, in which the stereogenic center is γ to the carboxylic acid.

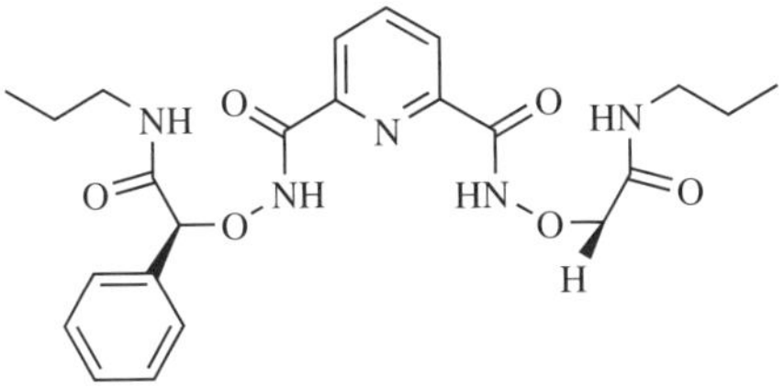

8.116 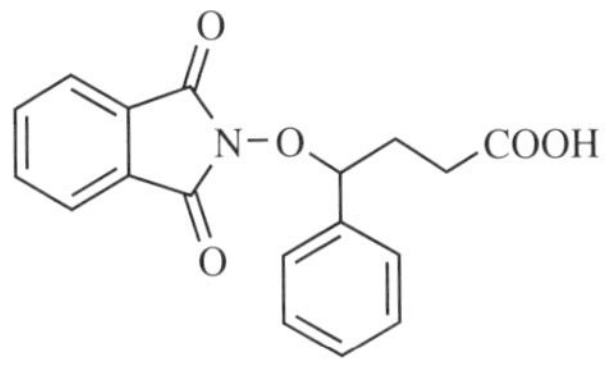   **8.117**

Compound **8.118** has carbonyl and NH groups that formed hydrogen bonds to and split the ^{1}H NMR spectra of α-substituted carboxylic acids.[1214]

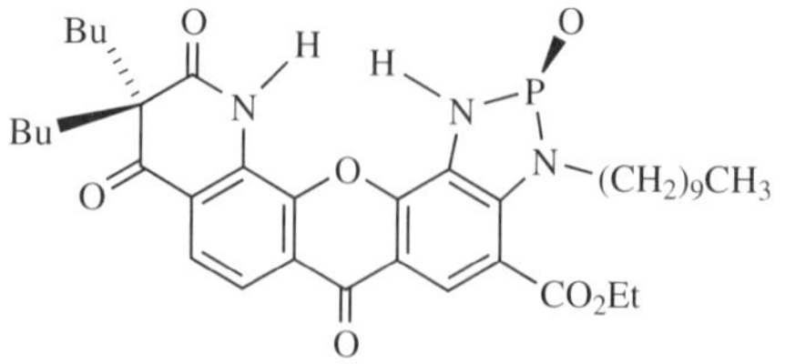

8.118

The anthryl-substituted **8.119** caused nonequivalence in the ^{1}H NMR spectrum of tetrabutyl ammonium mandelate in chloroform-d.[1215]

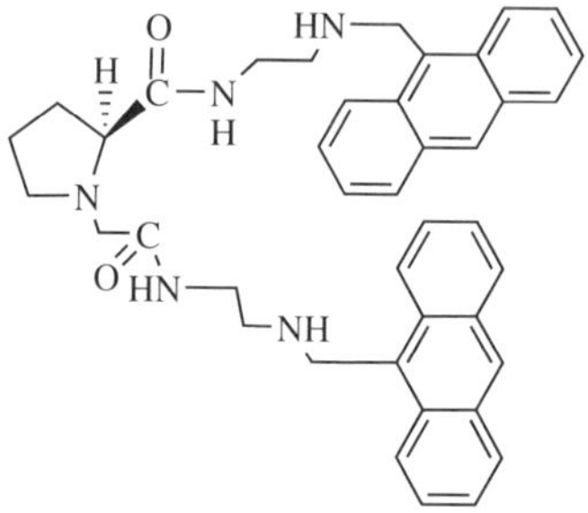

8.119

Proline-derived macrocyclic dioxopolyamines (**8.120**) with one or two proline units were evaluated as chiral solvating agents for carboxylic acids.[1216] The *bis*-proline derivative was generally more effective at causing enantiomeric discrimination in the ^{1}H NMR spectrum of substrates such as mandelic acid, naproxen, and dibenzoyl tartaric acid in chloroform-d.

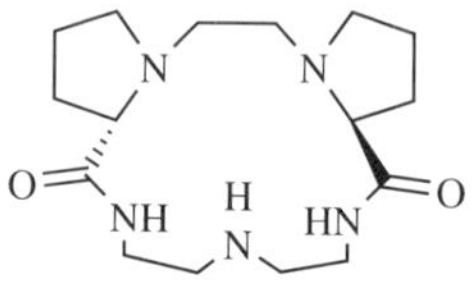

8.120

Dioxocyclens (**8.121**), which are prepared from an amino acid, caused a small amount of nonequivalence in the ^{1}H NMR spectra of mandelic acid, methoxyphenylacetic acid, and other mandelic acid derivatives.[1217]

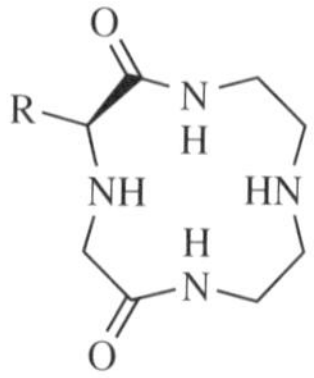

8.121

The host compound **8.122**, which is a derivative of an unnatural isoquinoline alkaloid, caused enantiomeric discrimination in the ^{1}H NMR spectrum of naproxen.[805]

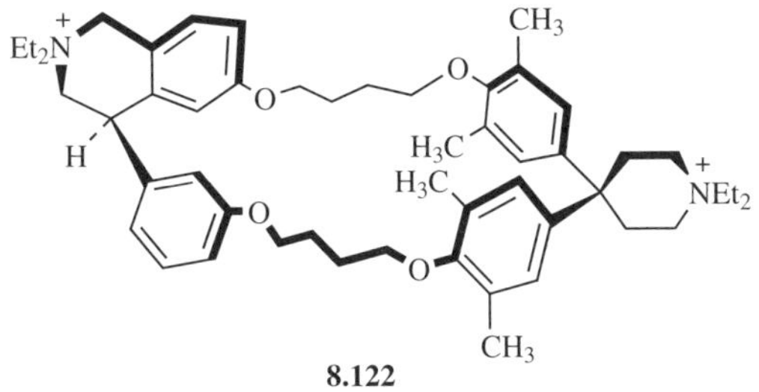

8.122

8.5. CYCLIC PEPTIDES

Several cyclic peptide and pseudopeptide receptors have been used to discriminate the NMR spectra of enantiomeric substrates. A cyclo(L-Pro-Gly)$_n$ (n = 3 or 4) caused nonequivalence of about 0.2 ppm of certain ^{13}C resonances of hydrochloride salts of the benzyl ester of proline and methyl esters of phenylalanine and valine in chloroform-d.[1218,1219] The amino acids bound to the cyclic peptide as shown in Figure 8.10.[1218]

The methyl, methine, and N-methyl resonances of trimethyl-1-phenethylammonium bromide (**8.123**) were distinguished in acetonitrile-d_3 in the presence of 24-, 27-, and 36-membered ring peptides containing glycine and N,N'-ethylene-bridged (S)-leucyl-(S)- leucine units.[1220] No enantiomeric discrimination was observed in chloroform-d or dimethylsulfoxide-d_6. This is one of the few systems capable of discriminating quaternary ammonium ions.

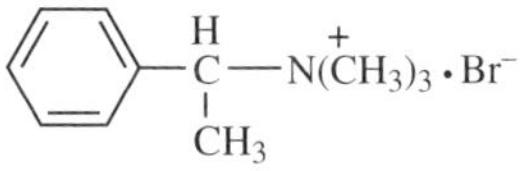

8.123

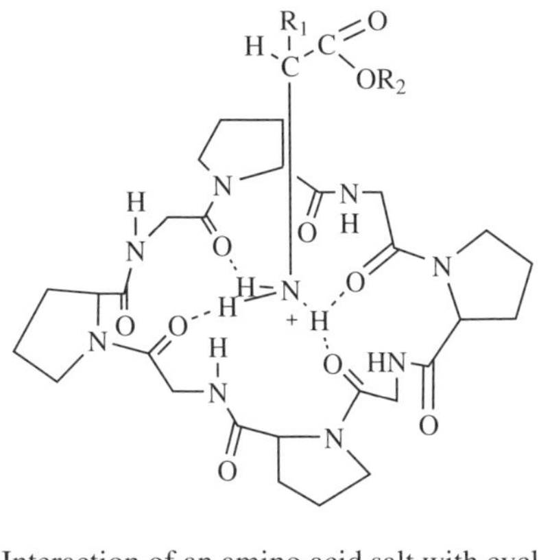

Figure 8.10. Interaction of an amino acid salt with cyclo(L-Pro–Gly)$_4$.

Similar 24-, 27-, and 36-membered cyclic pseudopeotides containing glycine and (2S,3'S)-4-methyl-2-(2'-oxo-3'-isobutyl-1-piperazinyl)pentanoic acid units (**8.124**) caused chiral discrimination in the ^{1}H and ^{13}C NMR spectra of species such as alanine-N-methylanilides (**8.125**), the hydrobromide salt of PEA, and several amino acid methyl esters in chloroform-d.[1221]

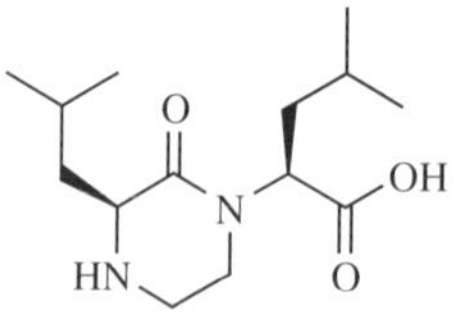

8.124

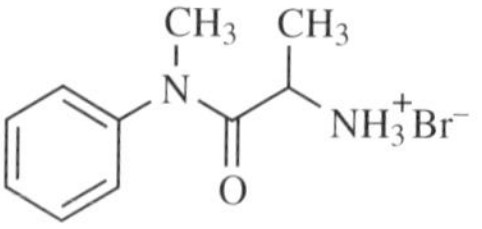

8.125

The cyclo(L-Phe-L-Pro)$_4$ species was able to enantiomerically discriminate D,L-phenylalanine methyl ester hydrochloride.[1222] The macrocyclic peptide receptor **8.126** exhibited chiral recognition with a range of amino acids and dipeptides.[1223]

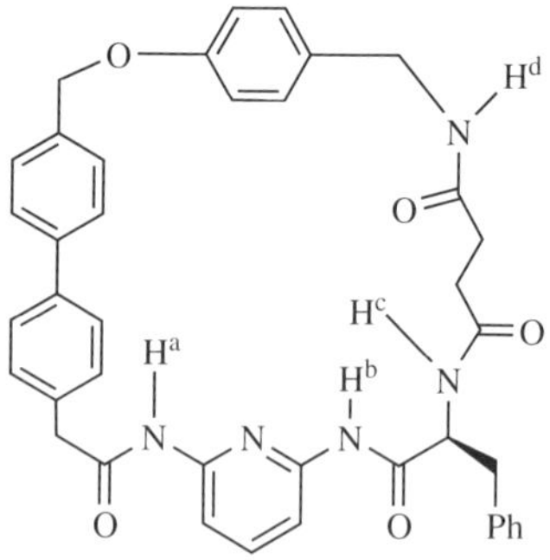

8.126

A cyclic hexapeptide with alternating L-proline and 3-aminobenzoic acid groups caused chiral discrimination in the ^{1}H NMR spectrum of N,N,N,-trimethyl-1-phenylethylammonium picrate (**8.123**).[1224] The spectrum was obtained in chloroform-d with 0.1% dimethylsulfoxide-d_6, and nonequivalence of 0.22 ppm for the methine and 0.06 ppm for the N-methyl resonance was observed.

8.6. MISCELLANEOUS RECEPTOR COMPOUNDS

8.6.1. Cyclic Tetraamidic Reagent

Compound **8.127** and similar analogs, which are patterned after chiral liquid chromatographic stationary phases, were examined using NMR spectroscopy.[1225] The spectra of substrates with a π-acid group such as dinitrobenzoyl derivatives of aryl carboxylic acid esters, arylamides, and dinitrophenyl ethyl sulfoxide exhibited nonequivalence of 0.01–0.13 ppm in the ^{1}H NMR spectrum in chloroform-d.

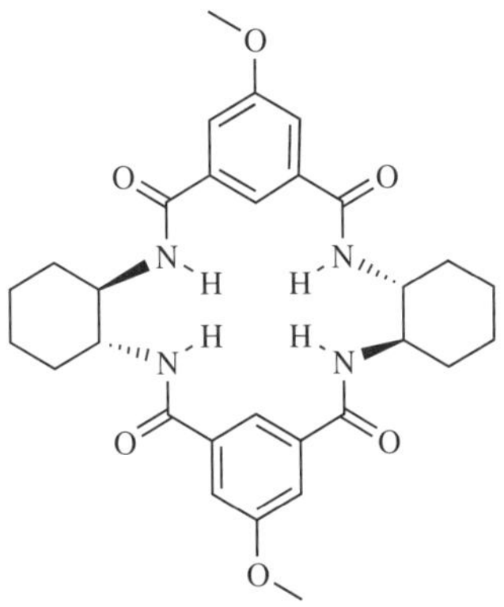

8.127

8.6.2. Cyclosophoraoses (Cyclic(1-2)-β-D-glucans)

The species *Rhizobium meliloti* produces a series of cyclodextrin-like cyclosophoraoses that contain from 17 to 27 (1-2)-β-D-glucan units. These compounds were shown to be effective chiral solvating agents for compounds such as propranolol (**8.4**),[1226] *N*-acetylphenylalanine, and catechin (**8.128**).[1227]

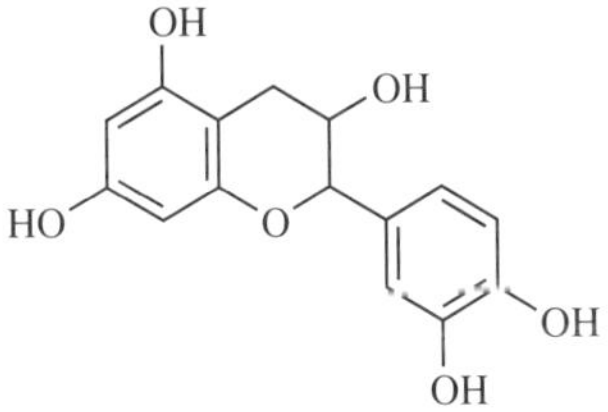

8.128

8.6.3. Glycophane Receptor

The glycophane (**8.129**), which has glucose and naphthyl units, exhibited chiral recognition in 1:1 mixtures of methanol-d_4/deuterium oxide toward 2,4-dinitrophenyl derivatives of amino acids.[1228] The reagent possessed both cyclodextrin-like and cyclophane-like properties.

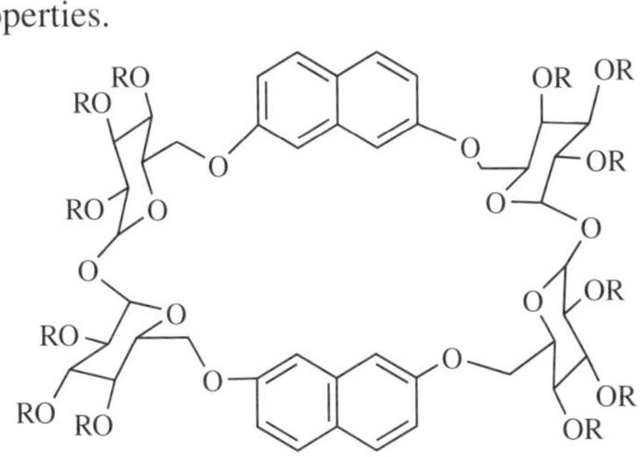

8.129

8.6.4. Cyclophane Receptors

Compound **8.130** is a water-soluble cyclophane that has been used as a chiral NMR solvating agent for carboxylic acids,[1229–1231] 1-arylethanols,[1230,1231] and arylamines.[1229,1330] The compound is a derivative of tartaric acid and contains a hydrophobic cavity. The ^{1}H NMR spectra of compounds such as mandelic acid, lactic acid, atrolactic acid, phenylglycine, phenylalanine, and NEA showed enantiomeric discrimination in the presence of **8.130**.[1229,1230] The aromatic rings in **8.130** caused shielding of the nuclei of the guest compounds. In some cases, small amounts of methanol-d_4 were added to solubilize the substrates.[1231]

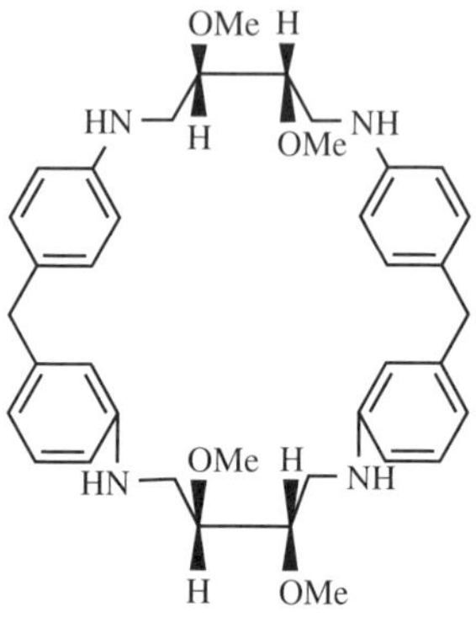

8.130

Other water-soluble cyclophane receptors (**8.131**) have been used as chiral solvating agents.[1232,1233] These caused nonequivalence in certain resonances of the ^{1}H NMR spectra of substrates such as menthol[1232,1233] and citronellol.[1233] The lipophillic portions of the substrates bound within the hydrophobic cavity of the cyclophanes.

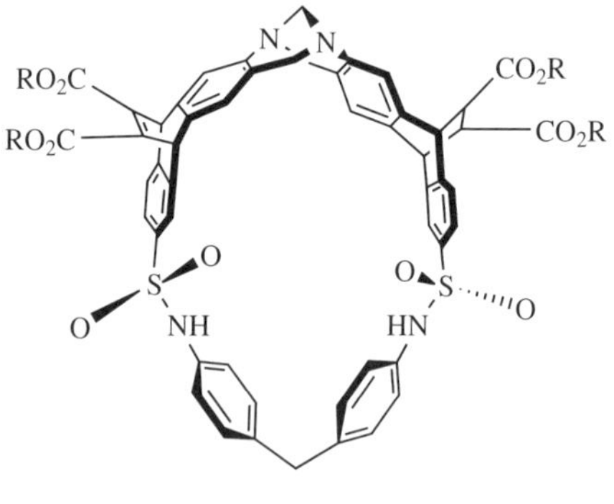

8.131

8.6.5. Cryptophane Receptor

The cryptophane host **8.132** was specifically designed to accommodate and enantiomerically discriminate the compound bromochlorofluoromethane (CHBrClF).[1234]

This was the first reported NMR enantiomeric discrimination of CHBrClF. The ^{1}H resonance was split in chloroform-d.

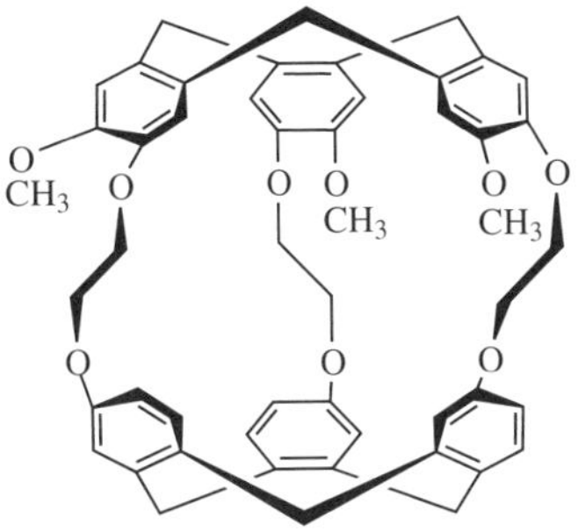

8.132

8.6.6. Receptor for Secondary Amines

The compound 2,6-bis(4,5-dihydro-4-phenylozazol-2-yl)pyridine (**8.133**) is an effective chiral NMR solvating agent for secondary amines.[1235] The amines were analyzed as their protonated cations and formed hydrogen bonds to the nitrogen atoms of **8.133**. Nonequivalence of 0.12–0.53 ppm was observed in the ^{1}H NMR spectra of compounds such as 2-methylpiperidine, 3-methylpiperidine, and proline methyl ester.

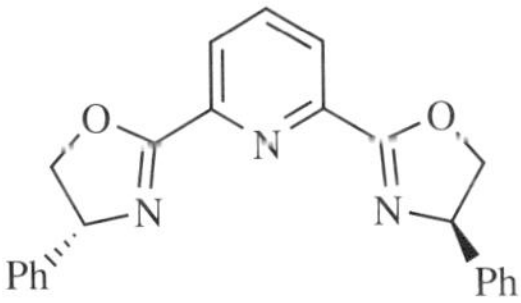

8.133

8.6.7. Spirobisindane Receptor

The spirobisindane host compound **8.134** formed chelate complexes with the di-cations of lysine and arginine.[1236] The resonance of the hydrogen *ortho* to the phosphonate group of **8.134** exhibited two separate peaks for the complexes of the D- and L-forms of the amino acids. No enantiomeric discrimination was observed for 1,2-diaminocyclohexane, L-histidine, or L-ornithine (**8.135**).

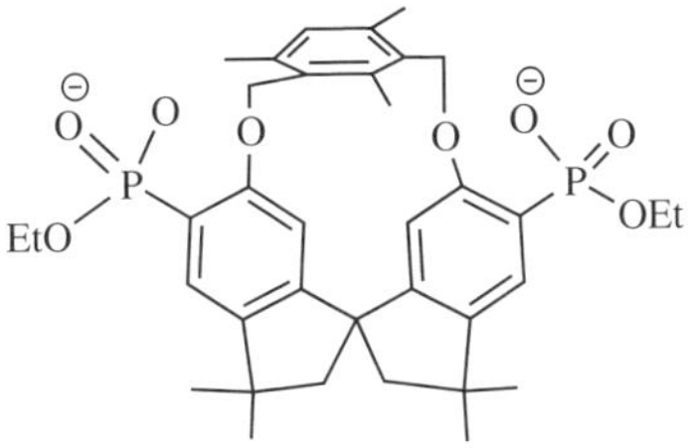

8.134

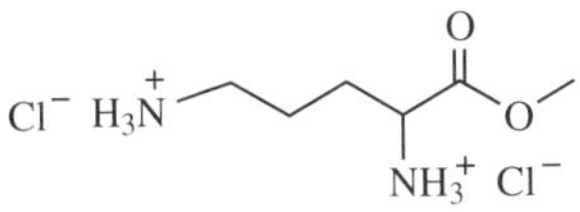

8.135

8.6.8. Receptors for Diols

Several resonances in the ^{1}H NMR spectrum of 2,2'-dihydroxy-1,1'-binaphthalene in chloroform-d exhibited enantiomeric discrimination in the presence of (−)-2,5-bis[4'(S)-isopropyloxazolin-2-yl]thiophene (**8.136**).[1237]

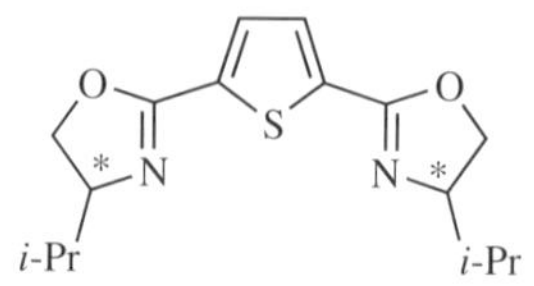

8.136

The twisted amidine (**8.137**) caused chiral discrimination in the ^{1}H NMR spectrum of 2,2'-dihydroxy-1,1'-binaphthalene[1238] and other diols such as 2,3-butanediol, 2,4-pentanediol, and 2,5-hexanediol.[1239] Figure 8.11 shows the nature of the hydrogen bonding interactions between diols and **8.136**. The resonances of the diastereotopic methyl groups of **8.137** were also resolved on binding with the substrate.

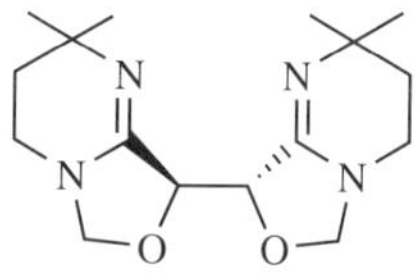

8.137

8.6.9. 1,1'-Binaphthalene-based Receptors

Host molecule **8.138** is based on a 1,1'-binaphthalene template.[1240] **8.138** was able to enantiomerically discriminate naproxen derivatives through a combination of hydrogen bonding, π–π interactions, and steric effects.

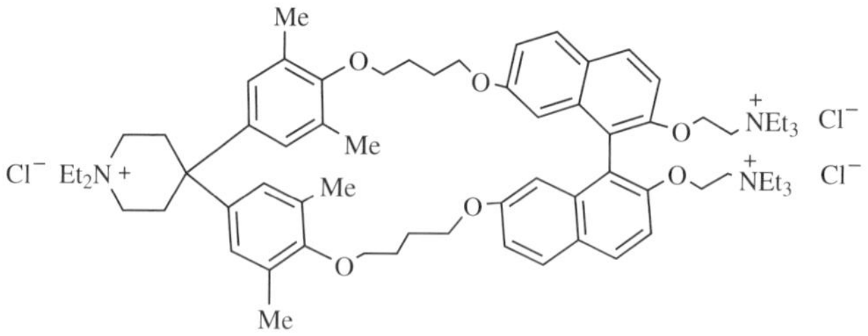

8.138

Host molecule **8.139** has been studied for its efficiency in preparing asymmetric barbiturate-isoxazoline conjugates.[1241] The receptor was able to distinguish the enantiomeric products so that their optical purity could be determined by recording the NMR spectrum of the crude reaction mixture.

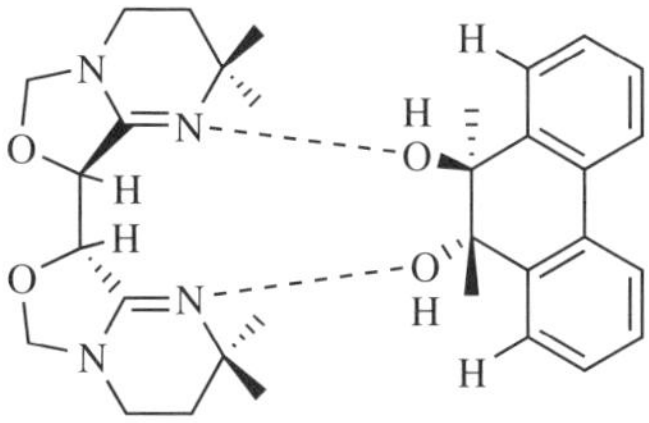

Figure 8.11. Hydrogen bonding interaction between diols and **132.8**

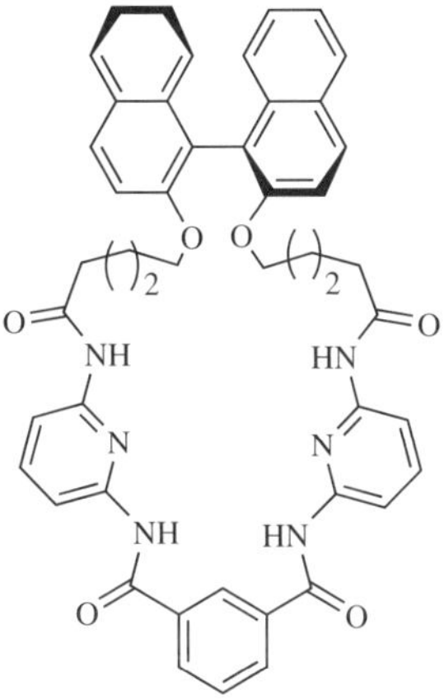

8.139

8.7. ENZYMES

8.7.1. Calmodulin

The compound (1-{β-[3-(p-methoxyphenyl)propyloxy]-p-methoxyphenethyl}-1H-imidazole hydrochloride (**8.140**) is the antagonist of mammalian receptor-operated calcium channels. A doubling of the ^{1}H NMR signal of the imidazolium H2 singlet was observed in the presence of calmodulin and was used to facilitate the study of certain aspects of this enzyme system.[1242]

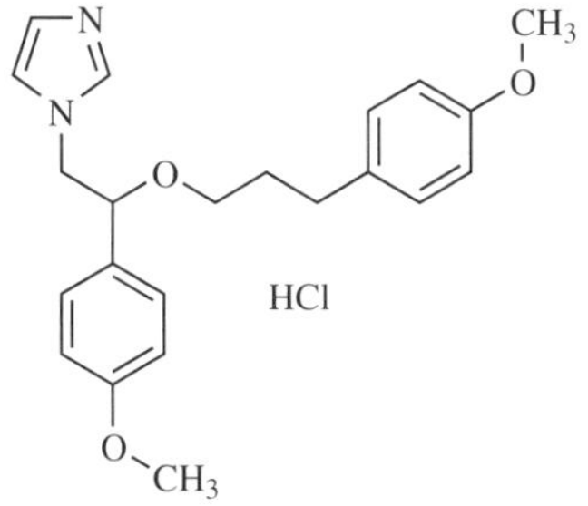

8.140

8.7.2. Chymotrypsin

The compound *N*-acetyl-DL-*p*-fluorophenylalaninal (**8.141**) binds to α-chymotrypsin. In the NMR spectrum of the mixture, separate signals were observed for the hemiacetal complex, bound aldehyde, free aldehyde, and free hydrate.[1243] The signals of the enantiomers were also distinct and could be used to determine that the L-aldehyde formed a strong hemiacetal complex.

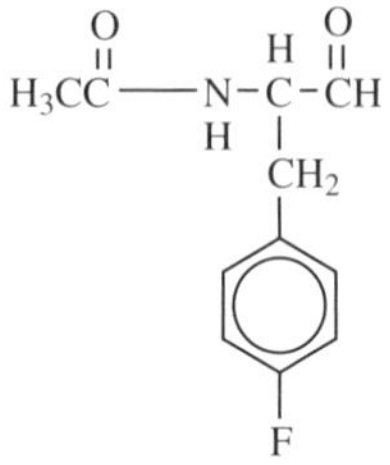

8.141

9

CHIRAL DISCRIMINATION WITH METAL-BASED REAGENTS

9.1. INTRODUCTION

The use of metal complexes to effect chiral recognition in NMR spectroscopy is a widely exploited strategy. Usually, the substrate functions as a ligand and displaces an existing ligand on the complex. With some reagents, the substrate can be thought of as binding to a coordinatively unsaturated metal.

One of the most significant developments in the entire field of NMR shift reagents was the discovery of the effect that paramagnetic lanthanide ions had on the spectra of bound substrates. Both achiral and chiral lanthanide species have found broad applicability, especially when high field NMR spectrometers were not commonly available. Dipolar shifts caused by the paramagnetism of the lanthanide ions were fundamental to the success of these reagents. Paramagnetism also leads to broadening in the NMR spectrum, and the magnitude of the broadening is larger at higher field strengths. As such, lanthanide complexes are not as widely used today as they once were. Nevertheless, the variety of substrate classes for which lanthanide-based reagents may be effective, combined with the range of solvents that different lanthanide species are soluble in, still makes lanthanide complexes important in the area of chiral NMR analysis.

A wealth of transition metal systems have also been studied for their applicability in chiral analysis via NMR spectroscopy. Complexes with palladium, platinum, rhodium, and silver are noteworthy for their ability to coordinate with soft Lewis

---

Discrimination of Chiral Compounds Using NMR Spectroscopy, by Thomas J. Wenzel
Copyright © 2007 John Wiley & Sons, Inc.

bases such as olefins, aromatics, phosphines, alkyl halides, and sulfur-, selenium-, and arsenic-containing compounds, among others. In certain cases, though, these metals are involved in complex formation and analysis of hard Lewis base substrates with oxygen- and nitrogen-containing moieties.

Cobalt, zinc, and ruthenium complexes, especially with porphyrin ligands, have been exploited for chiral recognition in NMR spectroscopy. The extensive shielding by the porphyrin ring of bound substrates causes exceptionally large shifts, and at times enantiomeric discrimination, of resonances. Specialized applications of nickel, cobalt, tin, titanium, tungsten, molybdenum, mercury, osmium, and uranium complexes have been reported.

In most cases other than the lanthanides, the metal is diamagnetic and the shifts and discrimination involve specific interactions from other ligands of the complex that are determined by the coordination geometry. Metal complexes are most commonly used for determining optical purity, but if the interaction is known specifically enough, absolute configurations can sometimes be assigned. Another common strategy is to look for trends in the relative shifts of chiral compounds with similar structures and use these empirical trends to assign absolute configurations.

9.2. LANTHANIDE COMPLEXES

9.2.1. Introduction

In 1969, Hinckley first showed that organic-soluble lanthanide tris β-diketonates were effective NMR shift reagents.[1244] The positive lanthanide ion in these complexes is coordinatively unsaturated, and as a hard Lewis acid, forms donor–acceptor complexes with hard Lewis bases. Paramagnetic lanthanide ions induce exceptionally large shifts in the NMR spectra of the donor compound. The shifts are primarily the result of dipolar (through-space) interactions of the magnetic field of the lanthanide ion. These through-space interactions are predictable according to the dipolar shift equation (Eq 1) initially derived by McConnell and Robertson.[1245] The term r is the distance between the lanthanide ion and the nucleus of interest, and θ is the angle between the principal magnetic axis of the donor–acceptor complex and the line connecting the lanthanide and the nucleus of interest.

$$\Delta\delta = K(3\cos^2\theta - 1)/r^3 \qquad (9.1)$$

In most cases, the distance term dominates such that the resonances of nuclei closer to the coordination site of the donor molecule exhibit the larger lanthanide-induced shifts. Except in unusual cases when θ is greater than 55.7° and the sign of the $(3\cos^2\theta - 1)$ term changes, the direction of the shifts is a function of the lanthanide ion. In a solvent such as chloroform, Eu(III) and Yb(III) are downfield shift reagents whereas Pr(III) is an upfield shift reagent. Certain other lanthanide ions induce larger shifts than Eu(III), Pr(III), or Yb(III) but these shifts are usually unnecessarily large. An ion such as Sm(III) induces smaller shifts, and the diamagnetic La(III) and Lu(III) only cause what are referred to as "complexation shifts." Given the scarcity of high-field NMR spectrometers at the time of Hinckley's discovery, interest in the development and use of lanthanide shift reagents mushroomed. The lanthanide tris β-diketonates are effective organic-soluble reagents for hard Lewis bases, which encompass almost all classes of nitrogen- and oxygen-containing compounds.

Organic-soluble shift reagents for soft Lewis bases such as olefins, aromatics, phosphines, and halogenated compounds were developed by using lanthanide tris β-diketonates [Ln(β-dik)$_3$] in conjunction with a silver β-diketonate [Ag(β-dik)].[1246,1247] These binuclear reagents involve the formation of a lanthanide tetrakis chelate anion that forms an ion pair with the Ag(I) as shown in Eq 9.2.

$$Ln(\beta\text{-dik})_3 + Ag(\beta\text{-dik}) = [Ln(\beta\text{-dik})_4]Ag \qquad (9.2)$$

The silver ion bonds to the soft Lewis base and the paramagnetic lanthanide causes shifts in the NMR spectrum. A range of lanthanide shift reagents for use in aqueous solution have also been developed over the years.

Not long after Hinckley's initial discovery, Whitesides reported the first application using an organic-soluble chiral lanthanide tris β-diketonate to discriminate the NMR spectra of chiral donors.[1248] The tris Eu(III) complex of 3-(*tert*-butylhydroxymethylene)-*d*-camphor (**9.1**) was shown to be effective for determining the optical purity of chiral amines. The reagent was less effective for other classes of hard Lewis bases.

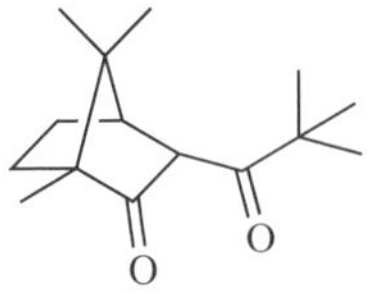

9.1

Soon after, the use of tris lanthanide complexes with 3-(trifluoroacetyl)-*d*-camphor (**9.2**) [H(tfc)],[1249] and 3-(heptafluorobutyryl)-*d*-camphor (**9.3**) [H(hfc)][1250] were introduced as chiral NMR shift reagents. These complexes were considerably more effective than those with 3-*tert*-butylhydroxymethylene-*d*-camphor, presumably because the electron-withdrawing properties of the trifluoromethyl and heptafluoropropyl groups enhanced the Lewis acidity of the lanthanide ion and caused stronger binding. In addition to amines, these early reports demonstrated the effectiveness of these chiral shift reagents for alcohols, epoxides, sulfoxides, aldehydes, ketones, and esters.[1249,1250] None of the complexes with **9.2** and **9.3** functioned most effectively for all compounds. Testing both may be warranted for a particular substrate.

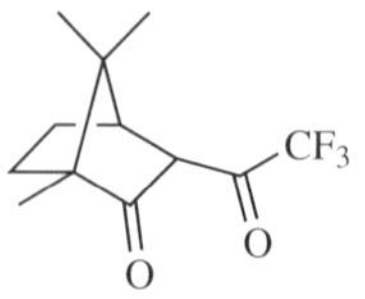

9.2

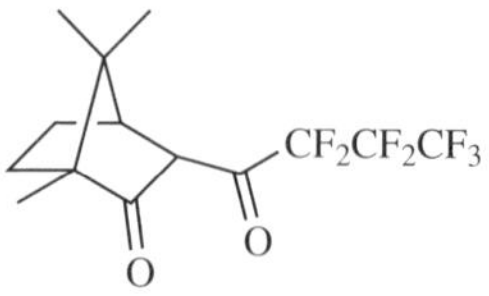

9.3

Organic-soluble lanthanide complexes with many other chiral ligands were tested for their effectiveness as chiral NMR shift reagents.[1251] The one noteworthy compound to surface from these studies was the Eu(III) complex of *d,d*-dicamphoyl methane (**9.4**).[1252] The chiral discrimination properties of Eu(dcm)$_3$ are truly impressive. For example, with 1-phenylethylamine (PEA), the methine and methyl resonances exhibited nonequivalence of 4.4 and 0.7 ppm, respectively, in a 1:1 mixture.[1252] The incredible effectiveness of this reagent was ascribed to the steric bulk of **9.4** relative to **9.2** and **9.3**. Presumably, the steric hindrance of lanthanide complexes with **9.4** caused a much more specific interaction of the enantiomeric pairs with the lanthanide than those which occurred with complexes of **9.2** and **9.3**, thereby enhancing the distinction in the NMR spectrum. Because the Eu(III) complex with **9.4** was not commercially available as early as those with **9.2** and **9.3**, it has not been used as much as those with **9.2** or **9.3**. However, the impressive discriminative ability of Eu(dcm)$_3$ recommends its use.

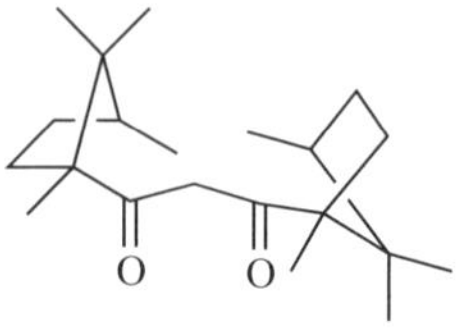

9.4

A primary attribute of chiral lanthanide shift reagents is the wide variety of compound classes they may potentially distinguish. The lanthanide tris β-diketonates can be used with almost any oxygen- or nitrogen-containing compound. The binuclear lanthanide–silver reagents extend the application to olefins, aromatics, phosphines, and halogenated compounds.

The lanthanide reagents are typically used for the determination of optical purity. The ligands in the lanthanide complexes are fluxional and alter their exact geometry to accommodate the binding of the substrate. As such, even with the bulky **9.4** ligand, there is no ability to predict the geometry of the lanthanide–donor complex with sufficient certainty to relate the magnitude of the lanthanide-induced shifts to absolute configuration. In some cases, as will be described, a group of similar compounds show empirical trends in the relative lanthanide-induced shifts that correlate with absolute configuration.

Lanthanide tris chelates of **9.2**, **9.3**, and **9.4** have the potential to bind either one or two substrates. The chiral discrimination in the 1:1 and 1:2 lanthanide–substrate complexes may be markedly different. A striking example was observed in the analysis of 2-phenyl-2-butanol with Eu(hfc)$_3$.[1253] In this case, adding increasing amounts of Eu(hfc)$_3$ caused the methyl resonance to show increasing enantiomeric discrimination. For the *ortho*-hydrogen resonance of the aromatic ring, the nonequivalence increased up to a lanthanide-to-substrate ratio of about 0.5 and then began to diminish. At a ratio above 1, the *ortho*-hydrogen resonances of the two enantiomers had recoalesced and then began to show a reverse sense of nonequivalence in the spectrum. A similar behavior was observed in the spectrum of 1-phenylethylamine with Eu(dcm)$_3$.[1252] The likely explanation was that chiral recognition in the 1:1 and 1:2 complexes was considerably different and the concentration of each complex was altered as the concentration ratio changed.

When using lanthanide shift reagents, it is best to obtain a series of spectra with increasing concentration of the lanthanide. The shift reagent can be conveniently added using small aliquots of a concentrated stock solution. This will facilitate tracking the resonances, which can be difficult because of the large shifts involved. It also enhances the likelihood of observing discrimination since it is impossible to know the optimal lanthanide-substrate ratio.

Two-dimensional NMR methods such as COSY spectra make it easier to identify the resonances of the minor enantiomer in a mixture. Another procedure for identifying the location of the minor enantiomer at levels as low as 0.3% has been reported.[1254] The method involved developing calibration plots of the lanthanide-induced shifts of a racemic mixture of the substrate. Knowing the relative slopes of the two lines, it was then possible to accurately measure the location of the major enantiomer and its overall shift and determine the expected location of the resonance of the minor enantiomer.

9.2.2. Peak Broadening with Lanthanide Shift Reagents

A concern when using lanthanide shift reagents is the peak broadening that occurs in the spectrum. The broadening is caused by two processes. The first is that the paramagnetic lanthanide ion reduces the relaxation time of the nucleus. The shortening of the excited state lifetime causes uncertainty broadening via the Heisenberg uncertainty principle. The second is that the substrate undergoes an exchange process between its bound and unbound form. The ideal circumstance is to have fast exchange so that the spectrum is a time-average of the bound and unbound form. The large size of the lanthanide–substrate complexes can slow down this exchange rate, causing broadening in the spectrum. The broadening that occurs is a function of the field strength and is more pronounced on higher field spectrometers. Lanthanide ions that cause larger shifts cause greater broadening as well. For chiral recognition studies with lanthanide shift reagents, it may be advisable to run the spectra at field strengths of 300 MHz or lower, although there are examples in the literature of successful discrimination at higher field strengths. There is no question that the broadening that occurs in the spectra with lanthanide shift reagents has reduced their usage in recent years. Nevertheless, for many substrates without reactive functional groups, there may not be other suitable options, and lanthanide shift reagents may represent the best chance of achieving enantiomeric discrimination.

One way of using chiral lanthanide shift reagents that may avoid unacceptable levels of broadening is to obtain proton decoupled ^{13}C NMR spectra. The shorter timescale of the ^{13}C nucleus compared to that of ^{1}H nucleus reduces the potential broadening from exchange effects. The sharp singlets of the decoupled ^{13}C signals may also be resolved without the onset of unacceptable broadening that complicates coupled hydrogen resonances. The ability of ^{13}C NMR as a way of monitoring optical purity with chiral lanthanide shift reagents was noted early after the development of chiral shift reagents for alcohols, amines, sulfoxides, ethers, and piperidine.[1255] The ^{1}H NMR spectrum of multistriatin (**9.5**) showed unacceptable broadening with Eu(hfc)$_3$, whereas the ^{13}C NMR spectrum showed excellent enantiomeric discrimination.[1256] A similar observation was made when studying chiral lactams with Eu(tfc)$_3$.[1257]

9.5

Another way to reduce line broadening is to employ a lanthanide ion such as Sm(III) that causes small shifts and therefore less broadening. The utility of Sm(tfc)$_3$ for the analysis of esters of amino acids has been demonstrated.[1258] The Hα (0.01–0.065 ppm) and methoxy (0.009–0.023 ppm) resonances exhibited nonequivalence. In many cases, resonances of hydrogen atoms on the aliphatic substituent group were also discriminated. The Hα and methoxy resonances for the D-isomer consistently showed the larger shift, whereas the shift order was reversed for resonances on the side chain. With appropriate caution, the shifts in the presence of Sm(tfc)$_3$ could be used to assign absolute configurations of these compounds.

Early work with lanthanide chelates on relatively low-field spectrometers was best done in nonpolar solvents to maximize the association and the enantiomeric discrimination. On high-field instruments, strong association will contribute to peak broadening. In some cases, then, better results may be obtained if more polar solvents are used. For example, the spectra of 1,2- and 1,3-dioxygenated compounds with Eu(tfc)$_3$ and Eu(hfc)$_3$ exhibited significant broadening in chloroform-d,[1259] presumably because of strong bidentate binding of the substrate to the lanthanide. The spectra in acetonitrile-d_3 had considerably less broadening and the optical purity of 16 of the 19 compounds examined could be determined. It was also found that small amounts of water in the solvent did not interfere with the analysis of these substrates.

Chiral discrimination in the ^{1}H NMR spectrum of thalidomide (**9.6**)[1260] and methoxamine (e*rythro*-α-(1-aminoethyl)-2,5-dimethoxybenzenemethanol –**9.7**)[1261] at 200 MHz in the presence of Eu(hfc)$_3$ was observed in acetonitrile-d_3.[1260] The coordinating ability of acetonitrile necessitated higher than usual europium-to-substrate ratios (10:1 in the case of thalidomide), but little broadening of the resonances was observed.

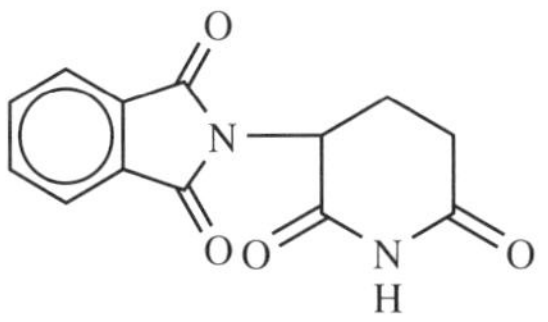

9.6

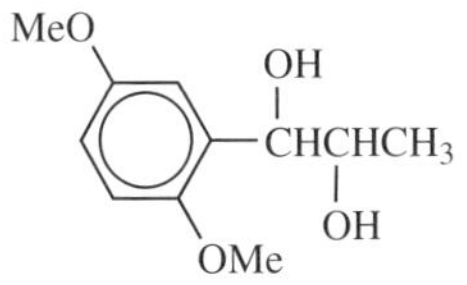

9.7

The magnitude of enantiomeric discrimination with lanthanide chelates was enhanced at lower temperatures because of more favorable association of the substrate with the lanthanide.[1252] On high-field spectrometers, warming the sample may diminish the broadening to acceptable levels by reducing the exchange broadening. By speeding up the exchange and reducing the association, smaller shifts and enantiomeric discrimination will be obtained, but the resonances may well be sharp enough to observe nonequivalence. Warming the probe to 50–70°C was used to advantage with systems in which a paramagnetic lanthanide ion was added to cyclodextrin derivatives.[1078] The ^{1}H NMR spectrum of ketotifen (**9.8**) in the presence of Eu(tfc)$_3$ exhibited significant broadening.[1262] Warming the sample to 75°C reduced the broadening to the extent that the enantiomeric discrimination was apparent in the spectrum.

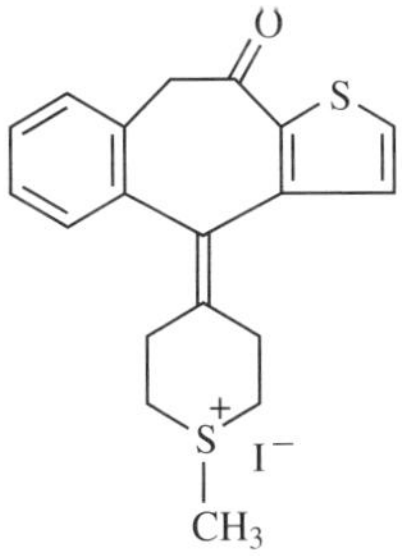

9.8

A general method of compensating for line broadening was described in the context of a study of β-alkoxy alcohols and diols using Eu(hfc)$_3$ and Yb(hfc)$_3$ at 300 MHz.[1263] In this study, the data were subjected to a rephrasing, baseline correction, and Gaussian line narrowing by as much as 10 Hz. The improvements made it possible to distinguish the resonances of the two enantiomers after the line narrowing procedures.

9.2.3. Catalytic Properties of Lanthanide Ions

Lanthanide ions, being Lewis acids, are known to catalyze certain types of reactions.[1264,1265,1266] In NMR studies using lanthanide tris β-diketonates for chiral discrimination, one must always be aware of the potential for a reaction to occur. One documented example involved a deracemization of dimethyl penta-2,3-dieneoates (**9.9**) in chloroform-d in the presence of Eu(hfc)$_3$.[1267,1268] Over nine days the ratio of the two enantiomers in a racemic mixture changed to 89:11, after which it did not change. The same phenomenon was observed in other 2,3-pentadienedioates as well, although the extent of deracemization varied.

$$\text{MeO}_2\text{C}\diagdown\text{C}=\text{C}\diagup\text{CO}_2\text{Me}$$

9.9

There are hundreds of published applications of chiral lanthanide shift reagents. Comprehensive reviews of the utilization of lanthanide shift reagents for chiral analysis were published in 1983[1269] and 1987.[1251] These reviews demonstrate the range of compounds for which chiral lanthanide shift reagents represent a potential means of determining enantiomeric purity. The relative order of binding of functional groups to lanthanides, and the binding of polyfunctional substrates to lanthanides is discussed in detail in an earlier review.[1251] Selected examples that demonstrate unique features, unusual substrates, or especially interesting applications are described herein.

9.2.4. Application of Lanthanide tris β-diketonates

A thorough investigation on the effectiveness of Eu(tfc)$_3$, Pr(tfc)$_3$, Eu(hfc)$_3$, and Pr(hfc)$_3$ in causing enantiomeric discrimination in the spectra of alcohols, esters, ketones, amines, sulfoxides, and sulfones was undertaken.[1253] Larger shifts were usually observed with the hfc chelates. However, the nonequivalence, which for some resonances was greater than 1 ppm, was quite variable with no discernible way of predicting whether better results would be obtained with a tfc or hfc chelate. Larger chiral discrimination was usually observed in carbon tetrachloride than chloroform-d. This was consistent with other reports of lanthanide shift reagents as the nonpolar solvent promoted association of the substrate with the lanthanide.[1252]

The effectiveness of Eu(III) and Pr(III) chelates of tfc, hfc, and dcm in causing chiral recognition in the spectra of amphetamine (**9.10**)[1270] and methamphetamine (**9.11**)[1271] has been compared. Eu(dcm)$_3$ provided the best discrimination. In an examination of ephedrine, methylephedrine, and 1-phenylethylamine, it was noted that the ^{13}C spin-lattice relaxation times of the pairs of enantiomers were different in the presence of Gd(dcm)$_3$.[1272] Relaxation effects exhibit only a distance dependence ($1/r^6$) and the findings indicated the different geometries of the diastereomeric Gd(dcm)$_3$ substrate complexes. A comparison of Eu(III) chelates of tfc, hfc, and dcm on α-amino acid derivatives indicated that the best results were obtained with Eu(hfc)$_3$.[1273] A number of solvents were examined and the largest distinctions were observed in chloroform-d and benzene-d_6.[1273]

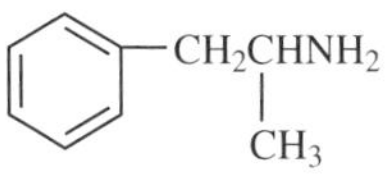

9.10: PhCH₂CH(NH₂)CH₃

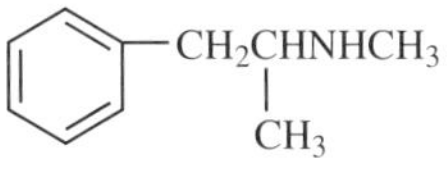

9.11: PhCH₂CH(NHCH₃)CH₃

At times it may be advantageous to simultaneously utilize chiral and achiral lanthanide chelates. For example, when examining the ^{1}H and ^{13}C NMR spectra of polycyclic compounds containing hydroxyl or acyl groups such as **9.12**, Eu(tfc)$_3$ and Eu(hfc)$_3$ did not cause enough large shifts at 200 or 300 MHz to create a first-order spectrum.[1274] Adding achiral Eu(fod)$_3$ (fod = 6,6,7,7,8,8,8-heptafluoro-2,2-dimethyl-3,5-octanedione), an especially effective achiral reagent, caused larger shifts and a first-order spectrum. Adding the chiral reagent then caused enantiomeric discrimination of some of the resonances.

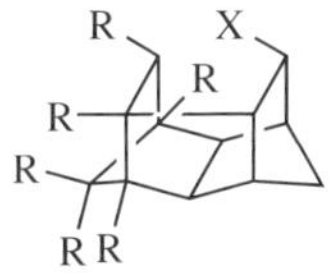

1: R = Cl, X = OH
2: R = Cl, X = OAc

9.12

9.2.1.1. Discrimination of Prochiral Nuclei

The ability of lanthanide chelates to distinguish prochiral groups in substrates such as dimethylsulfoxide, benzyl alcohol, 2-propanol, dibenzyl sulfoxide, 2-propanol, 2-propylamine, 2-methyl-2-butanol, and 2,2-dimethylpropanol, and dibenzyl sulfone that are enantiotopic by internal comparison was noted early in their development.[425,1250,1252] The chelates of hfc were more effective than those of tfc at discriminating prochiral nuclei.[1253] In the case of benzyl alcohol, the discrimination of the prochiral hydrogen atoms was confirmed by analyzing selectively deuterated species.[1275]

The ^{1}H NMR spectrum of *cis*- and *trans*-2,3-butylene oxide exhibited four sets of methyl and methine resonances in the presence of the chiral lanthanide chelates.[425] Similarly, the discrimination of enantiotopic hydrogen atoms in the ^{1}H NMR spectrum of dieldrin (**9.13**) demonstrated the *meso* stereochemistry of the compound.[425]

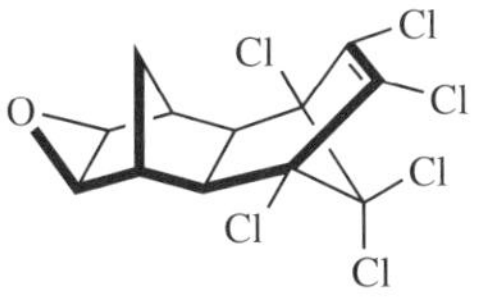

9.13

Eu(hfc)$_3$ has been shown to distinguish enantiotopic hydrogen atoms in allenic methyl esters such as **9.14**[1276] and cationic iminium species such as **9.15**.[1277] In the case of these cationic compounds, there are no donor groups that would be expected to bond to lanthanide ions. In this case it was likely that the anion of the organic salt associated with the Eu(hfc)$_3$ to create an anionic species ([Eu(hfc)$_3$Br]$^-$) that then formed an ion pair with the organic cation.

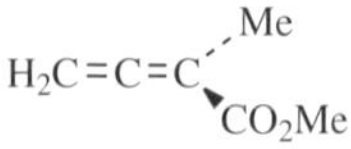

9.14

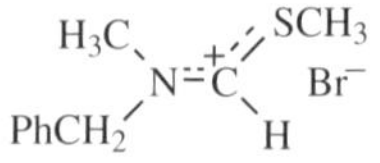

9.15

A procedure for discriminating the prochiral hydrogen atoms of 2-phenylethan-1-ol compounds using chiral lanthanide chelates has been described.[1278,1279] The alcohol was first converted into its corresponding N-(4-nitrophenylsulfonyl)-L-phenylalanyl ester (**9.16**). Addition of Yb(hfc)$_3$ caused separation of both the α- and β-prochiral protons.[1279] The shift of H$_S$ in **9.16** was always more sensitive than that of H$_R$.

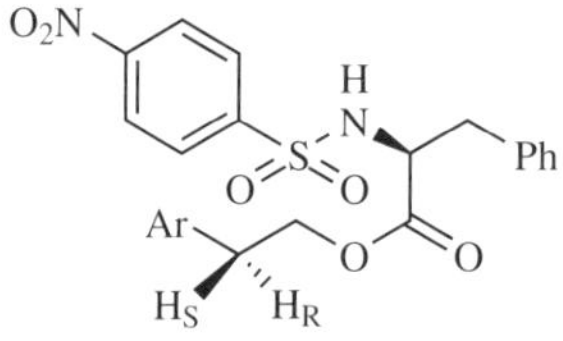

9.16

Chiral lanthanide shift reagents have been used with a variety of substrates to distinguish the *meso*-isomer from the corresponding pair of enantiomers.[1280–1285] For **9.17**, addition of Eu(tfc)$_3$ produced two methine singlets for the racemic pair that occurred with the *trans*-enantiomer, and two coupled doublets for the enantiotopic methine protons in the *cis*-(*meso*)-isomer.[1280] The *t*-butyl resonances of **9.18** were analyzed in the presence of Eu(tfc)$_3$ to distinguish the *cis*-(*meso*)-diol from the *trans*-d,l-racemate.[1282] The number of methyl resonances in the spectrum of **9.19** in the presence of Eu(tfc)$_3$ was used to establish the *cis*- and *trans*-structural configurations.[1283] A similar analysis of the stereochemistry of polycyclodiene pesticide metabolite diols (**9.20**) was performed using Eu(hfc)$_3$.[1281]

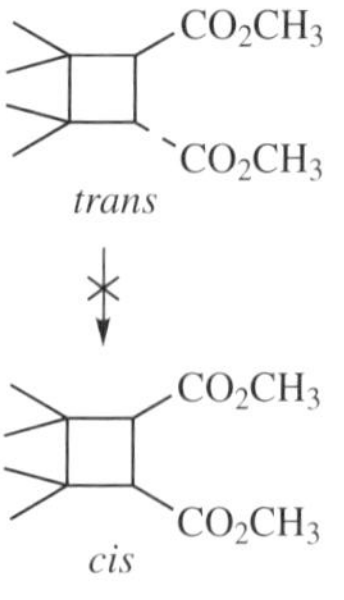

9.17

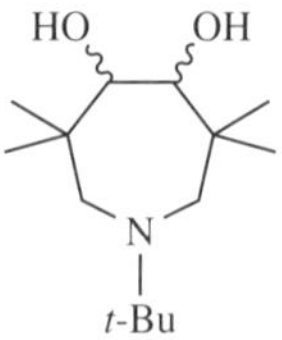

9.18

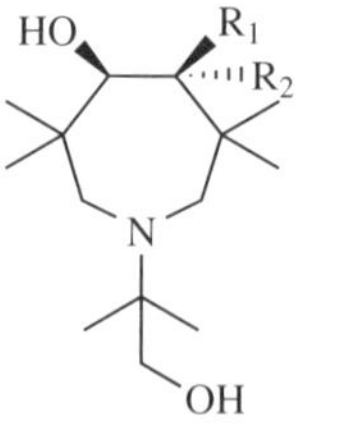

9.19
1: R$_1$ = OH, R$_2$ = H
2: R$_1$ = H, R$_2$ = OH

9.20

The analysis of *erythro-*, *threo-*, and *meso*-hydrobenzoins (**9.21**) was accomplished by analyzing the ^{1}H NMR spectrum in the presence of Eu(hfc)$_3$.[1285] The *threo-* and *erythro*-isomers were distinguishable on the basis of coupling constants as predicted by the Karplus equation. In the case of the *erythro*-isomer, four methine doublets were observed in the NMR spectrum, each representing a mixture of the two enantiomers.

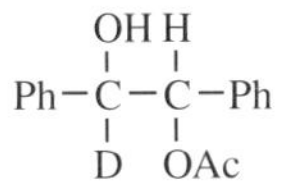

$$\begin{array}{c} \text{OH H} \\ | \quad | \\ \text{Ph}-\text{C}-\text{C}-\text{Ph} \\ | \quad | \\ \text{D} \quad \text{OAc} \end{array}$$

9.21

Eu(hfc)$_3$ was successfully used to distinguish the *meso* from racemic pair of bianthrone (**9.22**) compounds.[1284] The methine resonances at the point of attachment of the anthrone rings exhibited coupling in the *meso*-isomer in the presence of Eu(hfc)$_3$, but appeared as two singlets for the racemate. Other bianthrone derivatives have been examined as well with Eu(hfc)$_3$.[1286]

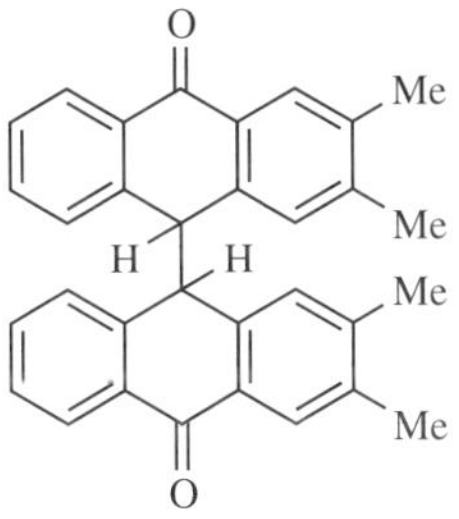

9.22

9.2.1.2. Analysis of Deuterium-Substituted Compounds

Lanthanide shift reagents have been used on several occasions to discriminate enantiomers that are chiral on the basis of deuterium[1287–1290] or tritium substitution.[1291] Sometimes ^{2}H or ^{3}H spectra were used for the analysis.

The enantiomers of primary 1-deuterio alcohols (RCHDOH) were distinguished in the presence of Eu(hfc)$_3$.[1287] For all six examples studied by using Eu(hfc)$_3$ with a ligand derived from *d*-camphor, the resonances of the (*R*)-enantiomer exhibited larger lanthanide-induced shifts than those of the (*S*)-enantiomer. Eu(hfc)$_3$ has been used to analyze similar tritiated alcohols.[1291] Proton decoupled tritium NMR spectra were used to analyze the enantiomeric purity.

The ^{1}H NMR spectrum of **9.23**, which is chiral by virtue of the deuterated methyl group, exhibited chiral discrimination in the presence of Eu(hfc)$_3$.[1288] The spectrum of benzhydrol **9.24**, which is chiral by virtue of the deuterated phenyl ring, had distinct resonances in the presence of Eu(dcm)$_3$.[1290] The resonances of the *ortho*-hydrogen exhibited nonequivalence of greater than 0.5 ppm, demonstrating the remarkable sensitivity of Eu(dcm)$_3$ to a rather small change in the compound. The optical purity of chiral 2,3-dideuterio-2-(methoxymethyl)spiro[cyclopropane-1,1'-indenes](**9.25**) was determined by analyzing the ^{1}H NMR spectrum in the presence of Eu(hfc)$_3$.[1289]

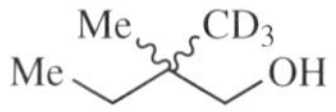

9.23

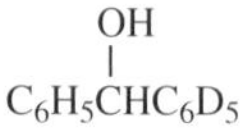

9.24

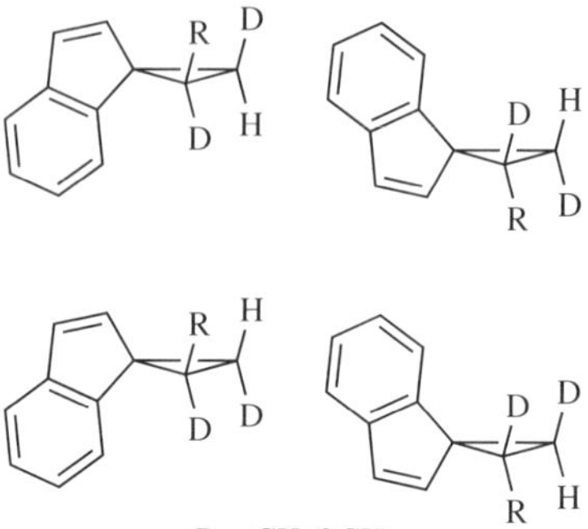

9.25

R = CH$_2$OCH$_3$

9.2.1.3. Analysis of Compounds Chiral by Virtue of Slow Rotation Lanthanide shift reagents have been used to study compounds that exhibit chirality on the basis of slow rotational properties.[1292–1298] Compound **9.26** exists as two racemic pairs of diastereomeric rotamers, and as such exhibited six methyl singlets in the ^{1}H NMR spectrum from the methyl and methoxy groups.[1292] Addition of Eu(tfc)$_3$ caused a doubling of each of these resonances so that 12 singlets were observed in the spectrum, indicating that each of the diastereomers consisted of a racemic pair.

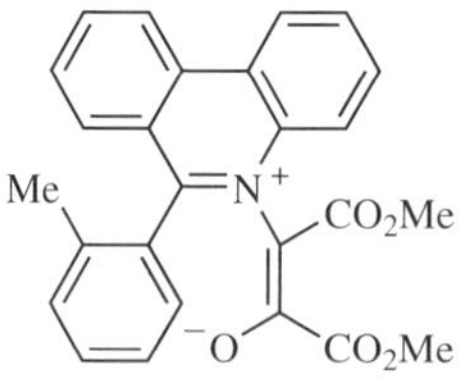

9.26

Eu(hfc)$_3$ has been used to study the rate of internal rotation in systems such as (E)-(2,6-dimethylphenyl)-benzylnitrosoamine (**9.27**),[1293] N,N-benzamides (**9.28**),[1297] and 1,3-dienes (**9.29**).[1296,1299] The discrimination of the enantiomers in the ^{1}H NMR spectrum in the presence of Eu(hfc)$_3$ was monitored in variable temperature studies to determine the barrier to rotation. Such findings must always be approached with caution, since it is possible that association of the lanthanide chelate may alter the

LANTHANIDE COMPLEXES

rotational properties of the substrate.

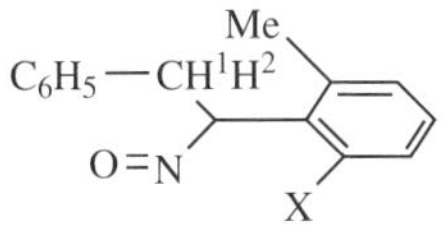

9.27, **9.28**, 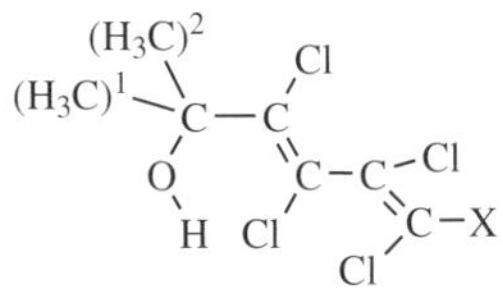 **9.29**

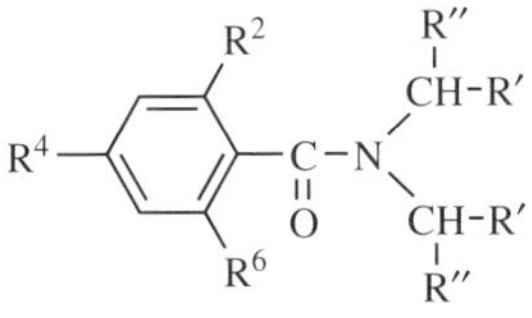

The racemization barrier in diene **9.30** was measured via a variable temperature study with Eu(tfc)$_3$.[1294] Restricted rotation about the ring-carbon bond in compounds of structure **9.31** and **9.32** causes a prochiral center. The barrier to rotation as a function of temperature was studied in the presence of the Pr(III) and Yb(III) chelates of tfc and hfc.[1295]

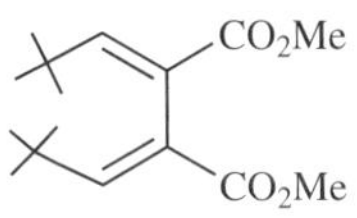

9.30, 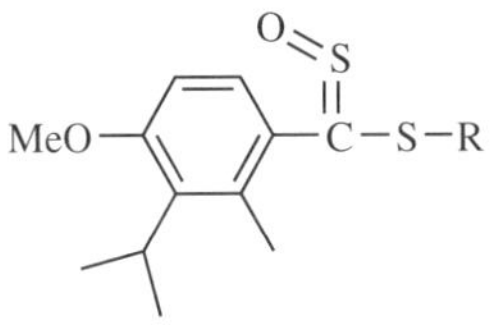 **9.31**, 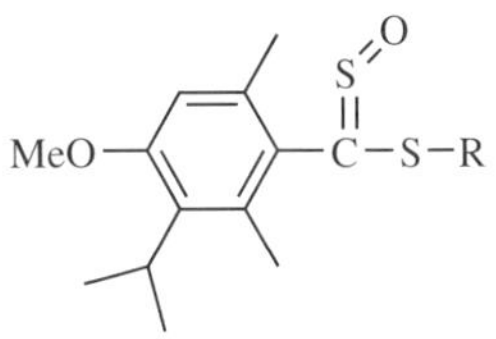 **9.32**

Slow rotation about the amide bond of N-boc-N-methyl amino acid derivatives (**9.33**) resulted in two N-methyl signals in the ^{1}H NMR spectrum because of the cis- and trans-conformations.[1298] Addition of Eu(hfc)$_3$ caused a doubling of both the N-methyl and N-Boc signals for the enantiomeric pairs.

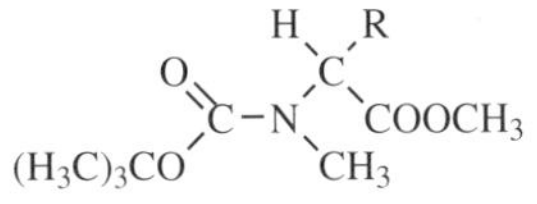

trans form 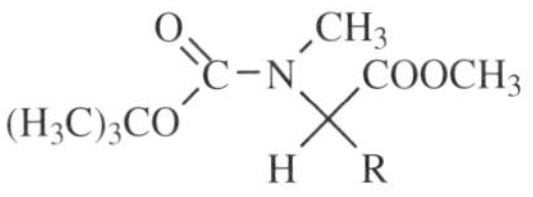   *cis* form

9.33

Compounds of general structure R–O–S–S–O–R are chiral by virtue of slow rotation about the disulfide (S–S) bond.[510] Addition of Eu(hfc)$_3$ in chloroform-*d* caused a splitting of the methylene resonances, confirming the chirality and enabling studies of the rotational barrier of compounds of this structure.

9.2.4.1. Analysis of Compounds Chiral by Virtue of Geometrical Constraints

Lanthanide shift reagents have been used on occasions to prove the chirality of compounds that are chiral by virtue of certain geometrical constraints. The chirality is demonstrated by a doubling of resonances in the NMR spectrum on addition of an appropriate lanthanide chelate.

The helical chirality of phenyl (**9.34**)[1300] and naphtha[9]annulenones (**9.35**)[1301] created by two chirally related, conformationally rigid forms was demonstrated by nonequivalence in the ^{1}H NMR spectrum in the presence of Eu(dcm)$_3$. The chirality of the spirophosphorane **9.36** was revealed by the ^{1}H NMR spectrum in the presence of Eu(tfc)$_3$.[1302]

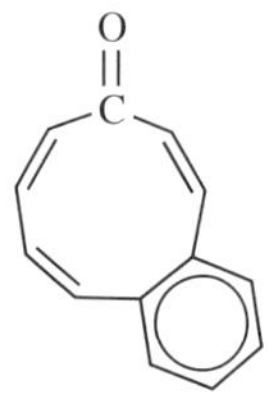

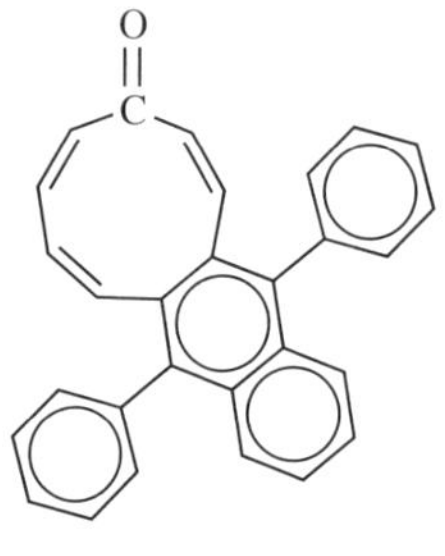

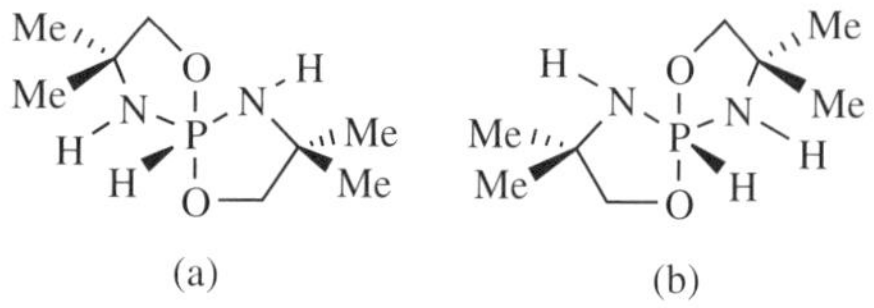

A doubling of resonances in the ^{1}H NMR spectrum of **9.37** and similar derivatives in the presence of Eu(tfc)$_3$ confirmed the chirality of these compounds.[1303] The chirality of the dimeric azamacrocycles (**9.38**) was confirmed on the basis of ^{1}H NMR spectra in the presence of Eu(dcm)$_3$.[1304] That 2,6,9-trioxabicyclo[3.3.1]-nona-3,7-diene-4,8-dicarbaldehyde (**9.39**) is chiral by virtue of a propeller-like structure was unequivocally demonstrated by a doubling of certain resonances in the presence of Eu(tfc)$_3$.[1305]

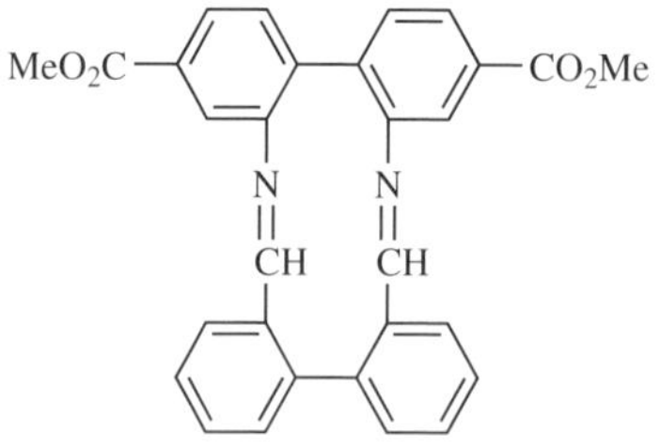

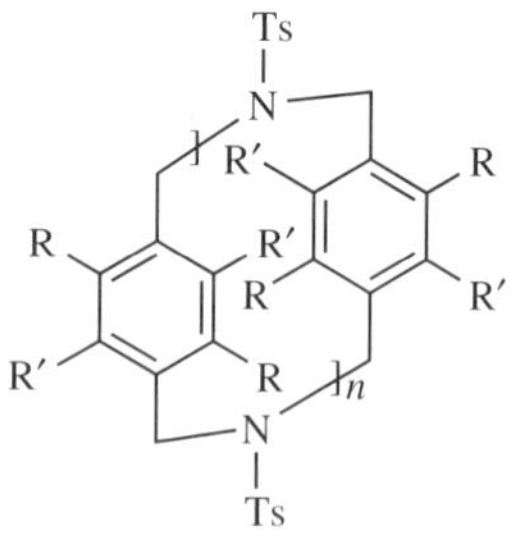

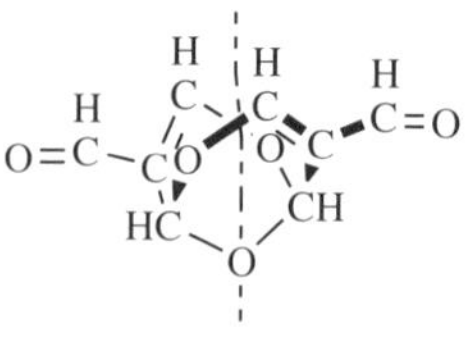

The chirality of the organoboron species **9.40** and similar compounds was monitored using ^{1}H NMR spectra in the presence of Eu(tfc)$_3$.[1306] It was believed that association of **9.40** with the Eu(III) occurred through one of the oxygen atoms.

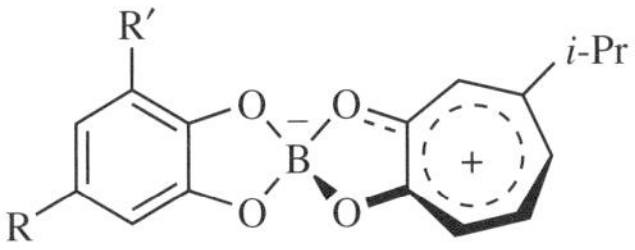

9.40

The NMR spectra of 2,2′6,6′-tetrasubstituted biphenyls (**9.41**) with hydroxyl, ester' or carboxylic acid substituent groups were enantiomerically discriminated in the presence of lanthanide chelates of tfc or hfc.[1307,1308] In one of these studies, it was noted that the resonances of the (S)-enantiomer consistently exhibited larger lanthanide-induced shifts than that of the (R)-enantiomer.[1308]

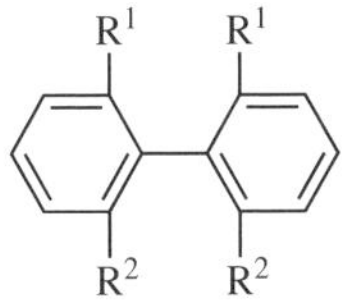

A: R^1 = OMe, R^2 = CO_2Me
B: R^1 = NO_2, R^2 = CO_2Me
C: R^1 = OMe, R^2 = CO_2OH

9.41

A series of singly bridged cyclophosphazene macrocyclics **9.42**[460] and **9.43**[459] showed duplication of resonances in the ^{31}P NMR spectrum in toluene-d_8 in the presence of Eu(hfc)$_3$. The doubling of resonances indicated that there was a 1:1 mixture of diastereomers from two racemic mixtures. The ^{31}P NMR spectrum of **9.42** with Eu(hfc)$_3$ first appeared to exhibit too many signals.[460] The extra peaks occurred because two Eu(hfc)$_3$ species bound to the *meso*-compound to form the species (Eu-*mR*-*mS*-Eu), thereby causing each *m* species to act independent of the other such that two ^{31}P signals result for the *meso*-isomer rather than only one.

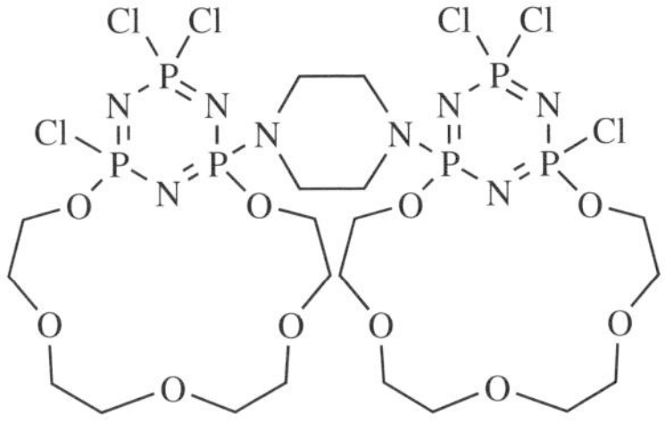

(*meso* + racemate)

9.42   **9.43**

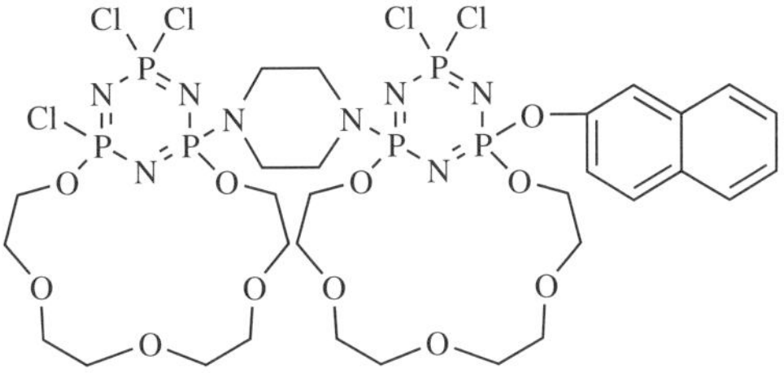

The chirality of the [2]-catenane **9.44**, which comprised two interlocked cyclic dimers, was demonstrated by a splitting of certain resonances in the presence of Eu(tfc)$_3$.[1309]

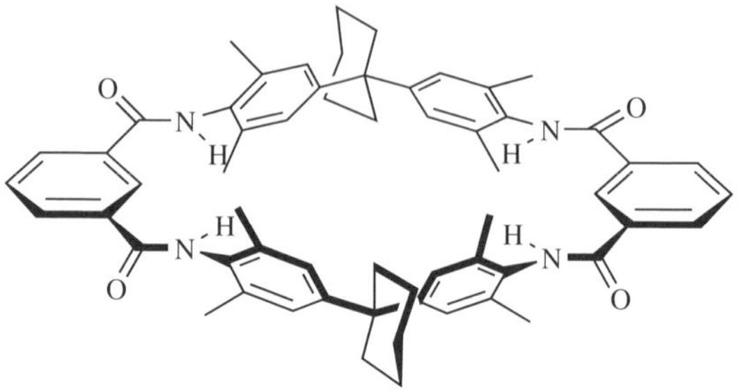

9.44

The optical purity of a chiral, isotactic propylene-carbon monoxide poly 1,4-ketone copolymer (**9.45**) was determined by monitoring the carbonyl carbon in the ^{13}C NMR spectrum in chloroform-d in the presence of Eu(tfc)$_3$.[1310]

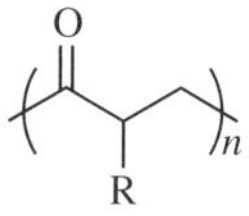

9.45

9.2.5. Miscellaneous Applications of Lanthanide tris β-diketonates

In some cases, derivatization of the substrate may be warranted prior to analysis with a lanthanide shift reagent. The optical purity of alcohols was reportedly easier to measure if they were first converted to their corresponding acetate.[1253] The methyl singlet of the esters showed enantiomeric discrimination in the presence of the lanthanide chelate. Similarly, carboxylic acids may be better analyzed if converted to methyl esters, although Eu(tfc)$_3$ has been used to analyze the optical purity of underivatized 2-(4-methylphenyl)butanoic acid.[1311] A procedure for the analysis of the enantiomeric purity of glycols, in which the 1,2-glycol was first converted to the corresponding 2-phenyl-1,3-dioxolanes using benzaldehyde as shown in Figure 9.1,

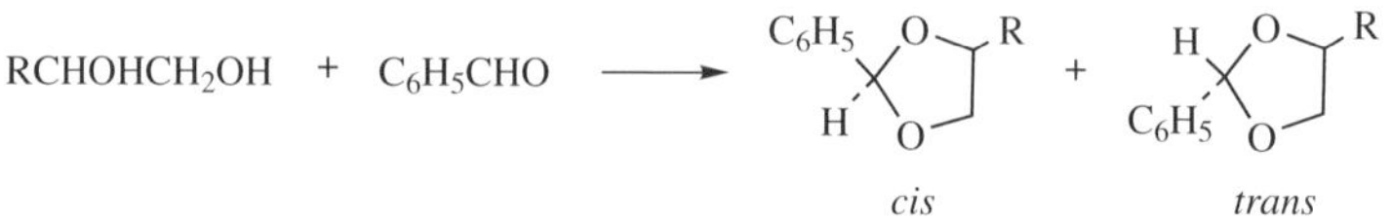

RCHOHCH$_2$OH + C$_6$H$_5$CHO ⟶ cis + trans

Figure 9.1. Conversion of 1,2-glycols to the corresponding 1-phenyl-1,3-dioxolanes using benzaldehyde.

has been described.[1312,1313] The *cis*- and *trans*-isomers were obtained in unequal amounts, but the benzylic proton was readily monitored and both isomers showed a doubling of the resonances in the presence of Eu(hfc)$_3$.

The use of Eu(hfc)$_3$ facilitated an analysis of a system that created the "chiralization" of a neutral spherical atom.[1314] A racemic mix of cryptophane A (**9.46**) bound xenon in its cavity in C$_2$D$_2$Cl$_4$. The ^{129}Xe NMR spectrum indicated the complexed nature of the xenon atoms. Addition of Eu(hfc)$_3$ caused the appearance of two ^{129}Xe signals, which arose from the xenon atoms within each of the enantiomers of **9.46**. Raising the concentration of Eu(hfc)$_3$ increases the nonequivalence in the ^{129}Xe NMR spectrum.

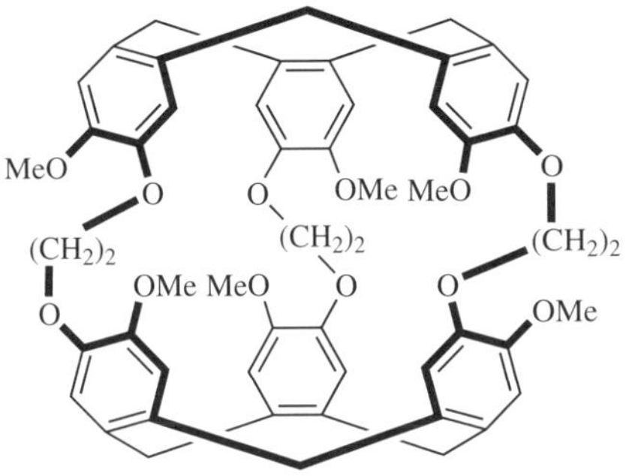

9.46

An exploratory study on the use of lanthanide shift reagents with *para*-hydrogen-induced polarization (PHIP) to examine aspects of asymmetric hydrogenations has been described.[1315] In the PHIP process, a hydrogenation is carried out using *para*-hydrogen. The resulting spin polarization effects caused enhancements of up to 10^5 in the NMR spectrum. The addition of the chiral lanthanide shift reagent served to distinguish the peaks of the enantiomers, enabling an analysis of the enantioselectivity of a chiral catalyst. The method was demonstrated on (*R*)- and (*S*)-1-phenyl-2-propen-1-ols. Eu(tfc)$_3$ was added to cause enantiomeric discrimination of the methyl resonances. The shift reagent could only be added at low concentrations (lanthanide-to-substrate ratio of 0.025) because higher concentrations eliminated the polarization effect. The method was also shown to work with 2,2,2-trifluoro-1-phenylethanol.

Lanthanide shift reagents have been used to examine enantiomeric discrimination by analyzing the spectra of less common nuclei than ^{1}H. The ^{17}O NMR resonances of the sulfonyl oxygen atoms of **9.47** and similar compounds were discriminated in the presence of Pr(hfc)$_3$.[1316] The enantiomeric purity of α-*C*-silylated amines and alcohols (**9.48**) was determined on the basis of the ^{29}Si NMR spectra in the presence of Eu(tfc)$_3$.[1317]

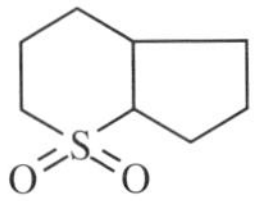

9.47 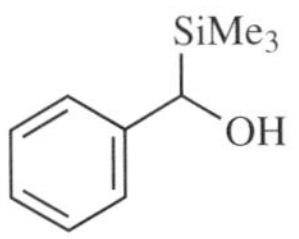   **9.48**

9.2.6. Analysis of Chiral Metal Complexes

Lanthanide tris chelates have been used to measure the optical purities of chiral metal complexes. In almost all of these studies, one or more ligands of the metal complex had a functionality that formed a donor–acceptor complex with the lanthanide ion.

Compound **9.49** consists of a mixture of a *meso* and a racemic pair.[1318] Addition of Eu(tfc)$_3$ caused discrimination of the methoxy signals of the racemic pair in the ^{1}H NMR spectrum, but did not resolve any of the enantiotopic hydrogen resonances in the *meso*-isomer. Association of **9.49** with the Eu(III) occurred through the carbonyl oxygen atom of the carboxy groups.

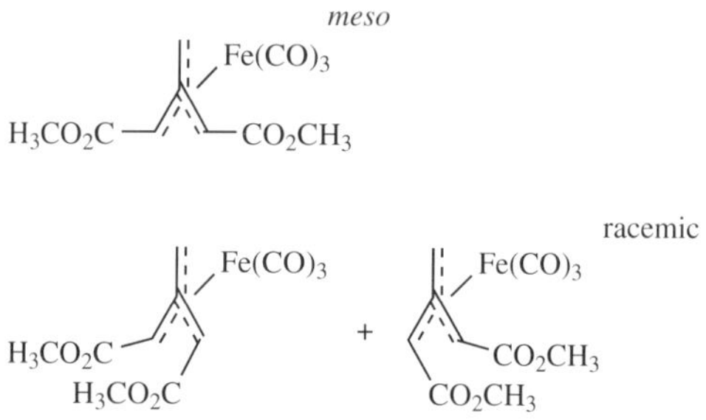

9.49

Other metal complexes that have an ester functionality in the ligand that associated with lanthanide ions in chiral β-diketonates and exhibited discrimination in the ^{1}H NMR spectrum included tricarbonyl(cyclohexadienyl)iron(I) salts (**9.50**)[1319–1321] and trimetal clusters of the form (μ^{3}CCO$_2$Me)CoMoNiCp$_2$(CO)$_5$.[1322] Other metal complexes have been shown to associate through sulfoxide (**9.51**),[1323] phosphine oxide (**9.52**),[1324] ketone (**9.53**),[1325] aldehyde (**9.54**),[1326] and hydroxyl groups (**9.55**).[1327] These studies used either Eu(tfc)$_3$ or Eu(hfc)$_3$ as the lanthanide chelate.

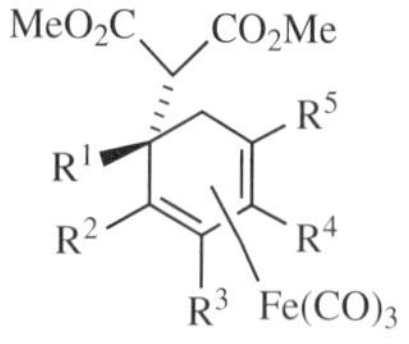

9.50 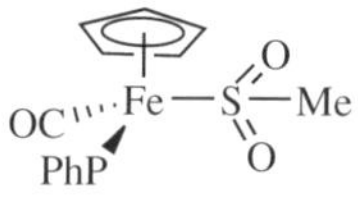 **9.51** 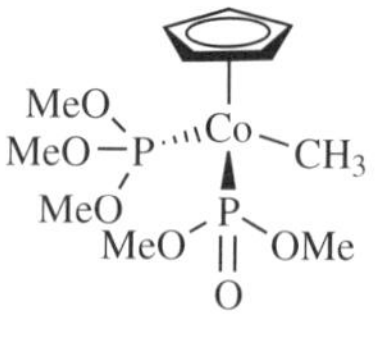 **9.52** 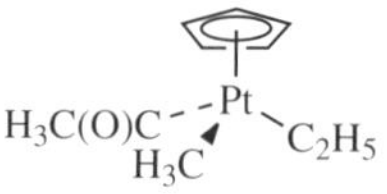 **9.53**

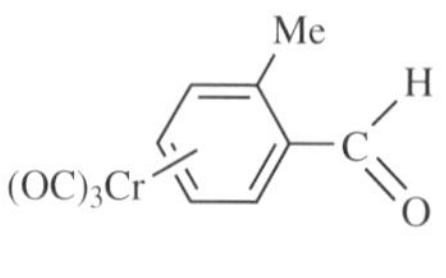

9.54 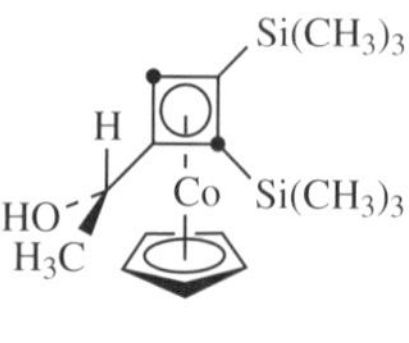 **9.55**

The nickel metalloporphyrin (**9.56**) forms a helical structure that is chiral. The chirality of **9.56** was unequivocally demonstrated in the ^{1}H NMR spectrum in the presence of Eu(tfc)$_3$ in methylene chloride-d_2.[1328] The nickel complex bound to the Eu(III) through the formyl group and the resonance of the formyl hydrogen exhibited nonequivalence.

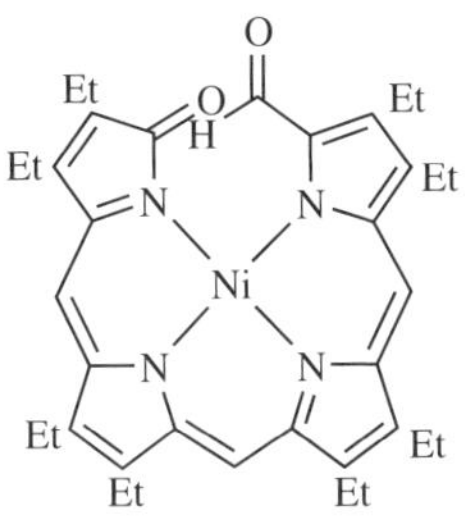

9.56

Carbonyl and cyano groups bonded to a metal ion can also bridge to a lanthanide ion. This interaction likely accounted for the chiral discrimination in the ^{1}H NMR spectrum of (η^5-C$_5$H$_5$)Fe(CO)(CN)(PPh$_3$)$_3$ in the presence of Yb(tfc)$_3$,[1329] β-[Ru$_4$(S$_2$CNMe$_2$)$_5$]$^+$ in the presence of Eu(hfc)$_3$,[1330] and ruthenium complexes such as [Ru(bpy)$_2$(CO)$_2$]Cl$_2$ with Eu(tfc)$_3$.[1331,1332] Discrimination of the latter ruthenium complexes was observed to occur in either chloroform-d or acetonitrile-d_3.

The spectra of the rhenium compounds **9.57**[1333] and **9.58**[1334,1335] exhibited chiral discrimination of the cyclopentadiene hydrogen resonances in the presence of Eu(hfc)$_3$. The site of association was not discussed but likely occurred through a bridging NO or CN group.

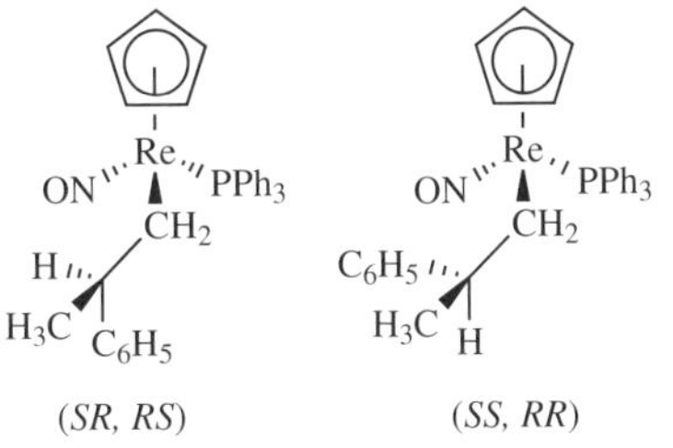

(SR, RS)   (SS, RR)   **9.58**

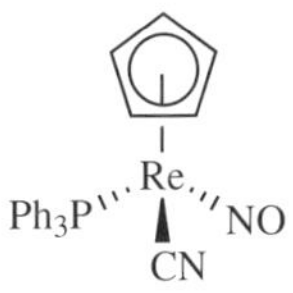

9.57

The enantiomers of the niobium complex (**4.42**) were distinguishable in the NMR spectrum in the presence of Eu(tfc)$_3$.[455] The niobium complex bound to the Eu(III) through the oxygen atom and the shifts were similar in magnitude to those in the spectra of sulfoxides.

Chiral discrimination in the presence of lanthanide tris β-diketonates has been observed in the spectra of metal complexes with ligands that do not have lanthanide-binding groups. Resonances of the tris-phenanthroline complex of ruthenium ([Ru(phen)$_3$]Cl$_2$) showed nonequivalence of up to 0.7 ppm in the ^{1}H NMR spectrum in the presence of Eu(tfc)$_3$ in nonpolar solvents.[1336] Presumably, the chloride anion associated with the Eu(tfc)$_3$ to form an anionic lanthanide species that interacted with the metal cation through ion pairing. It was important to have a small anion to facilitate formation of the anionic lanthanide species.

The enantiomeric purity of **9.59** was determined using Eu(tfc)$_3$ in methylene chloride-d_2 at 500 MHz.[1337] The association of **9.59** with the Eu(III) may have involved an ion pair as with compounds like [Ru(phen)$_3$]Cl$_2$, or may also have involved coordination through an amide carbonyl group.

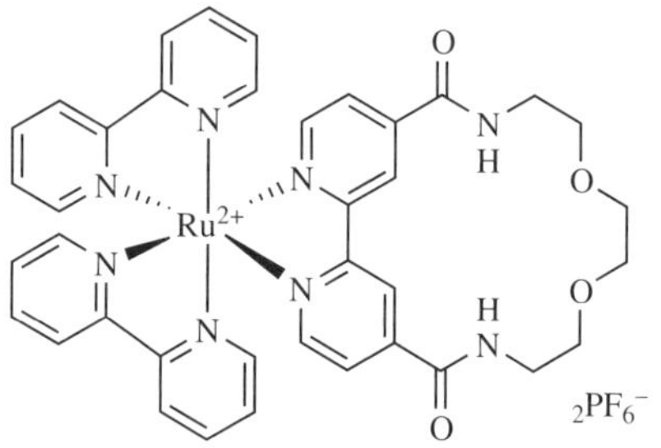

9.59

Transition metal complexes prepared from Schiff base ligands, an example of which is **9.60**, associate with Eu(III) in a bidentate manner through bridging oxygen atoms.[1338] The enantiomers of the nickel complex with a propylene diamine backbone (**9.60**) exhibited certain distinct resonances in the ^{1}H NMR spectrum with Eu(hfc)$_3$. The corresponding complex prepared from 2,3-diaminobutane consists of a racemic pair and *meso*-isomer. Addition of Eu(hfc)$_3$ to the racemic pair resulted in a duplication of certain peaks in the ^{1}H NMR spectrum. By contrast, addition of Eu(hfc)$_3$ to the *meso*-isomer resolved the enantiotopic hydrogens of the 2,3-diaminobutane group and they coupled to each other. The presence of the coupling confirmed the *meso*-configuration, whereas no coupling was observed between these hydrogens in the racemate.[1338]

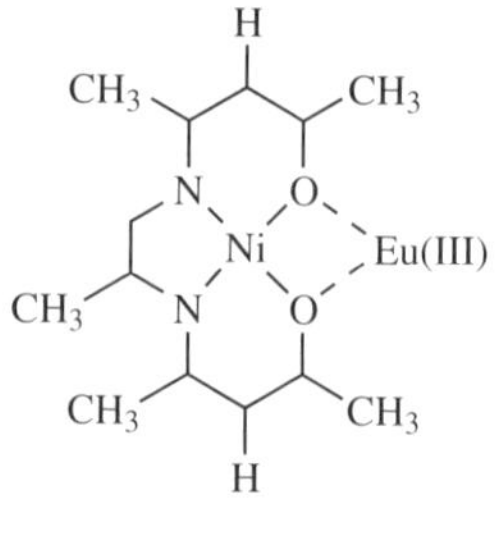

9.60

Eu(hfc)₃ was able to discriminate the prochiral olefin protons in the NMR spectrum of the *cis*- and *trans*-isomers of the iron tetracarbonyl complexes **9.61** and **9.62**.[1339] Binding to the Eu(III) occurred through the carbonyl oxygen atoms of the olefin ligand.

9.61   **9.62**

9.2.7. Assignment of Absolute Stereochemistry

Since the lanthanide tris β-diketonate complexes are fluxional in nature, it is not possible to predict the geometry of lanthanide–donor complexes with sufficient accuracy so as to use the magnitude of the lanthanide-induced shifts to assign absolute configurations. Instead, the magnitude of the lanthanide-induced shifts of an unknown must be compared to structurally similar compounds with known configurations, and the stereochemistry inferred based on the likelihood that the relative order of shifts of the (R)- and (S)-enantiomers will be the same. The studies described in the following section utilized lanthanide chelates with ligands prepared with d-camphor.

In a study of several secondary alkyl aryl alcohols, the resonances of the (R)-enantiomer consistently exhibited the larger lanthanide-induced shifts.[1340] However, in a subsequent study of 12 secondary carbinols with Eu(hfc)₃, consistent trends in the lanthanide-induced shifts were only observed among some of the derivatives.[1341] The relative magnitude of the shifts did not correlate with the steric effects of the substituent groups. It was recommended that using lanthanide shift data to assign the configuration of an unknown secondary carbinol required the analysis of a closely related series of known compounds.[1341] A later study of secondary carbinols with Pr(hfc)₃ also concluded that the lanthanide-induced shifts were not a reliable predictor of absolute stereochemistry.[1342]

More recently, however, a detailed analysis of the shifts of the ^{13}C resonances of secondary and tertiary carbinols with Pr(hfc)₃ were examined.[1343] The ^{13}C spectra were analyzed because of the reduced broadening. By analyzing a range of optically active secondary alcohols, an empirical rule was established. As shown in Figure 9.2, compounds with a hydroxyl group in the α-orientation had negative $\Delta\Delta\delta$ values ($\Delta\Delta\delta = \delta\Delta_{(R)-\text{Pr(tfc)}3} - \delta\Delta_{(S)-\text{Pr(tfc)}3}$), whereas those with the hydroxyl group in a β-orientation had positive $\Delta\Delta\delta$ values.[1343] For diols separated by two or more carbons, it was possible to treat each carbinol as an isolated unit and assign the stereochemistry about each stereocenter. The utility of the method was demonstrated by assigning the complete stereo structure of glisoprenin A (**9.63**) by using shifts in

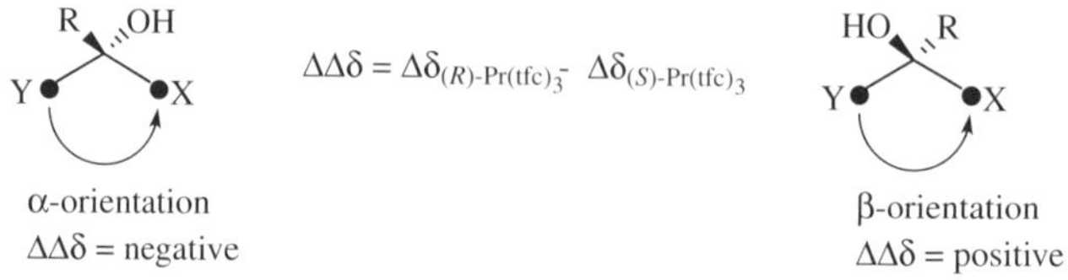

Figure 9.2. Correlation of $\Delta\Delta\delta$ values of ^{13}C resonances with absolute configuration of secondary and tertiary alcohols.

$\Delta\Delta\delta = \Delta\delta_{(R)\text{-Pr(tfc)}_3} - \Delta\delta_{(S)\text{-Pr(tfc)}_3}$

α-orientation
$\Delta\Delta\delta$ = negative

β-orientation
$\Delta\Delta\delta$ = positive

the ^{13}C NMR spectrum with Pr(hfc)$_3$.[1344] The validity of the assignment was confirmed by a total synthesis of **9.63**.[1345]

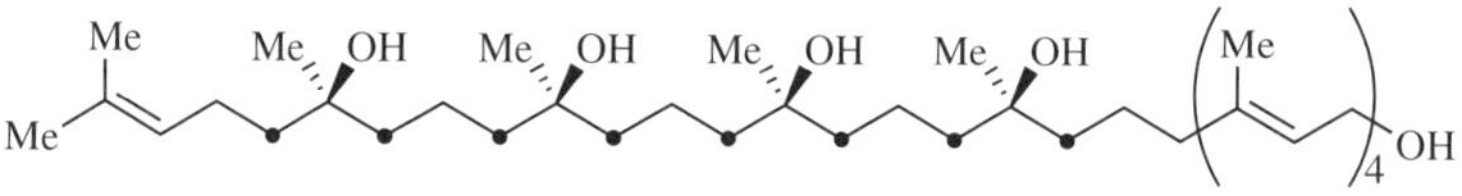

9.63

The lanthanide-induced shifts in the ^1H NMR spectra of a series of model *para-*, mono- or disubstituted benzhydrols (**9.64**) with Eu(dcm)$_3$ showed trends that correlated with absolute configurations, allowing the assignment of unknown compounds.[1346] The stereochemistries of **9.65** and analogous compounds were assigned by comparing shifts in the ^1H NMR spectrum with Eu(tfc)$_3$ to those of similar compounds with known configurations.[1347] The relative shifts in the ^1H NMR spectrum of the (*R*)- and (*S*)-enantiomers of 2-aryloxypropionyl derivatives (**9.66**) with Eu(hfc)$_3$ showed a consistent trend.[1348]

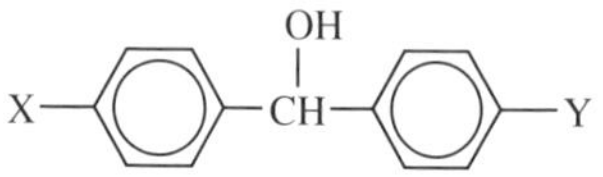

9.64

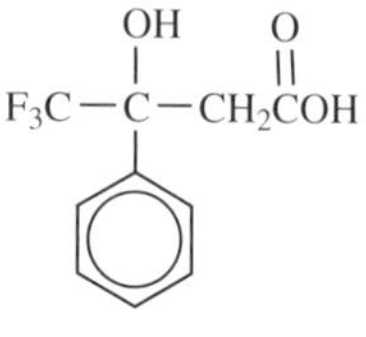

9.65

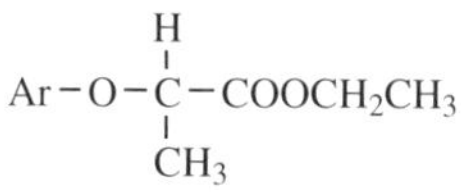

9.66

The shifts of the methoxy signals of the ^1H NMR spectra of α-amino acid methyl esters with Eu(tfc)$_3$ correlated with absolute configuration for eight of the nine amino acids studied.[1349] Only proline, a cyclic amino acid, showed the reverse trend. Similarly, the shifts of the methoxy resonances of *N*-acetyl or *N*-benzoyl-β-hetarylalanine methyl esters (**9.67**) in the presence of Eu(tfc)$_3$ correlated with the absolute stereochemistry.[1350] In all cases, the methoxy group of the (*R*)-enantiomer was more shielded in acetonitrile-d_3 and chloroform-*d*. Other resonances of the substrates did not show a consistent shift pattern with the stereochemical configuration.

ArCH$_2$CH(COOCH$_3$)(NHCOR)

9.67

The absolute configurations of menthyl butanoates (**4.62**) were determined on the basis of shift data with Eu(hfc)$_3$.[473] The assignment was made by comparing the lanthanide-induced shifts of the unknowns to similar compounds with known configurations. The shifts in the spectra of mono-, di-, and triglyceride derivatives with Eu(hfc)$_3$ showed consistent patterns that could be used to assign the absolute stereochemistry of an unknown by comparison to appropriate model substrates.[1351]

The relative shifts in the ^{1}H NMR spectra of N-phthaloyl-α-methylcyanoglycinates (**9.68**) with Eu(hfc)$_3$ showed patterns that correlated with the absolute configuration.[1352] The shifts of model compounds with known stereochemistries could then be used to assign the configuration of an unknown.

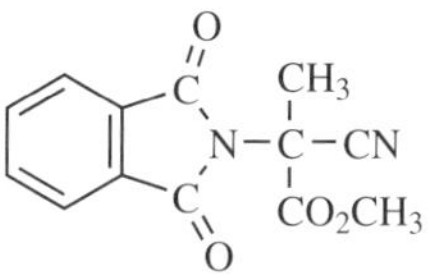

9.68

The stereochemistries of 2-methyl-5,6-dihydro-α-pyran 6,6-diacids (**9.69**) were assigned by first ozonizing the compounds to the corresponding lactic acids, and then comparing the shifts in the spectra with Eu(hfc)$_3$ to lactic acid derivatives with known configurations.[1353]

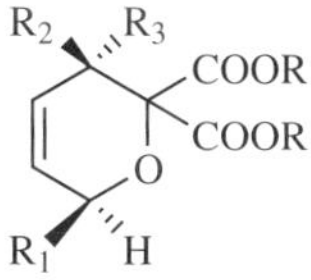

9.69

After showing that Eu(hfc)$_3$ could be used to discriminate the methyl resonances of enantiomers of lactones **9.70** at 400 MHz, the absolute configuration of the lactone produced in the degradation of C$_{34}$-Botryococcene, the major hydrocarbon constituent of *Botryococcus braunii*, was assigned by comparison to known compounds.[1354,1355] The stereochemistry of 3,11-dimethyl-2-nonacosanone (**9.71**), the sex pheromone of the German cockroach, was assigned by comparing the magnitude of the shifts with Eu(tfc)$_3$ to model compounds of known configuration.[1356]

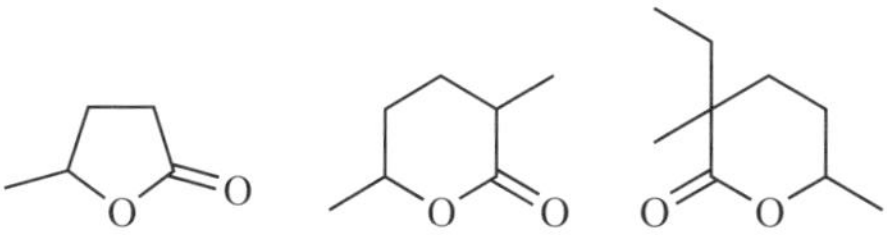

9.70

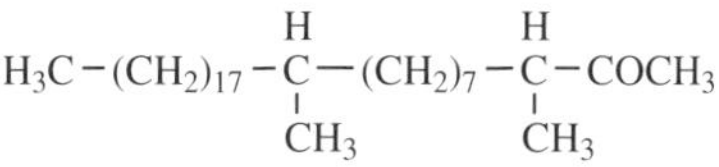

9.71

The potential of utilizing lanthanide-induced shifts to assign the absolute configurations of epoxides (**9.72**) and arene oxides (**9.73**) with Eu(tfc)$_3$ and Eu(hfc)$_3$ was examined.[1357] The shifts of the Hx and Hy resonances with Eu(tfc)$_3$ showed a consistent pattern in many examples that correlated with the absolute stereochemistry. Although the shifts in the spectra of acyclic epoxides also showed a consistent pattern, till now fewer of these were examined; hence the findings must be considered more tentative. The spectra of K-region arene oxides (**9.74**) showed a different pattern of lanthanide-induced shifts than **9.72** and **9.73**, but the pattern was consistent among this family, thus enabling a reliable assignment of absolute configuration based on shift data with Eu(tfc)$_3$.[1357] A caution when using lanthanide chelates with epoxides such as these was the observation that some of the arene oxides did aromatocize in the presence of the lanthanide chelate.

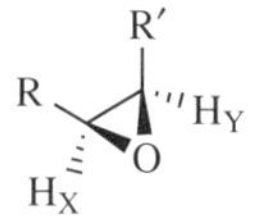

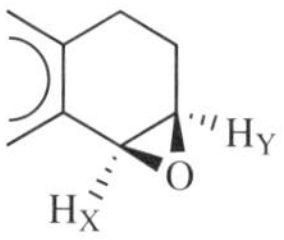

The shifts in the epoxide proton signals of 2'- and 4'-phenyl substituted chalcones (**9.75**) in the presence of Eu(hfc)$_3$ could be used to assign absolute stereochemistries.[1358] The resonances of the (+)-enantiomer consistently exhibited larger shifts in the spectrum.

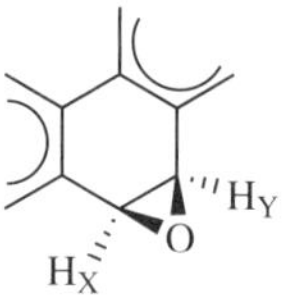

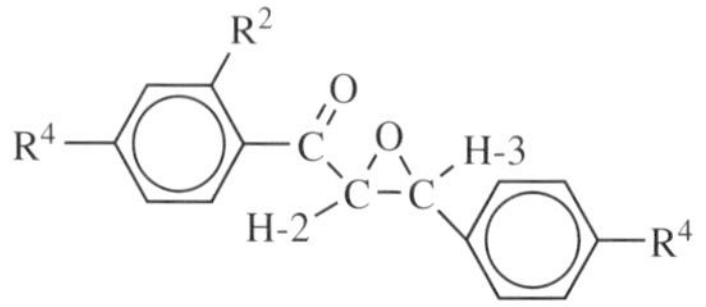

As these studies show, with proper caution, it is possible to assign the absolute configuration of a wide variety of substrates by comparison of the magnitudes of the lanthanide-induced shifts to appropriate model compounds.

9.2.8. Pr(III) Complex of Tetraphenylimidodiphosphinate—Pr(tpip)$_3$

The Pr(III) complex of tetraphenylimidodiphosphinate [Pr(tpip)$_3$] (**9.76**) has been shown to be an effective shift reagent for determining the optical purity of carboxylic acids in solvents such as chloroform-d, methylene chloride-d_2, and benzene-d_6.[1359,1360] The Pr(III) complex is achiral, but binding of carboxylate ions (R*CO$_2$) to Pr(tpip)$_3$ forms diastereomeric binuclear complexes of the formula (R*CO$_2$)$_2$Pr$_2$(tpip)$_4$. The dimeric complexes exhibited slow exchange on the NMR timescale.

Chiral discrimination in the spectra of compounds such as 2-phenylbutyric acid, 2-methylbutyric acid has been observed.[1359] The reagent caused nonequivalence in the ^{1}H spectrum of naproxen as well as an iron tricarbonyl complex that had an hexa-2,4-dienoic acid ligand (**9.77**).[1360] Pr(tpip)$_3$ has been shown to cause discrimination in the spectra of substrates where the chiral center is relatively remote from the carboxylic acid. For example, 5-(1,2-dithiolan-3-yl)pentanoic acid (**9.78**) has a chiral center five bonds removed from the carboxylate group, and certain resonances in the ^{1}H NMR spectrum exhibited nonequivalence in the presence of Pr(tpip)$_3$.[1360]

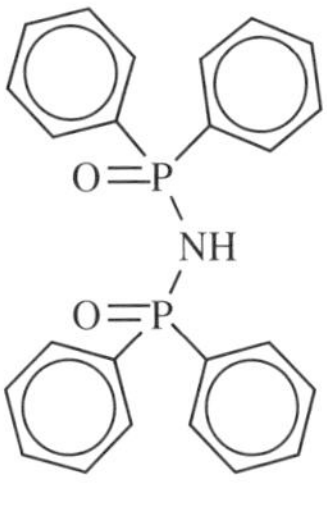

9.76

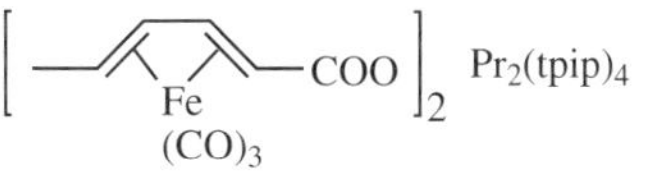

9.77

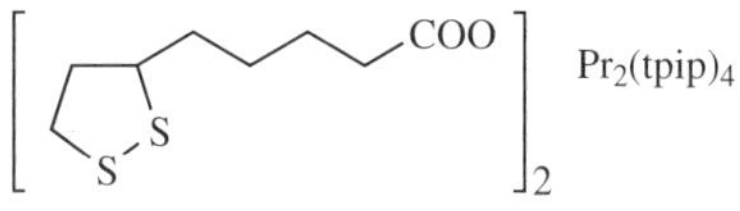

9.78

2-(5-{1-[(*tert*-Butoxycarbonyl)amino]-2-methylpropyl}thiophen-2-yl)propionic acid (**9.79**) has two chiral centers and formed six diastereomeric binuclear complexes with Pr(tpip)$_3$ that were distinguishable in the ^{1}H NMR spectrum.[1360] Adding an enantiomerically pure chiral carboxylate ion, such as (*S*)-2-chloropropionate (2-Cl-Prop) to Pr(tpip)$_3$, created a chiral complex that could be used to discriminate other chiral carboxylate ions. One example involved the discrimination of a compound such as 2-^{2}H-12-phenyldodecanoate that is chiral by virtue of deuterium substitution at the α-position. The enantiomers of the substrate acid (R*CO$_2$) were discriminated in the resulting mixed (R*CO$_2$)(2-Cl-prop)Pr$_2$(tpip)$_4$ binuclear complex.

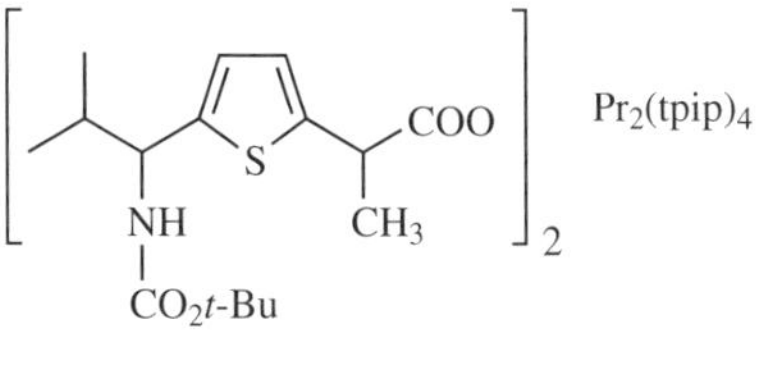

9.79

9.2.9. Binuclear Lanthanide–Silver Reagents

Organic-soluble binuclear lanthanide–silver complexes formed in solution by adding a lanthanide tris β-diketonate to a silver β-diketonate are effective NMR shift reagents for olefins, aromatics, halogenated compounds, and phosphines.[1246,1247] The mixture of the lanthanide and silver reagents leads to the formation of a lanthanide tetrakis chelate anion with the silver as an ion pair. The lanthanide ion causes the shifts in the

NMR spectrum, while the silver binds to the substrate. Chiral analogs of these reagents have been used to determine the enantiomeric purity of compounds with soft Lewis base functional groups.[1361–1367]

The lanthanide tris β-diketonate was typically a Eu(III), Pr(III), or Yb(III) complex of tfc or hfc, although complexes of hfc generally caused the larger enantiomeric discrimination. Binuclear complexes formed by using lanthanide chelates of dcm and 1,1-dicholylmethane were ineffective, presumably because these ligands were too sterically crowded to form the lanthanide tetrakis chelate anion.[1366] The use of Pr(III) and Yb(III), which causes larger shifts than Eu(III), was recommended in earlier studies on lower field instruments since the presence of the silver ion in the binuclear reagent caused a greater distance between the lanthanide and substrate than observed in conventional studies of hard Lewis bases that bind directly to the lanthanide ion in the tris β-diketonate.

A wide variety of silver β-diketonates with achiral and chiral ligands have been tested with the chiral lanthanide chelates. These include the silver complex of H(fod),[1361,1362] 1,1,1-trifluoro-2,4-pentanedione [H(tfa)],[1362] and H(tfc).[1362] Of these ligands, the Ag(tfa) was most effective with Pr(hfc)$_3$ and Ag(tfc) with Yb(hfc)$_3$.[1367] Also, β-diketonate ligands with functional groups such as phenyl, naphthyl, and thiophene rings that might provide other sites for association of the silver in the binuclear reagents were examined with the chiral Pr(III) and Yb(III) chelates of tfc and hfc. Some of these combinations were among the most effective chiral binuclear reagents studied.[1363,1366] The binuclear reagent formed by using achiral Eu(fod)$_3$ with the Ag(I) complex of d-camphor-10-sulfonate has also been shown to cause enantiomeric discrimination in the spectra of dienes containing bicyclo[3.3.1]nonanediene ring systems.[1365]

Even though a variety of silver β-diketonates are potential candidates for use in this application, today the only commercially available reagent is Ag(fod). The silver reagents exhibit light sensitivity and should be kept covered. Ag(fod) is one of the more stable of the silver reagents.

Exploratory studies on the potential applications of chiral binuclear reagents for enantiomeric discrimination demonstrated their effectiveness on compounds such as terpenes,[1361,1362] cyclohexenes,[1364] and 3,4,5,6-tetramethylphenanthrene.[1361] The binuclear reagents caused discrimination in both the ^{1}H and ^{13}C spectra of substrates.[1364,1367] The utility of Yb(hfc)$_3$/Ag(fod) in causing enantiomeric discrimination in the ^{1}H NMR spectra of 1,3-substituted allenes has been demonstrated.[1368–1370] These include derivatives with methyl,[1370] longer chain alkyl,[1369] phenyl,[1369] and trimethylsilyl groups.[1368] The combination Eu(tfc)$_3$/Ag(fod) was used to determine the enantiomeric purity of **9.80**.[1371] The resonance of the methyl group on the chiral carbon exhibited nonequivalence in the NMR spectrum. It was not discussed whether the silver bound at the alkyne or phenyl group.

$$H_3CC\equiv CCHCH_2-\text{C}_6H_5$$
$$\underset{CH_3}{|}$$

9.80

The enantiomeric purity of dienes such as **9.81** has been determined by using Yb(hfc)$_3$/Ag(fod).[1372] The binuclear reagents have also been applied to halogenated compounds. The combination Yb(hfc)$_3$/Ag(fod) was used to determine the enantiomic purity of dibromides **9.82** and **9.83**.[1373] A prior study showed that the silver ion in the binuclear reagents bound preferentially to the softest halogen, with a binding order of I > Br > Cl that was observed.[1247] In the case of **9.82** and **9.83**, the silver bound at the bromine atoms instead of the fluorine or chlorine atoms.

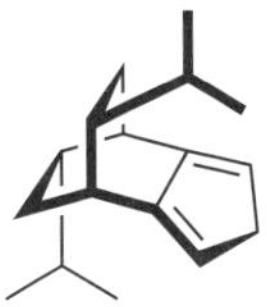

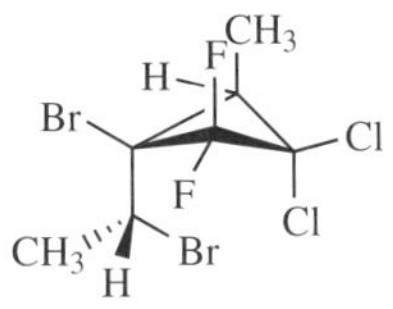

Compound **9.84** was shown to be enantiomerically pure by analyzing the ^{13}C NMR spectrum in the presence of Pr(hfc)$_3$/Ag(fod).[1374] Myrcene (**9.85**) and 3-methylpent-1-ene (**9.86**) were examined as model substrates and both exhibited nonequivalence of 0.1–0.3 ppm for certain resonances in the ^{13}C NMR spectrum with the binuclear reagent. The complete lack of any doubling of resonances in the ^{13}C NMR spectrum of **9.84** in the presence of Pr(hfc)$_3$/Ag(fod) was an evidence for its enantiomeric purity.

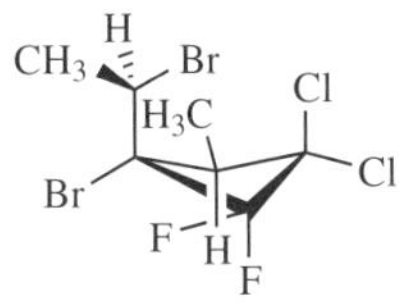

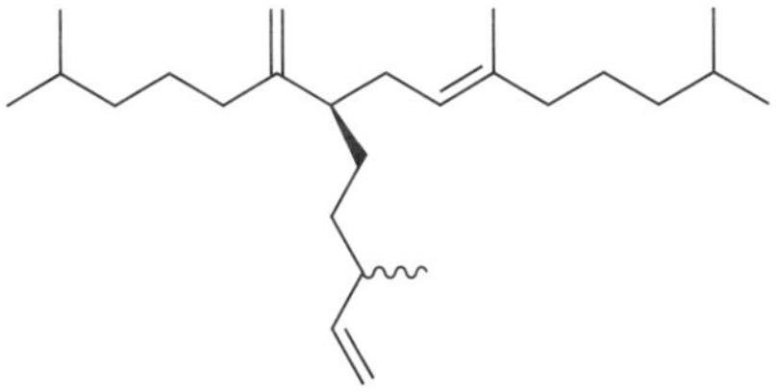

There are several examples of polyfunctional compounds with both hard and soft Lewis base moieties, in which a binuclear lanthanide–silver reagent was more effective for determining the enantiomeric purity than a lanthanide tris β-diketonate alone.[1375–1378]

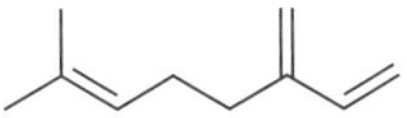

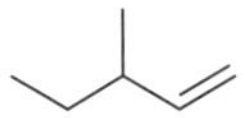

The enantiomeric purities of compounds such as **9.87** were determined by analyzing the methyl or methoxy resonance at 500 MHz in the presence of Yb(hfc)$_3$/Ag(fod).[1375] The nonequivalence in the spectrum was larger than that observed by using only Yb(hfc)$_3$. The addition of Eu(dcm)$_3$ did not cause enantiomeric discrimination in the spectrum of **9.88**, whereas Pr(hfc)$_3$/Ag(fod) did.[1376] The NMR spectra of photoadducts such as **9.89** exhibited chiral discrimination in the presence of Eu(hfc)$_3$/Ag(fod).[1377] There was no comment on whether lanthanide chelates by themselves were effective for the analysis.

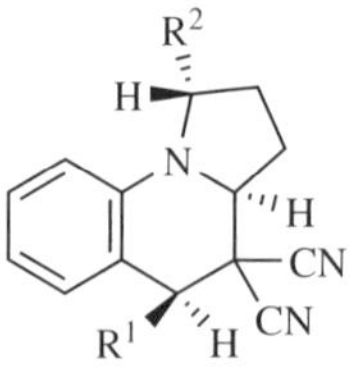

9.87, 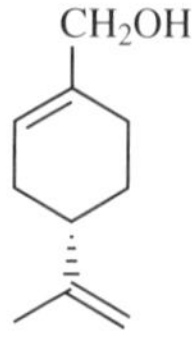 **9.88**, 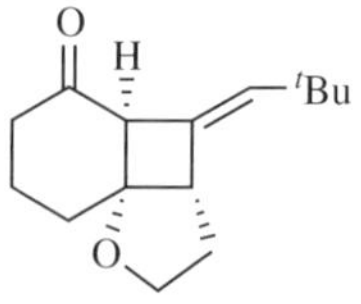 **9.89**

The helical chirality of the dinuclear zinc compound with dipyrromethene (**9.90**) was demonstrated by a doubling of certain resonances in the 200 MHz ^{1}H NMR spectrum in the presence of Eu(tfc)$_3$/Ag(fod) or Yb(hfc)$_3$/Ag(fod).[1378] Addition of Yb(hfc)$_3$ alone did not cause any splitting of the enantiomeric signals, whereas the further addition of Ag(fod) to the mixture did.

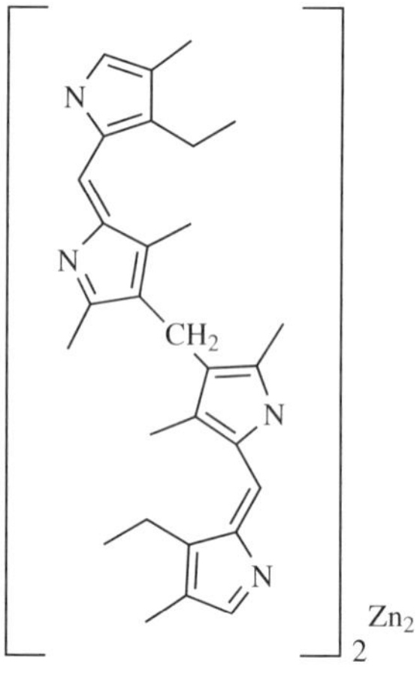

9.90

9.2.9.1. Analysis of Organic Salts The binuclear compounds are also effective chloroform-soluble NMR shift reagents for organic cations. The species Ag[Ln(β-dik)$_4$] is mixed with an organic salt with a halide anion. The silver halide precipitates from solution leaving an ion pair between the lanthanide tetrakis chelate anion and organic cation as shown in Eq 9.3.[1379] These reagents are effective for ammonium,[1379] sulfonium, and isothiouronium salts.[1380]

$$Ag[Ln(\beta\text{-dik})_4] + R^+X^- = R[Ln(\beta\text{-dik})_4] + AgX(s) \qquad (9.3)$$

The species [Eu(tfc)$_3$(fod)]$^-$, formed in situ from Eu(tfc)$_3$ and Ag(fod), caused enantiomeric discrimination in the ^{1}H NMR spectrum of *sec*-butylisothiouronium bromide (**9.91**).[1380] Enantiomeric discrimination was observed in the ^{1}H NMR spectra of alkylmethylphenyl sulfonium salts (**9.92**) (R = ethyl, *n*-butyl, *n*-octyl, and benzyl) in the presence of [Eu(tfc)$_3$(fod)]$^-$ and [Eu(tfc)$_4$]$^-$.[1381]

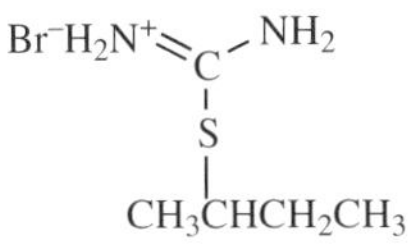

$$Br^-H_2N^+\!=\!\underset{\underset{CH_3CHCH_2CH_3}{|}}{\underset{S}{C}}\!-\!NH_2$$

9.91

Phenyl-S$^+$(CH$_3$)(R)

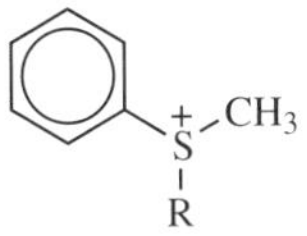

9.92

9.2.10. Aqueous Lanthanide Shift Reagents

Lanthanide species were first used for chiral discrimination in aqueous solutions in 1979.[1382] Optically pure carboxylate ligands were added to chloride salts of Eu(III), Pr(III), or Yb(III) to create a chiral lanthanide chelate in situ that was then used to discriminate another ligand. For example, addition of optically pure mandelate to EuCl$_3$ enabled the chiral discrimination of lactate. Addition of L-lactate to PrCl$_3$ allowed the discrimination of the enantiotopic protons of the glycolate ion (HOCH$_2$CO$_2^-$). Similarly, addition of D-mandelate to Eu(III) or Yb(III)chloride caused discrimination of the enantiotopic methyl groups of α-hydroxyisobutyrate.[1382] The success of this method was predicated on multiple carboxylate ions binding simultaneously to the lanthanide ion to create the steric crowding needed for chiral recognition.[1383] Another observation was that adding lanthanide ions to nonracemic mixtures of carboxylate ions sometimes caused a self-resolution of the enantiomers.[1382,1383]

Adding EuCl$_3$ to the *meso*-, (2*R*,3*R*)-, and (2*S*,3*S*)-isomers of tartaric acid caused chiral recognition in the ^{1}H NMR spectrum.[1384] Carnitine (Me$_3$N$^+$CH$_2$CH (OH)CH$_2$CO$_2^-$) formed a chelate bond with Eu(III) and in the presence of L-malic acid, enantiomeric discrimination was observed in the ^{1}H NMR spectrum.[1385] A pH of 9.1–9.3 was necessary to prevent precipitation, and the best ratio was 2:1 malate to carnitine.

Addition of a large excess of (*S*)-lactate to Pr(III) resulted in discrimination of the prochiral methylene resonances of citrate (**9.93**), causing four CH doublets that were all well separated.[484] The spectrum had to be recorded at a pH less than 6 to reduce the broadening to acceptable levels. A selectively deuterated derivative of citrate prepared enzymatically was used to confirm the assignment of resonances in the ^{1}H NMR spectrum.

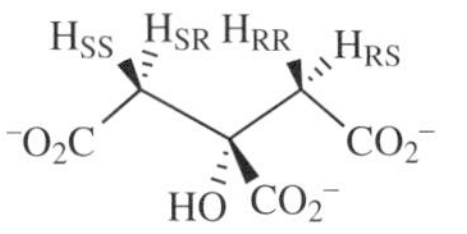

H$_{SS}$ H$_{SR}$ H$_{RR}$ H$_{RS}$
$^-$O$_2$C — C(HO)(CO$_2^-$) — CO$_2^-$

9.93

9.2.10.1. Complexes of Propylenediaminetetraacetate (pdta) The most studied aqueous chiral lanthanide shift reagent involves complexes with the ligand (R)-propylenediaminetetraacetic acid (pdta) (**9.94**). Ethylenediaminetetraacetic acid-like ligands (EDTA), of which pdta is one, do not thoroughly encapsulate the large lanthanide ions and leave an open coordination site for the bonding of carboxylate ions. Pdta provides a soluble lanthanide complex at basic pH, whereas free lanthanide ions would precipitate. In many cases, the relative shifts in the spectra of enantiomeric carboxylates in the presence of Eu(pdta) correlate with the absolute configuration.

$$HO_2C\diagdown\quad\diagup CO_2H$$
$$\quad\quad N\diagdown\quad\diagup N$$
$$HO_2C\diagup\quad\diagdown CO_2H$$

9.94

Eu(pdta) was effective in producing chiral discrimination in the ^{1}H NMR spectra of amino acids and hydroxy acids in deuterium oxide.[1386–1388] Discrimination of the enantiotopic groups in isobutyric acid was also observed with this reagent.[1386] It was important to adjust the pH so that the substrate was in its anionic form, which facilitated binding to the lanthanide ion. For α-amino acids at a pH of 8.5–10.5, the addition of the (R)-enantiomer of Eu(pdta) consistently shifted the Hα resonance of the L-isomer of the amino acids to lower frequency.[1387] The α-methyl signals of α-methylamino acids also showed a shift order that was consistent with the absolute configuration.[1389] In all 11 examples studied, the α-methyl resonance of the (S)-enantiomer shifted to lower frequency in the presence of the (R)-isomer of Eu(pdta).

α-Hydroxycarboxylic acids (**9.95**) likely bind to Eu(III) in a chelate manner involving the oxygen atoms of the carbonyl and hydroxyl groups.[1388] The shifts of the α-proton and α-methyl group exhibited a consistent trend in the presence of Eu(pdta) that correlated with the absolute stereochemistry.[1388] Using the (R)-isomer of Eu(pdta), the Hα resonance of the (S)-isomer of the substrate was consistently at a higher frequency than that of the (R)-isomer.

$$\begin{array}{c} OH \\ | \\ R-C-CO_2H \\ | \\ R' \end{array}$$

9.95

Aldonic acids (**9.96**) that have a hydroxy group α to the acid exhibited non-equivalence in their 400 MHz NMR spectra in the presence of Eu(pdta) under weak acidic conditions.[1390] The Cα configuration was important in determining the relative magnitude of the lanthanide-induced shifts. In the 17 examples studied, the (R)-isomer of Eu(pdta) consistently caused the Hα resonances of the (S)-isomer of the substrate to shift to higher frequency, whereas the Hβ and Hγ resonances of the (S)-isomer shifted to lower frequency. With some substrates, the relative order of the shifts exhibited a reversal at higher lanthanide ion concentrations, which was likely due to changes in the stoichiometry of the complex.[1390] This and other reports indicate how any studies with lanthanide species aimed in assigning absolute configurations must adhere to the concentration ranges over which established trends were observed.

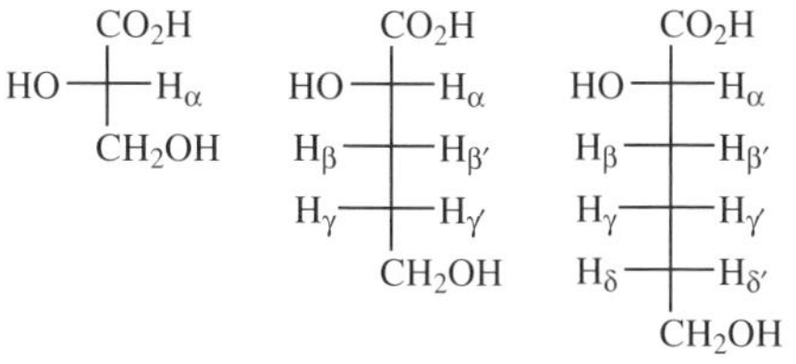

```
        CO₂H              CO₂H              CO₂H
   HO ──┼── Hα      HO ──┼── Hα       HO ──┼── Hα
        CH₂OH         Hβ ──┼── Hβ'      Hβ ──┼── Hβ'
                      Hγ ──┼── Hγ'      Hγ ──┼── Hγ'
                         CH₂OH          Hδ ──┼── Hδ'
                                           CH₂OH
```

9.96

The addition of Eu(pdta) caused nonequivalence in the ^{1}H NMR spectrum of β-amino acids at 400 MHz.[1391] The relative shifts of the α- and β-proton resonances showed a consistent trend with absolute configuration.

Eu(pdta) has been applied to the analyses of 2-hydroxyglutaric acid (**9.97**) and 5-oxoproline (**9.98**) in urine at 400 MHz.[1392] The absolute configuration of **9.97** and **9.98** can be used as a marker of inborn errors of metabolism. Enriched samples were used to confirm the assignment of the nonequivalent resonances. Deuterium oxide and Eu(pdta) were added directly to samples of urine, and either ^{1}H NMR or two-dimensional ^{1}H–^{13}C correlation spectroscopy was used to perform the analysis.

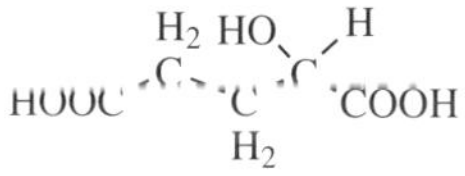

9.97 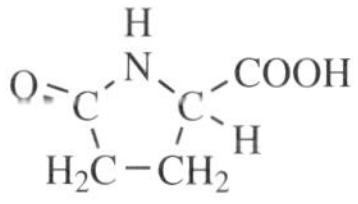   **9.98**

A potential problem with the use of Eu(pdta) is the extent of broadening in the spectrum on high field spectrometers. The Ce(III)[1393] or Sm(III)[1394] complexes of pdta reduced the broadening but still caused significant enough shifts (0.01–0.10 ppm for the Sm(III)complex) to discriminate enantiomers in the ^{1}H (400 MHz) or ^{13}C NMR spectra. As with Eu(pdta), the relative magnitude of the shifts of the enantiomers of α-amino acids with Ce(pdta) and Sm(pdta) correlated with the absolute configuration.

9.2.10.2. Complexes of N,N,N',N'-Tetrakis(pyridylmethyl)propylene Diamine (TPPN)

In contrast to pdta, N,N,N',N'-tetrakis(pyridylmethyl)propylene diamine (tppn) (**9.99**) is a neutral ligand. In contrast to the anionic lanthanide complexes with pdta, [Eu(III)(R)-tppnCl₂)]ClO₄ associated with underivatized α-amino acids at neutral rather than basic pH.[1395,1396] The binding of α-amino acids with Eu(tppn) was reportedly stronger than with Eu(pdta).[1396] Using the (R)-isomer of Eu(tppn), the resonance of the α-proton of the L-enantiomer of all of the amino acids tested shifted to higher frequency. The reverse correlation occurred for the α-methyl, β-methyl, and γ-hydrogen resonances. Eu(tppn) was also an effective chiral shift reagent for *N*-acyl amino acids.[1396]

9.99

Similar to lanthanide complexes with pdta, use of the Ce(III)[1397] or La(III)[1398] complex with tppn caused significant reductions in line broadening, while still causing chiral recognition in the ^{1}H NMR spectra (400 MHz) of α-amino acids, *N*-acyl amino acids, and carboxylic acids. The relative shifts correlated with the absolute configurations of the α-amino acids, but the nonequivalence was reversed when compared to the Eu(III) complex. The structure of the La(III) complex with α-amino acids shown in Figure 9.3 demonstrates how the Hα resonance of the D-enantiomer in the presence of the (*R*)-tppn complex occurred at higher frequencies relative to that of the L-enantiomer, and how the shift order was reversed for other protons of the substrate.

The utility of Eu(pdta) in analyzing the absolute configuration of 2-hydroxyglutaric acid was described earlier. The Eu(tppn) complex was equally effective in this application.[1392]

9.2.10.3. Complexes with Other Ligands Water-soluble chiral lanthanide complexes with other ligands have been reported and examined on a more limited basis as NMR discriminating agents. Eu(III) and Yb(III) form a 1:1 complex with (*S*)-carboxymethylsuccinic acid (**9.100**) that caused chiral discrimination in the

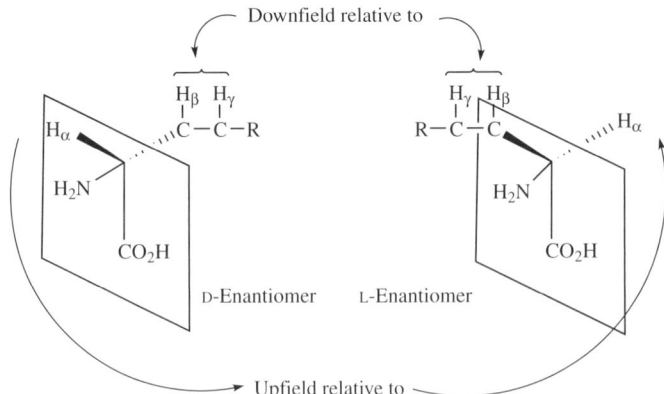

Figure 9.3. Relative positions of the protons in α-amino acid complexes with La(tppn).

^{1}H NMR spectrum of amino acids and (oxy)carboxylic acids.[1399]

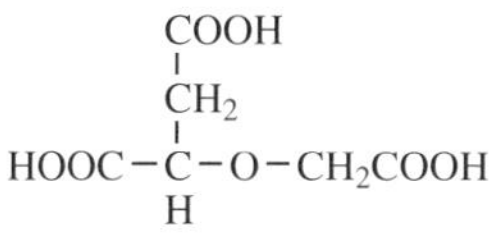

```
        COOH
         |
         CH₂
         |
HOOC — C — O — CH₂COOH
         |
         H
```

9.100

The Eu(III) complex of (*S,S*)-ethylenediamine-*N,N'*-disuccinic acid (**9.101**) produced enantiomeric discrimination in the ^{1}H NMR spectrum of phenyl glycine,[1400] alanine, serine, threonine, and phenylalanine.[1401] The best results were observed at pH values above eight. The Eu(III) complex with (*S,S*)-ethylenediamine *N,N'*-disuccinate was used to analyze the optical purity of tyrosine.[1402] The methine proton was resolved in the ^{1}H NMR spectrum. The effectiveness of Eu(III) complexes as chiral NMR shift reagents with two other *N,N'*-disuccinate ligands (**9.102** and **9.103**) was compared to **9.101**.[1403] The optimal pH for using these lanthanide complexes with α-amino acids was 9–10, as a precipitate formed at pH below 9 and above 10. A wide range of amino acids and α-alkylated amino acids were examined. Non-equivalence in the ^{1}H NMR spectrum with complexes of **9.103** ranged from 0.23–1.31 ppm and was considerably larger than that observed with the complexes of **9.101** and **9.102**.

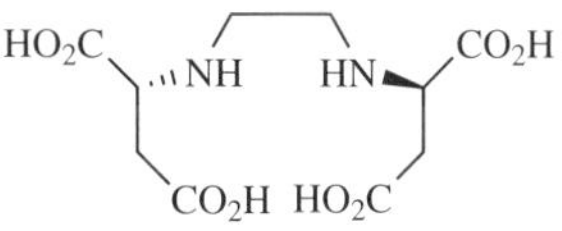

9.101 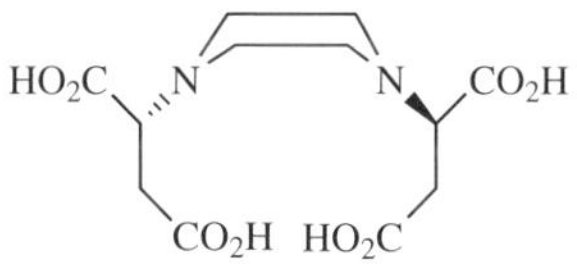 **9.102** 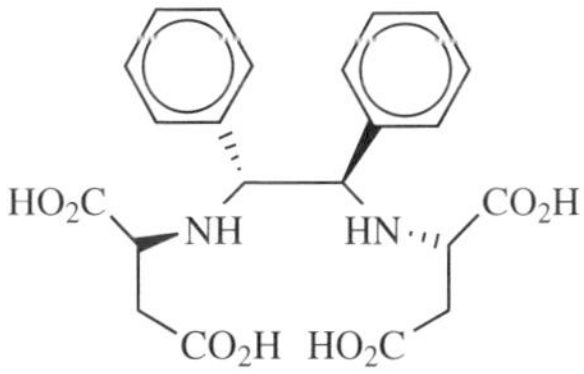 **9.103**

Eu(III) complexes with other EDTA-type ligands **9.104**, **9.105**, and **9.106** have been evaluated as aqueous chiral lanthanide shift reagents.[1404,1405] Better results were observed with the complexes of **9.105** and **9.106**, but enantiomeric discrimination was rather modest.

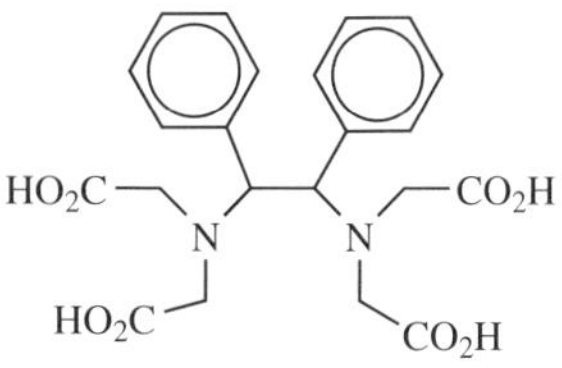

9.104 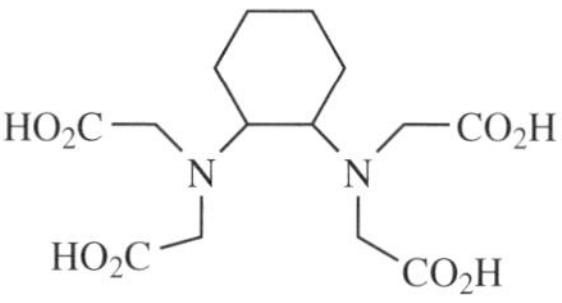 **9.105** 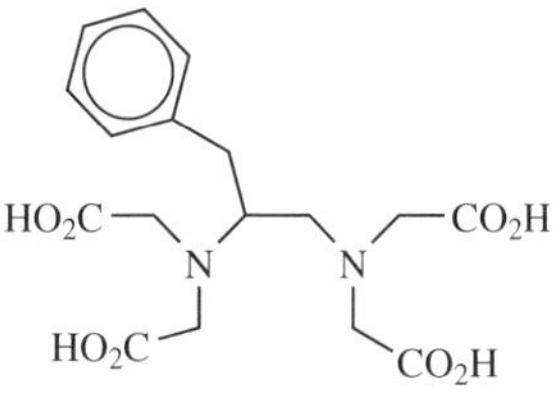 **9.106**

Eu(III) complexes with ligands that are bis-L-amino acid derivatives of either ethylenediamine[1406,1407] or diethylenetriamine[1408] have been examined as chiral lanthanide shift reagents. For the N,N'-ethylenediamine bis(L-amino acid) derivatives, the complex of a ligand with L-histidine residues (**9.107**) was judged most effective of those examined.[1406,1407] This complex caused enantiomeric discrimination in the ^{1}H and ^{13}C NMR spectra of α-amino acids in neutral solution. The shifts were not that large; however, nonequivalence on the order of 0.03–0.08 ppm and 0.05–0.2 ppm were observed in the ^{1}H (400 MHz) and ^{13}C (100 MHz) NMR spectra, respectively. The peaks were reportedly much sharper than those observed with other water-soluble lanthanide chelates.

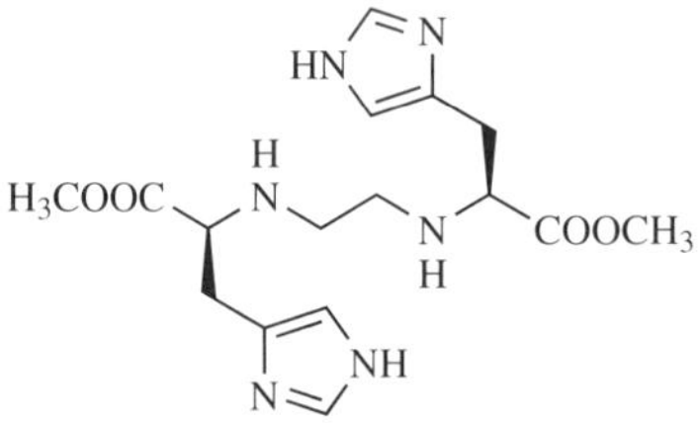

9.107

The Eu(III) complex with N,N-bis[2-{N-methyl((1S)-1-carboxy-3-methyl)butyl-amino}ethyl]glycine (**9.108**) was effective in causing chiral discrimination in the ^{1}H NMR spectra of α-amino acids and N-boc-Gly-Ala.[1408] The optimum pH for the analysis was 7–8, as the resonances were too broad at pH values above 8 and there was no enantiodistinction at pH values less than 7. Typical nonequivalence was on the order of 0.03–0.08 ppm.

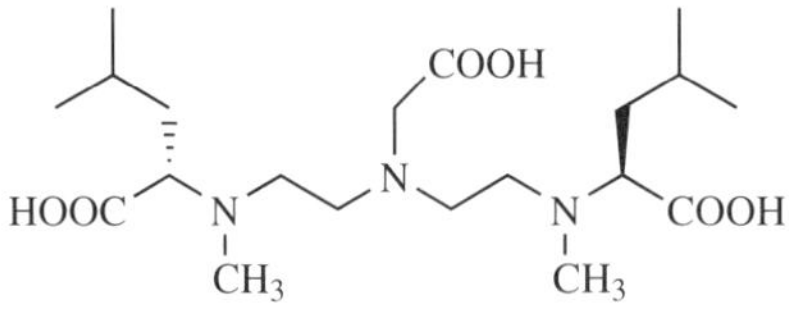

9.108

9.2.10.4. Complexes with Tetraazocyclodecane Macrocyclic Ligands Water-soluble lanthanide complexes suitable for use in chiral NMR studies have been prepared using chiral tetraazocyclodecane macrocyclic ligands.[213,1409,1410] The Yb(III) complex of **9.109** as its trifluoromethyl sulfonate salt provided a cationic species that produced chiral recognition in the ^{1}H NMR spectra of lactate and other alkyl and aryl α-hydroxy acids.[1409]

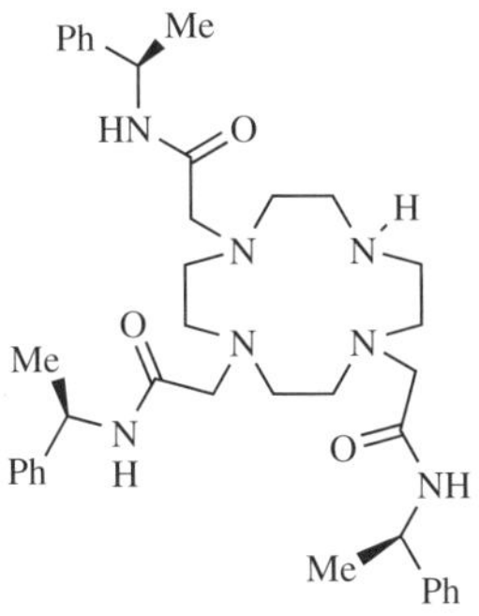

9.109

The chirality of anionic Tm(III) complex with DOTP (**9.110**) (Tm(dotp)$^{5-}$) was demonstrated by adding the cationic complex of La(III) with (S,S,S,S)-THP (**9.111**).[1410] Ion pairing of the Tm(III) and La(III) complexes caused a splitting of signals of the Tm(DOTP)$^{5-}$ species.

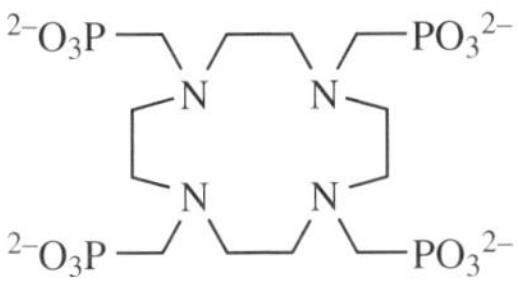

9.110 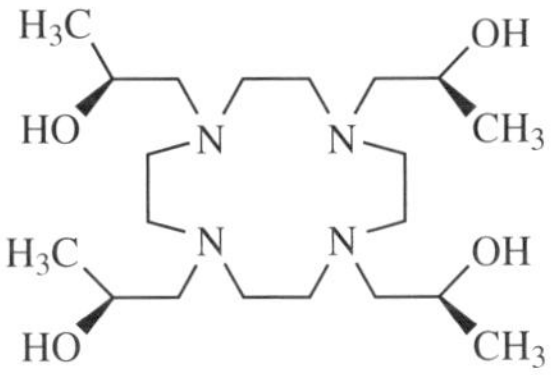 **9.111**

9.3. TRANSITION METAL COMPLEXES

9.3.1. Palladium Complexes

9.3.1.1. Bridged Dimers with Amine Ligands Bridged palladium dimers **9.112** and **9.113** can be prepared from enantiomerically pure N,N-dimethyl-(1-phenyl)ethylamine or N,N-dimethyl-1(1-naphthyl)ethylamine, respectively. Both ligands are commercially available and the palladium complexes are easily prepared and purified. The resulting dimers are a mixture of cis- and trans-complexes, but that is immaterial to their use for chiral NMR discrimination.[1411] As early as 1971, it was noted that the addition of a diphosphine to **9.112** displaced the bridging chloro ligand leading to the formation of a monopalladium complex as seen in Figure 9.4.[1412] The

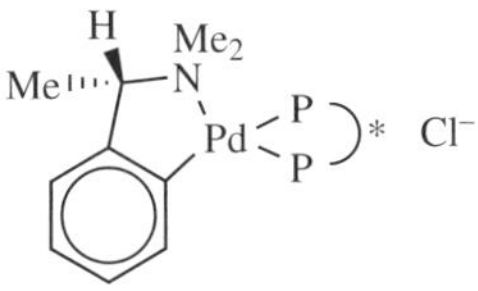

Figure 9.4. Monopalladium complex produced by reaction of a diphosphine with **111.9**.

potential utility of NMR spectroscopy to distinguish a pair of enantiomeric phosphines through the formation of their diastereomeric palladium complexes was noted in this first report, but not examined to a significant extent. Since then, compounds **9.112** and **9.113** have been used extensively for the analysis of a variety of chiral species that are capable of binding to palladium and forming the monopalladium species.[1413]

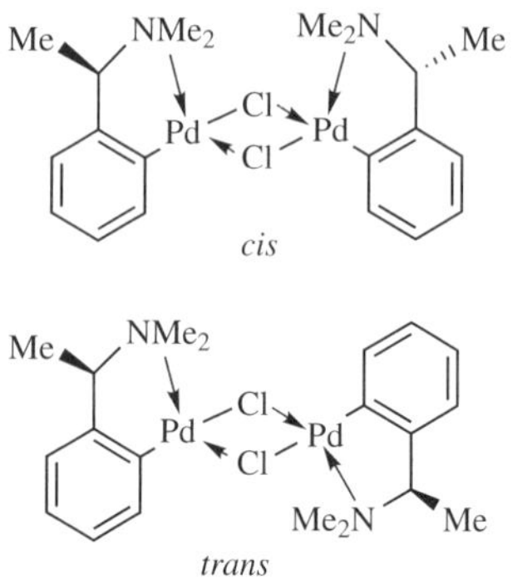

9.112

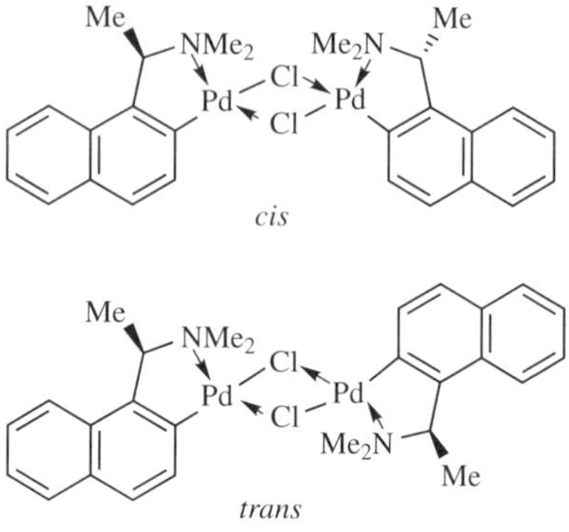

9.113

If the complexes are in slow exchange, it is possible to use resonances of either the substrate or phenyl/naphthyl amine ligand for the analysis. Comparisons of the effectiveness of **9.112** and **9.113** have usually noted that **9.113** caused a greater degree of discrimination in the NMR spectrum of the diastereomers. The greater effectiveness of **9.113** relative to **9.112** likely results because H8 of the naphthyl ring induces a conformational preference that forces the C-methyl group into an axial conformation in the five-membered metallocycle.[1414] A review article describing the applications of **9.112** and **9.113** for both the determination of optical purity and assignment of absolute configuration has been published.[1413]

Early applications of **9.112**[1415–1418] and **9.113**[1419,1420] involved the study of bidentate ligands. For example, **9.112** was used to enantiomerically discriminate 2-aminoethyl-n-butylphenylphosphine (**9.114**), a ligand that bound in a bidentate manner to palladium through the phosphorus and nitrogen atom.[1417] The diphosphine ligand o-phenylenebis(methylphenylphosphine) (**9.115**) exists in a *meso*-isomer [(R,S) and (S,R)] and pair of enantiomers [(R,R) and (S,S)]. Although not an NMR study, the chiral recognition properties of **9.112** were demonstrated through its ability to separate the three forms of **9.115** through a fractional crystallization procedure.[1416]

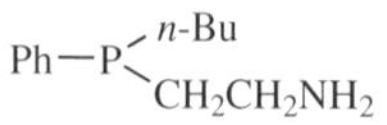

9.114

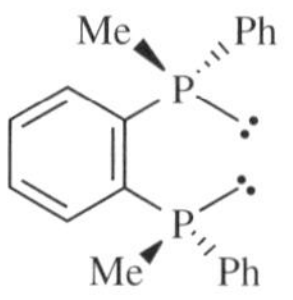

(*RS*)

9.115

The ability of **9.112** to cause nonequivalence in the ^{1}H and ^{31}P NMR spectra of diphosphine ligands such as **9.116** and **9.117** was examined.[1415] The utility of the ^{31}P NMR spectra for the analysis of optical purity was noted. The number of signals in the ^{31}P spectrum was determined by whether the phosphine ligand was in fast or slow exchange. **9.116** underwent fast exchange such that two singlets were observed in the ^{31}P NMR spectrum, one for each enantiomer. With **9.117**, slow exchange occurred and the ^{31}P NMR spectrum has two doublets for each enantiomer of the substrate. The two doublets corresponded to the phosphorus atoms *cis* and *trans* to the nitrogen atom, and doublets were observed because the two phosphorus atoms couple to each other. The doublets had different areas depending on the preference for the *cis*- or *trans*-isomer. Each of the doublets was then split into two resonances for the pair of enantiomers.

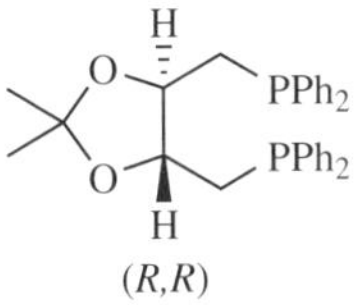

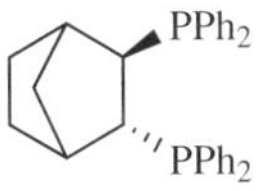

The utility of **9.113** in causing chiral recognition in the ^{1}H and ^{13}C NMR spectra of phosphine and arsine ligands such as methylphenyl(8-quinolyl)phosphine **9.118** was examined.[1419] **9.118** bound in a bidentate manner through the phosphorus (or arsenic) and nitrogen atom. Of particular importance was that the Hγ resonance (position adjacent to the metallated carbon) of the naphthyl ring shifted in a manner that correlated with the absolute stereochemistry of the phosphorus or arsenic atom.[1419] For the palladium complex with (*R*)-**9.113**, the (*S*)-enantiomer of **9.118** and similar compounds caused a pronounced shielding of the Hγ resonance that was not observed for the (*R*)-enantiomer.

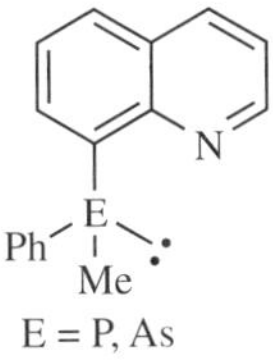

For the bidentate amino phosphine or arsine ligand **9.119**, the ^{1}H NMR spectrum with **9.113** showed distinct resonances for the two diastereomers.[1420] With **9.119**, use of (*R*)-**9.113** shielded the phenyl hydrogen atoms of the (*R*)-configuration of **9.119** more than those of the (*S*)-configuration. The shielding occurred because the naphthyl and phenyl rings were orthogonal to each other, such that the π electrons of the naphthyl ring were over the hydrogen atoms of the phenyl ring.

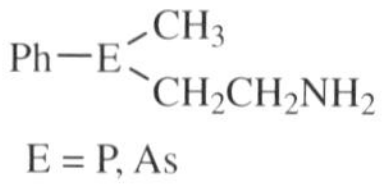

$$\text{Ph}-\text{E}\begin{smallmatrix}\diagup\text{CH}_3\\ \diagdown\text{CH}_2\text{CH}_2\text{NH}_2\end{smallmatrix}$$

E = P, As

9.119

9.112 has been used to determine the enantiomeric purity or assign the absolute configuration of diphosphines such as 2,5-bis(diphenylphosphine)hexane (**9.120**)[1421], a bis(phosphine) with phenyl and menthyl groups (**9.121**),[1422] 6,6′-dimethoxy-2,2′-diyl)bis(diphenylphosphines) (**9.122**),[1423] and bis(diphenylphosphino)butane (**9.123**).[1424] The assignment of absolute configuration was achieved by analyzing the shifts in the resonances of the alkyl group on the phenyl ligand. For example, the methine resonance of the (R,R)-substrate–palladium diastereomer of **9.122** was in the region from 5.2 to 5.35 ppm, whereas the corresponding resonance for the (S,R)-complex was at about 3.5 ppm.[1423] Other investigators have relied on the use of two-dimensional techniques, such as NOESY and ROESY, to show spatial proximity of nuclei on the phosphine and nuclei on the phenyl ligand as a means of assigning absolute configurations.[1422,1424,1425] Similarly, NOE connectivities were used to assign the stereochemistry of P-chiral phosphines **9.124** bonded to **9.113**.[1426] The specific configuration of the phenyl or naphthyl ligand in **9.112** and **9.113** constrains the phosphine ligand, thereby enabling these types of studies.

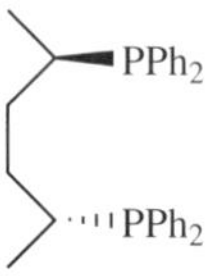

9.120

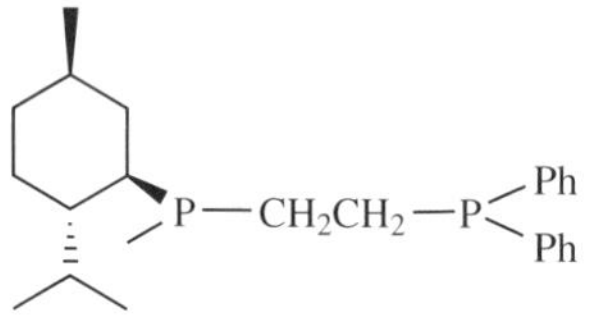

9.121

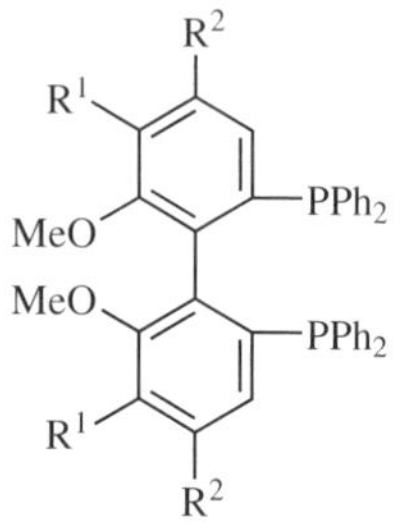

9.122

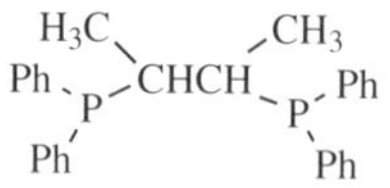

9.123

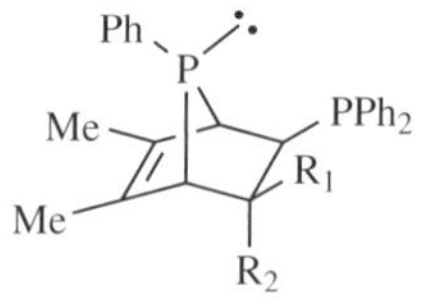

9.124

The use of NOE data with **9.113** has been used to assign the absolute configuration of monophosphines such as benzylisopropylphenylphosphine[1427] and benzylmesitylphenylphosphine.[1428] In the latter study, the palladium dimer with 1-(1-naphthyl) ethylamine (NEA) (**9.125**) was used for the analysis.[1428]

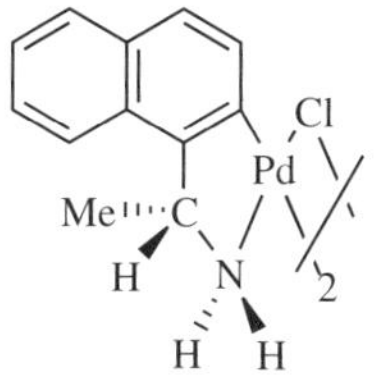

9.125

Dibenzophosphole 5-oxides (**9.126**) were reduced to their corresponding phosphines using trichlorosilane.[1429] Binding of the phosphine to **9.125** produced only the *cis*-isomer because of the destabilizing effect of having two soft ligands in a *trans*-arrangement. Nonequivalence was observed in the ^{1}H NMR spectrum of the diastereomers and the shifts of the C4 and C6 hydrogen resonances of the dibenzophosphole showed characteristic trends that could be used to assign the absolute configuration.[1429]

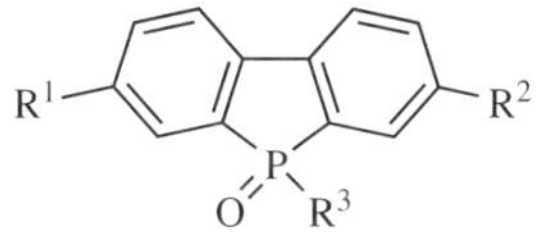

9.126

Other palladium compounds similar to **9.112** and **9.113**, that differ only in the nature of the substituent groups attached to the alkyl chain, have been examined for their utility as chiral NMR shift reagents. The palladium dimer with an imine ligand (**9.127**) caused nonequivalence in the ^{13}C or ^{31}P NMR spectra of diphenylphosphines, in which the alkyl substituent had a β-hydroxy group or hydroxyl-substituted cyclohexane ring such as **9.128**.[1430] The effectiveness of **9.127** was not compared to similar palladium reagents.

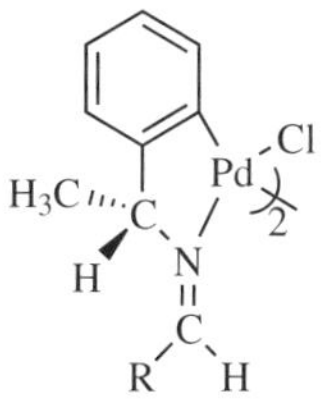

9.127 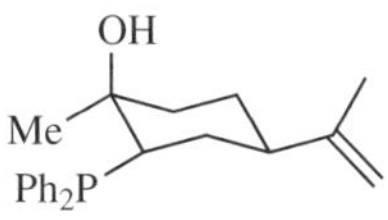   **9.128**

Palladium dimers with phenyl and naphthyl ligands, in which the alkyl portion is substituted with bulkier isopropyl (**9.129**) or *tertiary*-butyl (**9.130**) groups, have been evaluated.[1431–1433] Using ^{31}P NMR spectroscopy, the nonequivalence was often much greater with ligands that had bulkier groups as shown in Figure 9.5.[1432] The *t*-butyl ligand was subsequently applied to the analysis of

Figure 9.5. The magnitudes of diastereomeric peak separation (Δδ, ppm) obtained for the covalent and ionic platinum complexes. [Reproduced with permission from Elsevier.[1432]]

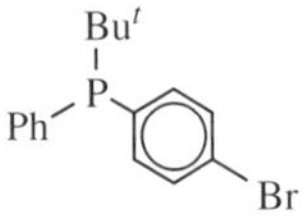

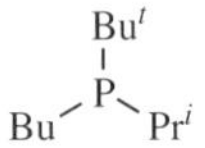

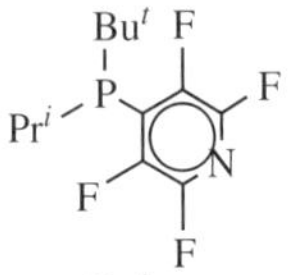

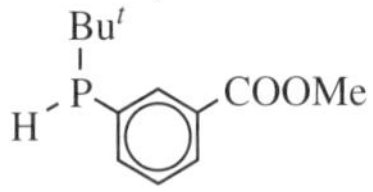

Side chain of CDA palladacycle:	(S$_c$) - 1a	(S$_c$R$_N$) - 1d	(S$_c$) - 1e
Bu^t-P(Me)(4-Br-C$_6$H$_4$)	0.052	0.390	0.281
Bu^t-P(Ph)(4-Br-C$_6$H$_4$)	0.289	0.437	0.654
Bu^t-P(Bu)(Pr^i)	0.505	1.077	2.567
Bu^t-P(Pr^i)(C$_5$F$_4$N)	1.521	2.520	2.510
Bu^t-P(H)(3-COOMe-C$_6$H$_4$)	1.507	2.342	5.355

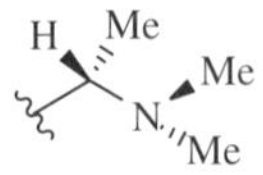

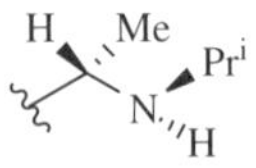

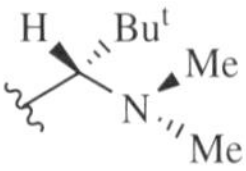

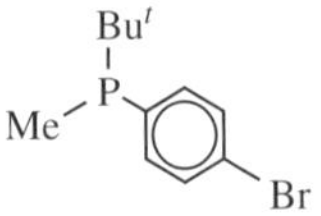

Data column values above:

Table header structures (side chain amine):
- (S$_c$) - 1a: H, Me, N(Me)(Me)
- (S$_c$R$_N$) - 1d: H, Me, N(Pr^i)(H)
- (S$_c$) - 1e: H, Bu^t, N(Me)(Me)

t-butylphenyl(4-bromophenyl)phosphine.[1433] NOE connectivities in the ^{1}H NMR spectrum were used to assign the absolute configuration of the phosphine.

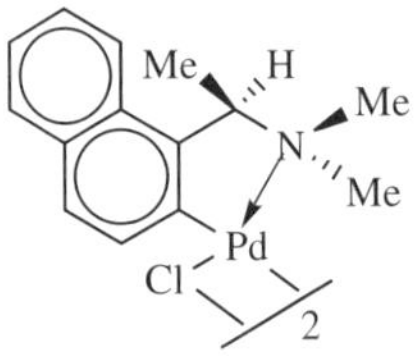

9.129 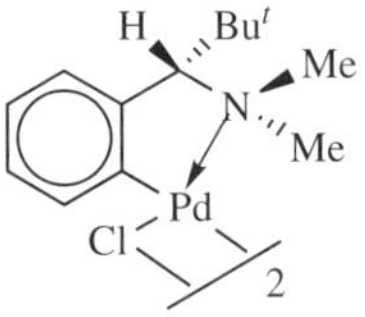 **9.130**

Palladium dimers of this type can be used for the chiral analysis of ligands other than phosphines. For example, **9.112** has been used to determine the optical purity of amino acids.[1434,1435] The α-amino acid bound to the palladium in a bidentate manner involving the nitrogen atom and oxygen atom of the carboxylate group.[1434] The palladium dimers were also effective for determining the optical purities of β- and γ-amino acids, as well as the methyl esters of β-amino acids.[1435] The effect of solvent, temperature, and aryl group of the ligand was examined in this report. The amino acids were in slow exchange and ^{1}H resonances were monitored. In some cases, nonequivalence up to 134.5 Hz at 300 MHz was obtained. Methanol-d_4 was typically the solvent of choice, although mixed solvents of methanol-d_4 with chloroform-d or benzene-d_6 sometimes resulted in greater distinction. In instances when the discrimination in the ^{1}H spectra was too small, the analysis was performed by cooling the sample or obtaining the ^{13}C NMR spectrum. The differences in the spectrum of the cis- and trans-complexes were used to assign the absolute configuration of β-amino acids.[1435] The assignment required a comparison of the shifts to those in the spectra of compounds with known configurations.[1435]

The palladium complex comparable to **9.112** in which the C-methyl group has been replaced with a trifluoromethyl group has been used to measure the optical purity of α-amino acids.[1436] The ^{19}F NMR was monitored and two signals, one for each diastereomer, were observed.

Two other palladium systems that in some ways are variations of the scheme employed with **9.112** and **9.113** have been used to analyze α-amino acids. One of these involved a palladium complex with a ferrocene ligand (**9.131**).[1437] Binding of the amino acid created a monopalladium species and the H3 resonance of the ferrocene ligand exhibited excellent nonequivalence in the spectra of the diastereomeric complexes.

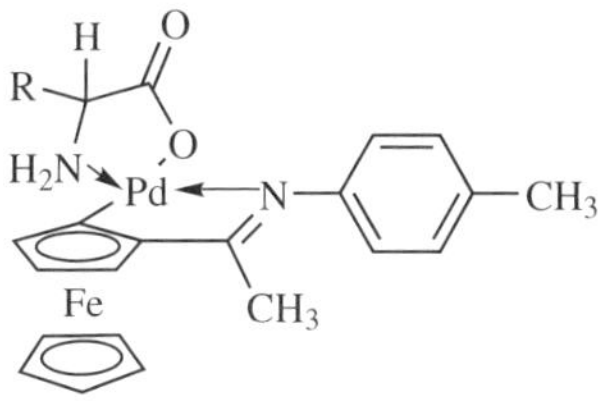

9.131

Similarly, binding of α-amino acids to a palladium dimer with a P*-chiral ligand (**9.132**) produced monopalladium species.[1438] The amino acids bound to **9.132** in an E/Z manner through the nitrogen and carboxylate oxygen atoms. Examination of the ^{31}P NMR showed four different signals for the four isomers with nonequivalence of up to 2 ppm.

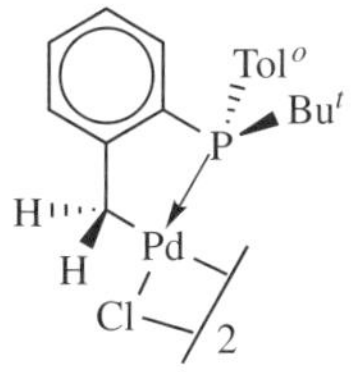

9.132

The compound [(methylsulfinyl)methyl]diphenylphosphine (**9.133**) binds to **9.113** in a bidentate manner through the phosphorus and sulfinyl oxygen atom.[1439] An analysis of the ^{1}H ROESY NMR data allowed the absolute stereochemistry of **9.133** to be determined. The diastereomeric complexes obtained by reacting a similar bidentate ligand [2-(methylsulfinyl)ethyl]amine (H$_2$NCH$_2$CH$_2$S(O)CH$_3$) with **9.113** produced two sets of *C*-methyl signals in the NMR spectrum.[1440]

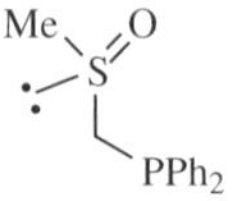

$$\text{9.133}$$

9.113 was shown to be an effective chiral derivatizing agent for the analysis of the optical purity of C_2-symmetric 1,2-diamines such as **9.134**.[1441] Both the Hγ and *C*-methyl resonances of the naphthyl ligand of **9.113** exhibited significant distinction for the diastereomeric complexes. The analysis was performed in solvents such as methylene chloride-d_2, chloroform-*d*, and deuterium oxide.

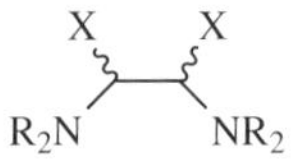

$$\text{9.134}$$

9.3.1.2. Miscellaneous Palladium Complexes   The palladium complex with the diphosphine ligand 2,2-dimethyl-4,5-bis(diphenylphosphinomethyl)-1,3-dioxolan (**9.135**) has been employed for the analysis of optical purity of alkenes and alkynes.[1442,1443] Alkene or alkyne substrates displaced the ethylene ligand in the original palladium complex and the ^{31}P NMR spectrum was conveniently monitored.

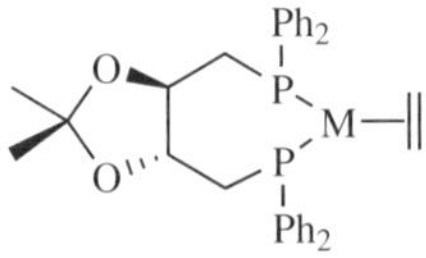

$$\text{9.135}$$

Palladium complexes with chiral 1,2-diamine ligands of structure **9.136** caused nonequivalence in the ^{1}H and ^{13}C NMR spectra of the diastereomeric complexes with α-amino acids and one β-amino acid.[1444] Complex formation was slow but addition of sodium deuteroxide facilitated the reaction. The complex in which R was a phenyl group typically gave the largest degree of enantiomeric discrimination, but it was only sparingly soluble in deuterium oxide. Therefore, the complex in which R was a methyl group was recommended.

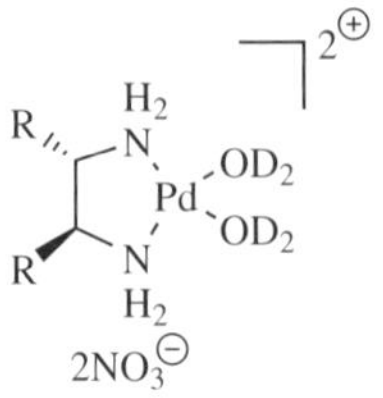

$$\text{9.136}$$

A palladium complex with a bis(diphenyl)phosphine(1,2-diaminocyclohexane) and 2,4-pentanedione ligands (**9.137**) was useful for analyzing the optical purity and assigning the absolute configuration of α-amino acids.[1445] Addition of the amino acid caused displacement of the 2,4-pentanedionate ligand. Four doublets were observed in the resulting ^{31}P NMR spectrum, two for each of the diastereomers. Fourteen α-amino acids were studied and the outer doublets were always from the L-isomer.[1445]

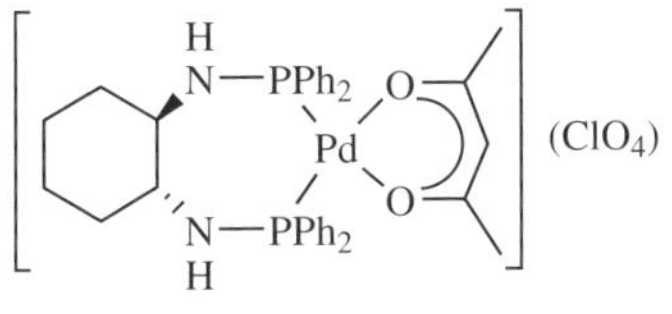

9.137

Palladium has also been used as a template to prepare complexes from chiral ligands that are then distinguishable through their NMR spectra. 1-Aminoalkylphosphonic acids (**9.138**) complexed with palladium in alkaline deuterium oxide to form PdL_2 complexes.[1446] The resulting ^{31}P NMR spectrum showed two peaks, one for the (R,R)- and (S,S)-complexes, the other for the (S,R)- and (R,S)-complexes. Eight of the ten derivatives that were analyzed showed nonequivalence in the ^{31}P NMR spectrum. This same method has been applied to the analysis of fluorinated aminomethane phosphonic acids.[1447]

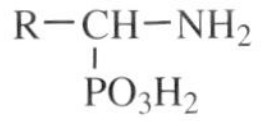

$$R-CH-NH_2$$
$$|$$
$$PO_3H_2$$

9.138

Addition of the chiral ligand (1-diphenylphosphino-3-benzyloxy)propane-2-thiol (**9.139**) to palladium formed the corresponding *cis*- and *trans*-[Pd(phosphinothiolato)$_2$] complexes.[1448] After 24 h, four resonances were observed in the ^{31}P NMR spectrum for the *cis*-[(R,R) and (S,S)], *trans*-[(R,R) and (S,S)], *cis*-(S,R), and *trans*-(S,R) species. The areas of these resonances were used to determine the enantiomeric purity of the thiol ligand.

9.139

The compound [3,7-(1′-phenylethyl)-3,7-diazabicyclo[3.3.1]nonane] (**9.140**) forms a rigid cavity that accommodates (π-allyl)palladium species with high selectivity.[1449] For example, the racemate of (π-allyl)palladium complexes such as (4-acetoxy-(1,2,3-η^3)cyclohexenyl)palladium (**9.141**) and 1-(S)-2-methylene-6,6-dimethylbicyclo[3.1.1]hept-2,3,10-η^3-enyl)palladium inserted one of the π-allyl

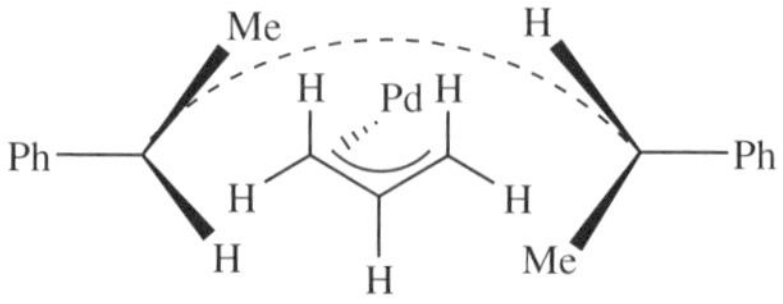

Figure 9.6. Insertion of a π-allyl ligand into the cavity of **137.9**.

ligands into the cavity of **9.140** as shown in Figure 9.6. The ^{1}H NMR spectrum showed signals for the (S,S)- and (R,R)-complexes. The association of the (π-allyl)-palladium with the cavity was specific enough such that NOE connectivity data could be used to assign the absolute configuration of the π-allyl ligand.[1449]

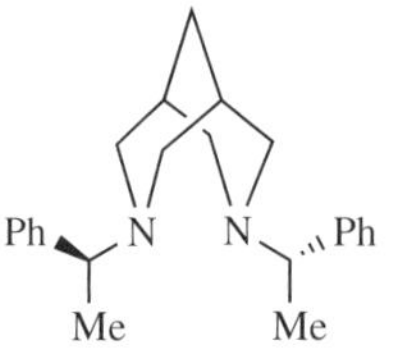

9.3.2. Platinum Complexes

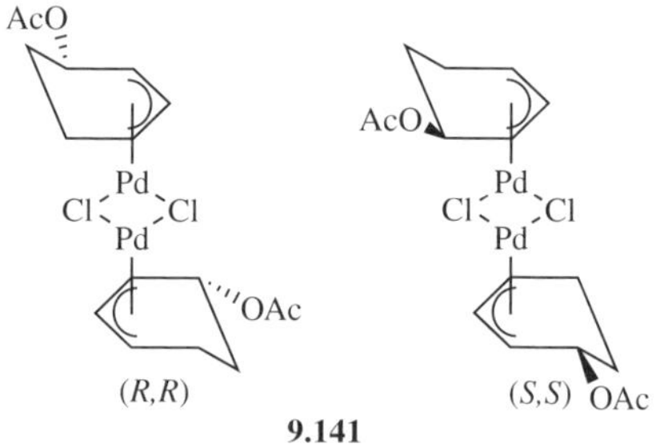

9.3.2.1. Complexes with Diphosphine Ligands Several chiral platinum complexes have been used as chiral NMR discriminating agents. Complex **9.142** has a chiral diphosphine ligand and bound ethylene group.[1442,1443,1450] Olefins, allenes, and alkynes displaced the ethylene, and ^{31}P NMR spectra in benzene-d_6 showed two signals for the resulting diastereomeric complexes. These included chiral substrates such as carvone, dimethyl *trans*-norbornene-2,3-dicarboxylate, cyclonona-1,2-diene, acrylamine, and 3-phenyl-3-hydroxy-prop-1-yne[1442,1443] and several cyclohexene derivatives.[1450] Carbonyl, hydroxyl, amide, and carboxy functional groups did not interfere with binding of the alkene moiety to the platinum.

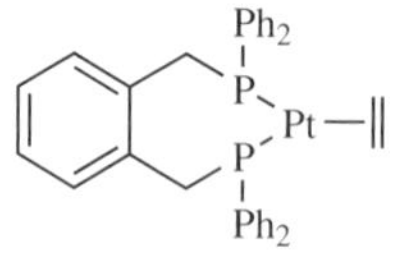

9.142

The effectiveness of platinum complexes of formula [Pt(P-P)CH$_3$(P*)]ClO$_4$, where P-P was five different chiral diphosphine ligands at distinguishing enantiomeric phosphines (P*) was examined.[1451] The ^{31}P NMR spectrum exhibited nonequivalence and the best results were obtained with (2S,3S)-bis(diphenylphosphino)butane, commonly known as (S,S)-chiraphos (**9.143**).

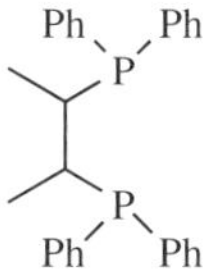

9.143

The enantiomers of 1,16-dihydroxytetraphenylene (**9.144**) were distinguished by examining the ^{1}H and ^{31}P NMR spectra of the palladium complex formed with (S)-2,2′-bis(diphenylphosphino)-1,1′binaphthyl-PtCO$_3$ (**9.145**).[1452]

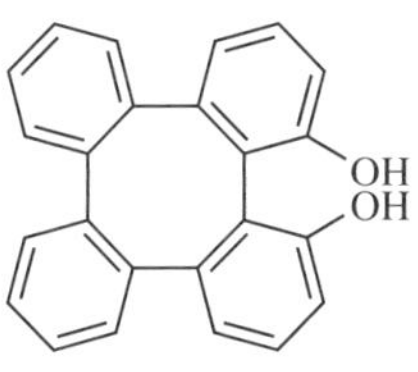

9.144 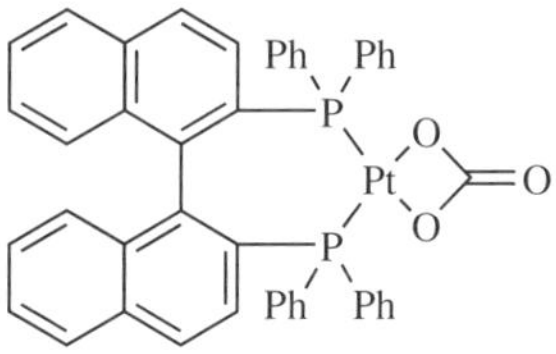   **9.145**

9.3.2.2. Complexes with Amine Ligands The use of ^{195}Pt NMR spectroscopy of platinum complexes incorporating chiral amine ligands to enantiomerically discriminate alkenes and allenes has been the focus of a series of reports.[1087,1453–1461] Two types of platinum complexes were used. In one, the amine was directly bonded to the platinum as in the *cis*- or *trans*-dichloro [(S)-α-methylbenzylamine][ethylene]platinum(II) complexes (**9.146**). The other involved an ionic complex of the chiral amine cation (AmH$^+$)with the anionic trichloro[ethylene]platinum(II) complex (**9.147**).

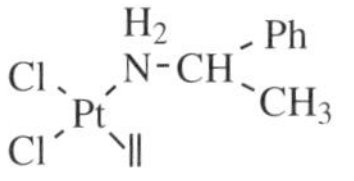

9.146 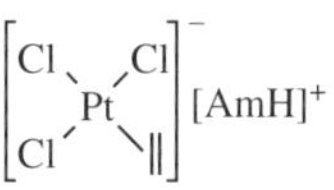   **9.147**

Olefins and allenes displaced the ethylene group and bound to the platinum. The ^{195}Pt NMR spectrum of the resulting diastereomeric complexes exhibited either two or four signals.[1453] The substrates have two prochiral faces. When four signals were observed, it indicated that each of the substrate enantiomers bound at the two prochiral faces. If only two signals were observed, binding occurred at only one of the prochiral faces.[1455,1456] Discrimination in the ^{195}Pt NMR spectrum of chiral allyl alcohols and allyl ethers was observed.[1453] A detailed study of 3-phenyl-3-methoxybut-1-ene (**9.148**), which only showed two ^{195}Pt signals, indicated that a hydrogen bond between the amide group of the amine ligand and oxygen of the substrate limited the binding of **9.148** to only one of the prochiral faces.[1455] A similar substrate without the methoxy group did not show as significant an extent of nonequivalence in the NMR spectrum.

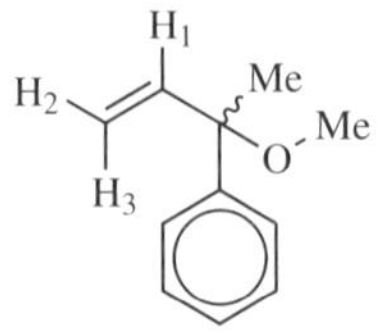

9.148

Allenes displaced the ethylene as well, although warming the sample to 30–40°C facilitated the reaction.[1087,1453] Many of the allenes caused four resonances in the ^{195}Pt spectrum, although in cases when the allene had a bulky *t*-butyl group, the steric hindrance resulted in binding of only one of the prochiral faces.[1089]

Ionic platinum complexes of structure **9.147** with 1-phenylethylamine, (*S,S*)-bis (α-methylbenzyl)amine (**9.149**),[1457] (1*S*,1′*S*)-bis[1-(1-naphthyl)ethyl]amine (**9.150**),[1458–1460] (1*S*,1′*S*)-*N*-[1′-(1-naphthyl)ethyl]-1-phenylethylamine (**9.151**),[1460] and (*S*)-3,5-dihydro-4*H*-dinaphthyl-[2,1-*c*:1′,2′-*e*]azepine (**9.152**)[1458] have been evaluated for their effectiveness as chiral NMR discriminating agents with alkenes and allenes. The effectiveness of these ionic complexes was also compared to the covalent amine complexes.

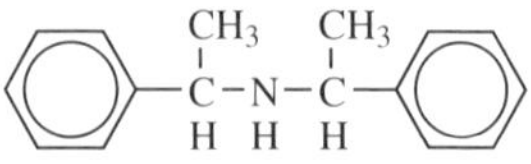

9.149

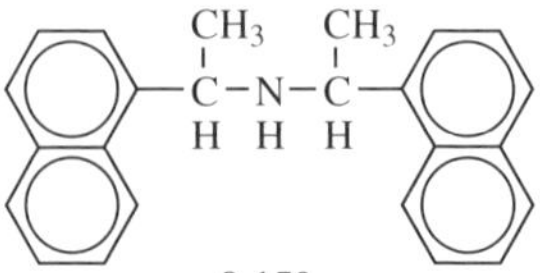

9.150

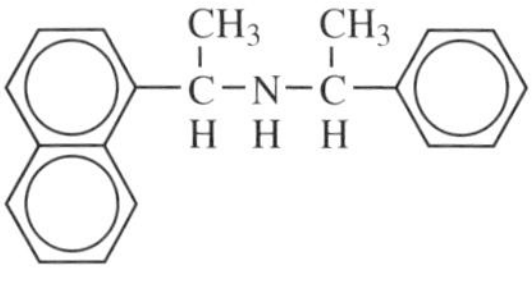

9.151

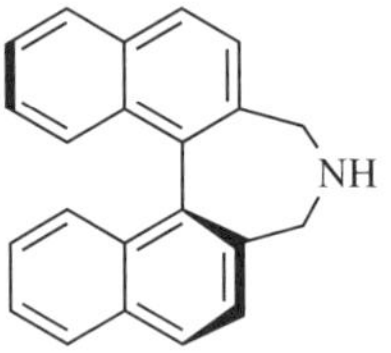

9.152

Spectra with the ionic complexes were typically obtained in benzene-d_6. The substrates underwent fast exchange in these systems. As with the covalent platinum complexes, either two or four platinum signals were observed depending on whether both prochiral faces of the substrate bound to the platinum. For substituted allenes with the platinum complex of **9.150**, there were always just two ^{195}Pt signals observed.[1459,1460] The complex with **9.150** was judged more versatile than that with **9.151** and the primary amines at it generally provided larger nonequivalence in the spectra of the diastereomeric products.[1460] For all 10 trisubstituted allenes that were examined with the platinum complex of **9.150**, the resonance for the derivative with the (*R*)-allene was at a higher frequency than that of the (*S*)-allene.[1459] The platinum complex with **9.150** was used to analyze the optical purity of vinyl ethers, allyl ethers, and olefins having stereogenic centers α or β to the double bond.[1460]

The platinum complex with **9.152** was effective for analyzing chiral alkenes and unsaturated ethers.[1458] Nonequivalence from 5 to 11 ppm was typically observed in the ^{195}Pt NMR spectra.[1458] For olefins with a chiral center α to the double bond, the *trans*-covalent complexes of form **9.146** were better than the ionic complexes, whereas for those with a chiral center β to the double bond, the ionic platinum complexes of form **9.147** were more effective.[1458] For open chain allyl ethers, the *trans*-form of **9.146** worked well, as did the ionic complex with **9.150**. For simple olefins, the *trans*-covalent and ionic complexes with **9.152** were preferred. For vinyl ethers the ionic platinum complex with **9.150** generally provided the largest discrimination.

A review of these systems has been published.[1461] In general, the ionic platinum complexes (**9.147**) are more effective since bulkier amine ligands may have trouble attaching directly to the platinum in the *trans*-complex (**9.146**).[1461] Figure 9.7 provides

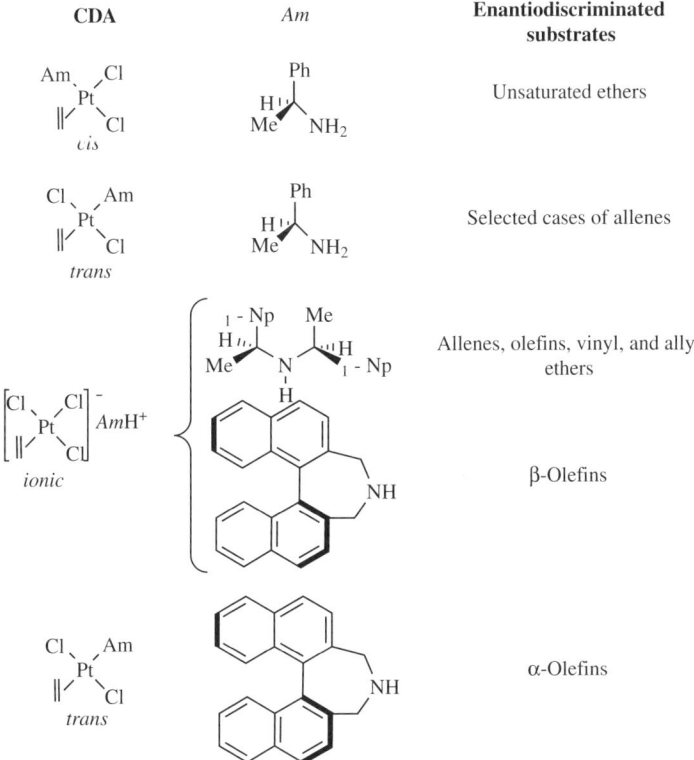

Figure 9.7. Summary of utility of covalent and ionic ethane–platinum(II) complexes for the analysis of unsaturated compounds. [Reproduced with permission.[1461]]

a summary of the recommended complexes to use for different categories of substrates. The platinum complexes themselves are not commercially available, although certain of the ligands are and reagents can be prepared by established methods.

9.3.2.3. Analysis of Geometrical Isomers The use of NMR spectroscopy to distinguish diastereomeric metal chelate complexes has been exploited for a long time.[1462] For example, the ^{1}H NMR spectra of Pt(IV) complexes with propylenediamine (pn) and ethylenediamine (en) ligands [(+) and (−)[Pt(S-pn)$_3$](C$_2$O$_4$)$_2$, [(+) and (−)[Pt(R-pn)$_2$en]Cl$_4$, and [(+) and (−)[Pt(R-pn)(en)$_2$]Cl$_4$ have distinct methyl signals for each pair.[1462] The signals could also be empirically related to the absolute configuration of the complexes.

9.3.3. Rhodium Complexes

9.3.3.1. Rhodium Dimer—[Rh$_2$(MTPA)$_4$] A dimeric rhodium complex with bridging α-methoxy-α-trifluoromethylphenylacetic acid (MTPA) ligands [Rh$_2$(MTPA)$_4$] (**9.153**) is an effective chiral NMR discriminating reagent for a broad range of compounds, most of which are characterized as soft Lewis bases. A review of the use of Rh$_2$(MTPA)$_4$ for chiral NMR discrimination was recently published.[1463] The rhodium centers in Rh$_2$(MTPA)$_4$ are coordinatively unsaturated and additional ligands can bind as illustrated in Figure 9.8. Since both 1:1 and 1:2 dimer–substrate complexes can form, adjusting the relative concentrations of the rhodium complex and substrate to optimal levels is often important with this system. Rh$_2$(MTPA)$_4$ is typically used for the analysis of enantiomeric purity, although with compounds such as phosphanes, phospholene chalcogenides, and spirochalcogenuranes, tentative trends observed for relative shifts may, with the study of more substrates, enable the assignment of absolute configurations. Rh$_2$(MTPA)$_4$ is not commercially available, although optically pure MTPA is, hence the complex can be prepared without too much difficulty.

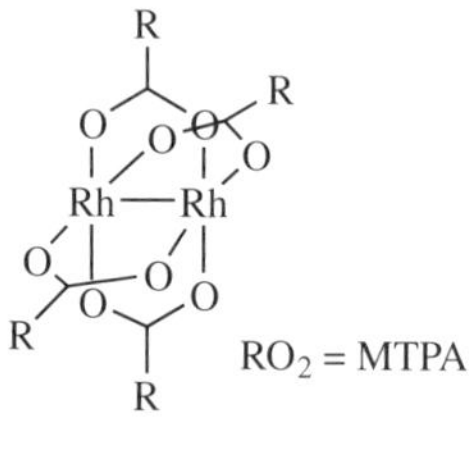

9.153

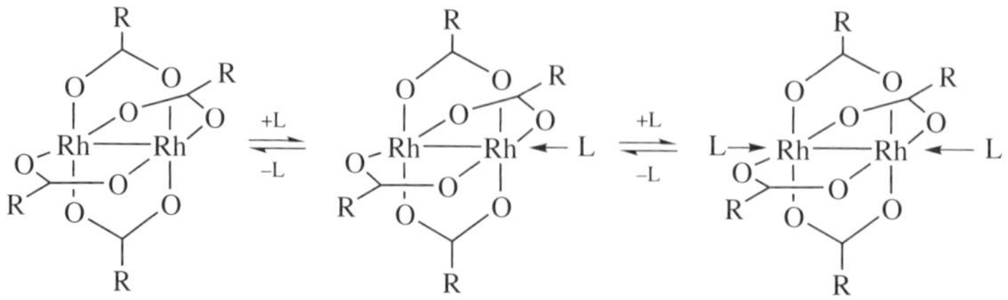

Figure 9.8. Binding of ligands to Rh$_2$(MTPA)$_4$.

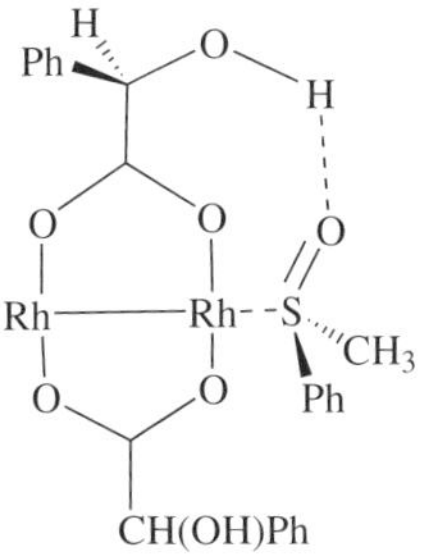

TRANSITION METAL COMPLEXES 379

Figure 9.9. Hydrogen bonding interaction of the sulfoxide oxygen with mandelic acid hydroxyl group in the corresponding rhodium-mandelic acid dimer.

There is only one report in which ligands other than MTPA were used in the dimeric rhodium complex.[1464] This involved an analysis of the chiral recognition in the ^{1}H and ^{13}C NMR spectra of methyl phenyl sulfoxide. The best results were observed using the dimer prepared from mandelic acid. A crystal structure confirmed that the sulfur atom bound to the rhodium and the hydroxyl group of the mandelic acid formed a hydrogen bond with the sulfoxide oxygen as shown in Figure 9.9. The stabilization that was observed in the mandelic acid complex is unlikely to occur and thereby enhances the discrimination with other classes of substrates. Rh$_2$(MTPA)$_4$ was therefore used in subsequent studies.

Rh$_2$(MTPA)$_4$ is especially useful for analyzing compounds with soft Lewis base donors such as olefin, phosphorus, sulfur, selenium, or iodine groups. The first report using Rh$_2$(MTPA)$_4$ for chiral discrimination in NMR spectroscopy involved an analysis of olefins such as α-pinene (**9.154**), limonene (**9.155**), 4-vinyl-1-cyclohexane, and carvone (**9.156**).[1465] Typical nonequivalence was 3–4 Hz on a 200 MHz spectrometer. No discrimination was observed in the ^{1}H NMR spectrum of camphene, suggesting that sterically crowded olefins were inhibited from binding to the rhodium.

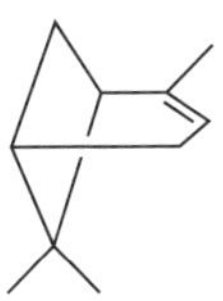

9.154 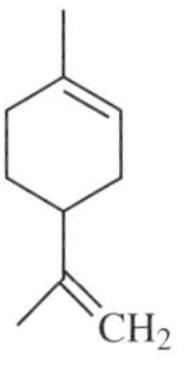   9.155 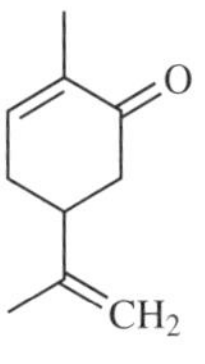   9.156

Phosphines bind to Rh$_2$(MTPA)$_4$ and the ^{1}H, ^{13}C, ^{31}P and, ^{103}Rh NMR spectra could potentially be used to determine enantiomeric purity.[1466,1467] The largest nonequivalence was typically observed in the ^{31}P NMR spectrum. The use of ^{31}P-detected ^{103}Rh HMQC NMR was the first use of ^{103}Rh NMR spectroscopy to recognize *P*-chirality in bonded ligands.[1466]

Rh$_2$(MTPA)$_4$ was useful in discriminating the enantiomers of aryl alkyl selenides such as 2-phenylselenylpentane (**9.157**),[1468] phenylselenenylmenthane (**9.158**), and menthene derivatives.[1469,1470] These compounds associated with the rhodium through the selenium atom rather than the phenyl ring. Chiral recognition was noted in both the ^{1}H and ^{13}C NMR spectra.[1469]

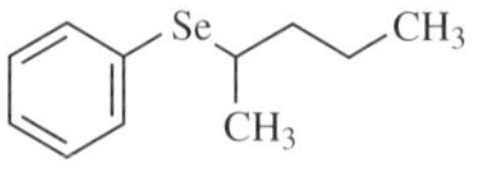

9.157

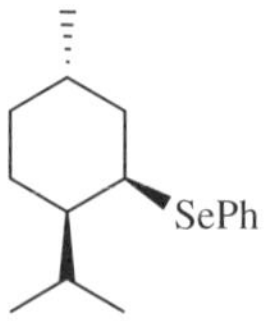

9.158

The potential of using $Rh_2(MTPA)_4$ to discriminate resonances in the ^{1}H, ^{13}C, ^{31}P, and ^{77}Se NMR spectra of phosphene selenides (P=Se) such as **9.159** has been examined.[1471] The ^{77}Se NMR spectrum was not useful because of severe broadening of the signals. Enantiomeric discrimination in the ^{1}H, ^{13}C, and ^{31}P spectra was noted.[1471] Chiral phosphorus thionate (P=S) compounds such as **9.160** and **9.161** bind to the rhodium atom. Chiral discrimination in the ^{1}H, ^{13}C, or ^{31}P spectra was observed for nine substrates with representative substituent groups.[1472]

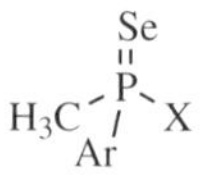

9.159

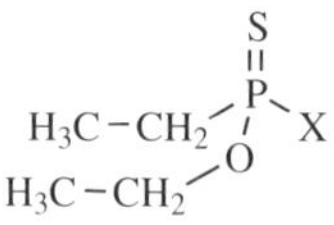

9.160

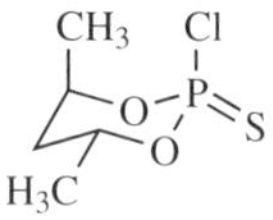

9.161

Nonequivalence was observed in the ^{1}H, ^{13}C, and ^{31}P NMR spectra of cyclic and bicyclic phospholene (**9.162**) and phospholane (**9.163**) chalcogenides with oxygen, selenium, and sulfur atoms in the presence of $Rh_2(MTPA)_4$.[1473] A later study showed that the ^{1}H NMR shifts of phospholene chalcogenides (**9.164** and **9.165**) exhibited trends that correlated with absolute configuration about the phosphorus atom.[1474] Examining similar model compounds with known configurations was recommended when performing a stereochemical analysis.

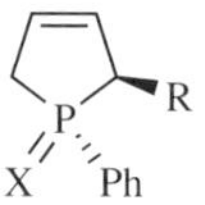

9.162 X = O, S, Se

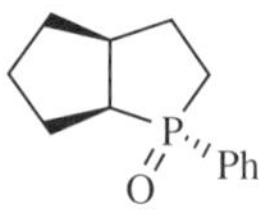

9.163

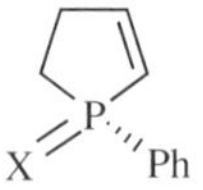

9.164 X=O, S, Se

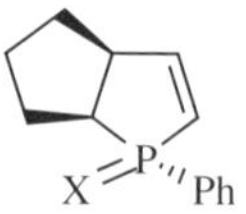

9.165 X=O, S, Se

$Rh_2(MTPA)_4$ was effective in discriminating spirochalcogenuranes of structure **9.166** with a Se, S, or Te atom.[1475] The shifts in the ^{1}H NMR spectra showed a consistent trend that correlated with absolute configuration, although the concentrations must be carefully adjusted for the analysis since 1:1 and 2:1 substrate–rhodium dimer complexes can form. Lower temperatures were best for studies aimed at assigning absolute stereochemistry. Analysis of optical purity was more readily accomplished at slightly elevated temperatures, where faster exchange caused less broadening of the resonances.

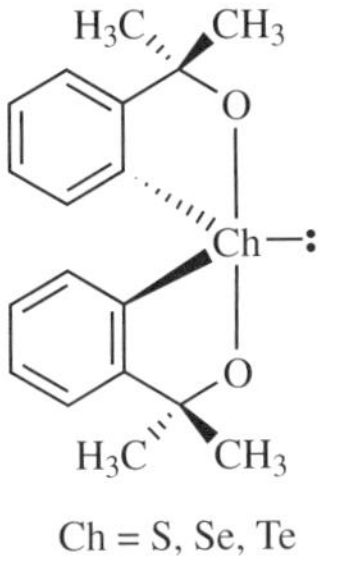

9.166

Ch = S, Se, Te

The ^{1}H and ^{13}C NMR spectra of bis(phosphaneoxides) such as **9.167** exhibited enantiomeric discrimination in the presence of Rh$_2$(MTPA)$_4$.[1476] The nonequivalence improved noticeably when the spectra were run at 213 K instead of ambient temperatures.

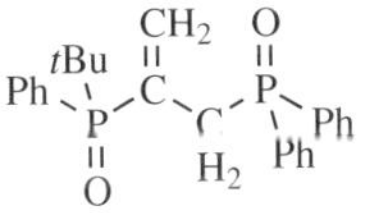

9.167

Secondary phosphine oxides exist as tautomers in solution as illustrated in Figure 9.10. In the absence of Rh$_2$(MTPA)$_4$, the phosphine oxide tautomer (P=O) is the preferred form. In the presence of Rh$_2$(MTPA)$_4$, the phosphine tautomer (P—OH) is favored over the phosphine oxide and binds to the rhodium through the phosphorus atom.[1477] Secondary phosphine oxides were examined, five of which were chiral and one of which had prochiral groups. All exhibited anisochronous signals in the ^{1}H NMR spectrum in the presence of Rh$_2$(MTPA)$_4$. Enantiomeric discrimination was also observed in the ^{13}C and ^{31}P NMR spectra, although obtaining sufficiently soluble compounds for ^{13}C NMR required using a mixed solvent system of acetone-d_6-chloroform-d.

Alkyl iodides such as 2-iodobutane bounded to Rh$_2$(MTPA)$_4$ and exhibited nonequivalence of up to 0.15 ppm in the ^{1}H NMR spectrum.[1468] The rhodium complex

Figure 9.10. Tautomers of secondary phosphine oxides.

was not effective in causing enantiomeric discrimination in the NMR spectra of the corresponding bromide derivatives. In certain cases, the chiral recognition was larger for the resonances of nuclei further from the binding site, which is rather atypical for many chiral discriminating agents. The enantiomeric discrimination with $Rh_2(MTPA)_4$ is caused by differences in the interaction of the substrate ligand with the MTPA ligand. Nuclei near the binding site with $Rh_2(MTPA)_4$ are often not close enough to the chiral center of the MTPA ligand to show significant chiral recognition, whereas those further removed interact more intimately with the MTPA ligands.

$Rh_2(MTPA)_4$ was effective in producing chiral discrimination in the NMR spectra of a number of atropisomeric diiodobiphenyls of structure **9.168**.[1478] For the analogs in which X was a hydrogen atom, chlorine atom, or methoxy group, only the iodine atom complexed with the rhodium. The NMR spectra of the *N,N*-dimethyl (**9.168f**) and bis(phenol) derivatives (**9.168g**) did not show enantiomeric recognition. The amino and thiol groups in **9.168e**, **9.168h**, and **9.168i** bound to the rhodium rather than the iodine atom. The distinction with **9.168e** was larger than in the corresponding diiodo (**9.168b**) compound.[1478]

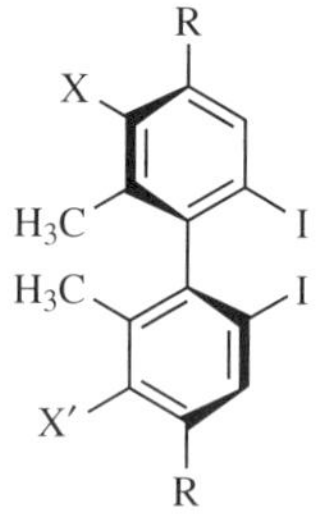

	a	b	c	d	e	f	g	h	i
R	H	CH_3	CH_3	CH_3	CH_3	CH_3	CH_3	CH_3	CH_3
X	H	H	OCH_3	Cl	H_2	$N(CH_3)_2$	OH	SH	SH
X'	H	H	OCH_3	Cl	H_2	$N(CH_3)_2$	OH	SH	H

9.168

Nitriles bind to $Rh_2(MTPA)_4$, although it is not clear whether it is through the lone pair of the nitrogen atom or π-orbital of the nitrile group. Even though nitriles exhibited only moderate shifts in the ^1H NMR spectrum with $Rh_2(MTPA)_4$, clear enantiomeric discrimination was often observed.[1479] As with other substrates of $Rh_2(MTPA)_4$, the resonances of nuclei that were somewhat remote from the cyano group exhibited greater chiral discrimination, presumably because they extended closer toward the MTPA ligands.

$Rh_2(MTPA)_4$ was effective for the enantiomeric analysis of mono- (**9.169**) and disubstituted oxirane derivatives.[1480] Annelated xanthine derivatives such as **9.170** bound to $Rh_2(MTPA)_4$ and exhibited nonequivalence in the resulting ^1H and ^{13}C NMR spectra.[1481,1482] The xanthines bound in a side-on manner involving π-electrons of the imidazole ring.[1482]

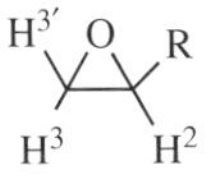

9.169

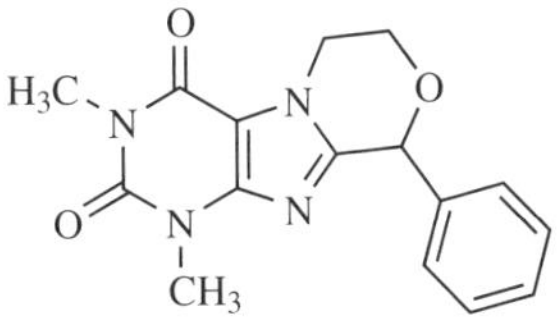

9.170

The utility of Rh$_2$(MTPA)$_4$ in producing nonequivalence in the ^{1}H and ^{13}C NMR spectra of chiral mesoionic oxatriazoles (**9.171**), thiatriazoles (**9.172**), and tetraazoles (**9.173**) has been demonstrated.[1483] The optimal rhodium–substrate ratio was 1:1, and the resonances of nuclei further from the binding site tended to have larger chiral recognition because they had greater interaction with the MTPA ligands.

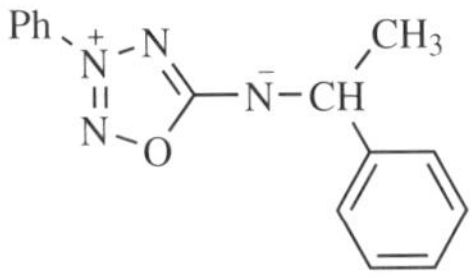

9.171

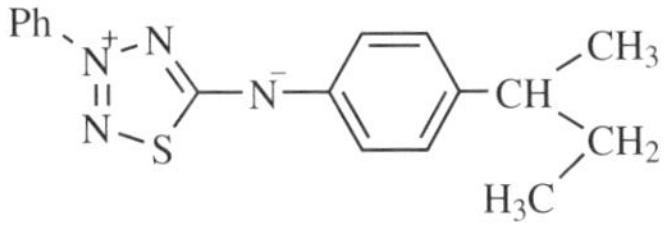

9.172

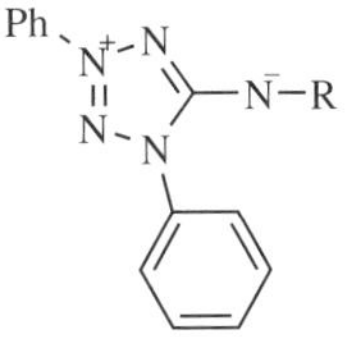

9.173

Whereas rhodium generally exhibits strong binding to soft Lewis bases, more recently, compounds that are hard Lewis bases have been shown to weakly bind to Rh$_2$(MTPA)$_4$. Enantiomeric discrimination was observed in the resulting NMR spectrum.[1484,1485] A series of chlorinated N-acylalanine methyl esters (metalaxyl derivatives) (**9.174**) bound to rhodium through either the amide or the ester carbonyl oxygen, depending on the degree of steric hindrance about the amide site.[1484] In most prior studies with Rh$_2$(MTPA)$_4$, a drop of acetone-d_6 was needed to aid solubility, but addition of acetone in this case significantly reduced or eliminated the chiral discrimination. Nonequivalence was observed in the spectra of the more dilute solutions in chloroform-d. Rh$_2$(MTPA)$_4$ was also an effective chiral discriminating agent for aliphatic ethers including **9.175**, acyclic dialkyl ethers, tetrahydrofurans, and tetrahydropyrans.[1485] Chiral recognition was observed in the ^{1}H and ^{13}C NMR spectra.

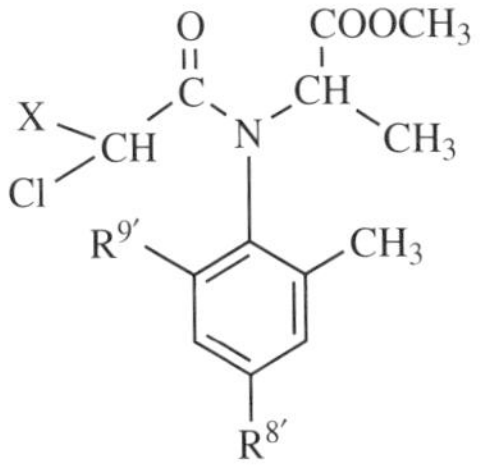

	1	2	3	4
X	H	H	Cl	Cl
R$^{8'}$	H	CH$_3$	Cl	CH$_3$
R$^{9'}$	CH$_3$	H	CH$_3$	H

9.175

9.174

9.3.3.2. Miscellaneous Rhodium Complexes Compound **9.176** was shown to be effective for the enantiomeric discrimination of a variety of aromatic compounds.[1486] Addition of methanol-d_4 to **9.176** rapidly hydrogenated the norbornadiene ligand, displacing it from the rhodium. Chiral aromatic compounds then displaced the methanol solvate. The aromatic ligand was in slow exchange, and nonequivalent signals were observed in the ^{31}P NMR spectrum of the two diastereomeric complexes. The system worked for compounds such as 2-phenylbutane, 1-phenylethanol, 1-(1-naphthyl)ethanol, 2-methylindoline, *N*-acetylphenylalanine, and 1,2-cyclohexylphenylethane. It was not effective with 1-phenylethylamine since the amine moiety preferentially bonded to the rhodium over the phenyl ring. Compounds in which the chiral center was α or β to the phenyl ring exhibited chiral discrimination in the ^{31}P NMR spectrum

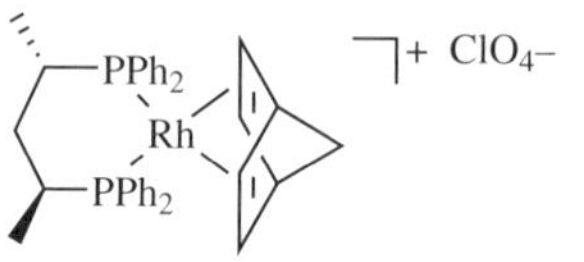

9.176

The ability of a chiral rhodium porphyrin complex with an αβαβ-tetramethylchiroporphyrin (**9.177**) to produce chiral recognition in the NMR spectra of compounds such as 2-aminopropanol[1487], glutamic acid dimethyl ester and aspartic acid dimethylester has been reported.[1488] The porphyrin ring induced substantial upfield shifts (as high as −5 ppm) in the ^{1}H NMR spectra of bound substrates. For the dimethyl esters of glutamic and aspartic acid, nonequivalence of up to 0.5 ppm was observed in the ^{1}H NMR spectrum.[1488] The rhodium complex was preferable to porphyrin complexes of cobalt since only 1:1 rhodium–substrate adducts were formed. The 2:1 complexes with cobalt led to the formation of a racemic [(*R*,*R*) and (*S*,*S*)] and *meso* [(*R*,*S*) and (*S*,*R*)] pair and produced a more complicated NMR spectrum. One drawback to the use of rhodium over cobalt was that over time, the system gradually achieved an equilibrium that favored the complexation of one isomer over the other. For 2-aminopropanol, this enrichment occurred over several days.[1487] For aspartic acid, it occurred over a matter of several minutes.[1488]

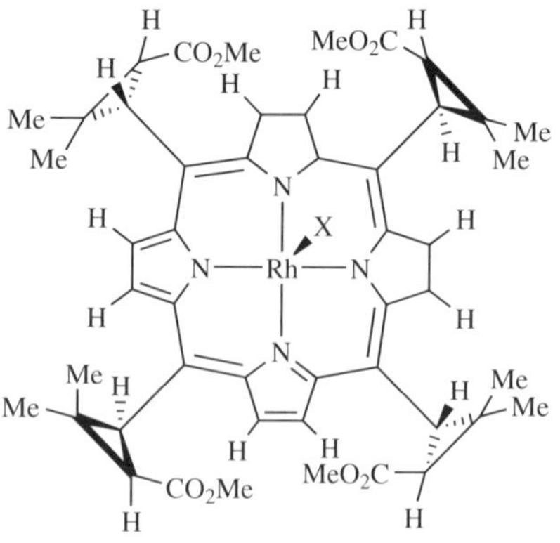

9.177

9.3.4. Cobalt Complexes

9.3.4.1. Complexes with Porphyrin Ligands The ability of a Co(III)-*meso*-tetraphenylporphyrin (**9.178**) to cause chiral recognition in the ^{1}H and ^{13}C NMR spectra in chloroform-*d* and dimethylsulfoxide-d_6 of nitrogenous bases such as 1-phenylethylamine (PEA), levamisole (**9.179**), and tetramisole has been described.[1489] **9.178** is achiral, but the binding of two amines to the porphyrin complex causes the enantiomeric [(*R,R*) and (*S,S*)] and pseudochiral [(*S,R*) and (*R,S*)] *exo* complexes. A doubling of the resonances occurred, and the intensities were related to the optical purity of the amine.

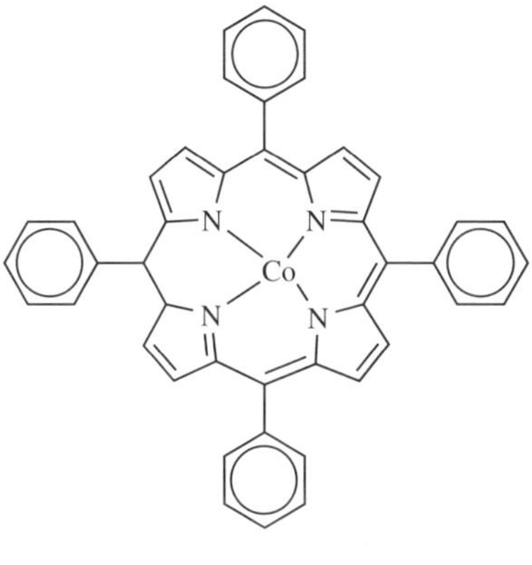

9.178

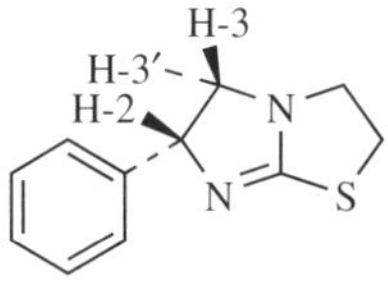

9.179

Binding two optically pure ligands such as L-PEA, L-phenylalanine methyl ester, L-phenylalanine benzyl ester, L-serine methyl ester to the Co(III) complex with a deuteroporphyrin dimethyl ester (**9.180**) unequivocally demonstrated the planar chirality of porphyrin complexes of this type.[1490] Protons of the two ligands in the complexes are rendered diastereotopic because of the planar chirality of the porphyrin ring and exhibited different chemical shifts.

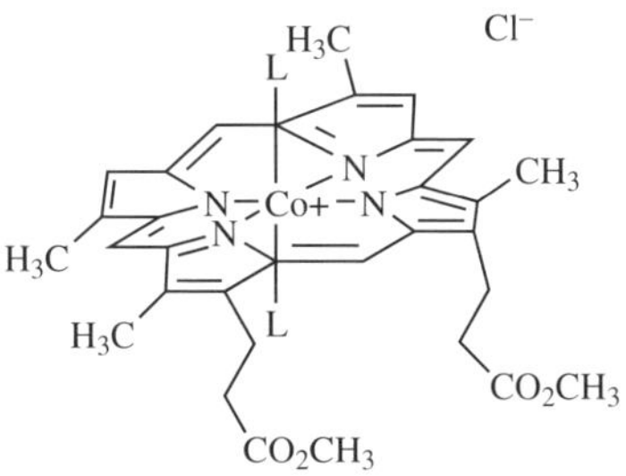

9.180

The Co(II) complex of a biconcave porphyrin (**9.181**) caused a doubling of resonances in the ^{1}H NMR spectrum of the dinitrile ligand **9.182**.[1491] The bridgehead proton resonance of **9.182** was split in the presence of **9.181** in benzene-d_6.

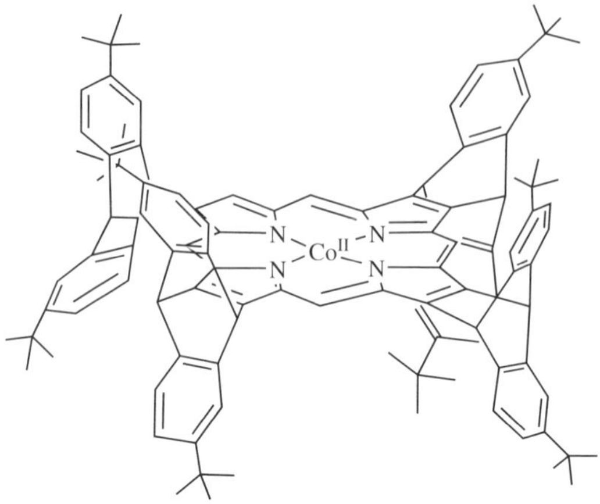

9.181

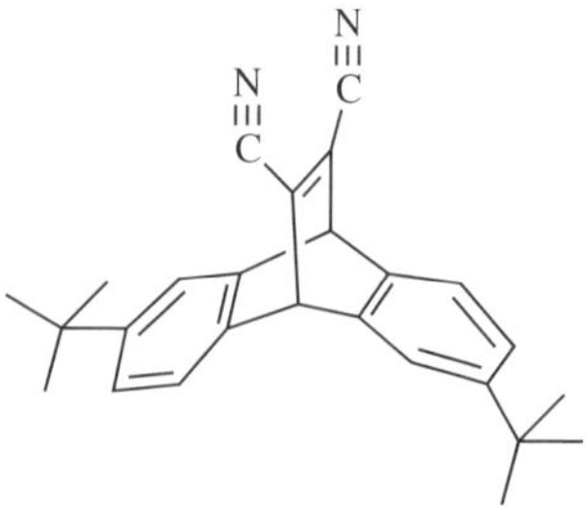

9.182

The chloro cobalt(III) complex with the αβαβ-tetramethylchiroporphyrin ligand (**9.177**) caused enantiomeric discrimination in the NMR spectra of a variety of chiral compounds.[1492–1497] Racemic primary amines such as 2-butylamine, 2-pentylamine, and 3-heptylamine bound in a 2:1 complex and three resonances of areas 1:1:2 were observed for the (R,R)-, (S,S)-, and (S,R)/(R,S)-complexes.[1492] The system was also effective for analyzing the optical purity of compounds such as 1-aminoindane, α-amino-γ-butyrolactone, 1-PEA, 2-aminomethoxypropane, 1-(1-naphthyl)ethylamine, 2-aziridine carboxylate (**9.183**), and 2-aziridine methanol.[1495] These amines underwent irreversible reactions to form bis-adducts with no detectable kinetic resolution. Chemical shifts of up to −6 ppm in the NMR spectrum were observed because of shielding from the porphyrin ring. The areas of the three resonances for the (R,R)-, (S,S)-, and (R,S)/(S,R)-isomers were used to determine optical purity.

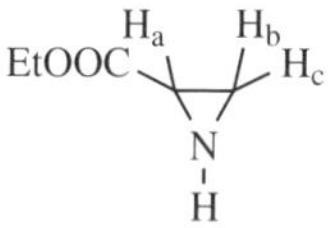

9.183

β-Amino alcohols such as pyrrolidinemethanol (**9.184**) and 2-amino-1-propanol exhibited nonequivalence of 0.1–0.3 ppm for some resonances in the ^{1}H NMR spectrum in the presence of **9.177**.[1493] If a stoichiometric excess of a racemic mixture of the β-aminoalcohols was added (3:1 substrate–cobalt ratio), the initial spectrum showed a pattern of peaks for the (R,R)-, (S,S)-, and (R,S)/(S,R)-isomers consistent with 1:1 binding of both enantiomers of the substrate. Over a period of time the intensity of the resonances changed to reflect preferential binding of the (R)-enantiomer. Therefore, a 2:1 stoichiometric ratio of substrate-to-cobalt should be used with these systems when determining optical purity.

9.184

Amino acid methyl esters bound to the cobalt in complexes with **9.177** in chloroform-d and nonequivalence of up to 0.5 ppm was observed in the ^{1}H NMR spectrum.[1495,1496] There was no detectable kinetic resolution with the amino acid methyl esters. The possibility of using **9.177** to analyze mixtures of several amino acid methyl esters was explored.[1497] For example, the ^{1}H NMR spectrum of a mixture of (S)-Leu, (S)-Asp, (R)-Asp, and (S)-Glu was analyzed. The simultaneous binding of two different amino acids to the cobalt did complicate the analysis. Using ^{1}H NMR one-dimensional TOCSY experiments of the mixture, it was possible to identify and assign the peaks to the different combinations and determine the concentration of each amino acid methyl ester in the mixture.

9.3.4.2. Analysis of Geometrical Isomers   As far back as 1967, the utility of NMR for analysis of chiral cobalt complexes was recognized for the (−)propylene-1,2-diamine (pn) dioxalato (C_2O_4) complex of cobalt [Co(pn)(C_2O_4)$_2$].[1498] The methyl resonance of the two isomers was distinct and based on the relationship of the methyl group to the pseudo C_3 axis, and the chemical shift correlated with the absolute configuration.

Metal complexes with polydentate ligands form various geometrical isomers depending on the spatial orientation of substituent groups on the ligand. Some of these geometrical isomers are chiral. NMR spectroscopy of cobalt complexes has been a useful tool in examining aspects of these geometrical isomers. For example, the bis Co(III) complex with the tridentate ligand ethylenediamine-N-acetate ($NH_2CH_2CH_2NHCO_2^-$)(edma) has six geometrical isomers, five of which are chiral.[1499] The ligand can be stereospecifically deuterated in deuterium oxide such that the (R)-configuration is obtained. The resulting ^{1}H NMR spectrum with the deuterated ligand showed different chemical shifts for resonances of the different isomers that were used to assign the absolute configurations.

The same methodology has been applied to the analysis of other cobalt complexes as well. The chiral isomers of the Co(III) complex μ-fac[Co(ida)Cs)(mdien)], in which ida is a selectively deuterated triidentate iminodiacetate ligand ((S)-HOOCCH$_2$NHCHD-COOH) and mdien (3-methyl-3-azapentane-1,5-diamine − $CH_3N(CH_2CH_2NH_2)_2$) a tridentate ligand, had distinct resonances that were assigned in the ^{1}H NMR spectrum.[1500] NOE data was used to confirm the assignments. Other complexes analyzed by the same method of selective deuteration include [Co(edma)(dien)]$^{2+}$ and [Co(edma) (mdien)] (dien = 3-azapentane-1,5-diamine) through the incorporation of a monodeuterated glycinate ligand,[1501] Co(mida)(edma) [mida = N-methyliminodiacetato − $CH_3N(CH_2COOH)_2$),[1502] and Co(edda)(gly) [edda = ethylenediamine-N,N'diacetate − ($-CH_2NHCH_2CO_2^-$) - a tetradentate ligand]. The edda ligand of the complex Co(edda)(en) (en = ethylenediamine) was selectively deuterated in deuterium oxide before preparation of the Co(edda)(gly) complex for analysis.[1503] ^{1}H and ^{13}C NMR spectra were used in the analysis for several of the complexes. Similarly, the di-2,4-pentanedioate 2-aminoethyl-n-butylphenylphosphine (**9.114**) complex of cobalt(III) produced a pair of diastereomeric (Δ and Λ) isomers that had distinct resonances in the ^{1}H NMR spectrum.[1417]

The diastereomers of the bis-dipeptide cobalt complex with L-phe-gly [Co(L-phe-gly)$_2$] have differences in the ^{1}H NMR spectrum that were used to assign the absolute configuration.[1504] Ring currents from the phenyl ring of the phenylalanine residue caused differential shielding that was used in the analysis.

The absolute configurations of Co(III) complexes with the tetradentate ethylene-bis-(S)-phenylalanine [(S,S)-EBPhe] (**9.185**) or ethylenebis-(S)-valine [(S,S)-EBV] ligands and an ethylenediamine or 1,10-phenanthroline ligand were assigned on the basis of the analysis of the ^{1}H and ^{13}C NMR spectra.[1505] The optically active (S,S)-EBPhe and (S,S)-EBV bound stereoselectively because only Δ diastereomers were formed.

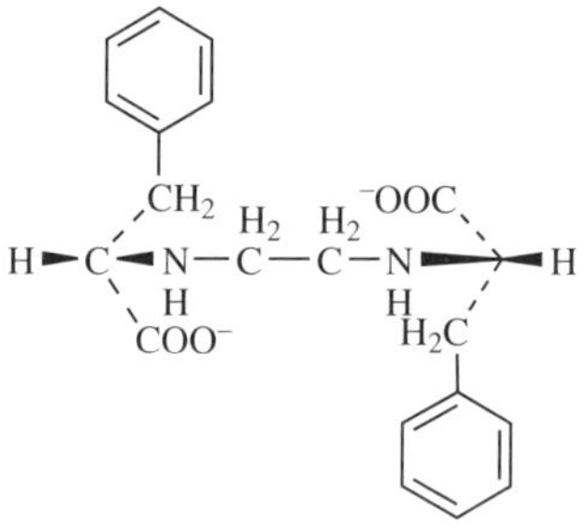

9.185

Cobalt has been used as a template for determining the absolute configurations of the phosphorus atoms in diphosphates as shown in Figure 9.11.[1506] A single ^{18}O atom was selectively incorporated into the diphosphate. The cobalt complexes of the S_P and

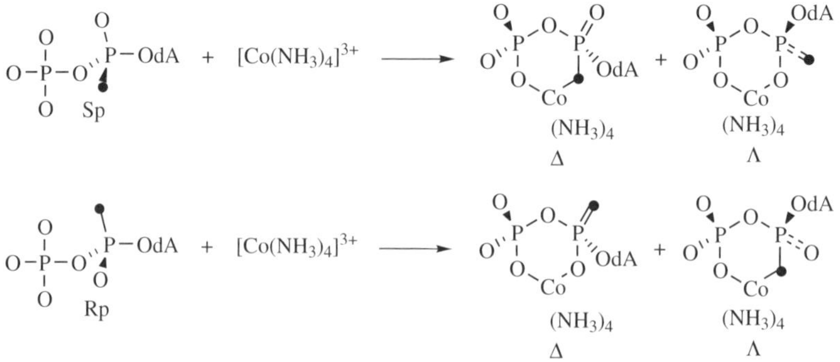

Figure 9.11. Configurational analysis of diphosphates on the basis of their cobalt complexes.

R_P diphosphates are in the Δ or Λ form, and ^{18}O perturbations in the ^{31}P NMR spectrum correlated with the absolute stereochemistry.

The Co(II) complex with adenosine-5′-triphosphate (ATP) was shown to function as an aqueous chiral shift reagent for L- and D-norepinephrines.[1507] ATP by itself associated with norepinephrine but did not provide any chiral discrimination in the ^1H NMR spectrum. The Co(II) did not hinder the association of ATP and norepinephrine and the dipolar shifts from the Co(II) caused nonequivalent resonances in the spectrum. The Co(III) complex with the tetradentate Schiff base ligand prepared from (R,R)-1,2-cyclohexanediamine and salicylaldehyde has been used as a chiral NMR shift reagent for the analysis of amino acids.[1508] The D-enantiomer generally exhibited greater association with the cobalt complex.

Another method of using Co(II) to determine the optical purity of amino acids is to prepare mixed ligand complexes.[1509] Preparation of the Co(II) complex with an enantiomerically pure amino acid allowed chiral discrimination of a racemic mixture of a second amino acid. The paramagnetic cobalt(II) ion induced differential shifts in the ^1H NMR spectra of the enantiomers.

A theoretical treatment of labile coordination of chiral ligands to a metal ion has been conducted by examining the binding of D- and L-histidine to Co(II).[1510] A pair of racemic complexes and a *meso*-complex is formed in such a system. The areas of the peaks for the different species can be used for the determination of optical purity.

9.3.4.3. Analysis of DNA Metal binding to DNA has been studied with the aid of chiral cobalt complexes. The racemic Co(III) complex with 1,10-phenanthroline, Co(phen)$_3^{3+}$, is in fast exchange between two enantiomeric forms. When added to double-stranded oligonucleotides, resonances in the ^1H NMR spectrum of the chiral Co(phen)$_3^{3+}$ complexes were distinct from each other.[1511] Of more interest was that the intensity of the two resonances was not equal as the Co(phen)$_3^{3+}$ rearranged to the enantiomer that preferentially bound to the oligonucleotide. Heating of the solution caused duplex melting and a re-equilibration of the Co(phen)$_3^{3+}$ to a racemic mixture, which indicated that the cobalt complex bound to the duplex but not to the single-thestranded oligonucleotides.

A similar study was done by analyzing the ^{59}Co NMR spectra of the binding of the Co(III) complex of ethylenediamine, [Co(en)$_3^{3+}$], with oligonucleotides.[1512,1513] The ^{59}Co signals exhibited large shifts to lower frequency on binding to the oligonucleotide. The signal that shifted further was also broader, reflecting differences in the relaxation times of the cobalt that occurred on binding. The Δ-Co(en)$_3^{3+}$ bound more tightly than the Λ-Co(en)$_3^{3+}$ to right-handed, guanine-rich DNA, as evidenced by an enhancement in its population. The reverse was true for left-handed DNA. Heating the sample to melt the duplex caused the Co(en)$_3^{3+}$ to rearrange to a racemate, indicating that binding occurred only with the double-stranded DNA.

9.3.5. Zinc Complexes

The chiral zinc porphyrin complex *trans*-5,15-bis(2-hydroxyphenyl)-10-{2,6-bis((methoxycarbonyl)methylphenyl}2,3,17,18-tetramethylporphyrinato]zinc(II)

(**9.186**) was evaluated for the analysis of amino acid methyl and benzyl esters.[1514–1516] **9.186** provided a metal ion for coordination and hydrogen bond donor and acceptor sites.[1516] The potential for steric repulsion existed as well. Chiral recognition was monitored in the ^{1}H NMR resonances of either the amino acid or porphyrin. Certain ^{1}H resonances exhibited nonequivalence from 0.1 to 0.15 ppm.[1514,1516]

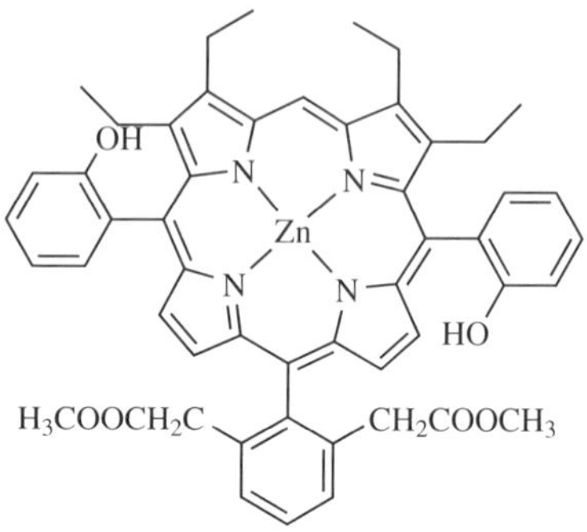

9.186

The chiral recognition properties of bis-zinc porphyrin complexes, or so-called zinc tweezers, have been analyzed through the use of chiral ligands.[1517–1520] Zinc tweezers are effective for studying the chirality of amine ligands using circular dichroism,[1518–1521] and NMR spectroscopy confirmed aspects of the binding of ligands.

Addition of L-histidine methyl and benzyl esters (**9.187**) to a bis-porphyrin zinc tweezer that was prepared from a Troger's base analog (**9.188**) favored binding with the (+)-**9.188** complex over the (−)-**9.188** complex.[1517] Resonances in the ^{1}H NMR spectrum of the histidine ligand had large shifts to lower frequency, with some appearing as high as −5 to −6 ppm. The study demonstrated the selective binding of amino acids to zinc tweezer complexes.

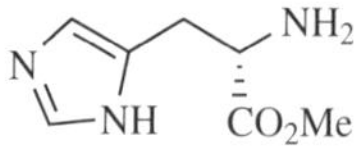

9.187

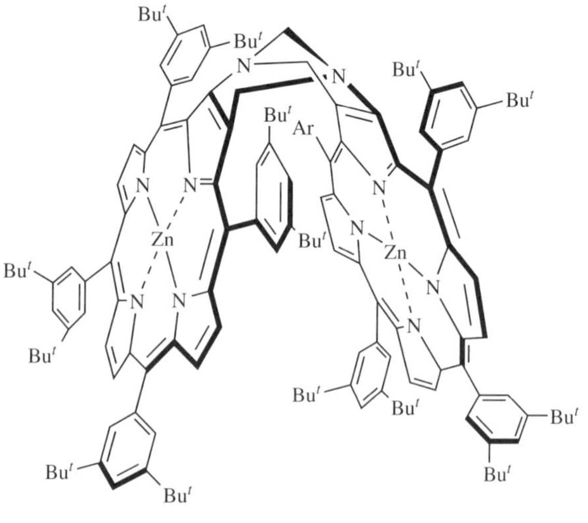

9.188

The inversion of chirality of a bis(zinc porphyrin) compound that occurred on the binding of optically pure amines was demonstrated by an analysis using ^{1}H NMR spectroscopy.[1518–1520] The spectra of the bis(zinc porphyrin) complex with either (R)- or (S)-1-phenylethylamine were identical. Monitoring of the NMR spectrum as the bis(zinc porphyrin)-(R)-amine was titrated with the (S)-amine showed an inversion from the left-handed (anti-(L)) to right-handed (anti-(R)) helice of the porphyrin, as evidenced by systematic alterations of the NMR spectrum.[1518] The spectrum with a racemic mixture of the amine was different than with either of the pure enantiomers and represented a 50:50 mixture of the two helical forms. The changes in the NMR spectrum occurred until the point at which a racemic mixture of the amine was present, and then reversed as the (S)-amine was in excess.

Zinc porphyrin tweezer compounds have been used as chiral NMR shift reagents, especially for bifunctional substrates that bound simultaneously to both zinc ions.[1522] The effectiveness of a zinc tweezer compound (**9.189**) in causing enantiomeric discrimination in the NMR spectra of diamines such as 1,2-diaminocyclohexane, an aziridine (**9.190**), and an isoxazoline (**9.191**) has been demonstrated. Shielding from the porphyrin ring caused exceptionally large shifts with resonances appearing at −7 ppm in some ^{1}H NMR spectra in some cases. Diamines have the potential to bridge the two zinc centers leading to exceptionally large binding constants.

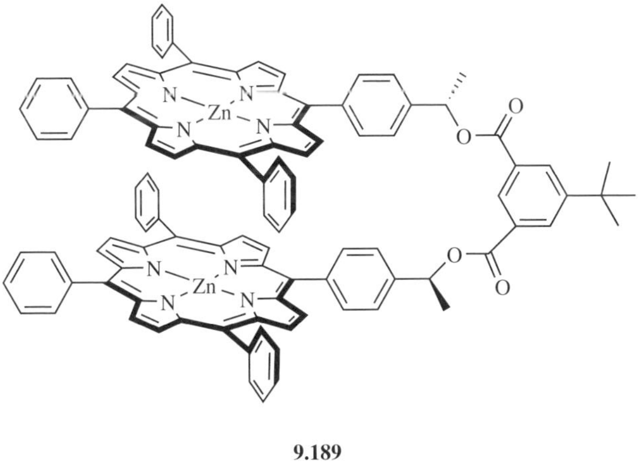

9.189

9.190

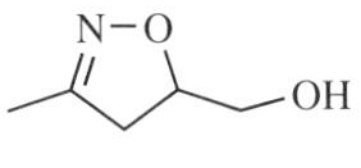

9.191

The homochiral helical dimer of a zinc complex of a linear tetapyrrole (**9.192**) acted as a host that caused enantiomeric discrimination in the NMR spectra of amino acid methyl esters of phenylalanine, leucine, and aspartic acid.[1523]

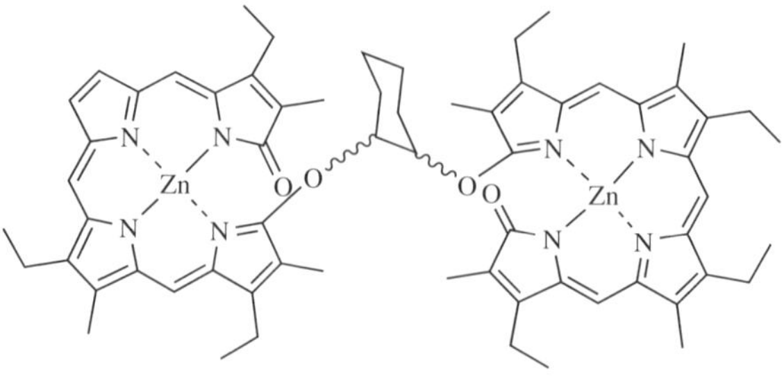

9.192

Zinc chelates with **9.193** formed propeller-like complexes that were effective chiral shift reagents for sulfoxides and sulfides.[1524] The analysis was done in acetone-d_6 and the chiral recognition was typically greater with the larger **9.193** than the corresponding ligand with a pyridyl group.

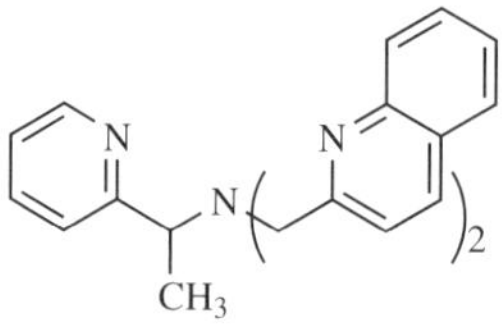

9.193

9.3.6. Ruthenium Complexes

A ruthenium picket fence porphyrin (**9.194**) with α-methoxy-α-trifluoromethylphenylacetic acid (MTPA) groups has been used for chiral recognition studies.[1525–1528] The symmetrical αβαβ-derivative was selected because it had fewer ^{19}F signals. Racemic phosphines bound to **9.194** to form 2:1 substrate–ruthenium complexes [(R,R), (S,S), and (R,S)/(S,R)] of areas 1:1:2 for a racemic mixture. No kinetic resolution was observed.[1527] The differences in the ^{19}F signals of the MTPA units were noted for the diastereomeric complexes. For benzylmethylphenylphosphine, there was a binding preference for only one diastereomer.[1525] This facilitated the analysis of the chiral recognition properties of the porphyrin when used as a catalyst.

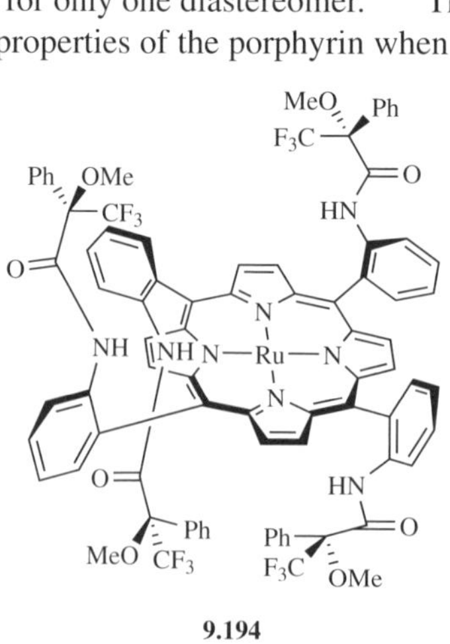

9.194

9.194 was also effective for the analysis of amino acid methyl esters of valine, leucine, alanine, *tert*-leucine, and tyrosine.[1526,1527] Nonequivalence in the ^{1}H and ^{19}F NMR spectra of the diastereomeric complexes was observed. When one enantiomer of the racemic mixture of the amino acid methyl ester bound preferably over the other, integration of the ^{19}F signals indicated the preference.

The ruthenium complex with **9.195** has been used to discriminate chiral isocyanides.[1528,1529] The porphyrin consisted of eight D-α-β-isopropylideneglycerol-γ-tosylate residues linked to the *ortho-meso* phenyl positions via ether bonds. With 1-phenylethylisocyanide, the (*R*)-enantiomer bound preferably over the (*S*)-enantiomer. The (*R,R*) signal was assigned by measuring the shifts with only the (*R*)-enantiomer. Integration of the intensity of the peaks for the (*S,S*), (*R,R*), and (*S,R*)/(*R,S*) indicated the binding preference in racemic mixtures of the isocyanides.

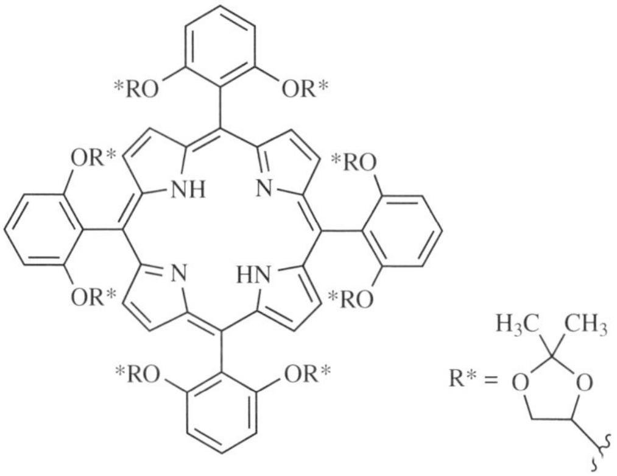

9.195

A ruthenium(II) chiroporphyrin (**9.177**) functioned as an enantioselective receptor for chiral aliphatic alcohols such as 2-octanol or 2-butanol.[1530] The sample needed to be cooled to −70°C to see the enantiomeric recognition, and the (*R*)-enantiomer of the alcohols bound preferentially to **9.177**. A biconcave ruthenium porphyrin (**9.181**) with dioxo-Ru(VI) and Ru(II) was evaluated for its ability to cause chiral discrimination in the ^{1}H NMR spectra of amines, alcohols, carboxylic acids, esters, nitriles, and nonpolar fullerene derivatives.[1531]

A ruthenium(II) complex with 2,2′-bipyridine or 1,10-phenanthroline ligands (L–L) and eilatin (**9.196**) of the form [Ru(L–L)$_2$(eilatin)]$^{2+}$ is a propeller-like molecule that dimerizes in acetonitrile-d_3.[1532] A clear preference was observed for the heterochiral over the homochiral dimer. This was reportedly the first observation of chiral recognition, solely on the basis of chiral stacking.

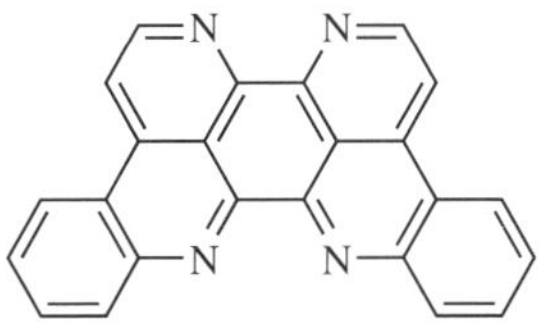

9.196

9.3.7. Nickel Complexes

The compound 1-diphenylphosphine-2-propanethiol (**9.197**) forms a bis-bidentate complex with Ni(II) involving the phosphorus and sulfur atom in a *trans*-arrangement.[1533] Depending on the chirality of the ligands, the nickel complexes can have either a *meso* (*R,S*) or enantiomeric [(*S,S*)/(*R,R*)] form. Discrimination was observed in the ^{31}P NMR spectrum and the area of the peaks indicated the optical purity of the ligand.

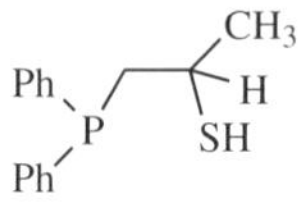

9.197

The [η^3-(+)-(1*R*,5*R*)-pinenyl]nickel bromide dimer forms 1:1 adducts with *t*-butylmethylphenylphosphine and methylphenylisopropylphosphine (**9.198**).[1534] Enantiomeric discrimination was observed in the ^{13}C and ^{31}P NMR spectra, enabling the determination of the optical purity of the phosphine. The extent of discrimination was much greater in the ^{13}C NMR spectra.

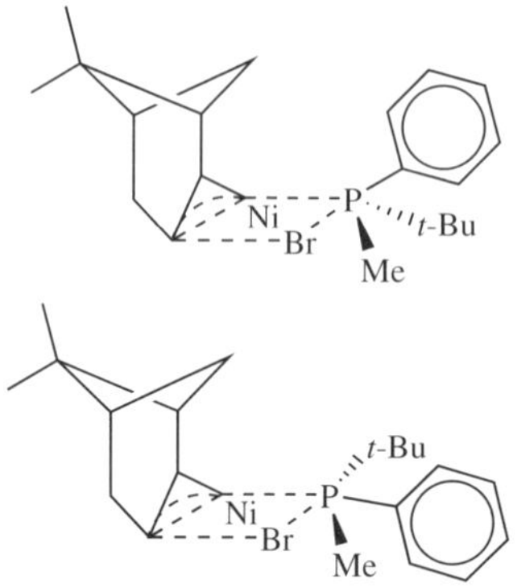

9.198

9.3.8. Copper Complexes

A copper derivative of a vaulted porphyrin with (*S*)-binaphthyl-L-alanine straps (**9.199**) was employed as a surrogate to study the chiral recognition properties of a catalytic Fe(III) complex that promotes olefin epoxidation.[1535] Spin-lattice relaxation times for each of the pure enantiomers of styrene oxide (**9.200**) were measured to determine which enantiomer fitted better into the pocket. Spin-lattice relaxation times exhibit a $1/r^6$ dependency that can be used to determine aspects of the geometry of the substrate–copper complex. The comparative relaxation times for the copper porphyrin were in agreement with the catalytic data for the Fe(III) porphyrin.

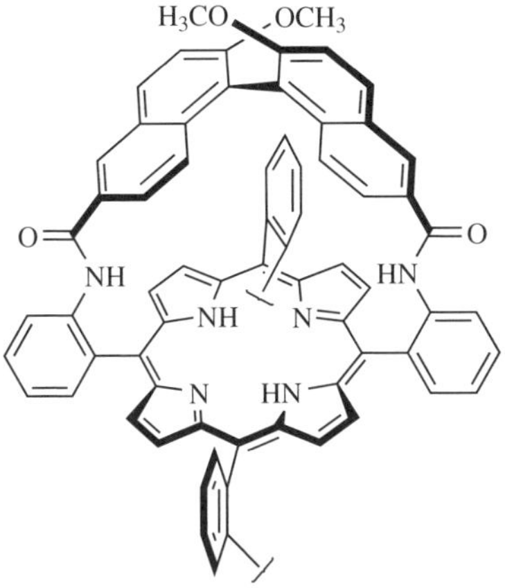

9.199

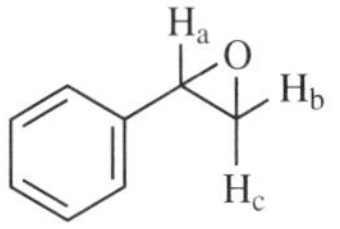

9.200

The binding strength of D- and L-amino acids was studied with a Cu(II) complex with a ligand consisting of a pinene-annulated derivative of 2,2′-bipyridine (**9.201**).[1536] The resonance of the proton of the α-carbon on the D-amino acid consistently exhibited greater broadening than observed for the resonance of the L-amino acid.

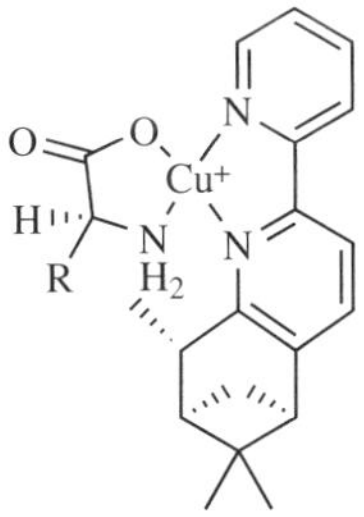

9.201

9.3.9. Silver Complexes

Silver complexes of *N,N′*-bis(mesitylmethyl)-1,2-diphenyl-1,2-ethanediamine (**9.202**) and the corresponding cyclohexyl-1,2-diamine derivative were evaluated as chiral NMR discriminating agents for olefins in chloroform-*d*, methylene chloride-d_2, and acetone-d_6.[1537] The compounds were effective for a range of olefins with a chiral center α to the double bond. Shielding from the mesityl rings caused large shifts to lower frequencies in the ^{1}H NMR spectrum of the olefin. The extent of enantiomeric discrimination was lower in acetone-d_6 than in chloroform-*d*. The silver salts could be prepared directly in the NMR tube by adding the amine ligand and the silver salt. The effect of the anion on the chiral recognition varied in the order triflate > nitrate > trifluoroacetate.

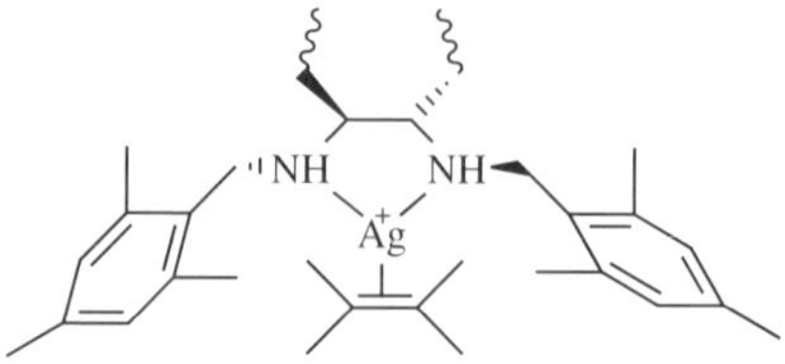

9.202

9.3.10. Tin Complexes

The ^{119}Sn NMR spectrum of tin compounds has been used to measure diastereomeric ratios of carboxylic acids.[1538] Reaction of Pb$_2$SnBr$_2$ with optically pure 2-phenylbutyric acid formed a complex with two bound carboxylate groups and one ^{119}Sn signal. The same complex with racemic 2-phenylbutyric acid had two ^{119}Sn signals, one for the (R,R) and (S,S) complexes and the other for the $(R,S)/(S,R)$ complexes. The ratio of the signals could be used to measure the extent of asymmetric induction by a tin catalyst.

A method for the analysis of the enantiomeric purity of alkyl chlorides involving the preparation of diastereomeric tetraorganostannanes and ^{119}Sn NMR spectroscopy has been described.[1539] The reaction of a racemic chiral alkyl halide with either diphenyltin dichloride (Ph$_2$SnCl$_2$) or dibutyltin dichloride produced a tetraorganosilane (Ph$_2$SnR*$_2$) that was a mix of the *meso* [$(R,S)/(S,R)$] or racemic [$(S,S)/(R,R)$] configuration. If the alkyl halide was racemic, two signals of equal area were observed in the ^{119}Sn NMR spectrum, and the ratio of the peaks could be used to determine optical purity. Nonequivalence was also observed in the ^{1}H and ^{13}C NMR spectra.

The enantiomeric purity of 1,2-diols can be determined by analysis of their corresponding 2,2-dibutyl-1,3,2-dioxastannolanes.[1540] The stannolanes form a dimer in chloroform-*d* as shown in Figure 9.12. The ^{1}H NMR spectra were generally too complex, but chiral discrimination in the ^{13}C NMR spectra was suitable for the analysis.

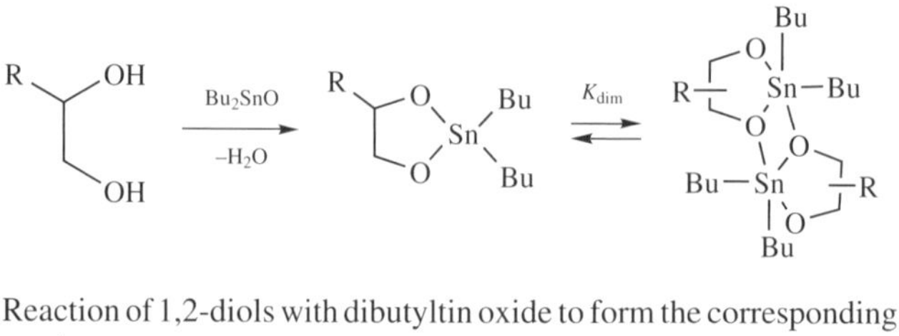

Figure 9.12. Reaction of 1,2-diols with dibutyltin oxide to form the corresponding 2,2-dibutyl-1,3,2-dioxastannolanes.

9.3.11. Titanium Complexes

The titanium dimer with (R,R)-diisopropyltartrate (DIPT) and isopropoxide groups (O^{i}Pr) [Ti$_2$DIPT$_2$(O^{i}Pr)$_4$] forms complexes with chiral α,N-disubstituted and N,N,α-trisubstituted β-amino alcohols.[1541] The resulting β-aminoalcohol complexes showed two sets of tartrate signals from the two diastereomeric complexes. The relative magnitude of the shifts was also used to assign the absolute configuration of the β-amino alcohol and study the mechanism of action of the titanium reagent.

9.3.12. Tungsten Complexes

The Ce(III) complex of the ammonium salt of a polyoxytungstate cluster [α1[P$_2$W$_{17}$O$_{61}$]$^{10-}$] was shown to be chiral through the addition of an optically pure amino acid.[1542] The carboxylate group of the optically pure amino acid bound to the Ce(III) and caused a splitting of the ^{31}P resonance due to the diastereomeric pairs of the monomeric polyoxometal amino acid complexes. Addition of achiral amino acids or racemic mixtures of amino acids did not cause a splitting of the ^{31}P resonances.

Similarly, the organotin-substituted α1-polyoxotungstate in which 1,2-diphenyl-1,2-diaminoethane was substituted onto the organotin moiety created diastereomeric species resulting in a splitting of the ^{31}P resonances.[1543] The two diastereomeric cluster species were distinguishable in acetonitrile-d_3, dimethylsulfoxide-d_6, and acetone-d_6.

9.3.13. Molybdenum Complexes

The molybdenum complex [C$_5$H$_5$Mo(CO)$_2$NN]PF$_6$ (**9.203**) in which NN is the chiral pyridine-2-carbaldimine ligand derived from pyridine-2-carbaldehyde and (S)-1-phenylethylamine exists as a pair of diastereomers that differ in the configuration about the molybdenum.[1544] The existence of the diastereomers was confirmed by the presence of two well-separated signals ($\Delta\Delta\delta = 14$ ppm) in the ^{95}Mo NMR spectrum. The ^{1}H NMR spectrum also showed two sets of peaks for the diastereomers.

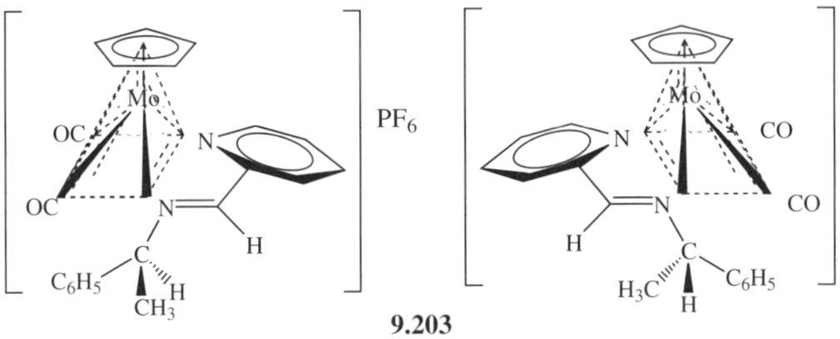

9.203

9.3.14. Mercury Complexes

The bis[O-n-butyl-P-phenylphosphonito] complex of mercury(I) (**9.204**) has two pairs of diastereomers because of the chiral phosphine ligand.[1545] The ^{199}Hg NMR spectrum in toluene or tetrahydrofuran exhibited two peaks for the two diastereomers.

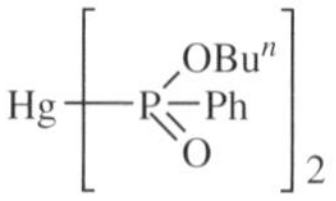

9.204

9.3.15. Osmium Complexes

Osmate(VI) esters of glycols with *trans-N,N,N′,N′*-tetramethyl-1,2-cyclohexanediamine (**9.205**) show enantiomerically discriminated ^{1}H NMR spectra of the glycol units.[1546] ^{1}H NMR shifts in the methyl resonances of the amine ligands of model glycols with the (*R,R*)-osmium complex and (*S,S*)-osmium complex were measured and used to assign the absolute configuration of unknown glycols obtained from the alkaline hydrolysis of two pyrrolizidine alkaloids.

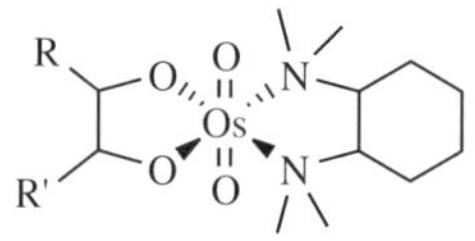

9.205

9.3.16. Uranium Complexes

The uranium complex of the salophen ligand (**9.206**) is chiral because the ligand deviates from planarity.[1547] Addition of optically pure 1-phenylethylamine, methyl-*p*-tolyl sulfoxide, or *N,N*-dimethyl-1-α-methylbenzyl ammonium chloride caused a splitting of signals for the racemic uranium complex. Addition of the racemate of the substrate caused only one set of signals for the uranium complex that were shifted from their original position. A set of equations were derived in which the shift data could be used to determine the association constants for the (*R*)- and (*S*)-isomers. While demonstrated on this uranium complex, the method of determining association constants is generally applicable to chiral host/guest systems.

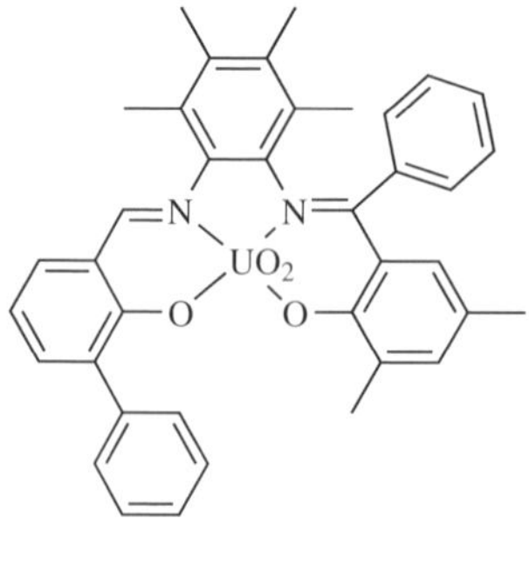

9.206

10

CHIRAL NMR DISCRIMINATION WITH HIGHLY ORDERED SYSTEMS

10.1. INTRODUCTION

The analysis of chiral compounds in liquid crystals, the most common of which has been employed is poly(γ-benzyl-L-glutamate) (PBLG), represents one of the most versatile techniques for determining the enantiomeric purity by NMR spectroscopy. Unlike the other systems described in earlier chapters, which required a specific interaction between the substrate and chiral reagent, studies with liquid crystals are only dependent on the enantiomers having different packing orders relative to the applied magnetic field. The liquid crystal itself adopts a particular orientation in a magnetic field. A pair of enantiomers dissolved in a liquid crystal and a cosolvent are likely to have different orientations relative to the magnetic field. Unlike solution-state studies where rapid tumbling occurs, the different orientations of the enantiomers in a liquid crystal can lead to differences in dipolar coupling and quadrupolar splitting. The doubling of peaks that occurs can be used to determine optical purity. Since no specific interactions are needed between the substrate and liquid crystal, essentially any compound is a candidate for analysis. This even includes aliphatic hydrocarbons.

Solid-state NMR studies of chiral recognition also depend on differences in the orientations of the two enantiomers in a solid lattice. Solid-state techniques have been far less explored, though, than the use of liquid crystals. Limited studies using micelles, ionic liquids, and polymers for chiral NMR discrimination have been

---

Discrimination of Chiral Compounds Using NMR Spectroscopy, by Thomas J. Wenzel
Copyright © 2007 John Wiley & Sons, Inc.

performed. In the case of a series of carbamate derivatives of cellulose polymers, these studies were aimed more at characterizing the nature of polymer-substrate interactions than determining optical purity, although nonequivalence in the NMR spectra of a number of substrates was noted.

A relatively recent set of findings involves the study of achiral compounds that aggregate in solution to form assemblies with chiral pockets. While these systems do not yet appear practical for routine NMR determination of optical purity or absolute configuration, they will no doubt be further explored in attempts to understand and exploit systems that assemble through noncovalent interactions.

10.2. LIQUID CRYSTALS

10.2.1. Introduction

Compounds capable of forming nematic liquid crystals become oriented in a magnetic field. Chiral liquid crystals form a cholesteric or helical structure and have the potential to be used for chiral discrimination. Unlike solution phase measurements, where rapid tumbling occurs, the ordered environment of a chiral liquid crystal presents additional opportunities, and at times complexities, when attempting the analysis of enantiomers. Generally the two enantiomers exhibit a different molecular order in the liquid crystal.

Several detailed analyses of the mechanism responsible for enantiomeric discrimination of compounds dissolved in poly(γ-benzyl-L-glutamate) (PBLG), the most common liquid crystal used for chiral NMR analysis, have been undertaken.[1548–1552] These typically involved a complete analysis of the ^{1}H and ^{13}C NMR spectra so as to obtain molecular ordering parameters. The analysis of the ^{1}H and ^{13}C NMR spectra of β-(trichloromethyl)-β-propiolactone (**10.1**) in PBLG-methylene chloride indicated that differences in the average molecular orientation of the two enantiomers were responsible for the chiral discrimination.[1549] The same conclusion was reached for the analysis of 3-butyn-2-ol in PBLG-chloroform.[1550]

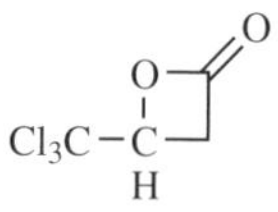

$$Cl_3C-\underset{H}{\overset{O}{\underset{|}{C}}}-O-\overset{O}{C}$$

10.1

A theoretical analysis of the full molecular order matrix for ethanol in PBLG confirmed the loss of symmetry for the prochiral groups.[1551] The sign and magnitude of the geminal and scalar couplings of the nonequivalent couplings of the two enantiotopic protons of ethanol could be determined. This procedure was then tested on a series of linear alcohols.

A thorough analysis of the orientational order parameter and conformational distribution functions of α-ethylhexanoic acid-d_{15} was undertaken.[1552] In the aggregate, these detailed analyses show that the recognition of enantiomers in oriented liquid crystals occurs because of different packing orders of the two enantiomers due to molecular shape, rather than specific and different intermolecular interactions between the enantiomers and the liquid crystal. The only situation where different packing orders were not the likely cause of chiral recognition occurred for compounds, such as 1-deutero-(2′,3′,4′,5′,6′-pentadeuterophenyl)phenylmethanol ($C_6H_5CD(OH)C_6D_5$), that are chiral by virtue of isotopic substitution.[1548]

The importance of packing order in influencing the chiral recognition has significance in the resulting NMR data. For most chiral solvating and derivatizing agents, the largest distinction tends to occur for nuclei that are close to the stereogenic center. With liquid crystals, the distance from the stereogenic center often has no influence on differences in packing order, such that the resonances of nuclei remote from the stereogenic center often exhibit large enantiomeric discrimination.

The different molecular ordering leads to three possible methods of distinguishing the enantiomers. One is straightforward chemical shift anisotropy that results in a splitting of particular resonances, as occur in solution-state measurements with chiral solvating and derivatizing agents. This is often the least useful method with liquid crystals because the differences in chemical shifts between enantiomers are often quite small.

A second involves differences in the dipolar coupling constants for nuclei of the enantiomers. For example, in an ordered environment, the ^1H–^{13}C coupling constant for a particular CH bond is likely to be different for a pair of enantiomers given their different orientation relative to the magnetic field. The ^{13}C signal would then appear in a proton-coupled spectrum as two doublets, both with the same chemical shifts (assuming there is no chemical shift anisotropy), but with different coupling constants. A comparison of the area of the doublets is used to determine optical purity.

The third, and most common method that has been used with liquid crystals, involves recording the NMR spectrum of a quadrupolar nucleus such as deuterium. In solution phase, quadrupolar splitting is not observed because of rapid tumbling. In the ordered environment of a liquid crystal, quadrupolar splitting of a ^2H resonance does occur and the magnitude of this splitting is often different for the two enantiomers because of their different orientation relative to the magnetic field. For a ^2H resonance, quadrupolar splitting causes the signal to appear as a doublet. The spectrum of a particular ^2H resonance for a pair of enantiomers then consists of two doublets (both with the same chemical shift assuming that there is no chemical shift anisotropy) as seen in Figure 10.1 for the spectrum of a mono-deuterated propionic acid.[1548] The areas of the two doublets can be used to determine the optical purity of the sample.

10.2.2. Early Studies with Liquid Crystals

The first demonstration that liquid crystals could be used for chiral NMR discrimination was in 1968 and involved the use of a cholesteryl chloride/cholesteryl myristate mixture. Chemical shift anisotropies of certain resonances in the ^1H NMR spectra of 3,3,3-trichloropropylene oxide and 2,3-epoxypropanal were observed.[1553]

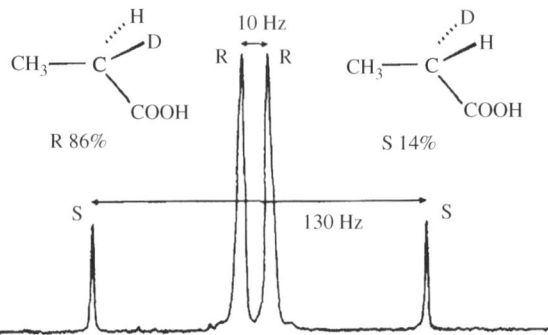

Figure 10.1. Proton-decoupled deuterium NMR spectrum of an (R)-enriched mixture of CH$_3$-CHD-COOH enantiomers (72% ee) dissolved in PBLG/CH$_2$Cl$_2$ liquid crystal at T = 300 K. Reproduced with permission from American Chemical Society.[1548]

A mixed liquid crystal with an active chiral component of optically pure sodium decyl-2-sulfate caused enantiomeric discrimination in the proton NMR spectrum of D/L-alanine.[1554] The ^{1}H methyl resonance of either D or L-alanine in the liquid crystal appeared as a doublet of triplets because of dipolar coupling within the methyl group (triplet) and scalar coupling to the methine group (doublet). A mixture of D/L-alanine showed a slight splitting of the outermost four resonances of the multiplet as shown in Figure 10.2. The center peak of the multiplet did not split because no chemical shift anisotropy was noticeable for the D- and L-enantiomers. The outer ones split because different ordering of the D- and L-enantiomers caused different dipolar ^{1}H–^{1}H coupling.

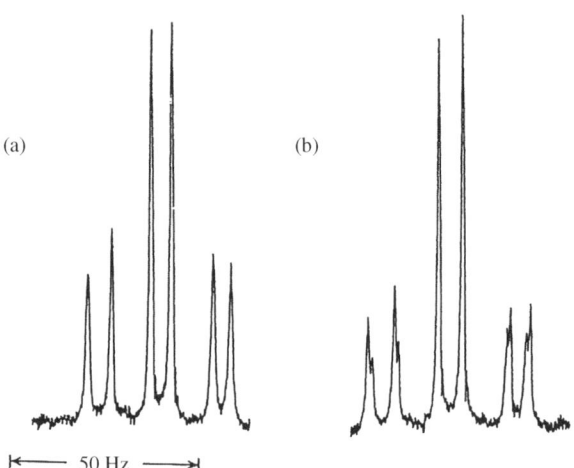

Figure 10.2. (a) The methyl resonance of D-alanine in sodium decyl-2-sulfate. (b) The methyl resonance in a similar phase but containing a mixture of D/L-alanine. The splitting of the outermost peaks of the multiplet occurs because of different dipolar coupling constants for the two enantiomers. Reproduced with permission from the Foreign Policy Research Institute.[1554]

The first use of differences in the quadrupolar splitting of the ^{2}H nucleus involved a study of benzyl-d_2-alcohol in a liquid crystal consisting of polybenzyl glutamate.[1555] Addition of the deuterated alcohol to liquid crystals from either polybenzyl-L-glutamate (PBLG) or polybenzyl-D-glutamate (PBDG) caused the enantiotopic deuterium atoms to become diastereotopic and appear as different resonances with different magnitudes of quadrupolar splitting. Use of a racemic mixture of PBLG/PBDG caused no doubling of the ^{2}H resonances.

The effectiveness of PBLG, poly-γ-ethyl-L-glutamate (PELG), and poly-ε-carbobenzyloxy-L-lysine (PCBLL) at causing enantiomeric discrimination in ^{2}H, ^{1}H, and ^{13}C NMR spectra was compared.[1556] All three were effective chiral recognition agents, and there was not enough difference to consistently recommend one over the others. PBLG has been overwhelmingly used in chiral NMR studies, although there are a few occasions that will be mentioned in subsequent sections in which PCBLL has proven to be more effective than PBLG.

10.2.3. ^{2}H NMR Analysis of Labeled Compounds

Since the first report on the use of ^{2}H NMR spectroscopy of substrates in PBLG, the scope of this system for chiral discrimination and the determination of optical purity has been extensively explored. The most impressive aspect of this system is the array of compound classes that can be analyzed, since the method only requires different ordering of the enantiomers in the liquid crystal and does not require any specific interaction between the substrate and liquid crystal. Initial reports focused on compounds that were deuterium labeled, both to provide adequate sensitivity and to simplify the ^{2}H spectra relative to that observed with natural abundance ^{2}H NMR spectroscopy.

Liquid crystals using PBLG as the chiral element always involve a cosolvent, the most common ones of which are methylene chloride, chloroform, or dimethylformamide (DMF). Sometimes it may be necessary to vary the cosolvents if solubility is an issue.

Incorporation of deuterium into the substrate is often achieved through the use of a suitable deuterium-labeled derivizating agent. Unlike other chiral NMR studies, the derivatizing reagents in these schemes are achiral, since the goal is only to incorporate a deuterium signal into each of the enantiomers. Therefore, the kinetic resolution and racemization that may occur with chiral derivatizing agents is not a concern.

Amino acids[1557] or the half-esters of lipase or esterase hydrolyses of dimethyl succinates (**10.2**)[1558] were esterified with methanol-d_4. The quadupolar splittings of the ^{2}H signal of the deuterated methoxy group were different for the two enantiomers. For neutral amino acid esters, the amine was protected as its benzophenone imine (**10.3**) to avoid decomposition during the NMR analysis.[1557]

Acetyl-d_3 chloride is an effective derivatizing agent for primary and secondary amines[1559,1560] and amino acids.[1560] For the amino acids, it was possible to prepare both the N-CD$_3$ and O-CD$_3$ derivatives and obtain two deuterium signals as probes. The amide derivative of the (R)-enantiomer consistently exhibited larger quadrupolar splittings than that of the (S)-enantiomer, although no consistent trend with absolute configuration was observed for the methoxy resonances.[1560] For the amide derivatives of secondary amines, rotational isomers were often observed. Under conditions of fast rotation, only two doublets were observed in the ^{2}H NMR spectrum of the two enantiomers. Under slow exchange, four doublets occurred in the ^{2}H NMR spectrum.[1559]

Another effective scheme for derivatizing amines, amino acids, and alcohols is to incorporate a perdeuterated benzoyl group into the compound.[1561] The o-, m-, and p-deuterium atoms of the benzoyl group provide three probe signals. The quadrupolar splittings were larger with amines than alcohols. Furthermore, the perdeutero benzoyl derivatives of primary and secondary amines always exhibited larger quadrupolar splitting with the (S)-enantiomer. For amino acids, the *para*-deuterium resonance was the optimal one to monitor and larger quadrupolar splits were consistently noted for the L-enantiomer of the amino acid.[1561]

Carboxylic acids were analyzed by conversion to their benzyl esters using perdeuterated benzyl alcohol.[1562] Using methylene chloride as the cosolvent was not that effective with these carboxylic acids, whereas good results were obtained with DMF. Unlike most systems for chiral discrimination, the stereogenic center does not need to be close to the deuterium label. So long as the two derivatives have different ordering in the liquid crystal, the quadrupolar splittings in the ^{2}H NMR spectrum are likely to be different. For example, the ^{2}H signal of the *para*-position of **10.4** was twelve bonds away from the chiral center but still showed different quadrupolar splitting that could be used to determine optical purity.

10.4

There are other examples of the use of PBLG liquid crystals to distinguish molecules in which the chirality only involves subtle changes to the substituent groups. For the homologous series of secondary alcohols of structure C_mCH(OH)C_n (m = n + 1), enantiomeric discrimination of the C5/C6 analog (6-dodecanol) was observed in the ^{2}H NMR spectrum.[1563] For the C_n–C_{13-n} series, which involves systematically moving the chiral site from the extremity toward the center of the alkyl chain, greater differences in the two quadrupolar splittings were observed the closer the stereogenic center was to the end of the chain.

An analogous study was done on a series of fluoroalkanes of structure C_mCFDC_n ($m = n + 1$).[1564] The ^2H signal appeared as a doublet of doublets and quadrupolar splitting of the ^2H signal was larger than the ^{19}F dipolar coupling. Chiral discrimination was not observed in the spectrum of [5-^2H]-5-fluorodecanes with PBLG, although the use of PCBLL/DMF was effective.

Compounds that are chiral by virtue of isotopic substitution, such as C_6H_5—CHD—OH and C_6H_5—CDOH—C_6D_5, exhibited enantiomerically distinguishable ^2H quadrupolar splitting in PBLG.[1548] The enantiotopic deuterium atoms in CHD$_2$OH are rendered diastereotopic with distinct resonances, as are all of the diastereotopic methylene groups of perdeutero octanol in PBLG.

A racemic mixture of the α,α'-dideuterated diol (**10.5**) exists in (*R,R*)-, (*S,S*)-, and (*R,S*)-(*meso*)-isomers.[1565] The ^2H NMR spectrum of **10.5** in PBLG consisted of four doublets of equivalent area, representing four distinct deuterium atoms. Two of these doublets, each of area two, corresponded to the (*R,R*)- and (*S,S*)-isomers, respectively. The other two doublets resulted from discrimination of the two enantiotopic deuterium atoms of the (*R,S*)-(*meso*)-isomer. Since the (*R,S*)-isomer is twice as probable in a racemate than either the (*R,R*)- or (*S,S*)-isomers, all four doublets were of equal intensity.

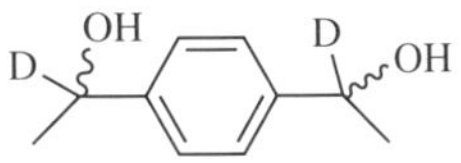

10.5

A similar pattern was observed in the ^2H spectrum of **10.6**.[1565] In this compound, the distinction occurred over two asymmetric carbon atoms that were nine bonds apart, demonstrating the effectiveness of PBLG at discriminating chiral compounds. Similarly, distinct ^2H resonances for the (*R,R,R*)-, (*R,R,S*)-, (*S,S,R*)-, and (*S,S,S*)-isomers of **10.7** were observed in PBLG, and the number of doublets and their corresponding intensities could be rationalized on the basis of the symmetry considerations and probability of occurrence.

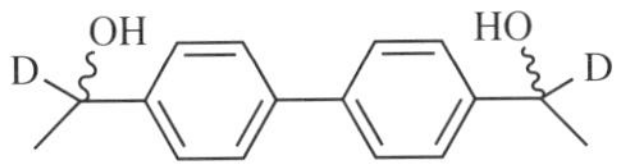

10.6 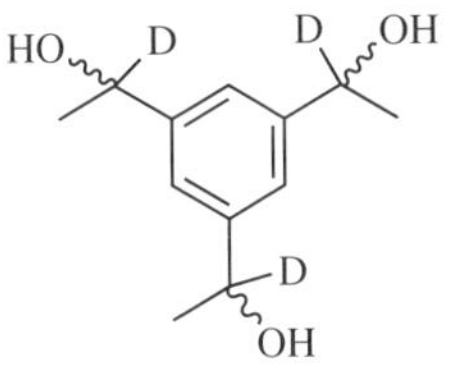   **10.7**

Other more complex alcohols have been analyzed using ^2H NMR spectroscopy with PBLG.[1566] In particular, the compound α,α'-dideutero-α,α'-1,3-benzenedimethanol monobenzoate (**10.8**) was analyzed to demonstrate the utility of liquid crystals for chiral analysis. With **10.8**, the presence of the *erythro*- and *threo*-diastereomers produced (*R,R*)-, (*R,S*)-, (*S,R*)-, and (*S,S*)-combinations, each with two nonequivalent deuterium atoms, for a total of eight possible doublets in the ^{2}H NMR spectrum. The use of a racemic mixture of PBLG and PBDG caused four ^{2}H doublets that could be used to measure the diastereomeric ratios. Use of only PBLG led to eight doublets, allowing the determination of optical purity.[1566]

10.8

Perdeuterated 4-pentyl-4′-cyanobiphenyl (**10.9**) is a flexible, prochiral molecule. Dissolving **10.9** in PBLG led to a reduction in symmetry that could be detected in the ^{2}H NMR spectrum.[1567] A detailed analysis indicated that the reduction in symmetry was caused by anisotropic solute–solvent interactions as opposed to changes in the conformational distribution.

10.9

The compound *cis*-decalin (**10.10**) can exist as two chiral invertomers. One- and two-dimensional ^{2}H NMR spectra of **10.10** with PBLG and PCBLL were used to study the inversion process.[1568] At high temperatures, deuterium nuclei of **10.10** that are enantiotopic under rapid kinetic averaging could be distinguished. At low temperature, the two chiral invertomers provided distinct quadrupolar splitting in the ^{2}H NMR spectrum. Other than a cyclodextrin system,[1070] the ability to study such a process in a hydrocarbon substrate using a chiral NMR discriminating agent has not been observed.

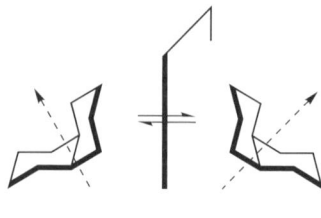

10.10

1-(2′,6′-Dideutero-4′-methylphenyl)naphthalene (**10.11**) is chiral under conditions of slow rotation.[1569] Under conditions of fast rotation, the deuterium atoms were equivalent and appeared as one doublet from quadrupolar splitting. Cooling the sample to 213 K caused two quadrupolar doublets, one for each of the enantiomers. The presence of enantiomers was confirmed by comparing the spectrum in a racemic mixture of PBLG and PBDG, where only one doublet was observed at low temperature. The quadrupolar splitting increased at the lower temperature. The rotational process was not significantly influenced in the liquid crystal environment.[1569]

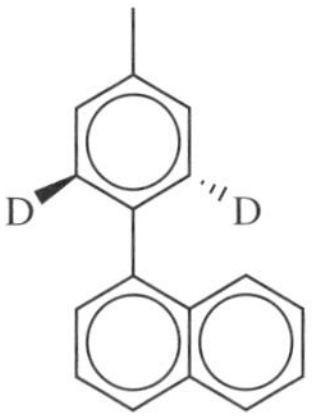

10.11

The enantiomeric and enantiotopic analysis of cone-shaped cyclotriveratrylenes (**10.12**) with C_3 and C_{3v} symmetries was undertaken using ^{2}H NMR spectroscopy in the presence of PBLG.[1570] In this case, DMF was needed as the cosolvent to obtain suitable results. These compounds do not have stereogenic or prostereogenic centers, but the substitution pattern can cause the compounds to be chiral (nona-substituted) or prochiral (hexa-substituted). Discrimination of these compounds occurred because of different ordering of the (*M*)- and (*P*)-isomers in PBLG.

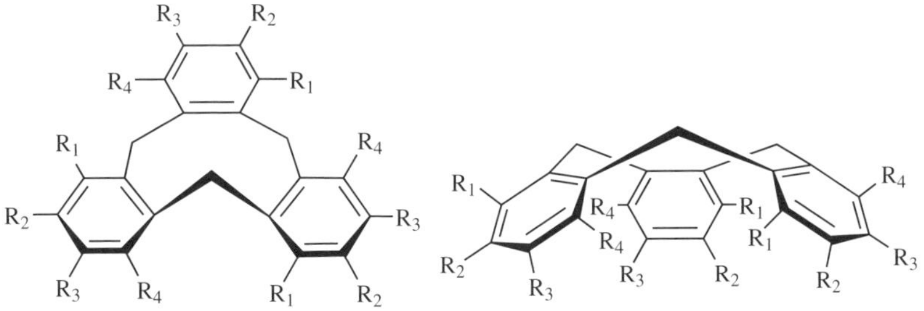

10.12

PBLG phases have also been used to demonstrate fundamental theoretical aspects of chirality via NMR spectroscopy. For example, using group theory arguments, it was shown that the effective symmetry of molecules with C_s, C_{2v}, S_4, or D_{2d} symmetry, which are compounds with prochiral faces, groups, or directions, should be reduced in a liquid crystal phase.[1571] The ^{2}H NMR spectrum of perdeuterated acenaphthene (**10.13**), which has C_{2v} symmetry, exhibited extra peaks when dissolved in PBLG that were indicative of a reduction in symmetry.

10.13

Differential quadrupolar splittings in the ^{2}H NMR signals of deuterium substituted compounds have been used to determine the optical purity of other compounds as well. These include various substituted benzylic alcohols.[1572,1573] The optical purities of a number of cyclic (**10.14**) and acyclic precursors to laurene and epilaurene were analyzed in PBLG-methylene chloride mixtures.[1574] The compounds 3-deutero-2-methylcyclohexanone and 3-deutero-2-methylcyclohexanol consist of *syn–syn*, *anti–syn*, *syn–anti*, and *anti–anti* forms. Using ^{2}H NMR spectroscopy in PBLG, it was possible to discriminate all of the enantiomeric forms of these compounds.[1575]

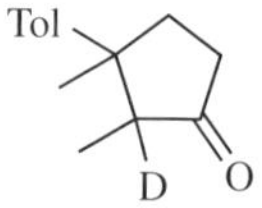

10.14

A thorough study of the use of PBLG showed how it was possible to discriminate enantiomers of alcohols, amines, carboxylic acids and esters, ethers, epoxides, tosylates, alkyl halides, and hydrocarbons.[1576] The analysis of alkyl halides and hydrocarbons is especially noteworthy since the range of chiral NMR reagents for these classes of compounds is quite limited.

Compounds that are axial chiral such as allenes (**10.15**), and methylene cyclohexanes either with polar or halogen substituent groups (**10.16**) have been analyzed by ^{2}H NMR spectroscopy with PBLG.[1577]

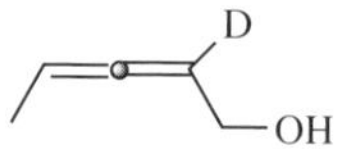

10.15 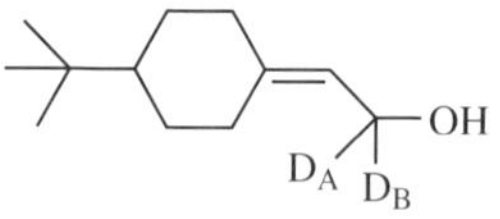 **10.16**

The aldehyde ligands in a series of chiral [(η^4-dienal)Fe(CO$_3$)] complexes were reduced to the corresponding deuterated alcohols using sodium borodeuteride.[1578] The reduction produced the ψ-endo and ψ-exo alcohols as shown in Figure 10.3.[1578] Four doublets were observed in the ^{2}H NMR spectrum, which enabled a determination of optical purity.

The mechanism and stereoselectivity of the SmI$_2$-mediated cyclization of δ-iodo-α,β-unsaturated esters (Fig. 10.4) was probed by ^{2}H NMR spectroscopy in PBLG.[1579]

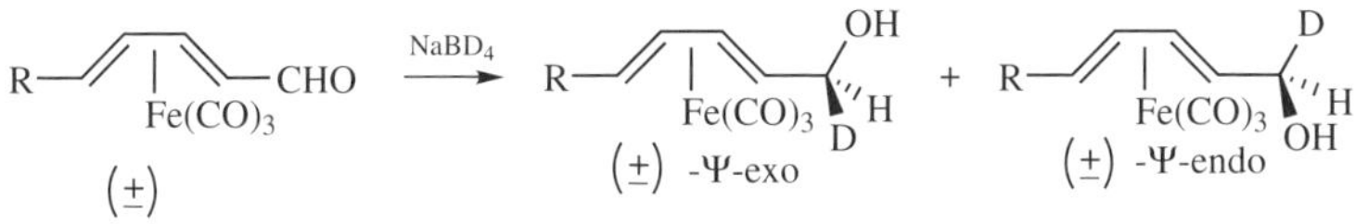

Figure 10.3. Ψ-exo and Ψ-endo alcohols produced by reduction of the corresponding aldehydes.

Incorporation of a CD_2 group into the synthesis facilitated the analysis. By recording ^2H NMR spectra in a racemic mixture of PBLG and PBDG, it was possible to distinguish the resonances of the *meso* and chiral pair since the *pro-(R)* and *pro-(S)* deuterium atoms in the *meso* compound are inequivalent. Obtaining the ^2H NMR spectrum in the presence of PBLG enabled a distinction of the enantiomeric pair. Similarly, the regiochemical and stereochemical outcomes of a palladium-catalyzed allylic alkylation of iso-cinnamyl-type substrates were assessed by ^2H NMR spectroscopy in PBLG-methylene chloride.[1580]

The ND group in amides undergoes slow exchange in PBLG liquid crystals. Dissolving amides in methanol-d_1 leads to an exchange to produce the deuterated amide, which can then be isolated and analyzed by ^2H NMR in PBLG.[1581]

The optical purity of [2-^2H]isopentenyl tosylate (**10.17**) was determined by analyzing the ^2H NMR spectrum in PBLG-methylene chloride at $-50°C$.[1582] The use of α-methoxy-α-trifluoromethylphenylacetic acid (MTPA) was not successful in performing the analysis. The optical purities of several γ-alkoxy secondary allyl carbamates (**10.18**) were determined by analyzing the ^2H NMR spectrum in PBLG.[1583]

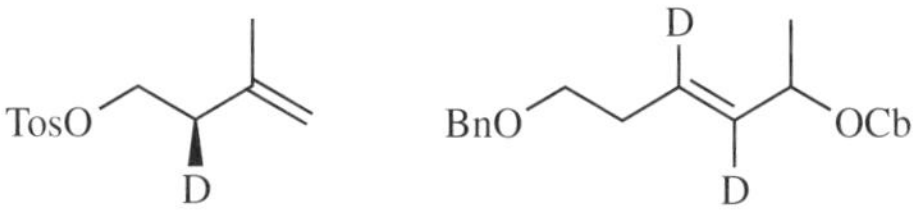

The hexyl ester of [$^2H_1$]-fluoroacetic acid exhibited different quadrupolar splittings for the two enantiomers in PBLG.[1584] The stereochemical course of fluorination during fluorometabolite biosynthesis in *Streptomyces cattleya* that had been given isotopically labeled precursors was subsequently analyzed using this NMR method.

Assignment of the ^2H resonances in a complex perdeuterated compound in PBLG can be challenging because each signal is split into a doublet through quadrupolar

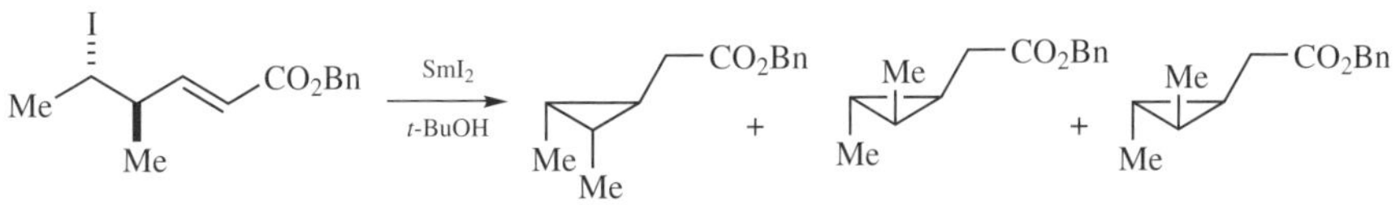

Figure 10.4. Reaction in the SmI_2-mediated cyclization of δ-iodo-α,β-unsaturated esters.

splitting. Furthermore, these signals will be doubled in the case of enantiomeric discrimination. A two-dimensional NMR method that used ^{2}H–^{13}C correlations to facilitate the assignment of the ^{2}H resonances has been described.[1585] The method also facilitated the determination of each enantiomer's set of quadrupolar peaks. The utility of this method was demonstrated on several examples of compounds.[1586]

10.2.4. Natural Abundance ^{2}H NMR Analysis

The methods described above using liquid crystals require deuterium incorporation into the compound, which in many cases may be impractical or undesirable. An alternative is to use natural abundance ^{2}H NMR spectroscopy, although this suffers from two limitations.

The first involves the low isotopic abundance of the ^{2}H nucleus, thereby requiring longer analysis times. The second and more significant limitation involves the complexity of the ^{2}H spectra caused by the presence of the quadrupolar splitting. With compounds that are deuterium labeled at a selective site, only one or a few signals are present in the spectrum and assignment of the resonances is usually straightforward. In natural abundance ^{2}H spectra, a doublet will now appear for every hydrogen resonance in the spectrum due to the quadrupolar splitting. For a mixture of enantiomers, every hydrogen has the potential to appear as two doublets. And since there is no way to easily predict a priori the magnitude of the quadrupolar splitting for a particular nucleus, assigning the resonances in the natural abundance ^{2}H NMR spectrum can be a challenging task. Nevertheless, several strategies that facilitate assigning natural abundance ^{2}H NMR spectra have been devised that render the analysis of nondeuterium-labeled compounds in liquid crystals a practical procedure.

For example, the compound but-3-en-2-ol has six different hydrogen atoms. A mixture of both enantiomers has the possibility of exhibiting 12 doublets in the natural abundance ^{2}H NMR spectrum in PBLG.[1587] A two-dimensional auto-correlation technique was devised to facilitate the assignment of the ^{2}H resonances, so that it was then possible to determine which ones exhibited enantiomeric discrimination on the basis of differential quadrupolar splittings. This method was subsequently applied to the analysis of the natural abundance ^{2}H NMR spectrum of 1-phenylethanol, 3-butyn-2-ol, and benzyl alcohol.[1588] Care must be exercised when examining a compound such as benzyl alcohol. Although the perdeuterated derivative of benzyl alcohol is prochiral, the natural abundance ^{2}H NMR will detect the compound that is mono-deuterated at the methylene group (CDH), which is a chiral species.

Assignment of the ^{2}H signals was further enhanced through the development of a two-dimensional Q-COSY pulse technique.[1589] The validity of the Q-COSY method was demonstrated on perdeuterated 1-pentanol, and then applied to the assignment of the natural abundance ^{2}H NMR spectra of 1-pentyn-3-ol and 3-butyn-2-ol.

The utility of the Q-COSY method in assigning natural abundance ^{2}H NMR spectra was further demonstrated with a series of chiral saturated alkanes including 3-methylhexane, 3-methylheptane, and 3-methyloctane in PBLG-chloroform.[1590] The use of a racemic mixture of PBLG and PBDG was used to facilitate assignment of the ^{2}H resonances as well, since no quadrupolar splitting occurs in this case. An overview of the use of the two-dimensional autocorrelation, Q-COSY and Q-resolved methods to facilitate the assignment of natural abundance ^{2}H NMR spectra have been published.[1591]

A further refinement has been the development of a modified z-gradient filter scheme that facilitates the collection and analysis of two-dimensional Q-COSY and Q-resolved natural abundance ^{2}H NMR spectra.[1592] A relatively simple procedure produces phased deuterium autocorrelation two-dimensional NMR spectra with significant gains in peak linewidth and resolution. The advantages of this method were demonstrated on the assignment of the enantiotopic hydrogens of perdeutero 1-butanol, but it is effective with the assignment of natural abundance ^{2}H NMR spectra as well.

The utility of natural abundance ^{2}H NMR spectroscopy to analyze the optical purity of **10.19** and **10.20** in PBLG was demonstrated.[1593] The methyl resonance at the stereogenic center was easily assigned and conveniently monitored. In addition, the ^{13}C resonances of the acetylenic carbons exhibited measurable chemical shift anisotropies for these compounds.

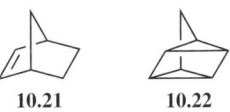

10.19   **10.20**

The use of natural abundance ^{2}H NMR spectroscopy for the analysis of chiral hydrocarbons was explored.[1594] The optical purity of 20 substrates encompassing rigid chiral bicyclic hydrocarbons (e.g., β-pinene), semi-rigid chiral aromatic, acetylenic or ethylenic derivatives (e.g., 3-methylpentene), flexible chiral alkanes (3-methylhexane), flexible chiral monocyclic hydrocarbons (e.g., (+/−)-*trans*-1, 3-dimethylcyclohexane), and chiral hydrocarbons without a stereogenic center (e.g., hepta-2,3-diene; 2,2′-dimethyl-1,1′-binaphthalene; *cis*-decalin(**10.10**)) was all successfully analyzed in PBLG using natural abundance ^{2}H NMR spectroscopy.

Natural abundance ^{2}H NMR analysis of bicyclo[2.2.1]hept-2-ene (norbornene – **10.21**) and quadricycane (tetracyclo[3.2.0.0.0]heptane–**10.22**), which have C_s and C_{2v} symmetry respectively, in PBLG enabled the assignment of nuclei to enantiotopic faces of a prochiral molecule.[1595] Both diastereotopic and enantiotopic nuclei could be assigned.

10.21   **10.22**

Finally, natural abundance ^{2}H NMR spectra in PBLG has been shown to be an effective way to analyze the H/D ratio of prochiral sites in naturally occurring compounds.[1596] Many natural systems show a stereoselective preference for incorporation of deuterium into prochiral methylene sites. The use of PBLG provides a straightforward means of distinguishing the *pro-(R)* and *pro-(S)* positions and assessing such kinetic isotope effects. The utility of the method was demonstrated on 1,1′-bis(phenylthio)hexane (C_5H_{11}-CH(S-C_6H_5)$_2$), a compound derived from the cleavage of methyl linoleate of safflower. The different quadrupolar splitting in the isochronous ^{2}H NMR spectrum of the signals was reportedly more significant than other solution phase methods for distinguishing pro-(R) and pro-(S) positions.[1596]

10.2.5. ^{13}C NMR Analysis

Two possibilities can cause enantiomeric discrimination in the ^{13}C NMR spectra of chiral compounds dissolved in liquid crystals. One is presence of chemical shift anisotropies of the ^{13}C resonances in the proton-decoupled spectrum. The other is the presence of different ^{1}H-^{13}C dipolar coupling for the two enantiomers because of their different packing orientations. The potential of both possibilities was demonstrated on 2-bromopropionic acid, β-(trichloromethyl)-β-propiolactone (**10.1**), and other substrates.[1597] A limitation occurs in complex molecules with many hydrogen atoms in that the proton-coupled ^{13}C spectra can be exceedingly complex.

A procedure to facilitate the assignment of proton-coupled ^{13}C NMR spectra and to determine the carbon-proton residual dipolar couplings for each enantiomer has been described.[1598] The process involved a two-dimensional carbon-proton selective refocusing that reduced the number of signals in the spectrum. Several example compounds were studied to show the applicability of the method.

A study of a variety of alcohols, carboxylic acids, heterocyclic, and axial chiral compounds such as 2,2′dihydroxy-1,1′-binaphthalene and o-methoxy-1,2-diphenyl-1,2-ethanediol in PBLG indicated that chemical shift anisotropy of 4–8 Hz was commonly observed in the proton-decoupled ^{13}C NMR spectra.[1599] With 2,2′-dihydroxy-1,1′-binaphthalene, DMF was used as the cosolvent to overcome solubility issues. For compounds with an isolated and easily identifiable methyl resonance, it was convenient to monitor this signal for differences in the dipolar coupling.

The optical purity of α,β-unsaturated δ-lactone **10.23** was determined by the presence of chemical shift anisotropy in the proton-decoupled ^{13}C spectrum in PBLG-chloroform.[1600] The antifungal compounds uniconazole (**10.24A**) and diniconazole (**10.24B**) exhibited differences in ^{13}C chemical shifts in PBLG-chloroform.[1601]

1-deutero-(2′,3′,4′,5′,6′-pentadeuterophenyl)phenylmethanol (C_6H_5–CDOH–C_6D_5) is chiral by virtue of isotopic substitution.[1602] It was possible to assign the pro-(R) and pro-(S) character of each of the rings by analyzing proton and deuterium-decoupled ^{13}C NMR spectra in PBLG-chloroform.

The optical purities of chiral monophosphine oxides (**10.25**) and boranes (**10.26**) were determined on the basis of the ^{13}C NMR spectrum in the presence of PBLG and PCBLL.[1603] Of the two liquid crystal reagents, better discrimination was observed with PCBLL. Chemical shift anisotropy of the ^{13}C resonances was one approach to performing the analysis. A second involved the difference in ^{31}P–^{13}C dipolar coupling for the ordered enantiomers that was observed in the proton-decoupled, phosphorus-coupled ^{13}C NMR spectrum.

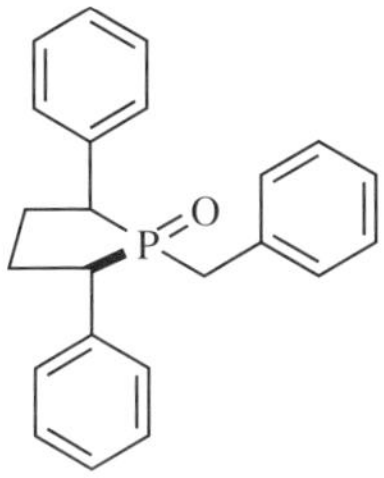

10.25

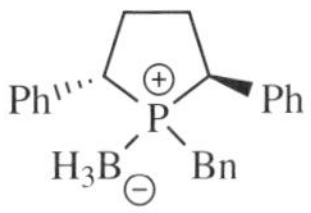

10.26

The optical purity of acyclic phosphonium salts such as **10.27** can be determined in PBLG/DMF and PCBLL/DMF liquid crystals.[1604] This represented the first enantiomeric discrimination of P-chirogenic phosphorus compounds in liquid crystals. Differential quadrupolar splitting in the ^{2}H NMR spectrum was used for deuterium-labeled compounds. For compounds that were not deuterium labeled, chemical shift anisotropy of up to 0.8 ppm in the proton-decoupled ^{13}C NMR spectrum was preferred for the analysis as no substantial discrimination was observed in the ^{31}P spectrum. Interpretation of some resonances in the ^{13}C spectra was complicated by ^{31}P coupling, although minimal differentiation in the dipolar ^{31}P–^{13}C coupling constants was observed for the enantiomers.

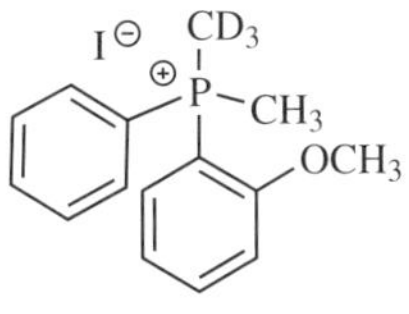

10.27

A rather elegant application of the use of liquid crystals for chiral discrimination involved the differentiation of enantiotopic directions in molecules such as *endo*-bicyclo[2.2.2]-oct-5-ene-2,3-dicarboxylic anhydride (**10.28**), bicyclo[2.2.1]hepta-2,5-diene (**10.29**), ethanol, and malononitrile (**10.30**).[1605] The ^{13}C and ^{2}H NMR spectra could be used to differentiate ^{1}H–^{1}H, ^{13}C–^{1}H, and ^{13}C–^{2}H enantiotopic directions in prochiral molecules with C_s and C_{2v} symmetry. For example, the ethylenic ^{13}C signals in the spectrum of **10.29** were distinct resonances when dissolved in PBLG. Similarly, the enantiotopic directions in ethanol were differentiated in PCBLL. Malononitrile is a molecule with C_2 symmetry and no prostereogenic carbon. However, the structure CXXYY is one in which the four X–Y relationships are ordered into two enantiotopic sets of two equivalent relationships. This relationship was first recognized by Mislow in 1967, and implies that a molecule such as $CH_2(CN)_2$ is prochiral in a chiral solvent.[1606] The proton-coupled ^{13}C NMR spectrum of malononitrile in PBLG provided the first spectral evidence for the differentiation of enantiotopic directions in such C_2 symmetric molecules.

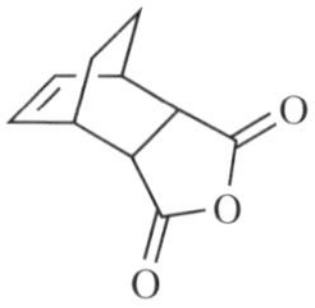

10.28  10.29 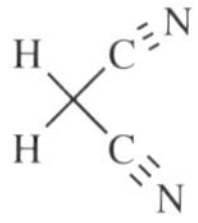 10.30

10.2.6. ^1H NMR Analysis

The use of proton NMR spectroscopy with liquid crystal systems is difficult because of the added complexities of differential dipolar coupling. Both chemical shift anisotropy and differential dipolar coupling were observed in the 400 MHz NMR spectrum of 2-bromopropionic acid, which has a relatively simple ^1H NMR spectrum, in PBLG.[1576]

The utility of the homonuclear Hartmann–Hahn (HOHAHA) two-dimensional correlation technique for assigning resonances in the ^1H NMR spectra of substrates in liquid crystals has been demonstrated.[1607] The compounds (+/−)-3,3,3-trichloroepoxypropane and 2-bromopropionic acid were used as examples.

More recently, a method to simplify proton spectra in liquid crystals by carrying out selective focusing through two-dimensional methods has been described.[1608] Using 1,2-dibromopropane in PBLG-chloroform as an example, it was possible to simplify the pattern of the methyl resonance by eliminating coupling to the methylene and methine hydrogen atoms, thereby leaving only dipolar coupling among the methyl hydrogen atoms. The utility of this method was demonstrated for the analysis of 3-methyl-4,4,4-trichlorobutyric-β-lactone (**10.31**) as well.[1608]

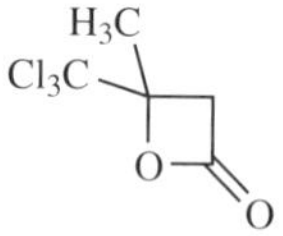

10.31

10.2.7. ^{19}F NMR Analysis

The ^{19}F NMR spectra of compounds with one or more fluorine atoms may exhibit chemical shift anisotropy or differential dipolar coupling in liquid crystals that can be used to analyze optical purity. Trifluoroacetic anhydride has been used as a derivatizing agent for amines, amino acids, and alcohols.[1609] Chemical shift anisotropy was frequently observed in the proton-decoupled ^{19}F NMR spectrum. The CF_3 group appeared as a 1:2:1 triplet in the proton-decoupled ^{19}F spectrum because of dipolar coupling among the fluorine atoms, and differences in the magnitude of the dipolar coupling caused a doubling of peaks that could be used to determine optical purity. The largest discrimination was observed for the amino acids and smallest for the alcohols.[1609] The ^{19}F NMR spectrum of 2,2,2-trifluoro-(α-methylbenzyl)acetamide (**10.32**) in PBLG-chloroform-d was used to analyze mixtures with high enantiomeric excess.[1581]

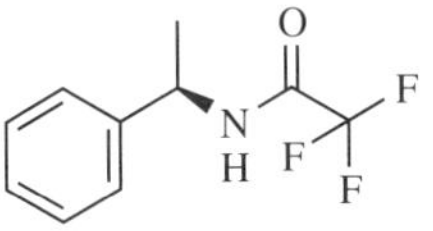

10.32

10.2.8. Analysis Using Other Nuclei

Other nuclei such as ^{10}B, ^{11}B, ^{133}Cs, and ^{14}N are quadrupolar and represent candidates for the analysis of optical purity of substrates dissolved in liquid crystals. The optical purities of a series of calcium and ammonium ions of borocryptates (**10.33**) were determined based on an analysis of their ^{10}B, ^{11}B, ^{133}Cs, and ^{14}N NMR spectra in PBLG-methylene chloride.[1610,1611] Differential quadrupolar splitting caused the ^{10}B signal to appear as two sextets, ^{11}B as two triplets, and ^{133}Cs as two septets. The areas of these resolved signals provided the optical purity.

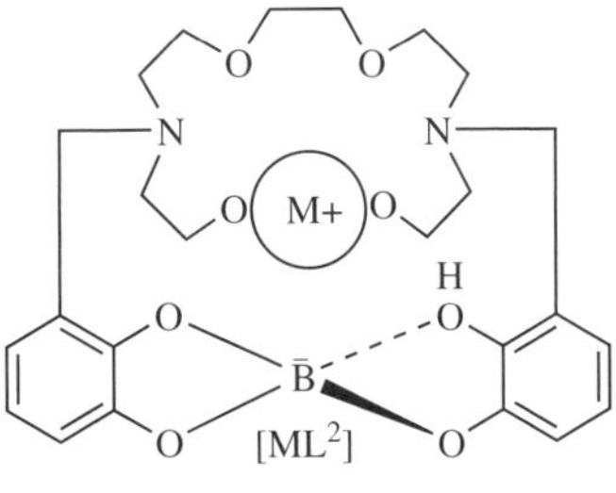

10.33

A thorough review of the use of natural abundance ^2H, ^1H, and ^{13}C NMR spectroscopy to determine the optical purity of compounds dissolved in liquid crystals has been published.[1591]

10.2.9. Miscellaneous Studies with Liquid Crystals

Surfactant molecules prepared with other amino acids have been used on occasions to form liquid crystals for the purpose of enantiomeric discrimination. For example, the ^{59}Co nucleus is quadrupolar and the two enantiomers of $[Co(en)_3]^{3+}$ exhibited different quadrupolar splittings of the ^{59}Co resonance when dissolved in a liquid crystal prepared from calcium N-dodecanoyl-L (or D)-alaninate.[1612] The difference in the quadrupolar splitting was more diagnostic than chemical shift anisotropy in the ^{59}Co spectrum.

L- and D-N-acyl-1-phenyl-d_5-2-aminopropane derivatives exhibited different ordering in a liquid crystal prepared by using cesium N-dodecanoyl-L-alaninate.[1613] The N-acyl group was comprised of linear chains of 1–10 carbons. The enantiomeric discrimination of these same substrates was also examined in a liquid crystal prepared using cesium N-dodecanoyl-L-threoninate.[1614]

Amino acid-based anionic surfactants consisting of a sulfonated amphiphile and L-phenylalanine or L-alanine units with penta- or tetra-decyl groups were evaluated as aqueous-based liquid crystal reagents.[1615] Aqueous solutions in the presence of chlorinated organic solvents provided homogeneous NMR solvents. Distinctions in the ^{1}H (different ^{1}H–^{1}H dipolar coupling) or ^{2}H (different quadrupolar splittings) NMR spectra of D- and L-alanine-$2d_1$ were observed.

A water-soluble liquid crystal prepared from a mixture of glucopon/hexanol and buffered water was shown to cause chiral discrimination in the ^{2}H NMR spectra of 2-deuteroalanine and 3,3-dideuterophenylalanine (**10.34**).[1616] Glucopon is a commercially available aqueous solution that consists of a mixture of mostly octyl- and decyl α- and β-mono and diglucosides.

$$D^A \quad D^B$$
$$\text{Ph-C(COOH)(NH}_2\text{)}$$

10.34

One last strategy for chiral analysis with liquid crystals involved the incorporation of a separate chiral discriminating agent into an achiral liquid crystal. The potential of this strategy was demonstrated by incorporating β-cyclodextrin or hydroxypropyl-β-cyclodextrin into cromolyn, an achiral nematic phase.[1617] The ^{2}H resonance of 1-deutero-1-phenylethanol in this system consisted of three doublets, one for the unbound substrate and the other two for the (R)- and (S)-enantiomers encaged within the cyclodextrin. The different quadrupolar splittings for the two encaged enantiomers indicated that they were not ordered the same within the cyclodextrin units.

10.3. MICELLES AND GELS

Several studies have examined micellar systems for evidence of chiral discrimination in the NMR spectra of substrates, although micelles have not yet been demonstrated to have widespread applicability for chiral NMR analysis.

Using amide **10.35** as a chiral probe molecule, no enantiomeric discrimination was observed in the ^{1}H NMR spectrum in the presence of micelles formed from **10.36**, **10.37**, or **10.38**.[1618] The addition of an achiral surfactant such as sodium dodecyl sulfate to **10.36** formed a mixed micelle with a more rigid structure, and enantiomeric discrimination was then observed in the NMR spectrum of **10.35**.

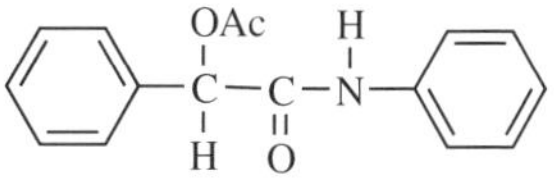

10.35

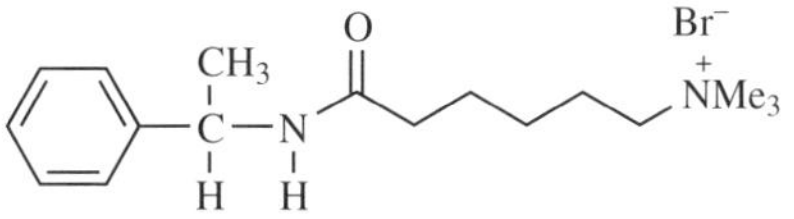

10.36

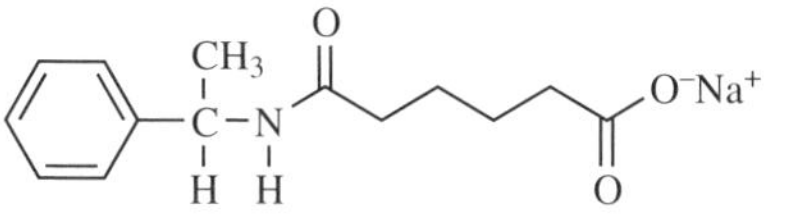

10.37

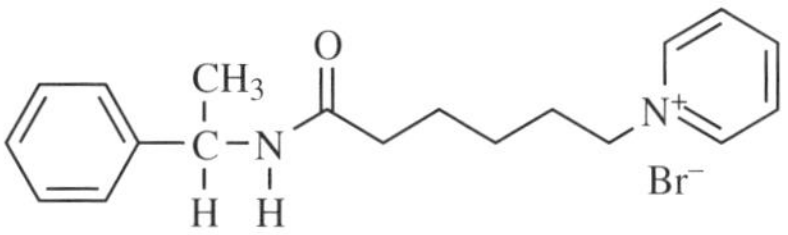

10.38

A common strategy has been used to form micelles with amino acid derivatives such as *N*-lauroyl-L-prolinate,[1619] *N*-dodecanoyl-L-prolinate,[1620,1621] *N*-dodecanoyl-L-alaninate and L-threolate,[1622] *N*-undecanoyl-L-isoleucinate,[1623] and *N*-undecanoyl-L-valinate-L-leucinate and L-leucinate-L-valinate.[1624] Enantiomeric discrimination in the ^{1}H and ^{13}C NMR spectra of *N,N'*-dibenzyl-2,2'-bipyridinium dibromide (**10.39**) was observed in the presence of sodium *N*-lauroyl-L-prolinate.[1619] The spectra of the LL- and DD-isomers, as well as the LD- and DD-isomers of ditryptophan and diphenylalanine were distinct in the presence of sodium *N*-dodecanoyl-L-prolinate.[1621]

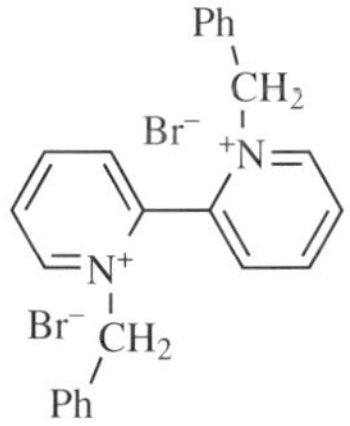

10.39

2,2'- and 2,3'-diamino-6-(octanoyloxymethyl)biphenyl (**10.40**) are chiral under conditions of slow rotation. The rotational barrier could be measured by dynamic ^{1}H NMR spectroscopy in the presence of sodium *N*-dodecanoyl-L-prolinate.[1620] The ^{59}Co signals of the Δ and Λ enantiomers of [Co(en)$_3$]$^{3+}$ were distinct in the presence of *N*-hexadecanoyl-L-prolinate.[1622] The enantiomeric discrimination with the prolinate derivative was larger than that observed with similar micelles incorporating alaninate or threonate moieties.

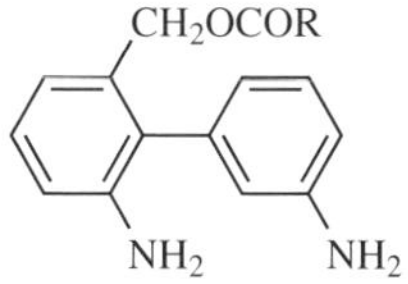

R = C$_7$H$_{15}$ or CH$_3$

10.40

Chiral recognition in the ^{1}H NMR spectrum of 1,1′-binaphthyl-2,2′-diylhydrogen phosphate (**10.41**) with micelles formed from sodium *N*-undecanoyl-L-isoleucinate[1623] and the corresponding L-valinate-L-leucinate and L-leucinate-L-valinate derivatives[1624] was observed and used to confirm relative retention rates in electrokinetic chromatographic separations with these micelles.

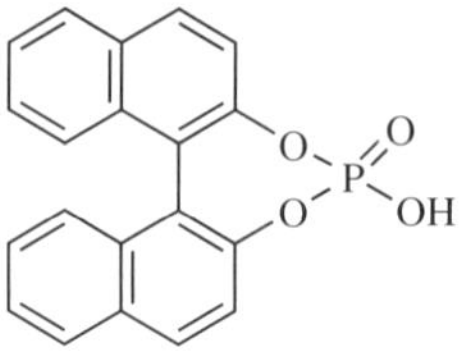

10.41

2-Carboxy-6-nitro-2′-dodecyloxybiphenyl (**10.42**) exhibits axial chirality. In the presence of *N*-hexadecyl-*N*-methyl-L-prolinolinium bromide (**10.43**), the compound deracemized and exhibited a diastereomeric ratio of 5:1.[1625] Warming the sample to high enough temperatures caused a fast interconversion such that only one set of signals was observed in the spectrum of **10.42**. Analysis of **10.42** using micelles prepared from sodium *N*-dodecanoyl-L-prolinate provided a ^{1}H NMR spectrum in which five of the seven aromatic signals exhibited chiral discrimination.[1626] ROESY data was used to study the interactions between **10.42** and the micelle. The micelle components existed in both the *E*- and *Z*-domains and the substrate exhibited greater binding to the *Z*-domain.[1626]

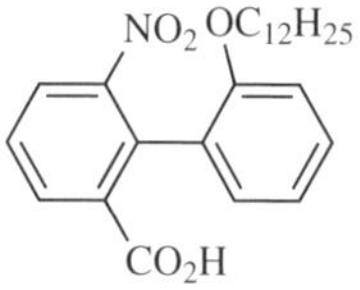

10.42 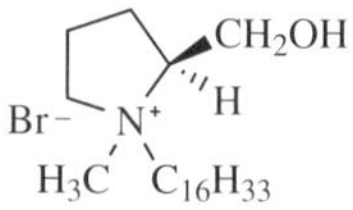   **10.43**

The ability of water-soluble micelles prepared from dodecylmaltopyranosides to discriminate the (*P*)- and (*M*)-enantiomers of [5] (**10.44**) and [7] thiaheterohelicenes (**10.45**) was examined.[1627] The micelle served to solubilize the helicene in water. The ^{1}H NMR spectrum of the [5]helicene was not enantiomerically discriminated, even at low temperatures, presumably due to a too fast inversion of the helix. The [7] helicene was enantiomerically discriminated and the (*M*)-enantiomer interacted more strongly with the micelle than with the (*P*)-enantiomer.

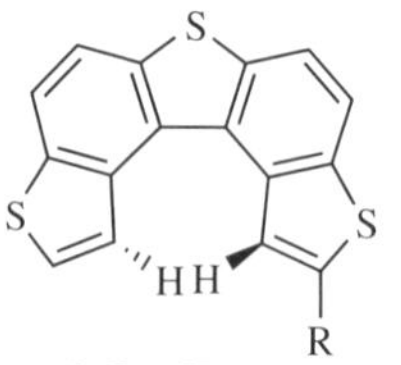

1: R = H
2: R = CH$_2$OH
3: R = COOH

10.44   **10.45**

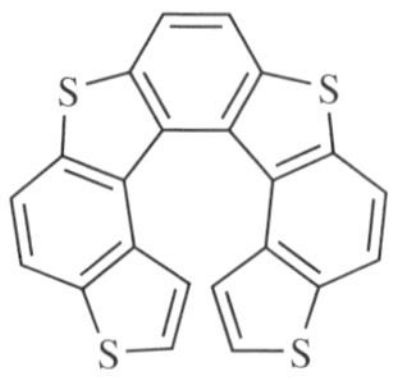

10.46 forms a gel in the presence of the coaggregating guest **10.47** and enantiomeric distinctions occurred in the NMR spectra in chloroform-d.[1628]

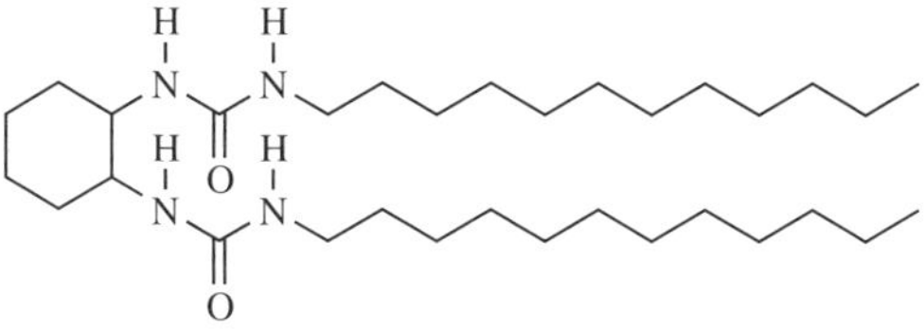

10.46

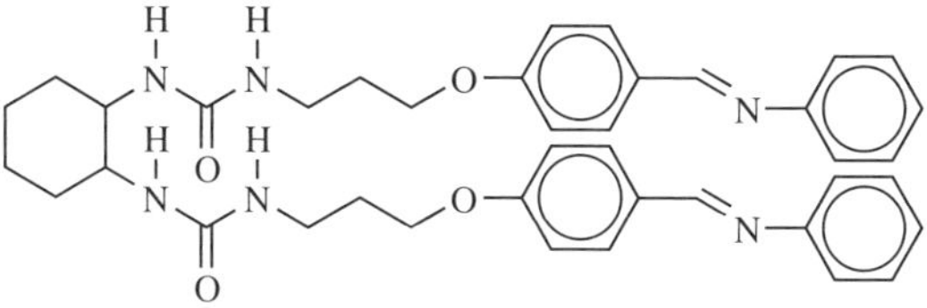

10.47

10.4. IONIC LIQUIDS

Recent reports have begun to explore the use of ionic liquids for chiral discrimination in NMR spectroscopy. Usually these reagents are not used as the solvent, but instead are dissolved in appropriate NMR solvents. The enantiomeric discrimination that has been observed in these studies is modest compared to that with many other chiral discriminating agents.

Using **10.48**, an ionic liquid prepared from ephedrine, a small nonequivalence was observed in the ^{19}F NMR spectrum of the sodium salt of α-methoxy-α-trifluoromethylphenylacetic acid (MTPA).[1629] Increasing the concentration of the ionic liquid in the NMR solvent increased the extent of enantiomeric discrimination. A small degree of nonequivalence was observed in the ^{19}F signal of the thiazolinium salt of MTPA (**10.49**) in benzene-d_6.[1630]

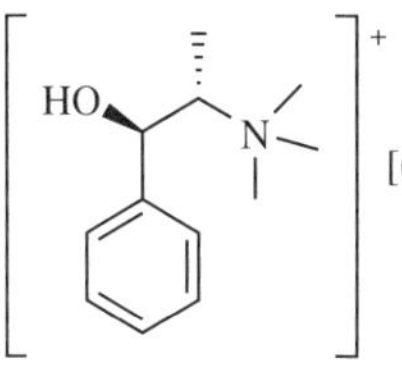

10.48 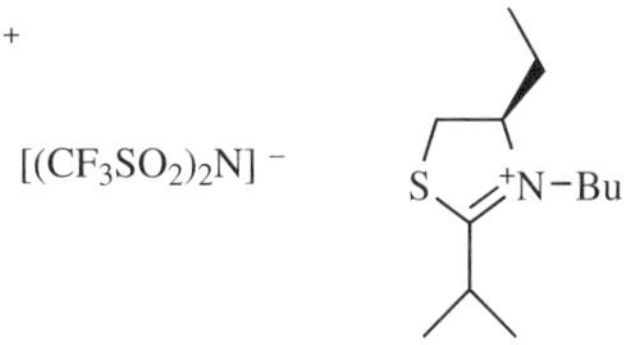   **10.49**

Chiral imidazolinium ionic liquids prepared from L-valine (**10.50**) caused chiral discrimination in the ^{1}H and ^{19}F NMR spectra of the potassium salt of MTPA in acetone-d_6.[1631] Replacement of the *t*-butyl group of the imadazolinium ion with a methyl group caused a signficant reduction in the chiral recognition. Addition of 18-crown-6 to trap the potassium ion increased the discrimination in the NMR spectrum.

Cyclophane-type imidazoliuim salts **10.51**[1632] and **10.52**[1633] form planar chiral ionic liquids. The chirality of **10.51** was confirmed by a splitting of certain resonances in the NMR spectrum on the addition of silver(I) (S)-(+)-10 camphor sulfonate.[1632] Similarly, addition of Eu(tfc)$_3$ or Eu(hfc)$_3$ (tfc = 3-trifluoroacetyl-d-camphor and hfc = 3-heptafluorobutyryl-d-camphor) to a racemic mixture of **10.52** caused a splitting of certain signals as well.[1633]

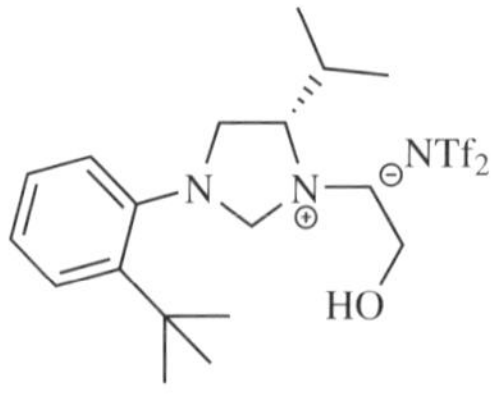

10.50

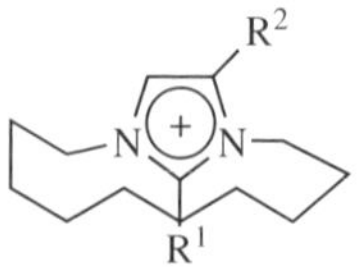

10.51

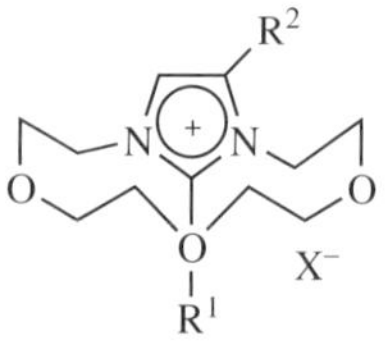

10.52

10.5. POLYMERS

10.5.1. Cellulose-carbamate Polymers

A series of carbamate derivatives of cellulose and amylose developed for use as stationary liquid chromatographic phases now account for the majority of liquid chromatographic separations of enantiomers.[1634] The three-dimensional structure of these polymers plays a significant role in their enantiodiscriminating properties. Certain of these cellulose carbamates are soluble in chloroform. Solubilization of these polymers leads to an opening of the structure and reduction in enantiodiscriminating abilities. Nevertheless, NMR studies of these polymers have been used to study the mechanism of substrate association. Also, modest enantiomeric recognition has often been observed in the ^{1}H NMR spectra for the substrates that were analyzed.

A *tris*(4-trimethylsilylphenylcarbamate) derivative of cellulose (**10.53**) caused a small extent of enantiomeric discrimination in certain resonances of *trans*-stilbene oxide (**10.54**) and 2-butanol.[1635] The enantiotopic methyl groups of 2-propanol exhibited distinct resonances in the presence of **10.53**. This same polymer caused chiral discrimination in the spectra of 2,2′-dihydroxy-1,1′-binaphthalene (BINOL), 1-phenylethanol, and other alkyl alcohols.[1636]

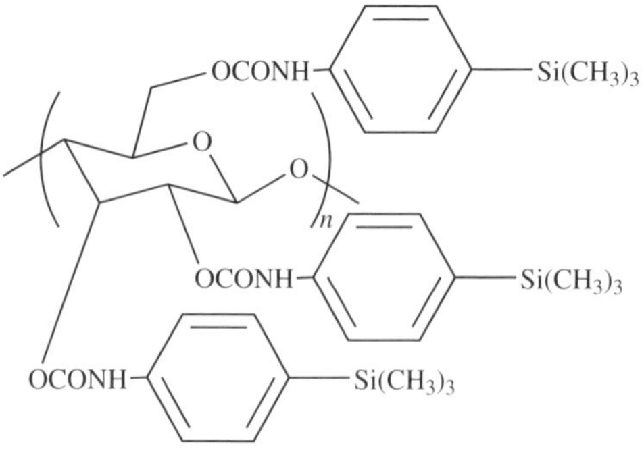

10.53

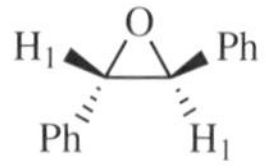

10.54

Cellulose or amylose derivatives with (5-fluoro-2-methylphenyl)carbamate,[1637,1638] (3,5-dimethylphenyl)- and (3,5-dichlorophenyl)carbamates,[1639,1640] and cyclohexylcarbamates[1641,1642] have been studied by NMR spectroscopy as well. Typical substrates examined in these studies include BINOL, 2,2′-dihydroxy-6,6′-dimethylbiphenyl (**10.55**), Troger's base (**10.56**), benzoin (**10.57**), mandelic acid, and *trans*-stilbene oxide (**10.54**). Discrimination was often noted in the ^{1}H or ^{13}C NMR spectra, but was usually quite modest relative to other chiral NMR discriminating agents.

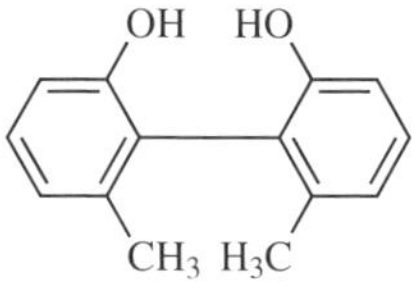

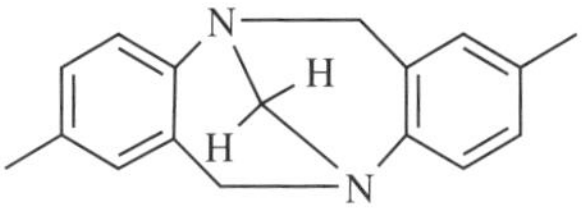

A cellulose derivative with 4-methylbenzoate was found to produce enantiomeric discrimination in the ^{13}C NMR spectrum of 1-phenylethanol in chloroform-d.[1643] No discrimination was observed in the ^{1}H NMR spectrum. In subsequent studies, mostly aimed at understanding the nature of the interaction of compounds with the cellulose polymer, chiral discrimination in the ^{13}C NMR spectra of 1-indanol (**10.58**), 1,2,3,4-tetrahydro-1-naphthol (**10.59**), glutethimide (**10.60**), and Troger's base (**10.56**) was noted.[1644]

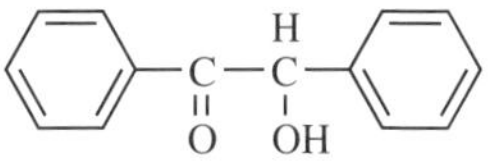

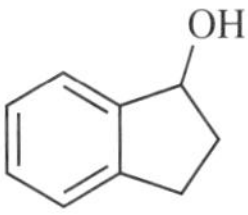

10.5.2. Poly 3- and 4-((S-(α-Methylbenzylcarbamoyl)phenyl Isocyanate

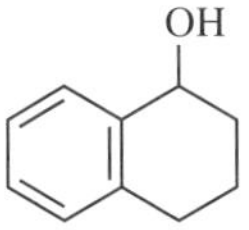

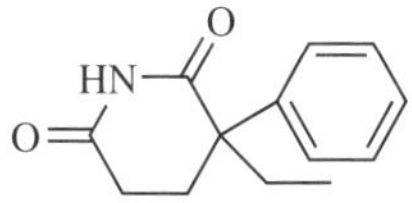

Chiral recognition in the ^{1}H NMR spectrum of BINOL and mandelic acid was observed in the ^{1}H NMR spectra with poly 3- and 4-((*S*-(α-methylbenzylcarbamoyl)-phenyl isocyanate in chloroform-d.[1645]

10.5.3. Poly(ester/β-sulfoxides)

Polymerization of ester and sulfoxide units into a polymer with structure **10.61** produces regions that have different stereochemical environments with an (*R,R*) or (*S,S*) structure and an (*R,S*) structure.[1646] The ^{13}C signals for positions a, b, c, and d are distinct for the two possibilities and integration of the signals enabled a determination of the enantiomeric excess that occurred in the polymerization reaction.

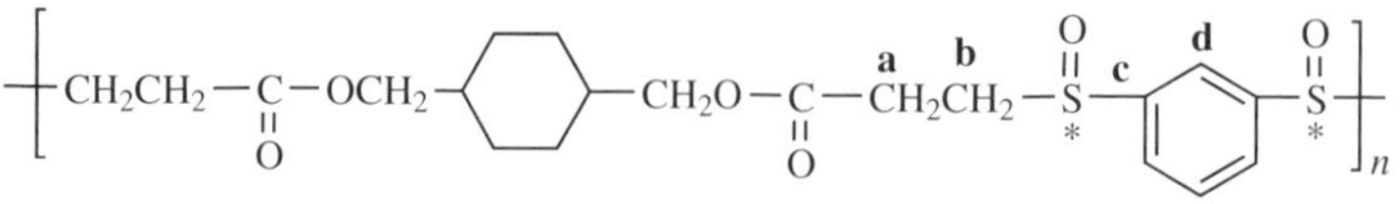

10.61

10.5.4. Chirasil-val

Chirasil-val is a polymer consisting of linked L-valine oligopeptides attached to a poly(β-methyl)siloxy α-methylpropanoic acid copolymer (**10.62**). This material was developed for use for the gas chromatographic separation of enantiomers, particularly N-trifluoroacetyl (TFA) amino acid methyl esters.[1647,1648] The mechanism of interaction with these substrates with the polymer was studied by NMR spectroscopy and a small extent of nonequivalence was observed in the ^{1}H and ^{19}F NMR spectra of N-TFA amino acid derivatives in carbon tetrachloride.[1647] No discrimination was observed in chloroform-d.

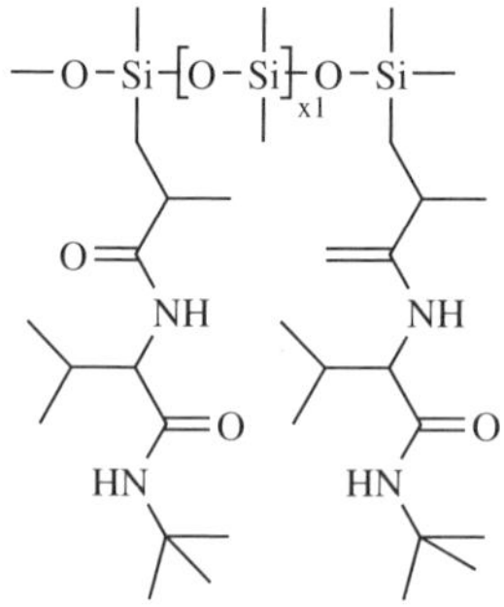

10.62

10.5.5. Poly(N-1-anthrylmaleimide)

The chiral recognition properties of a poly(N-1-anthrylmaleimide) (**10.63**) was studied using BINOL.[1649] A relatively modest nonequivalence was observed for certain resonances in the ^{1}H NMR spectrum.

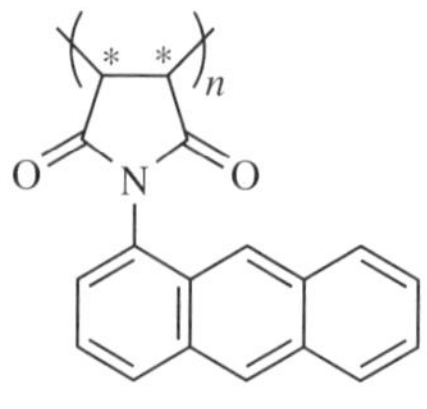

10.63

10.5.6. Poly[(1-6)-2,5-anhydro-3,4-di-*O*-methyl-D-glucitol]

The racemic protonated salt of phenyl glycine methyl ester underwent slow exchange with poly[(1-6)-2,5-anhydro-3,4-di-*O*-methyl-D-glucitol] (**10.64**) such that splitting of the C3, C4, and C6 ^{13}C resonances of the glucitol was observed in chloroform-*d*.[1650] The H-*ortho* aromatic resonance of the substrate also showed enantiomeric discrimination.

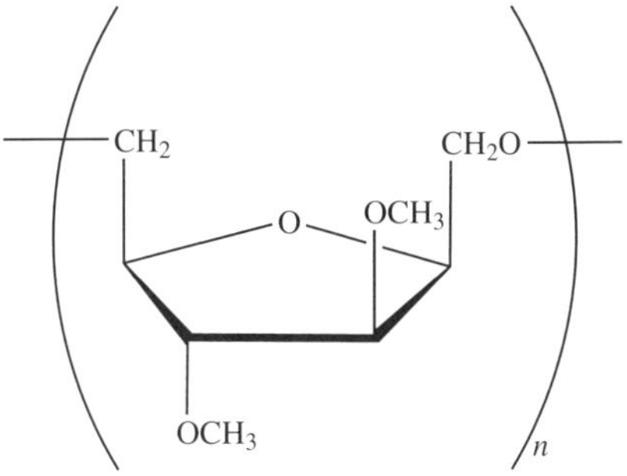

10.64

10.6. AGGREGATION OF ACHIRAL COMPOUNDS INTO CHIRAL ASSEMBLIES

Recent studies have examined systems in which achiral compounds can self-assemble to form chiral pockets capable of discriminating enantiomeric guest molecules. Some of these compounds were already described in the section on calix[4]resorcarenes in Chapter 8. Symmetrical molecules such as **10.65** can self-assemble into systems that have dissymmetric cavities.[1651,1652] Molecular recognition of chiral guests determined the nature of the cavity that was formed. The assembly occurred in solvents such as benzene-d_6 or toluene-d_8 and guests such as nopinone (**10.66**), camphor (**10.67**), camphor quinone (**10.68**), and pinanediol (**10.69**) were distinguished. The larger guests exhibited greater selectivity as to which enantiomer fit into the cavity. In some cases, the network of hydrogen bonds that held the cavity together was retained in solution for hours after removal of the guest compound.[1652]

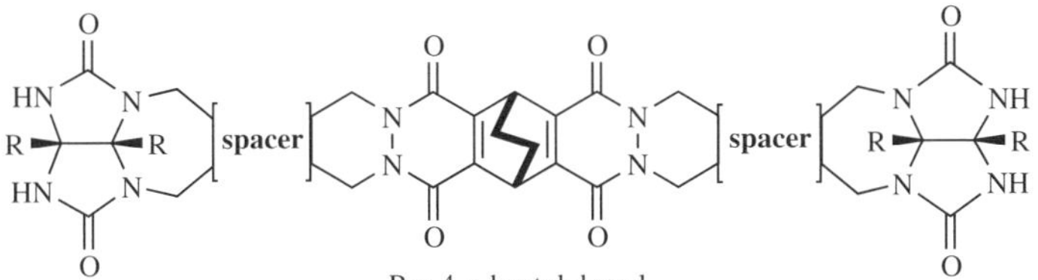

R = 4-*n*-heptylphenyl

10.65

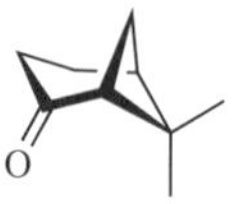

10.66 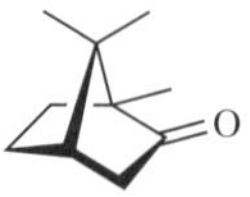 **10.67** 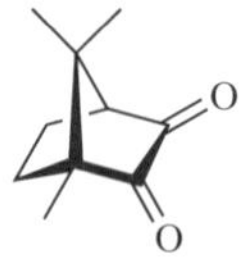 **10.68** 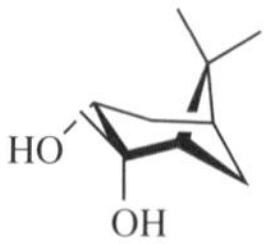 **10.69**

Compound **10.70** formed a cyclic tetramer with a chiral pocket capable of distinguishing chiral ketones such as 3-methylcyclohexanone, 3-methylcyclopentanone, and nor-camphor.[1653] A series of molecular clips of structure **10.71** formed dimers in chloroform-d.[1654] Mixing the $+/-$ pair produced only the hetero $+/-$ dimer in solution and none of the homodimer. The dimerization involved both hydrogen-bonding and π-π interactions. Mixing different clips, the $+/-$ heterodimer was still the only species that was formed.

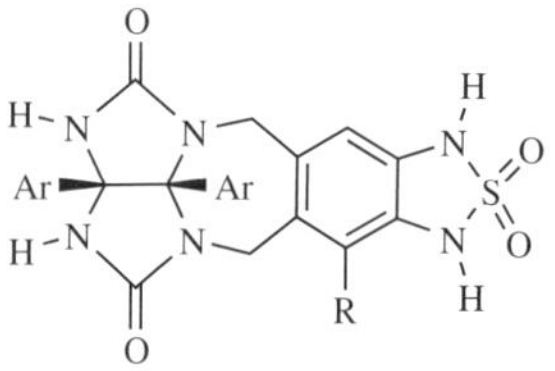

Ar = 4-dodecylphenyl
1: R = H
2: R = OH
10.70

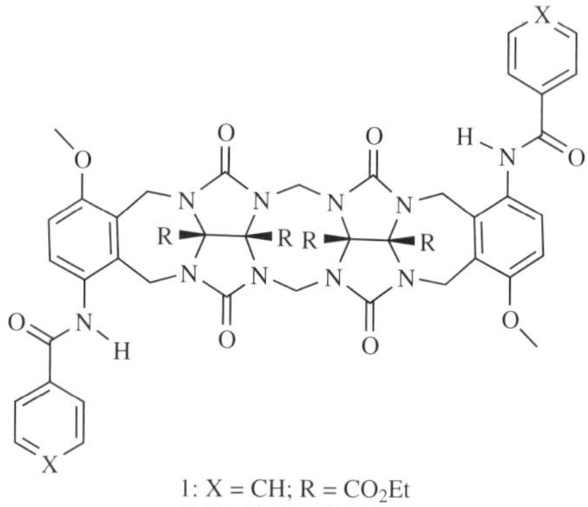

1: X = CH; R = CO$_2$Et
2: X = N; R = CO$_2$Et
10.71

Achiral constituent species consisting of 1,5-bis(2′,3′-dihydroxybenzamido)-naphthalene ligands (**10.72**) and Ga(III), Al(III), or Fe(III) assembled into M$_4$L$_6$ species with a chiral hydrophobic cavity.[1655] The enantiotopic protons of the prochiral species [CpRu(p-cymene)]$^+$ (**10.73**) were rendered diastereotopic in the ^{1}H NMR spectrum by encapsulation into the chiral cavity.

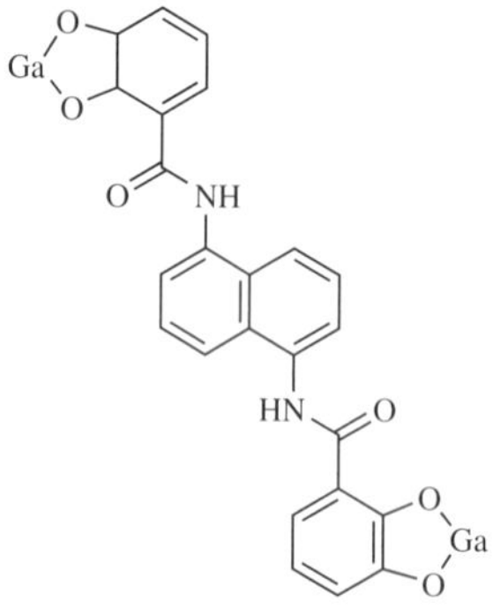

10.72

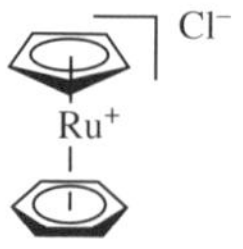

10.73

10.7. SOLID-STATE NMR SPECTROSCOPY

Several reports have described the use of solid-state NMR techniques for chiral analysis. One strategy is to examine the solid-state NMR spectra of crystals formed from optically pure or racemic mixtures. For example, the carbonyl and α-carbon signals in the optically pure, racemic, and *meso* crystal forms of tartaric acid exhibited distinct chemical shifts in their ^{13}C NMR spectra.[1656]

A similar phenomenon was observed in the solid-state ^{31}P NMR spectra of organo phosphorus compounds such as N-[(diphosphono)methyl]-3-*t*-butyl-thiomorpholine (**10.74**) and its corresponding 3-isobutyl derivative.[1657] The ^{31}P and ^{13}C chemical shielding tensors varied in racemic versus enantiomerically pure crystals because of different crystal symmetries. These differences could be used to determine optical purity.

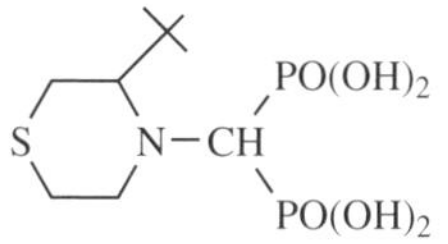

10.74

The utility of a solid-state technique known as one-dimensional exchange spectroscopy by sideband alteration (ODESSA) for the determination of the optical purity of chiral compounds has been demonstrated.[1658,1659] The method probes differences in internuclear distances between chemically equivalent sites and is based on chemical shielding tensors and a distant dependent Zeeman magnetization exchange between adjacent ^{31}P atoms. The ODESSA decay obtained through the use of a specific pulse sequence correlated with the optical purity of the compound. Separate samples of the racemate and pure enantiomers were needed to be able to perform the analysis, but the method worked even if there were identical isotropic chemical shifts. The method was demonstrated for cyclophosphamide **10.75** and L- and DL-phosphorylated amino acids.[1659] The different packing of the enantiomers of **10.75** and of the L-phosphorylated amino acids from the DL-phosphorylated amino acids caused different P–P distances that could be distinguished through the ODESSA technique.

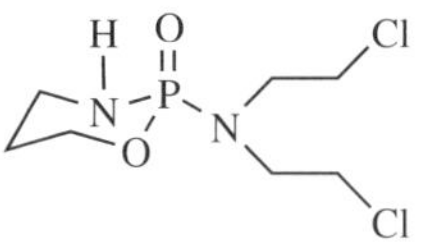

10.75

10.7.1. Tri-*O*-thymotide Clathrates

The cyclic triester of *O*-thymotic acid (**10.76**) forms a solid-state clathrate that crystallizes into two propeller forms with a (*P*)- or (*M*)-conformation.[1660]

Enantiomeric guest molecules co-crystallize into the clathrates and solid-state NMR spectroscopy has been used to monitor the discrimination. Enantiomeric discrimination in the ^{13}C NMR spectra of 2-hydroxy-, 2-chloro-, 2-bromo-, and 2-iodobutane was observed.[1661] The ^{2}H and ^{13}C NMR spectra of 2-bromo[1,1,1-^{2}H$_3$]butane also exhibited nonequivalence in solid-state complexes with **10.76**.[1662] Enantiotopic groups in prochiral compounds such as dimethylsulfoxide-d_6, 2-propanol-d_6, 2-bromopropane, and 2-propanol exhibited discrimination in the solid-state ^{13}C or ^{2}H NMR spectra of their clathrate complex.

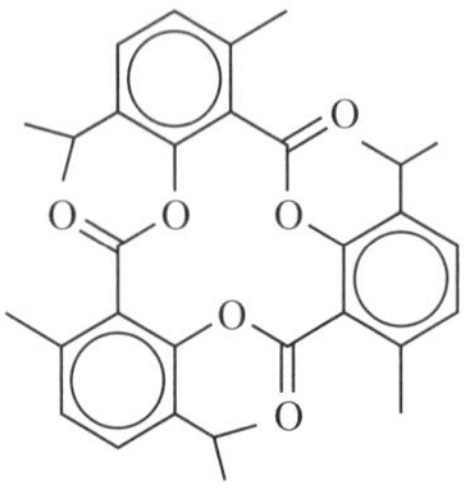

10.76

10.7.2. Cholic Acid Channels

Solid-state cholic acid (**10.77**) forms channels that can accommodate guest compounds. Solid-state ^{13}C NMR spectroscopy has been used to examine the complexation of γ-valerolactone (**10.78**) in a cholic acid host.[1663,1664] The methyl signal of the substrate was monitored in the analysis. The (S)-enantiomer of the substrate was more favored in the channels but the presence of four methyl resonances indicated that there was a preferential inclusion of a pair of enantiomers with an "inner" and "outer" species.[1664] The uptake of racemic aliphatic alcohols in the channels of nordeoxycholic acid (**10.79**) was studied by solid-state ^{13}C NMR spectroscopy.[1665]

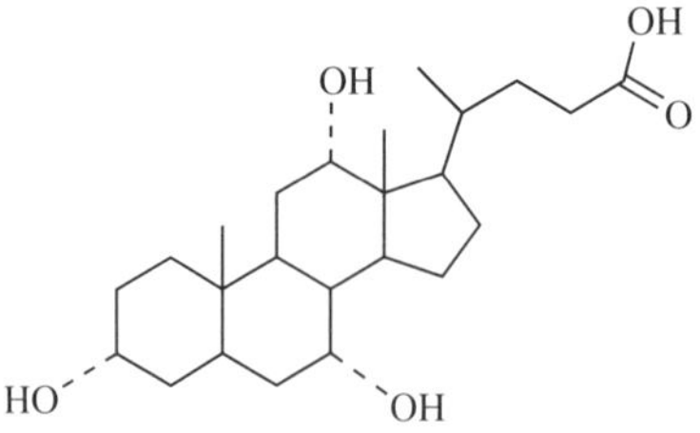

10.77 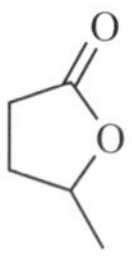   **10.78** 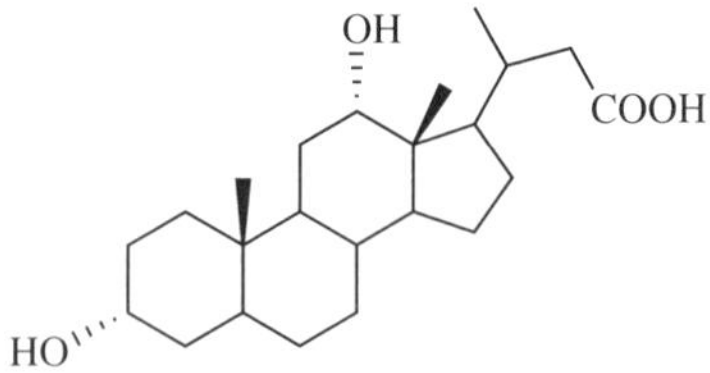   **10.79**

Methyl 7α,12α-dihydroxy-3α-phenylcarbamoyloxy-5β-cholan-24-oate (**10.80**), a soluble analog of a cholic-acid based liquid chromatographic phase, was used as an

NMR probe to study the complexation properties of BINOL.[1666] Nonequivalence of several resonances was noted in the ^{1}H and ^{13}C NMR spectra.

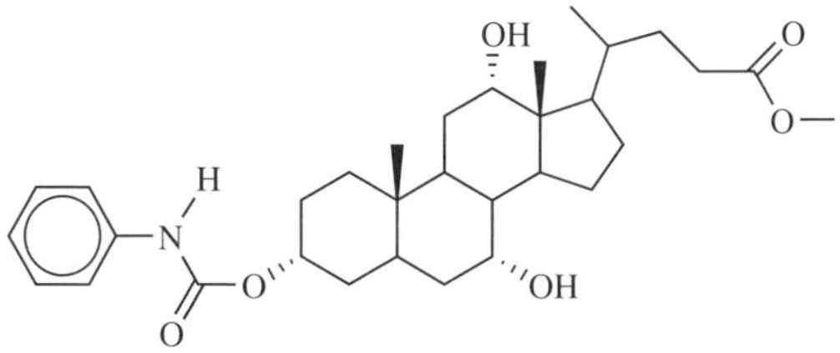

10.80

10.7.3. Analysis of Polymer-bound Compounds

The enantiomeric purity of polymer-bound compounds have been analyzed using MTPA derivatives.[1667,1668] MTPA derivatives of chloroalcohols such as **10.81** had broadened resonances but enough distinction between the signals of the diastereomers to determine enantiomeric purity.[1668] An on-bead analysis of the optical purity of polymer-bound olefins has been described. The olefins were bound to Wang-resin or TentaGel S-OH, subjected to a Sharpless asymmetric dehydroxylation producing compounds such as **10.82**, then subjected to derivatization with MTPA. The spectra were obtained using high resolution magic angle spinning (HRMAS) and the results were identical to those achieved by cleaving off the olefins and conducting a liquid chromatographic analysis.[1667]

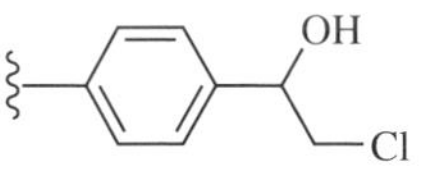

10.81 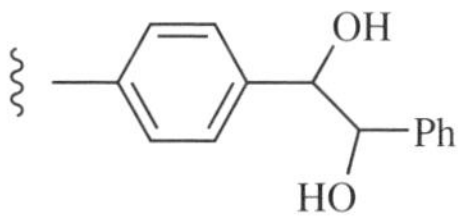   **10.82**

REFERENCES

1. Raban M, Mislow K. The determination of optical purity by nuclear magnetic resonance spectroscopy. *Tetrahedron Lett.* **1965**, 4249–4253.
2. Cawley A, Duxbury JP, Kee TP. NMR determination of enantiopurity via chiral derivatisation. *Tetrahedron: Asymmetry* **1998**, *9*, 1947–1949.
3. Vigneron JP, Dhaenens M, Horeau A. Nouvelle methode pour porter au maximum la purete optique d'un produit partiellement dedouble sans l'aide d'aucune substance chirale. *Tetrahedron* **1973**, *29*, 1055–1059.
4. Pirkle WH, Hoover DJ. NMR chiral solvating agents. *Top. Stereochem.* **1982**, *13*, 263–327.
5. Weisman GR. Nuclear magnetic resonance analysis using chiral solvating agents. *Asymmetric Synthesis* **1983**, *1*, 153–171.
6. Yamaguchi S. Nuclear magnetic resonance analysis using chiral derivatives. *Asymmetric Synthesis* **1983**, *1*, 125–152.
7. Rinaldi PL. The determination of absolute configuration using nuclear magnetic resonance techniques. *Prog. Nucl. Magn. Reson. Spectrosc.* **1982**, *15*, 291–352.
8. Takeuchi Y, Takahashi T. Flourine-containing agents for determining enantiomeric excess of chiral molecules. *Enantiocontrolled Synthesis of Fluoro-Organic Compounds*. Soloshonok VA, Ed. John Wiley & Sons, Ltd: New York, NY, **1999**, 497–534.
9. Parker D. NMR determination of enantiomeric purity. *Chem. Rev.* **1991**, *91*, 1441–1457.
10. Webb TH, Wilcox CS. Enantioselective and diastereoselective molecular recognition of neutral molecules. *Chem. Soc. Rev.* **1993**, 383–395.
11. Wenzel TJ, Wilcox JD. Chiral reagents for the determination of enantiomeric excess and absolute configuration using NMR spectroscopy. *Chirality* **2003**, *15*, 256–270.

Discrimination of Chiral Compounds Using NMR Spectroscopy, by Thomas J. Wenzel
Copyright © 2007 John Wiley & Sons, Inc.

12. Buckingham AD. Chirality in NMR spectroscopy. *Chem. Phys. Lett.* **2004**, *398*, 1–5.
13. Dale JA, Mosher HS. Nuclear magnetic resonance enantiomer reagents. Configurational correlations via nuclear magnetic resonance chemical shifts of diastereomeric mandelate, O-methylmandelate, and α-methoxy-α-trifluoromethylphenylacetate (MTPA) esters. *J. Am. Chem. Soc.* **1973**, *95*, 512–519.
14. Latypov SK, Seco JM, Quiñoá E, Riguera R. Conformational structure and dynamics of arylmethoxyacetates: DNMR spectroscopy and aromatic shielding effect. *J. Org. Chem.* **1995**, *60*, 504–515.
15. Seco JM, Latypov SK, Quiñoà E, Riguera R. Determining factors in the assignment of the absolute configuration of alcohols by NMR. The use of anisotropic effects on remote positions. *Tetrahedron* **1997**, *53*, 8541–8564.
16. Seco JM, Latypov S, Quiñoá E, Riguera R. New chirality recognizing reagents for the determination of absolute stereochemistry and enantiomeric purity by NMR. *Tetrahedron Lett.* **1994**, *35*, 2921–2924.
17. Kusumi T, Fujita Y, Ohtani I, Kakisawa H. Anomaly in the modified Mosher's method: absolute configurations of some marine cembranolides. *Tetrahedron Lett.* **1991**, *32*, 2923–2926.
18. Ohtani I, Kusumi T, Kashman Y, Kakisawa H. High-field FT NMR application of Mosher's method. The absolute configuration of marine terpenoids. *J. Am. Chem. Soc.* **1991**, *113*, 4092–4096.
19. Kusumi T, Ohtani II. Determination of the absolute configuration of biologically active compounds by the modified Mosher's method. *Biology-Chemistry Interface*, Dekker: New York, NY **1999**, 103–137.
20. Dale JA, Dull DJ, Mosher HS. α-Methoxy-α-trifluoromethylphenylacetic acid, a versatile reagent for the determination of enantiomeric composition of alcohols and amines. *J. Org. Chem.* **1969**, *34*, 2543–2549.
21. Dale JA, Mosher HS. Nuclear magnetic resonance nonequivalence of diastereomeric esters of α-substituted phenylacetic acids for the determination of stereochemical purity. *J. Am. Chem. Soc.* **1968**, *90*, 3732–3738.
22. Dale JA, Mosher HS. Enantiomeric purity of phenylethylene glycol and reliability of phenylglyoxylate asymmetric reductions in configurational assignments. *J. Org. Chem.* **1970**, *35*, 4002–4003.
23. Sullivan GR, Dale JA, Mosher HS. Correlation of configuration and ^{19}F chemical shifts of α-methoxy-α-trifluoromethylphenylacetate derivatives. *J. Org. Chem.* **1973**, *38*, 2143–2147.
24. Yang L, Andersen RJ. Absolute configuration of the anti-inflammatory sponge natural product contignasterol. *J. Nat. Prod.* **2002**, *65*, 1924–1926.
25. Mosandl A, Heusinger G, Gessner M. Analytical and sensory differentiation of 1-octen-3-ol enantiomers. *J. Agric. Food Chem.* **1986**, *34*, 119–122.
26. Seco JM, Quiñoá E, Riguera R. The assignment of absolute configurations by NMR of arylmethoxyacetate derivaties is this methodology being correctly used? *Tetrahedron: Asymmetry* **2000**, *11*, 2781–2791.
27. Latypov SK, Seco JM, Quiñoá E, Riguera R. MTPA vs MPA in the determination of the absolute configuration of chiral alcohols by ^{1}H NMR. *J. Org. Chem.* **1996**, *61*, 8569–8577.
28. Fernandez-Megia E, Ley SV. Recent studies on the use of MPA and MTPA in the determination of the absolute configuration of secondary alcohols by ^{1}H NMR. *Chemtracts*: *Org. Chem.* **1999**, *12*, 539–546.

29. Latypov SK, Galiullina NF, Aganov AV, Kataev VE, Riguera R. Determination of the absolute stereochemistry of alcohols and amines by NMR of the group directly linked to the chiral derivatizing reagent. *Tetrahedron* **2001**, *57*, 2231–2236.
30. Abdelmageed OH, Duclos RI, Abushanab E, Makriyannis A. Chirospecific syntheses of ^{2}H- and ^{13}C-labeled 1-O-alkyl-2-O-alkyl'-sn-glycero-3-phosphoethanolamines and 1-O-alkyl-2-O-alkyl'-sn-glycero-3-phosphocholines. *Chem. Phys. Lipids* **1990**, *54*, 49–59.
31. Riccio R, Finamore E, Santaniello M, Zollo F. Stereoselective synthesis of (24S)- and (24R)-24-(hydroxymethyl)cholesta-5,22(E)-dien-3β-ol: model compounds for stereochemical assignments of polyhydroxylated marine steroids. *J. Org. Chem.* **1990**, *55*, 2548–2552.
32. Hecht SS, Spratt TE, Trushin N. Absolute configuration of 4-(methylnitrosamino)-1-(3-pyridyl)-1-butanol formed metabolically from 4-(methylnitrosamino)-1-(3-pyridyl)-1-butanone. *Carcinogenesis* **1997**, *15*, 1851–1854.
33. Tamiaki H, Kitamoto H, Nishikawa A, Hibino T, Shibata R. Determination of 3^1- stereochemistry in synthetic bacteriochlorophyll-d homologues and self-aggregation of their zinc complexes. *Bioorg. Med. Chem.* **2004**, *12*, 1657–1666.
34. Zarevúcka M, Rejzek M, Šaman D, Wimmer Z, Vaněk T, Zhao Q, Legoy M. How to assign the absolute configuration of 2-substituted cyclopentanols. *Enantiomer* **1995**, *1*, 227–231.
35. Wimmer Z, Zarevúcka M, Rejzek M, Šaman D. Quantitative analysis of enzyme-mediated hydrolytic processes and their products using HPLC and NMR techniques. *Helv. Chim. Acta* **1997**, *80*, 818–831.
36. Sata NU, Wada S, Matsunaga S, Watabe S, van Soest RWM, Fusetani N. Rubrosides A-H, new bioactive tetramic acid glycosides from the marine sponge *Siliquariaspongia japonica*. *J. Org. Chem.* **1999**, *64*, 2331–2339.
37. Lemière GL, Willaert JJ, Dommisse RA, Lepoivre JA, Alderweireldt FC. Determination of the absolute configuration and enantiomeric purity of alcohols from the ^{13}C-NMR spectra of the corresponding MTPA esters. *Chirality* **1990**, *2*, 175–184.
38. Kubota T, Tsuda M, Kobayashi J. Absolute stereochemistry of amphidinolide C: synthesis of C-1–C-10 and C-17–C-29 segments. *Tetrahedron* **2003**, *59*, 1613–1625.
39. Fukushi Y, Yajima C, Mizutani J. Noe correlations in MTPA, MNCB and MBNC esters. *Tetrahedron Lett.* **1994**, *35*, 9417–9420.
40. Drescher M, Li Y, Hammerschmidt F. Enzymes in organic chemistry. 2. Lipase-catalysed hydrolysis of 1-acyloxy-2-arylethylphosphonates and synthesis of phosphonic acid analogues of L-phenylalanine and L-tyrosine. *Tetrahedron* **1995**, *51*, 4933–4946.
41. Hammerschmidt F, Li Y. Determination of absolute configuration of α-hydroxyphosphonates by ^{31}P NMR spectroscopy of corresponding Mosher esters. *Tetrahedron* **1994**, *50*, 10253–10264.
42. Li Y, Hammerschmidt F. Enzymes in organic chemistry. 1. Enantioselective hydrolysis of α-(acyloxy) phosphonates by esterolytic enzymes. *Tetrahedron: Asymmetry* **1993**, *4*, 109–120.
43. Grossmann G, Poncioni M, Bornand M, Jolivet B, Neuburger M, Séquin U. Bioactive butenolides from *Streptomyces antibioticus* TÜ 99: absolute configurations and synthesis of analogs. *Tetrahedron* **2003**, *59*, 3237–3251.
44. Joshi BS, Newton MG, Lee DW, Barber AD, Pelletier SW. Reversal of absolute stereochemistry of the pyrrolo[2,1-b]quinazoline alkaloids vasicine, vasicinone, vasicinol and vasicinolone. *Tetrahedron: Asymmetry* **1996**, *7*, 25–28.

45. Ohtani II, Hotta K, Ichikawa Y, Isobe M. Application of modified Mosher's method to α-aromatic secondary alcohols. Exception of the rule and conformational analyses. *Chem. Lett.* **1995**, 513–514.

46. Shi X, Attygalle AB, Liwo A, Hao M, Meinwald J, Dharmaratne HRW, Wanigasekera, MWAP. Absolute stereochemistry of soulattrolide and its analogues. *J. Org. Chem.* **1998**, *63*, 1233–1238.

47. Zhao G, Chao J, Zeng L, Rieser MJ, McLaughlin JL. The absolute configuration of adjacent bis-THF acetogenins and asiminocin, a novel highly potent asimicin isomer from *Asimina triloba*. *Bioorg. Med. Chem.* **1996**, *4*, 25–32.

48. Ohtani I, Kusumi T, Kashman Y, Kakisawa H. A new aspect of the high-field NMR application of Mosher's method. The absolute configuration of marine triterpene sipholenol-A. *J. Org. Chem.* **1991**, *56*, 1296–1298.

49. Ohtani I, Kusumi T, Ishitsuka MO, Kakisawa H. Absolute configurations of marine diterpenes possessing a xenicane skeleton. An application of an advanced Mosher's method. *Tetrahedron Lett.* **1989**, *30*, 3147–3150.

50. Kusumi T, Ohtani I, Inouye Y, Kakisawa H. Absolute configurations of cytotoxic marine cembranolides; consideration of Mosher's method. *Tetrahedron Lett.* **1988**, *29*, 4731–4734.

51. Kelly DR. A new method for the determination of the absolute stereochemistry of aromatic and heteroaromatic alkanets using Mosher's esters. *Tetrahedron: Asymmetry* **1999**, *10*, 2927–2934.

52. Arnone A, Nasini G, Vajna de Pava O. A hydroxytetradecatrienoic acid from *Mycosphaerella rubella*. *Phytochemistry* **1998**, *48*, 507–510.

53. Iorizzi M, Bryan P, McClintock J, Minale L, Palagiano E, Maurelli S, Riccio R, Zollo F. Chemical and biological investigation of the polar constituents of the starfish *Luidia clathrata*, collected in the Gulf of Mexico. *J. Nat. Prod.* **1995**, *58*, 653–671.

54. Searle PA, Molinski TF, Brzezinski LJ, Leahy JW. Absolute configuration of phorboxazoles A and B from the marine sponge *Phorbas* sp. 1. Macrolide and hemiketal rings. *J. Am. Chem. Soc.* **1996**, *118*, 9422–9423.

55. Bernart MW, Cardellina II, JH, Balaschak MS, Alexander MR, Shoemaker RH, Boyd MR. Cytotoxic falcarinol oxylipins from *Dendropanax arboreus*. *J. Nat. Prod.* **1996**, *59*, 748–753.

56. Baldwin BW, Morrow CJ. Differentiation and assignment of the proton NMR signals in the bis-MTPA ester of meso-α,α'-dimethyl-1,4-benzenedimethanol. *Tetrahedron: Asymmetry* **1996**, *7*, 2871–2878.

57. Guo Y, Gavagnin M, Trivellone E, Cimino G. Absolute stereochemistry of petroformynes, high molecular polyacetylenes from the marine sponge *Petrosia ficiformis*. *Tetrahedron* **1994**, *50*, 13261–13268.

58. Velten R, Steglich W, Anke T. Determination of the absolute configuration of a tetracyclic drimane sesquiterpenoid by Mosher's method. *Tetrahedron: Asymmetry* **1994**, *5*, 1229–1232.

59. Riccio R, Santaniello M, Greco OS, Minale L. Structure elucidation of (22E,24R,25R)-24-methyl-5α-cholest-22-ene-3β,4β,5,6α,8,14,15α,25,26-nonaol and (22E,24S)-24-methyl-5α-cholest-22-ene-3β,4β,5,6α,8,14,15α,25,28-nonaol, minor marine polyhydroxysteroids isolated from the starfish *Archaster typicus*. *J. Chem. Soc., Perkin Trans. 1* **1989**, 823–826.

60. Parker KA, Ledeboer MW. Asymmetric reduction. A convenient method for the reduction of alkynyl ketones. *J. Org. Chem.* **1996**, *61*, 3214–3217.
61. Queiroz EF, Kuhl C, Terreaux C, Mavi S, Hostettmann K. New dihydroalkylhexenones from *Lannea edulis*. *J. Nat. Prod.* **2003**, *66*, 578–580.
62. Sy L, Brown GD. Novel phenylpropanoids and lignans from *Illicium verum*. *J. Nat. Prod.* **1998**, *61*, 987–992.
63. Kusumi T, Hamada T, Ishitsuka MO, Ohtani I, Kakisawa H. Elucidation of the relative and absolute stereochemistry of lobatriene, a marine diterpene, by a modified Mosher method. *J. Org. Chem.* **1992**, *57*, 1033–1035.
64. Ichikawa A, Takahashi H, Ooi T, Kusumi T. Absolute configurations of some hydroxy-fatty acids produced by the insect genus *Laccifer*. *Biosci. Biotech. Biochem.* **1997**, *61*, 881–883.
65. Ichikawa A. Enantiomeric resolution of asymmetric glycol by a lipase-catalyzed transesterification reaction. *Chirality* **1999**, *11*, 338–342.
66. Ichikawa A. Application of the modified Mosher's method to acyclic glycols. *Enantiomer* **1997**, *2*, 327–331.
67. Ichikawa A. Application of the modified Mosher's method to acyclic glycols 2. Glycols possessing aromatic groups. *Enantiomer* **1998**, *3*, 255–261.
68. Boyd DR, Sharma ND, Boyle R, McMordie RAS, Chima J, Dalton H. A ^{1}H-NMR method for the determination of enantiomeric excess and absolute configuration of cis-dihydrodiol metabolites of polycyclic arenes and heteroarenes. *Tetrahedron Lett.* **1992**, *33*, 1241–1244.
69. Boyd DR, Sharma ND, Byrne B, Hand MV, Malone JF, Sheldrake GN, Blacker J, Dalton H. Enzymatic and chemoenzymatic synthesis and stereochemical assignment of cis-dihydrodiol derivatives of monosubstituted benzenes. *J. Chem. Soc., Perkin Trans.1* **1998**, 1935–1943.
70. Boyd DR, Dorrity MRJ, Hand MV, Malone JF, Sharma ND, Dalton H, Gray DJ, Sheldrake GN. Enantiomeric excess and absolute configuration determination of cis-dihydrodiols from bacterial metabolism of monocyclic arenes. *J. Am. Chem. Soc.* **1991**, *113*, 666–667.
71. Konno K, Fujishima T, Liu Z, Takayama H. Determination of absolute configuration of 1,3-diols by the modified Mosher's method using their di-MTPA esters. *Chirality* **2002**, *13*, 72–80.
72. Harada N, Saito A, Ono H, Gawronski J, Gawronska K, Sugioka T, Uda H, Kuriki T. A CD method for determination of the absolute stereochemistry of acyclic glycols. 1. Application of the CD exciton chirality method to acyclic 1,3-dibenzoate systems. *J. Am. Chem. Soc.* **1991**, *113*, 3842–3850.
73. Kouda K, Ooi T, and Kusumi T. Application of the modified Mosher's method to linear 1,3-diols. *Tetrahedron Lett.* **1991**, *40*, 3005–3008.
74. Lee J, Kobayashi Y, Tezuka K, Kishi Y. Toward creation of a universal NMR Database for the stereochemical assignment of acyclic compounds: proof of concept. *Org. Lett.* **1999**, *1*, 2181–2184.
75. MacMillan JB, Molinski TF. Caylobolide A, a unique 36-membered macrolactone from a Bahamian *Lyngbya majuscule*. *Org. Lett.* **2002**, *4*, 1535–1538.
76. Tan RX, Jensen PR, Williams PG, Fenical W. Isolation and structure assignment of rostratins A–D, cytotoxic disulfides produced by the marine-derived fungus *Exserohilum rostratum*. *J. Nat. Prod.* **2004**, *67*, 1374–1382.

77. Shimada H, Nishioka S, Singh S, Sahai M, Fujimoto Y. Absolute stereochemistry of non-adjacent bis-tetrahydrofuranic acetogenins. *Tetrahedron Lett.* **1994**, *35*, 3961–3964.
78. Gu Z, Zeng L, Fang X, Colman-Saizarbitoria T, Huo M, McLaughlin JL. Determining absolute configurations of stereocenters in annonaceous acetogenins through formaldehyde acetal derivatives and Mosher ester methodology. *J. Org. Chem.* **1994**, *59*, 5164–5172.
79. Shi G, He K, Liu X, Ye Q, MacDougal JM, McLaughlin JL. A novel application of Mosher's method to epimeric carbinols in acetogenins; absolute configurationspurities of 12-hydroxy-bullatacins A and B, new acetogenins from *Rollinia mucosa*. *Nat. Prod. Lett.* **1997**, *10*, 125–130.
80. Shi G, Gu Z, He K, Wood KV, Zeng L, Ye Q, MacDougal JM, McLaughlin JL. Applying Mosher's method to acetogenins bearing vicinal diols. The absolute configurations of muricatetrocin C and rollidecins A and B, new bioactive acetogenis from *Rollinia mucosa. Bioorg. Med. Chem.* **1996**, *4*, 1281–1286.
81. Rieser MJ, Hui Y, Rupprecht JK, Kozlowski JF, Wood KV, McLaughlin JL, Hanson PR, Zhuang Z, and Hoye TR. Determination of absolute configuration of stereogenic carbinol centers in annonaceous acetogenins by ^{1}H- and ^{19}F-NMR analysis of Mosher ester derivatives. *J. Am. Chem. Soc.* **1992**, *114*, 10203–10213.
82. Tsuda M, Toriyabe Y, Endo T, Kobayashi J. Application of modified Mosher's method for primary alcohols with a methyl group at C2 position. *Chem. Pharm. Bull.* **2003**, *51*, 448–451.
83. Yasuhara F, Yamaguchi S, Kasai R, Tanaka O. Assignment of absolute configuration of 2-substituted-1-propanols by ^{1}H NMR spectroscopy. *Tetrahedron Lett.* **1986**, *27*, 4033–4034.
84. De Rosa S, Milone A, Crispino A, Jaklin A, De Giulio A. Absolute configuration of 2, 6-dimethylheptyl sulfate and its distribution in ascidiacea. *J. Nat. Prod.* **1997**, *60*, 462–463.
85. Wimmer Z, Vašíčková S, Šaman D, Romaňuk M. Oxime ethers as potential juvenoids: synthesis of optical isomers. *Liebigs Ann. Chem.* **1990**, 847–852.
86. Ramón DJ, Guillena G, Seebach D. Nonreductive enantioselective ring opening of *N*-(methylsulfonyl)dicarboximides with diisopropoxytitanium $\alpha,\alpha,\alpha',\alpha'$-tetraaryl-1,3-dioxolane-4,5-dimethanolate. *Helv. Chim. Acta.* **1996**, *79*, 875–894.
87. Izumi S, Moriyoshi H, Hirata T. Identification of absolute configuration of tertiary alcohols by combination of Mosher's method and conformational analysis. *Bull. Chem. Soc. Jpn.* **1994**, *67*, 2600–2602.
88. Hoye TR, Renner MK. MTPA (Mosher) amides of cyclic secondary amines: conformational aspects and a useful method for assignment of amine configuration. *J. Org. Chem.* **1996**, *61*, 2056–2064.
89. Hoye TR, Renner MK. Applications of MTPA (Mosher) amides of secondary amines: assignment of absolute configuration in chiral cyclic amines. *J. Org. Chem.* **1996**, *61*, 8489–8495.
90. Kang C, Guo H, Qiu X, Bai X, Yao H, Gao L. Assignment of absolute configuration of cyclic secondary amines by NMR techniques using Mosher's method: a general procedure exemplified with (−)-isoanabasine. *Magn. Reson. Chem.* **2006**, *44*, 20–24.
91. Hietaniemi L, Pohjala E, Mälkönen P, Riekkola M. Assay of enantiomericpurities of β-blockers. Complications in the Mosher method. *Finn. Chem. Lett.* **1989**, *16*, 67–73.

92. Smith ECR, McQuaid LA, Paschal JW, DeHoniesto J. An enantioselective synthesis of D-(−)- and L-(+)-2-amino-3-phosphonopropanoic acid. *J. Org. Chem.* **1990**, *55*, 4472–4474.

93. Hull WE, Seeholzer K, Baumeister M, Ugi I. A modified synthesis of Mosher's acid and its use in a sensitive stereoisomer analysis of amino acid derivatives. *Tetrahedron* **1986**, *42*, 547–552.

94. Comba P, Hörmann A, Martin LL, Zipper L. Stereoselective coordination of chiral matrices: cobalt (III) chemistry of simple facially coordinating chiral triamines. *Helv. Chim. Acta* **1990**, *73*, 874–882.

95. Seco JM, Latypov SK, Quiñoá E, Riguera R. Choosing the right reagent for the determination of the absolute configuration of amines by NMR: MTPA or MPA? *J. Org. Chem.* **1997**, *62*, 7569–7574.

96. Kusumi T, Fukushima T, Ohtani I, Kakisawa H. Elucidation of the absolute configurations of amino acids and amines by the modified Mosher's method. *Tetrahedron Lett.* **1991**, *32*, 2939–2942.

97. Shi X, Attygalle AB, Meinwald J. Synthesis and absolute configuration of two defensive alkaloids from the Mexican bean beetle *Epilachna varivestis*. *Tetrahedron Lett.* **1997**, 38, 6479–6482.

98. Shi X, Attygalle AB, Xu S, Ahmad VU, Meinwald J. Synthesis and absolute configuration of 2-(12′-aminotridecyl)-pyrrolidine, a defensive alkaloid from the Mexican bean beetle, *Epilachna varivestis*. *Tetrahedron* **1996**, *52*, 6859–6868.

99. Kamiyama T, Umino T, Itezono Y, Nakamura Y, Satoh T, Yokose K. Sulfobacins A and B, novel von Willebrand factor receptor antagonists. II. Structural elucidation. *J. Antibiot.* **1995**, *48*, 929–936.

100. Buddrus J, Herzog H, Risch K. Determination of the enantiomeric ratio of organic ammonium halides or alkali carboxylates by NMR spectroscopy. *Anal. Chem.* **1994**, *66*, 40–42.

101. Baxter CAR, Richards HC. Substituted 1,2,3,4-tetrahydroquinolines. Measurement of optical purity by nuclear magnetic resonance spectroscopy. *Tetrahedron Lett.* **1972**, 3357–3358.

102. Navrátilová H. Use of S-Mosher acid as a chiral solvating agent for enantiomeric analysis of some trans-4-(4-fluor phenyl)-3-substituted-1-methylpiperidines by means of NMR spectroscopy. *Chirality*, **2001**, *13*, 731–735.

103. Navràtilovà H, de Gelder R, Kříž Z. Enantiodiscrimination in NMR spectra and X-ray structures of diastereomeric salts of trans-4-(4-fluorophenyl)-3-hydroxy-methyl-1-methylpiperidine with (S)-Mosher acid. *J. Chem. Soc., Perkin Trans. 2* **2002**, 2093–2099.

104. Benson SC, Cai P, Colon M, Haiza MA, Tokles M, Snyder JK. Use of carboxylic acids as chiral solvating agents for the determination of optical purity of chiral amines by NMR spectroscopy. *J. Org. Chem.* **1988**, *53*, 5335–5341.

105. Dyllick-Brenzinger R, Roberts JD. Chiral recognition by ^{15}N NMR spectroscopy. 8-Benzyl-5,6,7,8-tetrahydroquinoline. *J. Am. Chem. Soc.* **1980**, *102*, 1166–1167.

106. Villani, Jr. FJ, Costanzo MJ, Inners RR, Mutter MS, McClure DE. Determination of enantiomeric purity of tertiary amines by ^{1}H NMR of α-methoxy-α-(trifluoromethyl)-phenylacetic acid complexes. *J. Org. Chem.* **1986**, *51*, 3715–3718.

107. Oppolzer W, Kurth M, Reichlin D, Moffatt F. A reinvestigation of asymmetric induction in Diels-Alder reactions to chiral acrylates. *Tetrahedron Lett.* **1981**, *22*, 2545–2548.

108. Mori K, Nukada T, Ebata T. Synthesis of optically active forms of methyl (E)-2,4, 5-tetradecatrienoate, the pheromone of the male dried bean beetle, *Tetrahedron* **1981**, *37*, 1343–1347.

109. Van Os CPA, Vente M, Vliegenthart JFG. A NMR shift method for determination of the enantiomeric composition of hydroperoxides formed by lipoxygenase. *Biochim. Biophys. Acta* **1979**, *574*, 103–111.

110. Aareskjold K, Liaaen-Jensen S. Determination of enantiomeric composition of partly racemized carotenols. *Acta Chem. Scand.* **1982**, *36*, 499–504.

111. Yamaguchi S, Dale JA, Mosher HS. Asymmetric synthesis with (S)-(−)-*n*-butyl-*tert*-butylcarbinyl benzoylformate. *J. Org. Chem.* **1972**, *37*, 3174–3176.

112. Kabuto K, Yasuhara F, Yamaguchi S. Determination of enantiomeric purity of axially chiral biaryls. *Tetrahedron Lett.* **1980**, *21*, 307–308.

113. Yamaguchi S, Yasuhara F, Kabuto K. Use of shift reagent with diastereomeric MTPA esters for determination of configuration and enantiomeric purity of secondary carbinols in ^{1}H NMR spectroscopy. *Tetrahedron* **1976**, *32*, 1363–1367.

114. Yasuhara F, Yamaguchi S. Use of shift reagent with MTPA derivatives in ^{1}H NMR spectroscopy. III. Determination of absolute configuration and enantiomeric purity of primary carbinols with chiral center at the C-2 Position. *Tetrahedron Lett.* **1977**, 4085–4088.

115. Yasuhara F, Kabuto K, Yamaguchi S. Use of shift reagent with MTPA derivatives in ^{1}H NMR spectroscopy. IV. Determination of absolute configuration and enantiomeric purity of amino acid derivatives. *Tetrahedron Lett.* **1978**, 4289–4292.

116. Yasuhara F, Yamaguchi S. Determination of absolute configuration and enantiomeric purity of 2- and 3-hydroxycarboxylic acid esters. *Tetrahedron Lett.* **1980**, *21*, 2827–2830.

117. Kabuto K, Yasuhara F, Yamaguchi S. Determination of absolute configuration of axially chiral biaryls. *Tetrahedron Lett.* **1981**, 22, 659–662.

118. Yamaguchi S, Yasuhara F. Use of shift reagent with MTPA derivatives in ^{1}H NMR spectroscopy. II. Determination of absolute configuration and diastereomeric composition of secondary carbinols in epimeric mixture. *Tetrahedron Lett.* **1977**, 89–92.

119. Kabuto K, Yasuhara F, Yamaguchi S. A revised absolute configuration of 2-hydroxy-1,1′-binaphthyl. *Bull. Chem. Soc. Jpn.* **1983**, *56*, 1263–1264.

120. Merckx EM, Van de Wal AJ, Lepoivre JA, Alderweireldt FC. Use of shift reagent with MTPA derivatives in ^{19}F NMR spectroscopy: determination of absolute configuration of stereomeric 2- and 3- substituted cyclohexanols. *Bull. Soc. Chim. Belg.* **1987**, *87*, 21–25.

121. Merckx EM, Lepoivre JA, Lemière GL, Alderweireldt FC. Use of shift reagent with MTPA derivatives in ^{19}F NMR spectroscopy. IV. Determination of enantiomeric composition for a variety of secondary cycloalkanols. A survey. *Org. Magn. Reson.* **1983**, *21*, 380–387.

122. Vanhoeck L, Merckx EM, Bossaerts J, Lepoivre JA, Alderweireldt FC. A model for the complex of the (R)-MTPA ester of trans-4-tert-butylcyclohexanol with Eu(fod)$_3$, based on lanthanide induced shifts. *Org. Magn. Reson.* **1983**, *21*, 214–216.

123. Vanhoeck L, Bossaerts J, Dommisse RA, Lepoivre JA, Alderweireldt FC. The use of lanthanide-induced shifts for conformational and structural studies of (R)-MTPA (α-methoxy-α-trifluoromethyl-α-phenylacetic acid) esters. *Org. Magn. Reson.* **1984**, *22*, 24–28.

124. Van de Wal AJ, Merckx EM, Lemière GL, Van Osselaer TA, Lepoivre JA, Alderweireldt FC. Use of shift reagent with MTPA derivatives in ^{19}F NMR spectroscopy: III. Determination of absolute configuration and enantiomeric purity of secondary alcohols. *Bull. Soc. Chim. Belg.* **1978**, *87*, 545–552.

125. Sadozai CK, Merckx EM, Van de Wal AJ, Lemière GL, Esmans EL, Lepoivre JA, Alderweireldt FC. Enzymatic "in vitro" reduction of ketones. 9. Preparation of pure (1S,2S)-2-chlorocyclohexanol. *Bull. Soc. Chim. Belg.* **1982**, *91*, 163–170.

126. Lightner DA, Bouman TD, Gawroński JK, Gawrońska K, Chappuis JL, Crist BV, Hansen AE. Dissymmetric chromophores. 7. On the optical activity of conjugated cisoid dienes: an experimental-theoretical study of 5-alkyl-1,3-cyclohexadienes. *J. Am. Chem. Soc.* **1981**, *103*, 5314–5327.

127. Yasuhara F, Yamaguchi S, Takeda M, Abe T, Miyano S. Determination of the absolute configuration and enantiomeric purity of allylic and acetylenic alcohols. *Bull. Chem. Soc. Jpn.* **1991**, *64*, 3390–3394.

128. Yamaguchi S, Yasuhara F. Use of shift reagent with methoxy(trifluoromethyl) phenylacetic acid (MTPA) derivatives in proton NMR spectroscopy. II. Determination of absolute configuration and diastereomeric composition of secondary carbinols in epimeric mixtures. *Tetrahedron Lett.* **1977**, 89–92.

129. Kabuto K, Yasuhara F, Yamaguchi S. Determination of absolute configuration and enantiomeric purity of spirocyclic alcohols by ^{1}H NMR. *Tetrahedron Lett.* **1984**, *25*, 5801–5804.

130. Kalyanam N, Lightner DA. A convenient method of the determination of absolute configuration and enantiomeric excess of bicyclic secondary carbinols. *Tetrahedron Lett.* **1979**, 415–418.

131. Payan JP, Cossec B, Beydon D, Fabry JP, Ferrari E. Toxicokinetics and metabolism of 1, 2-diethylbenzene in male sprague dawley rats. 2. Evidence for in vitro and in vivo stereoselectivity of 1,2-diethylbenzene metabolism. *Drug Metab. Dispos.* **2001**, *29*, 868–876.

132. Toyota M, Omatsu I, Braggins J, Asakawa Y. New humulane-type sesquiterpenes from the liverworts *Tylimanthus tenellus* and *Marchantia emarginata* subsp. *tosana*. *Chem. Pharm. Bull.* **2004**, *52*, 481–484.

133. Sugimoto Y, Sakita T, Ikeda T, Moriyama Y, Murae T, Tsuyuki T, Takahashi T. Structure of a new ionone derivative, nigakialcohol from *Picrasma ailanthoides* PLANCHON. *Bull. Chem. Soc. Jpn.* **1997**, *52*, 3027–3032.

134. García R, Seco JM, Vázquez SA, Quiñoá E, Riguera R. Absolute configuration of secondary alcohols by ^{1}H NMR: in situ complexation of α-methoxyphenylacetic acid esters with barium(II). *J. Org. Chem.* **2002**, *67*, 4579–4589.

135. Omata K, Fujiwara T, Kabuto K. Use of diamagnetic lanthanide complex for extending the scope of NMR determination of absolute configuration by the modified Mosher's method. *Tetrahedron: Asymmetry* **2002**, *13*, 1655–1662.

136. Su B, Park EJ, Mbwambo ZH, Santarsiero BD, Mesecar AD, Fong HHS, Pezzuto JM, Kinghorn AD. New chemical constituents of *Euphorbia quinquecostata* and absolute configuration assignment by a convenient Mosher ester procedure carried out in NMR Tubes. *J. Nat. Prod.* **2002**, *65*, 1278–1282.

137. Lechner D, Stavri M, Oluwatuyi M, Pereda-Miranda R, Gibbons S. The antistaphylococcal activity of *Angelica dahurica* (Bai Zhi). *Phytochemistry* **2004**, *65*, 331–335.

138. Adamczyk M, Fishpaugh JR. Expeditious synthesis of Mosher using a solid supported carbodiimide. *Tetrahedron Lett.* **1996**, *37*, 7171–7172.

139. Porto S, Durán J, Seco JM, Quiñoà E, Riguera R. "Mix and shake" method for configurational assignment by NMR: application to chiral amines and alcohols. *Org. Lett.* **2003**, *5*, 2979–2982.

140. Queiroz EF, Wolfender J, Raoelison G, Hostettmann K. Determination of the absolute configuration of 6-alkylated α-pyrones from *Ravensara crassifolia* by LC-NMR. *Photochem. Anal.* **2003**, *14*, 34–39.

141. Kobayashi Y, Lee J, Tezuka K, Kishi Y. Toward creation of a universal NMR database for the stereochemical assignment of acyclic compounds: the case of two contiguous propionate units. *Org. Lett.* **1999**, *1*, 2177–2180.

142. Kobayashi Y, Tan C, Kishi Y. Stereochemical assignment of the C21-C38 portion of the desertomycin/oasomycin class of natural products by using universal NMR databases: prediction. *Angew. Chem., Int. Ed.* **2000**, *39*, 4279–4281.

143. Tan C, Kobayashi Y, Kishi Y. Stereochemical assignment of the C21-C38 portion of the desertomycin/oasomycin class of natural products by using universal NMR databases: proof. *Angew. Chem., Int. Ed.* **2000**, *39*, 4282–4284.

144. Kobayashi Y, Tan C, Kishi Y. Toward creation of a universal NMR database for stereochemical assignment: the case of 1,3,5-trisubstituted acyclic systems. *Helv. Chim. Acta* **2000**, *83*, 2562–2571.

145. Kobayashi Y, Tan C, Kishi Y. Toward creation of a universal NMR database for stereochemical assignment: complete structure of the desertomycin/oasomycin class of natural products. *J. Am. Chem. Soc.* **2001**, *123*, 2076–2078.

146. Raban M, Mislow K. The determination of optical purity by nuclear magnetic resonance spectroscopy. II. Compounds which owe their dissymmetry to deuterium substitution. *Tetrahedron Lett.* **1966**, 3961–3966.

147. Sandran DJ, Mislow K, Giddings WP, Dirlam J, Hanson GC. Stereochemistry of 2-benznorbornenone. *J. Am. Chem. Soc.* **1968**, *90*, 4877–4884.

148. Helmchen G, Otto R, Sauber K. Objective separation and absolute configuration of enantiomeric carboxylic acids and amines. 1. *Tetrahedron Lett.* **1972**, 3873–3878.

149. Trost BM, Belletire JL, Godleski S, McDougal PG, Balkovec JM. On the use of the O-methylmandelate ester for establishment of absolute configuration of secondary alcohols. *J. Org. Chem.* **1986**, *51*, 2370–2374.

150. Thomas HT, Mislow, K. Stereochemistry and chiroptical properties of the 3-phenyl-2-norbornanones. *J. Am. Chem. Soc.* **1970**, *92*, 6292–6298.

151. Lamshöft M, Schmickler H, Marner F. Determination of the absolute configuration of hydroxyiridals by chiroptical and NMR spectroscopic methods. *Eur. J. Org. Chem.* **2003**, 727–733.

152. Gross H, Wright AD, Beil W, König GM. Two new bicyclic cembranolides from a new *Sarcophyton* species and determination of the absolute configuration of sarcoglaucol-16-one. *Org. Biomol. Chem.* **2004**, *2*, 1133–1138.

153. Pehk T, Lippmaa E, Lopp M, Paju A, Borer BC, Taylor RJK. Determination of the absolute configuration of chiral secondary alcohols; new advances using ^{13}C- and 2D-NMR spectroscopy. *Tetrahedron: Asymmetry* **1993**, *4*, 1527–1532.

154. Adamczeski M, Quiñoá E, Crews P. Novel sponge-derived amino acids. 11. The entire absolute stereochemistry of the bengamides. *J. Org. Chem.* **1990**, *55*, 240–242.

155. Trost BM, Curran DP. On the stereochemistry of the bis-nor-Wieland-Miescher ketone. *Tetrahedron Lett.* **1981** *22*, 4929–4932.
156. Gupta S, Krasnoff SB, Renwick JAA, Roberts DW. Viridoxins A and B: novel toxins from the fungus *Metarhizium flavoviride*. *J. Org. Chem.* **1993**, *58*, 1062–1067.
157. Chen C, Reamer RA, Roy A, Chilenski JR. α-Methoxyphenyl acetic acid (MPA) for the configurational assignment of aromatic-heteroaromatic carbinols by ^{1}H NMR spectroscopy. *Tetrahedron Lett.* **2005**, *46*, 5593–5596.
158. Siegel C, Thornton ER. Diels-Alder reactions with dienes bearing a remote stereogenic center. Conformational model for diastereofacial selectivity. *Tetrahedron: Asymmetry* **1991**, *2*, 1413–1428.
159. Broom SJ, Wilkins AL, Lu Y, Ede RM. Novel nor-sesquiterpenoids in New Zealand honeys. The relative and absolute stereochemistry of the kamahines: an extension of the Mosher method to hemiacetals. *J. Org. Chem.* **1994**, *59*, 6425–6430.
160. Louzao I, Seco JM, Quiñoá E, Riguera R. The assignment of absolute configuration of cyanohydrins by NMR. *Chem. Commun.* **2006**, 1422–1424.
161. Kozlowski JK, Rath NP, Spilling CD. Determination of the enantiomeric purity and absolute configuration of α-hydroxy phosphonates. *Tetrahedron* **1995**, *51*, 6385–6396.
162. Skropeta D, Schmidt RR. Chiral, non-racemic α-hydroxyphosphonates and phosphonic acids via stereoselective hydroxylation of diallyl benzylphosphonates. *Tetrahedron: Asymmetry* **2003**, *14*, 265–273.
163. Wróblewski AE, Piotrowska DG. Enantiomeric phosphonate analogs of the paclitaxel C-13 side chain. *Tetrahedron: Asymmetry* **1999**, *10*, 2037–2043.
164. González-Morales A, Fernández-Zertuche M, Ordóñez M. Simultaneous separation and assignment of absolute configuration of γ-amino-β-hydroxyphosphonates by NMR using (S)-methoxyphenylacetic acid (MPA). *Rev. Soc. Quim. Méx.* **2004**, *48*, 239–245.
165. Greenland, GJ, Bowden, BF. Cembranoid diterpenes related to sarcophytol A from the soft coral *Sarcophyton trocheliophorum* (Alcyonacea). *Aust. J. Chem.* **1994**, *47*, 2013–2021.
166. Zhou B, Baj NJ, Glass TE, Malone S, Werkhoven MCM, van Troon F, David, Wisse JH, Kingston DGI. Bioactive labdane diterpenoids from *Renealmia alpinia* collected in the Suriname rainforest. *J. Nat. Prod.* **1997**, *60*, 1287–1293.
167. Falk H, Fröstl W, Schlögl K. Ferrocene derivatives. 56. Optically active, aromatically substituted spirans. 2. Synthesis, chiroptical properties, and absolute configuration of 7,7′-spiro-bi[3]ferrocenophane-6,6′-dione. *Monatsh. Chem.* **1971**, *102*, 1270–1278.
168. Rodríguez J, Riguera R. The natural polypropionate-derived esters of the mollusc *Onchidium* sp. *J. Org. Chem.* **1992**, *57*, 4624–4632.
169. Freire F, Seco JM, Quiñoá E, Riguera R. The prediction of the absolute stereochemistry of primary and secondary 1,2-diols by ^{1}H NMR spectroscopy: principles and applications. *Chem. Eur. J.* **2005**, *11*, 5509–5522.
170. Freire F, Seco JM, Quiñoá E, Riguera R. The assignment of the absolute configuration of 1,2-diols by low-temperature NMR of a single MPA derivative. *Org. Lett.* **2005**, *7*, 4855–4858.
171. Seco JM, Martino M, Quiñoá E, Riguera R. Absolute configuration of 1, *n*-diols by NMR: the importance of the combined anisotropic effects in bis-arylmethoxyacetates. *Org. Lett.* **2000**, *2*, 3261–3264.

REFERENCES 439

172. Freire F, Seco JM, Quiñoá E, Riguera R. Determining the absolute stereochemistry of secondary/secondary diols by ^{1}H NMR: basis and applications. *J. Org. Chem.* **2005**, *70*, 3778–3790.

173. Besada P, Terán C, Quezada E, Seco JM, Uriarte E. Resolution of racemic mixtures of carbocyclic analogues of nucleosides and assignment of their absolute configuration. *Nucleosides, Nucleotides Nucleic Acids* **2001**, *20*, 1359–1361.

174. Quezada E, Santana L, Uriarte E. Resolution of racemic carbonucleosides and assignment of the absolute configuration by NMR. *Tetrahedron: Asymmetry* **2001**, *12*, 2637–2639.

175. Jacobus J, Raban M, Mislow K. The preparation of (+)-N-methyl-1-(1-naphthyl) ethylamine and the determination of its optical purity by nuclear magnetic resonance. *J. Org. Chem.* **1968**, *33*, 1142–1145.

176. Jacobus J, Jones TB. Intrinsic and torsional diastereomers. The optical purity of (+)-(S)-deoxyephedrine. *J. Am. Chem. Soc.* **1970**, *92*, 4583–4585.

177. Ratajczak A, Żmuda H. Determination of optical purity of (+)-1-ferrocenyl-2-aminopropane by NMR spectroscopy. *Roczniki Chemii Ann. Soc. Chim. Polonorum* **1975**, *49*, 215–219.

178. Trost BM, Bunt RC, Pulley SR. On the use of O-methylmandelic acid for the establishment of absolute configuration of α-chiral primary amines. *J. Org. Chem.* **1994**, *59*, 4202–4205.

179. Latypov SK, Seco JM, Quiñoá E, Riguera R. Determination of the absolute stereochemistry of chiral amines by ^{1}H NMR of arylmethoxyacetic acid amides: the conformational model. *J. Org. Chem.* **1995**, *60*, 1538–1545.

180. Leiro V, Freire F, Quiñoá E, Riguera R. Absolute configuration of amino alcohols by ^{1}H-NMR. *Chem. Commun.* **2005**, 5554–5556.

181. Yabuuchi T, Kusumi T. NMR spectroscopic determination of the absolute configuration of chiral sulfoxides via N-(methoxyphenylacetyl)sulfoximines. *J. Am. Chem. Soc.* **1999**, *121*, 10646–10647.

182. Seco JM, Latypov S, Quiñoá E, Riguera R. Determination of the absolute configuration of alcohols by low temperature ^{1}H NMR of aryl(methoxy)acetates. *Tetrahedron: Asymmetry* **1995**, *6*, 107–110.

183. Andersson T, Borhan B, Berova N, Nakanishi K, Haugan JA, Liaaen-Jensen S. Absolute configurational assignment of 3-hydroxycarotenoids. *J. Chem. Soc., Perkin Trans. 1* **2000**, 2409–2414.

184. Latypov SK, Seco JM, Quiñoá E, Riguera R. Are both the (R)- and the (S)-MPA esters really needed for the assignment of the absolute configuration of secondary alcohols by NMR? The use of a single derivative. *J. Am. Chem. Soc.* **1998**, *120*, 877–882.

185. Simpson AF, Bodkin CD, Butts CP, Armitage MA, Gallagher T. Asymmetric reduction of prochiral cycloalkenones. The influence of exocyclic alkene geometry. *J. Chem. Soc., Perkin Trans. 1* **2000**, 3047–3054.

186. Scheid G, Kuit W, Ruijter E, Orru RVA, Henke E, Bornscheuer U, Wessjohann LA. A new route to protected acyloins and their enzymatic resolution with lipases. *Eur. J. Org. Chem.* **2004**, 1063–1074.

187. López B, Quiñoá E, Riguera R. Complexation with barium(II) allows the inference of the absolute configuration of primary amines by NMR. *J. Am. Chem. Soc.* **1999**, *121*, 9724–9725.

188. Garcia R, Seco JM, Vázquez SA, Quiñoá E, Riguera R. Role of barium(II) in the determination of the absolute configuration of chiral amines by ^{1}H NMR spectroscopy. *J. Org. Chem.* **2006**, *71*, 1119–1130.

189. Earle ME, Hultin PG, 4-Oxo-α-amino acids: a caution when determining absolute configurations by ^{1}H NMR using MPA derivatization and Ba^{2+} complexation. *Tetrahedron Lett.* **2000**, *41*, 7855–7858.

190. Zaitsev VP, Potapov VM, Dem'yanovich VM, Solov'eva LD. Stereochemical studies. XLIII. Certain features of the use of paramagnetic shift reagents during the determination of optical purity. *Zh. Org. Khim.* **1976**, *12*, 2326–2331.

191. Prestat G, Marchand A, Lebreton J, Guingant A, Pradère J. ^{1}H NMR evaluation of the enantiomeric purity of a series of heterocyclic β-dimethylamino esters and amides by using (S)-mandelic acid derivatives as chiral solvating agents. *Tetrahedron: Asymmetry* **1998**, *9*, 197–201.

192. Henrichs PM, Rodger CA, Caulfield T, Guo P. Differentiation of meso isomers from racemic mixtures with the combined use of chiral shift reagents and two-dimensional heteronuclear correlation NMR spectroscopy. *Magn. Res. Chem.* **1995**, *33*, 905–908.

193. Buist PH and Behrouzian B. Complexation of acyclic dialkyl sulfoxides with carboxylic acids: a stereochemical analysis of deshielding effects. *Magn. Res. Chem.* **1996**, *34*, 1013–1018.

194. Buist PH, and Marecak DM. Use of aromatic thia fatty acids as active site mapping agents for a yeast Δ^9 desaturase. *Can. J. Chem.* **1994**, *72*, 176–181.

195. Buist PH, Behrouzian B, MacIsaac KD, Cassel S, Rollin P, Imberty A, Gautier C, Pérez S, Genix P. Stereochemical analysis of D-glucopyranosyl-sulfoxides via a combined NMR, molecular modeling and X-ray crystallographic approach. *Tetrahedron: Asymmetry* **1999**, *10*, 2881–2889.

196. Buist PH, Marecak DM. Stereochemical analysis of a quasisymmetrical dialkyl sulfoxide obtained by a diverted biodehydrogenation reaction. *J. Am. Chem. Soc.* **1991**, *113*, 5877–5878.

197. Buist PH, Marecak D. (S)-α-Methoxyphenyl acetic acid: a new NMR chiral shift reagent for the stereochemical analysis of sulfoxides. *Tetrahedron: Asymmetry* **1995**, *6*, 7–10.

198. Buist PH, Behrouzian B, Cassel S, Lorin C, Rollin P, Imberty A, Perez S. Stereochemical analysis of D-galacto-sulfoxides using (S)-α-methoxyphenylacetic acid. *Tetrahedron: Asymmetry* **1997**, *8*, 1959–1961.

199. Gautier N, Noiret N, Nugier-Chauvin C, Patin H. NMR stereochemical analysis of chiral alkylsulfoxides with α-methoxyaryl acetic acids. *Tetrahedron: Asymmetry* **1997**, *8*, 501–505.

200. Buist PH, Marecak DM. Stereochemical analysis of sulfoxides obtained by derived desaturation. *J. Am. Chem. Soc.* **1992**, *114*, 5073–5080.

201. Yang Y, Kayser MM, Hooper D. Assignment of absolute configuration of 3-hydroxy β-lactams by the NMR "mix and shake" method. *Chirality* **2005**, *17*, 131–134.

202. Konishi K, Mori Y, and Taniguchi N. Quantitative analysis for enantiomeric purity of 2-alkanols by fluorine-19 magnetic resonance spectroscopy. *Analyst* **1969**, *94*, 1006–1009.

203. Whitesell JK, Reynolds D. Resolution of chiral alcohols with mandelic acid. *J. Org. Chem.* **1983**, *48*, 3548–3551.

204. Whitesell JK, Reynolds D. Resolution of chiral alcohols with mandelic acid. *J. Org. Chem.* **1983**, *48*, 3548–3551.

205. Chataigner I, Lebreton J, Durand D, Guingant A, Villiéras J. A new approach for the determination of the absolute configuration of secondary alcohols by ^{1}H NMR with O-substituted mandelate derivatives. *Tetrahedron Lett.* **1998**, *39*, 1759–1762.

206. Zhang B, Pionnier S. Natural stereospecific hydrogen isotope transfer in alcohol dehydrogenase-catalysed reduction. *Nukleonika* **2002**, *47*, S29–S31.

207. Rabiller C, Mesbahi M, Martin ML. ^{2}H-NMR resolution of the methylenic isotopomers of ethanol applied to the study of stereospecific enzyme-catalysed exchange. *Chirality* **1990**, *2*, 85–89.

208. Krois D, Lehner H. Resolution of chiral interconvertible diastereoisomers of a 2, 18-bridged biliverdin mediated by first-order asymmetric transformation. *J. Chem. Soc., Perkin Trans. 2* **1989**, 2085–2090.

209. Kataeva NA, Grishin YK, Dunina VV. Determination of the enantiomeric composition of *N,N*-dimethyl-α-ferrocenylethylamine by ^{1}H NMR spectroscopy. *Russ. Chem. Bull.* **2001**, *50*, 1323–1325.

210. Drandarov K, Guggisberg A, Hesse M. Asymmetric syntheses of the macrocyclic spermine alkaloids (−)-(*S*)-protoverbine, (−)-(*S*)-buchnerine, and their naturally occurring congenial alkaloids. *Helv. Chim. Acta* **2002**, *85*, 979–989.

211. Zingg SP, Arnett EM, McPhail AT, Bothner-By AA, Gilkerson WR. Chiral discrimination in the structures and energetics of association of stereoisomeric salts of mandelic acid with α-phenethylamine, ephedrine, and pseudoephedrine. *J. Am. Chem. Soc.* **1988**, *110*, 1565–1580.

212. Wang F, Wang Y, Polavarapu PL, Li T, Drabowicz J, Pietrusiewicz KM, Zygo K. Absolute configuration of tert-butyl-1-(2-methylnaphthyl)phosphine oxide. *J. Org. Chem.* **2002**, *67*, 6539–6541.

213. Terreno E, Botta M, Fedeli F, Mondino B, Milone L, Aime S. Enantioselective recognition between chiral α-hydroxy-carboxylates and macrocyclic heptadentate lanthanide(III) chelates. *Inorg. Chem.* **2003**, *42*, 4891–4897.

214. Parker D, Taylor RJ. Direct ^{1}H NMR assay of the enantiomeric composition of amines and β-amino alcohols using O-acetyl mandelic acid as a chiral solvating agent. *Tetrahedron* **1987**, *43*, 5451–5456.

215. Caira MR, Hunter R, Nassimbeni LR, Stevens AT. Resolution of albuterol acetonide, *Tetrahedron Asymmetry* **1999**, *10*, 2175–2189.

216. Giner J, Kiemle D, Zuniga DJ. Analysis of 2-deuterated isopentenyl alcohols by ^{1}H NMR of chiral esters. *Tetrahedron Lett.* **2002**, *43*, 1175–1177.

217. Abad J, Camps F. Arylacetic acid derivatization of 2,3- and internal erythro-squalene diols. Separation and absolute configuration determination. *Tetrahedron* **2004**, *60*, 11519–11525.

218. Itsuno S. Chiral polymer synthesis by means of repeated asymmetric reaction. *Prog. Polym. Sci.* **2005**, *30*, 540–558.

219. Itsuno S. Komura K. Highly stereoselective synthesis of chiral aldol polymers using repeated asymmetric Mukaiyama aldol reaction. *Tetrahedron* **2002**, *58*, 8237–8246.

220. Sureshan KM, Kiyosawa Y, Han F, Hyodo S, Uno Y, Watanabe Y. Resolution of synthetically useful myo-inositol derivatives using the chiral auxiliary O-acetylmandelic acid. *Tetrahedron: Asymmetry* **2005**, *16*, 231–241.

221. Sureshan KM, Miyasou T, Miyamori S, Watanabe Y, O-Acetylmandelic acid as a reliable chiral anisotropy reagent for the determination of absolute configuration of alcohols. *Tetrahedron: Asymmetry* **2004**, *15*, 3357–3364.

222. Sureshan KM, Miyasou T, Hayashi M, Watanabe Y. Is O-acetylmandelic acid a reliable chiral anisotropy reagent? *Tetrahedron: Asymmetry* **2004**, *15*, 3–7.

223. Huang S, Beale JM, Keller PJ, Floss HG. Synthesis of (R)- and (S)-[1-^{13}C$_1$,2-^{2}H$_1$] malonate and its stereochemical analysis by NMR spectroscopy. *J. Am. Chem. Soc.* **1986**, *108*, 1100–1101.

224. Panek JS, Sparks MA. Synthesis, resolution, and absolute stereochemical assignment of C1-oxygenated allysilanes and C3-oxygenated vinylsilanes. *Tetrahedron: Asymmetry* **1990**, *1*, 801–816.

225. Parve O, Aidnik M, Lille Ü, Martin I, Vallikivi I, Vares L, Pehk T. The tetrahydropyranyl-protected mandelic acid: a novel versatile chiral derivatising agent. *Tetrahedron: Asymmetry* **1998**, *9*, 885–896.

226. Aav R, Parve O, Pehk T, Claesson A, Martin I. Preparation of highly enantiopure stereoisomers of 1-(2,6-dimethylphenoxy)-2-aminopropane (mexiletine) *Tetrahedron: Asymmetry* **1999**, *10*, 3033–3038.

227. Haiza MA, Sanyal A, Snyder JK. O-Nitromandelic acid: a chiral solvating agent for the NMR determination of chiral diamine enantiomeric purity. *Chirality* **1997**, *9*, 556–562.

228. Bravo P, Ganazzoli F, Resnati G, De Munari S, Albinati A. Homochiral fluoro-organic compounds. 7. Determination of the absolute and relative configurations of fluorohydrins. *J. Chem. Res., Synop.* **1988**, 216–217.

229. Pabst A, Barron D, Sémon E, Schreier P. Two diastereomeric 3-oxo-α-ionol β-D-glucosides from raspberry fruit. *Phytochemistry* **1992**, *31*, 1649–1652.

230. Pabst A, Barron D, Sémon E, Schreier P. An α-ionol disaccharide glycoside from raspberry fruit. *Phytochemistry* **1992**, *31*, 2043–2046.

231. Arnone A, Bernardi R, Blasco F, Cardillo R, Resnati G, Gerus II, Kukhar VP. Trifluoromethyl vs. methyl ability to direct enantioselection in microbial reduction of carbonyl substrates. *Tetrahedron* **1998**, *54*, 2809–2818.

232. Bravo P, Frigerio M, Resnati G. Homochiral perfluoroalkyl-group-substituted secondary alcohols through stereoselective reduction of perfluoroalkyl 1-(p-tolylsulfinyl)alkyl ketones. *Synthesis* **1988**, 955–960.

233. Bravo P, Piovosi E, Resnati G, Fronza G. Asymmetric synthesis and structural analysis of 5-O-benzoyl-2,3-dideoxy-3-fluoro-α,β-D-ribofuranose and -xylofuranose from homochiral 1-fluoro-3-sulfinylacetone. *J. Org. Chem.* **1989**, *54*, 5171–5176.

234. Mosandl A, Günther C. Stereoisomeric flavor compounds. 20. Structure and properties of γ-lactone enantiomers. *J. Agric. Food Chem.* **1989**, *37*, 413–418.

235. Schellenberg A, Schmarr H, Eisenreich W, Engel K. Characterization of the enantiomers of γ- and δ-thiolactones. *Frontiers of Flavour Science* **1999**, 121–124.

236. Helmchen G, Schmierer R. Determination of the absolute configuration of chiral thiols by proton nuclear magnetic resonance spectroscopy of diastereomeric thiol esters. *Angew. Chem.* **1976**, *88*, 770–771.

237. Helmchen G, Schmierer R. Determination of the absolute configuration of chiral thiols by ^{1}H-NMR spectroscopy of diastereomeric thiol esters. *Angew. Chem. Int. Ed. Engl.* **1976**, *15*, 703–704.

238. Kowalczyk R, Skarżewski J. O-Methylatrolactic acid as a new reagent for determination of the enantiomeric purity and absolute configuration of chiral alcohols and amines. *Tetrahedron: Asymmetry* **2006**, *17*, 1370–1379.
239. Nadkarni PJ, Sawant MS, Trivedi GK. ^{1}H NMR correlation of chiral 3-phenylbutanoates for the determination of absolute stereochemistry of chiral alcohols. *Tetrahedron: Asymmetry* **1995**, *6*, 2001–2010.
240. Helmchen G. New method for the determination of the absolute configuration of chiral secondary alcohols and amines. NMR spectroscopy of diastereoisomeric esters and amides of α-phenylbutyric and hydratropic acid. *Tetrahedron Lett.* **1974**, 1527–1530.
241. Wang H, Gloer KB, Gloer JB, Scott JA, Malloch D. Anserinones A and B: new antifungal and antibacterial benzoquinones from the coprophilous fungus *Podospora anserina*. *J. Nat. Prod.* **1997**, *60*, 629–631.
242. Puder C, Zeeck A, Beil W. New biologically active rubiginones from *Streptomyces* sp. *J. Antibiot.* **2000**, *35*, 329–336.
243. Birnecker W, Wallnöfer B, Hofer O, Greger H. Relative and absolute configurations of two naturally occurring acetylenic spiroketal enol ether epoxides. *Tetrahedron* **1988**, *44*, 267–276.
244. Valente EJ, Pohl LR, Trager WF. Nuclear magnetic resonance configuration correlation of primary amine derivatives of α-methyl-α-methoxy(pentafluorophenyl)acetic acid. *J. Org. Chem.* **1980**, *45*, 543–546.
245. Pohl LR, Trager WF. A new chiral reagent for the determination of enantiomeric purity and absolute configuration of certain substituted β-arylethylamines. *J. Med. Chem.* **1973**, *16*, 475–479.
246. Takeuchi Y, Itoh N, Note H, Koizumi T, Yamaguchi K. α-Cyano-α-fluorophenylacetic acid (CFPA): a new reagent for determining enantiomeric excess that gives very large ^{19}F NMR Δδ values. *J. Am. Chem. Soc.* **1991**, *113*, 6318–6320.
247. Takeuchi Y, Itoh N, Satoh T, Koizumi T, Yamaguchi K. Chemistry of novel compounds with multifunctional carbon structure. 9. Molecular design, synthetic studies, and NMR investigation of several efficient chiral derivatizing reagents which give very large ^{19}F NMR Δδ values in enantiomeric excess determination. *J. Org. Chem.* **1993**, *58*, 1812–1820.
248. Takeuchi Y, Iwashita H, Yamada K, Gotaishi M, Kurose N, Koizumi T, Kabuto K, Kometani T. Chemistry of novel compounds possessing multifunctional carbon atoms. X. Synthetic studies of efficient and practical chiral derivatizing agents based on the α-cyano-α-fluorophenylacetic acid structure. *Chem. Pharm. Bull.* **1995**, *43*, 1668–1673.
249. Takeuchi Y, Konishi M, Hori H, Takahashi T, Kometani T, Kirk KL. Efficient synthesis of a new, highly versatile chiral derivatizing agent, α-cyano-α-fluoro-p-tolylacetic acid (CFTA). *Chem. Commun.* **1998**, 365–366.
250. Takahashi T, Fukuishima A, Tanaka Y, Takeuchi Y, Kabuto K, Kabuto C. CFTA, a new efficient agent for determination of absolute configurations of chiral secondary alcohols. *Chem. Commun.* **2000**, 788–789.
251. Fujiwara T, Omata K, Kabuto K, Kabuto C, Takahashi T, Segawa M, Takeuchi Y. Determination of the absolute configurations of α-amino esters from the ^{19}F NMR chemical shifts of their CFTA amide diastereomers. *Chem. Commun.* **2001**, 2694–2695.
252. Fujiwara T, Sasaki M, Omata K, Kabuto C, Kabuto K, Takeuchi Y. An efficient procedure for the resolution of α-cyano-α-fluoro-*p*-tolylacetic acid (CFTA) via the

diastereomeric N-carbobenzyloxy-cis-1-amino-2-indanol esters. *Tetrahedron: Asymmetry* **2004**, *15*, 555–563.

253. Takeuchi Y, Fujisawa H, Noyori R. A very reliable method for determination of absoute configuration of chiral secondary alcohols by ^{1}H NMR spectroscopy. *Org. Lett.* **2004**, *6*, 4607–4610.

254. Takeuchi Y, Itoh N, Koizumi T. The remarkably high reactivity of α-cyano-α-fluorophenylacetyl chloride (CFPA-Cl) towards hindered nucleophiles in enantiomeric excess determination. *J. Chem. Soc., Chem. Commun.* **1992**, 1514–1515.

255. Seco JM, Quiñoá E, Riguera R. Boc-phenylglycine: the reagent of choice for the assignment of the absolute configuration of α-chiral primary amines by ^{1}H NMR spectroscopy. *J. Org. Chem.* **1999**, *64*, 4669–4675.

256. Ito Y, Ishida K, Okada S, Murakami M. The absolute stereochemistry of anachelins, siderophores from the cyanobacterium *Anabaena cylindrical*. *Tetrahedron* **2004**, *60*, 9075–9080.

257. Pazos Y, Leiro V, Seco JM, Quiñoá E, Riguera R. Boc-phenylglycine: a chiral solvating agent for the assignment of the absolute configuration of amino alcohols and their ethers by NMR. *Tetrahedron: Asymmetry* **2004**, *15*, 1825–1829.

258. Harada K, Shimizu Y, Kawakami A, Fujii K. Determination of the absolute configuration of a secondary alcohol by NMR spectroscopy using difluorodinitrobenzene. *Tetrahedron Lett.* **1999**, *40*, 9081–9084.

259. Harada K, Shimizu Y, Fujii K. A chiral anisotropic reagent for determination of the absolute configuration of a primary amino compound. *Tetrahedron Lett.* **1998**, *39*, 6245–6248.

260. Harada K, Shimizu Y, Fujii K. A chiral anisotropic reagent for determination of the absolute configuration of a primary amino compound. *Tennen Yuki Kagobutsu Toronkai Koen Yoshishu* **1998**, *40*, 383–388.

261. Calmes M, Daunis J, Hanouneh A, Jacquier R. Determination of the chiral purity of benzylic amines using Marfey's reagent. *Tetrahedron: Asymmetry* **1993**, *4*, 2437–2440.

262. Takeuchi Y, Ogura H, Ishii Y, Koizumi T. Chemistry of novel compounds with multifunctional carbon structure. 4. Steric and electronic influences on the diastereoisomeric chemical shift differences in ^{19}F NMR spectra by introduction of fluorine, phenyl, and heteroatom groups into acetates. *J. Chem. Soc., Perkin Trans. 1* **1989**, 1721–1725.

263. Takeuchi Y, Ogura H, Ishii Y, Koizumi T. Chemistry of novel compounds with multifunctional carbon structure. VI. Synthetic studies and ^{19}F-nuclear magnetic resonance investigation of novel α,α-disubstituted fluoroacetates. *Chem. Pharm. Bull.* **1990**, *38*, 2404–2408.

264. Barrelle M, Hamman S. Substituted 2-fluoroacetic acids as chiral derivatizing agents. *J. Chem. Res., Synop.* **1995**, 316–317.

265. Hamman S, Barrelle M, Tetaz F, Beguin CG. Acide fluoro-2 phenyl-2 acetique: synthese, configuration absolue et emploi comme agent chiral de derivation. *J. Fluorine Chem.* **1987**, *37*, 85–94.

266. Hamman S. Acide-2-fluoro-2-phenylacetique, partie 3. correlation entre la configuration de ses amides d'amines chirales et les deplacements chimiques du proton et du fluor. *J. Fluorine Chem.* **1990**, *50*, 327–338.

267. Hamman S. Acide 2-fluoro-2-phényl acétique. Partie 5. RMN du fluor de ses amides formés avec des hydroxy amines et des fluoro amines. *J. Fluorine Chem.* **1993**, *62*, 5–13.

268. Hamman S. Acide 2-fluoro-2-phényl propanoïque: préparation et utilisation comme agent chiral de derivation. *J. Fluorine Chem.* **1993**, *60*, 225–232.

269. Barrelle M, Hamman S. Substituted 2-fluoroacetic acids as chiral derivatizing agents. *J. Chem. Res., Synop.* **1995**, 316–317.

270. Barrelle M, Boyer L, Chang-Fong J, Hamman S. 2-fluoro-2-phenylacetic acid as a chiral derivatizing agent. Part VI. Distinction of configurations of its esters with 2,2'-dihydroxy-1,1'-binaphthyl and some derivatives by ^{19}F NMR *Tetrahedron: Asymmetry* **1996**, *7*, 1961–1966.

271. Hamman S, Arnaud R, Barrelle M, Béguin CG. Solvation effects on the conformation of esters of chiral secondary alcohols and 2-fluoro-2-phenyl acetic acid used as a fluorine-substituted chiral derivatizing agent: an experimental and quantum mechanical study. *New J. Chem.* **1996**, *20*, 1113–1119.

272. Apparu M, Tiba YB, Léo P, Hamman S, Coulombeau C. Determination of the enantiometric purity and the configuration of β-aminoalcohols using (R)-2-fluorophenylacetic acid (AFPA) and fluorine-19 NMR: application to β-blockers. *Tetrahedron: Asymmetry* **2000**, *11*, 2885–2898.

273. Apparu M, Tiba YB, Léo P, Mathieu JP, Mauclaire L. Synthesis of and iodine-labelled analogue of practolol: (S)-3-[4-(4-iodobut-3-encarboxamido)phenoxy]-1-isopropylaminopropan-2-ol (AMI-9S). *J. Labelled Cpd. Radiopharm.* **1999**, *42*, 1195–1202.

274. Sonnet PE, Oliver JE, Waters RM, King G, Panicker S. A potential chiral derivatizing agent for 1,2-diglycerides. *Chem. Phys. Lipids* **1995**, *78*, 203–208.

275. Duret P, Waechter A, Figadère B, Hocquemiller R, Cavé A. Determination of absolute configurations of carbinols of annonaceous acetogenins with 2-naphthylmethoxyacetic acid esters. *J. Org. Chem.* **1998**, *63*, 4717–4720.

276. Takahashi H, Kato N, Iwashima M, Iguchi K. Determination of absolute configurations of tertiary alcohols by NMR spectroscopy. *Chem. Lett.* **1999**, 1181–1182.

277. Mada K, Ooi T, Kusumi T. NMR study of acutanol, a new cembrene alcohol, and sarcophytol A isolated from the soft coral *Sarcophyton acutangulum*. *Spectroscopy* **2001**, *15*, 177–182.

278. Ishii H, Krane S, Yasuhiro Itagaki Y, Berova N, Nakanishi K, Weldon PJ. Absolute configuration of a hydroxyfuranoid acid from the pelage of the genus *Bos*, 18-(6S,9R,10R)-bovidic acid. *J. Nat. Prod.* **2004**, *67*, 1426–1430.

279. Borde X, Nugier-Chauvin C, Noiret N, Patin H. (R)- and (S)-α-Methoxy-(1-naphthyl) acetic acids: resolution by fractional crystallization and use for the NMR stereochemical analysis of alkylsulfoxides. *Tetrahedron: Asymmetry* **1998**, *9*, 1087–1090.

280. Kusumi T, Takahashi K, Hashimoto T, Kan Y, Asakawa Y. Determination of the absolute configuration of ginnol, a long-chain aliphatic alcohol, by use of a new chiral anisotropic reagent. *Chem. Lett.* **1994**, 1093–1094.

281. Takahashi H, Iwashima M, Iguchi K. Determination of absolute configurations of β- or γ-methyl substituted secondary alcohols by NMR spectroscopy. *Tetrahedron Lett.* **1999**, *40*, 333–336.

282. Kusumi T, Takahashi H, Xu P, Fukushima T, Asakawa Y, Hashimoto T, Kan Y, Inouye Y. New chiral anisotropic reagents, NMR tools to elucidate the absolute configurations of long-chain organic compounds. *Tetrahedron Lett.* **1994**, *35*, 4397–4400.

283. Yamase H, Ooi T, Kusumi T. Determination of the absolute configuration of linear secondary alcohols adopting one enantiomer of the chiral anisotropic reagents, methoxy-(1- and 2-naphthyl)acetic acids. *Tetrahedron Lett.* **1998**, *39*, 8113–8116.
284. Clericuzio M, Toma L, Vidari G. Fungal Metabolites. 44. Isolation of a new caryophyllane ester from *Lactarius subumbonatus*: conformational analysis and absolute configuration. *Eur. J. Org. Chem.* **1999**. 2059–2065.
285. Arita S, Yabuuchi T, Kusumi T. Resolution of 1- and 2-naphthylmethoxyacetic acids, NMR reagents for absolute configuration determination, by use of L-phenylalaninol. *Chirality* **2003**, *15*, 609–614.
286. Vávra J, Vodička P, Streinz L, Buděšínský M, Koutek B, Ondráček J, Císařová I. Chiral derivatives of 2-(1-naphthyl)-2-phenylacetic acid. *Chirality* **2004**, *16*, 652–660.
287. Kasai Y, Naito J, Kuwahara S, Watanabe M, Ichikawa A, Harada N. Novel chiral molecular tools for preparation of enantiopure alcohols by resolution and simultaneous determination of their absolute configurations by the ^{1}H NMR anisotropy method. *J. Synth. Org. Chem., Jpn.* **2004**, *62*, 48–61.
288. Fujita T, Kuwahara S, Watanabe M, Harada N. Crystalline state conformation of 2-methoxy-2-(1-naphthyl)propionic acid ester. *Enantiomer* **2002**, *7*, 219–223.
289. Taji H, Watanabe M, Harada N, Naoki H, Ueda Y. Diastereomer method for determining % ee by ^{1}H NMR and/or MS spectrometry with complete removal of the kinetic resolution effect. *Org. Lett.* **2002**, *4*, 2699–2702.
290. Ichikawa A. Application of αMNPA to stereochemical studies of monoterpene alcohol. *Chirality* **1999**, *11*, 70–74.
291. Taji H, Kasai Y, Sugio A, Kuwahara S, Watanabe M, Harada N, Ichikawa A. Practical enantioresolution of alcohols with 2-methoxy-2-(1-naphthyl)propionic acid and determination of their absolute configurations by the ^{1}H NMR anisotropy method. *Chirality* **2002**, *14*, 81–84.
292. Kasai Y, Watanabe M, Harada N. Convenient method for determining the absolute configuration of chiral alcohols with racemic ^{1}H NMR anisotropy reagent, MαNP acid: use of HPLC-CD detector. *Chirality* **2003**, *15*, 295–299.
293. Kosaka M, Sekiguchi S, Naito J, Uemura M, Kuwahara S, Watanabe M, Harada N, Hiroi K. Synthesis of enantiopure phthalides including 3-butylphthalide, a fragrance component of celery oil, and determination of their absolute configurations. *Chirality* **2005**, *17*, 218–232.
294. Kosaka M, Sugito T, Kasai Y, Kuwahara S, Watanabe M, Harada N, Job GE, Shvet A, Pirkle WH. Enantioresolution and absolute configurations of chiral meta-substituted diphenylmethanols as determined by X-ray crystallographic and ^{1}H NMR anisotropy methods. *Chirality* **2003**, *15*, 324–328.
295. Kasai Y, Taji H, Fujita T, Yamamoto Y, Akagi M, Sugio A, Kuwahara S, Watanabe M, Harada N, Ichikawa A, Schurig V. MαNP acid, a powerful chiral molecular tool for preparation of enantiopure alcohols by resolution and determination of their absolute configurations by the ^{1}H NMR anisotropy method. *Chirality* **2004**, *16*, 569–585.
296. Harada N, Watanabe M, Kuwahara S, Sugio A, Kasai Y, Ichikawa A. 2-Methoxy-2-(1-naphthyl)propionic acid, a powerful chiral auxiliary for enantioresolution of alcohols and determination of their absolute configurations by the ^{1}H NMR anisotropy method. *Tetrahedron: Asymmetry* **2000**, *11*, 1249–1253.

297. Naito J, Kosaka M, Sugito T, Watanabe M, Harada N, Pirkle WH. Enantioresolution of fluorinated diphenylmethanols and determination of their absolute configurations by X-ray crystallographic and ^{1}H NMR anisotropy methods. *Chirality* **2004**, *16*, 22–35.

298. Kosaka M, Sugito T, Kasai Y, Kuwahara S, Watanabe M, Harada N, Job GE, Shvet A, Pirkle WH. Enantioresolution and absolute configurations of chiral meta-substituted diphenylmethanols as determined by X-ray crystallographic and ^{1}H NMR anisotropy methods. *Chirality* **2003**, *15*, 324–328.

299. Iwamoto H, Kobayashi Y, Kawatani T, Suzuki M, Fukazawa Y. Determination of absolute configuration of secondary alcohols using new chiral auxiliary and chemical shift calculation. *Tetrahedron Lett.* **2006**, *47*, 1519–1523.

300. Chinchilla R, Falvello LR, Nájera C. Determination of the absolute configuration of amines and α-amino acids by ^{1}H NMR of (R)-O-aryllactic acid amides. *J. Org. Chem.* **1996**, *61*, 7285–7290.

301. Gras J, Soto T, Heumann A. Remote anisotropic effects in diastereomeric esters of fluorinated O-aryllactic acids. *Eur. J. Org. Chem.* **2000**, 837–840.

302. Heumann A, Brunel JM, Faure R, Kolshorn H. Nuclear overhauser effects in diastereoisomeric vinyl ethers for the precise structure determination of chiral alcohols. *Chem. Commun.* **1996**, 1159–1160.

303. Tottie L, Moberg C, Heumann A. Derivatives of (R)-lactic acid for the analysis of chiral alcohols by ^{1}H NMR spectroscopy. *Acta Chem. Scand.* **1993**, *47*, 492–499.

304. Heumann A, Loutfi A, Ortiz B. New fluorinated O-aryl lactic acids: use as chiral derivatizing agents (CDAs) and determination of their enantiomeric purity with achiral diols. *Tetrahedron: Asymmetry* **1995**, *6*, 1073–1076.

305. Heumann A, Faure R. Fluorinated lactic acids: easily accessible reagents for the analysis of chiral compounds by ^{19}F NMR spectroscopy. ^{19}F NMR separation of the eight isomers of menthol. *J. Org. Chem.* **1993**, *58*, 1276–1279.

306. Chinchilla R, Foubelo F, Nájera C, Yus M. (R)-O-Aryllactic acids: convenient chiral solvating agents for direct ^{1}H NMR determination of the enantiomeric composition of amines and amino alcohols. *Tetrahedron: Asymmetry* **1995**, *6*, 1877–1880.

307. Kouda K, Kusumi T, Ping X, Kan Y, Hashimoto T, Asakawa Y. 2-anthrylmethoxyacetic acid, a new chiral anisotropic reagent for elucidating the absolute configuration of acyclic alcohols. *Tetrahedron Lett.* **1996**, *37*, 4541–4544.

308. Seco JM, Quiñoá E, Riguera R. 9-Anthrylmethoxyacetic acid esterification shifts-correlation with the absolute stereochemistry of secondary alcohols. *Tetrahedron* **1999**, *55*, 569–584.

309. Kouda K, Ooi T, Kaya K, Kusumi T. Absolute stereostructure of a 2,3,7,13-tetrahydroxyoctadecanoic acid, the framework of taurolipid B produced by a fresh-water protozoan, *Tetrahymena thermophila*. *Tetrahedron Lett.* **1996**, *37*, 6347–6350.

310. Seco JM, Tseng L, Godejohann M, Quiñoá E, Riguera R. Simultaneous enantioresolution and assignment of absolute configuration of secondary alcohols by directly coupled HPLC-NMR of 9-AMA esters. *Tetrahedron: Asymmetry* **2002**, *13*, 2149–2153.

311. Lenis LA, Ferreiro MJ, Debitus C, Jiménez C, Quiñoá E, Riguera R. The unusual presence of hydroxylated furanosesquiterpenes in the deep ocean tunicate *Ritterella rete*. Chemical interconversions and absolute stereochemistry. *Tetrahedron* **1998**, *54*, 5385–5406.

312. Abad J, Fabriás G, Camps F. Synthesis of dideuterated and enantiomers of monodeuterated tridecanoic acids at C-9 and C-10 positions. *J. Org. Chem.* **2000**, *65*, 8582–8588.

313. Ferreiro MJ, Latypov SK, Quiñoá E, Riguera R. Determination of the absolute configuration and enantiomeric purity of chiral primary alcohols by ^{1}H NMR of 9-anthrylmethoxyacetates. *Tetrahedron: Asymmetry* **1996**, *7*, 2195–2198.

314. Latypov SK, Ferreiro MJ, Quiñoá E, Riguera R. Assignment of the absolute configuration of β-chiral primary alcohols by NMR: scope and limitations. *J. Am. Chem. Soc.* **1998**, *120*, 4741–4751.

315. Seco JM, Quiñoá E, Riguera R. Incorrect procedure for the assignment of the absolute configuration of carbonucleosides by NMR: MPA must not be used with primary alcohols. *Tetrahedron: Asymmetry* **2002**, *13*, 919–921.

316. Pirkle WH, Simmons KA. Nuclear magnetic resonance determination of enantiomeric composition and absolute configuration of amines, alcohols, and thiols with α-[1-(9-anthryl)-2,2,2-trifluoroethoxy]acetic acid as a chiral derivatizing agent. *J. Org. Chem.* **1981**, *46*, 3239–3246.

317. Massad SK, Hawkins LD, Baker DC. A series of (2S)-2-O-protected-2-hydroxypropanals (L-lactaldehydes) suitable for use as optically active intermediates. *J. Org. Chem.* **1983**, *48*, 5180–5182.

318. Pirkle WH, Liu Y. Design, synthesis, resolution, determination of absolute configuration, and evaluation of a chiral naproxen selector. *J. Org. Chem.* **1994**, *59*, 6911–6916.

319. Ichikawa A, Ono H, Harada N. Synthesis and analytical properties of (*S*)-(+)-2-methoxy-2-(9-phenanthryl)propionic acid. *Tetrahedron: Asymmetry* **2003**, *14*, 1593–1597.

320. Ichikawa A, Ono H, Harada N. Stereochemical studies of chiral resolving agents, M9PP and H9PP acids. *Chirality* **2004**, *16*, 559–567.

321. Seco JM, Quiñoà E, Riguera R. A practical guide for the assignment of the absolute configuration of alcohols, amines and carboxylic acids by NMR. *Tetrahedron: Asymmetry* **2001**, *12*, 2915–2925.

322. Seco JM, Quiñoá E, Riguera R. The assignment of absolute configuration by NMR. *Chem. Rev.* **2004**, *104*, 17–117.

323. Takahashi H, Kato N, Iwashima M, Iguchi K. Determination of absolute configurations of tertiary alcohols by NMR spectroscopy. *Chem. Lett.* **1999**, 1181–1182.

324. Gerlach H, Zagalak B. Determination of the enantiomeric purity and absolute configuration of α-deuteriated primary alcohols. *J. Chem. Soc., Chem. Commun.* **1973**, 274–275.

325. Caspi E, Eck CR. Preparative scale synthesis of (1R) [1-^{2}H$_1$] or [1-^{3}H$_1$] primary alcohols of high optical purity. *J. Org. Chem.* **1977**, *42*, 767–768.

326. Kobayashi K, Jadhav PK, Zydowsky TM, Floss HG. A simple and efficient synthesis of chiral acetic acid of high optical purity. *J. Org. Chem.* **1983**, *48*, 3510–3512.

327. Furukawa J, Iwasaki S, Okuda S. Method for determination of enantiomeric composition and absolute configuration of 2,3-deuterated 3-alkylpropanols. *Tetrahedron Lett.* **1983**, *24*, 5257–5260.

328. Prabhakaran PC, Gould SJ, Orr GR, Coward JK. Synthesis of chirally deuteriated phthalimidopropanols and evaluation of their absolute stereochemistry. *J. Am. Chem. Soc.* **1988**, *110*, 5779–5784.

329. Schwab JM, Ray T, Ho C. Synthesis of (2R, 3R)- and (2S, 3S)-[2,3-^{2}H$_2$]oxirane and application of it to the synthesis of chirally labeled homoserine. *J. Am. Chem. Soc.* **1989**, *111*, 1057–1063.

330. Fretz H, Woggon W, Voges R. 49. The allylic oxidation of geraniol catalyzed by cytochrome P-450$_{Cath.}$, proceeding with retention of configuration. *Helv. Chim. Acta* **1989**, *72*, 391–400.

331. Ravn MM, Jin Q, Coates RM. Synthesis of allylic isoprenoid diphosphates by S$_N$2 displacement of diethyl phosphate. *Eur. J. Org. Chem.* **2000**, 1401–1410.

332. Brown JM, Parker D. Mechanism of asymmetric homogeneous hydrogenation. Rhodium-catalyzed reductions with deuterium and hydrogen deuteride. *Organometallics* **1982**, *1*, 950–956.

333. Martin GJ, Martin ML, Zhang B. Site-specific natural isotope fractionation of hydrogen in plant products studied by nuclear magnetic resonance. *Plant, Cell Environ.* **1992**, *15*, 1037–1050.

334. Armarego WLF, Milloy BA, Pendergast W. A highly stereospecific synthesis of (R)- and (S)-[2-^{2}H$_1$]glycine. *J. Chem. Soc., Perkin Trans. 1* **1976**, 2229–2237.

335. Pontoni G, Coward JK, Orr GR, Gould SJ. Stereochemical studies of enzyme-catalyzed alkyl-transfer reactions. An NMR method for distinguishing between the two prochiral hydrogens at C-1′ of spermidine. *Tetrahedron Lett.* **1983**, *24*, 151–154.

336. Parker D, Taylor RJ, Ferguson G, Tonge A. Origins of the proton NMR chemical shift non-equivalence in the diastereotopic methylene protons of camphanamides. *Tetrahedron* **1986**, *42*, 617–622.

337. Williams RM, Zhai D, Sinclair PJ. Asymmetric synthesis of (R)- and (S)-[2-^{2}H$_1$]glycine. *J. Org. Chem.* **1986**, *51*, 5021–5022.

338. Ramer SE, Cheng H, Vederas JC. Investigations of polypeptide biosynthesis: formation of peptide amides. *Pure Appl. Chem.* **1989**, *61*, 489–492.

339. Hameršak Z, Selestrin A, Lesac A, Šunjić V. Conformational study of α-arylethylamides of (−)-camphanic acid. *Tetrahedron: Asymmetry* **1998**, *9*, 1891–1897.

340. Brown JM, Parker D. Determination of optical purity at isotopically chiral sites by ^{2}H NMR. *Tetrahedron Lett.* **1981**, *22*, 2815–2818.

341. Parker D. ^{1}H and ^{2}H Nuclear magnetic resonance determination of the enantiomeric purity and absolute configuration of α-deuterated primary carboxylic acids, alcohols, and amines. *J. Chem. Soc., Perkin Trans. 2* **1983**, 83–88.

342. Sowinski JA, Toogood PL. Synthesis of an enantiomerically pure serine-derived thiazole. *J. Org. Chem.* **1996**, *61*, 7671–7676.

343. Jurczak J, Konowal A. Use of Eu(fod)$_3$ with diastereomeric (−)-ω-camphanic esters for determination of enantiomeric purity of alcohols by ^{1}H NMR spectroscopy. A convenient method for monitoring racemate resolution. *Pol. J. Chem.* **1978**, *52*, 1967–1971.

344. Jurczak J, Konowal A, Krawczyk Z, Ejchart A. Diastereomeric non-equivalence of (−)-ω-camphanic esters of chiral secondary alcohols in ^{13}C NMR spectroscopy. *Org. Magn. Reson.* **1981**, *15*, 193–196.

345. Jurczak J, Konowal A, Krawczyk Z, Ejchart A. Application of ^{1}H NMR-Eu(fod)$_3$-shifted spectra for the determination of the enantiomeric composition and absolute configuration of secondary alcohols, using (−)-ω-camphanic esters. *Org. Magn. Reson.* **1981**, *17*, 50–52.

346. Meier C, Laux WHG. Enantioselective synthesis of diisopropyl α-, β-, and γ-hydroxyarylalkylphosphonates from ketophosphonates: a study on the effect of the phosphonyl group. *Tetrahedron* **1996**, *52*, 589–598.
347. Tran HTN, Falk H. Concerning the chiral discrimination and helix inversion barrier in hypericinates and hypericin derivatives. *Monatsh. Chem.* **2002**, *133*, 1231–1237.
348. Williams RM, Sinclair PJ, Zhai D, Chen D. Practical asymmetric syntheses of α-amino acids through carbon-carbon bond constructions on electrophilic glycine templates. *J. Am. Chem. Soc.* **1988**, *110*, 1547–1557.
349. Arnold LD, Drover JCG, Vederas JC. Conversion of serine β-lactones to chiral α-amino acids by copper-containing organolithium and organomagnesium reagents. *J. Am. Chem. Soc.* **1987**, *109*, 4649–4659.
350. Kiefer M, Vogel R, Helmchen G, Nuber B. Resolution of (1,1'-binaphthalene)-2,2'-dithiol by enzyme catalysed hydrolysis of a racemic diacyl derivative. *Tetrahedron* **1994**, *50*, 7109–7114.
351. Kedzierski B, Thakker DR, Armstrong RN, Jerina DM. Absolute configuration of the K-region 4,5-dihydrodiols and 4,5-oxide of benzo[a]pyrene. *Tetrahedron Lett.* **1981**, *22*, 405–408.
352. Yagi H, Vyas KP, Tada M, Thakker DR, Jerina DM. Synthesis of the enantiomeric bay-region diol epoxides of benz[a]anthracene and chrysene. *J. Org. Chem.* **1982**, *47*, 1110–1117.
353. Yagi H, Thakker DR, Ittah Y, Croisy-Delcey M, Jerina DM. Synthesis and assignment of absolute configuration to the trans 3,4-dihydrodiols and 3,4-diol-1,2-epoxides of benzo[c]phenanthrene. *Tetrahedron Lett.* **1983**, *24*, 1349–1352.
354. Schollmeier M, Frank H, Oesch F, Platt KL. Assignment of absolute configuration to metabolically formed trans-dihydrodiols of dibenz[a,h]anthracene by two distinct spectroscopic methods. *J. Org. Chem.* **1986**, *51*, 5368–5372.
355. Lehr RE, Kumar S, Shirai N, Jerina DM. Synthesis of enantiomerically pure bay-region 3,4-diol 1,2-epoxide diastereomers and other derivatives of the potent carcinogen dibenz[c,h] acridine. *J. Org. Chem.* **1985**, *50*, 98–107.
356. Halpin RA, El-Naggar SF, McCombe KM, Vyas KP, Boyd DR, Jerina DM. Resolution and assignment of absolute configuration to the (+)- and (−)-cis and trans 3,4-diol metabolites of the anti-juvenile hormone precocene I. *Tetrahedron Lett.* **1982**, *23*, 1655–1658.
357. Galpin DR, Huitric AC. The use of nuclear magnetic resonance as a monitor in optical resolutions. *J. Org. Chem.* **1968**, *33*, 921–923.
358. Akhtar MN, Boyd DR, Hamilton JG. Synthesis of (+)- and (−)-naphthalene and anthracene 1,2-oxides. *J. Chem. Soc., Perkin Trans. 1* **1979**, 2437–2440.
359. D'Ambrosio M, Guerriero A, Pietra F. 168. Novel, racemic, or nearly-racemic antibacterial bromo- and chloroquinols and γ-lactams of the verongiaquinol and the cavernicolin type from the marine sponge *Aplysina* (= *Verongia*) *cavernicola*). *Helv. Chim. Acta* **1984**, *67*, 1484–1492.
360. Akasaka K, Imaizumi K, Sakakibara R, Terashima H, Ohrui H. Chiral discrimination of branched alkyl chain by labeling with chiral derivatization reagents. *Tennen Yuki Kagobutsu Toronkai Koen Yoshishu* **2000**, *42*, 565–570.
361. Ohrui H, Terashima H, Imaizumi K, Akasaka K. A solution of the "intrinsic problem" of diastereomer method in chiral discrimination. Development of a method for highly

efficient and sensitive discrimination of chiral alcohols. *Proc. Japan Acad. Ser. B Phys. Biol. Sci.* **2002**, *78*, 69–72.

362. Imaizumi K, Terasima H, Akasaka K, Ohrui H. Highly potent chiral labeling reagents for the discrimination of chiral alcohols. *Anal. Sci.* **2003**, *19*, 1243–1249.

363. Ohtaki T, Akasaka K, Kabuto C, Ohrui H. Chiral discrimination of secondary alcohols by both ^{1}H-NMR and HPLC after labeling with a chiral derivatization reagent, 2-(2,3-anthracenedicarboximide)cyclohexane carboxylic acid. *Chirality* **2005**, *17*, S171-S176.

364. Nishida Y, Itoh E, Abe M, Ohrui H, Meguro H. Synthesis of a series of fluorescent carboxylic acids with a 1,3-benzodioxole skeleton and their evaluation as chiral derivatizing reagents. *Anal. Sci.* **1995**, *11*, 213–220.

365. Matsugi M, Itoh K, Nojima M, Hagimoto Y, Kita Y. Determination of absolute configuration of trans-2-arylcyclohexanols using remarkable aryl-induced ^{1}H NMR shifts in diastereomeric derivatives. *Tetrahedron Lett.* **2001**, *42*, 6903–6905.

366. Matsugi M, Itoh K, Nojima M, Hagimoto Y, Kita Y. ^{1}H NMR determination of absolute configuration of 1- or 2-aryl-substituted alcohols and amines by means of their diastereomers: novel separation technique of diastereomeric derivatives of pyridyl alcohols by extraction. *Chem. Eur. J.* **2002**, *8*, 5551–5564.

367. Matsugi M, Itoh K, Nojima M, Hagimoto Y, Kita Y. A novel determination method of the absolute configuration of 1-aryl-1-alkylalcohols and amines by an intramolecular CH/π shielding effect in ^{1}H NMR. *Tetrahedron Lett.* **2001**, *42*, 8019–8022.

368. Matsugi M, Hagimoto Y, Itoh K, Nojima M, Kita Y. A simple determination method of the absolute configuration of 1-arylethanthiols by an intramolecular CH/π shielding effect in ^{1}H-NMR of diastereomeric thiol esters. *Chem Pharm. Bull.* **2003**, *51*, 460–462.

369. Fauconnot L, Nugier-Chauvin C, Noiret N, Patin H. Enantiomeric excess determination of some chiral sulfoxides by NMR: use of (S)-Ibuprofen® and (S)-Naproxen® as shift reagents. *Tetrahedron Lett.* **1997**, *38*, 7875–7878.

370. Blażewska K, Gajda T. (S)-Naproxen® and (S)-Ibuprofen®chlorides—convenient chemical derivatizing agents for the determination of the enantiomeric excess of hydroxyl and aminophosphonates by ^{31}P NMR. *Tetrahedron: Asymmetry* **2002**, *13*, 671–674.

371. Nishida Y, Kumagai M, Kamiyama A, Ohrui H, Meguro H. Determination of the absolute configuration of monosaccharides using (+) or (−) 2-tert-butyl-2-methyl-1,3-benzodioxole-4-carboxylic acid and high-resolution ^{1}H-n.m.r. spectroscopy. *Carbohydr. Res.* **1991**, *218*, 63–73.

372. Bai C, Liang B, Wang S, Zheng X, Lin Z, Huang W. Determination of the D- and L-configurations of amino deoxy sugars by (S)-TBMB carboxylic acid, a fluorescent chiral derivatization reagent. *Shengwu Huaxue Yu Shengwu Wuli Xuebao* **2002**, *34*, 615–618.

373. Wang Z, Hirose T, Shitara H, Goto M, Nohira H. Structure and chiral recognition ability of endo-3-benzamidonorborn-5-ene-2-carboxylic acid. *Bull. Chem. Soc. Jpn.* **2005**, *78*, 380–385.

374. Nagasawa K, Seto N, Hara C, Ito K. Coumarin-containing chiral discriminating agents. VII. New crystalline ^{1}H-NMR enantiomeric excess determination reagent for alcohols

and amines, (R)-(−)- and (S)-(+)-O-coumarinylmandelic acids. *Yakugaku Zasshi* **1997**, *117*, 786–799.

375. Nagasawa K, Yamashita A, Katoh S, Ito K, Wada K. N-Coumarinyl-L-proline, a novel chiral derivatizing agent for ^{1}H NMR determination of enantiomeric purities of alcohols and amines. *Chem. Pharm. Bull.* **1995**, *43*, 344–346.

376. Nagasawa K, Okazaki R, Yamashita A, Ito K, Wada K. New crystalline N-(coumarin-4-yl)-L-pyroglutamic acid. The first synthesis and application to ^{1}H NMR optical purity determination of alcohols and amines. *Heterocycles* **1997**, *45*, 1047–1050.

377. Nagasawa K, Seto N, Ito K. Coumarin-containing amino acids and oxy acids as chiral discriminating agents. Part III. Novel crystalline (R)-(+)- and (S)-(−)-O-coumarinyl lactic acids as chiral derivatizing agents for ^{1}H NMR inspection of optical purities of alcohols and amines. *Heterocycles* **1997**, *46*, 567–580.

378. Kim HC, Choi S, Kim H, Ahn K, Koh JH, Park J. Chiral 2,2-disubstituted cyclopropanecarboxylic acids: effective derivatizing agents for analysis of enantiomeric purity of alcohols and for resolution of 1,1′-bi-2-naphthol. *Tetrahedron Lett.* **1997**, *38*, 3959–3962.

379. Hirose T, Naito K, Shitara H, Nohira H, Baldwin BW. ^{1}H NMR study of chiral recognition of amines by chiral Kemp's acid diamide. *Tetrahedron: Asymmetry* **2001**, *12*, 375–380.

380. Hirose T, Naito K, Nakahara M, Shitara H, Aoki Y, Nohira H, Baldwin BW. New chiral Kemp's acid diamides for chiral amine recognition by ^{1}H NMR. *J. Inclusion Phenom. Macrocyclic Chem.* **2002**, *43*, 87–93.

381. Veinberga IG, Kazhoka KA, Turovskis IV, Dikovskaya KI, Freimanis YF. Determination of the enantiomeric composition of optically active derivatives of hydroxycyclopentenones using activated N-Boc-S-phenylalanine esters. *Zh. Org. Khim.* **1995**, *31*, 365–369.

382. Glowacki Z, Hoffmann M. ^{31}P NMR enantiomeric excess determination of 1-hydroxyalkylphosphonic acids via their diastereoisomeric phosphonodidepsipeptides. *Phosphorus, Sulfur Silicon Relat. Elem.* **1991**, *55*, 169–173.

383. Oba M, Ueno R, Fukuoka M, Kainosho M, Nishiyama K. Synthesis of L-threo- and L-erythro-[1-^{13}C, 2,3-^{2}H$_2$]amino acids: novel probes for conformational analysis of peptide side chains. *J. Chem. Soc., Perkin Trans. 1* **1995**, 1603–1609.

384. Brown JM, Field IP, Sidebottom PJ. Structural specificity in asymmetric charge-transfer complexation of helicenes. *Tetrahedron Lett.* **1981**, *22*, 4867–4970.

385. Kruizinga WH, Bolster J, Kellogg RM, Kamphuis J, Boesten WHJ, Meijer EM, Schoemaker HE. Synthesis of optically pure α-alkylated α-amino acids and a single-step method for enantiomeric excess determination. *J. Org. Chem.* **1988**, *53*, 1826–1827.

386. Moorlag H, Kruizinga WH, Kellogg RM. (S)-2-Chloropropanoyl chloride. A convenient reagent for the determination of the enantiomeric composition of α-substituted α-hydroxy acids. *Recl. Trav. Chim. Pays-Bas* **1990**, *109*, 479–480.

387. Andersson L, Kenne L. Synthesis and NMR studies of methyl 3-O-[(R)- and (S)-1-carboxyethyl]-α-D-gluco-, galacto- and manno-pyranosides. *Carbohydr. Res.* **1998**, *313*, 157–164.

388. Růžička J, Streinz L, Šaman D, Havlas Z, Wimmer Z, Zarevúcka M, Koutek B, Lešetický L. Chlorofluoroacetic acid as a highly versatile derivatizing agent: assignment of stereochemistry to esters of chiral alcohols. *Collect. Czech. Chem. Commun.* **2000**, *65*, 695–707.

389. Streinz L, Svatoš A, Vrkoč J, Meinwald J. Preparation of chlorofluoroacetic acid derivatives for the analysis of chiral alcohols. *J. Chem. Soc., Perkin Trans. 1* **1994**, 3509–3512.

390. Ishikawa N. The usefulness of optically active perfluorinated compounds. *J. Fluorine Chem.* **1984**, *25*, 17–20.

391. Kawa H, Yamaguchi F, Ishikawa N. Optically active perfluoro-2-propoxypropionic acid: a new chiral reagent for ^{19}F NMR study. *J. Fluorine Chem.* **1982**, *20*, 475–485.

392. Takahashi T, Okumoto H, Tsuji J, Harada N. Synthesis of a chiral steroid CD-ring synthon from D-leucine by means of diastereotopic face-selection. *J. Org. Chem.* **1984**, *49*, 948–950.

393. York WS, Hantus S, Albersheim P, Darvill AG. Determination of the absolute configuration of monosaccharides by ^{1}H NMR spectroscopy of their per-O-(S)-2-methylbutyrate derivatives. *Carbohydr. Res.* **1997**, *300*, 199–206.

394. Kubota T, Kanega J, Katagiri T. Application of (S)-trifluorolactic acid for chiral derivatizing agent. *J. Fluorine Chem.* **1999**, *97*, 213–221.

395. Kubota T, Kanega J, Katagiri T. Application of (S)-trifluorolactic acid for chiral derivatizing agent. *J. Fluorine Chem.* **1999**, *97*, 213–221.

396. Tanyeli C, Demir AS, Arkin AH, Akhmedov IM. PLE catalyzed enantiomeric separation of ($\pm$)-2-furylcarbinols. *Enantiomer* **1997**, *2*, 433–439.

397. Beugelmans R, Bigot A, Zhu J. A novel synthesis of K-13. *Tetrahedron Lett.* **1994**, *35*, 7391–7394.

398. Tanyeli C, Demir AS, Dikici E. New chiral synthon from the PLE catalyzed enantiomeric separation of 6-acetoxy-3-methylcyclohex-2-en-1-one. *Tetrahedron: Asymmetry* **1996**, *7*, 2399–2402.

399. Schwartz MA, Williams TH, Kolis SJ, Postma E, Sasso GJ. Biotransformation of prochiral 2-phenyl-1,3-di(4-pyridyl)-2-propanol to a chiral N-oxide metabolite. *Drug Metab. Dispos.* **1978**, *6*, 647–653.

400. Faigl F, Thurner A, Tárkányi G, Kovári J, Mordini A. Resolution and enantioselective rearrangements of amino group-containing oxiranyl ethers. *Tetrahedron: Asymmetry* **2002**, *13*, 59–68.

401. Miyano S, Okada S, Hotta H, Takeda M, Suzuki T, Kabuto C, Yasuhara F. Use of axially chiral 2'-methoxy-1,1'-binaphthyl-2-carboxylic acid as chiral derivatizing agent for discrimination of enantiomeric alcohols and amines by ^{1}H NMR. *Bull. Chem. Soc. Jpn.* **1989**, *62*, 3886–3891.

402. Fukushi Y, Yajima C, Mizutani J. A new method for establishment of absolute configurations of secondary alcohols by NMR spectroscopy. *Tetrahedron Lett.* **1994**, *35*, 599–602.

403. Latypov SK, Aganov AV, Tahara S, Fukushi Y. Conformational analysis of MNCB (MBNC) esters and amides: promising chiral reagents for stereoselective applications. *Tetrahedron* **1999**, *55*, 7305–7318.

404. Fukui H, Fukushi Y, Tahara S. NMR determination of the absolute configuration of β-chiral primary alcohols. *Tetrahedron Lett.* **2005**, *46*, 5089–5093.

405. Oi S, Ono H, Tanaka H, Matsuzaka Y, Miyano S. Investigation of the chiral discrimination mechanism using an axially asymmetric binaphthalene-based stationary phase for high-performance liquid chromatography. *J. Chromatogr. A* **1994**, *659*, 75–86.

406. Nabeta K, Ohkubo S, Hozumi R, Fukushi Y, Nakai H, Katoh K. Alloaromandendranes, bicyclogermacrane and 2,3-secoalloaromandendranes in cultured cells of the liverwort, *Heteroscyphus planus*. *Phytochemistry* **1996**, *43*, 83–93.

407. Ishikawa NK, Fukushi Y, Yamaji K, Tahara S, Takahashi K. Antimicrobial cuparene-type sesquiterpenes, enokipodins C and D, from a mycelial culture of *Flammulina velutipes*. *J. Nat. Prod.* **2001**, *64*, 932–934.
408. Pirkle WH, Burlingame TG, Beare SD. Optically active NMR solvents VI. The determination of optical purity and absolute configuration of amines. *Tetrahedron Lett.* **1968**, 5849–5852.
409. Pirkle WH, Beare SD. Optically active solvents in nuclear magnetic resonance spectroscopy. VII. Direct determination of optical purities and correlation of absolute configurations of sulfoxides. *J. Am. Chem. Soc.* **1968**, *90*, 6250–6251.
410. Pirkle WH, Pavlin MS. Determination of the enantiomeric purity of compound chiral by virtue of C-13 labelling. *J. Chem. Soc., Chem. Commun.* **1974**, 274–275.
411. Pirkle WH, Beare SD, Muntz RL. Assignment of absolute configuration of sulfoxides by NMR. A solvation model. *Tetrahedron Lett.* **1974**, 2295–2298.
412. Whitney TA, Pirkle WH. Determination of absolute configurations by NMR. Stereochemistry of the base-catalyzed decarboxylation of 2-methyl-2,3-dihydro-benzothiophene-2-carboxylic acid l-dioxide. *Tetrahedron Lett.* **1974**, 2299–2300.
413. Bucciarelli M, Forni A, Moretti I, Torre G. Configurational correlations for chiral episulphides and episulphoxides by ^{1}H NMR spectroscopy in optically active solvent. *Tetrahedron* **1977**, *33*, 999–1002.
414. Poje M, Nota O, Balenović K. Stereoselective oxidation of gem-disulphides with *Aspergillus niger*. *Tetrahedron* **1980**, *36*, 1895–1897.
415. Pirkle WH, Beare SD, Muntz RL. Optically active solvents for nuclear magnetic resonance. X. Enantiomeric nonequivalence of sulfinamides, sulfinates, sulfites, thiosulfinates, phosphine oxides, and amine oxides. *J. Am. Chem. Soc.* **1969**, *91*, 4575.
416. Pirkle WH, Hoekstra MS. Chiral nuclear magnetic resonance solvating agents. Resolution, determination of enantiomeric purity, and assignment of absolute configuration of cyclic and acyclic sulfinate esters. *J. Am. Chem. Soc.* **1976**, *98*, 1832–1839.
417. Lewis RA, Naumann K, DeBruin KE, Mislow K. t-Butylphosphonium salts: nucleophilic displacement as phosphorus with inversion of configuration. *Chem. Commun.* **1969**, 1010–1011.
418. Joesten MD, Smith HE, Vix VA. Effect of chiral solvents on ^{31}P nuclear magnetic resonance spectra of diastereoisomeric pyrophosphoramides. *J. Chem. Soc., Chem. Commun.* **1973**, 18–19.
419. Pirkle WH, Muntz RL, Paul IC. Chiral nuclear magnetic resonance solvents. XI. A method for determining the absolute configuration of chiral N,N-dialkylarylamine oxides. *J. Am. Chem. Soc.* **1971**, *93*, 2817–2819.
420. Pirkle WH, Beare SD. Optically active solvents in nuclear magnetic resonance spectroscopy. IX. Direct determinations of optical purities and correlations of absolute configurations of α-amino acids. *J. Am. Chem. Soc* **1969**, *91*, 5150–5155.
421. Moretti I, Taddei F, Torre G, Spassky N. Configurational correlations for chiral epoxides by nuclear magnetic resonance spectroscopy in optically active solvents. *J. Chem. Soc., Chem. Commun.* **1973**, 25–26.
422. Forni A, Moretti I, Torre G. Absolute configuration at chiral nitrogen in oxaziridines configurational correlations by NMR spectroscopy in optically active solvating agents. *Tetrahedron Lett.* **1978**, 2941–2944.

423. Bucciarelli M, Forni A, Moretti I, Torre G. Asymmetric synthesis at nitrogen by oxidation of imines with m-chloroperbenzoic acid in the presence of optically active carbinols. Absolute stereochemistry of chiral alcohol-imine-peracid solvates. *J. Chem. Soc., Perkin Trans. 1* **1980**, 2152–2161.

424. Häkli H, Mintas M, Mannschreck A. Preparative separations of enantiomeric diaziridines by liquid chromatography on triacetylcellulose. Racemizations monitored by polarimetry and by ^{1}H NMR. *Chem. Ber.* **1979**, *112*, 2028–2038.

425. Kainosho M, Ajisaka K, Pirkle WH, Beare SD. The use of chiral solvents or lanthanide shift reagents to distinguish meso from D or L diastereomers. *J. Am. Chem. Soc.* **1972**, *94*, 5924–5926.

426. Adzima LJ, Martin JC. Reactions of some new diaryldialkoxyspirosulfuranes. The barrier to cuneal inversion of configuration at sulfuranyl sulfur in diastereomeric spirosulfuranes. *J. Org. Chem.* **1977**, *42*, 4006–4016.

427. Kaspersen FM, Funke CW, Sperling EMG, van Rooy FAM, Wagenaars GN. Tritium nuclear magnetic resonance spectroscopy of [pyrrolidine-^{3}H] bepridil. *J. Chem. Soc., Perkin Trans. 2* **1986**, 585–591.

428. Shinoda S, Nishikimi T, Uchino S, Koie Y, Saito Y. Stereoselectivity in the substitution reaction of square-planar platinum(II) complexes determined in situ by nuclear magnetic resonance spectroscopy using a chiral solvent. *J. Chem. Soc., Dalton Trans.* **1984**, 2689–2693.

429. Küspert R, Mannschreck A. Application of NMR spectroscopy of chiral association complexes. 12. Rate determination of an intramolecular motion in a free molecule by means of dynamic NMR measurements in the presence of an auxiliary compound. *Org. Magn. Reson.* **1982**, *19*, 6–11.

430. Seri N, Simaan S, Botoshansky M, Kaftory M, Biali SE. A conformationally flexible tetrahydroxycalix[4]arene adopting the unusual 1,3-alternate conformation. *J. Org. Chem.* **2003**, *68*, 7140–7142.

431. Daligault F, Arboré A, Nugier-Chauvin C, Patin H. Chiral homoallylic and allylic sulfoxides as models for the stereochemical analysis of sulfoxide thiaoleates. *Tetrahedron: Asymmetry* **2004**, *15*, 917–924.

432. Pirkle WH, Sikkenga DL, Pavlin MS. Nuclear magnetic resonance determination of enantiomeric composition and absolute configuration of γ-lactones using chiral 2,2,2-trifluoro-1-(9-anthryl)ethanol. *J. Org. Chem.* **1977**, *42*, 384–387.

433. Pirkle WH, Sikkenga DL. The use of chiral solvating agents for nuclear magnetic resonance determination of enantiomeric purity and absolute configuration of lactones. Consequences of three-point interactions. *J. Org. Chem.* **1977**, *42*, 1370–1374.

434. Pirkle WH, Adams PE. Enantiomerically pure lactones. 3. Synthesis of and stereospecific conjugate additions to α,β-unsaturated lactones. *J. Org. Chem.* **1980**, *45*, 4117–4121.

435. Leborgne A, Moreau M, Spassky N. Determination of enantiomeric purity and absolute configuration of chiral β-propiolactones by nuclear magnetic resonance in optically active solvents. *Tetrahedron Lett.* **1983**, *24*, 1027–1030.

436. Latypov S, Franck X, Jullian J, Hocquemiller R, Figadère B. NMR determination of absolute configuration of butenolides of annonaceous type. *Chem. Eur. J.* **2002**, *8*, 5662–5666.

437. Pirkle WH, Robertson MR, Hyun MH. A liquid chromatographic method for resolving chiral lactams as their diastereomeric ureide derivatives. *J. Org. Chem.* **1984**, *49*, 2433–2437.

438. García-Martínez C, Taguchi Y, Oishi A, Hayamizu K. Configurational ^{1}H NMR study of optically active 7-(-1-phenylethyl)-2-oxa-7-azabicyclo[3.2.0]heptan-6-one derivatives using Pirkle's alcohols and a chiral shift reagent. *Magn. Reson. Chem.* **1998**, *36*, 429–435.

439. Pirkle WH, Rinaldi PL. Nuclear magnetic resonance determination of enantiomeric compositions of oxaziridines using chiral solvating agents. *J. Org. Chem.* **1977**, *42*, 3217–3219.

440. Pirkle WH, Rinaldi PL. Nuclear magnetic resonance determination of absolute configuration and enantiomeric compositions of chiral oxaziridines using chiral solvating agents. *J. Org. Chem.* **1978**, *43*, 4475–4480.

441. Bucciarelli M, Forni A, Moretti I, Prosyanik AV. Optically active *N*-alkyloxaziridine-3,3-dicarboxylic esters. *J. Chem. Res., Synop.* **1986**, 146–147.

442. Buccharelli M, Forni A, Moretti I, Torre G. Optically active trifluoromethylcarbinols as chiral solvating agents for asymmetric transformations at a ring-nitrogen atom. Synthesis of optically active *N*-chloroaziridines and stereochemical aspects of chiral solvent-aziridine solute complexes. *J. Org. Chem.* **1983**, *48*, 2640–2644.

443. Mintas M, Orhanović Z, Jakopčić K, Koller H, Stühler G, Mannschreck A. Liquid chromatography on triacetylcellulose. 9. Enantiomers of sterically hindered *N*-aryl-4-pyridones. Chromatographic enrichment and thermal interconversion. *Tetrahedron* **1985**, *41*, 229–233.

444. Clayden J, Kenworthy MN, Youssef LH, Helliwell M. Axial chirality in xanthene-4,5-dicarboxamides: 1,9-stereocontrol mediated by remote interactions between conformationally constrained amide groups. *Tetrahedron Lett.* **2000**, *41*, 5171–5175.

445. Garcia MB, Grilli S, Lunazzi L, Mazzanti A, Orelli LR. Conformational studies by dynamic NMR. 84. Structure, conformation, and stereodynamics of the atropisomers of *N*-aryl-tetrahydropyrimidines. *J. Org. Chem.* **2001**, *66*, 6679–6684.

446. Casarini D, Lunazzi L, Placucci G, Macciantelli D. Conformational studies by dynamic NMR. 32. Enantiomerization of chiral conformers in hindered naphthylamines and naphthyl nitroxides. *J. Org. Chem.* **1987**, *52*, 4721–4726.

447. Casarini D, Davalli S, Lunazzi L, Macciantelli D. Conformational studies of dynamic NMR. 37. Monitoring the stereomutations of symmetric amines by means of a chiral auxiliary agent. *J. Org. Chem.* **1989**, *54*, 4616–4619.

448. Casarini D, Lunazzi L, Pasquali F, Gasparrini F, Villani C. Conformation, stereodynamics, and chiral separation of the rotational enantiomers of hindered naphthyl ketones. *J. Am. Chem. Soc.* **1992**, *114*, 6521–6527.

449. Casarini D, Lunazzi L, Verbeek R. Conformational studies by dynamic NMR. 56. Enantiotopomerization and conformational analysis of hindered aryl alkyl ketones investigated by dynamic and solid state NMR. *Tetrahedron* **1996**, *52*, 2471–2480.

450. Guerra A, Lunazzi L. Conformational studies by dynamic NMR. 54. Trigonal nitrogen inversion and enantiomerization processes in the stereolabile chiral isomers of *N*-naphthylimines. *J. Org. Chem.* **1995**, *60*, 7959–7965.

451. Leardini R, Lunazzi L, Mazzanti A, McNab H, Nanni D. Conformational studies by dynamic NMR. 77. Stereomutation of the enantiomers of hindered O-substituted oximes. *Eur. J. Org. Chem.* **2000**, 3439–3446.

452. Kuttenberger M, Frieser M, Hofweber M, Mannschreck A. Axially chiral thioamides of acrylic acid: correlated and uncorrelated internal rotations. *Tetrahedron: Asymmetry* **1998**, *9*, 3629–3645.

453. Blanca MB, Yamamoto C, Okamoto Y, Biali SE, Kost D. Resolution and rotational barriers of quinolinone and acridone sulfenamide derivatives: demonstration of the S-N chiral axis. *J. Org. Chem.* **2000**, *65*, 8613–8620.

454. Takemura H, Shinmyozu T, Inazu T. Syntheses and properties of highly symmetrical cage compounds: pyridine analogues of hexa-m-xylylenetetraamine. *J. Am. Chem. Soc.* **1991**, *113*, 1323–1331.

455. Hoots JE, Lesch DA, Rauchfuss TB. Peracid oxidation of inorganic chalcogen ligands in transition metal complexes. *Inorg. Chem.* **1984**, *23*, 3130–3136.

456. Nabeshima T, Inaba T, Furukawa N, Ohshima S, Hosoya T, Yano Y. Selective recognition for heavy and transition metals by novel polyethers bearing bipyridines, and molecular chirality of pseudocrown structure in the Cu(I) complex. *Tetrahedron Lett.* **1990**, *31*, 6543–6546.

457. Chambron J, Mitchell DK, Sauvage J. Synthesis, characterization, and a proton NMR study of topologically chiral copper(I) [2]-catenates and achiral analogues. *J. Am. Chem. Soc.* **1992**, *114*, 4625–4631.

458. Tulloch AAD, Danopoulos AA, Tizzard GJ, Coles SJ, Hursthouse MB, Hay-Motherwell RS, Motherwell WB. Chiral 2,6-lutidinyl-biscarbene complexes of palladium. *Chem. Commun.* **2001**, 1270–1271.

459. Porwolik-Czomperlik I, Brandt K, Clayton TA, Davies DB, Eaton RJ, Shaw RA. Diastereoisomeric singly bridged cyclophosphazene-macrocyclic compounds. *Inorg. Chem.* **2002**, *41*, 4944–4951.

460. Beşil S, Coles SJ, Davies DB, Eaton RJ, Hursthouse MB, Kiliç A, Shaw RA, Çiftçi GY, Yeşilot S. Anomalous NMR behavior of meso compounds with remote stereogenic centers on addition of chiral shift reagent or chiral solvating agent. *J. Am. Chem. Soc.* **2003**, *125*, 4943–4950.

461. Beşli S, Coles SJ, Davies DB, Eaton RJ, Hursthouse MB, Kihe A, Shaw RA, Uslu A, Yeşilot S. Chirality in cyclotriphosphazenes with one stereogenic centre. *Inorg. Chem. Commun.* **2004**, *7*, 842–846.

462. Coles SJ, Davies DB, Eaton RJ, Hursthouse MB, Kiliç A, Shaw RA, Şahin Ş, Uslu A, Yeşilot S. Stereogenic properties of 1,3-disubstituted derivatives of cyclotriphosphazene: cis (meso) and trans (racemic) isomers. *Inorg. Chem. Commun.* **2004**, *7*, 657–661.

463. Uslu A, Coles SJ, Davies DB, Eaton RJ, Hursthouse MB, Kihç A, Shaw RA. Stereoisomerism in pentaerythritol-bridged cyclotriphosphazene tri-spiranes: spiro and ansa 1,3-propanediyldioxy disubstituted derivatives. *Eur. J. Inorg. Chem.* **2005**, 1042–1047.

464. Togni A, Pastor SD. Enantioselective synthesis of β-hydroxy-α-aminophosphonic acid precursors. *Tetrahedron Lett.* **1989**, *30*, 1071–1072.

465. Caccamese S, Principato G, Geraci C, Neri P. Resolution of inherently chiral 1,4-2,5-calix[8]bis-crown-4 derivatives by enantioselective HPLC. *Tetrahedron: Asymmetry* **1997**, *8*, 1169–1173.

466. Caccamese S, Bottino A, Cunsolo F, Parlato S, Neri P. Resolution of inherently chiral calix[4]arenes with AABB and CDCD substitution patters on the upper and lower rims, respectively. *Tetrahedron: Asymmetry* **2000**, *11*, 3103–3112.

467. Iki H, Kikuchi T, Shinkai S. Syntheses and spectral characterizations of tricarbonylchromium complexes of calix[4]arenes. *J. Chem. Soc., Perkin Trans. 1* **1993**, 205–210.

468. Aleksiuk O, Biali SE. Dixanthylene double calix[6]arene. *J. Org. Chem.* **1996**, *61*, 5670–5673.

469. Nam KC, Kim JM, Park YJ. Synthesis and structure identification of ABCH type calix[4]arenes: two step synthesis of asymmetrically substituted calix[4]arenes from monoalkylcalix[4]arenes. *Bull. Korean Chem. Soc.* **1998**, *19*, 770–776.

470. Tsantrizos YS, Ogilvie KK. Determination of the absolute stereochemistry of the fungal metabolite (R)-(−)-2-(4′-hydroxyphenyl)-2-hydroxyethanoic acid (pisolithin B). *Can. J. Chem.* **1991**, *69*, 772–778.

471. Hanna GM, Lau-Cam CA. Determination of the optical purity and absolute configuration of threo-methylphenidate by proton nuclear magnetic resonance spectroscopy with chiral solvating agent. *J. Pharm. Biomed. Anal.* **1993**, *11*, 665–670.

472. Jullian J, Franck X, Latypov S, Hocquemiller R, Figadère B. NMR determination of absolute configuration of α-acyloxy ketones. *Tetrahedron: Asymmetry* **2003**, *14*, 963–966.

473. Pausler MG, Rutledge PS. Experiments directed towards the synthesis of anthracyclinones. XXIV. Homochiral intermediates for vineomycin syntheses. *Aust. J. Chem.* **1994**, *47*, 2149–2160.

474. Stipanovic RD, McCormick JP, Schlemper EO, Hamper BC, Shinmyozu T, Pirkle, WH. Corroboration of techniques for assigning absolute configuration: lacinilene C methyl ether as an exemplary study. *J. Org. Chem.* **1986**, *51*, 2500–2504.

475. Pirkle WH, Boeder CW. Estimation of allene optical purities by nuclear magnetic resonance. *J. Org. Chem.* **1977**, *42*, 3697–3700.

476. Ang S, Low SH. Studies on the stereochemical characterization of N-methylated amino acids. *Aust. J. Chem.* **1991**, *44*, 1591–1601.

477. Bøgerø KP, Liljefors T, Arnt J, Hyttel J, Pedersen H. Octoclothepin enantiomers. A reinvestigation of their biochemical and pharmacological activity in relation to a new receptor-interaction model for dopamine D-2 receptor antagonists. *J. Med. Chem.* **1991**, *34*, 2023–2030.

478. Matthews JM, Dyatkin AB, Evangelisto M, Gauthier DA, Hecker LR, Hoekstra WJ, Liu F, Poulter BL, Sorgi KL, Maryanoff, BE. Synthesis, resolution, and absolute configuration of novel tricyclic benzodiazepines. *Tetrahedron: Asymmetry* **2004**, *15*, 1259–1267.

479. Gilmore J, Prowse W, Steggles D, Urquhart M, Olkowski J. Convenient synthesis of (2R)-and (2S)-2-(1-methylethyl)-5-oxo-2-phenylpentanenitrile, intermediates in the preparation of phenylalkylamine calcium channel blockers. *J. Chem. Soc., Perkin Trans. 1* **1996**, 2845–2850.

480. Krečmerová M, Buděšínský M, Masojídková M, Holý A. Synthesis of optically active N^6-alkyl derivatives of (R)-3-(adenine-9-yl)-2-hydroxypropanoic acid and related compounds. *Collect. Czech. Chem. Commun.* **2003**, *68*, 931–950.

481. Beaufour M, Merelli B, Menguy L, Cherton JC. Determination of the enantiomeric composition of chiral delta-2-oxazolines-1,3 by ^{1}H and ^{19}F NMR spectroscopy using chiral solvating agents. *Chirality* **2003**, *15*, 382–390.

482. Leonard J, Hewitt JD, Ouali D, Simpson SJ. Enantioselective opening of spiro epoxides derived from cis bicyclo[3.3.0.]octan-3,7-dione using chiral lithium amide bases. *Tetrahedron Lett.* **1990**, *31*, 6703–6706.

483. Tacke R, Reichel D, Gunther K Merget S. The first liquid-chromatographic separations of the (R)- and (S)-enantiomers of a chiral silanol, silane, and germane, *Z. Naturforsch* **1995**, *50b*, 568–572.

484. Anet FAL, Park J. Proton chemical shift assignments in citrate and trimethyl citrate in chiral media. *J. Am. Chem. Soc.* **1992**, *114*, 411–416.

485. Pirkle WH, Sikkenga DL. Use of achiral shift reagents to indicate relative stabilities of diastereomeric solvates. *J. Org. Chem.* **1975**, *40*, 3430–3434.

486. Jennison CPR, Mackay D. Lanthanide induced enhancement of enantiomeric shifts in chiral solvents and its use in the determination of optical purity. *Can. J. Chem.* **1973**, *51*, 3726–3732.

487. Young RN, Gauthier JY, Thérien M, Zamboni R. Asymmetric dithioacetals III. The preparation of the enantiomers of 3-((((3-(2-(7-chloroquinolin-2-yl-(E)-ethenyl) phenyl)-3-dimethylamino-3-oxopropyl-thio)methyl)thio)propionic acid (L-660,711) (MK-571), an antagonist of leukotriene. *Heterocycles* **1989**, *28*, 967–978.

488. Schacht J, Zugenmaier P, Horii F. Determination of the enantiomeric excess of ferroelectric liquid crystals derived from three natural amino acids. *Liq. Cryst.* **1999**, *26*, 525–533.

489. Anet FAL, Sweeting LM, Whitney TA, Cram DJ. Diastereomeric interactions in solution. *Tetrahedron Lett.* **1968**, 2617–2620.

490. Francotte E, Lang RW, Winkler T. New chiral fluoroanthryl derivatives: resolution of the enantiomers by chromatography on cellulose esters and their evaluation as chiral solvating agents in NMR spectroscopy. *Chirality* **1991**, *3*, 177–182.

491. de Moragas M, Cervelló E, Port A, Jaime C, Virgili A, Ancian B. Behavior of the 9-anthryl-tert-butylcarbinol as a chiral solvating agent. Study of diastereochemical association by intermolecular NOE and molecular dynamics calculations. *J. Org. Chem.* **1998**, *63*, 8689–8695.

492. Almer S, Cervelló E, Jaime C, Virgili A. Preparation of (*R*)-and (*S*)-1-adamantyl-9-anthrylmethanol. Conformational study and their behaviour as chiral solvating agents. *Tetrahedron: Asymmetry* **1999**, *10*, 3719–3725.

493. Pérez-Trujillo M, Virgili A, Molins E. Preparation, conformational analysis and behaviour as chiral solvating agents of 9-anthrylpentafluorophenylmethanol enantiomers: study of the diastereomeric association. *Tetrahedron: Asymmetry* **2004**, *15*, 1615–1621.

494. Pérez-Trujillo M, Maestre I, Jaime C, Alvarez-Larena A, Piniella JF, Virgili A. Enantioselective preparation and structural and conformational analysis of the chiral solvating agent α,α'-bis(trifluoromethyl)-1,8-anthracenedimethanol.*Tetrahedron: Asymmetry* **2005**, *16*, 3084–3093.

495. Pomares M, Sánchez-Ferrando F, Virgili A, Alvarez-Larena A, Piniella JF. Preparation and structural study of the enantiomers of α,α'-bis(trifluoromethyl)-9,10-anthracenedimethanol and its perdeuterated isotopomer, highly effective chiral solvating agents. *J. Org. Chem.* **2002**, *67*, 753–758.

496. Comelles J, Estivill C, Moreno-Mañas M, Virgili A, Vallribera A. (*R,R*)-α,α'-Bis (trifluoromethyl)-9,10-anthracenedimethanol: a chiral solvating agent for enantiomeric resolution of β-dicarbonyl compounds. *Tetrahedron* **2004**, *60*, 11541–11546.

497. Sánchez-Aris M, Estivill C, Virgili A. Synthesis and structural study of the enantiomers of α,α'-bis(trifluoromethyl)-10,10'-(9,9'-bianthryl)dimethanol as a chiral solvating agent. *Tetrahedron: Asymmetry* **2003**, *14*, 3129–3135.

498. Muñoz A, Virgili A. The use of perdeuterio 2,2,2-trifluoro-1-(9-anthryl) ethanol as chiral solvating agent. Chiral induction observed on ^{1}H and ^{13}C NMR. *Enantiomer* **2001**, *6*, 235–243.

499. Ferreiro MJ, Latypov SK, Quiñoá E, Riguera R. The use of ethyl 2-(9-anthryl)-2-hydroxyacetate for assignment of the absolute configuration of carboxylic acids by ¹H NMR. *Tetrahedron: Asymmetry* **1997**, *8*, 1015–1018.

500. Ferreiro MJ, Latypov SK, Quiñoá E, Riguera R. Assignment of the absolute configuration of α-chiral carboxylic acids by ¹H NMR spectroscopy. *J. Org. Chem.* **2000**, *65*, 2658–2666.

501. Akasaka K, Imaizumi K, Shichijyukari S, Ohrui H, Kabuto C. Enantiomeric discrimination of carboxylic acids having chiral centers remote from the carboxyl group by labeling with chiral fluorescent derivatization reagents. *Tennen Yuki Kagobutsu Toronkai Koen Yoshishu* **1998**, *40*, 79–84.

502. Muñoz A, Sánchez M, Junk T, Virgili A. A new case of chiral recognition between isotopomers. Preparation and study of (*R*) and (*S*) perdeuterio 2,2,2-trifluoro-1-(1-pyrenyl)ethanol. *J. Org. Chem.* **2000**, *65*, 5069–5071.

503. Muñoz A, Virgili A. Preparation and behavior of (*R*)- and (*S*)-2,2,2,-trifluoro-1-(1-pyrenyl)ethanol as chiral solvating agents: study of the diastereomeric association by Job's plots, intermolecular NOE and binding constants. *Tetrahedron: Asymmetry* **2002**, *13*, 1529–1534.

504. Brown E, Chevalier C, Huet F, Le Grumelec C, Lézé A, Touet J. Determination of the enantiomeric excesses of chiral acids by ¹⁹F NMR studies of their esters deriving from (*R*)-(+)-2-(trifluoromethyl)benzhydrol. *Tetrahedron: Asymmetry* **1994**, *5*, 1191–1194.

505. Jochims JC, Taigel G, Seeliger A. Proton resonance spectra of solvation diastereomers. *Tetrahedron Lett.* **1967**, 1901–1908.

506. Kainradl B, Langer E, Lehner H, Schlögl K. Stereochemistry of planar-chiral compounds. IV. Synthesis and absolute configuration of optically active, monosubstituted[2.2]metacyclophanes. *Justus Liebigs Ann. Chem.* **1972**, *766*, 16–31.

507. Takeuchi Y, Asahina M, Nagata K, Koizumi T. Chemistry of novel compounds with multifunctional carbon structure. 2. The first example of optically active multifunctional carbon compounds. *J. Chem. Soc., Perkin Trans. 1*, **1987**, 2203–2207.

508. Katayama M, Kato Y, Marumo S. Synthesis, absolute configuration and biological activity of both enantiomers of 2-(5,6-dichloro-3-indolyl)propionic acid: new dichloroindole auxins. *Biosci. Biotechnol. Biochem.* **2001**, *65*, 270–276.

509. Ute K, Kinoshita R, Matsui K, Miyatake N, Hatada K. Conformational asymmetry of a linear perfluoroalkyl chain in solution. Dynamic fluorine-19 NMR spectroscopy of the perfluoro-*n*-alkanes carrying a chiral end-group as a probe of magnetic nonequivalence. *Chem. Lett.* **1992**, 1337–1340.

510. Snyder JP, Nevins N, Tardif SL, Harpp DN. Inherently hindered rotation about a disulfide bond. *J. Am. Chem. Soc.* **1997**, *119*, 12685–12686.

511. Schwab JM, Klassen JB. Steric course of the allylic rearrangement catalyzed by β-hydroxydecanoylthioester dehydrase. Mechanistic implications. *J. Am. Chem. Soc.* **1984**, *106*, 7217–7227.

512. Au KG, Walsh CT. Stereochemical studies on a plasmid-coded fluoroacetate halidohydrolase. *Bioorg. Chem.* **1984**, *12*, 197–205.

513. Schwab JM, Klassen JB. B. Hydroxydecanoylthioester dehydrase. Steric course at substrate C-4 in the enzymes-catalyzed allylic rearrangement. *J. Chem. Soc., Chem. Commun.* **1984**, 296–297.

514. Schwab JM, Klassen JB. β-Hydroxydecanoylthioester dehydrase. Steric course at substrate C-2 and overall steric course of the enzyme-catalyzed allyic rearrangement. *J. Chem. Soc., Chem. Commun.* **1984**, 298–299.

515. Rawlings BJ, Reese PB, Ramer SE, Vederas JC. Comparison of fatty acid and polyketide biosynthesis: stereochemistry of cladosporin and oleic acid formation in *Cladosporium cladosporioides*. *J. Am. Chem. Soc.* **1989**, *111*, 3382–3390.

516. Tyrrell E, Tsang MWH, Skinner GA, Fawcett J. The application of ^{1}H nuclear magnetic resonance spectroscopy for the determination of the absolute configuration of chiral carboxylic acids. *Tetrahedron* **1996**, *52*, 9841–9852.

517. Noe CR, Knollmüller M, Wagner E, Völlenkle H. Chiral lactols. IV. Selectivities in acetal formation reactions of enantiomerically pure lactols using octahydro-8,9, 9-trimethyl-5,8-methano-2H-1-benzopyran-2-ol as a model. *Chem. Ber.* **1985**, *118*, 1733–1745.

518. Noe CR, Knollmüller M, Göstl G, Oberhauser B, Völlenkle H. Stereoelectronic effects and chiral recognition: a natural system of relationships between chiral compounds based on selectivities in the formation of acetals. *Angew. Chem. Int. Ed. Engl.* **1987**, *26*, 442–444.

519. Noe CR, Knollmüller M, Miculka C, Dungler K, Wagner E, Ettmayer P. Chiral lactols. XI. A method for the determination of the absolute configuration of chiral alkanols. *Chem. Ber.* **1994**, *127*, 887–892.

520. Noe CR, Knollmüller M, Oberhauser B, Steinbauer G, Wagner E. Chiral lactols. IV. A method for the determination of the absolute configuration of chiral α-hydroxy-substituted nitriles, alkynes, and aldehydes. *Chem. Ber.* **1986**, *119*, 729–743.

521. Noe CR, Knollmüller M, Gärtner P, Katikarides E, Gaischin L, Völlenkle H. Chiral lactols. XIII. On the determination of the absolute configuration of aromatic cyanohydrins and structurally related compounds. *Liebigs Ann. Chem.* **1995**, 1353–1360.

522. Girreser U, Noe CR. Synthesis of enantiopure (*S*)-α-hydroxy carboxanilides from (*S*)-α-amino acids. *Synthesis* **1995**, 1223–1224.

523. Mäurer M, Stegmann HB. Chiral recognition of diastereomeric esters and acetals by EPR and NMR investigations. *Chem. Ber.* **1990**, *123*, 1679–1685.

524. Stegmann HB, Mäurer M, Höfler U, Scheffler K, Hewgill F. Chiral recognition by ^{1}H-NMR, EPR, and ENDOR spectroscopy. *Chirality* **1993**, *5*, 282–287.

525. Girreser U, Haberhauer G, Noe CR. O,O-Acetal formation of exo-annelated octahydro-7,8,8-trimethyl-4,7-methanobenzofuran-2-ol with lactic acid and phenyllactic acid derivatives. *Monatsh. Chem.* **1998**, *129*, 281–289.

526. Lewis RA, Korpiun O, Mislow K. Configurational correlation of phosphinates by nuclear magnetic resonance and optical rotatory dispersion. *J. Am. Chem. Soc.* **1967**, *89*, 4786–4787.

527. Lewis RA, Korpiun O, Mislow K. Proton magnetic resonance spectra of menthyl phosphinates. *J. Am. Chem. Soc.* **1968**, *90*, 4847–4853.

528. Vitharana D, France JE, Scarpetti D, Bonneville GW, Majer P, Tsukamoto T. Synthesis and biological evaluation of (*R*)- and (*S*)-2-(phosphonomethyl)pentanedioic acids as inhibitors of glutamate carboxypeptides II. *Tetrahedron: Asymmetry* **2002**, *13*, 1609–1614.

529. Holt A, Jarvie AWP, Jervis GJ. The preparation and determination of optical purity of methylphenyl-1-naphthyl-(−)-menthoxysilanes. *J. Organomet. Chem.* **1970**, *21*, 75–77.

530. Holt A, Jarvie AWP, Jervis GJ. Preparation and spectroscopic properties of some new asymmetric organosilanes. *J. Chem. Soc., Perkin Trans. 2* **1973**, 114–122.
531. Corriu RJP, Lanneau GF. Synthèse et structure de nouveaux organosilanes chiraux bifonctionnels. *J. Organomet. Chem.* **1974**, *64*, 63–78.
532. Cole L, Löhr M, Rochin C. Reactions involving hexafluoropropylene oxide: novel ring opening reactions and resolution of a racemic mixture of a bromofluoro ester, ultrasound mediated Reformatsky reactions and stereoselectivity. *J. Chem. Soc., Perkin Trans. 1* **1998**, 2803–2811.
533. Sokolov VI, Reutov OA, Suleimanov GZ, Rozenberg VI, Petrovskii PV, Lutsenko AI, Fedin EI. L-Menthyl esters of α-bromomercuryphenylacetic acid: diastereoisomeric purity, symmetrization and reverse reaction. A stereochemical reinvestigation. *J. Organomet. Chem.* **1980**, *201*, 29–38.
534. Rahm A, Pereyre M. Hydrures organostanniques et syntheses asymetriques: obtention de quelques composes chiraux dans la serie des triorganostannyl-2 butanes et determination de leurs pouvoirs rotatoires maximaux. *J. Organomet. Chem.* **1975**, *88*, 79–92.
535. Katayama M, Kato Y, Marumo S. Synthesis, absolute configuration and biological activities of both enantiomers of 2-(5,7-dichloro-3-indolyl)propionic acid: a novel dichloroindole auxin and antiauxin. *Biosci. Biotechnol. Biochem.* **2004**, *68*, 1287–1292.
536. Ichikawa A, Hiradate S, Sugio A, Kuwahara S, Watanabe M, Harada N. Absolute configuration of 2-hydroxy-2-(1-naphthyl)propionic acid as determined by the ^{1}H NMR anisotropy method. *Tetrahedron: Asymmetry* **1999**, *10*, 4075–4078.
537. Ichikawa A, Hiradate S, Sugio A, Kuwahara S, Watanabe M, Harada N. Absolute configuration of 2-methoxy-2-(2-naphthyl)propionic acid as determined by the ^{1}H NMR anisotropy method. *Tetrahedron: Asymmetry* **2000**, *11*, 2669–2675.
538. Soncini P, Bonsignore S, Dalcanale E, Ugozzoli F. Cavitands as versatile molecular receptors. *J. Org. Chem.* **1992**, *57*, 4608–4612.
539. Tang H, Miura H, Kawakami Y. Enantiopure spiro[3.3]heptane-2,6-dicarboxylic acid. *Enantiomer* **2002**, *7*, 5–9.
540. Doyle TR, Vogl O. Haloaldehyde polymers. 45. Separation of fluorochlorobromoacetic acid into its antipodes; synthesis of optically active fluorochlorobromoacetaldehyde and its polymerization. *Polymer* **1991**, *32*, 751–760.
541. Eschler BM, Haynes RK, Ironside MD, Kremmydas S, Ridley DD, Hambley TW. A new resolution procedure for the preparation of both (*R*)-(+)- and (*S*)-(−)-4-tert-butoxycyclopent-2-enone from racemic 4-tert-butoxycyclopent-2-enone and conversion of (*R*)-(+)-4-tert-butoxycyclopent-2-enone into (*R*)-(+)-4-acetoxycyclopent-2-enone. A new method for the determination of the enantiomeric purities of the resolved enones. *J. Org. Chem.* **1991**, *56*, 4760–4766.
542. Barbeni M, Allegrone G, Cisero M, Guarda PA. GC and NMR enantiodiscrimination of 2-methyl substituted aliphatic acids via diastereomeric esterification with (*R*)-pantolactone. *Flavour Fragrance J.* **1992**, *7*, 163–167.
543. Toda F, Tanaka K, Ootani M, Hayashi A, Miyahara I, Hirotsu K. Structure study of host-guest molecular association in solution and in the solid state. *J. Chem. Soc., Chem. Commun.* **1993**, 1413–1415.
544. Tanaka K, Ootani M, Toda F. Optically active trans-bis(hydroxydiphenylmethyl)-2,2-dimethyl-1,3-dioxacyclopentane and its derivatives as chiral shift reagents for the

determination of enantiomeric purity and absolute configuration. *Tetrahedron: Asymmetry* **1992**, *3*, 709–712.

545. Tanaka K, Ootani M, Toda F. Optically active trans-bis(hydroxydiphenylmethyl)-2,2-dimethyl-1,3-dioxacyclopentane and its derivatives as chiral shift reagents for the determination of enantiomeric purity and absolute configuration. *Tetrahedron: Asymmetry* **1992**, *3*, 709–712.

546. Nishikawa K, Tsukada H, Abe S, Kishimoto M, Yasuoka N. Mode of molecular recognition during optical resolution: a structural study of the molecular complex involving both 3-hydroxypyrrolidines and (*R,R*)-(−)-trans-bis(hydroxydiphenylmethyl)-1,4-dioxaspiro[4,5]decane. *Chirality* **1999**, *11*, 166–171.

547. Tanaka N, Suemune H, Sakai K. Distinction of diastereofaces at the α-position of chiral cyclic acetals. *Tetrahedron: Asymmetry* **1992**, *3*, 1075–1086.

548. Vaida M, Shimon LJW, van Mil J, Ernst-Cabrera K, Addadi L, Leiserowitz L, Lahav M. Absolute asymmetric photochemistry using centrosymmetric single crystals. The host/guest system (E)-cinnamamide/(E)-cinnamic acid. *J. Am. Chem. Soc.* **1989**, *111*, 1029–1034.

549. Klaes M, Neumann B, Stammler H, Mattay J. Determination of the absolute configuration of inherently chiral resorc[4]arenes. *Eur. J. Org. Chem.* **2005**, 864–868.

550. Kasai R, Suzuo M, Asakawa J, Tanaka O. Carbon-13 chemical shifts of isoprenoid-β-D-glucopyranosides and -β-D-mannopyranosides. Stereochemical influences of aglycone alcohols. *Tetrahedron Lett.* **1977**, 175–178.

551. Tori K, Seo S, Yoshimura Y, Arita H, Tomita Y. Glycosidation shifts in carbon-13 NMR spectroscopy: carbon-13 signal shifts from aglycone and glucose to glucoside. *Tetrahedron Lett.* **1977**, 179–182.

552. Seo S, Tomita Y, Tori K, Yoshimura Y. Determination of the absolute configuration of a secondary hydroxy group in a chiral secondary alcohol using glycosidation shifts in carbon-13 nuclear magnetic resonance spectroscopy. *J. Am. Chem. Soc.* **1978**, *100*, 3331–3339.

553. Horibe I, Seo S, Yoshimura Y, Tori K. Glucosidation shifts of allylic and benzylic alcohols in ^{13}C NMR spectroscopy. *Org. Magn. Reson.* **1984**, *22*, 428–430.

554. Faghih R, Fontaine C, Horibe I, Imamura PM, Lukacs G, Olesker A, Seo S. Determination of the absolute configuration of chiral secondary alcohols from tetra-O-acetylglucosidation-induced ^{1}H nuclear magnetic resonance shifts. *J. Org. Chem.* **1985**, *50*, 4918–4920.

555. Seroka P, Plosiński M, Czub J, Sowiński P, Pawlak J. Monosaccharides as internal probes for the determination of the absolute configuration of 2-butanol. *Magn. Reson. Chem.* **2006**, *44*, 132–138.

556. Kurokawa-Nose Y, Shimada T, Wada H, Tanaka N, Murakami T, Saiki Y. Chemical and chemotaxonomical studies of ferns. LXXXIV. A novel 2-tetralol-type xyloside from *Asplenium wilfordii*. *Chem. Pharm. Bull.* **1993**, *41*, 930–932.

557. Otsuka H, Kamada K, Ogimi C, Hirata E, Takushi A, Takeda Y. Alangionosides A and B, ionol glycosides from leaves of *Alangium premnifolium*. *Phytochemistry* **1994**, *35*, 1331–1334.

558. Otsuka H, Hirata E, Shinzato T, Takeda Y. Stereochemistry of megastigmane glucosides from *Glochidion zeylanicum* and *Alangium premnifolium*. *Phytochemistry* **2003**, *62*, 763–768.

559. Otsuka H, Tamaki A. Platanionosides D–J: megastigmane glycosides from the leaves of *Alangium platanifolium* (SIEB. et ZUCC.) HARMS var. *platanifolium* SIEB. et ZUCC. *Chem. Pharm. Bull.* **2002**, *50*, 390–394.

560. Otsuka H, Kijima H, Hirata E, Shinzato T, Takushi A, Bando M, Takeda Y. Glochidionionosides A–D: megastigmane glucosides from leaves of *Glochidion zeylanicum* (GAERTN.) A. JUSS. *Chem. Pharm. Bull.* **2003**, *51*, 286–290.

561. Otsuka H, Yao M, Kamada K, Takeda Y. Alangionosides G-M: glycosides of megastigmane derivatives from the leaves of *Alangium premnifolium. Chem. Pharm. Bull.* **1995**, *43*, 754–759.

562. Řezanka T. Glycosides of polyenoic branched fatty acids from myxomycetes. *Phytochemistry.* **2002**, *60*, 639–646.

563. Řezanka T, and Dembitsky VM. Brominated oxylipins and oxylipin glycosides from Red Sea corals. *Eur. J. Org. Chem.* **2003**, 309–316.

564. Skouroumounis GK, Gunata YZ, Baumes RL. Determination of the absolute stereochemistry of chiral secondary megastigmane alcohols from shiraz leaves by NMR studies, *J. Essent. Oil Res.* **2000**, *12*, 661–666.

565. Trujillo M, Morales EQ, Vázquez JT. Tetra-O-benzoylglucosylation: a new ^{1}H nuclear magnetic resonance method for determination of the absolute configuration of secondary alcohols. *J. Org. Chem.* **1994**, *59*, 6637–6642.

566. Morales EQ, Padrón JI, Trujillo M, Vázquez JT. CD and ^{1}H NMR study of the rotational population dependence of the hydroxymethyl group in β glucopyranosides on the aglycon and its absolute configuration. *J. Org. Chem.* **1995**, *60*, 2537–2548.

567. Helm RF, Toikka M, Li K, Brunow G. Lignin model glycosides: preparation and optical resolution. *J. Chem. Soc., Perkin Trans. 1*, **1997**, 533–537.

568. Lancelin J, Beau J. Complete stereostructure of nystatin A_1: a proton NMR study. *Tetrahedron Lett.* **1989**, *30*, 4521–4524.

569. Pavia AA, Lacombe JM. ^{13}C NMR spectroscopy, a useful tool to determine the enantiomeric purity of synthetic threonine-containing glycopeptides. Spectra of diastereomeric α- and β-D-galactopyranosyl-L- and -D-threonine and -L- and -D-allothreonine. *J. Org. Chem.* **1983**, *48*, 2564–2568.

570. Fürstner A, Albert M, Mlynarski J, Matheu M, DeClercq E. Structure assignment, total synthesis, and antiviral evaluation of cycloviracin B_1. *J. Am. Chem. Soc.* **2003**, *125*, 13132–13142.

571. Shashkov AS, Usov AI, Knirel YA, Dmitriev BA, Kochetkov NK. Determination of the absolute and anomeric configurations of sugars in oligo- and polysaccharides by the effect of glycosylation in carbon-13 NMR spectra. *Russ. J. Bioorg. Chem.* **1981**, *7*, 1364–1371.

572. Nishizawa M, Kodama S, Yamane Y, Kayano K, Hatakeyama S, Yamada H. Synthesis and glycosylation shift of 1,1′-disaccharides. *Chem. Pharm. Bull.* **1994**, *42*, 982–984.

573. Nishida Y. Determination of the absolute configuration and the conformation of carbohydrate molecules based on the approach of analytical organic chemistry. *Tohoku J. Agric. Res.* **1995**, *46*, 73–91.

574. Lipkind GM, Shashkov AS, Nifant'ev NE, Kochetkov NK. Computer-assisted analysis of the structure of regular branched polysaccharides containing 2,3-disubstituted rhamnopyranose and mannopyranose residues on the basis of ^{13}C NMR data. *Carbohydr. Res.* **1992**, *237*, 11–22.

575. Kobayashi M. The fucofuranoside method, a new ^{1}H and ^{13}C nuclear magnetic resonance method to determine the absolute configuration of secondary alcohols. *Tetrahedron* **1997**, *53*, 5973–5994.

576. Ma Z, Hano Y, Qiu F, Chen Y, Nomura T. Determination of the absolute stereochemistry of lupine triterpenoids by fucofuranoside method and ORD spectrum. *Tetrahedron Lett.* **2004**, *45*, 3261–3263.
577. Kobayashi M. The fucofuranoside method for determining the absolute configuration of the tertiary alcohols substituted with methyl and two methylene groups. *Tetrahedron* **1998**, *54*, 10987–10998.
578. Kobayashi M. The ^{13}C NMR method for determining the absolute configuration of the 1,2-glycols consisting of secondary and tertiary hydroxyl groups. *Tetrahedron* **2000**, *56*, 1661–1665.
579. Kobayashi M. The arabinofuranoside method, a convenient substitute of the fucofuranoside method for determining the absolute configuration of the secondary alcohols. *Tetrahedron* **2002**, *58*, 9365–9371.
580. Shizuma M, Adachi H, Kawamura M, Takai Y, Takeda T, Sawada M. Chiral discrimination of fructo-oligosaccharides toward amino acid derivatives by induced-fitting chiral recognition. *J. Chem. Soc., Perkin Trans. 2* **2001**, 592–601.
581. Kano K, Negi S, Takaoka R, Kamo H, Kitae T, Yamaguchi M, Okubo H, Hirama M. Chiral recognition of tetrahelicene dicarboxylic acid by linear dextrins. *Chem. Lett.* **1997**, 715–716.
582. Toda F, Mori K, Okada J, Node M, Itoh A, Oomine K, Fuji K. New chiral shift reagents, optically active 2,2'-dihydroxy-1,1'-binaphthyl and 1,6-di(o-chlorophenyl)-1,6-diphenylhexa-2,4-diyne-1,6-diol. *Chem. Lett.* **1988**, 131–134.
583. Toda F, Mori K, Satō A. A determination method of absolute configuration. *Bull. Chem. Soc. Jpn.* **1988**, *61*, 4167–4169.
584. Drabowicz J, Duddeck H. ^{1}H NMR spectral nonequivalence of sulphoxide enantiomers in the presence of 2,2'-dihydroxy-1,1'-binaphthyl. *Sulfur Lett.* **1989**, *10*, 37–40.
585. Reynolds DP, Hollerton JC, Richards SA. Optical purity determination by ^{1}H NMR. *Analytical Applications of Spectroscopy*. Creaser CS, Davies AMC, Ed. Royal Society of Chemistry: London, England, **1988**, 346–348.
586. Dawson BA, Black DB, Lavoie A, LeBelle MJ. Nuclear magnetic resonance identification of the phenylalkylamine alkaloids of khat using a chiral solvating agent. *J. Forensic Sci.* **1994**, *39*, 1026–1038.
587. LeBelle MJ, Savard C, Dawson BA, Black DB, Katyal LK, Zrcek F, By AW. Chiral identification and determination of ephedrine, pseudoephedrine, methamphetamine, and metecathinone by gas chromatography and nuclear magnetic resonance. *Forensic Sci. Int.* **1995**, *71*, 215–223.
588. Nigmatov AG, Serebryakov EP. Catalytic asymmetric synthesis of 6-substituted derivatives of 1,3-cyclohexadienecarboxylic acid. *Russ. Chem. Bull.* **1993**, 233–234.
589. Azuzaya I, Okamoto I, Nakayama S, Tanatani A, Yamaguchi K, Shudo K, Kagechika H. A chiral *N*-methylbenzamide: spontaneous generation of optical activity. *Tetrahedron* **1999**, *55*, 11237–11246.
590. Michalik M, Döbler C. Determination of the chiral purity of aminoalcohols by ^{1}H NMR spectroscopy. *Tetrahedron* **1990**, *46*, 7739–7744.
591. Koy C, Michalik M, Döbler C, Oehme G. Chiral recognition of aminoalcohols by ^{1}H and ^{13}C NMR spectroscopy using binaphthyl derivatives as chiral solvating agents. *J. Prakt. Chem.* **1997**, *339*, 660–663.

592. Fukushi Y, Shigematsu K, Mizutani J, Tahara S. A new NMR chiral derivatizing reagent for determining the absolute configurations of carboxylic acids. *Tetrahedron Lett.* **1996**, *37*, 4737–4740.

593. Reeder J, Castro PP, Knobler CB, Martinborough E, Owens L, Diederich F. Chiral recognition of cinchona alkaloids at the minor and major grooves of 1,1′-binaphthyl receptors. *J. Org. Chem.* **1994**, *59*, 3151–3160.

594. Iuliano A, Bartalucci D, Uccello-Barretta G, Balzano F, Salvadori P. 3,5-Dinitrobenzoylphenylglycine analogues bearing the 1,1′-binaphthalene moiety—synthesis, conformational study, and application as chiral solvating agents. *Eur. J. Org. Chem.* **2001**, 2177–2184.

595. Toda F, Toyotaka R, Fukuda H. Optically active 4,4′,6,6′-tetrachloro-2,2′-bis(hydroxydiphenylmethyl)-biphenyl as a host for optical resolution and a chiral shift reagent. *Tetrahedron: Asymmetry* **1990**, *1*, 303–306.

596. Helder R, Arends R, Bolt W, Hiemstra H, Wynberg H. Alkaloid catalyzed asymmetric synthesis. III. The addition of mercaptans to 2-cyclohexene-1-one; determination of enantiomeric excess using carbon-13 NMR. *Tetrahedron Lett.* **1977**, 2181–2182.

597. Hiemstra H, Wynberg H. Determination of the enantiomeric purity of chiral ketones using ^{13}C NMR spectroscopy of their diastereomeric cyclic ketals. *Tetrahedron Lett.* **1977**, 2183–2186.

598. ten Hoeve W, Wynberg H. Chiral spiranes. Optical activity and nuclear magnetic resonance spectroscopy as a proof for stable twist conformations. *J. Org. Chem.* **1979**, *44*, 1508–1514.

599. Meyers AI, Williams DR, Erickson GW, White S, Druelinger M. Enantioselective alkylation of ketones via chiral, nonracemic lithioenamines. An asymmetric synthesis of α-alkyl and α,α′-dialkyl cyclic ketones. *J. Am. Chem. Soc.* **1981**, *103*, 3081–3087.

600. Lemière GL, Dommisse RA, Lepoivre JA, Alderweireldt FC, Hiemstra H, Wynberg H, Jones JB, Toone EJ. Determination of the absolute configuration of six-membered-ring ketones by ^{13}C NMR. *J. Am. Chem. Soc.* **1987**, *109*, 1363–1370.

601. Lange GL, Humber CC, Manthorpe JM. [2+2] Photoadditions with chiral 2,5-cyclohexadienone synthons. *Tetrahedron: Asymmetry* **2002**, *13*, 1355–1362.

602. Nagao T, Suenaga T, Ichihashi T, Fujimoto T, Yamamoto I, Kakehi A, Iriye R. Diastereoselective tandem Michael-intramolecular Wittig reactions of a cyclic phosphonium ylide with 8-phenylmenthyl enoates. *J. Org. Chem.* **2001**, *66*, 890–893.

603. Ridder D, Wunderlich H, Braun M. 1,1,2-Triphenyl-1,2-ethanediol: a host for carboxylic acids and amides in coordinatoclathrates. *Eur. J. Org. Chem.* **1998**, 1071–1076.

604. Shibata T, Ohkura T, Inayama S. Dithioacetal; as a chiroptical functional group to determine absolute and relative configurations by CD and ^{1}H NMR spectra. *Heterocycles* **1986**, *24*, 897–900.

605. von dem Bussche-Hünnefeld C, Beck AK, Lengweiler U, Seebach D. α,α,α′,α′-Tetraaryl-1,3-dioxolane-4,5-dimethanols (TADDOLs) for resolutions of alcohols and as chiral solvating agents in NMR spectroscopy. *Helv. Chim. Acta* **1992**, *75*, 438–441.

606. Schrader T. A chiral cyanohydrin phosphate for carbonyl umpolung—stereoselective synthesis of tertiary cyanohydrins. *Angew. Chem., Int. Ed. Engl.* **1995**, *34*, 917–919.

607. Pirkle WH. The nonequivalence of physical properties of enantiomers in optically active solvents. Differences in nuclear magnetic resonance spectra. I. *J. Am. Chem. Soc.* **1966**, *88*, 1837.

608. Burlingame TG, Pirkle WH. The nonequivalence of physical properties of enantiomers in optically active solvents. Differences in nuclear magnetic resonance spectra. II. *J. Am. Chem. Soc*. **1966**, *88*, 4294.

609. Pirkle WH, Burlingame TG. Nonequivalence of the nuclear magnetic resonance spectra of enantiomers in optically active solvents. III. *Tetrahedron Lett*. **1967**, 4039–4042.

610. Pirkle WH, Hoekstra MS. Carbon-13 nonequivalence of enantiomers in chiral solvents. *J. Magn. Reson*. 1975, *18*, 396–400.

611. Pirkle WH, Beare SD. Nonequivalence of the nuclear magnetic resonance spectra of enantiomers in optically active solvents. IV. Assignment of absolute configuration. *J. Am. Chem. Soc*. **1967**, *89*, 5485–5487.

612. Shore FL, Yuen GU. The absolute configuration of methyl 3-O-acetyl-2,3-dihydroxy-2-methylpropanoate by nuclear magnetic resonance and chemical determination. *J. Org. Chem*. **1972**, *37*, 3703–3707.

613. Kaehler H, Rehse K. Proton nuclear magnetic resonance of dihydroxysuccinic acid in optically active solvents. *Tetrahedron Letters* **1968**, 5019–5022.

614. Ejchart A, Jurczak J. NMR studies of diastereomeric dynamic systems. I. Chiral carboxylic acid. Chiral amine system. *Bull. Acad. Pol. Sci., Chim*. **1970**, *18*, 445–447.

615. Horeau A, Guette JP. Absolute configuration of α-isopropylphenylacetic acid, and a nuclear magnetic resonance determination of its optical purity. *C. R. Seances Acad. Sci. C* **1968**, *267*, 257–259.

616. Guette JP, Lacombe L. Horeau A. New examples of magnetic non-equivalence of enantiomers in a chiral medium. *C. R. Seances Acad. Sci. C* **1968**, *267*, 166–169.

617. Menezes PH, Gonçalves SMC, Hallwass F, Silva RO, Bieber LW, Simas AM. Efficient chiral discrimination by ^{77}Se NMR. *Org. Lett*. **2003**, *5*, 1601–1604.

618. Zemlicka J, Craine LE, Heeg M, Oliver JP. Enantioselective hydrolysis of dimethyl 2α,3α-[(dimethylmethylene)dioxy]-5β-hydroxy-1β,4β-cyclopentanedi-carboxylate with pig liver esterase. Stereoselective synthesis of methyl 2(*R*),3(*S*)-[(dimethylmethylene)dioxy]-5(*R*)-hydroxy-1(*S*)-carboxy-4(*R*)-cyclopentanecarboxylate. A cyclopentane synthone with all ring atoms chiral. *J. Org. Chem*. **1988**, *53*, 937–942.

619. Kühn M, Buddrus J. Analysis of chiral carboxylic acids by NMR using new optically active amines. *Tetrahedron: Asymmetry* **1993**, *4*, 207–210.

620. Reich-Rohrwig P. Schlögl K. Stereochemistry of the metallocenes. 20. Determination of the optical purity of methylferrocene-α-carboxylic acid. *Monatsh. Chem*. **1968**, *99*, 1752–1763.

621. Azzolina O, Vercesi D, Ghislandi V. Optical resolution and configuration of 2-(4-dimethylvinylphenyl)propionic acid. *Farmaco*, **1987**, *42*, 81–89.

622. Feringa B, Wynberg H. Asymmetric phenol oxidation. Stereospecific and stereoselective oxidative coupling of a chiral tetrahydronaphthol. *J. Org. Chem*. **1981**, *46*, 2547–2557.

623. Helmchen G, Ott R, Sauber K. Objective separation and absolute configuration of enantiomeric carboxylic acids and amines. 1. *Tetrahedron Lett*. **1972**, 3873–3878.

624. Hoye TR, Koltun DO. An NMR strategy for determination of configuration of remote stereogenic centers: 3-methylcarboxylic acids. *J. Am. Chem. Soc*. **1998**, *120*, 4638–4643.

625. Hoye TR, Hamad AS, Koltun DO, Tennakoon MA. An NMR method for determination of configuration of β-substituted carboxylic acids. *Tetrahedron Lett*. **2000**, *41*, 2289–2293.

626. Blažević N, Z—inič M, Kovač T, Šunjić V, Kajfež F. Identification methods for (±)-alpha-(3-benzoylphenyl)-propionic acid (ketoprofen) and its enantiomers based on physico-chemical properties. *Acta Pharm. Jugoslav.* **1975**, *25*, 155–163.

627. Ejchart A, Jurczak J, Bańkowski K. NMR studies of diastereomeric dynamic systems with α-amino acid derivatives. *Bull. Acad. Pol. Sci., Chim.* **1971**, *19*, 731–734.

628. Klys A, Nazarski RB, Zakrzewski J. Diastereomeric amides from rac-3,3',4,4'-tetramethyl-1,1'-diphosphaferrocene-2-carboxylic acid and (S)-α-phenylethylamine: separation and determination of absolute configuration. *J. Organomet. Chem.* **2001**, *627*, 135–138.

629. De Munari S, Marazzi G, Faustini F, Villa V, Carluccio L. Directed resolution of 2-fluoro-2-methylhexanoic acid. *J. Fluorine Chem.* **1986**, *34*, 157–166.

630. Köhler H, Gerlach H. Synthese des dysidins. *Helv. Chim. Acta* **1984**, *57*, 1783–1792.

631. Ghislandi V, La Manna A, Azzolina O, Gazzaniga A, Vercesi D. Configurational relationships in antiphlogistic hydratropic acids. *Farmaco* **1982**, *37*, 81–93.

632. De Munari S, Marazzi G, Forgione A, Longo A, Lombardi P. Determination of absolute configuration of the derivative from 2-[4-(1-oxo-2-isoindolinyl)-phenyl]-propionic acid and R-(+)-1-phenylethylamine by ^{1}H-NMR spectroscopy; use of shift reagent with diastereoisomeric amides. *Tetrahedron Lett.* **1980**, *21*, 2273–2276.

633. Pirkle WH, Beare SD. Nuclear magnetic resonance spectroscopy in optically active solvents. V. The determination of the absolute configuration of (+)-α-hydroxy-α-trifluoromethylphenylacetic acid. *Tetrahedron Lett.* **1968**, 2579–2582.

634. Korver O. Optical rotatory dispersion and circular dichroism of substituted mandelic acids. *Tetrahedron* **1970**, *26*, 5507–5518.

635. Mikolajczyk M, Para M, Ejchart A, Jurczak J. Correlation of the absolute configuration of phosphorus thio-acids by nuclear magnetic resonance spectroscopy. *Chem. Commun.* **1970**, 654.

636. Mikolajczyk M, Omelańczuk J, Leitloff M, Drabowicz J, Ejchart A, Jurczak J. ^{1}H, ^{31}P, and ^{13}C nuclear magnetic resonance nonequivalence of diastereomeric salts of chiral phosphorus thio acids with optically active amines. A method for determining the optical purity and configuration of chiral phosphorus thio acids. *J. Am. Chem. Soc.* **1978**, *100*, 7003–7008.

637. Glowacki Z, Hoffman M, Rachoń J. Direct ^{31}P NMR determination of the enantiomeric composition of 1-hydroxyalkylphosphonates using 1-(1-naphthyl)ethylamine as a chiral agent. *Phosphorus, Sulfur Silicon Relat. Elem.* **1993**, *82*, 39–47.

638. Hirashima A, Eto M. Synthesis of optically active cyclic phosphorothionates and phosphoramidothionates with insecticidal activity by using a chiral phosphorylating agent. *Agric. Biol. Chem.* **1983**, *47*, 2831–2839.

639. Wu S, Casida JE. Asymmetric synthesis of (R_p)- and (S_p)-2-ethyl-, (R_p)-2-pentyloxy-, (S_p)-2-pentylthio- and (S_p)-2-pentylamino-4H-1,3,2-benzodioxaphosphorin 2-oxides. *Phosphorus, Sulfur Silicon Relat. Elem.* **1994**, *88*, 129–137.

640. Hoyer G, Rosenberg D, Rufer C, Seeger A. Bestimmung der optischen reinheit und absoluten konfiguration von enantiomeren aralkylaminen. *Tetrahedron Lett.* **1972**, 985–988.

641. Takano S, Kasahara C, Ogasawara K. A simultaneous determination of enantiomeric purity and chirality of a certain cyclic enone relating to the gibbane framework. *Heterocycles* **1981**, *16*, 1669–1671.

642. Mamlok L, Marquet A, Lacombe L. Feasibility of determining the optical purity of ketones by NMR. *Tetrahedron Lett.* **1971**, 1039–1042.

643. Reiner J, Dagnino Jr. R, Goldman E, Webb TR. Facile determination of the optical purity of α-N-boc-amino aldehydes. *Tetrahedron Lett.* **1993**, *34*, 5425–5428.

644. Nowick JS, Powell NA, Nguyen TM, Noronha G. An improved method for the synthesis of enantiomerically pure amino acid ester isocyanates. *J. Org. Chem.* **1992**, *57*, 7364–7366.

645. Nikiforov TT, Simeonov MF. Synthesis and absolute configuration of diastereomeric 3-substituted 1-[1′-(S)-phenylethyl]-2-pyrrolidinones. *Dolk. Bulg. Akad. Nauk.* **1986**, *39*, 73–76.

646. Grotjahn DB, Joubran C. Coordination of chiral amines to coordinatively unsaturated Cp*Ir-amino acid complexes allows determination of enantiomeric purity. *Tetrahedron: Asymmetry* **1995**, *6*, 745–752.

647. Abraham RJ, Plant JE, Bedford GR. The NMR spectra of the porphyrins. The NMR spectra of some tetraarylchlorins. *Org. Magn. Reson.* **1983**, *21*, 745–756.

648. Jörss E, Schuler P, Maichle-Mössmer C, Abram S, Stegmann HB. Chiral recognition by EPR and ENDOR spectroscopy via diastereomers. *Enantiomer* **1997**, *2*, 5–16.

649. Kollár L, Berente Z, Forintos H, Keglevich G. Detection of the enantiomers of P-stereogenic pentacoordinated phosphorus compounds: ^{31}P NMR of oxaphosphetes in optically active solvents. *Tetrahedron: Asymmetry* **2000**, *11*, 4433–4436.

650. Pappalardo S, Parisi MF. Inherently chiral calix[4]crown ethers. *Tetrahedron Lett.* **1996**, *37*, 1493–1496.

651. Wang Y, Mosher HS. ^{19}F NMR non-equivalence of diastereomeric amides of 2,2,2-trifluoro-1-phenylethylamine. *Tetrahedron Lett.* **1991**, *32*, 987–990.

652. You T, Mosher HS. Synthesis and study of a new chiral agent—R-(−)-β-trifluoromethyl-β-methoxy-β-phenylethylamine. *Youji Huaxue* **1989**, *9*, 518–520.

653. You T, Hu X, Zhang D. Use of chiral reagent TMPEA for measuring enantiomeric purity of chiral ketones. *Gaodeng Xuexiao Huaxue Xuebao* **1997**, *18*, 132–133.

654. Yao L, Su C, Qi L, Chen X, Zhang L, Hu Y, Huang S. Synthesis of the diastereomeric imines of MTPEA and their spectral characteristics. *Huaxue Yanjiu Yu Yingyong* **1997**, *9*, 510–513.

655. Koos M, Mosher HS. α-Amino-α-trifluoromethyl-phenylacetonitrile: a potential reagent for ^{19}F NMR determination of enantiomeric purity of acids. *Tetrahedron* **1993**, *49*, 1541–1546.

656. Hamman S. Fluoro-2 phenyl-2 amino-1 ethane: obtention des enantiomeres, configuration absolue et emploi comme agent chiral de derivation. *J. Fluorine Chem.* **1989**, *45*, 377–387.

657. Port A, Virgili A, Jaime C. Preparation of enantiomers of 9-(1-amino-2,2-dimethylpropyl)-9,10-dihydroanthracene. Conformational study and their behaviour as chiral solvating agents. *Tetrahedron: Asymmetry.* **1996**, *7*, 1295–1302.

658. Pirkle WH, Sowin TJ. Synthesis and separation of diastereomeric imino alcohol derivatives of chiral phthalides: a method for assignment of phthalide absolute configurations. *J. Org. Chem.* **1987**, *52*, 3011–3017.

659. Karl V, Kaunzinger A, Gutser J, Steuer P, Angles-Angel J, Mosandl A. Stereoisomeric flavour compounds LXVII. 2-, 3-, and 4-Alkyl-branched acids, part 1: general approach to the synthesis of the enantiopure acids. *Chirality* **1994**, *6*, 420–426.

660. Bartschat D, Kuntzsch C, Heil M, Schittrigkeit A, Schumacher K, Mang M, Mosandl A, Kaiser R. Chiral compounds of essential oils XXI: (E,Z)-2,3-dihydrofarnesals-chirospecific analysis and structure elucidation of the stereoisomers. *Phytochem. Anal.* **1997**, *8*, 159–166.
661. Castellani L, Flieger M, Mannina L, Sedmera P, Segre AL, Sinibaldi M. A nuclear magnetic resonance study of the diastereoisomer complexes formed between a terguride derivative and naproxen. *Chirality* **1994**, *6*, 543–548.
662. Enders D, Thomas CR, Runsink J. 5-Amino-4-aryl-2,2-dimethyl-1,3-dioxins: application as chiral NMR shift reagents and derivatizing agents for acidic compounds. *Tetrahedron: Asymmetry* **1999**, *10*, 323–326.
663. Halpern B, Nitecki DE, Weinstein B. The steric purity of model peptides by NMR spectroscopy. *Tetrahedron Lett.* **1967**, 3075–3078.
664. Halpern B, Chew LF, Weinstein B. Measurement of racemization in peptide synthesis by nuclear magnetic resonance spectroscopy. *J. Am. Chem. Soc.* **1967**, *89*, 5051–5052.
665. Weinstein B, Pritchard AE. Amino acids and peptides. XXVIII. Determination of racemization in peptide synthesis by nuclear magnetic resonance spectroscopy. *J. Chem. Soc., Perkin Trans. 1* **1972**, 1015–1020.
666. Davies JS, Thomas RJ, Williams MK. Nuclear magnetic resonance spectra of benzoyl-dipeptide esters. Convenient test for racemization in peptide synthesis. *J. Chem. Soc., Chem. Commun.* **1975**, 76–77.
667. Benoiton NL, Kuroda K, Chen FMF. A series of lysyldipeptide derivatives for racemization studies in peptide synthesis. *Int. J. Pept. Protein Res.* **1979**, *13*, 403–408.
668. Davies JS, Mohammed AK. Assessment of racemization in *N*-alkylated amino acid derivatives during peptide coupling in a model dipeptide system. *J. Chem. Soc., Perkin Trans. 1* **1981**, 2982–2990.
669. Davies JS, Thomas RJ. Studies on the diastereoisomeric and conformational aspects of benzoyl dipeptide esters, as a means of assessing racemization using nuclear magnetic resonance spectroscopy. *J. Chem. Soc., Perkin Trans. 1* **1981**, 1639–1646.
670. Xing X, Fichera A, Kumar K. A novel synthesis of enantiomerically pure 5,5,5,5′,5′,5′-hexafluoroleucine. *Org. Lett.* **2001**, *3*, 1285–1286.
671. Babu IR, Hamill EK, Kumar K. A highly stereospecific and efficient synthesis of homopentafluoro-phenylalanine. *J. Org. Chem.* **2004**, *69*, 5468–5470.
672. Dewey RS, Schoenewaldt EF, Joshua H, Paleveda WJ, Schwam H, Barkemeyer H, Arison BH, Veber DF, Denkewalter RG, Hirschmann R. Synthesis of peptides in aqueous medium. V. Preparation and use of 2,5-thiazolidinediones (NTA's). Use of the ^{13}C-H nuclear magnetic resonance signal as internal standard for quantitative studies. *J. Am. Chem. Soc.* **1968**, *90*, 3254–3255.
673. Nagai Y, Kusumi T. New chiral anisotropic reagents for determining the absolute configuration of carboxylic acids. *Tetrahedron Lett.* **1995**, *36*, 1853–1856.
674. Yabuuchi T, Ooi T, Kusumi T. Application of phenylglycine methyl ester (PGME) to determination of the absolute configuration of carboxylic acids having phenylalkyl group. *Chirality* **1997**, *9*, 550–555.
675. Yabuuchi T, Kusumi T. Phenylglycine methyl ester, a useful tool for absolute configuration determination of various chiral carboxylic acids. *J. Org. Chem.* **2000**, *65*, 397–404.

676. Ichikawa A, Ono H, Hiradate S, Watanabe M, Harada N. Absolute configurations of 2-methoxy-2-(1-naphthyl)propionic acid and 2-methoxy-2-(2-naphthyl)propionic acid as determined by the phenylglycine methyl ester (PGME) method. *Tetrahedron: Asymmetry* **2002**, *13*, 1167–1172.

677. Sasaki K, Satake M, Yasumoto T. Identification of the absolute configuration of pectenotoxin-6, a polyether macrolide compound, by NMR spectroscopic method using a chiral anisotropic reagent, phenylglycine methyl ester. *Biosci., Biotechnol., Biochem.* **1997**, *61*, 1783–1785.

678. D'Ambrosio M, Guerriero A, Deharo E, Debitus C, Munoz V, Pietra F. New types of potentially antimalarial agents: epidioxy-substituted norditerpene and norsesterpenes from the marine sponge *Diacarnus levii*. *Helv. Chim. Acta* **1998**, *81*, 1285–1292.

679. Singh IP, Milligan KE, Gerwick WH. Tanikolide, a toxic and antifungal lactone from the marine cyanobacterium *Lyngbya majuscule*. *J. Nat. Prod.* **1999**, *62*, 1333–1335.

680. Su B, Takaishi Y, Yabuuchi T, Kusumi T, Tori M, Takaoka S. The structure and absolute configurations of macrophyllic acids A-E, five new rearranged eudesmane sesquiterpene acids from the bark of *Inula macrophylla*. *Tetrahedron Lett.* **2000**, *41*, 2395–2400.

681. Sasaki K, Onodera H, Yasumoto T. Applicability of the phenylglycine methyl ester to the establishment of the absolute configuration of dinophysistoxin-1, a polyether toxin responsible for diarrhetic shellfish poisoning. *Enantiomer* **1998**, *3*, 59–63.

682. Van Klink JW, Baek S, Barlow AJ, Ishii H, Nakanishi K, Berova N, Perry NB, Weavers RT. Determination of the absolute configuration of *Anisotome* irregular diterpenes: application of CD and NMR methods. *Chirality* **2004**, *16*, 549–558.

683. Schmidt EW, Faulkner DJ. Absolute configuration of methyl (2Z,6R,8R,9E)-3,6-epoxy-4,6,8-triethyl-2,4,9-dodecatrienoate from the sponge *Plakortis halichondrioides*. *Tetrahedron Lett.* **1996**, *37*, 6681–6684.

684. Morohashi A, Satake M, Nagai H, Oshima Y, Yasumoto T. The absolute configuration of gambieric acids A–D, potent antifungal polyethers, isolated from the marine dinoflagellate *Gambierdiscus toxicus*. *Tetrahedron* **2000**, *56*, 8995–9001.

685. Hanuš LO, Řezanka T, Dembitsky VM. A trinorsesterterpene glycoside from the North American fern *Woodwardia virginica* (L.) Smith. *Phytochemistry* **2003**, *63*, 869–875.

686. Su B, Takaishi Y, Yabuuchi T, Kusumi T, Tori M, Takaoka S, Honda G, Ito M, Takeda Y, Kodzhimatov OK, Ashurmetov O. Sesquiterpenes and monoterpenes from the bark of *Inula macrophylla*. *J. Nat. Prod.* **2001**, *64*, 466–471.

687. Yasumoto T. The chemistry and biological function of natural marine toxins. *Chem. Rec.* **2001**, *1*, 228–242.

688. Restelli A, Annunziata R, Pellacini F, Ferrario F. NMR determination of absolute configurations in 2-alkylthiazolidine-4-carboxylic acids. *J. Heterocycl. Chem.* **1990**, *27*, 1035–1039.

689. Yang X, Wang G, Zhong C, Wu X, Fu E. Novel NMR chiral solvating agents derived from (1R,2R)-diaminocyclohexane: synthesis and enantiodiscrimination for chiral carboxylic acids. *Tetrahedron: Asymmetry* **2006**, *17*, 916–921.

690. Port A, Virgili A, Alvarez-Larena A, Piniella JF. Preparation of enantiomers of 1-(1-naphthyl)-2,2-dimethylpropylamine and their behaviour as chiral solvating agents: study of diastereochemic association by Job's plots and intermolecular NOE measurements. *Tetrahedron: Asymmetry* **2000**, *11*, 3747–3757.

691. Chi Y, Peelan TJ, Gellman SH. A rapid ^{1}H NMR assay for enantiomeric excess of α-substituted aldehydes. *Org. Lett.* **2005**, *7*, 3469–3472.
692. Meyers AI, Brich Z. ^{1}H NMR determination of enantiomeric purities of 2-substituted aldehydes via the aldimine of 2-amino-1-methoxymenth-8-ene. *J. Chem. Soc., Chem. Commun.* **1979**, 567–568.
693. Mikolajczyk M, Omelańczuk J. Configuration of the optically active phosphorus thioacids. III. Synthesis and optical resolution of O-methyl isopropylphosphonothioic acid. *Phosphorus* **1973**, *3*, 47–50.
694. Glowacki Z, Hoffman M, Rachoń J. ^{31}P NMR determination of the enantiomeric composition of N-protected 1-aminoalkylphosphonates using 1-(1-naphthyl) ethylamine and ephedrine as the chiral agents. *Phosphorus, Sulfur Silicon Relat. Elem.* **1995**, *104*, 21–32.
695. Glowacki Z, Hoffman M. ^{31}P NMR determination of the enantiomeric composition of N-phthaloyl-1-aminoalkyl-phosphonate monoesters using 1-(1-naphthyl)ethylamine and ephedrine as the chiral agents. *Phosphorus, Sulfur Silicon Relat. Elem.* **1998**, *134/135*, 279–285.
696. Mikolajczyk M, Lyżwa P, Drabowicz J. A new efficient procedure for asymmetric synthesis of α-aminophosphonic acids via addition of lithiated bis(diethylamino) phosphine borane complex to enantiopure sulfinimines. *Tetrahedron: Asymmetry* **2002**, *13*, 2571–2576.
697. Boulet CA, Tregear SJ, Hansen AS. Use of 2D NMR for the assignment of structure of 1,3,2-oxaza-phospholidin-2-ones. *Heterocycles* **1991**, *32*, 2105–2110.
698. Bernhard W, Brügger P, Daly JJ, Schönholzer P, Weber RH, Hansen H. 47. First optically active heptalenes and their absolute configuration. *Helv. Chim. Acta* **1985**, *68*, 415–428.
699. Agami C, Meynier F, Berlan J, Besace Y, Brochard L. Chiral 3-substituted aldehydes: determination of absolute configurations and enantiomeric excesses by NMR analysis of derived oxazolidines. *J. Org. Chem.* **1986**, *51*, 73–75.
700. Anet FAL, Kopelevich M. Detection and assignments of diastereotopic chemical shifts in partially deuteriated methyl groups of a chiral molecule. *J. Am. Chem. Soc.* **1989**, *111*, 3429–3431.
701. Anet FAL, O'Leary DJ, Beale JM, Floss HG. Stereogenic (chiral) methyl groups: determination of configuration by direct tritium nuclear magnetic resonance spectroscopy. *J. Am. Chem. Soc.* **1989**, *111*, 8935–8936.
702. Allen BD, O'Leary DJ. Fomenting proton anisochronicity in the CH$_2$D group. *J. Am. Chem. Soc.* **2003**, *125*, 9018–9019.
703. Allen BD, Cintrat J, Faucher N, Berthault P, Rousseau B, O'Leary DJ. An isosparteine derivative for stereochemical assignment of stereogenic (chiral) methyl groups using tritium NMR: theory and experiment. *J. Am. Chem. Soc.* **2005**, *127*, 412–420.
704. Vignon SA, Wong J, Tseng H, Stoddart JF. Helical chirality in donor-acceptor catenanes. *Org. Lett.* **2004**, *6*, 1095–1098.
705. Aime S, Botta M, Crich SG, Terreno E, Anelli PL, Uggeri F. Spectral discrimination of chiral macrocyclic paramagnetic metal complexes by NMR techniques. *Chem. Eur. J.* **1999**, *5*, 1261–1266.
706. Bailey DJ, O'Hagan D, Tavasli M. A short synthesis of (S)-2-(diphenylmethyl)pyrrolidine, a chiral solvating agent for NMR analysis. *Tetrahedron: Asymmetry* **1997**, *8*, 149–153.

707. Webster FX, Zeng X, Silverstein RM. Following the course of resolution of carboxylic acids by ^{13}C NMR spectrometry of amine salts. *J. Org. Chem.* **1982**, *47*, 5225–5226.
708. Rosini C, Uccello-Barretta G, Pini D, Abete C, Salvadori P. Quinine: An inexpensive chiral solvating agent for the determination of enantiomeric composition of binaphthyl derivatives and alkylarylcarbinols by NMR spectroscopy. *J. Org. Chem.* **1988**, *53*, 4579–4581.
709. Uccello-Barretta G, Balzano F, Salvadori P. Rationalization of the multireceptorial character of chiral solvating agents based on quinine and its derivatives: overview of selected NMR investigations. *Chirality* **2005**, *17*, S243–S248.
710. Uccello-Barretta G, Pini D, Rosini C, Salvadori P. Stereochemical features of the stereoselective interaction between (R)- or (S)-7,7′-bis(1-propen-3-oxy)-2,2′-dihydroxy-1,1′-binaphthyl and quinine. *J. Chromatogr. A* **1994**, *666*, 541–548.
711. Salvadori P, Rosini C, Pini D, Bertucci C, Altemura P, Uccello-Barretta G, Raffaelli A. A novel application of *Cinchona* alkaloids as chiral auxiliaries: preparation and use of a new family of chiral stationary phases for the chromatographic resolution of racemates. *Tetrahedron* **1987**, *43*, 4969–4978.
712. Klein J, Hartenstein H, Sicker D. First discrimination of enantiomeric cyclic hemiacetals and methyl acetals derived from hydroxamic acids and lactams of *Gramineae* by means of ^{1}H NMR using carious chiral solvating agents. *Magn. Reson. Chem.* **1994**, *32*, 727–731.
713. Uccello-Barretta G, Pini D, Mastantuono A, Salvadori P. Direct NMR assay of the enantiomeric purity of chiral β-hydroxy esters by using quinine as chiral solvating agent. *Tetrahedron: Asymmetry* **1995**, *6*, 1965–1972.
714. Abid M, Török B. *Cinchona* alkaloid induced chiral discrimination for the determination of the enantiomeric composition of α-trifluoro-methylated-hydroxyl compounds by ^{19}F NMR spectroscopy. *Tetrahedron: Asymmetry* **2005**, *16*, 1547–1555.
715. Winter-Werner B, Diederich F, Gramlich V. 113. Analogs of *Cinchona* alkaloids incorporating a 9,9′-spirobifluorene moiety. *Helv. Chim. Acta* **1996**, *79*, 1338–1360.
716. Klein J, Borsdorf R. NMR spectroscopic determination of enantiomers via their triphenyltin derivatives: a new application of chiral solvating agents. *Fresenius J. Anal. Chem.* **1994**, *350*, 644–646.
717. Knoll S, Klein J, Ristau T, Borsdorf R. NMR-spectroscopic investigations of chiral carboxylic acids via their tributyltin derivatives. *J. Prakt. Chem./Chem.-Ztg.* **1996**, *338*, 374–375.
718. Klein J, Lebelt I, Borsdorf R. Triphenyltin (IV) derivatives of *N*-acetyl amino acids, *N*-acetyl dipeptides: preparation, NMR investigations and NMR spectroscopic resolution of the enantiomeric composition. *J. Prakt. Chem./Chem.-Ztg.* **1996**, *338*, 544–548.
719. Żymańczyk-Duda E, Skwarczyński M, Lejczak B, Kafarski P. Accurate assay of enantiopurity of 1-hydroxy- and 2-hydroxyalkylphosphonate esters. *Tetrahedron: Asymmetry* **1996**, *7*, 1277–1280.
720. Maly A, Lejczak B, Kafarski P. Quinine as chiral discriminator for determination of enantiomeric excess of diethyl 1,2-dihydroxyalkanephosphonates. *Tetrahedron: Asymmetry* **2003**, *4*, 1019–1024.
721. Kotoris CC, Wen W, Lough A, Taylor SD. Preparation of chiral α-monofluoroalkylphosphonic acids and their evaluation as inhibitors of protein tyrosine phosphatase 1B. *J. Chem. Soc., Perkin Trans. 1*, **2000**, 1271–1281.

722. Meyer F, Uziel J, Papini AM, Jugé S. Triphenylphosphonium salts bearing an L-alanyl substituent: short synthesis and enantiomeric analysis by NMR. *Tetrahedron Lett.* **2001**, *42*, 3981–3984.

723. Sweeting LM, Anet FAL. Magnetic resonance study of diastereomeric interactions of nitroaromatic sulfoxides with alcohols in solution. *Org. Magn. Reson.* **1984**, *22*, 539–542.

724. Uccello-Barretta G, Balzano F, Quintavalli C, Salvadori P. Different enantioselective interaction pathways induced by derivatized quinines. *J. Org. Chem.* **2000**, *65*, 3596–3602.

725. Uccello-Barretta G, Bardoni S, Balzano F, Salvadori P. Versatile chiral auxiliaries for NMR spectroscopy based on carbamoyl derivatives of dihydroquinine. *Tetrahedron: Asymmetry* **2001**, *12*, 2019–2023.

726. Uccello-Barretta G, Mirabella F, Balzano F, Salvadori P. C11 versus C9 carbamoylation of quinine: a new class of versatile polyfunctional chiral solvating agents. *Tetrahedron: Asymmetry* **2003**, *14*, 1511–1516.

727. Maier NM, Schefzick S, Lombardo GM, Feliz M, Rissanen K, Lindner W, Lipkowitz KB. Elucidation of the chiral recognition mechanism of *Cinchona* alkaloid carbamate-type receptors for 3,5-dinitrobenzoyl amino acids. *J. Am. Chem. Soc.* **2002**, *124*, 8611–8629.

728. Hellriegel C, Skogsberg U, Albert K, Lämmerhofer M, Maier NM, Lindner W. Characterization of a chiral stationary phase by HR/MAS NMR spectroscopy and investigation of enantioselective interaction with chiral ligates by transferred NOE. *J. Am. Chem. Soc.* **2004**, *126*, 3809–3816.

729. Shapiro M, Lin M, Versace R, Petter RC. Application of HMQC in determination of chiral purity using quinine. *Tetrahedron: Asymmetry* **1996**, *7*, 2169–2172.

730. Wilen SH, Qi JZ, Williard PG. Resolution, asymmetric transformation, and configuration of Tröger's base. Application of Tröger's base as a chiral solvating agent. *J. Org. Chem.* **1991**, *56*, 485–487.

731. Toda F, Tanaka K, Mori K. Novel 1,2- and 1,3-double chiral recognitions. Optical resolution of α- and β-haloacetylenic alcohols by complexation with brucine. *Chem. Lett.* **1983**, 827–830.

732. Crassous J, Jiang Z, Schurig V, Polavarapu PL. Preparation of (+)-chlorofluoroiodomethane, determination of its enantiomeric excess and of its absolute configuration. *Tetrahedron: Asymmetry* **2004**, *15*, 1995–2001.

733. Mistry N, Roberts AD, Tranter GE, Francis P, Barylski I, Ismail IM, Nicholson JK, Lindon JC. Directly coupled chiral HPLC-NMR and HPLC-CD spectroscopy as complementary methods for structural and enantiomeric isomer identification: application to atracurium besylate. *Anal. Chem.* **1999**, *71*, 2838–2843.

734. Fulwood R, Parker D. A chiral solvating agent for direct NMR assay of the enantiomeric purity of carboxylic acids. *Tetrahedron: Asymmetry* **1992**, *3*, 25–28.

735. Fulwood R, Parker D. 1,2-Diphenylethane-1,2-diamine: an effective NMR chiral solvating agent for chiral carboxylic acids. *J. Chem. Soc., Perkin Trans. 2* **1994**, 57–64.

736. Guo J, Knapp DW, Boegeman S. A practical preparation of enantiomerically pure (*R*)- and (*S*)-2-bromohexadecanoic acids. *Tetrahedron: Asymmetry* **2000**, *11*, 4105–4111.

737. Huang G, Li C. Effective NMR chiral solvating agent for 2-(2,4,5-trichlorophenoxy) propanoic acid. *Fenxi Ceshi Xuebao* **2003**, *22*, 48–51.

738. Rajamoorthi K, Stockton GW. Direct NMR determination of optical purity of nicotinic and quinolinic carboxylic acid compounds using 1,2-diphenylethane-1,2-diamine as a chiral solvating agent. *Spectrosc. Lett.* **2001**, *34*, 255–266.

739. Yoo S, Kim S. (1S,2S)-(−)-1,2-Diphenylethylenediamine; a good chiral solvating agent for the determination of enantiomeric purity of chiral alcohols by ^{1}H NMR. *Bull. Korean Chem. Soc.* **1996**, *17*, 673–674.

740. Alexakis A, Frutos JC, Mangeney P. An easy and fast way to determine the enantiomeric purity of substituted cyclanones. *Tetrahedron: Asymmetry* **1993**, *4*, 2431–2434.

741. Mangeney P, Alexakis A, Normant JF. Resolution and determination of enantiomeric excesses of chiral aldehydes via chiral imidazolidines. *Tetrahedron Lett.* **1988**, *29*, 2677–2680.

742. Cuvinot D, Mangeney P, Alexakis A, Normant J, Lellouche J. Chiral trifluoro diamines as convenient reagents for determining the enantiomeric purity of aldehydes by use of ^{19}F NMR spectroscopy. *J. Org. Chem.* **1989**, *54*, 2420–2425.

743. Lapitskaya MA, Zatonsky GV, Pivnitsky KK. Enantiomeric NMR analysis of organic acids with the Corey chiral controller. *Mendeleev Commun.* **1999**, 149–151.

744. Kobayashi Y, Hayashi N, Tan C, Kishi Y. Toward the creation of NMR databases in chiral solvents for assignments of relative and absolute stereochemistry: proof of concept. *Org. Lett.* **2001**, *3*, 2245–2248.

745. Hayashi N, Kobayashi Y, Kishi Y. Toward the creation of NMR databases in chiral solvents for assignments of relative and absolute stereochemistry: scope and limitation. *Org. Lett.* **2001**, *3*, 2249–2252.

746. Fidanze S, Song F, Szlosek-Pinaud M, Small PLC, Kishi Y. Complete structure of the mycolactones. *J. Am. Chem. Soc.* **2001**, *123*, 10117–10118.

747. Kobayashi Y, Hayashi N, Kishi Y. Toward the creation of NMR databases in chiral solvents for assignments of relative and absolute stereochemistry: NMR desymmetrization of meso compounds. *Org. Lett.* **2001**, *3*, 2253–2255.

748. Kobayashi Y, Hayashi N, Kishi Y. Toward the creation of NMR databases in chiral solvents: bidentate chiral NMR solvents for assignment of the absolute configuration of acyclic secondary alcohols. *Org. Lett.* **2002**, *4*, 411–414.

749. Higashibayashi S, Kishi Y. Assignment of the relative and absolute configurations of acyclic secondary 1,2-diols. *Tetrahedron* **2004**, *60*, 11977–11982.

750. Kobayashi Y, Hayashi N, Kishi Y. Application of chiral bidentate NMR solvents for assignment of the absolute configuration of alcohols: scope and limitation. *Tetrahedron Lett.* **2003**, *44*, 7489–7491.

751. Kobayashi Y, Czechtizky W, Kishi Y. Complete stereochemistry of tetrafibricin. *Org. Lett.* **2003**, *5*, 93–96.

752. Dambruoso P, Bassarello C, Bifulco G, Appendino G, Battaglia A, Fontana G, Gomez-Paloma L. Advances in the universal NMR database approach. 2′-substituted taxanes as probes for an improved protocol of diastereomeric differentiation. *Org. Lett.* **2005**, *7*, 983–986.

753. Zhang Q, Carrera G, Gomes MJS, Aires-de-Sousa J. Automatic assignment of absolute configuration from 1D NMR data. *J. Org. Chem.* **2005**, *70*, 2120–2130.

754. Deshmukh M, Duñach E, Juge S, Kagan HB. A convenient family of chiral shift reagents for measurement of enantiomeric excesses of sulfoxides. *Tetrahedron Lett.* **1984**, *25*, 3467–3470.

755. Buist PH, Marecak DM, Partington ET, Skala P. Enantioselective sulfoxidation of a fatty acid analogue by Bakers' yeast. *J. Org. Chem.* **1990**, *55*, 5667–5669.

756. Cho H, Ramaswamy S, Plapp BV. Flexibility of liver alcohol dehydrogenase in stereoselective binding of 3-butylthiolane 1-oxides. *Biochemistry* **1997**, *36*, 382–389.

757. El Ouazzani H, Khiar N, Fernández I, Alcudia F. General method for asymmetric synthesis of α-methylsulfinyl ketones: application to the synthesis of optically pure oxisuran and bioisosteres. *J. Org. Chem.* **1997**, *62*, 287–291.

758. Morrow RJ, Millership JS, Collier PS. Diastereotopic analysis of mesoridazine besylate (serentil). *Chirality* **2004**, *16*, 534–540.

759. Duñach E, Kagan HB. A simple chiral shift reagent for measurement of enantiomeric excesses of phosphine oxides. *Tetrahedron Lett.* **1985**, *26*, 2649–2652.

760. Peyronel J, Samuel O, Fiaud J. New chiral bicyclic phosphoramides derived from L-glutamic acid. *J. Org. Chem.* **1987**, *52*, 5320–5325.

761. Tambuté A, Begos A, Lienne M, Caude M, Rosset R. New chiral stationary phases containing a phosphorus atom as an asymmetric centre. I. Synthesis and first chromatographic results. *J. Chromatogr.* **1987**, *396*, 65–81.

762. Pakulski Z, Demchuk OM, Kwiatosz R, Osiński PW, Świerczyńska W, Pietrusiewicz KM. The classical Kagan's amides are still practical NMR chiral shift reagents: determination of enantiomeric purity of P-chirogenic phospholene oxides. *Tetrahedron: Asymmetry* **2003**, *14*, 1459–1462.

763. Jursic BS, Zdravkovski Z. Molecular mechanics calculations and comparison of proton, fluorine, and carbon NMR diastereomer discrimination via nonbonding interactions between fluorine-labeled enantiomeric amides and enantiomerically pure chiral solvating agents. *J. Org. Chem.* **1993**, *58*, 5245–5250.

764. Jursic BS. Spectroscopic and molecular mechanics calculations of discrimination between enantiomers possessing an electron rich aromatic group directly attached to the chiral carbon atom with optically pure benzoyl derivatives. *J. Chem. Soc., Perkin Trans. 2*, **1994**, 961–969.

765. Jursic BS, Zdravkovski Z, Zuanic M. Determination of enantiomeric composition of 2-phenyl-2-(2-piperidyl)acetamide. A routine method for evaluation of enantiomeric purity of primary amides. *Tetrahedron: Asymmetry* **1994**, *5*, 1711–1716.

766. Pirkle WH, Pochapsky TC. Considerations of chiral recognition relevant to the liquid chromatography separation of enantiomers. *Chem. Rev.* **1989**, *89*, 347–362.

767. Pirkle WH, Tsipouras A. 3,5-Dinitrobenzoyl amino acid esters. Broadly applicable chiral solvating agents for NMR determination of enantiomeric purity. *Tetrahedron Lett.* **1985**, *26*, 2989–2992.

768. Pirkle WH, Pochapsky TC. Intermolecular ^{1}H{^{1}H} nuclear overhauser effects in diastereomeric complexes: support for a chromatographically derived chiral recognition model. *J. Am. Chem. Soc.* **1986**, *108*, 5627–5628.

769. Pirkle WH, Pochapsky TC. Chiral molecular recognition in small bimolecular systems: a spectroscopic investigation into the nature of diastereomeric complexes. *J. Am. Chem. Soc.* **1987**, *109*, 5975–5982.

770. Wenzel TJ, Morin CA, Brechting AA. Lanthanide-chiral resolving agent mixtures as chiral NMR shift reagents. *J. Org. Chem.* **1992**, *57*, 3594–3599.

771. Wenzel TJ, Bean AC, Dunham SL. Chiral NMR shift reagents: mixtures of lanthanide tris (β-diketonates) with chiral carboxylate anions. *Magn. Reson. Chem.* **1997**, *35*, 395–402.
772. Wenzel TJ, Brogan KL. Lanthanide-chiral carboxylate and chiral ester mixtures as NMR shift reagents. *Enantiomer* **2000**, *5*, 293–302.
773. Pirkle WH, Welch CJ. Chromatographic and ^{1}H NMR support for a proposed chiral recognition model. *J. Chromatogr. A* **1994**, *683*, 347–353.
774. Pirkle WH, Selness SR. Chiral recognition studies: intra- and intermolecular ^{1}H{^{1}H}-nuclear overhauser effects as effective tools in the study of bimolecular complexes. *J. Org. Chem.* **1995**, *60*, 3252–3256.
775. Pirkle WH, Gan KZ. A direct chromatographic separation of enantiomers chiral by virtue of isotopic substitution. *Tetrahedron: Asymmetry* **1997**, *8*, 811–814.
776. Job GE, Shvets A, Pirkle WH, Kuwahara S, Kosaka M, Kasai Y, Taji H, Fujita K, Watanabe M, Harada N. The effects of aromatic substituents on the chromatographic enantioseparation of diarylmethyl esters with the Whelk-O1 chiral stationary phase. *J. Chromatogr. A* **2004**, *1055*, 41–53.
777. Koscho ME, Pirkle WH. Investigation of a broadly applicable chiral selector used in enantioselective chromatography (Whelk-O 1) as a chiral solvating agent for NMR determination of enantiomeric composition. *Tetrahedron: Asymmetry* **2005**, *16*, 3345–3351.
778. Wenzel TJ, Miles RD, Weinstein SE. Chiral nuclear magnetic resonance shift reagents: lanthanide mixtures with 1-(1-naphthyl)ethylurea derivatives of amino acids. *Chirality* **1997**, *9*, 1–9.
779. Skogsberg U, Händel H, Sanchez D, Albert K. Comparisons of the interactions between two analytes and two structurally similar chiral stationary phases using high-performance liquid chromatography, suspended-state high-resolution magic angle spinning nuclear magnetic resonance and solid-state nuclear magnetic resonance spectroscopy. *J. Chromatogr. A* **2004**, *1023*, 215–223.
780. Skogsberg U, Allenmark S. Determination of enantiomer separation factors by nuclear magnetic resonance spectroscopy and by chiral liquid chromatography. *J. Chromatogr. A* **2001**, *921*, 161–167.
781. Allenmark S, Skogsberg U, Thunberg L. Chiral selectors based on C_2-symmetric dicarboxylic acids. *Tetrahedron: Asymmetry* **2000**, *11*, 3527–3534.
782. Allenmark S, Skogsberg U. A new chiral selector based on trans-acenaphthen-1, 2-dicarboxylic acid. *Enantiomer* **2000**, *5*, 451–455.
783. Skogsberg U, Thunberg L, Allenmark S. Solvent-dependent enantioselective interaction of some bis-allylamide chiral selectors studied by NMR. *Chirality* **2001**, *13*, 272–278.
784. Im SH, Ryoo JJ, Lee K, Choi S, Jeong YH, Jung YS, Hyun MH. NMR studies of chiral discrimination relevant to the enantioseparation of *N*-acylarylalkylamines by an (*R*)-phenylglycinol-derived chiral selector. *Chirality* **2002**, *14*, 329–333.
785. Hyun MO, Han SC, Jeong KK. A new NMR chiral solvating agent derived from (*R*)-4-hydroxyphenylglycine. *Bull. Korean Chem. Soc.* **1998**, *19*, 595–598.
786. Uccello-Barretta G, Rosini C, Pini D, Salvadori P. A spectroscopic study of the interaction of D- and L-*N*-(3,5-dinitrobenzoyl)valine methyl ester with *n*-butylamide of (*S*)-2-[(phenylcarbamoyl)oxy]propionic acid: direct evidence for a chromatographic chiral recognition rationale. *J. Am. Chem. Soc.* **1990**, *112*, 2707–2710.

787. Pini D, Uccello-Barretta G, Rosini C, Salvadori P. N-(n-Butylamide) of (S)-2-(phenylcarbamoyloxy)propionic acid: a new chiral solvating agent, derived from L-lactic acid, for the enantiomeric purity determination of derivatized amino acids. *Chirality* **1991**, *3*, 174–176.

788. Jursic BS, Goldberg, SI. Enantiomerically pure 2,2'-oxybis[N-(1-phenylethyl)acetamide]. An especially effective chiral solvating agent for determinations of enantiomer compositions by NMR spectroscopy. *J. Org. Chem.* **1992**, *57*, 7370–7372.

789. Hanyu M, Yanagihara R, Miyazawa T, Yamada T. Chiral recognition ability of quaternary ammonium salts by tetrapeptides. *Peptide Science*. Aoyagi H, Ed. The Japanese Peptide Society: Osaka, Japan, **2002**, 373–376.

790. Spisni A, Corradini R, Marchelli R, Dossena A. Chiral recognition of amino acid derivatives: an NMR investigation of the selector and the diastereomeric complexes. *J. Org. Chem.* **1989**, *54*, 684–688.

791. Bauer A, Bach T. Assignment of the absolute configuration of 7-substituted 3-azabicyclo[3.3.1]nonan-2-ones by NMR-titration experiments. *Tetrahedron: Asymmetry* **2004**, *15*, 3799–3803.

792. Bergmann H, Grosch B, Sitterberg S, Bach T. An enantiomerically pure 1,5,7-trimethyl-3-azabicyclo[3.3.1]nonan-2-one as ^{1}H NMR shift reagent for the ee determination of chiral lactams, quinolones, and oxazolidinones. *J. Org. Chem.* 2004, *69*, 970–973.

793. Bach T, Bergmann H, Grosch B, Harms K. Highly enantioselective intra- and intermolecular [2+2] photocycloaddition reactions of 2-quinolones mediated by a chiral lactam host: host-guest interactions, product configuration, and the origin of the stereoselectivity in solution. *J. Am. Chem. Soc.* **2002**, *124*, 7982–7990.

794. Murray RS, Boyd VA, Lynch VM, Negrete GR. Pyrimidinone conjugate for the assignment of the absolute configuration of α-chiral carboxylic acids by ^{1}H-NMR. *ARKIVOC* **2001**, 58–73.

795. Sugimoto H, Yamane Y, Inoue S. Enantiomeric discrimination by novel optically active isocyanurates having peripheral amino acid units. *Tetrahedron: Asymmetry* **2000**, *11*, 2067–2075.

796. Tachibana Y, Ando M, Kuzuhara H. Asymmetric synthesis of α-deuterated α-amino acids through nonenzymatic transamination reaction and the determination of their enantiomeric excesses. *Bull. Chem. Soc. Jpn.* **1983**, 3652–3656.

797. Fukui H, Fukushi Y, Tahara S. NMR determination of the absolute configuration of chiral 1,2- and 1,3-diols. *Tetrahedron Lett.* **2003**, *44*, 4063–4065.

798. Chin J, Kim DC, Kim H, Panosyan FB, Kim KM. Chiral shift reagent for amino acids based on resonance-assisted hydrogen bonding. *Org. Lett.* **2004**, *6*, 2591–2593.

799. Kim KM, Park H, Kim H, Chin J, Nam W. Enantioselective recognition of 1,2-amino alcohols by reversible formation of imines with resonance-assisted hydrogen bonds. *Org. Lett.* **2005**, *7*, 3525–3527.

800. Harada T, Kurokawa H, Kagamihara Y, Tanaka S, Inoue A, Oku A. Stereoselective acetalization of 1,3-alkanediols by l-menthone: application to the resolution of racemic 1,3-alkanediols and to the determination of the absolute configuration of enantiomeric 1,3-alkanediols. *J. Org. Chem.* **1992**, *57*, 1412–1421.

801. Meyers AI, White SK, Fuentes LM. Determination of enantiomeric composition of chiral 2-substituted-1,2-glycols via ^{13}C-NMR and HPLC. *Tetrahedron Lett.* **1983**, *24*, 3551–3554.

REFERENCES

802. Rice KC, Brossi A. Expedient synthesis of racemic and optically active N-norreticuline and N-substituted and 6′-bromo-N-norreticulines. *J. Org. Chem.* **1980**, *45*, 592–601.
803. Dumont R, Brossi A, Silverton JV. Facile conversion of natural colchicine into (±)-congeners and (+)-enantiomers including 2-demethyl analogues. *J. Org. Chem.* **1986**, *51*, 2515–2521.
804. Greiner E, Folk JE, Jacobson AE, Rice KC. A novel and facile preparation of bremazocine enantiomers through optically pure N-norbremazocines. *Bioorg. Med. Chem.* **2004**, *12*, 233–238.
805. Dharanipragada R, Diederich F. Diastereomeric complex formation between a novel optically active host and naproxen in aqueous solution. *Tetrahedron Lett.* **1987**, *28*, 2443–2446.
806. Elman B, Moberg C. The synthesis of (+)- and (−)-1-(1-isoquinolinyl)-1-(2-pyridyl)ethanol. A chiral ligand with useful chelating properties. *Tetrahedron* **1986**, *42*, 223–228.
807. Kim KH, Kim HJ, Kim J, Lee JH, Lee SC. Determination of the optical purity of (R)-terbutaline by ^{1}H-NMR and RP-LC using chiral derivatizing agent, (S)-(−)-α-methylbenzyl isocyanate. *J. Pharm. Biomed. Anal.* **2001**, *25*, 947–956.
808. Pomares M, Grabuleda X, Jaime C, Virgili A, Álvarez-Larena Á, Piniella JF. Configurational and conformational NMR study of enantiopure 2,2-dimethyl-1-(1-naphthyl)-propanol via its carbamate derivatives. *Magn. Reson. Chem.* **1999**, *37*, 885–890.
809. Pirkle WH, Hauske JR. Broad spectrum methods for the resolution of optical isomers. A discussion of the reasons underlying the chromatographic separability of some diastereomeric carbamates. *J. Org. Chem.* **1977**, *42*, 1839–1844.
810. Pirkle WH, Simmons KA, Boeder CW. Dynamic NMR studies of diastereomeric carbamates: implications toward the determination of relative configuration by NMR. *J. Org. Chem.* **1979**, *44*, 4891–4896.
811. Sonnet PE, Dudley RL, Osman S, Pfeffer PE, Schwartz D. Configuration analysis of unsaturated hydroxyl fatty acids. *J. Chromatogr.* **1991**, *586*, 255–258.
812. Robertson DW, Krushinski JH, Fuller RW, Leander JD. Absolute configurations and pharmacological activities of the optical isomers of fluoxetine, a selective serotonin-uptake inhibitor. *J. Med. Chem.* **1988**, *31*, 1412–1417.
813. García-Martínez C, Taguchi Y, Oishi A, Hayamizu K. Ring current effects of phenyl and naphthyl groups: internal probes for determining the absolute configuration of chiral azetidin-2-ones by ^{1}H NMR. *Tetrahedron: Asymmetry* **1998**, *9*, 955–965.
814. Jeon DJ, Kim JS, Lee JN, Kim HR, Ryu EK. (R)-[1-(1-Naphthyl)ethyl] isothiocyanate and (S)-1-phenylethyl isothiocyanate. New chirality recognizing reagents for the determination of enantiomeric purity of chiral amines by NMR. *Chem. Lett.* **2000**, 40–41.
815. Nabeya A, Endo T. α-Methoxy-α-(trifluoromethyl)benzyl isocyanate. A convenient reagent for the determination of the enantiomeric composition of primary and secondary amines. *J. Org. Chem.* **1988**, *53*, 3358–3361.
816. Kram TC, Lurie IS. The determination of enantiomeric composition of methamphetamine by ^{1}H-NMR spectroscopy. *Forensic Sci. Int.* **1992**, *55*, 131–137.
817. Vodička P, Streinz L, Koutek B, Buděšínský M, Ondráček J, Císařova I. (S)-2-Chloro-2-fluoroethanoyl isocyanate, a chiral derivative of trichloroacetyl isocyanate. *Chirality* **2003**, *15*, 472–478.

818. Dana G, Danechpajouh H. Carbon-13 NMR study of the stereoisomerism of 4, 5-disubstituted 2,2-dimethyl-1,3-dioxolanes. *Bull. Soc. Chim. Fr.* **1980**, 395–399.
819. Solladié G, Hanquet G, Rolland C. Stereoselective sulfoxide directed reduction of 1, 2-diketo-derivatives to enantiomerically pure syn and anti 1,2-diols. *Tetrahedron Lett.* **1997**, *38*, 5847–5850.
820. Oh H, Shin H, Oh, G-S, Pae H-O, Chai K-Y, Chung H-T, Lee H-S. The absolute configuration of prunioside A from Spiraea prunifolia and biological activities of related compounds. *Phytochem.* **2003**, *64*, 1113–1118.
821. Wu F, Gu Z, Zeng L, Zhao G, Zhang Y, McLaughlin JL, Sastrodihardjo S. Two new cytotoxic monotetrahydrofuran annonaceous acetogenins, annomuricins A and B, from the leaves of *Annona muricata. J. Nat. Prod.* **1995**, *58*, 830–836.
822. Wróblewski AE, Glowacka IE. Synthesis of (1R,2S)- and (1S,2S)-3-azido-1,2-dihydroxypropylphosphonates. *Tetrahedron: Asymmetry* **2002**, *13*, 989–994.
823. Devijver C, Salmoun M, Daloze D, Braekman JC, De Weerdt WH, De Kluijver MJ, Gomez R. (2R,3R,7Z)-2-aminotetradec-7-ene-1,3-diol, a new amino alcohol from the caribbean sponge *Haliclona vansoesti. J. Nat. Prod.* **2000**, *63*, 978–980.
824. Szechner B. Synthesis and absolute configuration of two diastereoisomeric (1R, 2S, 3R)- and (1S, 2S, 3R)-2-amino-1-(2-furyl)-1,3-butandiols. *Tetrahedron* **1981**, *37*, 949–952.
825. Lakshman MK, Chaturvedi S, Kole PL, Windels JH, Myers MB, Brown MA. Resolution of polycyclic aromatic hydrocarbon dihydrodiols via diastereomeric formaldehyde acetals. *Tetrahedron: Asymmetry* **1997**, *8*, 3375–3378.
826. Pirkle WH, Simmons KA. Improved chiral derivatizing agents for the chromatographic resolution of racemic primary amines. *J. Org. Chem.* **1983**, *48*, 2520–2527.
827. Song CE, Lee SG, Lee KC, Kim IO, Jeong JH. Chromatographic resolution of racemic α-halocarboxylic acids and O-substituted α-hydroxycarboxylic acids via diastereomeric *N*-acyloxazolidinones. *J. Chromatogr. A* **1993**, *654*, 303–308.
828. Smith MB, Dembofsky BT, Son YC. 5(R)-Methyl-1-(chloromethyl)-2-pyrrolidinone: a new reagent for the determination of enantiomeric composition of alcohols, *J. Org. Chem.* **1994**, *59*, 1719–1725.
829. Chen P, Tao S, Smith MB. 5R-Methyl-1-chloromethyl-2-pyrrolidinone: determining enantiomeric excess in chiral nonracemic amines. *Synth. Commun.* **1998**, *28*, 1641–1648.
830. Domínguez BE, García PE, Cárdenas J. Determining the absolute configuration of secondary alcohols by means of a chiral auxiliary and NOESY. *Tetrahedron: Asymmetry* **2005**, *16*, 3976–3985.
831. Kawai T, Kodama K, Ooi T, Kusumi T. 1,3-Dipolar addition of nitrones to symmetrically substituted allenes: for the determination of absolute configuration of chiral allenes by NMR spectroscopy. *Tetrahedron Lett.* **2004**, *45*, 4097–4099.
832. Rodriguez-Escrich S, Popa D, Jimeno C, Vidal-Ferran A, Pericàs MA. (*S*)-2-[(*R*)-Fluoro(phenyl)methyl]oxirane: a general reagent for determining the ee of α-chiral amines. *Org. Lett.* **2005**, *7*, 3829–3832.
833. Uccello-Barretta G, Iuliano A, Menicagli R, Peluso P, Pieroni E, Salvadori P. 1,3,5-Triazine multiselector systems: new tools for chiral discrimination. *Chirality* **1997**, *9*, 113–121.

834. Uccello-Barretta G, Iuliano A, Franchi E, Balzano F, Salvadori P. Di- and tri-1-(1-naphthyl)ethylamino-substituted 1,3,5-triazine derivatives: a new class of versatile chiral auxiliaries for NMR spectroscopy. *J. Org. Chem.* **1998**, *63*, 9197–9203.

835. Uccello-Barretta G, Samaritani S, Menicagli R, Salvadori P. 2,4,6-Tri[(S)-1′-methylbenzylamino]-1,3,5-triazine: a new NMR chiral solvating agent for 3,5-dinitrophenyl derivatives; an attempt at a chiral discrimination rationale. *Tetrahedron: Asymmetry* **2000**, *11*, 3901–3912.

836. Iuliano A, Uccello-Barretta G, Salvadori P. 1-Arylethylamino-substituted s-triazine derivatives as chiral solvating agents for the determination of the enantiomeric composition of chiral compounds. *Tetrahedron: Asymmetry* **2000**, *11*, 1555–1563.

837. Nemoto H, Tsutsumi H, Yuzawa S, Peng X, Zhong W, Xie J, Miyoshi N, Suzuki I, Shibuya M. Highly efficient chiral resolution and determination of absolute configuration of 2-alkanols by using a cyclopenta[b]furan derivative. *Tetrahedron Lett.* **2004**, *45*, 1667–1670.

838. Hoyer GA, Rosenberg D, Rufer C, Seeger A. Determination of the optical purity and absolute configuration of enantiomeric aralkylamines. *Tetrahedron Lett.* **1972**, 985–988.

839. Christoffers J, Schulze Y, Pickardt J. Synthesis, resolution, and absolute configuration of trans-1-amino-2-dimethylaminocyclohexane. *Tetrahedron* **2001**, *57*, 1765–1769.

840. Dehmlow EV, Westerheide R. (S)-3-aminothiolane: a new chiral building block. *Synthesis* **1992**, 947–949.

841. Maier NM, Zoltewicz JA. Dynamic equilibration of diastereomeric salts of atropisomers. Proton NMR spectra of 1,8-di(3′-pyridyl)naphthalene in the presence of R-camphorsulfonic acid. *Tetrahedron* **1997**, *53*, 465–468.

842. Mehmandoust M, Marazano C, Singh R, Gillet B, Césario M, Fourrey J, Das BC. Synthesis of chiral isoquinuclidines and determination of their absolute configuration. *Tetrahedron Lett.* **1988**, *29*, 4423–4426.

843. Chow H, Wan C, Ng M. A versatile method for the resolution and absolute configuration assignment of substituted 1,1′-bi-2-naphthols. *J. Org. Chem.* **1996**, *61*, 8712–8714.

844. Bourgeois J, Seiler P, Fibbioli M, Pretsch E, Diederich F, Echegoyen L. Cyclophane-type fullerene-dibenzo[18]crown-6 conjugates with trans-1, trans-2, and trans-3 addition patterns: regioselective templated synthesis, X-ray crystal structure, ionophoric properties, and cation-complexation-dependent redox behavior. *Helv. Chim. Acta*, **1999**, *82*, 1572–1595.

845. Komatsuzaki N, Uno M, Shirai K, Takai Y, Tanaka T, Sawada M, Takahashi S. An anion receptor based on poly-substituted cobalticinium complexes. *Bull. Chem. Soc. Jpn.* **1996**, *69*, 17–24.

846. Rees L, Suckling CJ, Wood HCS. A simple and effective method for preparation of the 6(R)- and 6(S)-diastereoisomers of 5-formyltetrahydrofolate (leucovorin). *J. Chem. Soc., Chem. Commun* **1987**, 470–472.

847. Towers MDKN, Woodgate PD, Brimble MA. Addition of 2-[(trimethylsilyloxy)]furan to 2-acetyl-1,4-benzoquinone using chiral non-racemic copper(II)-pybox catalysts. *ARKIVOC* **2003**, 43–55.

848. Preite MD, González-Sierra M, Zinczuk J, Rúveda EA. (N-Methylphenylsulfoximidoyl)methyllithium: a versatile reagent for the determination of absolute configuration of six-membered ring ketones. *Tetrahedron: Asymmetry* **1994**, *5*, 503–506.

849. Saigo K, Sekimoto K, Yonezawa N, Ishii F, Hasegawa M. Optically active anti head-to-head coumarin dimer. A new agent for the determination of enantiomeric excess of amines and alcohols. *Bull. Chem. Soc. Jpn.* **1985**, *58*, 1006–1009.

850. Adegawa Y, Kashima T, Saigo K. Determination of enantiomeric excesses of chiral amines by using an enantiomerically pure anti head-to-head coumarin dimer derivative. *Tetrahedron: Asymmetry* **1993**, *4*, 1421–1424.

851. Ohki A, Miyashita R, Naka K, Maeda S. Enantioselective extraction of di-O-benzoyl-tartrate anion by ion-pair extractant having binaphthyl-unit. *Bull. Chem. Soc. Jpn.* **1991**, *64*, 2714–2719.

852. Toşa M, Paizs C, Majdik C, Novák L, Kolonits P, Irimie F, Poppe L. Synthesis of optically active 3-substituted-10-alkyl-10H-phenothiazine-5-oxides by enantioselective biotransformations. *Tetrahedron: Asymmetry* **2002**, *13*, 211–221.

853. Torrens A, Castrillo JA, Frigola J, Salgado L, Redondo J. Optical resolution and enantiomeric purity determination of the analgesic cizolirtine. *Chirality* **1999**, *11*, 63–69.

854. Kolasa T, Miller MJ. A simple method for distinguishing optical isomers of chiral amines, hydroxylamines, amino acids, and peptides. *J. Org. Chem.* **1986**, *51*, 3055–3058.

855. van Boeckel CAA, Visser GM, Hegstrom RA, van Boom JH. Amplification of chirality based upon the association of nucleic acid strands of opposite handedness. *J. Mol. Evol.* **1987**, *25*, 100–105.

856. Honzawa S, Okubo H, Anzai S, Yamaguchi M, Tsumoto K, Kumagai I. Chiral recognition in the binding of helicenediamine to double strand DNA: interactions between low molecular weight helical compounds and a helical polymer. *Bioorg. Med. Chem.* **2002**, *10*, 3213–3218.

857. Nabeta K, Yamamoto M, Koshino H, Fukui H, Fukushi Y, Tahara S. Absolute configuration of kelsoene and prespatane. *Biosci. Biotechnol. Biochem.* **1999**, *63*, 1772–1776.

858. Fukui H, Fukushi Y, Tahara S. NMR determination of the absolute configuration of cyclic chiral alkenes. *Tetrahedron Lett.* **1999**, *40*, 325–328.

859. Fukui H, Fukushi Y, Ohashi S, Oikawa H, Tahara S. NMR determination of the absolute configuration of a macrophomate synthase inhibitor by using an axial chiral reagent. *Biosci. Biotechnol. Biochem.* **2001**, *65*, 1920–1923.

860. Jassbi AR, Fukushi Y, Tahara S. Determination of absolute configuration of decipinone, a diterpenoid ester with a myrsinane-type carbon skeleton, by NMR spectroscopy. *Helv. Chim. Acta*, **2002**, *85*, 1706–1713.

861. Fedin EI, Davankov VA. NMR investigations of the enantiomeric excess effects in solutions with weak intermolecular association. *Chirality* **1995**, *7*, 326–330.

862. Williams T, Pitcher RG, Bommer P, Gutzwiller J, Uskoković M. Diastereomeric solute-solute interactions of enantiomers in achiral solvents. Nonequivalence of the nuclear magnetic resonance spectra of racemic and optically active dihydroquinine. *J. Am. Chem. Soc.* **1969**, *91*, 1871–1872.

863. Thong CM, Marraud M, Neel J. Estimation of the optical purity of chiral peptide derivatives by proton NMR spectroscopy. *C. R. Seances Acad. Sci. C* **1975**, *281*, 691–694.

864. Ghosh SK. Chiral recognition in dipeptides containing 1-aminocyclopropane carboxylic acid or α-aminoisobutyric acid: NMR studies in solution. *J. Peptide Res.* **1999**, *53*, 261–274.

865. Jursic BS, Goldberg SI. Enantiomer discrimination arising from solute-solute interactions in partially resolved chloroform solutions of chiral carboxamides. *J. Org. Chem.* **1992**, *57*, 7172–7174.

866. Tait A, Colorni E, Bella MD. Stereospecific synthesis of 2-[(2H-1,2,4-benzothiadiazine-1,1-dioxide-3-yl)thio]propanoic acids: enantiomeric excess evaluation by ^{1}H NMR. *Tetrahedron: Asymmetry* **1997**, *8*, 2199–2207.

867. Navrátilová H, Potáček M. Enantiomeric enrichment of (+)-(3R,4S)-4-(4-fluorophenyl)-3-hydroxymethyl-1-methylpiperidine by crystallization. *Enantiomer* **2001**, *6*, 333–337.

868. Claessens CG, Torres T. Chiral self-discrimination in a M_3L_2 subphthalocyanine cage. *J. Am. Chem. Soc.* **2002**, *124*, 14522–14523.

869. Bergman SD, Kol M. π-Stacking induced NMR spectrum splitting enantiomerically enriched Ru(II) complexes: evaluation of enantiomeric excess. *Inorg. Chem.* **2005**, *44*, 1647–1654.

870. Evans MA, Morken JP. Isotopically chiral probes for in situ high-throughput asymmetric reaction analysis. *J. Am. Chem. Soc.* **2002**, *124*, 9020–9021.

871. Reetz MT, Eipper A, Tielmann P, Mynott R. A practical NMR-based high-throughput assay for screening enantioselective catalysts and biocatalysts. *Adv. Synth. Catal.* **2002**, *344*, 1008–1016.

872. Reetz MT, Tielmann P, Eipper A, Ross A, Schlotterbeck G. A high-throughput NMR-based ee-assay using chemical shift imaging. *Chem. Commun.* **2004**, 1366–1367.

873. Harger MJP. Nuclear magnetic resonance non-equivalence of the enantiomers in optically active samples of phosphinic amides. *J. Chem. Soc., Chem. Commun.* **1976**, 555–556.

874. Harger MJP. Proton magnetic resonance non-equivalence of the enantiomers of alkylphenylphosphinic amides. *J. Chem. Soc., Perkin Trans. 2* **1977**, 1882–1887.

875. Ouryupin AB, Kadyko MI, Petrovskii PV, Fedin EI. Chiral discrimination in nonracemic mixtures of methanephosphonic acid, N,N'-bis(1-phenylethyl)diamides. *Chirality* **1994**, *6*, 1–4.

876. Uryupin BA, Kadyko MI, Petrovskii PV, Fedin EI, Sanshe M, Wolf R, Mastryukova TA, Kabachnik MI. NMR determination of enantiomeric composition in the case of statistically controlled associate-diastereomer anisochrony. *Zh. Obshch. Khim.* **1995**, *65*, 18–22.

877. Harger MJP. Chemical shift non-equivalence of enantiomers in the proton magnetic resonance spectra of partly resolved phosphinothioic acids. *J. Chem. Soc., Perkin Trans. 2* **1978**, 326–331.

878. Kuchen W, Kutter J. (−)D- and (+)D-4-Methoxyphenylmethylthiophosphinic acid check of the resolution by ^{31}P NMR spectroscopy. *Z. Naturforsch., B: Chem. Sci.* **1979**, *34B*, 1332–1333.

879. Harger MJP. The proton magnetic resonance spectra of chiral phosphinate esters. chemical shift non-equivalence of enantiomers induced by optically active phosphinothioic acids. *J. Chem. Soc., Perkin Trans. 2* **1980**, 1505–1511.

880. Freeman R, Haynes RK, Loughlin WA, Mitchell C, Stokes JV. Synthetic utilization of highly stereoselective conjugate addition reactions of phosphorus and sulfur stabilized allylic carbanions. *Pure Appl. Chem.* **1993**, *65*, 647–654.

881. Drescher M, Felsinger S, Hammerschmidt F, Kählig H, Schmidt S, Wuggenig F. Enzymes in organic chemistry 7. Evaluation of homochiral t-butyl(phenyl)phosphinothioic acid for the determination of enantiomeric excesses and absolute configurations of α-substituted phosphonates. *Phosphorus, Sulfur Silicon Relat. Elem.* **1998**, *140*, 79–93.

882. Haynes RK, Au-Yeung T, Chan W, Lam W, Li Z, Yeung L, Chan ASC, Li P, Koen M, Mitchell CR, Vonwiller SC. Reaction of metallated tert-butyl(phenyl)phosphane oxide with electrophiles as a route to functionalized tertiary phosphane oxides: alkylation reactions. *Eur. J. Org. Chem.* **2000**, 3205–3216.

883. Bentrude WG, Moriyama M, Mueller H, Sopchik AE. The reactions of ethoxy radicals with optically active tertiary phosphines. Stereochemistry of the substitution process and the question of permutation modes for the possible phosphoranyl radical intermediates. *J. Am. Chem. Soc.* **1983**, *105*, 6053–6061.

884. Moriyama M, Bentrude WG. Optically active phosphines. Facile preparation of the optically active *n*-propylmethylbenzyl- and methylphenylbenzylphosphine oxides as precursors to the corresponding tertiary phosphines. *J. Am. Chem. Soc.* **1983**, *105*, 4727–4733.

885. Omelanzcuk J, Sopchik AE, Lee S, Akutagawa K, Cairns SM, Bentrude WG. Photo-Arbuzov rearrangements of benzyl phosphites. *J. Am. Chem. Soc.* **1988**, *110*, 6908–6909.

886. Haynes RK, Freeman RN, Mitchell CR, Vonwiller SC. Preparation of enantiomerically pure tertiary phosphine oxides from, and assay of enantiomeric purity with, (R_p)- and (S_p)-tert-butylphenylphosphinothioic acids. *J. Org. Chem.* **1994**, *59*, 2919–2921.

887. Hammerschmidt F, Schmidt S. Deprotonation of secondary benzylic phosphates. Configurationally stable benzylic carbanions with a diethoxyphosphoryloxy substituent and their rearrangement to optically active tertiary α-hydroxyphosphonates. *Chem. Ber.* **1996**, *129*, 1503–1508.

888. Drabowicz J, Dudziński B, Mikolajczyk M. Chiral t-butylphenylphosphinothioic acid: a new NMR solvating agent for determination of enantiomeric excesses of sulfoxides. *Tetrahedron: Asymmetry* **1992**, *3*, 1231–1234.

889. Drabowicz J, Dudziński B, Mikolajczyk M, Colonna S, Gaggero N. Chiral t-butylphenylphosphinothioic acid in NMR analysis of tertiary amine oxides with a stereogenic nitrogen atom. *Tetrahedron: Asymmetry* **1997**, *8*, 2267–2270.

890. Antkowiak R, Antkowiak WZ, Nowak-Wydra B. Studies on a chirality of orellanine. Spectral nonequivalence of atropisomers of tetra-O-methylorellanine and related compounds induced by chiral solvating agents. *Heterocycles* **2002**, *58*, 137–146.

891. Omelańczuk J, Mikolajczyk M. Chiral t-butylphenylphosphinothioic acid: a useful chiral solvating agent for direct determination of enantiomeric purity of alcohols, thiols, amines, diols, aminoalcohols and related compounds. *Tetrahedron: Asymmetry* **1996**, *7*, 2687–2694.

892. Gunderson KG, Shapiro MJ, Doti RA, Skiles JW. Simultaneous determination of enantiomeric purity and erythro/threo relationship in chiral 1,2-aminoalcohols by NMR using (R)-(+)-t-butylphenylphosphinothioic acid. *Tetrahedron: Asymmetry* **1999**, *10*, 3263–3266.

893. Markowicz SW, Pokrzeptowicz K, Karolak-Wojciechowska J, Czylkowski R, Omelańczuk J, Sobczak A. Enantiomerically pure α-pinene derivatives from material

of 65% enantiomeric purity. 1. Di[3α-(2α-hydroxy)pinane]amine. *Tetrahedron: Asymmetry* **2002**, *13*, 1981–1991.

894. Szawkalo J, Zawadzka A, Wojtasiewicz K, Leniewski A, Drabowicz J, Czarnocki Z. First enantioselective synthesis of the antitumour alkaloid (+)-crispine A and determination of its enantiomeric purity by ^{1}H NMR. *Tetrahedron: Asymmetry* **2005**, *16*, 3619–3621.

895. Johnson CR, Elliott RC, Penning TD. Determination of enantiomeric purities of alcohols and amines by a ^{31}P NMR technique. *J. Am. Chem. Soc.* **1984**, *106*, 5020–5022.

896. Cullis PM, Iagrossi A, Rous AJ. Thiophosphoryl transfer reactions: a general synthesis and configurational analysis of O-substituted [^{16}O,^{18}O]thiophosphates. *J. Am. Chem. Soc.* **1986**, *108*, 7869–7870.

897. Cullis PM, Iagrossi A, Rous AJ, Schilling MB. Epimerisations and non-stereospecific reactions of 1,3,2-oxazaphospholidin-2-ones and -2-thiones. *J. Chem. Soc., Chem. Commun.* **1987**, 996–998.

898. Dehmlow EV, Sauerbier C. Note on carbonyl derivatives with a chiral phosphorylhydrazine. *Z. Naturforsch., B: Chem. Sci.* **1989**, *44b*, 240–242.

899. Anderson RC, Shapiro MJ. 2-Chloro-4(R),5(R)-dimethyl-2-oxo-1,3,2-dioxaphospholane, a new chiral derivatizing agent. *J. Org. Chem.* **1984**, *49*, 1304–1305.

900. Brown HC, Imai, T, Desai MC, Singaram B. Chiral synthesis via organoboranes 3. Conversion of boronic esters of essentially 100% optical purity to aldehydes, acids, and homologated alcohols of very high enantiomeric purities. *J. Am. Chem. Soc.* **1985**, *107*, 4980–4983.

901. Brown HC, Naik RG, Bakshi RK, Pyun C, Singaram B. Chiral synthesis via organoboranes. 4. Synthetic utility of boronic esters of essentially 100% optical purity. Synthesis of homologated boronic acids and esters of very high enantiomeric purities. *J. Org. Chem.* **1985**, *50*, 5586–5592.

902. Brown HC, Kim K-W, Cole TE, Singaram B. Chiral synthesis via organoboranes. 8. Synthetic utility of boronic esters of essentially 100% optical purity. Synthesis of primary amines of very high enantiomeric purities. *J. Am. Chem. Soc.* **1986**, *108*, 6761–6764.

903. Hägele G, Boisnière B, Helie I, Rabiller C, Martin GJ, Martin ML. Monitoring site-specific natural hydrogen isotope fractionation using a novel phosphorus-containing chiral discriminating reagent. *Magn. Reson. Chem.* **1991**, *29*, 813–822.

904. ven Hoeve W, Wynberg H. The design of resolving agents. Chiral cyclic phosphoric acids. *J. Org. Chem.* **1985**, *50*, 4508–4514.

905. Hulst R, Zijlstra RWJ, Feringa BL, de Vries NK, ten Hoeve W, Wynberg H. A new ^{31}P NMR method for the enantiomeric excess determination of alcohols, amines, and amino acid esters. *Tetrahedron Lett.* **1993**, *34*, 1339–1342.

906. Hulst R, Zijlstra RWJ, de Vries NK, Feringa BL. (S)-2H-2-oxo-5,5-dimethyl-4(R)-phenyl-1,3,2-dioxaphosphorinane, a new reagent for the enantiomeric excess determination of unprotected amino acids using ^{31}P NMR. *Tetrahedron: Asymmetry* **1994**, *5*, 1701–1710.

907. Hulst R, de Vries NK, Feringa BL. The ^{31}P-NMR spectroscopic determination of the enantiomeric excess of unprotected amino acids. *Angew. Chem. Int. Ed. Engl.* **1999**, *31*, 1092–1093.

908. Hulst R, de Vries NK, Feringa BL. The rational design and application of new chiral phosphonates for the enantiomeric excess determination of unprotected amino acids.

Remarkable pH dependency of the diastereomeric shift differences. *Tetrahedron* **1994**, *50*, 11721–11728.

909. Oshikawa T, Yamashita M, Kumagai S, Seo K, Kobayashi J. Optically active diazacyclophosphamides: novel efficient reagents for the determination of the absolute configuration of amines and alcohols. *J. Chem. Soc., Chem. Commun.* **1995**, 435–436.

910. Feringa BL, Smaardijk AA, Wynberg H, Strijtveen B, Kellogg RM. Alkylphosphonic dichlorides, new reagents for the enantiomeric excess determination of chiral alcohols and thiols by ^{31}P NMR. *Tetrahedron Lett.* **1986**, *27*, 997–1000.

911. Vigneron JP, Dhaenens M, Horeau A. Maximizing optical purity of a partially resolved product without using a chiral compound. *Tetrahedron* **1973**, *29*, 1055–1059.

912. Strijtveen B, Feringa BL, Kellogg RM. Methylphosphonic dichloride as reagent for the determination of the enantiomeric excess of chiral thiols. Scope and limitations. *Tetrahedron* **1987**, *43*, 123–130.

913. Feringa BL, Strijtveen B, Kellogg RM. Enantiomeric excess determination without chiral auxiliary compounds. A new ^{31}P NMR method for amino acid esters and primary amines. *J. Org. Chem.* **1986**, *51*, 5484–5486.

914. Ouryupin AB, Kadyko MI, Petrovskii PV, Fedin EI. Chiral discrimination in nonracemic mixtures of methanephosphonic acid, *N,N'*-bis(1-phenylethyl)diamides. *Chirality* **1994**, *6*, 1–4.

915. Speelman JC, Talma AG, Kellogg RM, Meetsma A, de Boer JL, Beurskens PT, Bosman WP. Molecular structure of a chiral 3,5-bridged pyridine and the effect of structure on circular dichroic spectra. *J. Org. Chem.* **1989**, *54*, 1055–1062.

916. Strijtveen B, Kellogg RM. Synthesis of (racemization prone) optically active thiols by S_N2 substitution using cesium thiocarboxylates. *J. Org. Chem.* **1986**, *51*, 3664–3671.

917. Feringa BL. ^{31}P NMR nonequivalence of diastereoisomeric O,O-dialkyl phosphorodithioates. *J. Chem. Soc., Chem. Commun.* **1987**, 695–696.

918. Strijtveen B, Kellogg RM. Synthesis and determination of enantiomeric excesses of non-racemic tert-thiols derived from chiral secondary α-mercaptocarboxylic acids. *Tetrahedron* **1987**, *43*, 5039–5054.

919. Dimukhametov MN, Ismaev IE. Simple ^{31}P NMR method of determination of enantiomeric purity of chiral O-alkyl alkylthiophosphonic acids. *Izv. Akad. Nauk SSSR, Ser. Khim.* **1989**, 2225–2228.

920. Shapiro MJ, Archinal AE, Jarema MA. Chiral purity determination by ^{1}H NMR spectroscopy. Novel use of 1,1'-binaphthyl-2,2'-diylphosphoric acid. *J. Org. Chem.* **1989**, *54*, 5826–5828.

921. Höök BB, Johansson AM, Hjorth S, Sundell S, Hacksell U. Synthesis of (+)-(R)- and (−)-(S)-5-hydroxy-2-methyl-2-dipropylaminotetralin: effects on rat hippocampal output of 5-HT, 5-HIAA, and DOPAC as determined by in vivo microdialysis. *Chirality* **1993**, *5*, 112–119.

922. Ravard A, Crooks PA. Chiral purity determination of tobacco alkaloids and nicotine-like compounds by ^{1}H NMR spectroscopy in the presence of 1,1'-binaphthyl-2,2'-diylphosphoric acid. *Chirality* **1996**, *8*, 295–299.

923. Ansari AM, Craig JC. Kinetic resolution of 7-chloro-4-{4-[ethyl(2-hydroxyethyl)amino]-1-methyl-butylamino}quinoline (hydroxychloroquine) by an atropisomeric resolving agent. *J. Chem. Soc., Perkin Trans. 2* **1994**, 1731–1733.

924. Dietrich-Buchecker C, Rapenne G, Sauvage J, De Cian A, Fischer J. A dicopper(I) trefoil knot with m-phenylene bridges between the ligand subunits: synthesis, resolution, and absolute configuration. *Chem. Eur. J.* **1999**, *5*, 1432–1439.

925. Alexakis A, Mutti S, Normant JF, Mangeney P. A new reagent for a very simple and efficient determination of enantiomeric purity of alcohols by ^{31}P NMR. *Tetrahedron: Asymmetry* **1990**, *1*, 437–440.

926. Alexakis A, Mutti S, Mangeney P. A new reagent for the determination of the optical purity of primary, secondary, and tertiary chiral alcohols and of thiols. *J. Org. Chem.* **1992**, *57*, 1224–1237.

927. Alexakis A, Frutos JC, Mutti S, Mangeney P. Chiral diamines for a new protocol to determine the enantiomeric composition of alcohols, thiols, and amines by ^{31}P, ^{1}H, ^{13}C, and ^{19}F NMR. *J. Org. Chem.* **1994**, *59*, 3326–3334.

928. Nowacki J, Szychowski J, Wojtasiewicz K, Matuszewska I, Jaremko M, Czarnocki Z. The determination of enantiomeric excess of the products of Baker's yeast-mediated reduction of δ-alkoxycarbonyl-β-keto esters, via the derivation to Alexakis' alkoxydiazaphosphoridine. *Heterocycles* **2002**, *57*, 705–714.

929. Alexakis A, Frutos JC, Mangeney P, Meyers AI, Moorlag H. Determination of enantiomeric purity of hydroxy biaryls using ^{1}H and ^{31}P-NMR on their diazaphospholidine derivatives. *Tetrahedron Lett.* **1994**, *35*, 5125–5128.

930. Alexakis A, Chauvin A. Determination of the enantiomeric excess of chiral carboxylic acids by ^{31}P NMR with C_2 symmetrical diamines. *Tetrahedron: Asymmetry* **2001**, *12*, 1411–1415.

931. Hulst R, de Vries NK, Feringa BL. α-Phenylethylamine based chiral phospholidines; new agents for the determination of the enantiomeric excess of chiral alcohols, amines and thiols by means of ^{31}P NMR. *Tetrahedron: Asymmetry* **1994**, *5*, 699–708.

932. de Parrodi CA, Moreno GE, Quintero L, Juaristi E. Application of phosphorylated reagents derived from *N,N'*-di-[(*S*)-α-phenylethyl]-cyclohexane-1,2-diamines in the determination of the enantiomeric purity of chiral alcohols. *Tetrahedron: Asymmetry* **1998**, *9*, 2093–2099.

933. Moreno GE, Mastranzo VM, Quintero L, de Parrodi CA, Juaristi E. The Use of *N,N'*-di[α-phenylethyl]-diamines as phosphorylated chiral derivatizing agents for the determination of the enantiomeric purity of chiral secondary alcohols. *Rev. Soc. Quim. Mex.* **2003**, *47*, 127–129.

934. Devitt PG, Mitchell MC, Weetman JM, Taylor RJ, Kee TP. Accurate in situ ^{31}P{^{1}H} assay of enantiopurity in α-hydroxyphosphonate esters using a diazaphospholidine derivatizing agent. *Tetrahedron: Asymmetry* **1995**, *6*, 2039–2044.

935. Davies SR, Mitchell MC, Cain CP, Devitt PG, Taylor RJ, Kee TP. Phospho-transfer catalysis. On the asymmetric hydrophosphonylation of aldehydes. *J. Organomet. Chem.* **1998**, *550*, 29–57.

936. Chauvin A, Bernardinelli G, Alexakis A. Determination of the absolute configuration of chiral aryl-alkyl carbinols using organophosphorus diamine derivatizing agents by ^{31}P NMR spectroscopy. *Tetrahedron: Asymmetry* **2004**, *15*, 1857–1879.

937. Reymond S, Brunel JM, Buono G. New chiral organophosphorus derivatizing agent for the determination of enantiomeric composition of chloro- and bromohydrins by ^{31}P NMR spectroscopy. *Tetrahedron: Asymmetry* **2000**, *11*, 1273–1278.

938. Brunel J, Pardigon O, Maffei M, Buono G. The use of (4*R*,5*R*)-dicarboalkoxy 2-chloro 1,3,2-dioxaphospholanes as new chiral derivatizing agents for the determination of enantiomeric purity of alcohols by ^{31}P NMR. *Tetrahedron: Asymmetry* **1992**, *3*, 1243–1246.

939. Che C, Zhang ZN, Huang GL, Wang XX, Qin ZH. Using the chiral organophosphorus derivatizing agents for determination of the enantiomeric composition of chiral carboxylic acids by ^{31}P NMR spectroscopy. *Chin. Chem. Lett.* **2004**, *15*, 675–678.

940. Bredikhin AA, Bredikhina ZA, Nigmatzyanov FF. On the use of seven-membered phosphorus heterocycles based on 2,2′-dihydroxy-1,1′-binaphthalene and 1,4:3,6-dianhydro-D-mannitol in the ^{31}P NMR analysis of the enantiomeric composition of chiral alcohols. *Russ. Chem. Bull.* **1998**, *47*, 411–416.

941. Kato N. Direct chirality determination of secondary carbinol by chirality recognition ability of C$_2$ symmetry 1,1′-binaphthyl-2,2′-diyl phosphoryl chloride. *J. Am. Chem. Soc.* **1990**, *112*, 254–257.

942. Li KY, Zhou ZH, Yeung CH, Tang CC. A chiral phosphorus derivatizing agent for the determination of the enantiomeric excess of chiral alcohols, amines by ^{31}P NMR. *Chin. Chem. Lett.* **2001**, *12*, 907–908.

943. Li KY, Zhou ZH, Chan ASC, Tang CC. An efficient chiral phosphorus derivatizing agent for the determination of the enantiomeric excess of chiral alcohols and amines by ^{31}P NMR spectroscopy. *Heteroat. Chem.* **2002**, *13*, 93–95.

944. Bredikhin AA, Bredikhin ZA, Gaisina LM, Strunskaya EI, Azancheev NM. Cyclic phosphorochloridites (chloridates) based on chiral butane-2,3-diol and dihydrobenzoin as reagents for the analysis of enantiomeric compositions of alcohols by ^{31}P NMR spectroscopy. *Russ. Chem. Bull.* **2000**, *49*, 310–313.

945. Alexakis A, Chauvin A. TADDOL organophosphorus derivatising agents for the determination of the enantiomeric excess of chiral alcohols and carboxylic acids by ^{31}P and ^{1}H NMR spectroscopy. *Tetrahedron: Asymmetry* **2000**, *11*, 4245–4248.

946. Bredikhin AA, Strunskaya EI, Azancheev NM, Bredikhina ZA. (4R,5R)-Bis(*N*,*N*-dimethylaminocarbonyl)-2-chloro-1,3,2-dioxaphospholane: a convenient reagent for control of enantiomeric composition of chiral alcohols by ^{31}P NMR spectroscopy. *Russ. Chem. Bull.* **1998**, *47*, 174–176.

947. Bredikhina ZA, Novikova VG, Azancheev NM, Bredikhin AA. Regiochemistry of cleavage of monosubstituted oxiranes by phosphorochloridites. Enantiomeric composition of oxiranes by ^{31}P NMR spectral data. *Russ. J. Gen. Chem.* **2002**, *72*, 1207–1214.

948. Bredikhin AA, Strunskaya EI, Novikova VG, Azancheev NM, Sharafutdinova DR, Bredikhina ZA. Jacobsen-type enantioselective hydrolysis of aryl glycidyl ethers. ^{31}P NMR analysis of the enantiomeric composition of oxiranes. *Russ. Chem. Bull.* **2004**, *53*, 213–218.

949. Kolodiazhnyi OI, Demchuk OM, Gerschkovich AA. Application of the dimenthyl chlorophosphite for the chiral analysis of amines, amino acids and peptides. *Tetrahedron: Asymmetry* **1999**, *10*, 1729–1799.

950. Brunel JM, Faure B. A new ^{31}P NMR method for the enantiomeric excess determination of diols and secondary diamines with C$_2$ symmetry. *Tetrahedron: Asymmetry* **1995**, *6*, 2353–2356.

951. Garner CM, McWhorter C, Goerke AR. Menthyldichlorophosphate: a chiral derivatizing agent for symmetrical diols. *Tetrahedron Lett.* **1997**, *38*, 7717–7720.

952. Feringa BL, Smaardijk A, Wynberg H. Simple ^{31}P NMR method for the determination of enantiomeric purity of alcohols not requiring chiral auxiliary compounds. *J. Am. Chem. Soc.* **1985**, *107*, 4798–4799.

953. Chaloner PA, Perera SAR. Enantioselective addition of diethylzinc to benzaldehyde in the presence of ephedrine derivatives. *Tetrahedron Lett.* **1987**, 3013–3014.

954. Welch CJ. The measurement of enantiopurity using phosphorus-NMR. *Tetrahedron: Asymmetry* **1991**, *2*, 1127–1132.

955. Dimukhametov MN, Garifzyanova GG, Azancheev NM, Al'fonsov VA. Hexaalkylphosphorous triamides as enantiodifferentiating agents for chiral alcohols. *Russ. J. Gen. Chem.* **1999**, *69*, 1965–1966.

956. Liu X, Ilankumaran P, Guzei IA, Verkade JG. P[(*S*,*S*,*S*)-PhHMeCNCH$_2$CH$_2$]$_3$N: a new chiral ^{31}P and ^{1}H NMR spectroscopic reagent for the direct determination of ee values of chiral azides. *J. Org. Chem.* **2000**, *65*, 701–706.

957. Weix DJ, Dreher SD, Katz TJ. [5]HELOL phosphite: a helically grooved sensor of remote chirality. *J. Am. Chem. Soc.* **2000**, *122*, 10027–10032.

958. Wang DZ, Katz TJ. A [5]HELOL analogue that senses remote chirality in alcohols, phenols, amines, and carboxylic acids. *J. Org. Chem.* **2005**, *70*, 8497–8502.

959. Lacour J, Ginglinger C, Favarger F, Torche-Haldimann S. Application of TRISPHAT anion as NMR chiral shift reagent. *Chem. Commun.* **1997**, 2285–2286.

960. Bergman SD, Frantz R, Gut D, Kol M, Lacour J. Effective chiral recognition among ions in polar media. *Chem. Commun.* **2006**, 850–852.

961. Frantz R, Pinto A, Constant S, Bernardinelli G, Lacour J. Fluorinated TRISPHAT anions: spectroscopic probes for detailed asymmetric ion pairing studies. *Angew. Chem., Int. Ed.* **2005**, *44*, 5060–5064.

962. Brissard M, Gruselle M, Malézieux B, Thouvenot R, Guyard-Duhayon C, Convert O. An anionic {[MnCo(ox)$_3$]$^-$}$_n$ network with appropriate cavities for the enantioselective recognition and resolution of the hexacoordinated monocation [Ru(bpy)$_2$(ppy)]$^+$ (bpy = bipyridine, ppy = phenylpyridine). *Eur. J. Inorg. Chem.* **2001**, 1745–1751.

963. Brissard M, Amouri H, Gruselle M, Thouvenot R. Dédoublement et determination des excès énantiomériques des complexes de ruthénium^{II} hexacoordinés [Ru(bpy)$_2$ppy]$^+$ et [Ru(bpy)$_2$Quo]$^+$ (bpy = bipyridine, ppy = phénylpyridine, Quo = 8-quinolate) utilisés comme templates dans la construction de réseaux anioniques tridimensionnels. *C. R. Chim.* **2002**, *5*, 53–58.

964. Brissard M, Convert O, Gruselle M, Guyard-Duhayon C, Thouvenot R. Enantiospecific synthesis of Δ and Λ [Ru(bpy)$_2$ppy]$^+$ and [Ru(bpy)$_2$quo]$^+$ (bpy = 2,2′-bipyridine, ppy = phenylpyridine-H$^+$, quo = 8-hydroxyquinolate): ^{1}H and ^{13}C NMR studies and X-ray structure determination of rac-[Ru(bpy)$_2$quo]PF$_6$. *Inorg. Chem.* **2003**, *42*, 1378–1385.

965. Monchaud D, Lacour J, Coudret C, Fraysse S. TRISPHAT salts. Efficient NMR chiral shift and resolving agents for substituted cyclometallated ruthenium bis(diimine) complexes. *J. Organomet. Chem.* **2001**, *624*, 388–391.

966. Casper R, Amouri H, Gruselle M, Cordier C, Malezieux B, Duval R, Leveque H. Efficient asymmetric synthesis of Δ- and Λ-enantiomers of (bipyridyl)ruthenium complexes and crystallographic analysis of Δ -bis(2,2′-bipyridine) (2,2′-bipyridine-4,4′-dicarboxylato)ruthenium: diastereoselective homo- and heterochiral ion pairing revisited. *Eur. J. Inorg. Chem.* **2003**, 499–505.

967. Bruylants G, Bresson C, Boisdenghien A, Pierard F, Mesmaeker AK, Lacour J, Bartik K. Comparison of the NMR enantiodifferentiation of a chiral ruthenium(II) complex of C_2 symmetry using the TRISPHAT anion and a lanthanide shift reagent. *New J. Chem.* **2003**, *27*, 748–751.

968. Chavarot M, Ménage S, Hamelin O, Charnay F, Pécaut J, Fontecave M. "Chiral-at-metal" octahedral ruthenium(II) complexes with achiral ligands: a new type of enantioselective catalyst. *Inorg. Chem.* **2003**, *42*, 4810–4816.

969. Desvergnes-Breuil V, Hebbe V, Dietrich-Buchecker C, Sauvage J, Lacour J. NMR evaluation of the configurational stability of Cu(I) complexes. *Inorg. Chem.* **2003**, *42*, 255–257.

970. Amouri H, Thouvenot R, Gruselle M. Differentiation of planar chiral enantiomers of [Cp*M(2-alkyl-phenoxo)] [BF$_4$] {M = Rh, Ir} by the trisphat anion. *C. R. Chim.* **2002**, *5*, 257–262.

971. Gruselle M, Thouvenot R, Malézieux B, Train C, Gredin P, Demeschik TV, Troitskaya LL, Sokolov VI. Enantioselective self-assembly of bimetallic [Mn^{II} (Δ)-Cr^{III}(C$_2$O$_4$)$_3$]$^-$ and [Mn^{II} (Λ)-Cr^{III}(C$_2$O$_4$)$_3$]$^-$ layered anionic networks templated by the optically active (Rp)- and (Sp)-[1-CH$_2$N(n-C$_3$H$_7$)$_3$-2-CH$_3$–C$_5$H$_3$Fe–C$_5$H$_5$]$^+$ ions. *Chem. Eur. J.* **2004**, *10*, 4763–4769.

972. Planas JG, Prim D, Rose E, Rose-Munch F, Monchaud D, Lacour J. TRISPHAT anion. An efficient NMR chiral shift counterion for cationic tricarbonyl manganese complexes with planar chirality. *Organometallics* **2001**, *20*, 4107–4110.

973. Ratni H, Jodry JJ, Lacour J, Kündig EP. [n-Bu$_4$N][Δ-TRISPHAT] salt, an efficient NMR chiral shift reagent for neutral planar chiral tricarbonylchromium complexes. *Organometallics* **2000**, *19*, 3997–3999.

974. Planas JG, Prim D, Rose-Munch F, Rose E, Thouvenot R, Vaissermann J. Synthesis of new chiral heterobimetallic chromium-ruthenium complexes by regioselective insertion of ruthenium into the C-S bond of tricarbonyl-η^6-[(thiophenyl)arene] chromium complexes. *Organometallics* **2002**, *21*, 4385–4389.

975. Berger A, Djukic J, Pfeffer M, Lacour J, Vial L, de Cian A, Kyritsakas-Gruber N. Chloride-promoted synthesis of cis bis-chelated palladium(II) complexes from ortho-mercurated tricarbonyl(η^6-arene)chromium complexes. *Organometallics* **2003**, *22*, 5243–5260.

976. Amouri H, Casper R, Gruselle M, Guyard-Duhayon C, Boubekeur K, Lev DA, Collins LSB, Grotjahn DB. Chiral recognition and resolution mediated by π-π interactions: synthesis and X-ray structure of trans-[(Sp,Sp)-bis(Cp*Ru)-carbazolyl][Δ-trisphat]. *Organometallics* **2004**, *23*, 4338–4341.

977. Amouri H, Thouvenot R, Gruselle M, Malézieux B, Vaissermann J. Synthesis, NMR, and X-ray molecular structure of a unique chiral propeller-like cobalt complex [(Co$_2$(CO)$_4$)μ,η^2,η^2-(–H$_2$CC$\equiv$CCH$_2$–)(–dppm)$_2$][BF$_4$]$_2$ and differentiation of its enantiomers by the trisphat anion. *Organometallics* **2001**, *20*, 1904–1906.

978. Jodry JJ, Lacour J. Efficient resolution of a dinuclear triple helicate by asymmetric extraction/precipitation with TRISPHAT anions as resolving agents. *Chem. Eur. J.* **2000**, *6*, 4297–4304.

979. Lacour J, Jodry JJ, Monchaud D. Configurational ordering of a cationic dinuclear triple helicate by chiral TRISPHAT anions. *Chem. Commun.* **2001**, 2302–2303.

980. Mimassi L, Guyard-Duhayon C, Rager MN, Amouri H. Chiral recognition and resolution of the enantiomers of supramolecular triangular hosts: synthesis, circular dichroism, NMR, and X-ray molecular structure of [Li⊂(R,R,R)-{Cp*Rh(5-chloro-2, 3-dioxopyridine)}$_3$][Δ-trisphat]. *Inorg. Chem.* **2004**, *43*, 6644–6649.

981. Bark T, von Zelewsky A, Rappoport D, Neuburger M, Schaffner S, Lacour J, Jodry J. Synthesis and stereochemical properties of chiral square complexes of iron(II). *Chem. Eur. J.* **2004**, *10*, 4839–4845.

982. Ginglinger C, Jeannerat D, Lacour J, Jugé S, Uziel J. ^{1}H and ^{31}P NMR determination of the enantiomeric purity of quaternary phosphonium cations using TRISPHAT as chiral shift agent. *Tetrahedron Lett.* **1998**, *39*, 7495–7498.

983. Hebbe V, Londez A, Goujon-Ginglinger C, Meyer F, Uziel J, Jugé S, Lacour J. NMR enantiodifferentiation of triphenylphosphonium salts by chiral hexacoordinated phosphate anions. *Tetrahedron Lett.* **2003**, *44*, 2467–2471.

984. Vial L, Lacour J. Conformational preference and configurational control of highly symmetric spirobi[dibenzazepinium] cation. *Org. Lett.* **2002**, *4*, 3939–3942.

985. Pasquini C, Desvergnes-Breuil V, Jodry JJ, Cort AD, Lacour J. Chiral anion-mediated asymmetric induction onto chiral diquats. *Tetrahedron Lett.* **2002**, *43*, 423–426.

986. Lacour J, Londez A, Goujon-Ginglinger C, Buss V, Bernardinelli G. Configurational ordering of cationic chiral dyes using a novel C_2-symmetric hexacoordinated phosphate anion. *Org. Lett.* **2000**, *2*, 4185–4188.

987. Lacour J, Vial L, Herse C. Efficient NMR enantiodifferentiation of chiral quats with BINPHAT anion. *Org. Lett.* **2002**, *4*, 1351–1354.

988. Herse C, Bas D, Krebs FC, Bürgi T, Weber J, Wesolowski T, Laursen BW, Lacour J. A highly configurationally stable [4]heterohelicenium cation. *Angew. Chem., Int. Ed.* **2003**, *42*, 3162–3166.

989. Pasquato L, Herse C, Lacour J. NMR enantiodifferentiation of thiiranium cations by chiral hexacoordinated phosphate anions. *Tetrahedron Lett.* **2002**, *43*, 5517–5520.

990. Djukic J, Berger A, Duquenne M, Pfeffer M, de Cian A, Kyritsakas-Gruber N. Synthesis of nonracemic ortho-mercurated and ortho-ruthenated complexes of 2-[tricarbonyl(η^6-phenyl)chromium]pyridine. *Organometallics* **2004**, *23*, 5757–5767.

991. Martínez-Viviente E, Pregosin PS, Vial L, Herse C, Lacour J. ^{1}H, ^{19}F, and ^{31}P PGSE NMR diffusion studies on chiral organic salts: ion pairing and the dependence of a diffusion value on diastereomeric structure. *Chem. Eur. J.* **2004**, *10*, 2912–2918.

992. Lacour J, Vial L, Bernardinelli G. One-step stereoselective synthesis and NMR chiral shift properties of a novel hexacoordinated phosphorus cation. *Org. Lett.* **2002**, *4*, 2309–2312.

993. Lacour J, Hebbe-Viton V. Recent developments in chiral anion mediated asymmetric chemistry. *Chem. Soc. Rev.* **2003**, *32*, 373–382.

994. Gerlt JA, Coderre JA. Oxygen chiral phosphodiesters. 1. Synthesis and configurational analysis of cyclic [^{18}O]-2'-deoxyadenosine 3',5'-monophosphate. *J. Am. Chem. Soc.* **1980**, *102*, 4531–4533.

995. Webb MR, Trentham DR. Analysis of chiral inorganic [^{16}O, ^{17}O, ^{18}O]thiophosphate and the stereochemistry of the 3-phosphoglycerate kinase reaction. *J. Biol. Chem.* **1980**, *255*, 1775–1779.

996. Buchwald SL, Knowles JR. Determination of the absolute configuration of [^{16}O, ^{17}O, ^{18}O]phosphate monoesters by using ^{31}P NMR. *J. Am. Chem. Soc.* **1980**, *102*, 6601–6602.

997. Arnold JRP, Lowe G. Synthesis and stereochemical analysis of chiral inorganic [^{16}O, ^{17}O, ^{18}O]thiophosphate. *J. Chem. Soc., Chem. Commun.* **1986**, 865–867.

998. Bruzik K, Tsai M. Phospholipids chiral at phosphorus. 5. Synthesis and configurational analysis of chiral [^{17}O,^{18}O]phosphatidylethanolamine. *J. Am. Chem. Soc.* **1984**, *106*, 747–754.

999. Lowe G. Chiral [^{16}O, ^{17}O, ^{18}O]phosphate esters. *Acc. Chem. Res.* **1983**, *16*, 244–251.

1000. Speckhard DC, Pecoraro VL, Knight WB, Cleland WW. Determination of the absolute configurations of the isomers of triamminecobalt(III) adenosine triphosphate. *J. Am. Chem. Soc.* **1986**, *108*, 4167–4171.

1001. Coderre JA, Mehdi S, Demou PC, Weber R, Traficante DD, Gerlt JA. Oxygen chiral phosphodiesters. 3. Use of ^{17}O NMR spectroscopy to demonstrate configurational differences in the diastereomers of cyclic 2′-deoxyadenosine 3′,5′-[^{17}O, ^{18}O]monophosphate. *J. Am. Chem. Soc.* **1981**, *103*, 1870–1872.

1002. Bentrude WG, Sopchik AE, Gajda T. Stereo- and regiochemistries of the oxidations of 2-methoxy-5-tert-butyl-1,3,2-dioxaphosphorinanes and the cyclic methyl 3′,5′-phosphite of thymidine by H$_2$O/I$_2$ and O$_2$/AIBN to P-chiral phosphates. ^{17}O NMR assignment of phosphorus configuration to the diastereomeric thymidine cyclic methyl 3′,5′-monophosphates. *J. Am. Chem. Soc.* **1989**, *111*, 3981–3987.

1003. Jankowski S, Olejniczak AB, Lesnikowski ZJ. P-Chiral oligonucleotides. 2D ROESY NMR assignment of absolute configuration at phosphorus and conformational analysis of 5′-O-monomethoxytrityl-(2′-O-deoxyribonucleoside) 3′-O-[O-(4-nitrophenyl)]-methanephosphonates. *Nucleosides, Nucleotides Nucleic Acids* **2002**, *21*, 177–190.

1004. Olejniczak S, Sobczak M, Potrzebowski MJ, Polak M, Plavec J, Nawrot B. Assignment of absolute configuration at phosphorus of P-chiral diastereomers of deoxyribonucleoside methanephosphonamidates by means of NMR spectroscopy. *Tetrahedron* **2004**, *60*, 3979–3986.

1005. Hulst R, Kellogg RM, Feringa BL. New methodologies for enantiomeric excess (ee) determination based on phosphorus NMR. *Recl. Trav. Chim. Pays-Bas* **1995**, *114*, 115–138.

1006. Silks LA, Dunlap RB, Odom JD. Quantitative detection of remotely disposed chiral centers using selenium-77 nuclear magnetic resonance spectroscopy. *J. Am. Chem. Soc.* **1990**, *112*, 4979–4982.

1007. Silks LA, Peng J, Odom JD, Dunlap RB. Synthesis of (4S,5R)-(−)-4-methyl-5-phenyloxazolidine-2-selone: a chiral auxiliary reagent capable of detecting the enantiomers of (R,S)-lipoic acid by ^{77}Se nuclear magnetic resonance spectroscopy. *J. Chem. Soc., Perkin Trans. 1* **1991**, 2495–2498.

1008. Peng J, Odom JD, Dunlap RB, Silks LA. Use of a selone chiral derivatizing agent for the absolute configurational assignment of stereogenic centers. *Tetrahedron: Asymmetry* **1994**, *5*, 1627–1630.

1009. Peng J, Barr ME, Ashburn DA, Lebioda L, Garber AR, Martinez RA, Odom JD, Dunlap RB, Silks LA. Synthesis and characterization of acylated chiral oxazolidine-2-selones: selone chiral derivatizing agents for the detection and quantitation of remotely disposed chiral centers. *J. Org. Chem.* **1995**, *60*, 5540–5549.

1010. Smith AB, Friestad GK, Barbosa J, Bertounesque E, Duan JJ, Hull KG, Iwashima M, Qiu Y, Spoors PG, Salvatore BA. Total synthesis of (+)-calyculin A and (−)-calyculin

B: cyanotetraene construction, asymmetric synthesis of the C(26–37) oxazole, fragment assembly, and final elaboration. *J. Am. Chem. Soc.* **1999**, *121*, 10478–10486.

1011. Wu R, Odom JD, Dunlap RB, Silks LA. Simple enantiomeric excess determination of alcohols using chiral selones and ^{77}Se NMR spectroscopy. *Tetrahedron: Asymmetry* **1995**, *6*, 833–834.

1012. Wu R, Odom JD, Dunlap RB, Silks LA. Reaction of alcohols (via the Mitsunobu reaction) and alkyl halides with chiral selone derivatizing agents. *Tetrahedron: Asymmetry* **1999**, *10*, 1465–1470.

1013. Wu R, Hernández G, Odom JD, Dunlap RB, Silks LA. Simple enantiomeric excess determination of amines using chiral selones: unusual N-H···Se bonding detected by HMQC ^{1}H/^{77}Se NMR spectroscopy. *Chem. Commun.* **1996**, 1125–1126.

1014. House KL, O'Connor MJ, Silks LA, Dunlap RB, Odom JD. (*R,R*)-*N,N'*-Dimethylcyclohexyl-1,2-diazaselenophospholidine as a chiral derivatizing agent for the evaluation of chiral alcohols. *Chirality* **1994**, *6*, 196–201.

1015. Wu R, Silks LA, Odom JD, Dunlap RB. Highly sensitive spectroscopic detection of chiral molecular structures by ^{77}Se NMR spectroscopy. *Spectroscopy* **1996**, *11*, 37–42.

1016. Wu R, Barr ME, Hernandez G, Orji CC, Silks LA. Recent progress in the synthesis and application of chiral selones: exploitation of the selenocarbonyl group. *Recent Res. Devel. In Org. & Bioorg. Chem.* **1998**, *2*, 29–42.

1017. Michelsen P, Annby U, Gronowitz S. On the use of optically active phenylselenopropionic acid in enantiomer analysis. *Chem. Scripta* **1984**, *24*, 251–252.

1018. Duddeck H, Hotopp T. ^{77}Se and ^{13}C NMR spectroscopy of some 1,2,3-selenadiazoles. *Magn. Reson. Chem.* **1995**, *33*, 490–492.

1019. Burgess K, Porte AM. A reagent for determining optical purities of diols by formation of diastereomeric arylboronate esters. *Angew. Chem., Int. Ed. Engl.* **1994**, *33*, 1182–1184.

1020. Resnick SM, Torok DS, Gibson DT. Chemoenzymatic synthesis of chiral boronates for the ^{1}H NMR determination of the absolute configuration and enantiomeric excess of bacterial and synthetic cis-diols. *J. Org. Chem.* **1995**, *60*, 3546–3549.

1021. Resnick SM, Gibson DT. Oxidation of 6,7-dihydro-5H-benzocycloheptene by bacterial strains expressing naphthalene dioxygenase, biphenyl dioxygenase, and toluene dioxygenase yields homochiral monol of cis-diol enantiomers as major products. *Appl. Environ. Microbiol.* **1996**, *62*, 1364–1368.

1022. Boyd DR, Sharma ND, Hempenstall F, Kennedy MA, Malone JF, Allen CCR, Resnick SM, Gibson DT. bis-cis-Dihydrodiols: a new class of metabolites resulting from biphenyl dioxygenase-catalyzed sequential asymmetric cis-dihydroxylation of polycyclic arenes and heteroarenes. *J. Org. Chem.* **1999**, *64*, 4005–4011.

1023. Boyd DR, Sharma ND, Agarwal R, Resnick SM, Schocken MJ, Gibson DT, Sayer JM, Yagi H, Jerina DM. Bacterial dioxygenase-catalysed dihydroxylation and chemical resolution routes to enantiopure *cis*-dihydrodiols of chrysene. *J. Chem. Soc., Perkin Trans. 1* **1997**, 1715–1723.

1024. Browne CE, Dobbs TK, Hecht SS, Eisenbraun EJ. Stereochemical assignment of (E)- and (Z)-2-(1-naphthyl)-1-phenylpropene and their photocyclization to 5-methylchrysene. *J. Org. Chem.* **1978**, *43*, 1656–1660.

1025. Fernández-Gadea F, Rodríguez B. Complexation of ent-8,13β-epoxylabdane-12,14-diols with boric acid. A method for establishing the C(14) configuration. *J. Org. Chem.* **1984**, *49*, 4721–4724.

1026. Tokles M, Snyder JK. Camphanylboronic acid, a chiral derivatizing agent for optical purity determination of diols. *Tetrahedron Lett.* **1988**, *29*, 6063–6066.

1027. Brown JM, Leppard SW, Lloyd-Jones GC. A rapid assay for the enantiomeric purity of secondary alcohols using 4*S*,5*R*-4-methyl,5-phenyl-1,3,2-oxazaborolidine (ephedrine-borane). *Tetrahedron: Asymmetry* **1992**, *3*, 261–266.

1028. Caselli E, Danieli C, Morandi S, Bonfiglio B, Forni A, Prati F. (*S*)-(+)-*N*-Acetylphenylglycineboronic acid: a chiral derivatizing agent for ee determination of 1,2-diols. *Org. Lett.* **2003**, *5*, 4863–4866.

1029. Morandi S, Caselli E, Forni A, Bucciarelli M, Torre G, Prati F. Enantiomeric excess of 1,2-diols by formation of cyclic boronates: an improved method. *Tetrahedron: Asymmetry* **2005**, *16*, 2918–2926.

1030. Pérez-Fuertes Y, Kelly AM, Johnson AL, Arimori S, Bull SD, James TD. Simple protocol for NMR analysis of the enantiomeric purity of primary amines. *Org. Lett.* **2006**, *8*, 609–612.

1031. Kelly AM, Pérez-Fuertes Y, Arimori S, Bull SD, James TD. Simple protocol for NMR analysis of the enantiomeric purity of diols. *Org. Lett.* **2006**, *8*, 1971–1974.

1032. Nozaki K, Yoshida M, Takaya H. Chiral bimetallic boronic esters: A donor-acceptor coexisting receptor for amines. *Bull. Chem. Soc. Jpn.*, **1996**, *69*, 2043–2052.

1033. Ollivault M, Monnier L, Carboni B. Facile determination of the diastereoisomeric purity of 2,3-pinanediol (1-chloroalkyl)boronates. Isolation of boronic esters containing a configurationally stable boron atom. *Tetrahedron: Asymmetry* **1997**, *8*, 1955–1958.

1034. Terunuma D, Kato M, Kamei M, Uchida H, Nohira H. Application of optically active organosilane to the determination of optical purities of amines, alcohols, and carboxylic acids. *Chem. Lett.* **1985**, 13–14.

1035. Terunuma D, Kato M, Kamei M, Uchida H, Ueno S, Nohira H. Syntheses of new optically active organosilanes and their applications for the optical-purity-determining agents. *Bull. Chem. Soc. Jpn.* **1986**, *59*, 3581–3587.

1036. Chan TH, Peng Q, Wang D, Guo JA. Chiral silyl acetals as convenient reagents for determining enantiomeric purity of alcohols. *J. Chem. Soc., Chem. Commun.* **1987**, 325–326.

1037. Kaye PT, Learmonth RA. Chiral organosilicon compounds. II. Comparison of gas-liquid chromatographic and NMR analyses of enantiomeric alcohols. *J. Chromatogr.* **1990**, *503*, 437–441.

1038. Weibel DB, Walker TR, Schroeder FC, Meinwald J. Chiral silylation reagents for the determination of absolute configuration by NMR spectroscopy. *Org. Lett.* **2000**, *2*, 2381–2383.

1039. Schroeder FC, Weibel DB, Meinwald J. Chiral silylation reagents: determining configuration via NMR-spectroscopic coanalysis. *Org. Lett.* **2004**, *6*, 3019–3022.

1040. Smedley SR, Schroeder FC, Weibel DB, Meinwald J, Lafleur KA, Renwick JA, Rutowski R, Eisner T. Mayolenes: labile defensive lipids from the glandular hairs of a caterpillar (*Pieris rapae*). *Proc. Natl. Acad. Sci. U.S.A.* **2002**, *99*, 6822–6827.

1041. Williamson RT, Sosa ACB, Mitra A, Seaton PJ, Weibel DB, Schroeder FC, Meinwald J, Koehn FE. New silyl ether reagents for the absolute stereochemical determination of secondary alcohols. *Org. Lett.* **2003**, *5*, 1745–1748.

1042. Wang X. Determination of enantiomeric purity of alcohols using achiral diphenyldichlorosilane. *Tetrahedron Lett.* **1991**, *32*, 3651–3654.

1043. Dodziuk H, Koźmiński W, Ejchart A. NMR studies of chiral recognition by cyclodextrins. *Chirality* **2004**, *16*, 90–105.

1044. Chankvetadze B. Combined approach using capillary electrophoresis and NMR spectroscopy for an understanding of enantioselective recognition mechanisms by cyclodextrins. *Chem. Soc. Rev.* **2004**, *33*, 337–347.

1045. MacNicol DD. Cyclodextrins as NMR shift reagents for hydrocarbons. *Tetrahedron Lett.* **1975**, 3325–3326.

1046. MacNicol DD, Rycroft DS. Cyclodextrins as chiral nuclear magnetic resonance shift reagents. *Tetrahedron Lett.* **1977**, 2173–2176.

1047. Uekama K, Imai T, Hirayama F, Otagiri M, Hibi T, Yamasaki M. ^{1}H-NMR spectroscopic evidence on chiral discrimination of d1-pirprofen by β-cyclodextrin complexation. *Chem. Lett.* **1985**, 61–64.

1048. Greatbanks D, Pickford R. Cyclodextrins as chiral complexing agents in water, and their application to optical purity measurements. *Magn. Reson. Chem.* **1987**, *25*, 208–215.

1049. Casy AF, Mercer AD. Application of cyclodextrins to chiral analysis by ^{1}H NMR spectroscopy. *Magn. Reson. Chem.* **1988**, *26*, 765–774.

1050. Bernabé M, Jiménez-Barbero J, Martin-Lomas M, Penadés S, Vicent C. Chiral recognition of 1-O-allyl- and 1-O-benzyl-D- and -L- myo-inositol by cyclomalto-hexaose and -heptaose (α- and β-cyclodextrin). *Carbohydr. Res.* **1990**, *208*, 255–259.

1051. Richards JJ, Webb ML. Determination of enantiomeric excess by inclusion complexation using nuclear magnetic resonance spectroscopy. *Anal. Proc.* **1992**, *29*, 251–253.

1052. Amato ME, Pappalardo GC, Perly B. High-field NMR techniques and molecular modelling in the study of the inclusion complex of the cognition activator suronacrine (HP-128) with cyclodextrins. *Magn. Reson. Chem.* **1993**, *31*, 455–460.

1053. Casy AF. Use of cyclodextrins in chiral discrimination by NMR spectroscopy: an example of atropisomerism. *Magn. Reson. Chem.* **1993**, *31*, 416–417.

1054. Owens PK, Fell AF, Coleman MW, Kinns M, Berridge JC. Use of ^{1}H-NMR spectroscopy to determine the enantioselective mechanism of neutral and anionic cyclodextrins in capillary electrophoresis. *J. Pharm. Biomed. Anal.* **1997**, *15*, 1603–1619.

1055. Redenti E, Ventura P, Fronza G, Selva A, Rivara S, Plazzi PV, Mor M. Experimental and theoretical analysis of the interaction of (±)-cis-ketoconazole with β-cyclodextrin in the presence of (+)-L-tartaric acid. *J. Pharm. Sci.* **1999**, *88*, 599–607.

1056. Redondo J, Blázquez M, Torrens A. Chiral discrimination of the analgesic cizolirtine by using cyclodextrins: a ^{1}H NMR study on the solution structures of their host-guest complexes. *Chirality* **1999**, *11*, 694–700.

1057. Aga DS, Heberle S, Rentsch D, Hany R, Müller SR. Sulfonic and oxanilic acid metabolites of acetanilide herbicides: separation of diastereomers and enantiomers by capillary zone electrophoresis and identification by ^{1}H NMR spectroscopy. *Environ. Sci. Technol.* **1999**, *33*, 3462–3468.

1058. Fogassy K, Harmat V, Böcskei Z, Tárkányi G, Tőke L, Faigl F. Efficient synthesis and resolution of (±)-1-[2-carboxy-6-(trifluoromethyl)phenyl]pyrrole-2-carboxylic acid. *Tetrahedron: Asymmetry* **2000**, *11*, 4771–4780.

1059. Kano K, Negi S, Kamo H, Kitae T, Yamaguchi M, Okubo H, Hirama M. Recognition of helicity by native cyclodextrins. Highly enantioselective complexation of tetrahelicene dicarboxylic acid with β-cyclodextrin. *Chem. Lett.* **1998**, 151–152.

1060. Kano K, Kamo H, Negi S, Kitae T, Takaoka R, Yamaguchi M, Okubo H, Hirama M. Chiral recognition of an anionic tetrahelicene by native cyclodextrins. Enantioselective dominated by location of a hydrophilic group of the guest in a cyclodextrin cavity. *J. Chem. Soc., Perkin Trans. 2*, **1999**, 15–21.

1061. Matthews DB, Hinton RH, Wright B, Wilson ID, Stevenson D. Bioanalysis of p-trifluoromethylmandelic acid and Mosher's acid by chiral gas chromatography and fluorine nuclear magnetic resonance to study chiral inversion: application to rat urine samples. *J. Chromatogr. B* **1997**, *695*, 279–285.

1062. Brown SE, Coates JH, Lincoln SF, Coghlan DR, Easton CJ. Chiral molecular recognition: a ^{19}F nuclear magnetic resonance study of the diastereoisomer inclusion complexes formed between fluorinated amino acid derivatives and α-cyclodextrin in aqueous solution. *J. Chem. Soc., Faraday Trans.* **1991**, *87*, 2699–2703.

1063. Li S, Purdy WC. Circular dichroism, ultraviolet, and proton nuclear magnetic resonance spectroscopic studies of the chiral recognition mechanism of β-cyclodextrin. *Anal. Chem.* **1992**, *64*, 1405–1412.

1064. Yamashoji Y, Ariga T, Asano S, Tanaka M. Chiral recognition and enantiomeric separation of alanine β-naphthylamide by cyclodextrins. *Anal. Chim. Acta* **1992**, *168*, 39–47.

1065. Kahle C, Deubner R, Schollmayer C, Scheiber J, Baumann K, Holzgrabe U. NMR spectroscopic and molecular modelling studies on cyclodextrin-dipeptide inclusion complexes. *Eur. J. Org. Chem.* **2005**, 1578–1589.

1066. Jursic BS, Patel PK. Cyclomaltooligosaccharide (cyclodextrin)-assisted enantiomeric recognition of benzo[lmn][3,8]phenanthroline-derived amino acids. *Carbohydr. Res.* **2005**, *340*, 1413–1418.

1067. Jursic BS, Patel PK. Cyclodextrin assisted enantiomeric recognition of benzo[de] isoquinoline-1,3-dione derived amino acids. *Tetrahedron* **2005**, *61*, 919–926.

1068. Berlicki L, Rudzińska E, Kafarski P. Enantiodifferentiation of aminophosphonic and aminophosphinic acids with α- and β-cyclodextrins. *Tetrahedron: Asymmetry* **2003**, *14*, 1535–1539.

1069. Berlicki L, Rudzińska E, Mucha A, Kafarski P. Cyclodextrins as NMR probes in the study of the enantiomeric compositions of *N*-benzyloxycarbonylamino-phophonic and phosphinic acids. *Tetrahedron: Asymmetry* **2004**, *15*, 1597–1602.

1070. Dodziuk H, Sitkowski J, Stefaniak L, Jurczak J, Sybilska D. ^{13}C NMR differentiation of diastereoisomeric complexes of cis-decalin with β-cyclodextrin. *J. Chem. Soc., Chem. Commun.* **1992**, 207–208.

1071. Botsi A, Yannakopoulou K, Hadjoudis E, Perly B. NMR differentiation of enantiomeric (+)- and (−)-α-pinene via complexation with cyclodextrins in water. *J. Chem. Soc., Chem. Commun.* **1993**, 1085–1086.

1072. Dodziuk H, Sitkowski J, Stefaniak L, Sybilska D. Structure of cyclodextrins and their complexes. 7. Unusual NMR manifestation of chiral recognition of (1*R*,5*R*)- and (1*S*,5*S*)-α-pinenes by α-cyclodextrin in different solvents. *Pol. J. Chem.* **1996**, *70*, 1361–1364.

1073. Botsi A, Perly B, Hadjoudis E. (+)- and (−)-α-pinene as chiral recognition probes with natural cyclodextrins and their permethylated derivatives. An aqueous NMR study. *J. Chem. Soc., Perkin Trans. 2* **1997**, 89–94.

1074. Dodziuk H, Koźmiński W, Lukin O, Sybilska D. NMR manifestations and molecular dynamics modeling of chiral recognition of α-pinenes by α-cyclodextrin. *J. Mol. Struct.* **2000**, *523*, 205–212.

1075. Dodziuk H, Ejchart A, Lukin O, Vysotsky MO. ^{1}H and ^{13}C NMR and molecular dynamics study of chiral recognition of camphor enantiomers by α-cyclodextrin. *J. Org. Chem.* **1999**, *64*, 1502–1507.

1076. Abrams SR, Reaney MJT, Abrams GD, Mazurek T, Shaw AC, Gusta LV. Ratio of (*S*)- to (*R*)-abscisic acid from plant cell cultures supplied with racemic aba. *Phytochemistry* **1989**, *28*, 2885–2889.

1077. Aime S, Botta M, Parker D, Williams JAG. Nuclear magnetic resonance studies of neutral lanthanide(III) complexes with tetraaza-macrocyclic ligands containing three phosphinate and one carboxamide co-ordinating arms. *J. Chem. Soc., Dalton Trans.* **1995**, 2259–2266.

1078. Wenzel TJ, Bogyo MS, Lebeau EL. Lanthanide-cyclodextrin complexes as probes for elucidating optical purity by NMR spectroscopy. *J. Am. Chem. Soc.* **1994**, *116*, 4858–4865.

1079. Wenzel TJ, Miles RD, Zomlefer K, Frederique DE, Roan MA, Troughton JS, Pond BV, Colby AL. Dysprosium(III)-diethylenetriaminepentaacetate complexes of aminocyclodextrins as chiral NMR shift reagents. *Chirality* **2000**, *12*, 30–37.

1080. Kean SD, Easton CJ, Lincoln SF, Parker D. A preparative and solution study of a modified β-cyclodextrin and its europium(III) complex, and their interactions with racemic amino acid anions. *Aust. J. Chem.* **2001**, *54*, 535–539.

1081. Taylor A, Williams DAR, Wilson ID. Derivatized β-cyclodextrins combined with high field NMR for enantiomer analysis: application to ICI 185,282 (5(Z)-7-([2RS,4RS,5SR]-4-o-hydroxyphenyl-2-trifluoromethyl-1,3-dioxan-5-yl) heptenoic acid). *J. Pharm. Biomed. Anal.* **1991**, *9*, 493–496.

1082. Kano K, Kato Y, Kodera M. Mechanism for chiral recognition of binaphthyl derivatives by cyclodextrins. *J. Chem. Soc., Perkin Trans. 2*, **1996**, 1211–1217.

1083. Park KK, Park JM. ^{1}H NMR study of the inclusion complexes of chiral aromatic guests with β-cyclodextrin and its derivatives: discrimination of aromatic protons and chiral recognition. *Bull. Korean Chem. Soc.* **1996**, *17*, 1052–1056.

1084. Chankvetadze B, Burjanadze N, Pintore G, Strickmann D, Bergenthal D, Blaschke G. Chiral recognition of verapamil by cyclodextrins studied with capillary electrophoresis, NMR spectroscopy, and electrospray ionization mass spectrometry. *Chirality* **1999**, *11*, 635–644.

1085. Kano K, Hasegawa H. Chiral recognition of helical metal complexes by modified cyclodextrins. *J. Am. Chem. Soc.* **2001**, *123*, 10616–10627.

1086. Uccello-Barretta G, Balzano F, Caporusso AM, Salvadori P. Direct determination of the enantiomeric purity of chiral trisubstituted allenes by using permethylated cyclodextrin as a chiral solvating agent. *J. Org. Chem.* **1994**, *59*, 836–839.

1087. Uccello-Barretta G, Balzano F, Caporusso AM, Iodice A, Salvadori P. Permethylated β-cyclodextrin as chiral solvating agent for the NMR assignment of the absolute configuration of chiral trisubstituted allenes. *J. Org. Chem.* **1995**, *60*, 2227–2231.

1088. Uccello-Barretta G, Balzano F, Salvadori P, Lazzaroni R, Caporusso AM, Menicagli R. Different NMR approaches to the chiral analysis of trisubstituted allenes devoid of polar functional groups and aromatic hydrocarbons. *Enantiomer* **1996**, *1*, 365–375.

1089. Uccello-Barretta G, Balzano F, Cuzzola A, Menicagli R, Caporusso AM, Salvadori P. Permethylated cyclodextrins as chiral solvating agents for the determination of the enantiomeric composition of apolar substrates by NMR. *Gazz. Chim. Ital.* **1997**, *127*, 383–386.

1090. Uccello-Barretta G, Balzano F, Menicagli R, Salvadori P. NMR chiral analysis of aromatic hydrocarbons by using permethylated β-cyclodextrin as chiral solvating agent. *J. Org. Chem.* **1996**, *61*, 363–365.

1091. Sørbye K, Tautermann C, Carlsen P, Fiksdahl A. *N,N*-1,2-Benzenedisulfonylimide, a new cyclic leaving group for the stereoselective nucleophilic substitution of amines. *Tetrahedron: Asymmetry* **1998**, *9*, 681–689.

1092. Schwarz-Barać S, Ritter H, Schollmeyer D. Cyclodextrins in polymer synthesis: enantiodiscrimination in free-radical polymerization of cyclodextrin-complexed racemic *N*-methacryloyl-D,L-phenylalanine methyl ester. *Macromol. Rapid Commun.* **2003**, *24*, 325–330.

1093. Uccello-Barretta G, Cuzzola A, Balzano F, Menicagli R, Iuliano A, Salvadori P. A new stereochemical model from NMR for benzoylated cyclodextrins, promising new chiral solvating agents for the chiral analysis of 3,5-dinitrophenyl derivatives. *J. Org. Chem.* **1997**, *62*, 827–835.

1094. Uccello-Barretta G, Cuzzola A, Balzano F, Menicagli R, Salvadori P. Benzoylated and benzylated cyclodextrins: a new class of chiral solvating agents for chiral recognition of 3,5-dinitrophenyl derivatives by ^{1}H-NMR spectroscopy. *Eur. J. Org. Chem.* **1998**, 2009–2012.

1095. Kuroda Y, Suzuki Y, He J, Kawabata T, Shibukawa A, Wada H, Fujima H, Go-oh Y, Imai E, Nakagawa T. Enantiorecognition of a new chiral selector, β-cyclodextrin perphenylcarbamate, as studied by NMR spectroscopy and molecular energy calculation. *J. Chem. Soc., Perkin Trans. 2* **1995**, 1749–1759.

1096. Uccello-Barretta G, Ferri L, Balzano F, Salvadori P. Partially versus exhaustively carbamoylated cyclodextrins: NMR investigation on enantiodiscriminating capabilities in solution. *Eur. J. Org. Chem.* **2003**, 1741–1748.

1097. Uccello-Barretta G, Balzano F, Sicoli G, Scarselli A, Salvadori P. NMR enantiodiscrimination of polar and apolar substrates by multifunctional cyclodextrins. *Eur. J. Org. Chem.* **2005**, 5349–5355.

1098. Schurig V, Grosenick H, Juza M. Enantiomer separation of chiral inhalation anesthetics (enflurane, isoflurane and desflurane) by gas chromatography on a γ-cyclodextrin derivatives. *Recl. Trav. Chim. Pays-Bas 1995*, *114*, 211–219.

1099. Grosenick H, Juza M, Klein J, Schurig V. NMR spectroscopic investigation of the enantioselective complexation between the inhalation anesthetics enflurane and isoflurane and a γ-cyclodextrin derivative. *Enantiomer* **1996**, *1*, 337–349.

1100. Taylor A, Blackledge CA, Nicholson JK, Williams DAR, Wilson ID. Use of cyclodextrins and their derivatives for chiral analysis by high field nuclear magnetic resonance spectroscopy. *Anal. Proc.* **1992**, *29*, 229–231.

1101. Kim KH, Park YH. Enantioselective inclusion between terbutaline enantiomers and hydroxypropyl-β-cyclodextrin. *Int. J. Pharm.* **1998**, *175*, 247–253.

1102. Branch SK, Holzgrabe U, Jefferies TM, Mallwitz H, Matchett MW. Chiral discrimination of phenethylamines with β-cyclodextrin and heptakis(2,3-di-O-acetyl)β-cyclodextrin by capillary electrophoresis and NMR spectroscopy. *J. Pharm. Biomed. Anal.* **1994**, *12*, 1507–1517.

1103. Branch SK, Holzgrabe U, Jefferies TM, Mallwitz H, Oxley FJR. Effect of β-cyclodextrin acetylation on the resolution of phenethylamines with capillary electrophoresis and nuclear magnetic resonance spectroscopy. *J. Chromatogr. A* **1997**, *758*, 277–292.

1104. Thunhorst M, Holzgrabe U. Utilizing NMR spectroscopy for assessing drug enantiomeric composition. *Magn. Reson. Chem.* **1998**, *36*, 211–216.

1105. Holzgrabe U, Mallwitz H, Branch SK, Jefferies TM, Wiese M. Chiral discrimination by NMR spectroscopy of ephedrine and *N*-methylephedrine induced by β-cyclodextrin, heptakis(2,3-di-O-acetyl)β-cyclodextrin, and heptakis(6-O-acetyl)β-cyclodextrin. *Chirality* **1997**, *9*, 211–219.

1106. Süss F, Kahle C, Holzgrabe U, Scriba GKE. Studies on the chiral recognition of peptide enantiomers by neutral and sulfated β-cyclodextrin and heptakis-(2,3-di-O-acetyl)-β-cyclodextrin using capillary electrophoresis and nuclear magnetic resonance. *Electrophoresis* **2002**, *23*, 1301–1307.

1107. Murakami T, Harata K, Morimoto S. Synthesis and chiral recognition property of 3-acetylamino-3-deoxy-β-cyclodextrin. *Chem. Lett.* **1988**, 553–556.

1108. Murakami T, Harata K, Morimoto S. Synthesis of mono-acetylamino-β-cyclodextrins and ^{1}H-NMR study on their inclusion complexes with chiral guests. *Chem. Express* **1989**, *4*, 645–648.

1109. Gosnat M, Djedaïni-Pilard F, Perly B. Étude par RMN de la reconnaissance chirale par des cyclodextrines modifiées. *J. Chim. Phys.* **1995**, *92*, 1777–1781.

1110. Djedaini-Pilard F, Gosnat M, Brucato-Mauclaire V, Creminon C, Dalbiez JP, Pilard S, Luijten W, Perly B. New asymmetric β-cyclodextrin derivatives designed for chiral recognition. *Proceedings of the International Symposium on Cyclodextrins, Spain, 1998.* Kluwer: Dordrecht, Netherlands **1999**, 625–628.

1111. Köhler JEH, Hohla M, Richters M, König WA. Cyclodextrin derivatives as chiral selectors. Investigation of the interaction with (*R,S*)-methyl-2-chloropropionate by enantioselective gas chromatography, NMR spectroscopy, and molecular dynamics simulation. *Angew. Chem., Int. Ed. Engl.* **1992**, *31*, 319–320.

1112. Ramig K, Krishnaswami A, Rozov LA. Chiral interactions of the fluoroether anesthetics desflurane, isoflurane, enflurane, and analogues with modified cyclodextrins studied by capillary gas chromatography and nuclear magnetic resonance spectroscopy: a simple method for column-suitability screening. *Tetrahedron* **1996**, *52*, 319–330.

1113. Aturki Z, Desiderio C, Mannina L, Fanali S. Chiral separations by capillary zone electrophoresis with the use of cyanoethylated-β-cyclodextrin as chiral selector. *J. Chromatogr. A* **1998**, *817*, 91–104.

1114. Bates PS, Kataky R, Parker D. Chiral sensors based on lipophilic cyclodextrins: interrogation of enantioselectivity by combined NMR, structural correlation and electrode response studies. *J. Chem. Soc., Perkin Trans. 2* **1994**, 669–675.

1115. Hickel A, Gradnig G, Schall M, Zangger K, Griengl H. Determination of enantiomeric purity of mandelonitrile with derivatized cyclodextrins in water. *Spectrochim. Acta, Part A* **1997**, *53*, 451–455.

1116. Reuben J, Rao CT, Pitha J. Distribution of substituents in carboxymethyl ethers of cyclomaltoheptaose. *Carbohydr. Res.* **1994**, *258*, 281–285.

1117. Dignam CF, Randall LA, Blacken RD, Cunningham PR, Lester SG, Brown MJ, French SC, Aniagyei SE, Wenzel TJ. Carboxymethylated cyclodextrin derivatives as chiral NMR discriminating agents. *Tetrahedron: Asymmetry* **2006**, *17*, 1199–1208.

1118. Endresz G, Chankvetadze B, Bergenthal D, Blaschke G. Comparative capillary electrophoretic and nuclear magnetic resonance studies of the chiral recognition of racemic metomidate with cyclodextrin hosts. *J. Chromatogr. A* **1996**, *732*, 133–142.

1119. Owens PK, Fell AF, Coleman MW, Berridge JC. Screening of cyclodextrins by nuclear magnetic resonance for the design of chiral capillary electrophoresis separations. *J. Chromatogr. A* **1998**, *797*, 149–164.

1120. Chankvetadze B, Schulte G, Bergenthal D, Blaschke G. Comparative capillary electrophoresis and NMR studies of enantioseparation of dimethindene with cyclodextrins. *J. Chromatogr. A* **1998**, *798*, 315–323.

1121. Chankvetadze B, Pintore G, Burjanadze N, Bergenthal D, Strickmann D, Cerri R, Blaschke G. Capillary electrophoresis, nuclear magnetic resonance and mass spectrometry studies of opposite chiral recognition of chlorpheniramine enantiomers with various cyclodextrins. *Electrophoresis* **1998**, *19*, 2101–2108.

1122. Park K, Kim KH, Jung S, Lim H, Hong C, Kang J. Enantioselective stabilization of inclusion complexes of metoprolol in carboxymethylated β-cyclodextrin. *J. Pharm. Biomed. Anal.* **2002**, *27*, 569–576.

1123. Lee S, Yi D, Jung S. NMR spectroscopic analysis on the chiral recognition of noradrenaline by β-cyclodextrin (β-CD) and carboxymethyl-β-cyclodextrin (CM-β-CD). *Bull. Korean Chem. Soc.* **2004**, *25*, 216–220.

1124. Chankvetadze B, Endresz G, Schulte G, Bergenthal D, Blaschke G. Capillary electrophoresis and ^{1}H NMR studies on chiral recognition of atropisomeric binaphthyl derivatives by cyclodextrin hosts. *J. Chromatogr. A* **1996**, *732*, 143–150.

1125. Smith KJ, Wilcox JD, Mirick GE, Wacker LS, Ryan NS, Vensel DA, Readling R, Domush HL, Amonoo EP, Shariff SS, Wenzel TJ. Calix[4]arene, calix[4]resorcarene, and cyclodextrin derivatives and their lanthanide complexes as chiral NMR shift reagents. *Chirality* **2003**, *15*, S150–S158.

1126. Wenzel TJ, Amoono EP, Shariff SS, Aniagyei SE. Sulfated and carboxymethylated cyclodextrins and their lanthanide complexes as chiral NMR discriminating agents. *Tetrahedron: Asymmetry* **2003**, *14*, 3099–3104.

1127. Kitae T, Nakayama T, Kano K. Chiral recognition of α-amino acids by charged cyclodextrins through cooperative effects of Coulomb interaction and inclusion. *J. Chem. Soc., Perkin Trans. 2*, **1998**, 207–212.

1128. Kano K, Hasegawa H. Chiral recognition of Ru(phen)$_3^{2+}$ by anionic cyclodextrins. *Chem. Lett.* **2000**, 698–699.

1129. Kano K, Hasegawa H. Interactions with charged cyclodextrins and chiral recognition. *J. Inclusion Phenom. Macrocyclic Chem.* **2001**, *41*, 41–47.

1130. Chankvetadze B, Endresz G, Bergenthal D, Blaschke G. Enantioseparation of mianserine analogues using capillary electrophoresis with neutral and charged cyclodextrin buffer modifiers. ^{13}C NMR study of the chiral recognition mechanism. *J. Chromatogr. A* **1995**, *717*, 245–253.

1131. Owens PK, Fell AF, Coleman MW, Berridge JC. Complexation of voriconazole stereoisomers with neutral and anionic derivatised cyclodextrins. *J. Inclusion Phenom. Macrocyclic Chem.* **2000**, *38*, 133–151.

1132. Chankvetadze B, Lomsadze K, Bergenthal D, Breitkreutz J, Bergander K, Blasche G. Mechanistic study on the opposite migration order of clenbuterol enantiomers in capillary electrophoresis with β-cyclodextrin and single-isomer heptakis(2,3-diacetyl-6-sulfo)-β-cyclodextrin. *Electrophoresis* **2001**, *22*, 3178–3184.

1133. Chankvetadze B, Burjanadze N, Maynard DM, Bergander K, Bergenthal D, Blaschke G. Comparative enantioseparations with native β-cyclodextrin and heptakis-(2-O-methyl-3,6-di-O-sulfo)-β-cyclodextrin in capillary electrophoresis. *Electrophoresis* **2002**, *23*, 3027–3034.

1134. Zhou Z, Thompson R, Reamer RA, Lin Z, French M, Ellison D, Wyvratt J. Mechanistic study of the enantiomeric recognition of a basic compound with negatively charged single-isomer γ-cyclodextrin derivatives using capillary electrophoresis, nuclear magnetic resonance spectroscopy, and infrared spectroscopy. *Electrophoresis* **2003**, *24*, 2448–2455.

1135. Brown SE, Coates JH, Duckworth PA, Lincoln SF, Easton CJ, May BL. Substituent effects and chiral discrimination in the complexation of benzoic, 4-methylbenzoic and (RS)-2-phenylpropanoic acids and their conjugate bases by β-cyclodextrin and 6^A-amino-6^A-deoxy-β-cyclodextrin in aqueous solution: potentiometric titration and ^{1}H nuclear magnetic resonance spectroscopic study. *J. Chem. Soc. Faraday Trans.* **1993**, *89*, 1035–1040.

1136. Kano K, Kitae T, Takashima H. Use of electrostatic interaction for chiral recognition. Enantioselective complexation of anionic binaphthyls with protonated amino-β-cyclodextrin. *J. Inclusion Phenom. Mol. Recognit. Chem.* **1996**, *25*, 243–248.

1137. Kitae T, Takashima H, Kano K. Chiral recognition of phenylacetic acid derivatives by aminated cyclodextrins. *J. Inclusion Phenom. Macrocyclic Chem.* **1999**, *33*, 345–359.

1138. Rekharsky M, Yamamura H, Kawai M, Inoue Y. Critical difference in chiral recognition of N-Cbz-D/L-aspartic and –glutamic acids by mono- and bis(trimethylammonio)-β-cyclodextrins. *J. Am. Chem. Soc.* **2001**, *123*, 5360–5361.

1139. Park KK, Lim HS, Park JW. Chiral discrimination of phenylacetic acid derivatives by xylylenediamine-modified β-cyclodextrins. *Bull. Korean Chem. Soc.* **1999**, *29*, 211–213.

1140. Kyba EB, Koga K, Sousa LR, Siegel MG, Cram DJ. Chiral recognition in molecular complexing. *J. Am. Chem. Soc.* **1973**, *95*, 2692–2693.

1141. Lovely AE, Wenzel TJ. Chiral NMR discrimination of secondary amines using (18-crown-6)-2,3,11,12-tetracarboxylic acid. *Org. Lett.* **2006**, *8*, 2823–2826.

1142. Machida Y, Nishi H, Nakamura K. Nuclear magnetic resonance studies for the chiral recognition of the novel chiral stationary phase derived from 18-crown-6 tetracarboxylic acid. *J. Chromatogr. A* **1998**, *810*, 33–41.

1143. Machida Y, Nishi H, Nakamura K. Crystallographic studies for the chiral recognition of the novel chiral stationary phase derived from (+)-(R)-18-crown-6 tetracarboxylic acid. *Chirality* **1999**, *11*, 173–178.

1144. Wenzel TJ, Thurston JE. (+)-(18-Crown-6)-2,3,11,12-tetracarboxylic acid and its ytterbium(III) complex as chiral NMR discriminating agents. *J. Org. Chem.* **2000**, *65*, 1243–1248.

1145. Wenzel TJ, Thurston JE. Enantiomeric discrimination in the NMR spectra of underivatized amino acids and α-methyl amino acids using (+)-(18-crown-6)-2,3,11, 12-tetracarboxylic acid. *Tetrahedron Lett.* **2000**, *41*, 3769–3772.

1146. Wenzel TJ, Freeman BE, Sek DC, Zopf JJ, Nakamura T, Yongzhu J, Hirose K, Tobe Y. Chiral recognition in NMR spectroscopy using crown ethers and their ytterbium(III) complexes. *Anal. Bioanal. Chem.* **2004**, *378*, 1536–1547.

1147. Machida Y, Kagawa M, Nishi H. Nuclear magnetic resonance studies for the chiral recognition of (+)-(R)-18-crown-6-tetracarboxylic acid to amino compounds. *J. Pharm. Biomed. Anal.* **2003**, *30*, 1929–1942.

1148. Lee W, Bang E, Baek C, Lee W. Chiral discrimination studies of (+)-(18-crown-6)-2,3,11,12-tetracarboxylic acid by high-performance liquid chromatography and NMR spectroscopy. *Magn. Reson. Chem.* **2004**, *42*, 389–395.

1149. Curtis WD, Laidler DA, Stoddart JF, Jones GH. To enzyme analogues by lock and key chemistry with crown compounds. 1. Enantiomeric differentiation by configurationally chiral cryptands synthesised from L-tartaric and D-mannitol. *J. Chem. Soc., Perkin Trans. 1* **1977**, 1756–1769.

1150. Curtis WD, Laidler DA, Stoddart JF, Jones GH. Chiral recognition by configurationally chiral cryptands. *J. Chem. Soc., Chem. Commun.* **1975**, 835–837.

1151. Dietl F, Merz A, Tomahogh R. Synthesis of (+)-2R,3R,11R,12R- and (−)-2S,3S,11S,12S-tetraphenyl-18-crown-6. *Tetrahedron Lett.* **1982**, *23*, 5255–5258.

1152. Nakatsuji Y, Nakahara Y, Muramatsu A, Kida T, Akashi M. Novel C_2-symmetric chiral 18-crown-6 derivatives with two aromatic sidearms as chiral NMR discriminating agents. *Tetrahedron Lett.* **2005**, *46*, 4331–4335.

1153. Weinstein SE, Vining MS, Wenzel TJ. Lanthanide-crown ether mixtures as chiral NMR shift reagents for amino acid esters, amines and amino alcohols. *Magn. Reson. Chem.* **1997**, *35*, 273–280.

1154. Kyba EP, Timko JM, Kaplan LJ, de Jong F, Gokel GW, Cram DJ. Host-guest complexation. 11. Survey of chiral recognition of amine and amino ester salts by dilocular bisdinaphthyl hosts. *J. Am. Chem. Soc.* **1978**, *100*, 4555–4568.

1155. Peacock SC, Domeier LA, Gaeta FCA, Helgeson RC, Timko JM, Cram DJ. Host-guest complexation. 13. High chiral recognition of amino esters by dilocular hosts containing extended steric barriers. *J. Am. Chem. Soc.* **1978**, *100*, 8190–8202.

1156. Lingenfelter DS, Helgeson RC, Cram DJ. Host-guest complexation. 23. High chiral recognition of amino acid and ester guests by hosts containing one chiral element. *J. Org. Chem.* **1981**, *46*, 393–406.

1157. Stoddart JF. Tate and Lyle lecture, from carbohydrates to enzyme analogs. *Chem. Soc. Rev.* **1979**, *8*, 85–142.

1158. Joly J, Nazhaoui M, Dumont B. Synthesis and complexation behavior of some crown ethers derived from D-hexopyranosides and D-mannitol towards racemic phenylglycine salts. *Bull. Soc. Chim. Fr.* **1994**, *131*, 369–380.

1159. Wenzel TJ, Thurston JE, Sek DC, Joly J. Utility of crown ethers derived from methyl β-D-galactopyranoside and their lanthanide couples as chiral NMR discriminating agents. *Tetrahedron: Asymmetry* **2001**, *12*, 1125–1130.

1160. Morin FG, Grant DM. The preparation of new chiral diphenyl-substituted macrocyclic polyether-diester compounds and their enantiomeric recognition of chiral organic ammonium salts. *J. Heterocyclic Chem.* **1984**, *21*, 897–901.

1161. Davidson RB, Bradshaw JS, Jones BA, Dalley NK, Christensen JJ, Izatt RM, Morin FG, Grant DM. Enantiomeric recognition of organic ammonium salts by chiral crown ethers based on the pyridino-18-crown-6 structure. *J. Org. Chem.* **1984**, *49*, 353–357.

1162. Li Y, Echegoyen L, Martínez-Díaz MV, de Mendoza J, Torres T. Enantiomeric recognition of organic ammonium salts by chiral dialkyl-substituted triazole-18-crown-6 ligands. *J. Org. Chem.* **1991**, *56*, 4193–4196.

1163. Bradshaw JS, Huszthy P, McDaniel CW, Oue M, Zhu CY, Izatt RM. Enantiomeric recognition of organic ammonium salts by chiral pyridino-18-crown-6 ligands: a short review. *J. Coord. Chem.* **1992**, *27*, 105–114.

1164. Wang T, Bradshaw JS, Huszthy P, Izatt RM. Various aspects of enantiomeric recognition of (S,S)-dimethylpyridino-18-crown-6 by several organic ammonium salts. *Supramol. Chem.* **1996**, *6*, 251–255.

1165. Habata Y, Bradshaw JS, Young JJ, Castle SL, Huszthy P, Pyo T, Lee ML, and Izatt RM. New pyridino-18-crown-6 ligands containing two methyl, two tert-butyl, or two allyl substituents on chiral positions next to the pyridine ring. *J. Org. Chem.* **1996**, *61*, 8391–8396.

1166. Hellier PC, Bradshaw JS, Young JJ, Zhang XX, Izatt RM. Chiral pyridine-based macrobicyclic clefts: synthesis and enantiomeric recognition of ammonium salts. *J. Org. Chem.* **1996**, *61*, 7270–7275.

1167. Reetz MT, Rudolph J, Mynott R. Enantiotopic group recognition: direct evidence for selective complexation of enantiotopic groups by a chiral host. *J. Am. Chem. Soc.* **1996**, *118*, 4494–4495.

1168. Zhang XX, Izatt RM, Zhu CY, Bradshaw JS. Thermodynamic and NMR studies of solvent effect on enantiomeric recognition for chiral organic ammonium guests by chiral diketopyridino-18-crown-6 type ligands at 25.0 °C. *Supramol. Chem.* **1996**, *6*, 267–274.

1169. Samu E, Huszthy P, Horváth G, Szöllősy Á, Neszmélyi A. Enantiomerically pure chiral pyridino-crown ethers: synthesis and enantioselectivity toward the enantiomers of α-(1-naphthyl)ethylammonium perchlorate. *Tetrahedron: Asymmetry* **1999**, *10*, 3615–3626.

1170. Redd JT, Bradshaw JS, Huszthy P, Izatt RM. Pyrimidino-and proton-ionizable pyrimidino-crown ether ligands: synthesis and preliminary complexation studies. *J. Inclusion Phenom. Mol. Recognit. Chem.* **1997**, *29*, 301–308.

1171. Farkas V, Szalay L, Vass E, Hollósi M, Horváth G, Huszthy P. Probing the discriminating power of chiral crown hosts by CD spectroscopy. *Chirality* **2003**, *15*, S65–S73.

1172. Lee C, Teng P, Wong W, Kwong H, Chan ASC. New C_2-symmetric 2,2′-bipyridine crown macrocycles for enantioselective recognition of amino acid derivatives. *Tetrahedron* **2005**, *61*, 7924–7930.

1173. Li Y, Echegoyen L, Martínez-Díaz MV, de Mendoza J, Torres T. Enantiomeric recognition between chiral triazole-18-crown-6 ligands and organic ammonium cations assessed by ^{13}C and ^{1}H NMR relaxation times. *J. Org. Chem.* **1994**, *59*, 6539–6542.

1174. Naemura K, Ueno K, Takeuchi S, Tobe Y, Kaneda T, Sakata Y. Azophenolic acerands having chiral 1-phenyl-cis-1,2-cyclohexanediol units: a correlation between enantiorecognitive coloration and host-guest complementarity. *J. Am. Chem. Soc.* **1993**, *115*, 8475–8476.

1175. Naemura K, Mizo-oku T, Kamada K, Hirose K, Tobe Y, Sawada M, Takai Y. Preparation of homochiral crown ether containing (S)-1-(1-adamantyl)ethane-1,2-diol as a chiral subunit and its enantioselective complexation with an organic ammonium cation. *Tetrahedron: Asymmetry* **1994**, *5*, 1549–1558.

1176. Naemura K, Tobe Y, Kaneda T. Preparation of chiral and meso-crown ethers incorporating cyclohexane-1,2-diol derivatives as a steric barrier and their complexation with chiral and achiral amines. *Coord. Chem. Rev.* **1996**, *148*, 199–219.

1177. Hirose K, Fuji J, Kamada K, Tobe Y, Naemura K. Temperature dependent inversion of enantiomer selectivity in the complexation of optically active azophenolic crown ethers containing alkyl substituents as chiral barriers with chiral amines. *J. Chem. Soc., Perkin Trans. 2* **1997**, 1649–1657.

1178. Naemura K, Nishikawa Y, Fuji J, Hirose K, Tobe Y. Preparation of homochiral phenolic crown ethers containing para-substituted phenol moiety and chiral subunits derived from (S)-1-phenylethane-1,2-diol: their chiral recognition behaviour in complexation with neutral amines. *Tetrahedron: Asymmetry* **1997**, *8*, 873–882.

1179. Naemura K, Matsunaga K, Fuji J, Ogasahara K, Nishikawa Y, Hirose K, Tobe Y. Temperature dependence of enantioselectivity in complexes of optically active phenolic crown ethers with chiral amines in solution. *Anal. Sci.* **1998**, *14*, 175–182.

1180. Hirose K, Ogasahara K, Nishioka K, Tobe Y, Naemura K. Enantioselective complexation of phenolic crown ethers with chiral aminoethanol derivatives: effects of substituents of aromatic rings of hosts and guests on complexation. *J. Chem. Soc., Perkin Trans. 2* **2000**, 1984–1993.

1181. Sawada M, Takai Y, Yamada H, Hirayama S, Kaneda T, Tanaka T, Kamada K, Mizooku T, Takeuchi S, Ueno K, Hirose K, Tobe Y, Naemura K. Chiral recognition in host-guest complexation determined by the enantiomer-labeled guest method using fast atom bombardment mass spectrometry. *J. Am. Chem. Soc.* **1995**, *117*, 7726–7736.

1182. Chadwick DJ, Cliffe IA, Sutherland IO, Newton RF. Chiral diaza-18-crown-6 derivatives. *J. Chem. Soc., Chem. Commun.*. **1981**, 992–994.

1183. Chadwick DJ, Cliffe IA, Sutherland IO, Newton RF. The formation of complexes between aza derivatives of crown ethers and primary alkylammonium salts. 7. Chiral derivatives of aza crown ethers. *J. Chem. Soc., Perkin Trans. 1* **1985**, 1707–1717.

1184. Galán A, Andreu D, Echavarren AM, Prados P, de Mendoza J. A receptor for the enantioselective recognition of phenylalanine and tryptophan under neutral conditions. *J. Am. Chem. Soc.* **1992**, *114*, 1511–1512.

1185. Lu G, Liu F, Song W, He W, Prados P, de Mendoza J. Enantioselective recognition of a pair of novel receptors for amino acid zwitterions. *Mater. Sci. Eng., C* **1999**, *10*, 29–32.

1186. Yanagihara R, Tominaga M, Aoyama Y. Chiral host-guest interaction. A water-soluble calix[4]resorcarene having L-proline moieties as a non-lanthanide chiral NMR shift reagent for chiral aromatic guests in water. *J. Org. Chem.* **1994**, *59*, 6865–6867.

1187. Dignam CF, Richards CJ, Zopf JJ, Wacker LS, Wenzel TJ. An enantioselective NMR shift reagent for cationic aromatics. *Org. Lett.* **2005**, *7*, 1773–1776.

1188. Dignam CF, Zopf JJ, Richards CJ, Wenzel TJ. Water-soluble calix[4]resorcarenes as enantioselective NMR shift reagents for aromatic compounds. *J. Org. Chem.* **2005**, *70*, 8071–8078.

1189. Yoon J, Cram DJ. Chiral recognition properties in complexation of two asymmetric hemicarcerands. *J. Am. Chem. Soc.* **1997**, *119*, 11796–11806.

1190. Saito S, Nuckolls C, Rebek. J. New molecular vessels: synthesis and chiroselective recognition. *J. Am. Chem. Soc.* **2000**, *122*, 9628–9630.

1191. Yamanaka M, Rebek J. Constellational diastereomers in encapsulation complexes. *Chem. Commun.* **2004**, 1690–1691.

1192. Scarso A, Shivanyuk A, Hayashida O, Rebek J. Asymmetric environments in encapsulation complexes. *J. Am. Chem. Soc.* **2003**, *125*, 6239–6243.

1193. Ito K, Kida A, Ohba Y, Sone T. Synthesis of chiral calixarene analog incorporating amino acid residues: molecular recognition for a racemic ammonium ion by the macrocycles. *Chem. Lett.* **1998**, 1221–1222.

1194. Ito K, Noike M, Kida A, Ohba Y. Synthesis of chiral homoazacalix[4]arenes incorporating amino acid residues: molecular recognition for racemic quaternary ammonium ions. *J. Org. Chem.* **2002**, *67*, 7519–7522.

1195. Ito K, Ohba Y. Synthesis and properties of chiral homoazacalixarenes possessing amino acid residues. *Recent Res. Devel. Pure & Applied Chem.* **1999**, *3*, 255–281.
1196. Ito K, Ohta T, Ohba Y, Sone T. Syntheses of chiral bishomodiazacalix[4]arenes incorporating amino acid residues: molecular recognition for racemic ammonium ions by the macrocycles possessing tyrosine residues. *J. Heterocyclic Chem.* **2000**, *37*, 79–85.
1197. He Y, Xiao Y, Meng L, Zeng Z, Wu X, Wu C. New type chiral calix[4](aza)crowns: synthesis and chiral recognition. *Tetrahedron Lett.* **2002**, *43*, 6249–6253.
1198. Narumi F, Hattori T, Matsumura N, Onodera T, Katagiri H, Kabuto C, Kameyama H, Miyano S. Synthesis of an inherently chiral O,O′-bridged thiacalix[4]crowncarboxylic acid and its application to a chiral solvating agent. *Tetrahedron* **2004**, *60*, 7827–7833.
1199. Tsubaki K, Otsubo T, Kinoshita T, Kawada M, Fuji K. The first example for cycloenantiomeric hexahomooxacalix[3]arenes. *Chem. Pharm. Bull.* **2001**, *49*, 507–509.
1200. Liu S, He Y, Qing G, Xu K, Qin H. Fluorescent sensors for amino acid anions based on calix[4]arenes bearing two dansyl groups. *Tetrahedron: Asymmetry* **2005**, *16*, 1527–1534.
1201. Qing G, He Y, Zhao Y, Hu C, Liu S, Yang X. Calix[4]arene-based chromogenic chemosensor for the α-phenylglycine anion: synthesis and chiral recognition. *Eur. J. Org. Chem.* **2006**, 1574–1580.
1202. Garrier E, Le Gac S, Jabin I. First enantiopure calix[6]aza-cryptand: synthesis and chiral recognition properties towards neutral molecules. *Tetrahedron: Asymmetry* **2005**, *16*, 3767–3771.
1203. Barea E, Navarro JAR, Salas JM, Quirós M, Willermann M, Lippert B. Chiral pyrimidine metallacalixarenes: synthesis, structure and host-guest chemistry. *Chem. Eur. J.* **2003**, *9*, 4414–4421.
1204. Prins LJ, Huskens J, de Jong F, Timmerman P, Reinhoudt DN. Complete asymmetric induction of supramolecular chirality in a hydrogen-bonded assembly. *Nature* **1999**, *398*, 498–502.
1205. Ishi-i T, Crego-Calama M, Timmerman P, Reinhoudt DN, Shinkai S. Self-assembled receptors for enantioselective recognition of chiral carboxylic acids in a highly cooperative manner. *Angew. Chem., Int. Ed.* **2002**, *41*, 1924–1929.
1206. Castellano RK, Kim BH, Rebek Jr. J. Chiral capsules: asymmetric binding in calixarene-based dimers. *J. Am. Chem. Soc.* **1997**, *119*, 12671–12672.
1207. Fernandes SA, Nachtigall FF, Lazzarotto M, Fujiwara FY, Marsaioli AJ. 'Non-covalent synthesis' of a chiral host of calix[6]arene and enantiomeric discrimination. *Magn. Reson. Chem.* **2005**, *43*, 398–404.
1208. Gleich A, Schmidtchen FP, Mikulcik P, Müller G. Enantiodifferentiation of carboxylates by chiral building blocks for abiotic anion receptors. *J. Chem. Soc., Chem. Commun.* **1990**, 55–57.
1209. Echavarren A, Galán A, Lehn J, de Mendoza J. Chiral recognition of aromatic carboxylate anions by an optically active abiotic receptor containing a rigid guanidinium binding subunit. *J. Am. Chem. Soc.* **1989**, *111*, 4994–4995.
1210. Convery MA, Davis AP, Dunne CJ, MacKinnon JW. Synthesis and properties of enantiopure bicyclic amidines. *Tetrahedron Lett.* **1995**, *36*, 4279–4282.
1211. Stork T, Helmchen G. A new chiral solvating agent for carboxylic acids based on directed hydrogen bonding. *Recl. Trav. Chim. Pays-Bas* **1995**, *114*, 253–254.

1212. Bilz A, Stork T, Helmchen G. New chiral solvating agents for carboxylic acids: discrimination of enantiotopic nuclei and binding properties. *Tetrahedron: Asymmetry* **1997**, *8*, 3999–4002.

1213. Yang D, Li X, Fan Y, Zhang D. Enantioselective recognition of carboxylates: a receptor derived from α-aminoxy acids functions as a chiral shift reagent for carboxylic acids. *J. Am. Chem. Soc.* **2005**, *127*, 7996–7997.

1214. Martín M, Raposo C, Almaraz M, Crego M, Caballero C, Grande M, Morán JR. Efficient recognition of chiral carbamoyl-α-hydroxyacids with a cleft-type receptor. *Angew. Chem. Int. Ed. Engl.* **1999**, *35*, 2386–2388.

1215. Xu K, Wu X, He Y, Liu S, Qing G, Meng L. Synthesis and chiral recognition of novel chiral fluorescence receptors bearing 9-anthryl moieties. *Tetrahedron: Asymmetry* **2005**, *16*, 833–839.

1216. Yang X, Wu X, Fang M, Yuan Q, Fu E. Novel rigid chiral macrocyclic dioxopolyamines derived from L-proline as chiral solvating agents for carboxylic acids. *Tetrahedron: Asymmetry* **2004**, *15*, 2491–2497.

1217. Yuan Q, Fu E, Wu X, Fang M, Xue P, Wu C, Chen J. A convenient synthesis of chiral dioxocyclens and application as chiral solvating agents. *Tetrahedron Lett.* **2002**, *43*, 3935–3937.

1218. Deber CM, Blout ER. Amino acid-cyclic peptide complexes. *J. Am. Chem. Soc.* **1974**, *96*, 7566–7568.

1219. Bartman B, Deber CM, Blout ER. ^{13}C NMR relaxation studies of complexes between cyclo(L-Pro-Gly)$_3$ and amino acids. Conformational aspects of stepwise binding. *J. Am. Chem. Soc.* **1977**, *99*, 1028–1033.

1220. Miyake H, Shibata K, Kojima Y, Yamashita T, Ohsuka A. Macrocyclic peptides. 5. Chiral recognition of (R)- and (S)-trimethyl-1-phenethylammonium bromides by 24-, 27- and 36-membered ring peptides containing glycine and N,N'-ethylene-bridged (S)-leucyl-(S)-leucine. *Makromol. Chem., Rapid Commun.* **1990**, *11*, 667–671.

1221. Miyake H, Kojima Y, Yamashita T, Ohsuka A. Enantiomeric differentiations of various (R)- and (S)- ammonium and (R)- and (S)-α-amino acid ester salts by macrocyclic pseudopeptides. *Makromol. Chem.* **1993**, *194*, 1925–1933.

1222. Ishizu T, Noguchi S. Enantiomer-differentiating ability of cyclo(L-Phe-L-Pro)$_4$ having a rigid skeleton for phenylalanine methylester hydrochloride. *Chem. Pharm. Bull.* **1997**, *45*, 1202–1204.

1223. Dowden J, Edwards PD, Flack SS, Kilburn JD. Synthesis and binding properties of a macrocyclic peptide receptor. *Chem. Eur. J.* **1999**, *5*, 79–89.

1224. Heinrichs G, Vial L, Lacour J, Kubik S. Enantioselective recognition of a chiral quaternary ammonium ion by C$_3$ symmetric cyclic hexapeptides. *Chem. Commun.* **2003**, 1252–1253.

1225. Uccello-Barretta G, Balzano F, Martinelli J, Berni M, Villani C, Gasparrini F. NMR enantiodiscrimination by cyclic tetraamidic chiral solvating agents. *Tetrahedron: Asymmetry* **2005**, *16*, 3746–3751.

1226. Lee S, Choi Y, Lee S, Jeong K, Jung S. Chiral recognition based on enantioselective interactions of propranolol enantiomers with cyclosophoraoses isolated from *Rhizobium meliloti*. *Chirality* **2004**, *16*, 204–210.

1227. Lee S, Jung S. ^{13}C NMR spectroscopic analysis on the chiral discrimination of N-acetylphenylalanine, catechin and propranolol induced by cyclic (1→2)-β-D-glucans (cyclosophoraoses). *Carbohydr. Res.* **2002**, *337*, 1785–1789.

1228. Coterón JM, Vicent C, Bosso C, Penadés S. Glycophanes, cyclodextrin-cyclophane hybrid receptors for apolar binding in aqueous solutions. A stereoselective carbohydrate-carbohydrate interaction in water. *J. Am. Chem. Soc.* **1993**, *115*, 10066–10076.

1229. Takahashi I, Odashima K, Koga K. Diastereomeric host-guest complex formation by an optically active paracyclophane in water. *Tetrahedron Lett.* **1984**, *25*, 973–976.

1230. Takahashi I, Odashima K, Koga K, Kitajima H. Utility of an optically active paracyclophane as a new type of shift reagent for nuclear magnetic resonance (NMR) spectra based on the diastereomeric host-guest complex formation. *Chem. Express* **1992**, *7*, 653–656.

1231. Takahashi I, Aoyagi Y, Nakamura I, Kitagawa A, Matsumoto K, Kitajima H, Isa K, Odashima K, Koga K. Studies on water-soluble artificial receptors containing chiral side chains derived from carbohydrates. 2. Formation of diastereomeric inclusion complexes between optically active cyclophane TCP44 and chiral aromatic guests in acidic aqueous solutions. *Heterocycles* **1999**, *51*, 1371–1388.

1232. Webb TH, Suh H, Wilcox CS. Enantioselective and diastereoselective molecular recognition of alicyclic substrates in aqueous media by a chiral, resolved synthetic receptor. *J. Am. Chem. Soc.* **1991**, *113*, 8554–8555.

1233. Wilcox CS, Webb TH, Zawacki FJ, Glagovich N, Suh H. Selectivity in molecular recognition of steroids, alkanes and alicyclic substrates in aqueous media. *Supramol. Chem.* **1993**, *1*, 129–137.

1234. Canceill J, Lacombe L, Collet A. Analytical optical resolution of bromochlorofluoromethane by enantioselective inclusion into a tailor-made "cryptophane" and determination of its maximum rotation. *J. Am. Chem. Soc.* **1985**, *107*, 6993–6996.

1235. Sada K, Tateishi Y, Shinkai S. A chiral pybox ligand as a new chiral shift reagent for secondary dialkylammonium cations. *Chem. Lett.* **2004**, *33*, 582–583.

1236. Wehner M, Schrader T, Finocchiaro P, Failla S, Consiglio G. A chiral sensor for arginine and lysine. *Org. Lett.* **2000**, *2*, 605–608.

1237. Gao M, Wang B, Liu H, Xu Z. Synthesis of chiral 2,5-bis(oxazolinyl)thiophenes and their application as chiral shift reagents for 1,1′-bi-2-naphthol. *Chin. J. Chem.* **2002**, *20*, 85–89.

1238. Dobashi Y, Kobayashi K, Sato N, Dobashi A. Chiral recognition of 1,1′-bi-2-naphthol enantiomers by twisted amidine derivatives: rational design of a highly enantioselective receptor. *Tetrahedron Lett.* **1998**, *39*, 2985–2988.

1239. Dobashi Y, Kobayashi K, Dobashi A. A new twisted dual hydrogen bond acceptor featuring amidine functionality for chiral recognition: a high level of enantioselection of diol derivatives. *Tetrahedron Lett.* **1998**, *39*, 93–96.

1240. Castro PP, Georgiadis TM, Diederich F. Chiral recognition in clefts and cyclophane cavities shaped by the 1,1′-binaphthyl major groove. *J. Org. Chem.* **1989**, *54*, 5835–5838.

1241. Rasmussen BS, Elezcano U, Skrydstrup T. Synthesis and binding properties of chiral macrocyclic barbiturate receptors: application to nitrile oxide cyclizations. *J. Chem. Soc., Perkin Trans. 1* **2002**, 1723–1733.

1242. Reid DG, MacLachlan LK, Robinson SP, Camilleri P, Dyke CA, Thorpe CJ. Calmodulin discriminates between the two enantiomers of the receptor-operated calcium channel blocker SK&F 96365: a study using ^{1}H-NMR and chiral HPLC. *Chirality* **1990**, *2*, 229–232.

1243. Shah DO, Gorenstein DG. Fluorine nuclear magnetic resonance spectroscopy of "transition state analogue" complexes of the D and L enantiomers of *N*-acetyl-p-fluorophenylalaninal and α-chymotrypsin. *Biochemistry* **1983**, *22*, 6096–6101.

1244. Hinckley CC. Paramagnetic shifts in solutions of cholesterol and the dipyridine adduct of trisdipivalomethanatoeuropium(III). A shift reagent. *J. Am. Chem. Soc.* **1969**, *91*, 5160–5162.

1245. McConnell HM, Robertson RE, Isotropic nuclear resonance shifts. *J. Chem. Phys.* **1958**, *29*, 1361–1365.

1246. Wenzel TJ, Bettes TC, Sadlowski JE, Sievers RE. New binuclear lanthanide NMR shift reagents effective for aromatic compounds. *J. Am. Chem. Soc.* **1980**, *102*, 5903–5904.

1247. Wenzel TJ, Sievers RE. Binuclear complexes of lanthanide(III) and silver(I) and their function as shift reagents for olefins, aromatics, and halogenated compounds. *Anal. Chem.* **1981**, *53*, 393–399.

1248. Whitesides GM, Lewis DW. Tris[3-(tert-butylhydroxymethylene)-d-camphorato]europium(III). A reagent for determining enantiomeric purity. *J. Am. Chem. Soc.* **1970**, *92*, 6979–6980.

1249. Goering HL, Eikenberry JN, Koermer GS. Tris[3-(trifluoromethylhydroxymethylene)-d-camphorato]europium(III). A chiral shift reagent for direct determination of enantiomeric compositions. *J. Am. Chem. Soc.* **1971**, *93*, 5913–5914.

1250. Fraser RR, Petit MA, Saunders JK. Determination of enantiomeric purity by an optically active nuclear magnetic resonance shift reagent of wide applicability. *Chem. Commun.* **1971**, 1450–1451.

1251. Wenzel TJ. *NMR Shift Reagents*. CRC Press, Inc: Boca Raton, FL, **1978**.

1252. McCreary MD, Lewis DW, Wernick DL, Whitesides GM. The determination of enantiomeric purity using chiral lanthanide shift reagents. *J. Am. Chem. Soc.* **1974**, *96*, 1038–1054.

1253. Goering HE, Eikenberry JN, Koermer GS, Lattimer CJ. Direct determination of enantiomeric compositions with optically active nuclear magnetic resonance lanthanide shift reagents. *J. Am. Chem. Soc.* **1974**, *96*, 1493–1501.

1254. Gupta AK, Kazlauskas RJ. Calibration plots to aid determination of high enantiomeric purity using chiral lanthanide shift reagents. *Tetrahedron: Asymmetry* **1992**, *3*, 243–246.

1255. Fraser RR, Stothers JB, Tan CT. Determination of optical purity with chiral shift reagents and ^{13}C magnetic resonance. *J. Magn. Reson.* **1973**, *10*, 95–97.

1256. Pearce GT, Gore WE, Silverstein RM. Synthesis and absolute configuration of multistriatin. *J. Org. Chem.* **1976**, *41*, 2797–2803.

1257. Belleney J, Bui C, Carrière FJ. Determination of the enantiomeric purity of disubstituted β-lactams by NMR spectroscopy using a chiral shift reagent. *Magn. Reson. Chem.* **1990**, *28*, 606–611.

1258. Omata K, Aoyagi S, Kabuto K. Observing the enantiomeric ^{1}H chemical shift non-equivalence of several α-amino ester signals using tris[3-(trifluoromethylhydroxy-methylene)-(+)-camphorato]samarium(III): a chiral lanthanide shift reagent that causes minimal line bordering. *Tetrahedron: Asymmetry* **2004**, *15*, 2351–2356.

1259. Sweeting LM, Crans DC, Whitesides GM. Determination of enantiomeric purity of polar substrates with chiral lanthanide NMR shift reagents in polar solvents. *J. Org. Chem.* **1987**, *52*, 2273–2276.

1260. Rothchild R, Sanders K, Venkatasubban KS. NMR studies of drugs. Use of lanthanide shift reagents in polar solvent with thalidomide. *Spectrosc. Lett.* **1993**, *26*, 597–619.

1261. DeArment M, Eastabrooks M, Venkatasubban KS, Benshafrut R, Rothchild R, Wyss H. NMR studies of drugs. Application of a chiral lanthanide shift reagent to methoxamine in chloroform-d or acetonitrile-d_3. *Spectrosc. Lett.* **1994**, *27*, 533–555.

1262. Polívka Z, Buděšínský M, Holubek J, Schneider B, Šedivý Z, Svátek E, Matoušová O, Metyš J, Valchář M, Souček R, Protiva M. 4H-Benzo[4,5]cyclohepta[1,2-b]thiophenes and 9,10-dihydro derivatives - sulfonium analogues of pizotifen and ketotifen; chirality of ketotifen; synthesis of the 2-bromo derivative of ketotifen. *Collect. Czech. Chem. Commun.* **1989**, *54*, 2443–2469.

1263. Peterson PE, Stepanian M. Methodology for the analysis of products from asymmetric syntheses using chiral NMR shift reagents. Relative complexion constants of enantiomers. *J. Org. Chem.* **1988**, *53*, 1907–1911.

1264. Wenzel TJ, Ciak JM. Europium, tris(6,6,7,7,8,8,8-heptafluoropropyl-2,3dimethyl-3,5-octanedianato. *Electronic Encyclopedia of Reagents for Organic Synthesis*. John Wiley & Sons, Ltd: Sussex, UK. **2004**, 1–26.

1265. Wenzel TJ, Ciak JM. Europium, tris[3-(2,2,3,3,4,4,4-heptafluoro-1-(oxo-kO)butyl]-1,7,7-trimethyl[2.2.1]heptan-2-onato-kO]. *Electronic Encyclopedia of Reagents for Organic Synthesis*. John Wiley & Sons, Ltd: Sussex, UK. **2004**, 1–8.

1266. Wenzel TJ, Ciak JM. Europium, tris[1,7,7-trimethyl-3-(trifluoroacetyl-kO)bicyclo[2.2.1]heptan-2-onato-kO]. *Electronic Encyclopedia of Reagents for Organic Synthesis*. John Wiley & Sons, Ltd: Sussex, UK. **2004**, 1–6.

1267. Naruse Y, Watanabe H, Ishiyama Y, Yoshida T. Enantiomeric enrichment of allenedicarboxylates by a chiral organoeuropium reagent. *J. Org. Chem.* **1997**, *62*, 3862–3866.

1268. Naruse Y, Watanabe H, Inagaki S. A facile access to chiral allenedicarboxylates by deracemization process. *Tetrahedron: Asymmetry* **1992**, *3*, 603–606.

1269. Fraser RR. Nuclear magnetic resonance analysis using chiral shift reagents. *Asymmetric Synthesis* **1983**, *1*, 173–196.

1270. Liu JH, Tsay JT. Use of chiral lanthanide shift reagents for the nuclear magnetic resonance spectrometric determination of amphetamine enantiomers. *Analyst* **1982**, *107*, 544–549.

1271. Liu JH, Ramesh S, Tsay JT, Ku WW, Fitzgerald MP, Angelos SA, Lins CLK. Approaches to drug sample differentiation. II: Nuclear magnetic resonance spectrometric determination of methamphetamine enantiomers. *J. Forensic Sci.* **1981**, *26*, 656–663.

1272. Hofer E, Keuper R, Renken H. Chiral relaxation reagents (CRR) in ^{13}C NMR spectroscopy. *Tetrahedron Lett.* **1984**, *25*, 1141–1142.

1273. Calmes M, Daunis J, Jacquier R, Verducci J. A general and accurate NMR determination of the enantiomeric purity of α-aminoacids and α-aminoacid derivatives. *Tetrahedron* **1987**, *43*, 2285–2292.

1274. Axt M, Alifantes J, Costa VEU. Use of chiral lanthanide shift reagents in the elucidation of NMR signals from enantiomeric mixtures of polycyclic compounds. *J. Chem. Soc., Perkin Trans. 2* **1999**, 2783–2788.

1275. Fraser RR, Petit MA, Miskow M. Separation of nuclear magnetic resonance signals of internally enantiotropic protons using a chiral shift reagent. The deuterium isotope effect on geminal proton-proton coupling constants. *J. Am. Chem. Soc.* **1972**, *94*, 3253–3254.

1276. Lang RW, Hansen HJ. Proton NMR spectroscopic analysis of prochiral allenic esters using optically active europium shift reagents. *Helv. Chim. Acta* **1979**, *64*, 1458–1465.

1277. Lefevre F, Rabiller C, Mannschreck A, Martin GJ. Cation-anion association in iminium salts. Use of europium complexes as ^{1}H N.M.R. auxiliary compounds. *J. Chem. Soc. Chem. Commun.* **1979**, 942–943.

1278. Coxon JM, Cambridge JRA, Nam SGC. NMR separation of β-prochiral protons to the ether oxygen of chiral esters with lanthanide shift reagents. *Org. Lett.* **2001**, *3*, 4225–4227.

1279. Coxon JA, Cambridge JRA, Nam SGC. Identification of β-prochiral protons to the ether oxygen of chiral esters of 2-arylethan-1-ols with d-Yb(hfc)$_3$ shift reagents. *Synlett* **2004**, *8*, 1422–1424.

1280. Goe GL. Photochemical addition of dimethyl maleate to 2,3-dimethyl-2-butene. Use of a chiral shift agent. *J. Org. Chem.* **1973**, *38*, 4285–4288.

1281. McKinney JD, Matthews HB, Wilson NK. Determination of optical purity and prochirality of chlorinated polycyclodiene pesticide metabolites. *Tetrahedron Lett.* **1973**, 1895–1898.

1282. Johnson PY, Jacobs I, Kerkman DJ. The chemistry of hindered systems. Syntheses and properties of tetramethylazacycloheptanes and related acyclic amines. *J. Org. Chem.* **1975**, *40*, 2710–2720.

1283. Johnson PY, Kerkman DJ. The chemistry of hindered systems. 2. The acyloin reaction– an approach to regiospecifically hydroxylated tetramethylazacycloheptane systems. *J. Org. Chem.* **1976**, *41*, 1768–1773.

1284. Cameron DW, Edmonds JS, Feutrill GI, Hoy AE. The use of chiral NMR shift reagents in a study of the stereochemistry of bianthrones. *Aust. J. Chem.* **1976**, *29*, 2225–2262.

1285. Joseph-Nathan P, Román LU. ^{1}H NMR study of hydrobenzoins with a chiral shift reagent. *Spectroscopy* **1991**, *9*, 47–54.

1286. Agranat I, Tapuhi Y, Lallemand JY. Manifestation of chirality in dynamic stereoisomers lacking asymmetric centers by the LIS technique. The chirality of bianthrones. *Nouv. J. Chim.* **1979**, *3*, 59–61.

1287. Reich CJ, Sullivan GR, Mosher HS. Chiral 1-deuterio alcohols. Synthesis and determination of enantiomeric purity by chiral lanthanide NMR shift reagents. *Tetrahedron Lett.* **1973**, 1505–1508.

1288. Harrison DM, Quinn P. The stereospecific synthesis of (*S*)-2-[^{2}H$_3$]methyl 2-methylbutanol. Characterisation of the (*R*) and (*S*) enantiomers of the racemic [^{2}H$_3$]alcohol by ^{2}H-NMR in the presence of a chiral shift reagent. *Tetrahedron Lett.* **1983**, *24*, 831–834.

1289. Baldwin JE, Black KA. Complete kinetic analysis of thermal stereomutations among the eight 2,3-dideuterio-2-(methoxymethyl)spiro[cyclopropane-1,1'-indenes]. *J. Am. Chem. Soc.* **1984**, *106*, 1029–1040.

1290. Makino T, Orfanopoulos M, You T, Wu B, Mosher CW, Mosher HS. Chiral benzhydrol-2,3,4,5,6-d$_5$. *J. Org. Chem.* **1985**, *50*, 5357–5360.

1291. Elvidge JA, Evans EA, Jones JR, Zhang LM. Determination of optical purity by ^{3}H NMR spectroscopy. *Proceedings of the Second International Symposium, Kansas City, MO, 1985.* Muccino RR, Ed. Elsevier Science Publishers: Amsterdam, **1985**, 401–408.

1292. Acheson RM, Selby IA. Direct observation of enantiomeric rotational isomers using ^{1}H nuclear magnetic resonance spectroscopy. *J. Chem. Soc. Chem. Comm.* **1973**, 537–539.

1293. Mannschreck A, Jonas V, Kolb B. Determination of rate of internal rotation by ^{1}H-NMR spectroscopy in the presence of optically active auxiliary compounds. *Angew. Chem., Int. Ed.* **1973**, *12*, 909–910.

1294. Jelinski LW, Kiefer EF. Dissymmetric 1,3-dienes. IV. Syntheses and enantiomerization barriers of some acyclic vic-dialkylidene compounds. *J. Am. Chem. Soc.* **1976**, *98*, 282–283.

1295. Tangerman A, Zwanenburg B. Recognition of atropisomerism and thermally labile prochirality, and measurement of barriers to rotation sulfines (thione oxides) by use of chiral shift reagents. *Recl. Trav. Chim. Pays-Bas* **1997**, *96*, 196–199.

1296. Becher G, Burgemeister T, Henschel H, Mannschreck A. Applications of NMR spectroscopy of chiral association complexes. *Org. Magn. Reson.* **1978**, *11*, 481–486.

1297. Holík M, Mannschreck A. Application of NMR spectroscopy of chiral association complexes. *Org. Magn. Reson.* **1979**, *12*, 223–228.

1298. Calmes M, Daunis J, Elyacoubi R, Jacquier R. Control of the enantiomeric purity of N-BOC N-methylamino acid building blocks by a convenient NMR method. *Tetrahedron: Asymmetry* **1990**, *1*, 329–334.

1299. Mannschreck A, Jonas V, Bödecker H, Elbe H, Köbrich G. Diastereomeric association complexes of chiral butadienes. *Tetrahedron Lett.* **1974**, 2153–2156.

1300. Anastassiou AG, Hasan M. Evidence for skeletal enantiomerism in benzo[9]annulenone. *Tetrahedron Lett.* **1983**, *24*, 4279–4282.

1301. Anastassiou AG, Hasan M. The existence of helical chirality in a sterically congested naphtho[9]annulenone. *Helv. Chim. Acta* **1982**, *65*, 2526–2528.

1302. Houalua D, Sanchez M, Wolf R. Mise en evidence de la chiralité de l'édifice spirophosphoranique. *Org. Magn. Reson.* **1973**, *5*, 451–452.

1303. Behnam BA, Hall DM. 2,2′-Bridged biphenyls with 12-membered heterocyclic bridging rings. 1. Tetrabenzo[b,d,h,j][1,6]diazacyclododecines. *J. Chem. Soc., Perkin Trans. 1* **1980**, 107–112.

1304. Bottino F, Di Grazia M, Finocchiaro P, Fronczek FR, Mamo A, Pappalardo S. Reaction of tosylamide monosodium salt with bis(halomethyl) compounds: an easy entry to symmetrical *N*-tosyl aza macrocycles. *J. Org. Chem.* **1988**, *53*, 3521–3529.

1305. Arnold Z, Buděšínský M. 2,6,9-Trioxabicylo[3.3.1.]nona-3,7-diene-4,8-dicarbaldehyde, a dissymmetric propeller-like molecule. The structure and chirality proof. *J. Org. Chem.* **1988**, *53*, 5352–5353.

1306. Uncuţa C, Bally I, Drăghici C, Chiraleu F, Balaban AT. Heterocyclic organoboron compounds. XVII. Axial chirality of boron chelates evidenced by ^{1}H-NMR spectroscopy in the presence of chiral additives. *Rev. Roum. Chim.* **1984**, *29*, 121–129.

1307. Behnam BA, Hall DM, Modarai B. Detection of enantiomers in dissymmetric biaryls with chiral shift reagents. *Tetrahedron Lett.* **1979**, 2619–2620.

1308. Diaz E, Rojas-Dávila E, Guzmán A, Joseph-Nathan P. The use of chiral shift reagents in an NMR study of 2,2′,6,6′-tetrasubstituted biphenyls. *Org. Magn. Reson.* **1980**, *14*, 439–443.

1309. Hunter CA. Synthesis and structure elucidation of a new [2]-catenane. *J. Am. Chem. Soc.* **1992**, *114*, 5303–5311.

1310. Jiang Z, Sen A. Palladium(II)-catalyzed isospecific alternating copolymerization of aliphatic α-olefins with carbon monoxide and isospecific alternating isomerization cooligomerization of a 1,2-disubstituted olefin with carbon monoxide. Synthesis of novel, optically active, isotactic 1,4- and 1,5-polyketones. *J. Am. Chem. Soc.* **1995**, *117*, 4455–4467.

1311. Tanaka Y, Chiyo T, Iijima S, Shimizu T, Kusano T. Asymmetric decarboxylation of 2-ethyl-2-(4-methylphenyl)-propane-1,3-dioic acid in cholesteric liquid crystals. *Mol. Cryst. Liq. Cryst.* **1983**, *99*, 255–266.

1312. Eliel EL, Ko K. Determination of enantiomeric purity of glycols RCHOHCH$_2$OH. *Tetrahedron Lett.* **1983**, *24*, 3547–3550.

1313. Ko K, Frazee WJ, Eliel EL. Asymmetric reactions based on 1,3-oxathianes. 3. Secondary α-hydroxyacids, RCHOHCO$_2$H and glycols RCHOHCH$_2$OH. *Tetrahedron* **1984**, *40*, 1333–1343.

1314. Bartik K, El Haouaj M, Luhmer M, Collet A, Reisse J. Can monoatomic xenon become chiral? *ChemPhysChem* **2000**, *1*, 221–224.

1315. Ulrich C, Bargon J. Parahydrogen-induced polarization in the presence of NMR shift reagents. *Magn. Reson. Chem.* **2000**, *38*, 33–37.

1316. Barbarella G, Bongini A, Ceré V, Paolucci C, Pollicino S. Differentiation of sulfonyl oxygens in cyclic systems by ^{17}O NMR lanthanide induced shifts. *J. Org. Chem.* **1993**, *58*, 6499–6502.

1317. Fortis F, Barbe B, Pétraud M, Picard J. Unprecedented use of ^{29}Si NMR spectroscopy for a convenient determination of enantiomeric excesses of chiral α-C-silylated amines and alcohols. *Chem. Commun.* **1999**, 527–528.

1318. Krull IS. The mechanism of ring opening in the reactions between cis- and trans-FEIST's esters with diiron nonacarbonyl. *J. Organomet. Chem.* **1973**, *57*, 363–372.

1319. Stephenson GR. Transition-metal mediated asymmetric synthesis. IV Chelation of malonate esters by a chiral lanthanide shift reagent: a general method to measure the enantiomeric excess of tricarbonyl(cyclohexadienyl)iron(1+) salts. *Aust. J. Chem.* **1982**, *35*, 1939–1943.

1320. Stephenson GR. Transition metal mediated asymmetric synthesis. II Complete resolution of (2R,5S)-(−)-tricarbonyl-[1–5-η-(2-methoxy-5-methylcyclohexadienyl)]iron(1+) hexafluorophosphate(1−) by selective crystallization of enantiomers. *Aust. J. Chem.* **1981**, *34*, 2339–2345.

1321. Astley ST, Meyer M, Stephenson GR. Hippeastrine synthesis: a combined bio-dioxygenation/organoiron chirality relay approach. *Tetrahedron Lett.* **1993**, *34*, 2035–2038.

1322. Blumhofer R, Vahrenkamp H. Chiral μ3-alkylidyne trimetal clusters: preparation and investigation of the optical activity. *Chem. Berg.* **1986**, *119*, 683–698.

1323. Flood TC, DiSanti FJ, Miles DL. General synthesis of optically pure chiral iron primary alkyls via a metal-activated alkylation reagent. *J. Chem. Soc. Chem. Comm.* **1975**, 336–337.

1324. Kläui W, Neukomm H. Unterscheidung enantiotoper/diastereotoper protonen und regioselektive spinentkopplung in metallorganischen komplexen mit hilfe von shiftreagenzien. *Org. Magn. Reson.* **1977**, *10*, 126–131.

1325. Hamer G, Shaver A. Cyclopentadienyl platinum(IV) complexes: ^{1}H, ^{13}C nuclear magnetic resonance and optically active shift reagent study. *Can. J. Chem.* **1980**, *58*, 2011–2015.

1326. Solladié-Cavallo A, Suffert J. Enantiomeric purity of chiral metal carbonyl complexes using Eu(hfc)$_3$. *Magn. Reson. Chem.* **1985**, *23*, 739–743.

1327. Ville G, Vollhardt KPC, Winter MJ. On the reversibility of η^4-cyclobutadiene-metal formation from complexed alkynes: unimolecular isomerization of labeled racemic and enantiomerically enriched η^5-cyclopentadienyl- η^4-cyclobutadiene-cobalt complexes. *J. Am. Chem. Soc.* **1981**, *103*, 5267–5269.

1328. Koerner R, Olmstead MM, Ozarowski A, Phillips SL, Van Calcar PM, Winkler K, Balch AL. Possible intermediates in biological metalloporphyrin oxidative degradation. Nickel, copper, and cobalt complexes of octaethylformybiliverdin and their conversion to a verdoheme. *J. Am. Chem. Soc.* **1998**, *120*, 1274–1284.

1329. Reger DL. Cyanide, isocyanide, and nitrile derivatives of cyclopentadienyliron. Interaction of chiral metal complexes with an optically active shift reagent. *Inorg. Chem.* **1975**, *14*, 660–664.

1330. Hendrickson AR, Hope JM, Martin RL. Tris-and pentakis-dialkyldithiocarbamates of ruthenium, [Ru(S$_2$CNR$_2$)$_3$]^n and [Ru$_2$(S$_2$CNR$_2$)$_5$]^n ($n = +1, 0,$ and -1): chemical and electrochemical interrelations. *J. Chem. Soc., Dalton Trans.* **1975**, 2032–2039.

1331. Rutherford TJ, Pellegrini PA, Aldrich-Wright J, Junk PC, Keene FR. Isolation of enantiomers of a range of tris(bidentate)ruthenium(II) species using chromatographic resolution and stereoretentive synthetic methods. *Eur. J. Inorg. Chem.* **1998**, 1677–1688.

1332. Rutherford TJ, Quagliotto MG, Keene FR. Chiral [Ru(pp)$_2$(CO$_2$)]$^{2+}$ species (pp = bibidentate polypyridyl ligand) and their use in the stereoselective synthesis of ligand-bridged dinuclear complexes. *Inorg. Chem.* **1995**, *34*, 3857–3858.

1333. Peng T, Gladysz JA. Regiospecific, diastereospecific and enantiospecific nucleophilic additions to chiral monosubstituted alkene complexes of the formula [η^5-C$_5$H$_5$)Re(NO)(PPh$_3$)(H$_2$C=CHR)]$^+$ BF$_4^-$. *Tetrahedron Lett.* **1990**, *31*, 4417–4420.

1334. Dewey MA, Gladysz JA. Mechanism of equilibration of diastereomeric rhenium amide complexes of the formula (η^5-C$_5$H$_5$)Re(NO)(PPh$_3$)(NHCHRR′): rhenium vs carbon epimerization. *Organometallics* **1990**, *9*, 1351–1353.

1335. Dewey MA, Stark GA, Gladysz JA. Configurational equilibria in amido and lithioamido complexes of formulas (η^5-C$_5$H$_5$)Re(NO)(PAr$_3$)(NHCHRR′) and (η^5-C$_5$H$_5$)Re(NO)(PAr$_3$)(NLiR″): epimerization occurs at rhenium via phosphine dissociation. *Organometallics* **1996**, *15*, 4798–4807.

1336. Barton JK, Nowick JS. Application of chiral lanthanide-induced shift reagents to optically active cations: the use of tris[3-(trifluoromethylhydroxymethylene)-(+)-camphorato]europium(III) to determine the enantiomeric purity of tris(phenanthroline) ruthenium (II) dichloride. *J. Chem. Soc., Chem. Commun.* **1984**, 1650–1652.

1337. Uppadine LH, Keene FR, Beer PD. Approaches towards the enantioselective recognition of anionic guest species using chiral receptors based on rhenium(I) and ruthenium(II) with amide bipyridine ligands. *J. Chem. Soc., Dalton Trans.* **2001**, 2188–2198.

1338. Lindoy LF, Moody WE. Nuclear magnetic resonance studies of metal complexes using lanthanide shift reagents. Lanthanide-induced shifts in the ^{1}H (and ^{13}C) spectra of diamagnetic metal complexes of quadridentate ligands incorporating oxygen-nitrogen donor atoms. *J. Am. Chem. Soc.* **1977**, *99*, 5863–5870.

1339. Schurig V. NMR-spektroskopischer nachweis der chiralität π-komplexierter prochiraler olefine. *Tetrahedron Lett.* **1976**, 1269–1272.

1340. Červinka O, Maloň P, Trška P. Asymmetric reactions. XLVII. NMR-shift differences of enantiotopic proton signals in some alkylarylmethanols induced by tris[3-trifluoroace-tyl-(+)-camphorato]europium(III). *Collection Czechoslov. Chem. Commun.* **1973**, *38*, 3299–3301.

1341. Sullivan GR, Ciavarella D, Mosher HS. Correlation of configuration of chiral secondary carbinols by use of a chiral lanthanide nuclear magnetic resonance shift reagent. *J. Org. Chem.* **1974**, *39*, 2411–2412.

1342. Yoshida N, Morita H, Oyama K, Lee H-H. Enzymatic syntheses of chiral sec-alcohols and their NMR chemical shifts induced by a chiral lanthanide shift reagent. *Chem. Express* **1989**, *4*, 721–724.

1343. Ghosh I, Zeng H, Kishi Y. Application of chiral lanthanide shift reagents for assignment of absolute configuration of alcohols. *Org. Lett.* **2004**, *6*, 4715–4718.

1344. Ghosh I, Kishi Y, Tomoda H, Omura S. Use of a chiral praseodymium shift reagent in predicting the complete stereostructure of glisoprenin A. *Org. Lett.* **2004**, *6*, 4719–4722.

1345. Adams CM, Ghosh I, Kishi Y. Validation of lanthanide chiral shift reagents for determination of absolute configuration: total synthesis of glisoprenin A. *Org. Lett.* **2004**, *6*, 4723–4726.

1346. Capillon J, Lacombe L. Effets des substituants dans les specters de rmn de benzhydrols substitués en présence de Eu (dcm)$_3$. *Can. J. Chem.* **1979**, *57*, 1446–1450.

1347. Mioskowski C, Solladie G. Enantiomeric purity and absolute configuration of β-hydroxy-β-(trifluoromethyl)-β-phenylpropionic acid obtained by asymmetric synthesis. *Tetrahedron* **1973**, *29*, 3669–3674.

1348. Azzolina O, Ghislandi V. Antiphlogistic aryloxypropionic acids: configurational study. *Farmaco* **1993**, *48*, 713–724.

1349. Ajisaka K, Kamisaku M, Kainosho M. Enantiomeric shift difference induced by the lanthanide shift reagent: its correlation with absolute configuration of α-amino acids. *Chem. Lett.* **1972**, 857–858.

1350. Cativiala C, Diaz de Villegas MD, Garcia JI, Mayoral JA, Meléndez E. Determination of the enantiomeric purity by ^{1}H NMR with Eu(tfc)$_3$ of β-hetarylalanine derivatives. Correlation of the enantiomeric shift difference with absolute configuration. *Bull. Soc. Chim. Belg.* **1984**, *93*, 479–482.

1351. Bus J, Lok CM, Groenewegen A. Determination of enantiomeric purity of glycerides with a chiral PMR shift reagent. *Chem. Phys. Lipids* **1976**, *16*, 123–132.

1352. Hudhomme P, Duguay G. Control of the enantiomeric purity and correlation with the absolute configuration of *N*-protected 2-cyanoglycinates. *Tetrahedron: Asymmetry* **1993**, *4*, 1891–1900.

1353. Jankowski K, Rabczenko A. Application of an optically active nuclear magnetic resonance shift reagent to configurational problems. *J. Org. Chem.* **1975**, 960–961.

1354. White JD, Somers TC, Reddy GN. Degradation and absolute configurational assignment to C_{34}-botryococcene. *J. Org. Chem.* **1992**, *57*, 4991–4998.

1355. White JD, Somers TC, Reddy GN. Absolute configuration of (−)-botryococcene. *J. Am. Chem. Soc.* **1986**, *108*, 5352–5353.

1356. Nishida R, Kuwahara Y, Fukami H, Ishii S. Female sex pheromone of the German cockroach, *Blattella germanica* (L.) (Orthoptera: Blattellidae), responsible for male wing-raising. IV. The absolute configuration of the pheromone, 3,11-dimethyl-2-nonacosanone. *J. Chem. Ecol.* **1979**, *5*, 289–297.

1357. Yeh HJC, Balani SK, Yagi H, Greene RME, Sharma ND, Boyd DR, Jerina DM. Use of chiral lanthanide shift reagents in the determination of enantiomer composition and absolute configuration of epoxides and arene oxides. *J. Org. Chem.* **1986**, *51*, 5439–5443.

1358. Langin-Lantéri M, Fonbonne C, Huet J, Petit-Ramel M. Chiral phenyl-substituted chalcone epoxides. I. Correlative determination of absolute configuration by ^{1}H NMR lanthanide-induced shifts. *Magn. Reson. Chem.* **1987**, *25*, 216–218.

1359. Alvarez C, Barkaoui L, Goasdoue N, Daran JC, Platzer N, Rudler H, Vaissermann J. A new method for the determination of the enantiomeric purity of carboxylic acids: reaction of carboxylates with tris(tetraphenylimidodiphosphinato)praseodymium and X-ray structure of a dinuclear dicarboxylato adduct. *J. Chem. Soc., Chem. Commun.* **1989**, 1507–1509.

1360. Platzer N, Rudler H, Alvarez C, Barkaoui L, Denise B, Goasdoué N, Rager M, Vaissermann J, Daran J. Praseodymium NMR-shift reagents for carboxylic acids and their carboxylates: synthesis, X-ray structure and applications. *Bull. Soc. Chim. Fr.* **1995**, *132*, 95–113.

1361. Offermann W, Mannschreck A. Chiral recognition of alkene and arene hydrocarbons by ^{1}H and ^{13}C NMR. Determination of enantiomeric purity. *Tetrahedron Lett.* **1981**, *22*, 3227–3230.

1362. Wenzel TJ, Sievers RE. Nuclear magnetic resonance studies of terpenes with chiral and achiral lanthanide(III)-silver(I) binuclear shift reagents. *J. Am. Chem. Soc.* **1982**, *104*, 382–388.

1363. Wenzel TJ, Lalonde DR. New binuclear NMR shift reagents for olefins and aromatics. *J. Org. Chem.* **1983**, *48*, 1951–1954.

1364. Offermann W, Mannschreck A. Application of NMR spectroscopy of chiral association complexes. 14. Chiral recognition of terpene and cyclohexene hydrocarbons by ^{1}H and ^{13}C NMR. *Org. Magn. Reson.* **1984**, *22*, 355–363.

1365. Krasutskii PA, Rodionov VN, Tikhonov VP, Yurchenko AG. Silver d-camphor-10-sulfonate as a new chiral reagent for enantiomeric resolution of olefins. *Teor. Eksp. Khim.* **1984**, *20*, 58–63.

1366. Wenzel TJ, Ruggles AC, Lalonde DR. Binuclear lanthanide(III)—silver(I) NMR shift reagents: investigations of new achiral and chiral analogs. *Magn. Reson. Chem.* **1985**, *23*, 778–783.

1367. Baldovini N, Tomi F, Casanova J. Enantiomeric differentiation of terpenic olefins by carbon-13 NMR using chiral binuclear shift reagents. *Magn. Reson. Chem.* **2001**, *39*, 621–624.

1368. Peterson PE, Jensen BL. The preparation of 1,3-bis(trimethylsilyl)allene and the observation of the enantiomers using olefin complexing chiral shift reagents. *Tetrahedron Lett.* **1984**, *25*, 5711–5714.

1369. Mannschreck A, Munninger W, Burgemeister T, Gore J, Cazes B. Chiral recognition of allenic hydrocarbons by ^{1}H NMR. *Tetrahedron* **1986**, *42*, 399–408.

1370. Pasto DJ, Sugi KD. On the maximum rotation and the solvobromination and – mercuration of enantioenriched 1,3-dimethylallene. *J. Org. Chem.* **1991**, *56*, 4157–4160.

1371. Wells MB, McConathy JE, White PS, Templeton JL. Regioselective and stereoselective reactions of 2-butyne bound to a resolved chiral tungsten(II) center. *Organometallics* **2002**, *21*, 5007–5020.

1372. Chen Z, Eriks K, Halterman RL. Asymmetric synthesis and metalation of C_2-symmetric annulated bicyclooctylcyclopentadienes. *Organometallics* **1991**, *10*, 3449–3458.

1373. Pasto DJ, Sugi KD. Stereochemical features of the (2 + 2) cycloaddition reactions of chiral allenes. 3. The cycloaddition of enantioenriched 1,3-dimethylallene with the symmetrically 1,1-disubstituted alkenes 1,1-dichloro-2,2-difluoroethene and 1,1-diphenylethene. *J. Org. Chem.* **1992**, *57*, 12–17.

1374. Johns L, Belt S, Lewis CA, Rowland S, Massé G, Robert J, König WA. Configurations of polyunsaturated sesterterpenoids from the diatom, *Haslea ostrearia*. *Phytochemistry* **2000**, *53*, 607–611.

1375. Nijhuis WHN, Verboom W, El-Fadl AA, van Hummel GJ, Reinhoudt DN. Stereochemical aspects of the "tert-amino effect". 2. Enantio-and diastereoselectivity in the synthesis of quinolines, pyrrolo[1,2-*a*]quinolines, and [1,4]oxazino [4,3-a]quinolines. *J. Org. Chem.* **1989**, *54*, 209–216.

1376. Wang Q, Fan SY, Wong HNC, Li Z, Fung BM, Twieg RJ, Nguyen HT. Enantioselective synthesis of chiral liquid crystalline compounds from monoterpenes. *Tetrahedron* **1993**, *49*, 619–638.

1377. Carreira EM, Hastings CA, Shepard MS, Yerkey LA, Millward DB. Asymmetric induction in intramolecular [2 + 2]- photocycloadditions of 1,3-disubstituted allenes with enones and enoates. *J. Am. Chem. Soc.* **1994**, *116*, 6622–6630.

1378. Thompson A, Dolphin D. Nuclear magnetic resonance studies of helical dipyrromethene-zinc complexes. *Org. Lett.* **2000**, *2*, 1315–1318.

1379. Wenzel TJ, Zaia J. Lanthanide tetrakis (β-diketonates) as effective NMR shift reagents for organic salts. *J. Org. Chem.* **1985**, *50*, 1322–1324.

1380. Wenzel TJ, Zaia J. Organic-soluble lanthanide nuclear magnetic resonance shift reagents for sulfonium and isothiouronium salts. *Anal. Chem.* **1987**, *59*, 562–567.

1381. Green TK, Whetstine JR, Son E. Enantiomeric purity of alkylmethylphenylsulfonium ions with chiral NMR shift reagents: racemization by pyramidal inversion as observed by ^{1}H NMR spectroscopy. *Tetrahedron: Asymmetry* **1997**, *8*, 3175–3181.

1382. Reuben J. Chiral interactions in aqueous solution mediated by lanthanoid ions. NMR spectral resolution of enantiomeric nuclei. *J. Chem. Soc. Chem. Commun.* **1979**, 68–69.

1383. Reuben J. Aqueous lanthanide shift reagents. 8. Chiral interactions and stereochemical assignments of chemically and isotopically chiral ligands. *J. Am. Chem. Soc.* **1980**, *102*, 2232–2237.

1384. Pfenninger VA, Hochrainer A. NMR-spektroskopische methoden als reinheitskriterien isomerer weinsäuren und weinsäureester. *Helv. Chim. Acta* **1981**, *64*, 1558–1562.

1385. Bounoure J, Souppe J. Direct ^{1}H NMR determination of the enantiomeric excess of carnitine. *Analyst* **1988**, *113*, 1143–1144.

1386. Kabuto K, Sasaki Y. The europium(III)-(*R*)-propylenediaminetetra-acetate ion: a promising chiral shift reagent for ^{1}H NMR spectroscopy in aqueous solution. *J. Chem. Soc., Chem. Commun.* **1984**, 316–318.

1387. Kabuto K, Sasaki Y. Highly consistent correlation between absolute configuration of α-amino acids and their shift induced by the NMR chiral shift reagent propylenediaminetetraacetatoeuropium(III) in aqueous solution. *J. Chem. Soc., Chem. Commun.* **1987**, 670–671.

1388. Kabuto K, Sasaki Y. Consistent correlation between absolute configuration of α-hydroxy carboxylic acids and lanthanoid induced shift by an NMR chiral shift reagent in aqueous solution. *Chem. Lett.* **1989**, 385–388.

1389. Kabuto K, Sasaki Y. A facile NMR method for assigning absolute configuration of underivatized α-methyl-α-amino acids using a chiral lanthanoid shift reagent for aqueous solution. *Tetrahedron Lett.* **1990**, *31*, 1031–1034.

1390. Kabuto K, Sasaki K, Sasaki Y. Absolute configuration of aldonic acids and lanthanoid induced shift by the chiral shift reagent propylenediaminetetraacetatoeuropium(III) in aqueous solution. *Tetrahedron: Asymmetry* **1992**, *3*, 1357–1360.

1391. Sasaki M, Omata K, Kabuto K, Sasaki Y. Enantiomer signal separation of β-amino acids induced by Eu(III)-pdta; relation between the absolute configurations and relative shifts of enantiomer signals. *Kidorui* **2003**, *42*, 198–199.

1392. Bal D, Gradowska W, Gryff-Keller A. Determination of the absolute configuration of 2-hydroxyglutaric acid and 5-oxoproline in urine samples by high-resolution NMR spectroscopy in the presence of chiral lanthanide complexes. *J. Pharm. Biomed. Anal.* **2002**, *28*, 1061–1071.

1393. Ogasawara K, Omata K, Kabuto K, Jin H, Sasaki Y. Optically active Ce(III) - propylenediaminetetracetate complex: a chiral shift reagent for aqueous solution causing less signal broadening. *Kidorui* **1999**, *34*, 152–153.

1394. Inamoto A, Ogasawara K, Omata K, Kabuto K, Sasaki Y. Samarium(III)–propylenediaminetetraacetate complex: a water-soluble chiral shift reagent for use in high-field NMR. *Org. Lett.* **2000**, *2*, 3543–3545.

1395. Kabuto K, Hazama R, Umakoshi K, Sasaki Y. A new lanthanide complex, [EuIII((R)-tppn)Cl2]ClO4, as a chiral NMR shift reagent for aqueous solution. *Kidorui* **1995**, *26*, 244–245.

1396. Hazama R, Umakoshi K, Kabuto C, Kabuto K, Sasaki Y. A europium(III)-N,N,N',N'-tetrakis(2-pyridylmethyl)-(R)-propylenediamine complex as a new chiral lanthanide NMR shift reagent for aqueous neutral solution. *Chem. Commun.* **1996**, 15–16.

1397. Sato J, Omata K, Kabuto K, Jin H, Umakoshi K, Sasaki Y. Properties of chiral Ce(III)-tppn complex as a chiral shift reagent in aqueous solution. *Kidorui*, **1998**, *32*, 58–59.

1398. Sato J, Jin H, Omata K, Kabuto K, Sasaki Y. Resolution of enantiomer signals by diamagnetic lanthanum(III)-N,N,N',N'-tetrakis(2-pyridinylmethyl)-(R)-propylenediamine complex in ^{1}H NMR. *Enantiomer*, **1999**, *4*, 147–150.

1399. Peters JA, Vijverberg CAM, Kieboom APG, van Bekkum H. Lanthanide(III) salts of (S)-carboxymethyloxysuccinic acid: chiral lanthanide shift reagents for aqueous solution. *Tetrahedron Lett.* **1983**, *24*, 3141–3144.

1400. Kido J, Okamoto Y, Brittain HG. A new chiral shift reagent for aqueous solutions: Eu{(S,S)-ethylenediamine-N,N'-disuccinate}, Eu(EDDS). *J. Coord. Chem.* **1990**, *21*, 107–110.

1401. Kido J, Okamoto Y, Brittain HG. Europium (S,S)-ethylenediamine-N,N'-disuccinate as a chiral lanthanide shift reagent for aqueous solutions. *J. Org. Chem.* **1991**, *56*, 1412–1415.

1402. Makriyannis A, El Kheir AA, Ayad MM, Metwally MF, Hassin WES. Proton nuclear magnetic resonance spectral resolution of enantiomeric amino acid tyrosine. *Chin. Pharm. J.* **2001**, *53*, 1–5.

1403. Hulst R, de Vries NK, Feringa BL. Asymmetric synthesis of new chiral europium N,N'-disuccinate complexes: shift reagents for aqueous solutions and application in the enantiomeric excess determination of amino acids. *J. Org. Chem.* **1994**, *59*, 7453–7458.

1404. Asano H, Omata K, Kabuto K, Sasaki Y. Eu(III) complexes of optically active EDTA analogs as NMR chiral shift reagents. *Kidorui* **1996**, *28*, 352–353.

1405. Asano H, Omata K, Kabuto K, Sasaki Y. Properties of some optically active Eu(III)-EDTA analogs as NMR chiral shift reagents in methanol. *Kidorui* **1997**, *30*, 352–353.

1406. Takemura M, Yamato K, Doe M, Watanabe M, Miyake H, Kikunaga T, Yanagihara N, Kojima Y. Europium(III)-N,N'-ethylenebis(L-amino acid) complexes as new chiral NMR lanthanide shift reagents for unprotected α-amino acids in neutral aqueous solution. *Bull. Chem. Soc. Jpn.* **2001**, *74*, 707–715.

1407. Kojima Y, Watanabe M, Seki Y, Yamato K, Miyake H. Syntheses of novel structurally constrained (S)- histidyl-(S)-histidine derivatives and their copper (II) complexes. *Chem. Lett.* **1995**, 797–798.

1408. Watanabe M, Hasegawa T, Miyake H, Kojima Y. New chiral NMR shift reagent for α-amino acids and *N*-protected oligopeptides in aqueous solution. *Chem. Lett.* **2001**, 4–5.

1409. Dickins RS, Love CS, Puschmann H. Bidentate lactate binding in aqueous solution in a cationic, heptadentate lanthanide complex: an effective chiral derivatising agent. *Chem. Commun.* **2001**, 2308–2309.

1410. Corsi DM, van Bekkum H, Peters JA. Ion-pair interactions of lanthanide(III) complexes in aqueous solutions. *Inorg. Chem.* **2000**, *39*, 4802–4808.

1411. Hockless DCR, Gugger PA, Leung P, Mayadunne RC, Pabel M, Wild SB. Facile interconversions between diastereomers of chloro-bridged palladium(II) dimers of orthometallated (±)-dimethyl[1-(1-naphthyl)ethyl]amine. *Tetrahedron* **1997**, *53*, 4083–4094.

1412. Otsuka S, Nakamura A, Kano T, Tani K. Partial resolution of racemic tertiary phosphines with an asymmetric palladium complex. *J. Am. Chem. Soc.* **1971**, *93*, 4301–4303.

1413. Wild SB. Resolutions of tertiary phosphines and arsines with orthometallated palladium(II)-amine complexes. *Coord. Chem. Rev.* **1997**, *166*, 291–311.

1414. Berens U, Brown JM, Long J, Selke R. Synthesis and resolution of 2,2′-bis-diphenylphosphino [3,3′]biindolyl; a new atropisomeric ligand for transition metal catalysis. *Tetrahedron: Asymmetry* **1996**, *7*, 285–292.

1415. Kyba EP, Rines SP. A spectroscopic method for the determination of optical purities of chiral, chelating diphosphines. *J. Org. Chem.* **1982**, *47*, 4800–4801.

1416. Roberts NK, Wild SB. Resolutions using metal complexes. Synthesis, separation into diastereoisomers, and resolution of o-phenylenebis(methylphenylphosphine) using palladium complexes containing optically active ortho-metalated dimethyl(α-methylbenzyl)amines. *J. Am. Chem. Soc.* **1979**, *101*, 6254–6260.

1417. Kashiwabara K, Kinoshita I, Fujita J. Optical resolution of 2-aminoethyl-*n*-butylphenylphosphine and the bis(acetylacetonato) cobalt(III) complex containing the resolved aminophosphine. *Chem. Lett.* **1978**, 673–676.

1418. Kyba EP, Rines SP. A spectroscopic method for the determination of optical purities of chiral, chelating diphosphines. *J. Org. Chem.* **1982**, *47*, 4800–4801.

1419. Allen DG, McLaughlin GM, Robertson GB, Steffen WL, Salem G, Wild SB. Resolutions involving metal complexation. Preparation and resolution of (*R,S*)-methylphenyl(8-quinolyl)phosphine and its arsenic analogue. Crystal and molecular structure of (+)$_{589}$-[(*R*)-dimethyl(1-ethyl-α-naphthyl)aminato-C^2,N]-[(*S*)-methylphenyl(8-quinolyl)phosphine]palladium(II) hexafluorophosphate. *Inorg. Chem.* **1982**, *21*, 1007–1014.

1420. Martin JWL, Palmer JAL, Wild SB. Resolutions involving metal complexation. Resolution of (±)-1-amino-2-(methylphenylarsino)ethane and its phosphorus analogue. Stereochemistry and stability of square-planar bis(bidentate) complexes of bivalent palladium and platinum. *Inorg. Chem.* **1984**, *23*, 2664–2668.

1421. Wiesauer C, Kratky C, Weissensteiner W. Synthesis, determination of enantiomeric purity and structural characterisation of enantiopure (2R,5R)-(+)-2,5-bis-(diphenylphosphino)-hexane, a chiral DPPB analogue. *Tetrahedron: Asymmetry* **1996**, *7*, 397–398.

1422. Jiang Q, Rüegger H, Venanzi LM. Determination of the absolute configurations of the epimers of the P-chiral phosphine Ph$_2$PCH$_2$CH$_2$P*Ph(L-(−)-menthyl) by use of two-dimensional NMR spectroscopy in combination with a palladium(II) "reporter complex". *J. Organomet. Chem.* **1995**, *488*, 233–240.

1423. Schmid R, Foricher J, Cereghetti M, Schönholzer P. 35. Axially dissymmetric diphosphines in the biphenyl series: synthesis of (6,6'-dimethoxybiphenyl-2,2'-diyl)bis(diphenylphosphine) ('MeO-BIPHEP') and analogues via an ortho-lithiation/iodination Ullmann-reaction approach. *Helv. Chim. Acta* **1991**, *74*, 370–389.

1424. McFarlane W, Swarbrick JD, Bookham JL. An NMR study of the solution conformations of di(tertiary phosphine) complexes of orthopalladated 1-phenyl-1-(*N,N*-dimethylamino)ethane. *J. Chem. Soc., Dalton Trans.* **1998**, 3233–3238.

1425. Bookham JL, McFarlane W. Chiral ligands: unambiguous assignment of absolute configuration by NMR spectroscopy. *J. Chem. Soc., Chem. Commun.* **1993**, 1352–1354.

1426. Aw B, Hor TSA, Selvaratnam S, Mok KF, White AJP, Williams DJ, Rees NH, McFarlane W, Leung P. Asymmetric synthesis of P-chiral diphosphines. Steric effects on the palladium-complex-promoted asymmetric Diels-Alder reaction between a dimethylphenylphosphole and (E/Z)-methyl-substituted diphenylvinylphosphines. *Inorg. Chem.* **1997**, *36*, 2138–2146.

1427. Albert J, Bosque R, Cadena JM, Granell JR, Muller G, Ordinas JI. Synthesis and resolution of benzylisopropylphenylphosphine, a monodentate P-chiral ligand, *Tetrahedron: Asymmetry* **2000**, *11*, 3335–3343.

1428. Albert J, Bosque R, Cadena JM, Delgado S, Granell J, Muller G, Ordinas JI, Bardia MF, Solans X. Synthesis, resolution, and reactivity of benzylmesitylphenylphosphine: a new P-chiral bulky ligand. *Chem. Eur. J.* **2002**, *8*, 2249–2287.

1429. Durán E, Gordo E, Granell J, Font-Bardía M, Solans X, Velasco D, López-Calahorra F. Enantiomeric resolution and determination of the absolute configuration of dibenzophosphole 5-oxides. *Tetrahedron: Asymmetry* **2001**, *12*, 1987–1997.

1430. Albert J, Granell J, Muller G, Sainz D, Font-Bardia M, Solans X. Chiral cyclopalladated compounds for enantiomeric purities of functionalized phosphines by means of multinuclear NMR. *Tetrahedron: Asymmetry* **1995**, *6*, 325–328.

1431. Dunina VV, Golovan EB, Gulyukina NS, Buyevich AV. Enantiomeric discrimination in the complexation of ortho-palladated α-arylalkylamines with the racemic tert-butylmethylphenylphosphine. *Tetrahedron: Asymmetry* **1995**, *6*, 2731–2746.

1432. Dunina VV, Kuz'mina LG, Kazakova MY, Grishin YK, Veits YA, Kazakova EI. Ortho-palladated α-phenylalkylamines for enantiomeric purity determination of monodentate P*-chiral phosphines. *Tetrahedron: Asymmetry* **1997**, *8*, 2537–2545.

1433. Dunina VV, Kuz'mina LG, Rubina MY, Grishin YK, Veits YA, Kazakova EI. A resolution of the monodentate P*-chiral phosphine PBu^{t}C^{6}H$_4$Br-4 and its NMR-deduced absolute configuration. *Tetrahedron: Asymmetry* **1999**, *10*, 1483–1497.

1434. Bohm A, Hauck T, Beck W. Metal complexes of biologically important ligands. 124. NMR-spectroscopic determination of the enantiomeric ratio of α-amino acids by use of a chiral orthometallated palladium complex. *Z. Naturforsch., B: Chem. Sci.* **1999**, *54*, 1360–1362.

1435. Böhm A, Seebach D. Determination of enantiomer purity of β- and γ-amino acids by NMR analysis of diastereoisomeric palladium complexes. *Helv. Chim. Acta*, **2000**, *83*, 3262–3278.

1436. Levrat F, Stoeckli-Evans H, Engel N. Enantiomeric excess determination of α-amino acids by ^{19}F NMR spectroscopy of their *N,N*-dimethyl-(2,2,2-trifluoro-1-phenyleth-yl)amine-C,N)palladium complexes. *Tetrahedron: Asymmetry* **2002**, *13*, 2335–2344.

1437. Cui XY, Wu YJ, Yang LR. Enantiomeric excess determination of amino acids with chiral cyclopalladated ferrocenylimine. *Chin. Chem. Lett.* **1999**, *10*, 127–128.

1438. Dunina VV, Gorunova ON, Livantsov MV, Grishin YK. P*-Chiral phosphapalladacycle as derivatizing agent for enantiomeric purity determination of α-amino acids by means of ^{31}P NMR spectroscopy. *Tetrahedron: Asymmetry* **2000**, *11*, 2907–2961.

1439. Leung P, Quek GH, Lang H, Liu AM, Mok KF, White AJP, Williams DJ, Rees NH, McFarlane W. Optical resolution and the stereoelectronic properties of chelating (±)-[(methylsulfinyl)methyl]diphenylphosphine. *J. Chem. Soc., Dalton Trans.* **1998**, 1639–1643.

1440. Lim CC, Leung P, Sim KY. Resolution and enantiomeric purities of [2-(methylsulfi-ny)ethyl]amine. *Tetrahedron: Asymmetry* **1994**, *5*, 1883–1886.

1441. Chooi SYM, Leung P, Lim CC, Mok KF, Quek GH, Sim KY, Tan MK. Versatile chiral palladium(II) complexes for enantiomeric purities of 1,2-diamines. *Tetrahedron: Asymmetry* **1992**, *3*, 529–532.

1442. Parker D, Taylor RJ. Simple organometallic chiral derivatising agents for the ^{31}P N.M.Rs. assay of the enantiomeric purity of certain η^2-donors. *J. Chem. Soc., Chem. Commun*, **1987**, 1781–1783.

1443. Parker D, Taylor RJ. Scope and limitations of organometallic chiral derivatising agents for the ^{31}P NMR determination of the enantiomeric purity of chiral η^2 donors. *Tetrahedron* **1988**, *44*, 2241–2248.

1444. Staubach B, Buddrus J. Determination of the enantiomeric ratio of unprotected amino acids by NMR spectroscopy with C$_2$-chiral palladium compounds. *Angew. Chem., Int. Ed. Engl.* **1996**, *35*, 1344–1346.

1445. Bravo J, Cativiela C, Chaves JE, Navarro R, Urriolabeitia EP. ^{31}P NMR spectroscopy as a powerful tool for the determination of enantiomeric excess and absolute configurations of α-amino acids. *Inorg. Chem.* **2003**, *42*, 1006–1013.

1446. Glowacki Z, Topolski M, Matczak-Jon E, Hoffmann M. ^{31}P NMR enantiomeric purity determination of free 1-aminoalkylphosphonic acids via their diastereoisomeric Pd(II) complexes. *Magn. Reson. Chem.* **1989**, *27*, 922–924.

1447. Vaubaillon D, Giraud V, Rabiller C, Gruss U, Haas A, Hägele G. Resolution of fluorinated aminomethanphosphonic acids catalysed by penicillin G acylase III. *Phosphorus, Sulfur Silicon Relat. Elem.* **1997**, *126*, 177–183.

1448. Brugat N, Duran J, Polo A, Real J, Álvarez-Larena Á, Piniella JF. Synthesis and characterization of a new chiral phosphinothiol ligand and its palladium(II) complexes. *Tetrahedron: Asymmetry* **2002**, *13*, 569–577.

1449. Gogoll A, Johansson C, Axén A, Grennberg H. Determination of absolute configuration of (π-allyl)palladium complexes by NMR spectroscopy and stereoselective complexation. *Chem. Eur. J*, **2001**, *7*, 396–403.

1450. Fulwood R, Parker D, Ferguson G, Kaltner B. Crystal structure and further applications of the chiral derivatising agent η^2-ethene platinum-DIOP. *J. Organomet. Chem.* **1991**, *419*, 269–276.

1451. Payne NC, Stephan DW. Studies in enantiomeric discrimination I. Chiral phosphine complexes of platinum. *J. Organomet. Chem.* **1981**, *221*, 223–230.

1452. Wen J, Hong W, Yuan K, Mak TCW, Wong HNC. Synthesis, resolution, and application of 1,16-dihydroxytetraphenylene as a novel building block in molecular recognition and assembly. *J. Org. Chem.* **2003**, *68*, 8918–8931.

1453. Salvadori P, Uccello-Barretta G, Bertozzi S, Settambolo R, Lazzaroni R. Determination of the enantiomeric purity of chiral allyl alcohols and allyl ethers by ^{195}Pt NMR Spectroscopy. *J. Org. Chem.* **1988**, *53*, 5678–5770.

1454. Salvadori P, Uccello-Barretta G, Lazzaroni R, Caporusso AM. New method for the enantiomeric excess determination of chiral trisubstituted allenes by ^{195}Pt NMR of trans-dichloro[(S)-α-methylbenzylamine](allene)platinum(II) complexes. *J. Chem. Soc., Chem. Commun.* **1990**, 1121–1123.

1455. Uccello-Barretta G, Bernardini R, Balzano F, Lazzaroni R, Salvadori P. On the high efficiency of cis-dichloro[(S)-α-methylbenzylamine](ethylene)platinum(II) as chiral derivatizing agent for the determination of the enantiomeric composition of chiral unsaturated ethers by ^{195}Pt-NMR spectroscopy: a spectroscopic conformational and configurational characterization in solution of diastereoisomeric complexes cis-dichloro[(S)-α-methylbenzylamine][(S)- and (R)-3-phenyl-3-methoxybut-1-ene]-platinum(II). *J. Organomet. Chem.* **2000**, *605*, 68–73.

1456. Uccello-Barretta G, Bernardini R, Lazzaroni R, Salvadori P. Use of cis-dichloro[(S)-α-methylbenzylamine](ethylene)Pt(II) as a chiral derivatizing agent for the determination by ^{195}Pt-NMR of the enantiomeric composition of unsaturated ethers or alcohols having a quaternary chiral carbon atom. *J. Organomet. Chem.* **2000**, *598*, 174–178.

1457. Uccello-Barretta G, Bernardini R, Lazzaroni R, Salvador P. [PtCl$_3$(C$_2$H$_4$)]$^-$ [(S,S)-(PhMeCH)$_2$NH$_2$]$^+$: a new and versatile ionic organometallic chiral derivatizing agent for the determination of the enantiomeric composition of chiral unsaturated compounds by ^{195}Pt NMR spectroscopy. *Org. Lett.* **2000**, *2*, 1795–1798.

1458. Uccello-Barretta G, Bernardini R, Balzano F, Venturelli F, Caporusso AM, Salvadori P. trans-[PtCl$_2$(Am)(C$_2$H$_4$)] and [PtCl$_3$(C$_2$H$_4$)]$^-$[AmH]$^+$, containing binaphthyl secondary amines. Efficient chiral derivatizing agents for the enantiodiscrimination of chiral olefins by ^{195}Pt NMR spectroscopy. *Eur. J. Org. Chem.* **2001**, 3651–3655.

1459. Uccello-Barretta G, Bernardini R, Balzano F, Caporusso AM, Salvadori P. ^{195}Pt NMR determination of the enantiomeric purity and absolute configuration of trisubstituted allenes by using [PtCl$_3$(C$_2$H$_4$)]$^-$[(S,S)-(1-NpMeCH)$_2$NH$_2$]$^+$ as CDA. *Org. Lett.* **2001**, *3*, 205–207.

1460. Uccello-Barretta G, Bernardini R, Balzano F, Salvadori P. [PtCl$_3$(C$_2$H$_4$)]$^-$[AmH]$^+$ complexes containing chiral secondary amines: use as chiral derivatizing agents for the enantiodiscrimination of unsaturated compounds by ^{195}Pt NMR spectroscopy and NMR stereochemical investigation. *J. Org. Chem.* **2001**, *66*, 123–129.

1461. Uccello-Barretta G, Bernardini R, Balzano F, Salvadori P. Overall view of the use of chiral platinum(II) complexes as chiral derivatizing agents (CDAs) for the enantiodiscrimination of unsaturated compounds by ^{195}Pt NMR. *Chirality* **2002**, *14*, 484–489.

1462. Froebe LR, Brushmiller JG. A PMR method of assigning the absolute configurations of platinum(IV) propylenediamine coordination complexes. *Inorg. Chim. Acta* **1970**, *4*, 481–483.

1463. Duddeck H. Rh$_2$[MTPA]$_4$, a dirhodium complex as NMR auxiliary for chiral recognition. *The Chemical Record* **2005**, *5*, 1–14.

1464. Meyer C, Duddeck H. Chiral discrimination of methyl phenyl sulfoxide using dirhodium complexes. *Magn. Reson. Chem.* **2000**, *38*, 29–32.

1465. Wypchlo K, Duddeck H. Chiral recognition of olefins by ^{1}H NMR spectroscopy in the presence of a chiral dirhodium complex. *Tetrahedron: Asymmetry* **1994**, *5*, 27–30.

1466. Magiera D, Baumann W, Podkorytov IS, Omelanczuk J, Duddeck H. Stable Rh$_2$(MTPA)$_4$ phosphane adducts. The first example of P-chirality recognition by ^{103}Rh NMR spectroscopy. *Eur. J. Inorg. Chem.* **2002**, 3252–3257.

1467. Magiera D, Omelanczuk J, Dziuba K, Pietrusiewicz KM, Duddeck H. Phosphine-Rh$_2$[(R)-MTPA]$_4$ adducts in solution: characterization by NMR spectroscopy and chiral discrimination. *Organometallics* **2003**, *22*, 2464–2471.

1468. Hameed S, Ahmad R, Duddeck H. Chiral recognition of selenides and iodides by ^{1}H NMR spectroscopy in the presence of a chiral dirhodium complex. *Heteroat. Chem.* **1998**, *9*, 471–474.

1469. Malik S, Moeller S, Duddeck H, Choudhary MI. Phenylselenenylmethane derivatives and their enantiomeric discrimination by ^{1}H and ^{13}C NMR spectroscopy in the presence of a chiral dirhodium complex. *Magn. Reson. Chem.* **2002**, *40*, 659–665.

1470. Duddeck H, Malik S, Moeller S, Gáti T, Tóth G, Rozwadowski Z. Chirality recognition of selenium compounds by NMR spectroscopy in the presence of a chiral dirhodium complex. *Phosphorus, Sulfur Silicon Relat. Elem.* **2005**, *180*, 993–1000.

1471. Malik S, Duddeck H, Omelanczuk J, Choudhary MI. First direct discrimination of chiral phosphine selenide (P=Se) derivatives by multinuclear magnetic resonance spectroscopy in the presence of a chiral dirhodium complex. *Chirality* **2002**, *14*, 407–411.

1472. Rockitt S, Duddeck H, Omelanczuk J. First direct discrimination of chiral phosphorus thionate (P=S) derivatives by multinuclear magnetic resonance spectroscopy in the presence of a chiral dirhodium complex. *Chirality* **2001**, *13*, 214–223.

1473. Magiera D, Moeller S, Drzazga Z, Pakulski Z, Pietrusiewicz KM, Duddeck H. Chiral phospholene and phospholane chalcogenides: stereochemistry and chiral recognition by multinuclear NMR spectroscopy of their Rh$_2$[(R)-MTPA]$_4$ adducts. *Chirality* **2003**, *15*, 391–399.

1474. Moeller S, Drzazga Z, Pakulski Z, Pietrusiewicz KM, Duddeck H. The dirhodium-method in the determination of absolute configurations of phospholene chalcogenides. *Chirality* **2006**, *18*, 395–397.

1475. Gáti T, Tóth G, Drabowicz J, Moeller S, Hofer E, Polavarapu P, Duddeck H. Effective enantiodifferentiation of spirochalcogenuranes by the dirhodium method: towards the determination of absolute configurations? *Chirality* **2005**, *17*, S40–S47.

1476. Gáti T, Simon A, Tóth G, Szmigielska A, Maj AM, Pietrusiewicz KM, Moeller S, Magiera D, Duddeck H. Bis(phosphane oxide) adducts of $Rh_2(MTPA)_4$. Kinetics and chirality discrimination. *Eur. J. Inorg. Chem.* **2004**, 2160–2166.

1477. Magiera D, Szmigielska A, Pietrusiewicz KM, Duddeck H. Secondary phosphine oxides: tautomerism and chiral recognition monitored by multinuclear NMR spectroscopy of their $Rh_2[(R)-MTPA]_4$ adducts. *Chirality* **2004**, *16*, 57–64.

1478. Moeller S, Albert D, Duddeck H, Simon A, Tóth G, Demchuk OM, Pietrusiewicz KM. Chiral atropisomeric diiodobiphenyls. Enantiodifferentiation by the dirhodium method. *Tetrahedron: Asymmetry* **2004**, *15*, 3609–3616.

1479. Hameed S, Ahmad R, Duddeck H. Chiral recognition of nitriles by ^1H NMR spectroscopy in the presence of a chiral dirhodium complex. *Magn. Reson. Chem.* **1998**, *36*, S47–S53.

1480. Wypchlo K, Duddeck H. Chiral recognition of epoxides by ^1H NMR spectroscopy in the presence of a chiral dirhodium complex. *Chirality* **1997**, *9*, 601–603.

1481. Rockitt S, Duddeck H, Drabczynska A, Kiec-Kononowicz K. Chiral discrimination of some annelated xanthine derivatives by the dirhodium method. *Eur. J. Org. Chem.* **2000**, 3489–3496.

1482. Rockitt S, Wartchow R, Duddeck H, Drabczynska A, Kiec-Kononowicz K. Modes of xanthine complexation to dirhodium tetrakis[(R)-α-methoxy-α-(trifluoromethyl)-phenylacetate] in solution and in the solid state. *Z. Naturforsch., B: Chem. Sci.* **2001**, *56b*, 319–324.

1483. Jaźwiński J, Rozwadowski Z, Magiera D, Duddeck H. Multinuclear magnetic resonance study of chiral mesoionic oxa- and thiatriazole and tetrazole derivatives: adducts with dirhodium complexes and chiral recognition. *Magn. Reson. Chem.* **2003**, *41*, 315–323.

1484. Diaz Gomez E, Jios J, Della Vedova CO, March HD, Di Loreto HE, Toth G, Simon A, Albert D, Moeller S, Wartchow R, Duddeck H. N-(Chloroacetyl)- and N-(dichloroacetyl)-N-(xylyl)alanine esters: assignment of the absolute configurations and enantiodifferentiation by the dirhodium method. *Tetrahedron: Asymmetry* **2005**, *16*, 2285–2293.

1485. Gomez ED, Albert D, Duddeck H, Kozhushkov SI, de Meijere A. Enantiodifferentiation of aliphatic ethers by the dirhodium method. *Eur. J. Org. Chem.* **2006**, 2278–2280.

1486. Buriak JM, Osborn JA. A simple in situ ^{31}P NMR method for the determination of the enantiomeric purity of aromatic substrates. *J. Chem. Soc., Chem. Commun.*, **1995**, 689–690.

1487. Simonato J, Pécaut J, Marchon J. Iodorhodium(III) tetramethylchiroporphyrin: potential reagent for chiral selection and analysis of amino acids. *Inorg. Chim. Acta* **2001**, *315*, 240–244.

1488. Claeys-Bruno M, Pécaut J, Bardet M, Marchon J. An improved chiroporphyrin shift reagent for the enantiomeric assay of amino acid derivatives by ^1H NMR spectroscopy: evaluation of iodorhodium(III) tetramethylchiroporphyrin. *J. Porphyrins Phthalocyanines* **2002**, *6*, 396–402.

1489. Abraham RJ, Plant J, Bedford GR, Wright B. A novel use of cobalt(III) meso-tetraphenyl-porphyrin as a chiral shift reagent. *Org. Magn. Reson.* **1984**, *22*, 57–60.

1490. Gaudemer A, Gaudemer F, Merienne C. Structural studies of metalloporphyrins. VII. NMR evidence for planar chirality in natural porphyrins. *Org. Magn. Reson.* **1983**, *21*, 83–85.
1491. Schwenninger R, Ramondenc Y, Wurst K, Schlögl J, Kräutler B. A highly enantiopure biconcave porphyrin with effective D_4-structure. *Chem. Eur. J.* **2000**, *6*, 1214–1223.
1492. Toronto D, Sarrazin F, Pécaut J, Marchon J, Shang M, Scheidt WR. A cobalt(III) chiroporphyrin and its amine adducts. A potential chiral NMR shift reagent for amines. *Inorg. Chem.* **1998**, *37*, 526–532.
1493. Simonato J, Pécaut J, Marchon J. Kinetic and structural factors governing chiral recognition in cobalt(III) chiroporphyrin-amino alcohol complexes. *J. Am. Chem. Soc.* **1998**, *120*, 7363–7364.
1494. Claeys-Bruno M, Toronto D, Pécaut J, Bardet M, Marchon J. Three-point versus two-point attachment of (*R*)- and (*S*)-amino acid methyl esters to a cobalt(III) chiroporphyrin: implications for the analysis of amino acid enantiomers by ^{1}H NMR spectroscopy. *J. Am. Chem. Soc.* **2001**, *123*, 11067–11068,
1495. Simonato J, Chappellet S, Pécaut J, Baret P, Marchon J. Evaluation of cobalt(III) tetramethylchiroporphyrin as an analytical reagent for the determination of enantiomer composition of primary amines and aziridines by ^{1}H NMR spectroscopy. *New J. Chem.* **2001**, *25*, 714–720.
1496. Claeys-Bruno M, Bardet M, Marchon J. Détermination de la composition énantiomérique d'esters d'acides aminés par spectroscopie de RMN à l'aide d'une chiroporphyrine de cobalt(III) comme agent chiral de déplacement chimique. *C. R. Chim.* **2002**, *5*, 21–25.
1497. Claeys-Bruno M, Bardet M, Marchon J. One-dimensional TOCSY ^{1}H NMR experiments on adducts of amino acid esters with cobalt(III) tetramethylchiroporphyrin. Prospects for the analysis of amino acid mixtures. *Magn. Reson. Chem.* **2002**, *40*, 647–652.
1498. Brushmiller JG, Stadtherr LG. The use of proton magnetic resonance in assigning the absolute configuration of coordination complexes containing optically active propylenediamine ligands. *Inorg. Nucl. Chem. Lett.* **1967**, *3*, 525–529.
1499. Ama T, Kawaguchi H, Yasui T, Matsumoto K, Ooi S. The stereochemistry of the six geometrical isomers of the bis(ethylenediamine-N-acetato)cobalt(III) ion, $[Co(edma)_2]^{+1}$. *Bull. Chem. Soc. Jpn.* **1985**, *58*, 2561–2568.
1500. Ama T, Kaizaki S, Kawaguchi H, Yasui T. Determination of the absolute configuration of u-fac-$[Co(ida)(mdien)]^+$ by ^{1}H NMR spectroscopy. *Chem. Lett.* **1987**, 441–444,
1501. Yasui T, Shikiji T, Koine N, Ama T, Kawaguchi H. Preparation and stereochemistry of cobalt(III) complexes with ethylenediamine-N-acetate and 3-azapentane-1,5-diamine or 3-methyl-3-azapentane-1,5-diamine. *Bull. Chem. Soc. Jpn.* **1987**, *60*, 595–601.
1502. Ama T, Niiyama R, Kawaguchi H, Yasui T. Stereochemistry of the (iminodiaceto or *N*-methyliminodiacetato)-(ethylenediamine-*N*-acetato)cobalt(III) complexes. *Bull. Chem. Soc. Jpn.* **1987**, *60*, 119–123.
1503. Ama T, Kuwamura K, Kawaguchi H, Yasui T. Determination of the absolute configuration of the chiral (ethylenediamine-*N,N'*-diacetate)(glycinato)cobalt(III) complex by ^{1}H NMR spectroscopy. *Kochi Daigaku Rigakubu Kiyo, Kagaku* **1991**, *12*, 1–8.
1504. Gillard RD, Mitchell PR. Determination of the absolute configuration of bis(dipeptidato)-cobaltate(III) complexes by ^{1}H nuclear magnetic resonance spectroscopy. *J. Chem. Soc., Chem. Comm.* **1978**, 428–429.

1505. Strašák M, Novomeský P, Butvin P. Steric and hydrophobic effects on stereoselectivity in ternary cobalt(III) complexes of amino acid derivatives. *Collect. Czech. Chem. Commun.* **1990**, *55*, 1518–1528.

1506. Coderre JA, Gerlt JA. Oxygen chiral phosphodiesters. 2. Enzymatic synthesis and configurational analysis of [α-^{18}O]-2′-deoxyadenosine 5′-diphosphate. *J. Am. Chem. Soc.* **1980**, *102*, 6594–6597.

1507. Grant J, Reuben J. Nuclear magnetic resonance manifestation of the configurational differences between the L- and D-norepinephrine complexes with cobaltous adenosine 5′-triphosphate, an aqueous chiral shift reagent. *J. Am. Chem. Soc.* **1978**, *100*, 5209–5210.

1508. Fujii Y, Yoshikawa Y, Syoji M, Shinohara H. The stereochemistry and reactivity of metal-Schiff base complexes. VII. Contribution of hydrophobic interligand interaction to chiral recognition of phenylalaninate and tryptophanate with (1R,2R)-*N*,*N*′-disalicylidene-1,2-cyclohexanediaminecobalt(III) complex. *Bull. Chem. Soc. Jpn.* **1990**, *63*, 138–146.

1509. Tikhonov VP. Study of mixed ligand cobalt(II) complexes with α-amino acids. Resolution of enantiomers in the NMR spectra. *Koord. Khim.* **1991**, *7*, 1094–1100.

1510. Marty W, Pasquier ML, Gampp H. 164. Horeau's coupling of enantiomers revisited. The reversible coupling of enantiomers to form diastereoisomers in kinetically labile metal complexes. *Helv. Chim. Acta* **1987**, *70*, 1774–1785.

1511. Rehmann JP, Barton JK. ^{1}H NMR studies of tris(phenanthroline) metal complexes bound to oligonucleotides: characterization of binding modes. *Biochemistry* **1990**, *29*, 1701–1709.

1512. Xu Q, Shoemaker RK, Braunlin WH. Induction of B Λ transitions of deoxyoligonucleotides by multivalent cations in dilute aqueous solution. *Biophys. J.* **1993**, *65*, 1039–1049.

1513. Xu Q, Jampani SRB, Deng H, Braunlin WH. Chiral recognition of deoxyoligonucleotides by Δ-and Λ-tris(ethylenediamine)cobalt(III). *Biochemistry* **1995**, *34*, 14059–14065.

1514. Mizutani T, Ema T, Tomita T, Kuroda Y, Ogoshi H. Design and synthesis of a trifunctional chiral porphyrin with C_2 symmetry as a chiral recognition host for amino acid esters. *J. Am. Chem. Soc.* **1994**, *116*, 4240–4250.

1515. Ogoshi H, Ema T, Kato Y, Mizutani T, Kuroda Y. Molecular recognition: new chiral metalloporphyrins as receptor models. *Supramol. Chem.* **1995**, *6*, 115–124.

1516. Mizutani T, Ema T, Tomita T, Kuroda Y, Ogoshi H. Design and synthesis of a trifunctional chiral porphyrin with C_2 symmetry as a chiral recognition host for amino acid esters. *J. Am. Chem. Soc.* **1994**, *116*, 4240–4250.

1517. Crossley MJ, Mackay LG, Try AC. Enantioselective recognition of histidine and lysine esters by porphyrin chiral clefts and detection of amino acid conformations in the bound state. *J. Chem. Soc., Chem. Commun.* **1995**, 1925–1927.

1518. Borovkov VV, Lintuluoto JM, Inoue Y. Elucidation of the mechanism of supramolecular chirality inversion in bis(zinc porphyrin) by dynamic approach using CD and ^{1}H NMR spectroscopy. *J. Phys. Chem. A* **2000**, *104*, 9213–9219.

1519. Borovkov VV, Lintuluoto JM, Inoue Y. Supramolecular chirogenesis in zinc porphyrins: mechanism, role of guest structure, and application for the absolute configuration determination. *J. Am. Chem. Soc.* **2001**, *123*, 2979–2989.

1520. Borovkov VV, Lintuluoto JM, Sugeta H, Fujiki M, Arakawa R, Inoue Y. Supramolecular chirogenesis in zinc porphyrins: equilibria, binding properties, and thermodynamics. *J. Am. Chem. Soc.* **2002**, *124*, 2993–3006.

1521. Huang X, Fujioka N, Pescitelli G, Koehn FE, Williamson RT, Nakanishi K, Berova N. Absolute configurational assignments of secondary amines by CD-sensitive dimeric zinc porphyrin host. *J. Am. Chem. Soc.* **2002**, *124*, 10320–10335.

1522. Ema T, Ouchi N, Doi T, Korenaga T, Sakai T. Highly sensitive chiral shift reagent bearing two zinc porphyrins. *Org. Lett.* **2005**, *7*, 3985–3988.

1523. Yagi S, Sadachi H, Kashiwagi Y, Takagishi T, Mizutani T, Kitagawa S, Ogoshi H. Chiral recognition of α-amino esters on the chiral helical surface of zinc bilinone. *Chem. Lett.* **2000**, *2*, 1054–1055.

1524. Canary JW, Allen CS, Castagnetto JM, Wang Y. Conformationally driven, propeller-like chirality in labile coordination complexes. *J. Am. Chem. Soc.* **1995**, *117*, 8484–8485.

1525. Le Maux P, Bahri H, Simonneaux G. Molecular recognition of racemic phosphines by a chiral ruthenium porphyrin. *J. Chem. Soc., Chem. Commun.*, **1991**, 1350–1352.

1526. Morice C, Le Maux P, Simonneaux G, Toupet L. Chiral recognition of amino esters by ruthenium porphyrin complexes and crystal structure of {5,10,15,20-tetrakis[o-(3,3,3-trifluoro-2-methoxy-2-phenylpropanoylamino)phenyl]porphyrin}-bis(L-valine methyl ester)ruthenium(II) (α,α,β,β isomer). *J. Chem. Soc., Dalton Trans.*, **1998**, 4165–4171.

1527. Galardon E, Le Maux P, Bondon A, Simonneaux G. Chiral recognition of amino esters by a ruthenium porphyrin complex: kinetics of the exchange process determined by ^{1}H NMR. *Tetrahedron: Asymmetry* **1999**, *10*, 4203–4210.

1528. Simonneaux G, Le Maux P. Optically active ruthenium porphyrins: chiral recognition and asymmetric catalysis. *Coord. Chem. Rev.* **2002**, *228*, 43–60.

1529. Galardon E, Lukas M, Le Maux P, Simonneaux G. Synthesis and characterisation of a new chiral ruthenium picket-fence porphyrin and its use in chiral recognition of racemic isocyanides. *Tetrahedron Lett.* **1999**, *40*, 2753–2756.

1530. Mazzanti M, Veyrat M, Ramasseul R, Marchon J, Turowska-Tyrk I, Shang M, Scheidt WR. A new ruthenium(II) chiroporphyrin containing a multipoint recognition site: enantioselective receptor of chiral aliphatic alcohols. *Inorg. Chem.* **1996**, *35*, 3733–3734.

1531. Schwenninger R, Schlögl J, Maynollo J, Gruber K, Ochsenbein P, Bürgi H, Konrat R, Kräutler B. Metal complexes of a biconcave porphyrin with D_4-structure. Versatile chiral shift agents. *Chem. Eur. J.* **2001**, *7*, 2676–2686.

1532. Gut D, Rudi A, Kopilov J, Goldberg I, Kol M. Pairing of propellers: dimerization of octahedral ruthenium(II) and osmium(II) complexes of eilatin via π-π stacking featuring heterochiral recognition. *J. Am. Chem. Soc.* **2002**, *124*, 5449–5456.

1533. Pasquier ML, Marty W. A new method for determining enantiomeric excess, where the sample serves as its own reference. *Angew. Chem., Int. Ed. Engl.* **1985**, *24*, 315–316.

1534. Mynott R, Richter WJ, Wilke G. Determination of the enantiomeric excess of Horner phosphines by ^{13}C NMR spectroscopy. A ^{13}C and ^{31}P NMR study of the diastereomeric complexes formed with [η^3-(+)(1R,5R)-pinenyl]nickel bromide dimer. *Z. Naturforsch., B: Chem. Sci.* **1986**, *41b*, 85–88.

1535. Groves JT, Crowley SJ, Shalyaev KV. Paramagnetic ^{1}H-NMR relaxation probes of stereoselectivity in metalloporphyrin catalyzed olefin epoxidation. *Chirality* **1998**, *10*, 106–119.

1536. Seymour JL, Tureček F, Malkov AV, Kočovský P. Chiral recognition in solution and the gas phase. Experimental and theoretical studies of aromatic D- and L-amino acid-Cu (II)-chiragen complexes. *J. Mass Spectrom.* **2004**, *39*, 1044–1052.

1537. Cucciolito ME, Flores G, Vitagliano A. Chiral recognition in silver(I) olefin complexes with chiral diamines. Resolution of racemic alkenes and NMR discrimination of enantiomers. *Organometallics* **2004**, *23*, 15–17.

1538. Otera J, Yano T. ^{119}Sn NMR spectroscopic determination of diastereomeric ratio of some optically active organotin compounds. *Bull. Chem. Soc. Jpn.* **1985**, *58*, 387–388.

1539. Klein J, Neels S, Borsdorf R. Investigations on diastereoisomeric tetraorganotin compounds: the use of ^{119}Sn NMR spectroscopy for the direct determination of the diastereoisomeric composition. *J. Chem. Soc., Perkin Trans. 2* **1994**, 2523–2524.

1540. Luchinat C, Roelens S. Enantiomeric purity determination of 1,2-diols through NMR spectroscopy without chiral auxiliaries. *J. Am. Chem. Soc.* **1986**, *108*, 4873–4878.

1541. Potvin PG. A microscale NMR method of determining absolute stereochemistries in β-amino alcohols by enantioselective complexation and the mode of action of their oxidative resolution. *J. Org. Chem.* **1992**, *57*, 3272–3274.

1542. Sadakane M, Dickman MH, Pope MT. Chiral polyoxotungstates. 1. Stereoselective interaction of amino acids with enantiomers of $[Ce^{III}(\alpha_1\text{-}P_2W_{17}O_{61})(H_2O)_x]^{7-}$. The structure of DL-$[Ce_2(H_2O)_8(P_2W_{17}O_{61})_2]^{14-}$. *Inorg. Chem.* **2001**, *40*, 2715–2719.

1543. Bareyt S, Piligkos S, Hasenknopf B, Gouzerh P, Lacôte E, Thorimbert S, Malacria M. Efficient preparation of functionalized hybrid organic/inorganic wells–Dawson-type polyoxotungstates. *J. Am. Chem. Soc.* **2005**, *127*, 6788–6794.

1544. Minelli M, Rockway TW, Enemark JH, Brunner H, Muschiol M. Direct observation of diastereomers with opposite Mo configurations by ^{95}Mo NMR. *J. Organomet. Chem.* **1981**, *217*, C34–C36.

1545. Eichbichler J, Peringer P. Observation of diastereomers by ^{199}Hg NMR. *J. Organomet. Chem.* **1982**, *231*, 95–96.

1546. Resch JF, Meinwald J. Use of osmate(VI) ester trans-N,N,N′,N′-tetramethyl-1,2-cyclohexanediamine complexes for determination of glycol stereochemistry. *Tetrahedron Lett.* **1981**, *22*, 3159–3162.

1547. Cort AD, Murua JIM, Pasquini C, Pons M, Schiaffino L. Evaluation of chiral recognition ability of a novel uranyl-salophen-based receptor: an easy and rapid testing protocol. *Chem. Eur. J.* **2004**, *10*, 3301–3307.

1548. Meddour A, Canet I, Loewenstein A, Péchiné JM, Courtieu J. Observation of enantiomers, chiral by virtue of isotopic substitution, through deuterium NMR in a polypeptide liquid crystal. *J. Am. Chem. Soc.* **1994**, *116*, 9652–9656.

1549. Lesot P, Gounelle Y, Merlet D, Loewenstein A, Courtieu J. Measurement and analysis of the molecular ordering tensors of two enantiomers oriented in a polypeptide liquid crystalline system. *J. Phys. Chem.* **1995**, *99*, 14871–14875.

1550. Lesot P, Merlet D, Courtier J, Emsley JW, Rantala TT, Jokisaari J. Calculation of the molecular ordering parameters of (±)-3-butyn-2-ol dissolved in an organic solution of poly(γ-benzyl-L-glutamate). *J. Phys. Chem. A* **1997**, *101*, 5719–5724.

1551. Merlet D, Loewenstein A, Smadja W, Courtieu J, Lesot P. Quantitative description of the facial discrimination of molecules containing a prochiral group by NMR in a chiral liquid crystal. *J. Am. Chem. Soc.* **1998**, *120*, 963–969.

1552. Emsley JW, Lesot P, Merlet D. The orientational order and conformational distributions of the two enantiomers in a racemic mixture of a chiral, flexible molecule dissolved in a chiral nematic liquid crystalline solvent. *Phys. Chem. Chem. Phys.* **2004**, *6*, 522–530.

1553. Sackmann E, Meiboom S, Snyder LC. The nuclear magnetic resonance spectra of enantiomers in optically active liquid crystals. *J. Am. Chem. Soc.* **1968**, *90*, 2183–2184.

1554. Tracey AS, Diehl P. The interaction of D- and L-alanine with an optically active model membrane system. *FEBS Lett.* **1975**, *59*, 131–132

1555. Czarniecka K, Samulski ET. Polypeptide liquid crystals: a deuterium NMR study. *Mol. Cryst. Liq. Cryst.* **1981**, *63*, 205–214.

1556. Aroulanda C, Sarfati M, Courtieu J, Lesot P. Investigation of the enantioselectivity of three polypeptide liquid-crystalline solvents using NMR spectroscopy. *Enantiomer* **2001**, *6*, 281–287.

1557. Canet I, Meddour A, Courtieu J, Canet JL, Salaün J. New, and accurate method to determine the enantiomeric purity of amino acids based on deuterium NMR in a cholesteric lyotropic liquid crystal. *J. Am. Chem. Soc.* **1994**, *116*, 2155–2156.

1558. Chevtchouk T, Ollivier J, Salaün J, Merlet D, Courtieu J. Determination of the region- and enantioselectivity of the enzymatic hydrolyses of succinates based on H,C COLOC analysis and deuterium NMR in a chiral liquid crystal. *Tetrahedron: Asymmetry* **1997**, *8*, 999–1003.

1559. Canet J, Canet I, Courtieu J, Da Silva S, Gelas J, Troin Y. Acetyl-d_3 chloride: a convenient nonchiral derivatizing agent (NCDA) for a facile enantiomeric excess determination of amines through deuterium NMR. *J. Org. Chem.* **1996**, *61*, 9035–9037.

1560. Chalard P, Bertrand M, Canet I, Théry V, Remuson R, Jeminet G. Determination of absolute configurations of amines and amino acids using nonchiral derivatizing agents (NCDA) and deuterium NMR. *Org. Lett.* **2000**, *2*, 2431–2432.

1561. Meddour A, Loewenstein A, Péchiné J, Courtieu J. An achiral deuterated derivatizing agent for enantiomeric analysis through NMR in a liquid crystalline solvent. *Tetrahedron: Asymmetry* **1997**, *8*, 485–494.

1562. Meddour A, Courtieu J. Achiral deuterated derivatizing agent for enantiomeric analysis of carboxylic acids by NMR in a chiral liquid crystalline solvent. *Tetrahedron: Asymmetry* **2000**, *11*, 3635–3644.

1563. Meddour A, Atkinson D, Loewenstein A, Courtieu J. Enantiomeric analysis of homologous series of secondary alcohols by deuterium NMR spectroscopy in a chiral nematic liquid crystal: influence of molecular geometry on chiral discrimination. *Chem. Eur. J.* **1988**, *4*, 1142–1147.

1564. Tavasli M, Courtieu J, Goss RJM, Meddour A, O'Hagan D. Extreme enantiomeric discrimination of fluoroalkanes using deuterium NMR in chiral liquid crystalline media. *Chem. Commun.* **2002**, 844–845.

1565. Meddour A, Canlet C, Blanco L, Courtieu J. Diastereomeric shape recognition using NMR spectroscopy in a chiral liquid crystalline solvent. *Angew. Chem., Int. Ed.* **1999**, *38*, 2391–2393.

1566. Canlet C, Merlet D, Lesot P, Meddour A, Lowenstein A, Courtieu J. Deuterium NMR stereochemical analysis of threo-erythro isomers bearing remote stereogenic centers in racemic and non-racemic liquid crystalline solvents. *Tetrahedron: Asymmetry* **2000**, *11*, 1911–1918.

1567. Emsley JW, Lesot P, Courtieu J, Merlet D. The effect of a chiral nematic solvent on the orientational order and conformational distribution of a flexible prochiral solute. *Phys. Chem. Chem. Phys.* **2004**, *6*, 5331–5337.

1568. Sarfati M, Aroulanda C, Courtieu J, Lesot P. Enantiomeric recognition of chiral invertomers through NMR in chiral oriented phases: a study of cis-decalin. *Tetrahedron: Asymmetry* **2001**, *12*, 737–744.

1569. Lesot P, Lafon O, Kagan HB, Fan C. Study of molecular rotational isomerism using deuterium NMR in chiral oriented solvents. *Chem. Commun.* **2006**, 389–391.

1570. Lesot P, Merlet D, Sarfati M, Courtieu J, Zimmermann H, Luz Z. Enantiomeric and enantiotopic analysis of cone-shaped compounds with C_3 and C_{3v} symmetry using NMR spectroscopy in chiral anisotropic solvents. *J. Am. Chem. Soc.* **2002**, *124*, 10071–10082.

1571. Merlet D, Emsley JW, Lesot P, Courtieu J. The relationship between molecular symmetry and second-rank orientational order parameters for molecules in chiral liquid crystalline solvents. *J. Chem. Phys.* **1999**, *111*, 6890–6896.

1572. Lafontaine E, Pechine JM, Courtieu J, Mayne CL. Visualization of enantiomers in cholesteric solvents through deuterium NMR. *Liq. Cryst.* **1990**, *7*, 293–298.

1573. Bayle J, Courtieu J, Gabetty E, Loewenstein A, Péchiné JM. Enantiomeric analysis in a polypeptide lyotropic liquid crystal through proton decoupled deuterium NMR. *New J. Chem.* **1992**, *16*, 837–838.

1574. Canet J, Fadel A, Salaün J, Canet-Fresse I, Courtieu J. Enantiomeric excess analysis of sesquiterpene precursors through proton decoupled deuterium NMR in cholesteric lyotropic liquid crystal. *Tetrahedron: Asymmetry* **1993**, *4*, 31–34.

1575. Canet I, Courtieu J, Dauphin G, Gourcy J, Veschambre H. Enantiomeric excess analysis of (2R,3S)-3-deuterio-2-methylcyclohexanone and (1S,2R,3S)-3-deuterio-2-methylcyclohexanol, through deuterium NMR in a polypeptide lyotropic liquid crystal. *Tetrahedron: Asymmetry* **1995**, *6*, 333–336.

1576. Canet I, Courtieu J, Loewenstein A, Meddour A, Pechine JM. Enantiomeric analysis in a polypeptide lyotropic liquid crystal by deuterium NMR. *J. Am. Chem. Soc.* **1995**, *117*, 6520–6526.

1577. Smadja W, Auffret S, Berdagué P, Merlet D, Canlet C, Courtieu J, Legros J, Boutros A, Fiaud J. Visualisation of axial chirality using ^{2}H-{^{1}H} NMR in poly(γ-benzyl L-glutamate), a chiral liquid crystal solvent. *Chem. Commun.* **1997**, 2031–2032.

1578. Canet J, Canet I, Gelas J, Ripoche I, Troin Y. Facile and performant enantiomeric excess analysis of diene iron tricarbonyl complexes through deuterium NMR. *Tetrahedron: Asymmetry* **1997**, *8*, 2447–2451.

1579. Villar H, Guibé F, Aroulanda C, Lesot P. Investigation of SmI_2-mediated cyclisation of δ-iodo-α,β-unsaturated esters by deuterium 2D NMR in oriented solvents. *Tetrahedron: Asymmetry* **2002**, *13*, 1465–1475.

1580. Gouriou L, Lloyd-Jones GC, Vyskočil Š, Kočovský P. ^{2}H-quadrupolar coupling-based analysis of stereochemical and regiochemical memory in the Pd-catalysed allylic alkylation of iso-cinnamyl type substrates employing the chiral monophosphine ligands 'MOP' and 'MAP'. *J. Organomet. Chem.* **2003**, *687*, 525–537.

1581. Phillips AR, Sharman GJ. The measurement of high enantiomeric excesses in chiral liquid crystals using ^{19}F NMR and exchangeable protons in ^{2}H NMR. *Chem. Commun.* **2004**, 1330–1331.

1582. Leyes AE, Poulter CD. Synthesis of (R)-[2-^{2}H]isopentenyl diphosphate and determination of its enantiopurity by ^{2}H NMR spectroscopy in a lyotropic medium. *Org. Lett.* **1999**, *1*, 1067–1070.

1583. Razon P, Dhulut S, Bezzenine-Lafollée S, Courtieu J, Pancrazi A, Ardisson J. Secondary allyltitanium(IV) reagents in aldehyde allylation I. Extension of the Hoppe reaction to γ-alkoxy secondary allyl carbamates. *Synthesis* **2005**, 102–108.

1584. Goss RJM, Meddour A, Courtieu J, O'Hagan D. [^{2}H$_1$]-Fluoroacetic acid. Resolution and assignment of the absolute configuration by ^{2}H-NMR in chiral liquid crystalline solvent. *Synthesis and Applications of Isotopically Labelled Compounds* **2001**, *7*, 163–166.

1585. Lesot P, Sarfati M, Merlet D, Ancian B, Emsley JW, Timimi BA. 2D-NMR strategy dedicated to the analysis of weakly ordered, fully deuterated enantiomers in chiral liquid crystals. *J. Am. Chem. Soc.* **2003**, *125*, 7689–7695.

1586. Lafon O, Berdagué P, Lesot P. Use of two-dimensional correlation between ^{2}H quadrupolar splittings and ^{13}C CSA's for assignment of NMR spectra in chiral nematics. *Phys. Chem. Chem. Phys.* **2004**, *6*, 1080–1084.

1587. Merlet D, Ancian B, Smadja W, Courtieu J, Lesot P. Analysis of natural abundance deuterium NMR spectra of enantiomers in chiral liquid crystals via 2D auto-correlation experiments. *Chem. Commun.* **1998**, 2301–2302.

1588. Lesot P, Merlet D, Lowenstein A, Courtieu J. Enantiomeric visualization using proton-decoupled natural abundance deuterium NMR in poly (γ-benzyl-L-glutamate) liquid crystalline solutions. *Tetrahedron: Asymmetry* **1998**, *9*, 1871–1881.

1589. Merlet D, Ancian B, Courtieu J, Lesot P. Two-dimensional deuterium NMR spectroscopy of chiral molecules oriented in a polypeptide liquid crystal: applications for the enantiomeric analysis through natural abundance deuterium NMR. *J. Am. Chem. Soc.* **1999**, *121*, 5249–5258.

1590. Sarfati M, Courtieu J, Lesot P. First successful enantiomeric discrimination of chiral alkanes using NMR spectroscopy. *Chem. Commun.* **2002**, 1113–1114.

1591. Sarfati M, Lesot P, Merlet D, Courtieu J. Theoretical and experimental aspects of enantiomeric differentiation using natural abundance multinuclear NMR spectroscopy in chiral polypeptide liquid crystals. *Chem. Commun.* **2000**, 2069–2081.

1592. Lafon O, Lesot P, Merlet D, Courtieu J. Modified z-gradient filtering as a mean to obtain phased deuterium autocorrelation 2D NMR spectra in oriented solvents. *J. Magn. Reson.* **2004**, *171*, 135–142.

1593. Parenty A, Campagne J, Aroulanda C, Lesot P. Routine use of natural abundance deuterium NMR in a polypeptidic chiral oriented solvent for the determination of the enantiomeric composition of chiral building blocks. *Org. Lett.* **2002**, *4*, 1663–1666.

1594. Lesot P, Sarfati M, Courtieu J. Natural abundance deuterium NMR spectroscopy in polypeptide liquid crystals as a new and incisive means for the enantiodifferentiation of chiral hydrocarbons. *Chem. Eur. J.* **2003**, *9*, 1724–1745.

1595. Aroulanda C, Lesot P, Merlet D, Courtieu J. Structural ambiguities revisited in two bridged ring systems exhibiting enantiotopic elements, using natural abundance deuterium NMR in chiral liquid crystals. *J. Phys. Chem. A.* **2003**, *107*, 10911–10918.

1596. Lesot P, Aroulanda C, Billault I. Exploring the analytical potential of NMR spectroscopy in chiral anisotropic media for the study of the natural abundance deuterium distribution in organic molecules. *Anal. Chem.* **2004**, *76*, 2827–2835.

1597. Lesot P, Merlet D, Meddour A, Courtieu J, Loewenstein A. Visualization of enantiomers in a polypeptide liquid-crystal solvent through carbon-13 NMR spectroscopy. *J. Chem. Soc., Faraday Trans.* **1995**, *91*, 1371–1375.

1598. Farjon J, Baltaze J, Lesot P, Merlet D, Courtieu J. Heteronuclear selective refocusing 2D NMR experiments for the spectral analysis of enantiomers in chiral oriented solvents. *Magn. Reson. Chem.* **2004**, *42*, 594–599.

1599. Meddour A, Berdague P, Hedli A, Courtieu J, Lesot P. Proton-decoupled carbon-13 NMR spectroscopy in a lyotropic chiral nematic solvent as an analytical tool for the measurement of the enantiomeric excess. *J. Am. Chem. Soc.* **1997**, *119*, 4502–4508.

1600. Dirat O, Kouklovsky C, Langlois Y, Lesot P, Courtieu J. Double diastereoselection in asymmetric [2+3] cycloaddition of chiral oxazoline *N*-oxides: application to the kinetic resolution of a racemic α, β-unsaturated δ-lactone. *Tetrahedron: Asymmetry* **1991**, *10*, 3197–3207.

1601. Sugiura M, Kimura A, Fujiwara H. Discrimination of enantiomers by means of NMR spectroscopy using chiral liquid crystalline solution: application to triazole fungicides, uniconazole and diniconazole. *Magn. Reson. Chem.* **2006**, *44*, 121–126.

1602. Lesot P, Lafon O, Courtieu J, Berdagué P. Analysis of the ^{13}C NMR spectra of molecules, chiral by isotopic substitution, dissolved in a chiral oriented environment: towards the absolute assignment of the pro-*R*/pro-*S* character of enantiotopic ligands in prochiral molecules. *Chem. Eur. J.* **2004**, *10*, 3741–3746.

1603. Rivard M, Guillen F, Fiaud J, Aroulanda C, Lesot P. Efficient enantiodiscrimination of chiral monophosphine oxides and boranes by phosphorus coupled ^{13}C NMR spectroscopy in the presence of chiral ordering agents. *Tetrahedron: Asymmetry* **2003**, *14*, 1141–1142.

1604. Meddour A, Uziel J, Courtieu J, Juge S. Enantiodifferentiation of acyclic phosphonium salts in chiral liquid crystalline solutions. *Tetrahedron: Asymmetry* **2006**, *17*, 1424–1429.

1605. Aroulanda C, Merlet D, Courtieu J, Lesot P. NMR experimental evidence of the differentiation of enantiotopic directions in C_s and C_{2v} molecules using partially oriented, chiral media. *J. Am. Chem. Soc.* **2001**, *123*, 12059–12066.

1606. Mislow K, Raban M. Stereoisomeric relations of groups in molecules. *Top. Stereochem.* **1967**, *1*, 1–38.

1607. Lesot P, Merlet D, Courtieu J, Emsley JW. Discrimination and analysis of the NMR spectra of enantiomers dissolved in chiral liquid solvents through 2D correlation experiments. *Liq. Cryst.* **1996**, *21*, 427–435.

1608. Farjon J, Merlet D, Lesot P, Courtieu J. Enantiomeric excess measurements in weakly oriented chiral liquid crystal solvents through 2D ^{1}H selective refocusing experiments. *J. Magn. Reson.* **2002**, *158*, 169–172.

1609. Jakubcova M, Meddour A, Péchiné J, Baklouti A, Courtieu J. Measurement of the optical purity of fluorinated compounds using proton decoupled ^{19}F NMR spectroscopy in a chiral liquid crystal solvent. *J. Fluorine Chem.* **1997**, *86*, 149–153.

1610. Graf E, Graff R, Hosseini MW, Huguenard C, Taulelle F. Probing peristaltic chirality of alkaline cations: NMR study of alkaline borocryptates. *Chem. Commun.* **1997**, 1459–1460.

1611. Huguenard C, Taulelle F, Graf E, Hosseini MW. NMR in a liquid crystal solvent: study of the chirality of borocryptates. *J. Chim. Phys.* **1998**, *95*, 341–349.

1612. Iida M, Mizuno Y, Koine N. Chiral discrimination in the interactions between [Co(en)$_3$]$^{3+}$ and cholesteric liquid crystals by ^{59}Co NMR spectroscopy. *Chem. Lett.* **1994**, 481–484.

1613. Weiss-López BE, Azocar M, Montecinos R, Cassels BK, Araya-Maturana R. Differential incorporation of L- and D-*N*-acyl-1-phenyl-d$_5$-2-aminopropane in a cesium *N*-dodecanoyl-l-alaninate cholesteric nematic lyomesophase. *Langmuir* **2001**, *17*, 6910–6914.

1614. Ahumada H, Montecinos R, Alegria S, Araya-Maturana R, Weiss-López BE. Chiral discrimination of L- and D-N-acyl-1-phenyl-d$_5$-2-aminopropanes in a cesium *N*-dodecanoyl-L-threoninate cholesteric nematic lyomesophase. *J. Chil. Chem. Soc.* **2005**, *50*, 421–425.

1615. Baczko K, Larpent C, Lesot P. New amino acid-based anionic surfactants and their use as enantiodiscriminating lyotropic liquid crystalline NMR solvents. *Tetrahedron: Asymmetry* **2004**, *15*, 971–982.

1616. Solgadi A, Meddour A, Courtieu J. Enantiomeric discrimination of water soluble compounds using deuterium NMR in a glucopon/buffered water/*n*-hexanol chiral lyotropic liquid crystal. *Tetrahedron: Asymmetry* **2004**, *15*, 1315–1318.

1617. Péchiné JM, Meddour A, Courtieu J. Monitoring the differential ordering of enantiomers included into cyclodextrins through deuterium NMR in lyotropic liquid crystals. *Chem. Commun.* **2002**, 1734–1735.

1618. Jursic BS. An enantiomeric discrimination in aqueous mixed chiral micelles through hydrogen bonding. *Tetrahedron Lett.* **1993**, *34*, 963–966.

1619. Belogi G, Croce M, Mancini G. Discrimination of enantiomers by chiral micelles observed by NMR spectroscopy. *Langmuir* **1997**, *13*, 2903–2904.

1620. Andreani R, Bombelli C, Borocci S, Lah J, Mancini G, Mencarelli P, Vesnaver G, Villani C. New biphenylic derivatives: synthesis, characterization and enantiodiscrimination in chiral aggregates. *Tetrahedron: Asymmetry* **2004**, *15*, 987–994.

1621. Bombelli C, Borocci S, Lupi F, Mancini G, Mannina L, Segre AL, Viel S. Chiral recognition of dipeptides in a biomembrane model. *J. Am. Chem. Soc.* **2004**, *126*, 13354–13362.

1622. Iida M, Mizuno Y, Koine N. ^{59}Co NMR spectroscopic chiral discrimination of [Co(en)$_3$]$^{3+}$ enantiomers in ionic interacting systems. *Bull. Chem. Soc. Jpn.* **1995**, *68*, 1337–1344.

1623. Yarabe HH, Rugutt JK, McCarroll ME, Warner IM. Capillary electrophoretic separation of binaphthyl enantiomers with two polymeric chiral surfactants: ^{1}H-nuclear magnetic resonance and fluorescence spectroscopy study. *Electrophoresis* **2000**, *21*, 2025–2032.

1624. Rugutt JK, Billiot E, Warner IM. NMR study of the interaction of monomeric and polymeric chiral surfactants with (*R*)- and (*S*)-1,1′-binaphthyl-2,2′-diyl hydrogen phosphate. *Langmuir* **2000**, *16*, 3022–3029.

1625. Borocci S, Erba M, Mancini G, Scipioni A. Deracemization of an axially chiral biphenylic structure in chiral micellar aggregates. *Langmuir* **1998**, *14*, 1960–1962.

1626. Bella J, Borocci S, Mancini G. Recognition in organized aggregates formed by a chiral amidic surfactant. *Langmuir* **1999**, *15*, 8025–8031.

1627. Nakagawa H, Gomi K, Yamada K. Chiral recognition of thiaheterohelicenes by alkyl β-D-pyranoside micelles. Influence of extension of helix. *Chem. Pharm. Bull.* **2001**, *49*, 49–53.

1628. de Loos M, van Esch J, Kellogg RM, Feringa BL. Chiral recognition in bis-urea-based aggregates and organogels through cooperative interactions. *Angew. Chem., Int. Ed.* **2001**, *40*, 613–616.

1629. Wasserscheid P, Bösmann A, Bolm C. Synthesis and properties of ionic liquids derived from the 'chiral pool'. *Chem. Commun.* **2002**, 200–201.

1630. Levillain J, Dubant G, Abrunhosa I, Gulea M, Gaumont A. Synthesis and properties of thiazoline based ionic liquids derived from the chiral pool. *Chem. Commun.* **2003**, 2914–2915.

1631. Clavier H, Boulanger L, Audic N, Toupet L, Mauduit M, Guillemin J. Design and synthesis of imidazolinium salts derived from (L)-valine. Investigation of their potential in chiral molecular recognition. *Chem. Commun.* **2004**, 1224–1225.

1632. Ishida Y, Miyauchi H, Saigo K. Design and synthesis of a novel imidazolium-based ionic liquid with planar chirality. *Chem. Commun.* **2002**, 2240–2241.

1633. Ishida Y, Sasaki D, Miyauchi H, Saigo K. Design and synthesis of novel imidazolium-based ionic liquids with a pseudo crown-ether moiety: diastereomeric interaction of a racemic ionic liquid with enantiopure europium complexes. *Tetrahedron Lett.* **2004**, *45*, 9455–9459.

1634. Yamamoto C, Okamoto Y. Optically active polymers for chiral separation. *Bull. Chem. Soc. Jpn.* **2004**, *77*, 227–257.

1635. Yashima E, Yamada M, Okamoto Y. An NMR study of chiral recognition relevant to the liquid chromatographic separation of enantiomers by a cellulose derivative. *Chem. Lett.* **1994**, 579–582.

1636. Okamoto Y, Yashima E. Preparation of polysaccharide derivatives and their chiral recognition mechanism. *Macromol. Symp.* **1995**, *99*, 15–23.

1637. Yashima E, Yamamoto C, Okamoto Y. NMR studies of chiral discrimination relevant to the liquid chromatographic enantioseparation by a cellulose phenylcarbamate derivative. *J. Am. Chem. Soc.* **1996**, *118*, 4036–4048.

1638. Okamoto Y, Yashima E, Yamamoto C. NMR studies of chiral discrimination by phenylcarbamate derivatives of cellulose. *Macromol. Symp.* **1997**, *120*, 127–137.

1639. Yashima E, Yamada M, Yamamoto C, Nakashima M, Okamoto Y. Chromatographic enantioseparation and chiral discrimination in NMR by trisphenylcarbamate derivatives of cellulose, amylose, oligosaccharides, and cyclodextrins. *Enantiomer* **1997**, *2*, 225–240.

1640. Yamamoto C, Yashima E, Okamoto Y. Structural analysis of amylose tris(3,5-dimethylphenylcarbamate) by NMR relevant to its chiral recognition mechanism in HPLC. *J. Am. Chem. Soc.* **2002**, *124*, 12583–12589.

1641. Kubota T, Yamamoto C, Okamoto Y. Tris(cyclohexylcarbamate)s of cellulose and amylose as potential chiral stationary phases for high-performance liquid chromatography and thin-layer chromatography. *J. Am. Chem. Soc.* **2000**, *122*, 4056–4059.

1642. Kubota T, Yamamoto C, Okamoto Y. Chromatographic enantioseparation by cycloalkylcarbamate derivatives of cellulose and amylose. *Chirality* **2002**, *14*, 372–376.

1643. Oguni K, Matsumoto A, Isokawa A. ^{13}C NMR study on diastereomeric interactions between cellulose tris(4-methylbenzoate) and 1-phenylethanol enantiomers. *Polym. J.* **1994**, *26*, 1257–1261.

1644. Oguni K, Ito M, Isokawa A, Matsumoto A. ^{13}C nuclear spin-spin relaxation times (T_{2C}s) of enantiomers in the presence of column packing material and solvents for chiral discrimination HPLC. *Chirality* **1996**, *8*, 372–376.

1645. Maeda K, Okamoto Y. Synthesis and conformation of optically active poly(phenylisocyanate)s bearing an ((S)-(α-methylbenzyl)carbamoyl) group. *Macromolecules* **1998**, *31*, 1046–1052.

1646. Angeloni AS, Laus M, Caretti D, Chiellini E, Galli G. Chiral liquid-crystalline poly(ester/β-sulfoxide)s by asymmetric oxidation of prochiral nematic poly(ester/β-sulfide)s. *Makromol. Chem.* **1990**, *191*, 2787–2793.

1647. Koppenhoefer B, Hummel M. Chemical shift nonequivalence of enantiomers of amino acid derivatives in ^{1}H NMR and ^{19}F NMR spectroscopy induced by L-Chirasil-Val. *Z. Naturforsch., B: Chem. Sci.* **1992**, *47b*, 1034–1036.

1648. Lohmiller K, Bayer E, Koppenhoefer B. Recognition of enantiomers by Chirasil-Val and oligopeptide analogues as studied by gas-phase calorimetry and ^{1}H NMR spectroscopy in solution. *J. Chromatogr.* **1993**, *634*, 65–77.

1649. Isobe Y, Onimura K, Tsutsumi H, Oishi T. Asymmetric polymerization of N-1-anthrylmaleimide with diethylzinc-chiral ligand complexes and optical resolution using the polymer. *Polym. J.* **2002**, *34*, 18–24.

1650. Kakuchi T, Harada Y, Satoh T, Yokota K, Hashimoto H. New macromolecular ionophore: enantioselective membrane transport of racemic amino acid by poly[(1→6)-2,5-anhydro-3,4-di-O-methyl-D-glucitol. *Polymer* **1994**, *35*, 204–206.

1651. Rivera JM, Martin T, Rebek J. Chiral spaces: dissymmetric capsules through self-assembly. *Science* **1998**, *279*, 1021–1023.

1652. Rivera JM, Craig SL, Martín T, Rebek, J. Chiral guests as their ghosts in reversibly assembled hosts. *Angew. Chem., Int. Ed.* **2000**, *39*, 2130–2132.

1653. Nuckolls C, Hof F, Martín T, Rebek J. Chiral microenvironments in self-assembled capsules. *J. Am. Chem. Soc.* **1999**, *121*, 10281–10285.

1654. Wu A, Chakraborty A, Fettinger JC, Flowers RA, Isaacs L. Molecular clips that undergo heterochiral aggregation and self-sorting. *Angew. Chem., Int. Ed.* **2002**, *41*, 4028–4031.

1655. Fiedler D, Pagliero D, Brumaghim JL, Bergman RG, Raymond KN. Encapsulation of cationic ruthenium complexes into a chiral self-assembled cage. *Inorg. Chem.* **2004**, *43*, 846–848.

1656. Hill HDW, Zens AP, Jacobus J. Solid-state NMR spectroscopy. Distinction of diastereomers and determination of optical purity. *J. Am. Chem. Soc.* **1979**, *101*, 7090–7091.

1657. Andersen KV, Bildsøe H, Jakobsen HJ. Determination of enantiomeric purity from solid-state ^{31}P MAS NMR of organophosphorus compounds. *Magn. Reson. Chem.* **1990**, *28*, S47–S51.

1658. Potrzebowski MJ, Tadeusiak E, Misiura K, Ciesielski W, Bujacz G, Tekely P. A new method for distinguishing between enantiomers and racemates and assignment of enantiomeric purity by means of solid-state NMR. Examples from oxazaphosphorinanes. *Chem. Eur. J.* **2002**, *8*, 5007–5011.

1659. Potrzebowski MJ, Assfeld X, Ganicz K, Olejniczak S, Cartier A, Gardiennet C, Tekely P. An experimental and theoretical study of the ^{13}C and ^{31}P chemical shielding tensors in solid O-phosphorylated amino acids. *J. Am. Chem. Soc.* **2003**, *125*, 4223–4232.

1660. Facey GA, Ripmeester JA. Solid state NMR as a tool to differentiate between the enantiotopic methyl groups of prochiral guest molecules in tri-O-thymotide clathrates. *Mol. Cryst. Liq. Cryst.* **1992**, *211*, 167–175.

1661. Ripmeester JA, Burlinson NE. Chiral discrimination and solid-state ^{13}C NMR. Application to tri-o-thymotide clathrates. *J. Am. Chem. Soc.* **1985**, *107*, 3713–3714.

1662. Facey G. Ripmeester JA. Probing molecular motion and chemical reactions inside the chiral tri-o-thymotide clathrate cavity by solid state NMR techniques. *J. Chem. Soc., Chem. Commun.* **1990**, 1585–1587.

1663. Imashiro F, Kuwahara D, Terao T. ^{13}C Solid-state NMR study on populations, conformations, and molecular motions of γ-valerolactone enantiomers enclathrated in the chiral cholic acid host. *J. Chem. Soc., Perkin Trans. 2* **1993**, 1759–1763.

1664. Nakamura S, Imashiro F, Takegoshi K, Terao T. Sequential arrangement of γ-valerolactone enantiomers enclathrated in cholic acid channels as studied by ^{13}C solid-state NMR: elucidation of the optical resolution mechanism. *J. Am. Chem. Soc.* **2004**, *126*, 8769–8776.

1665. Kato K, Aoki Y, Sugahara M, Tohnai N, Sada K, Miyata M. Interpretation of enantioresolution in nordeoxycholic acid channels based on the four-location model. *Chirality* **2003**, *15*, 53–59.

1666. Vaton-Chanvrier L, Oulyadi H, Combret Y, Coquerel G, Combret JC. Chiral recognition of binaphthyl derivatives: a chiral recognition model on the basis of chromatography, spectroscopy, and molecular mechanistic calculations for the enantioseparation of 1,1'-binaphthyl derivatives on cholic acid-bonded stationary phases. *Chirality* **2001**, *13*, 668–674.

1667. Riedl R, Tappe R, Berkessel A. Probing the scope of the asymmetric dihydroxylation of polymer-bound olefins. Monitoring by HRMAS NMR allows for reaction control and on-bead measurement of enantiomeric excess. *J. Am. Chem. Soc.* **1998**, *120*, 8994–9000.

1668. Antonsson T, Jacobsson U, Moberg C, Rákoś L. Preparation of polymer-supported (*R*)- and (*S*)-styrene oxide. *J. Org. Chem.* **1989**, *54*, 1191–1194.

INDEX

Abscisic acid, analysis of, 285
N-Acetal-DL-p-fluorophenylalaninal, analysis of, 330
Acetals, analysis of, 176
Acetic acid, monodeutero, monotritio, analysis of, 173–174
Acetutolol, analysis of, 27
Acryloins, analysis of, 47
Aggregating systems, 319
   achiral compounds, 423–424
   calix[4]arene dimelamines, 318
   calix[4]resorcarenes, 314–315
   metal complexes, 424
Albuterol, analysis of, 54
Alcohols
   analysis of using
      amine-based reagents, 161–162, 176–177
      calixarenes, 314
      carboxylic acid-based reagents, 18, 40, 90, 95, 98
      cyclodextrins, 290
      hydroxyl-containing reagents, 124, 145, 149, 152
      lanthanide complexes, 333, 336, 338–339, 346
      liquid crystals, 404–405, 408, 412, 414–415
      miscellaneous reagents, 193–194, 196–197, 206, 214, 216–217, 219–221
      phosphorus-containing reagents, 234, 237–240, 245, 248–249
      selenium-containing reagents, 266–267
      silyl-containing reagents, 274–275
      solid-state NMR spectroscopy, 426
      transition metal complexes, 393
   as chiral discriminating agents, 102–152
Diols, analysis of using
   boron-containing reagents, 269, 273–274
   carboxylic acid-based reagents, 19–25, 33, 42–43, 54, 75, 86–87
   metal complexes, 337, 340, 351, 398
   miscellaneous reagents, 212, 328, 423
   liquid crystals, 405, 412

Discrimination of Chiral Compounds Using NMR Spectroscopy, by Thomas J. Wenzel
Copyright © 2007 John Wiley & Sons, Inc.

Alcohols (*Continued*)
    phosphorus-containing reagents, 234
    1,2-diols, analysis of using
        amine-based reagents, 187
        boron-containing reagents, 268–272
        calixarenes/calix[4]resorcarenes, 313–314, 318
        carboxylic acid-based reagents, 21, 24, 43, 77
        hydroxyl-containing reagents, 124, 127, 142
        miscellaneous reagents, 203, 206, 211–212
        phosphorus-containing reagents, 243, 248
        metal complexes, 336, 346, 396
    1,3-diols, analysis of using
        amine-based reagents, 185–187
        boron-containing reagents, 268, 272
        carboxylic acid-based reagents, 22–23, 37–38, 43, 55, 77
        miscellaneous reagents, 203, 206, 211–212, 336
        phosphorus-containing reagents, 243, 248
    1,4-diols, analysis of, 24, 43, 77, 97, 187, 272
    1,5-diols, analysis of, 24, 43, 77, 187
    erythro/threo configuration, 21–23, 25, 43, 77, 187, 211–212, 269
    phenols, analysis of, 243, 250
    polyols, analysis of, 37, 42, 270
        annonaceous acetogenins, 17, 24–25, 68, 111
        1,3,5-triols, 37, 185–186
    primary, analysis of using
        carboxylic acid-based reagents, 25, 31, 43–44, 71, 73, 77, 79, 82–83, 85, 88, 95, 100
        liquid crystals, 403, 410–411, 413
        metal complexes, 339, 340–341
        miscellaneous reagents, 140, 218, 266, 275–276, 316
        phosphorus-containing reagents, 234, 236, 243, 246–247, 249
    secondary, analysis of
        acetylenic, 32, 131, 180, 216, 243–244, 400

acyloins, 47
allylic, 32, 46, 138, 249, 275, 375
cyclohexanols, 32
    2-substituted, 15, 32, 70, 89, 408
    3-substituted, 15, 32, 58
cyclooctenols, 32
cyclopentanols, 14, 32, 46
glycerols, 13, 66, 267, 353
hemiacetal esters, 41
α-hydroxy aldehydes, 131
α-hydroxy carboxanilides, 131
oxetanes, 97
using
    amine-based reagents, 154, 175, 177, 180, 182, 187, 189
    aryl-containing carboxylic acids, 51, 54, 56–59, 63, 66–68, 70–73, 75, 78–80
    carboxylic acid-based reagents, 85, 87–89, 91, 94, 97, 100–101
    CFTA, 60–62
    glycosidation shifts, 138–144
    isocyanates, 207–208, 210–211
    liquid crystals, 401, 404–405, 408, 410, 412, 416
    metal complexes, 339, 347, 351–352, 393
    miscellaneous reagents, 127, 131–132, 214, 216–217, 266, 272, 289, 313, 326, 420–421, 426
    MPA, 38–42, 46, 48
    MTPA, 8, 11–19, 30–31
    phosphorus-containing reagents, 234, 236, 243–247, 249–250
    silyl-containing reagents, 275–277
tertiary, analysis of, 26, 69, 73, 80, 87, 89, 142–144, 177, 180, 187, 243, 246, 351
Aldehydes
    analysis of, 153, 319, 333
    3-alkyl, 173
    3-aryl, 173
    α-amino, 159
    using amine-based reagents, 159, 163, 168, 170–171, 173, 183
    as chiral discriminating agents, 203–205

INDEX 539

Alkanes, analysis of
  aliphatic, 410–411
  cyclic, 284, 285, 319, 406, 411
Alkenes, analysis of, 223–224
  using
    cyclodextrins, 285, 287, 293
    lanthanide-silver complexes, 332–334, 355–357
    liquid crystals, 408, 411
    palladium complexes, 372
    platinum complexes, 374–378
    rhodium complexes, 379
    silver complexes, 395–396
Alkyl halides, analysis of, 266, 314–315, 396, 426
  using
    lanthanide-silver complexes, 332–334, 355, 357
    liquid crystals, 405, 408
    rhodium complexes, 379, 381–382
Alkynes, analysis of, 356, 372, 374
Allenes, analysis of, 215
  using
    cyclodextrins, 289, 291
    lanthanide-silver complexes, 356
    platinum complexes, 374–378
    TFAE, 120–121
Alphaprodine, analysis of, 281
Amides
  analysis of, 193–194, 196, 200–201, 324, 409, 415
  as chiral discriminating agents, 191–200
Amine oxides, analysis of, 105, 149, 233
Amines,
  analysis of using
    boron-containing reagents, 272
    carboxylic acid-based reagents, 90, 91, 94, 97–98
    cyclodextrins, 290–291
    hydroxyl-containing reagents, 124, 131, 145, 148–149
    lanthanide complexes, 333, 336, 338
    liquid crystals, 404, 408, 414–415
    miscellaneous reagents, 194–195, 197, 206, 217
    phosphorus-containing reagents, 234, 237–239, 245, 250

    selenium-containing reagents, 266
    silyl-containing reagents, 274
    transition metal complexes, 372, 391, 393
  diamines, analysis of, 52, 57, 248, 391
  heterocyclic, analysis of, 29, 96, 313
  piperidines, analysis of, 26, 29, 136, 327, 336
  pyrrolidines, analysis of, 26, 136
  primary, analysis of using
    carboxylic acid-based reagents, 12, 28–29, 38–39, 44–45, 48, 52, 54, 59–60, 62, 64–65, 70, 73–74, 78, 80–81, 84–85, 91, 272
    crown ethers, 300–310
    cyclodextrins, 288, 290
    hydroxyl-containing reagents, 103, 124–125, 136, 145–146
    isocyanate reagents, 207, 209–210
    liquid crystals, 404
    metal complexes, 334, 347, 385–386, 398
    miscellaneous reagents, 179, 196, 204, 213–218, 221–222, 316
    phosphorus-containing reagents, 234, 236, 243, 246
    receptor compounds, 324, 326
  quaternary amines, 200, 257–258, 315, 323–324, 358, 398
  secondary, analysis of using
    carboxylic acid-based reagents, 26, 27, 44, 52, 54, 58, 60, 66, 74, 81, 84
    crown ethers, 301–303
    hydroxyl-containing reagents, 121, 136, 145–146
    isocyanate reagents, 207, 209–210
    lanthanide complexes, 339
    liquid crystals, 404, 416
    miscellaneous reagents, 199, 215–216, 219–220,
    phosphorus-based reagents, 236, 240–241, 243, 248
    receptor compounds, 327
  tertiary, analysis of using
    carboxylic acid-based reagents, 27, 30, 49, 52
    hydroxyl-containing reagents, 108, 121

Amines (*Continued*)
  phosphorus-containing reagents, 240–241
  as chiral discriminating agents, 153–189
Amino acids
  analysis of, 156, 194, 283–284, 289, 291, 313, 324, 340, 343, 404, 415–416, 425
    N-acetyl, 177, 275, 292, 299, 320, 325, 352, 384
    N-acyl, 362, 383
    α-alkylated, 93, 301, 360
    β-amino acids, 106, 361, 370–372
    γ-amino acids, 370
    N,N-dimethyl, 121, 129
    N-(3,5-dinitrobenzoyl), 179, 199, 216, 290
    dinitrophenyl, 283, 290, 325
    esters, 214, 216, 237, 296, 313, 316, 323–324, 423
      using
        carboxylic acid-based reagents, 28, 31, 32, 73–74, 85
        crown ethers, 301–302, 304–309
        hydroxyl-containing reagents, 106, 108, 136, 144
        metal complexes, 336, 352, 384
    isotopically substituted, 84, 92, 203
    γ-oxo-α-amino acids, 48
    N-trifluoroacetyl, 200, 422
    underivatized, 84, 141, 204, 222, 225, 317, 320, 324, 326–327, 389, 395
      using
        amine-based reagents, 165, 169
        crown ethers, 301–302, 310
        cyclodextrins, 286–287
        lanthanide complexes, 338, 360–364
        liquid crystals, 402–404, 416
        palladium complexes, 370–373
        phosphorus-containing reagents, 237, 248
  as chiral discriminating agents, 62–63, 92–93, 193–195, 197
Amino alcohols, analysis of, 205, 216, 221, 268, 313
    using
      aryl-containing carboxylic acids, 28, 45, 54, 63, 66, 74
      axial chiral alcohols, 145–148
      crown ethers, 301–302, 304, 306
      cyclodextrins, 290, 293
      phosphorus-containing reagents, 234, 237, 243
      transition metal complexes, 384, 386, 397
α-Amino esters, analysis of, 61–62, 94, 216, 236, 239, 272
β-Amino esters, analysis of, 272
β-Amino ethers, analysis of, 272
Aminoglutethimide, analysis of, 302
Aminophosphines, analysis of, 367
Aminophosphinic acids, analysis of, 284
Aminophosphonic acids, analysis of, 284, 373
Amlodipine maleate, analysis of, 282, 295, 297
Amphetamine, analysis of, 338
Amphidinolide C, analysis of, 15
Analgesics, analysis of, 281
Anions, analysis of, 250, 259, 298–300, 365
Annonaceous acetogenins, analysis of, 17, 24–25, 68, 111
Antihistamines, analysis of, 281
Aromatic hydrocarbons, analysis of using
    cyclodextrins, 289
    lanthanide-silver complexes, 332–334, 355–356
    liquid crystals, 407–408, 411
    rhodium complexes, 384
Arsenoxides, analysis of, 148
Arsineoxides, analysis of, 148
Arsines, analysis of, 367
Aryl-containing carboxylic acids as chiral discriminating agents, 8–81
Atenolol, analysis of, 290
Atracurium besylate, analysis of, 181
Axial chiral compounds, analysis of, 112–113
  biaryls, 31, 85, 176, 233, 244, 283, 288, 299, 345, 412, 417–418, 421
  BINOL, 202, 216, 218, 328, 412, 420–422, 427
Azides, analysis of, 249

Aziridines, analysis of, 112, 386, 391

Baclofen, analysis of, 302
Barbiturate-isoxazoline conjugates, analysis of, 328
Barbituric acids, analysis of, 318
Barium method, 34, 47–48, 50, 80–81
Benzodiazepinones, analysis of, 194
Betulinic acid, analysis of, 143
ent-Bicyclogermacrene, analysis of, 101
Boron-containing compounds as chiral discriminating agents, 268–274
Boron-10 NMR spectra, 415
Boron-11 NMR spectra, 415
Broadening with lanthanide complexes, 335–337
  carbon-13 spectra, 336, 351–352, 361
  Ce(III), 361–362
  La(III), 362
  Sm(III), 336, 361
Bromochlorofluoromethane, analysis of, 326–327
Brompheniramine, analysis of, 294, 313
Bullatin, analysis of, 17
Butenolides, analysis of, 16

Calixarenes
  analysis of, 108–109, 118–119, 160–161
  as chiral discriminating agents, 311–312, 315–320
Calixresorcarenes
  analysis of, 134, 137
  as chiral discriminating agents, 311–315, 319–320
Caltechin, analysis of, 325
Carbamates, analysis of, 409
Carbinoxamine maleate, analysis of, 281, 286, 296, 298, 313
Carbohydrates, analysis of
  amino deoxy sugars, 90
  arabinose derivatives, 14, 90, 94
  disaccharides, 141–142
  fructose, 90
  galactose, 84, 90, 94
  glucose, 84, 90, 94
  mannitol, 84
  mannose, 90, 94
  myo-inositol, 55, 281
  polysaccharides, 141–142
  rhamnose, 90
  ribose, 90
  xylose, 90, 94, 138
Carbon-13
  isotopic chirality, 104, 227–228
  NMR spectra, assignment of resonances in liquid crystals, 412
  satellite signals, 241
Carboxylic acids
  analysis of, 197–198, 216, 265, 274, 290, 324, 421
  α-chiral, 126, 128, 130, 137, 147, 155–156, 163, 169, 202, 213–214, 299, 320–321
  β-chiral, 128, 130, 156, 163, 166, 168
  γ-chiral, 163, 321
  α-deuterated, 129, 130
  using
    amine-based reagents, 153–157, 161–162, 164–168, 170, 173, 175, 177, 179, 181–182, 184
    calixarenes, 315–316
    cyclodextrins, 283, 288, 292–293, 299
    hydroxyl-containing reagents, 123, 126–136, 150–151
    lanthanide complexes, 345–346, 354–355, 359–364
    liquid crystals, 400, 403–404, 408–409, 412, 414
    phosphorus-containing reagents, 244, 246–247, 249–250
    specialized receptors, 318, 320–323, 326
    transition metal complexes, 393, 396
  as chiral discriminating agents, 82–101
Carnitine, analysis of, 359
Cathinone, analysis of, 145
Cationic substrates, analysis of, 250–259, 299, 358–359
  ammonium ions, 281–282, 294–296, 298, 358
  bipyridinium ions, 417
  helicene cation, 258
  isothiouronium ions, 358–359
  quaternary amines, 200, 257–258, 315, 323–324, 358, 398
  metal complexes, 250–254, 259, 288, 296
  monomethenium dyes, 258

Cationic substrates (*Continued*)
  phosphonium salts, 178, 221–222, 257, 413
  pyrrolidinium ions, 200
  spirobi[dibenzazepinium], 257
  sulfonium ions, 358–359
  thiiranium ions, 258
Caylobolide A, analysis of, 23
Cesium-133 NMR spectra, 415
Chlorofluoroacetic acid, analysis of, 135
Chlorofluoroiodoacetic acid, analysis of, 181
Chlorpheniramine, analysis of, 294–295, 313
Circular dichroism, 23, 43, 86–87, 390
  exciton chirality method, 23, 76
Citrate, analysis of, 122, 359
Citronellol, analysis of, 275
Cizolirtine, analysis of, 221, 282
Clenbuterol, analysis of, 297
Cobalt-59 NMR spectra, 389, 415, 417
Cocaine, analysis of, 128
Conformational preference, of
  amine-based reagents, 156, 163–166
  aryl-containing carboxylic acid reagents, 51, 54–56, 61–62, 69–75, 77–78, 80–81
  carboxylic acid-based reagents, 97–98, 100–101
  hydroxyl-containing reagents, 126, 131, 140
  miscellaneous reagents, 208, 210–211, 213, 276, 366
  MPA derivatives, 39, 41, 43–45, 47
  MTPA derivatives, 8–14, 26–28
Crispine A, analysis of, 234
Crown ethers, as chiral discriminating agents, 300–310
Cyanohydrins, analysis of, 41, 131, 136, 152, 293
Cyclodextrins, as chiral discriminating agents, 279–300
  anionic, 293–298
  cationic, 298–299
  neutral, 280–293
Cyclohexanols, analysis of, 32
  2-substituted, 15, 32, 70, 89, 408
  3-substituted, 15, 32, 58
Cyclooctenols, analysis of, 32
Cyclopentanols, analysis of, 14, 32, 46

Cyclophosphazenes, analysis of, 116–117, 345

Database methods, 23, 35–38, 153, 184–189
  BMBA-pMe, 187–189
  DMBA, 184–186
  DMBA/MTPA, 189
  carbon-13, 35–37, 153, 186–187
  proton NMR spectra, 153, 186
Decipinone, analysis of, 224
Deoxyephedrine, analysis of, 44
Deracemization, 338
Desertomycin, analysis of, 37
Desflurane, analysis of, 291–292
Desymmetrization, 228
Deuterium
  isotopic chirality, 38, 76, 157, 196, 342, 387–388
  carboxylic acids, 129–130, 182
  alcohols, 83, 142, 341–342, 401, 405–406, 412
  α-deuterated primary alcohols, 4, 38, 51, 54, 62, 82–83, 236, 265, 341
  sulfoxides, 50, 104
NMR spectra
  assignment methods in liquid crystals, 409–411
  with
    carboxylic acid-based reagents, 51–52, 62, 84
    hydroxyl-containing reagents, 104, 129–130
    lanthanide complexes, 341
    liquid crystals, 401, 403–413, 416
    phosphorus-containing reagents, 236
    solid-state systems, 426
Diamines, as chiral discriminating agents, 182–184, 187–189
Diaziridines, analysis of, 106
Diffusion measurements, 259
2,3-Dihydrofarnesal, analysis of, 163
Dihydroquinine, analysis of, 225
Dimethindene maleate, analysis of, 281, 295, 298
Diniconazole, analysis of, 412
Diols, analysis of using
  boron-containing reagents, 269, 273–274
  carboxylic acid-based reagents, 19–25, 33, 42–43, 54, 75, 86–87

INDEX **543**

metal complexes, 337, 340, 351, 398
miscellaneous reagents, 212, 328, 423
liquid crystals, 405, 412
phosphorus-containing reagents, 234
1,2-diols, analysis of using
  amine-based reagents, 187
  boron-containing reagents, 268–272
  calixarenes/calix[4]resorcarenes, 313–314, 318
  carboxylic acid-based reagents, 21, 24, 43, 77
  hydroxyl-containing reagents, 124, 127, 142
  miscellaneous reagents, 203, 206, 211–212
  phosphorus-containing reagents, 243, 248
  metal complexes, 336, 346, 396
1,3-diols, analysis of using
  amine-based reagents, 185–187
  boron-containing reagents, 268, 272
  carboxylic acid-based reagents, 22–23, 37–38, 43, 55, 77
  miscellaneous reagents, 203, 206, 211–212, 336
  phosphorus-containing reagents, 243, 248
1,4-diols, analysis of, 24, 43, 77, 97, 187, 272
1,5-diols, analysis of, 24, 43, 77, 187
Diols, *erythro/threo* configuration, 21–23, 25, 43, 77, 187, 211–212, 269
Dipeptides, analysis of, 164–165, 177, 225, 248, 283, 291, 298, 324, 364, 417
Diphosphines, analysis of, 365–368, 374
Diphosphates, analysis of, 388
Disulfides, analysis of, 343
DNA, analysis of, 223, 389
Dothiepine, analysis of, 292
Doxylamine succinate, analysis of, 281, 286, 296, 298, 313

Enantiotopic faces, groups, and directions, 407–408, 411, 413–414
  acenaphthene, 408
  bicycloheptadiene, 413–414
  malononitrile, 413–414
  quadricycane, 411
  norbornene, 411

Enflurane, analysis of, 291–292
Enokipodin C, analysis of, 101
Enones, analysis of, 135
Ephedrine, analysis of, 52, 293, 338
3-Epibetulinic acid, analysis of, 143
Epimerization, 150, 170, 204, 235, 265
Episulfides, 104
Episulfoxides, 104
Epoxides, analysis of, 86, 106, 196, 248, 315
using
  liquid crystals, 401, 408
  metal complexes, 333, 354, 382, 394
  polymers, 420–421
Esters, analysis of, 146, 193, 393
using
  lanthanide complexes, 333, 338–339, 346, 352–353
  liquid crystals, 408–409
Ethers, analysis of, 120, 122, 124, 206, 209, 336, 383, 408, 414
  allyl ethers, 375, 377
  fluoro ethers, 291–292
  vinyl ethers, 209, 377

Fenaldopam, analysis of, 281
Fluorine-19 NMR spectra, with
  amine-based reagents, 154, 165, 176–178, 181, 183
  carboxylic acid-based reagents, 10, 12–14, 17–18, 24–25, 29, 32–33, 60–62, 66, 73, 80–81, 94–95
  cyclodextrins, 280, 283, 287, 291, 292, 298
  hydroxyl-containing reagents, 121, 128, 129, 133, 135
  ionic liquids, 419
  liquid crystals, 414, 415
  metal complexes, 371, 392, 393
  miscellaneous reagents, 193, 210, 216
  phosphorus-containing reagents, 243, 252, 259
Fluoxetine, analysis of, 124–125, 208

Gambieric acid, analysis of, 168
Geometric enantiomers, analysis of, 128, 134, 137, 181, 219, 241, 259, 344–347, 375, 415
Glisoprenin A, analysis of, 351
Glutethimide, analysis of, 421
Glycerols, analysis of, 13, 66, 267, 353

Glycosidation shifts, 138–144

Halogenated alkanes, analysis of, 266, 314–315, 396, 426
  using
    lanthanide-silver complexes, 332–334, 355, 357
    liquid crystals, 405, 408
    rhodium complexes, 379, 381–382
Halohydrins, analysis of, 87, 245–246
Helically chiral compounds, 93, 223, 258, 283
  calix[4]arene dimelamines, 318
  catenanes, 174, 260
  1,12-dimethylbenzo[c]phenanthrene-5,8-dicarboxylic acid, 144
  hypericin, 85
  naphtha[9]annulenones, 344
  [7]thiaheterohelicenes, 418
  zinc complexes, 358, 391
Hemiacetals, analysis of, 41, 176
Heterocyclic amines, analysis of, 29, 96, 313
High-throughput optical purity, 227–228
Hydrins
  cyano, 41, 131, 136, 152, 293
  halo, 87, 245–246
  sulfenylfluoro, 58
Hydrocarbons
  alkanes, analysis of
    aliphatic, 410–411
    cyclic, 284–285, 319, 406, 411
  alkenes, analysis of, 223–224
    using
      cyclodextrins, 285, 287, 293
      lanthanide-silver complexes, 332–334, 355–357
      liquid crystals, 408, 411
      palladium complexes, 372
      platinum complexes, 374–378
      rhodium complexes, 379
      silver complexes, 395–396
  alkynes, analysis of, 356, 372, 374
  allenes, analysis of, 215
    using
      cyclodextrins, 289, 291
      lanthanide-silver complexes, 356
      platinum complexes, 374–378
      TFAE, 120–121

  aromatic, analysis of using
    cyclodextrins, 289
    lanthanide-silver complexes, 332–334, 355–356
    liquid crystals, 407–408, 411
    rhodium complexes, 384
  fullerenes, 393
α-Hydroxy aldehydes, analysis of, 131
α-Hydroxy carboxanilides, analysis of, 131
Hydroxychloroquine, analysis of, 241
Hydroxyl-containing reagents as chiral discriminating agents, 102–152
3-Hydroxy-β-ionine, analysis of, 68

Ibuprofen, analysis of, 281
Imidazolidinones, analysis of, 197, 318
Imines, analysis of, 107
Indoprofen, analysis of, 157
Ionic liquids as chiral discriminating agents, 419–420
Ionol glycosides, analysis of, 139
Isoanabasane, analysis of, 27
Isocyanates
  analysis of, 159, 213
  as chiral discriminating agents, 206–211
Isocyanides, analysis of, 393
Isoflurane, analysis of, 291–292
Isothiocyanates, as chiral discriminating agents, 210–211
Isothiouronium salts, analysis of, 358–359
Isotopic chirality
  carbon-13, 104, 227–228
  deuterium, 38, 76, 157, 196, 342, 387–388
    carboxylic acids, 129–130, 182
    alcohols, 83, 142, 341–342, 401, 405–406, 412
    α-deuterated primary alcohols, 4, 38, 51, 54, 62, 82–83, 236, 265, 341
    sulfoxides, 50, 104
  oxygen-16, 234–235
  oxygen-18, 234–235
Isoxazolines, analysis of, 391

Kamahines, analysis of, 41
cis-Ketoconozole, analysis of, 282
Ketones
  analysis of, 220, 285, 319, 408, 423–424
  cyclic, 128, 149–150, 183–184, 220, 408, 424

INDEX 545

using
  amine-based reagents, 153, 159, 161, 183–184
  hydroxyl-containing reagents, 114, 120, 128, 145, 149, 150–151,
  lanthanide complexes, 333, 338, 344, 346
  phosphorus-containing reagents, 234–236
as chiral discriminating agents, 205–206
Ketoprofen, analysis of, 156
Ketotifen, analysis of, 337
Kinetic isotope effects, 411–412
Kinetic resolution, 149, 170, 214, 216, 228, 403
  with
    boron-containing reagents, 268, 271
    carboxylic acid-based reagents, 2, 12, 60, 81
    phosphorus-containing reagents, 243–246, 248
    selenium-containing reagent, 265
    silyl-containing reagent, 274
    transition metal complexes, 386–387, 392

Lacinilene C, analysis of, 120, 154
Lactams
  analysis of, 196, 336
    β-lactams, 112
    γ-lactams, 111–112, 159, 201
  as chiral discriminating agents, 201–202
Lactones, analysis of, 57, 111, 122, 179, 194, 196, 400, 412, 426
  β-lactones, 414
  δ-lactones, 110–111, 353, 412
  γ-lactones, 110, 353
  phthalides, 162
Lanthanide complexes
  coupled to other reagents, 157, 194–195, 197, 209, 420
  carboxylic acid-based, 48, 83, 85, 97
  crown ethers, 302, 305–306, 309
  cyclodextrins, 286–287, 296, 298–299
  hydroxyl-containing, 122–123, 133, 145
  MTPA, 30–35
  as chiral discriminating agents, 332–365
Levamisole, analysis of, 385
Lipoic acid, analysis of, 265

Liquid chromatography-NMR spectroscopy, 35, 75, 181
Liquid crystals as chiral discriminating agents, 399–416
  chemical shift anisotropy, 401, 412–415
  dipolar coupling, 401, 412–415
  mechanism of discrimination, 400–401
  quadrupolar splitting, 401, 403

Macrophomate synthase inhibitor, analysis of, 224
Mercaptoalcohols, analysis of, 234
Mercury-199 NMR spectra, 73, 397
meso-Isomers, distinguishing from racemic pair, 181
  alcohols, 19, 185, 239, 249, 277, 340–341, 405, 420
  amines, 49, 307
  bianthrones, 341
  carboxylic acids, 150, 359, 425
  cyclophosphazenes, 116, 117, 345
  2,3-diaminosuccinate, 107
  epoxides, 339
  esters, 340, 409
  metal complexes, 348, 350, 389, 394, 396
  nitrogen heterocycles, 218
  phosphines, 366
  phosphoramides, 105
  porphyrin-like compounds, 52
  quaternary amines, 257
  thiols, 239
  using proton-carbon heteronuclear two-dimensional methods, 49
Mesoridazine, analysis of, 192
Metacalixarenes, analysis of, 318
Metal complexes, analysis of, 13
  bimetallic, 241, 254–255, 259, 358, 397
  chromium, 254, 348
  clusters, 256, 348
  cobalt, 219, 296, 348, 387–389, 415–417
  copper, 115, 253
  geometric isomers, 378, 387–389, 397–398
  iridium, 159, 253
  iron, 43–44, 52, 155–156, 251, 296, 348–349, 351, 355, 408–409
  lanthanide, 53, 95, 175, 285
  manganese, 254
  mercury, 133, 397–398

Metal complexes (*Continued*)
  molybdenum, 397
  nickel, 349–350
  niobium, 115, 349
  palladium, 115, 373–374
  platinum, 348, 378
  rhodium, 253, 296
  ruthenium, 251–253, 288, 296, 349–350
  tin, 133
  uranium, 398
  zinc, 159–160, 296
Metalaxyl derivatives, analysis of, 383
Methadone hydrochloride, analysis of, 281
Methamphetamine, analysis of, 146, 211, 338
Methcathinone, analysis of, 146
Methoxamine, analysis of, 336
α-Methoxyphenylacetic acid as chiral discriminating agent, 38–51
α-Methoxy-α-trifluoromethylphenylacetic acid as chiral discriminating agent, 11–38
Methylephedrine, analysis of, 338
Metolachor ESA, analysis of, 282
Metomidate, analysis of, 294, 297
Metoprolol, analysis of, 27, 295
Mianserine, analysis of, 296
Micelles as chiral discriminating agents, 416–419
Microginin, analysis of, 64
Mix-and-shake derivatization, 35, 50–51, 62
Modified Mosher method, 10, 18
Molybdenum-95 NMR spectra, 397
Mosher's reagent as chiral discriminating agent, 11–38
MPA as chiral discriminating agent, 38–51
MTPA as chiral discriminating agent, 11–38
Multistriatin, analysis of, 336

Naproxen, analysis of, 157, 163, 244, 292, 320–323, 328, 355
Negastignane-3,9-diol, analysis of, 139
Neobenodine hydrochloride, analysis of, 281
Nicotine, analysis of, 241
Nicotinic acids, analysis of, 182
Nitriles, analysis of, 382, 385–386, 393
Nitrogen-14 NMR spectra, 415
Nitrogen-15 NMR spectra, 52, 257
Noradrenaline, analysis of, 295
Norepinephrines, analysis of, 145, 389

Norpseudoephedrin, analysis of, 145
Nuclear Overhauser effect (NOE), with
  amine-based reagents, 164–165, 168, 172, 176–177, 179
  carboxylic acid-based reagents, 15, 25–26, 28, 64, 73, 97, 100–101
  cyclodextrins, 279
  hydroxyl-containing reagents, 138–139, 141
  miscellaneous, 194, 196, 203, 207, 214, 224, 418
  phosphorus-containing reagents, 263
  transition metal complexes, 368, 370–371, 374, 387
Nystatin A, analysis of, 141

Oasomycins, analysis of, 23, 36–37, 184–185
Oxadiazines, analysis of, 123
Oxamiquine, analysis of, 295, 297
Oxatriazoles, analysis of, 383
Oxaziridines, analysis of, 106, 112
Oxazolidinones, analysis of, 197, 201
Oxazolines, analysis of, 121, 265
Oxetanes, analysis of, 97
Oxygen-16, isotope effects, 235, 260–261
Oxygen-17
  isotope effects, 260–261
  NMR spectra, 262–263, 347
Oxygen-18, isotope effects, 235, 260–262, 388–389

para-Hydrogen-induced polarization, 347
threo-Phenidates, analysis of, 119

Pheniramine hydrochloride, analysis of, 281, 286, 294, 298, 313
Phenols, analysis of, 243, 250
Phosphamides, analysis of, 192, 381, 425
Phosphane oxides, analysis of, 53, 233
Phosphanes, analysis of, 378
Phosphates, analysis of, 288, 295, 299, 418
  configurational analysis, 260–263
  oxygen isotope methods, 260–263
  ROESY studies, 263
  diphosphates, 388
Phosphene selenides, analysis of, 380
Phosphinate esters, analysis of, 232
Phosphinates, analysis of, 132–133, 148–149
Phosphine-boranes, analysis of, 413

INDEX 547

Phosphine oxides, analysis of, 53, 105, 117, 149, 192–193, 232, 369, 381, 413
Phosphine thiols, analysis of, 394
Phosphines, analysis of, 332–334, 355, 365–371, 379, 392, 394
    diphosphines, 365–368, 374
Phosphonic acids, analysis of, 92, 158, 172
Phosphinic amides, analysis of, 230–231
Phospholane chalcogenides, analysis of, 380
Phospholene chalcogenides, analysis of, 378, 380
Phosphonates, analysis of, 117, 133, 178, 232, 240
    amino alkyl, 89, 172
    α-hydroxy, 15, 42, 85, 89, 151, 158, 232–233, 245
    β-hydroxy, 42, 85, 89, 177, 233
    γ-hydroxy, 85
Phosphonic acid diamides, analysis of, 231
Phosphonium salts, analysis of, 178, 221–222, 257, 413
Phosphoramides, analysis of, 105
Phosphoroalkyldichloridates, analysis of, 172
Phosphorus-containing chiral discriminating agents, 230–264
Phosphorus-31 NMR spectra, with
    amine-based reagents, 157–158, 160–161, 169, 172, 175, 177–178
    carboxylic acid-based reagents, 5, 15, 42, 85, 89, 92
    cyclodextrins, 284–285, 288
    hydroxyl-containing reagents, 105, 116–117, 133, 150
    liquid crystals, 413
    metal complexes, 345, 367, 369, 371–375, 379–381, 384, 389, 394, 397
    miscellaneous reagents, 192–193, 221
    phosphorus-containing reagents, 229–230, 233–240, 243–250, 257, 259, 261–262
    selenium-containing reagents, 267
    solid-state systems, 425
Phosphorus chlorothionates, analysis of, 158, 380
Phosphorus thioacids, analysis of, 157, 171, 178, 231–232
Phosphorus thionates, analysis of, 380
Pinoresorcinol, analysis of, 275

Piperidines, analysis of, 26, 29, 136, 327, 336
Pirkle's alcohol as chiral discriminating agent, 109–123
Pirprofen, analysis of, 280
Pisolithin B, analysis of, 119
Phthalides, analysis of, 162
Planotriol monoacetate, analysis of, 101
Platinum-195 NMR spectra, 375–378
Polyamine alkaloids, analysis of, 52, 54
Polymers
    analysis of, 55, 346, 421–422
    as chiral discriminating agents, 420–423
Polyols, analysis of, 37, 42, 270
    annonaceous acetogenins, 17, 24–25, 68, 111
    1,3,5-triols, 37, 185–186
Primaquine, analysis of, 302
Prochiral compounds, analysis of, 280, 406
    acetic acid, monodeutero, monotritio, 173–174
    alcohols, 1, 4, 51–52, 82–84, 138, 187, 236, 265, 280, 339–341, 400, 426
    alkenes, 375, 411, 413–414
    allenes, 375
    amines, 84–85, 339
    amino acids, 92–93
    bromoalkanes, 426
    carboxylic acids, 129–130, 173–174, 182, 321, 355, 360
    citrate, 122, 359
    cyclotriveratrylenes, 407
    α-deuterated primary alcohols, 4, 38, 51, 54, 62, 82–83, 236, 265, 341
    epoxides, 339
    esters, 340, 409
    imines, 114
    iminium species, 340
    metal complexes, 351, 424
    phosphinates, 132–133
    phosphine oxides, 381
    sulfones, 339
    sulfoxides, 339, 426
    thiols, 411–412
Propoxyphene hydrochloride, analysis of, 281
Propranolol, analysis of, 27, 280, 286, 298, 313, 325
Proton NMR spectra, assignment methods in liquid crystals, 414

Pseudoephedrine, analysis of, 52
Pyrrolidines, analysis of, 26, 136

Quinidine, analysis of, 147
Quinine, analysis of, 147
Quinolinic acids, analysis of, 182
Quinolones, analysis of, 201
Quinone methides, analysis of, 160

Racemization, with,
 amine-based reagents, 174, 182
 carboxylic acid-based reagents, 2, 12, 38, 40, 45, 51, 55–56, 59, 71, 73, 94, 95
 hydroxyl-containing reagents, 133, 149–150
 liquid crystals, 403
 phosphorus-containing reagents, 245
 selenium-containing reagents, 265
Receptor compounds as chiral discriminating agents, 320–329
 for amines, 327
 for carboxylic acids, 320–323
 for diols, 328
Remote chiral centers, analysis of
 alcohols, 61, 67, 88, 250, 266–267, 405
 amines, 250, 272
 carboxylic acids, 127, 250, 265
 esters, 404
 phenols, 250, 275
Rhodium-103 NMR spectra, 379
Rostratin A, analysis of, 23
Rotational enantiomers, analysis of, 108, 113–114, 146, 257, 342–343, 407, 417
 amides, 113, 342
 calixarenes, 108–109
 1,3-dienes, 342–343
 disulfides, 129, 343
 imines, 114
 naphthyl ketones, 114
 nitrosamines, 342
 oximes, O-substituted, 114
 thioamides, 114

Sarcoglaucol-16-one, analysis of, 39
Sarcophytol, analysis of, 67
Selenides, analysis of, 379, 380

Selenium-containing compounds as chiral discriminating agents, 264–268
Selenium-77 NMR spectra, 5, 155, 229, 265–268, 380
Selenoxides, analysis of, 145, 149
Self discrimination, 224–227, 230–231
 carboxamides, 225
 carboxylate ions, 359
 dipeptides, 225
 phosphinic amides, 230
 ruthenium complexes, 226–227, 393
 thiophosphinic acids, 231
Silanes, analysis of, 55, 133
Silicon-29 NMR spectra, 347
Silyl-containing compounds as chiral discriminating agents, 274–277
Solid-state NMR spectroscopy, 425–427
 alkyl halides, 426
 cyclophosphamide, 425
 lactone, 426
 MTPA derivatives, 427
 one-dimensional exchange spectroscopy by side band alteration (ODESSA), 425
 organophosphorus compounds, 425
Soulattrolide, analysis of, 17
Spin-lattice relaxation
 copper complex, 394–395
 phosphorus-31 NMR spectra, 246
 lanthanide complexes, 335, 338
Spiroacetals, analysis of, 206
Spiroalcohols, analysis of, 33
Spirochalcogenuranes, analysis of, 107, 123, 378, 380–381
Spirocyclic oxaphosphates, analysis of, 160
Spirophosphoranes, analysis of, 344
Spirosulfuranes, analysis of, 107, 123, 381
Steroids, analysis of, 13, 20, 33, 140, 143, 187
Subphthalocyanines, analysis of, 226
Sulfides, analysis of, 392
Sulfinamides, analysis of, 104, 114
Sulfinate esters, analysis of, 104
Sulfinates, analysis of, 104
Sulfites, analysis of, 104
Sulfonamide esters, analysis of, 25, 164
Sulfonamides, analysis of, 25, 164
Sulfonates, analysis of, 124
Sulfones, analysis of, 104, 338–339, 347

INDEX   549

Sulfonic acids, analysis of, 184
Sulfonium salts, analysis of, 358–359
Sulfonyl chlorides, analysis of, 158
Sulfoxides, analysis of, 178, 221, 233, 314, 319, 324
   using
      amide reagents, 191–192, 194–197
      carboxylic acid-based reagents, 45–46, 49–50, 68–69, 89
      hydroxyl-containing reagents, 103–104, 109–110, 122, 124, 127, 145, 148–149
      metal complexes, 333, 336, 338–339, 371–372, 379, 392, 398
Sulfoximines, analysis of, 57, 149
Suprofen, analysis of, 157

Taxanes, analysis of, 188
Telenzepine, analysis of, 281
Temperature, variable
   effect on conformational preference, 10, 54, 69, 75, 126
      of MPA derivatives, 3, 13, 43, 46–47, 79
      systems under slow exchange, 113–114, 146, 253, 257, 260, 343, 389, 406–407, 418
   to enhance enantiomeric discrimination, 30, 196, 239, 371, 381, 393, 409
   to reduce broadening, 194, 337
Terbutaline, analysis of, 207
Tetraazoles, analysis of, 383
Tetrafibricin, analysis of, 187
Tetrahydrofolate, analysis of, 219, 220
Tetramisole, analysis of, 385
TFAE as chiral discriminating agent, 109–123
Thalidomide, analysis of, 336
Thiatriazoles, analysis of, 383
Thiazoline-2-thione, analysis of, 128
N-Thiocarboxyanhydrides, analysis of, 165
Thiolactones, analysis of, 57
α-Thiolamides, analysis of, 239

α-Thiolcarboxylic acids, analysis of, 239, 245
β-Thiolcarboxylic acids, analysis of, 239
α-Thiolesters, analysis of, 245
Thiols
   analysis of using
      carboxylic acid-based reagents, 57, 78, 85–86, 89
      Noe's reagent, 131
      phosphorus-containing reagents, 234, 237, 239, 243, 245
      transition metal complexes, 373, 382
   as chiral discriminating agents, 135, 149–151
Thiophosphates, configuration analysis of, 234–235, 260–263
Thiophosphinates, analysis of, 232
Thiophosphoranes, analysis of, 169
Thiosulfinates, analysis of, 104
Timolol, analysis of, 27
Tin-119 NMR spectra, 396
Tosylates, analysis of, 408, 409
Transition metal complexes as chiral discriminating agents, 365–398
2,2,2-Trifluoro-1-(9-anthryl)ethanol as chiral discriminating agent, 109–123
Tritium NMR spectroscopy, 84, 108, 173–174, 341
Troger's base, analysis of, 421

Uniconazole, analysis of, 412

Vasicinone alkaloids, analysis of, 16
Verapamil, analysis of, 288
Vericonazole, analysis of, 297
Viridoxins, analysis of, 40

Xanthene-4,5-dicarboxamides, analysis of, 113
Xanthines, analysis of, 382
Xenon-129 NMR spectra, 347

Zeaxanthin, analysis of, 47